RNA INTERFERENCE TECHNOLOGY

RNA Interference (RNAi) technology has rapidly become one of the key methods used in functional genomics. RNAi is used to block the expression of genes and create phenotypes that can potentially yield clues about the function of these genes. In the postgenomic era, the elucidation of the physiological function of genes has become the rate-limiting step in the quest to develop "gene-based drugs" and RNAi could potentially play a pivotal role in the validation of such novel drugs. In this cutting-edge overview, the basic concepts of RNAi biology are discussed, as well as the current and potential applications. Leading experts from both academia and industry have contributed to this invaluable reference for graduate students, postdocs, and researchers from academia wanting to initiate RNAi research in their own labs, as well as for those working in research and development in biotech and pharmaceutical companies who need to understand this emerging technology.

Krishnarao Appasani is the Founder and Chief Executive Officer of Gene-Expression Systems, a gene discovery company focusing on functional genomics in cancer research.

RNA Interference Technology

FROM BASIC SCIENCE TO DRUG DEVELOPMENT

Edited by

Krishnarao Appasani
GeneExpression Systems, Inc.

Forewords by

Andrew Fire
Stanford University, co-discoverer of RNAi
and
Marshall Nirenberg
National Institutes of Health
Winner of the Nobel Prize in Physiology or Medicine, 1968

DAMAGED

CAMBRIDGE
UNIVERSITY PRESS

PUBLISHED BY THE PRESS SYNDICATE OF THE UNIVERSITY OF CAMBRIDGE
The Pitt Building, Trumpington Street, Cambridge, United Kingdom

CAMBRIDGE UNIVERSITY PRESS
The Edinburgh Building, Cambridge CB2 2RU, UK
40 West 20th Street, New York, NY 10011-4211, USA
477 Williamstown Road, Port Melbourne, VIC 3207, Australia
Ruiz de Alarcón 13, 28014 Madrid, Spain
Dock House, The Waterfront, Cape Town 8001, South Africa

http://www.cambridge.org

First published 2005

Printed in the United States of America

Typefaces ITC Stone Serif 9/13.5 pt. and Poppl-Laudatio   *System* LaTeX 2_ε [TB]

A catalogue record for this book is available from the British Library.

Library of Congress Cataloguing in Publication Data
RNA interference technology : from basic science to drug development / edited by
   Krishnarao Appasani.
       p. cm.
   Includes bibliographical references and index.
   ISBN 0-521-83677-8 (alk. paper)
   1. Small interfering RNA.   2. Gene silencing.   I. Appasani, Krishnarao.
   QP623.5.S63R63   2004
   572.8'8 – dc22   2004054562

ISBN 0 521 83677 8 hardback

In memory of my parents

For my teachers, family members
and especially my wife Shyamala and sons Raakish and Raghu

Contents

Foreword

Andrew Fire

It has been a privilege to watch the growth of RNA interference technology over the last ten years. Starting with a mixture of curiosity and chagrin, the field has grown into a substantial enterprise which impacts (and utilizes resources from) virtually every field of biomedical research. Research in RNAi derives from a set of apparently unconnected observations: strange pigment patterns in plants, unexpected failures and successes in antisense and overexpression studies, small regulatory RNAs in bacteria. If there is an underlying and recurring scientific lesson, it has been: "Pursue the unexpected." Basic and applied research each advance as a consequence of this pursuit; certainly this has been no better illustrated than in the last ten years of RNAi.

The work of hundreds of researchers in different fields that is reported in this book should provide the reader with both solid information (needed for experimental design and evaluation) and a lively and hopeful scientific story (needed to keep us all going through the long haul of scientific research). Our knowledge of the realm of genetic regulation by small RNAs has grown with remarkable speed. Starting in 1981 with a single known example of a modulatory short RNA (regulating copy number of the ColE1 plasmid), small RNAs are now known to regulate genetic activity at virtually every level: DNA and chromosome structure, transcription, RNA structure and stability, translation, and protein stability. Likewise, our ability to experimentally alter cells using this system has advanced at an unprecedented rate. As recently as 1990, the known examples of experimentally-induced silencing were a few unusual and accidental plant pigmentation patterns; now there are extensive menus of silencing-based methods as part of the "standard" molecular biology toolkit.

Work in this field is by no means finished. We still don't understand all of the modalities of RNA-triggered genetic regulation, why these modalities exist, and how they interact with each other. We don't have a clear picture the full extent of RNA-based regulation. As these questions are further investigated and understood, and as the underlying mechanisms are understood in detail, it will become possible to carry out more and more sophisticated experimental manipulations of genetic function. More questions: How do some organisms encapsulate

RNA triggers to produce a systemic response? How are long term RNAi effects perpetuated? What is the link between RNAi and immunity? What biological effects will come from the selective or global inactivation or augmentation of the RNAi pathway? How can we best use RNAi to discover the most sensitive and critical targets for biological investigation and drug development? Can we cure diseases by specifically triggering the RNAi pathway to attack errant genes? Can we treat other diseases by up- or down-regulating components of the RNAi machinery itself in specific cell types? How will cells and organisms respond in the long term to continuous modulation or use of the RNAi machinery?

We'll all be busy for quite a while in addressing these questions. Based on the first years of the field, one thing that can certainly be expected is a few more surprises.

Stanford, California, USA
August 2, 2004

Foreword

Marshall Nirenberg

RNA interference is a powerful tool that has been used to inhibit gene function either by increasing the destruction of mRNA corresponding to the gene, or in some cases, by inhibiting the transcription of the gene or the translation of mRNA to the corresponding protein. Exploring gene function by the classical approach of generating mutants of a gene often is much more laborious and time consuming then silencing gene function by RNAi using double-stranded RNA or double-stranded oligoribonucleotides about twenty two nucleotide residues in length. This book edited by Krishnarao Appasani is a timely and comprehensive compendium of information on RNAi and will be useful to experts on RNAi as well as investigators in many fields of research who may be interested in using RNAi to explore problems they are studying.

The RNAi field is only six years old. Research on RNAi has been expanding at an extraordinarily rapid rate, yet the field is in its infancy. There is great interest in using RNAi as a means of exploring gene function during embryonic development and in the adult in many organisms. Many aspects of RNAi remain to be explored. For example, the reactions and the molecules required for RNAi targeted destruction of mRNA are incompletely known. Similarly, the mechanisms of RNAi targeted modification of DNA, which regulates, transcription of DNA, as well as RNA targeted inhibition of mRNA translation are only partially known. Also, the functions of most micro RNA genes have not yet been explored. Since RNAi also can be used to regulate gene expression in specific cell types, the possibility that RNAi can be used therapeutically to treat diseases or certain viral infections by targeted gene silencing is an exciting, challenging possibility. However, difficult problems have to be overcome such as the problem of delivery of appropriate double-stranded oligoribonucleotides into cells, the stability, concentration, and toxicity of the oligoribonucleotides, and the length of time the oligoribonucleotides remain in the cells. These are challenging research problems. Nevertheless, the use of oligoribonucleotides as therapeutic agents to silence gene expression has great potential for the future. Libraries of small interfering RNAs (siRNAs) or short hairpin RNAs (shRNA) have been constructed and have been screened in cultured cells. In addition, methods have been devised for high throughput screening of siRNA or shRNA libraries. RNAi has been used to inhibit replication of viruses in cultured cells such as HIV, hepatitis C virus, and hepatitis B virus. The oncogenic fusion protein p210 in chronic myelogenous leukemia cells promotes cell division in these cells. Both siRNA and a lentivirus vector containing shRNA have been shown to reduce the levels of p210 protein in cell lines and thereby inhibit cell division. In addition, RNAi has been used in intact mice to reduce the function of a mutant gene which results in the movement disorder, spinocerebellar ataxia type one. Treatment of mice by RNAi resulted in improved motor coordination and the cellular changes in the brain characteristic of the disease were no longer visible. RNAi also is being investigated as a therapy for ocular diseases.

It is too early to say how successful RNAi therapy will be. However, it is clear that RNAi is a powerful tool that has revolutionized basic research and that the ability of RNAi to down-regulate almost any gene affords remarkable opportunities to explore the use of duplex oligoribonucleotides as therapeutic agents for many diseases.

Laboratory of Biochemical Genetics
National Heart, Lung, and Blood Institute
National Institutes of Health
Bethesda, MD

Contributors

Jeff Aalfs
Wyeth Research
35 Cambridge Park Drive
Cambridge, MA 02140
USA

Hideo Akashi
Department of Chemistry and Biotechnology
School of Engineering
The University of Tokyo
Hongo, Tokyo 113-8656
Japan

Ahmad Z. Amin
Biology Department
616 Fordham Hall, CB#3280
University of North Carolina
Chapel Hill, NC 27599-3280
USA

Krishnarao Appasani, PhD., MBA
GeneExpression Systems, Inc.
P.O. Box 540170
Waltham, Massachusetts 02454-0170
USA
E-mail: DrAppasani@expressgenes.com

Greg Arndt, PhD.
Johnson & Johnson Research
Level 4, 1 Central Avenue
Eveleigh, NSW 1430
Sydney, Australia
E-mail: garndt@medau.jnj.com

Thomas Baeriswyl
University of Zurich, Institute of Zoology
Winterthurerstrasse 190, CH-8057
Zurich, Switzerland

Mehdi Banan
Ambion, Inc.
2130 Woodward Street
Austin, Texas 78744
USA

John E. Bisi
Cellular Genomics
GlaxoSmithKline R&D
Stevenage, Herts
UK

Peter Blume-Jensen, PhD.
Department of Molecular Oncology
Serono Reproductive Biology Institute
One Technology Place
Rockland, MA 02370
USA

Queta Boese, PhD.
Dharmacon, Inc.
2650 Crescent Dr, Suite #100
Lafayette, CO 80026
USA
E-mail: boese.q@dharmacon.com

Dimitris Bourikas
University of Zurich, Institute of Zoology
Winterthurerstrasse 190, CH-8057
Zurich, Switzerland

Michael Boutros, PhD.
German Cancer Research Center (DKFZ/B110)
Im Neuenheimer Feld 580
69120 Heidelberg
Germany
E-mail: m.boutros@dkfz.de

David Brown, PhD.
Ambion, Inc.
2130 Woodward Street
Austin, Texas 78744
USA
E-mail: dbrown@ambion.com

Frank Buchholz
Max Plank Institute of Molecular Cell Biology and Genetics
Pfotenhauer Strasse 108 Dresden
Germany
E-mail: buchholz@mpi-cbg.de

Mike Byrom
Ambion, Inc.
2130 Woodward Street
Austin, Texas 78744
USA

Federico Calegari
Max Plank Institute of Molecular Cell Biology and Genetics
Pfotenhauer Strasse 108 Dresden
Germany
E-mail: Calegari@mpi-cbg.de

Howard Y. Chang
Departments of Biochemistry and Dermatology
Stanford University School of Medicine
Stanford, CA 94305
USA

Padmanabhan Chellappan, PhD.
International Laboratory for Tropical
  Agricultural Biotechnology
Donald Danforth Plant Science Center
975 N. Warson Rd.
St Louis, MO 63132
USA
E-mail: iltab@danforthcenter.org

Jen-Tsan Chi
Departments of Biochemistry and Dermatology
Stanford University School of Medicine
Stanford, CA 94305
USA
E-mail: chi@pmgm2.stanford.edu

Chris Childs
Wyeth Research
35 Cambridge Park Drive
Cambridge, MA 02140
USA

Neil J. Clarke, PhD.
Cellular Genomics
GlaxoSmithKline R&D
Stevenage, Herts
UK
E-mail: neil.j.clarke@gsk.com

Sandra Clauder-Münster
Anadys Pharmaceuticals Europe GmbH
Meyerhofstr.1
69117 Heidelberg
Germany

Caretha L. Creasy
Cellular Genomics
GlaxoSmithKline R&D
Stevenage, Herts
UK

Ye Ding, PhD.
New York State Health Department
Wodsworth Center
Division of Molecular Medicine, Room C-660
Empire State Plaza
Albany, NY 12201-0509
USA
E-mail: yding@wadsworth.org

Graeme Doran
Department of Human Anatomy and Genetics
South Parks Road
University of Oxford
Oxford OX1 3QU
UK
E-mail: graeme.doran@st-edmund-hall.oxford.ac.uk

Mark E. Drew, PhD.
Dept. of Mol. Microbiology, Rm. 9210
Washington University School of Medicine
Box 8230, 4940 Parkview Place
St. Louis, MO 63110
USA
E-mail: drew@borcim.wustl.edu

Nathaniel R. Dudley, PhD.
Biology Department
616 Fordham Hall, CB#3280
University of North Carolina
Chapel Hill, NC 27599-3280
USA
E-mail: ndudley@unc.edu

Michael K. Dush, PhD.
Cellular Genomics
GlaxoSmithKline R&D
Stevenage, Herts
UK

Derek M. Dykxhoorn, PhD.
The Center for Blood Research
Harvard Medical School
800 Huntington Ave
Boston, MA 02151
USA
E-mail: dykxhoor@cbr.med.harvard.edu

Mark R. Edbrooke, PhD.
Cellular Genomics
GlaxoSmithKline R&D
Stevenage, Herts
UK
E-mail: mark.r.edbrooke@gsk.com

Paul T. Englund, PhD.
Department of Biological Chemistry
Johns Hopkins Medical School
725 N. Wolfe St.
Baltimore, MD 21205
USA
E-mail: penglund@jhmi.edu

Andrew Farmer, D.Phil.
BD Biosciences Clontech
1020 East Meadow Circle
Palo Alto, CA 94303
USA
E-mail: Andrew_Farmer@bd.com; aafarmer@clontech.com

Claude M. Fauquet, PhD.
International Laboratory for Tropical Agricultural Biotechnology
Donald Danforth Plant Science Center
975 N. Warson Rd.
St Louis, MO 63132
USA
E-mail: iltab@danforthcenter.org

James E. Ferrell, Jr., PhD.
Department of Molecular Pharmacology
Stanford University School of Medicine
269 Campus Drive, CCSR Rm 3160
Stanford, CA 94305-5174
USA

Andrew Fire, PhD.
Departments of Pathology and Genetics
Stanford University School of Medicine
300 Pasteur Drive, Room L235
Stanford, CA 94305-5324
USA
E-mail: afire@stanford.edu

Kris J. Fisher
Cellular Genomics
GlaxoSmithKline R&D
Stevenage, Herts
UK

Lance Ford, PhD.
Ambion, Inc.
2130 Woodward Street
Austin, Texas 78744
USA

Takashi Futami
Department of Chemistry and Biotechnology
School of Engineering
The University of Tokyo
Hongo, Tokyo 113-8656
Japan

Marc Gentzel
European Molecular Biology Organization
Meyerhofstr.1
69117 Heidelberg
Germany

Bob Goldstein, PhD.
Biology Department
616 Fordham Hall, CB#3280
University of North Carolina
Chapel Hill, NC 27599-3280
USA
E-mail: bobg@unc.edu

Alla Grishok, PhD.
Center for Cancer Research
Massachusetts Institute of Technology
40 Ames Street
Cambridge, MA
USA
E-mail: agrishok@mit.edu

Philipp Hadwiger
Research and Development
Alnylam Europe AG
Fritz-Hornschuch-Strasse 9
95326 Kulmbach
Germany
E-mail: phadwiger@alnylam.de; hpvornlocher@alnylam.de

Steven A. Haney
Wyeth Research
35 Cambridge Park Drive
Cambridge, MA 02140
USA
E-mail: shaney@wyeth.com

Marc Hild, PhD.
Novartis Institute for Biomedical Research
100 Technology Square
Cambridge, MA 02139
USA

Wieland B. Huttner
Max Plank Institute of Molecular Cell Biology and Genetics
Pfotenhauer Strasse 108 Dresden
Germany
E-mail: huttner@mpi-cbg.de

John M. Johnson III
Cellular Genomics
GlaxoSmithKline R&D
Stevenage, Herts
UK

Takashi Kadowaki
Department of Internal Medicine
Graduate School of Medicine
The University of Tokyo
Hongo, Tokyo 113-8655
Japan

Shih-Chu Kao
Department of Neurology and Center for Neurologic Diseases
Brigham and Women's Hospital
Harvard Medical School
4 Blackfan Circle, HIM 760
Boston, MA 02115
USA

Mark A. Kay, MD., PhD.
Stanford University School of Medicine
Departments of Pediatrics and Genetics
Program in Human Gene Therapy
Stanford, CA 94305
USA

Anastasia Khvorova, PhD.
Dharmacon, Inc.
2650 Crescent Dr, Suite #100
Lafayette, CO 80026
USA
E-mail: khvorova.a@dharmacon.com

Ralf Kittler
Max Plank Institute of Molecular Cell Biology and Genetics
Pfotenhauer Strasse 108 Dresden
Germany

Kenneth S. Kosik, MD.
Department of Neurology and Center for Neurologic Diseases
Brigham and Women's Hospital
Harvard Medical School
4 Blackfan Circle, HIM 760
Boston, MA 02115
USA

Anna M. Krichevsky, PhD.
Department of Neurology and Center for Neurologic Diseases
Brigham and Women's Hospital
Harvard Medical School
4 Blackfan Circle, HIM 760
Boston, MA 02115
USA
E-mail: krichevsky@cnd.bwh.harvard.edu; akrichevsky@rics.bwh.harvard.edu

Po-Tsan Ku
Ambion, Inc.
2130 Woodward Street
Austin, Texas 78744
USA

Stefan Kubicka, M.D.
Department of Gastroenterology
Medical School of Hannover
Carl-Neuberg-Str. 1
30623 Hannover
Germany
E-mail: kubicka.Stefan@mh-hannover.de

Patricia E. Kuwabara, PhD.
Department of Biochemistry
University of Bristol
The School of Medical Sciences
University Walk, Bristol BS8 1TD
UK
E-mail: p.kuwabara@bristol.ac.uk

Eric Lader, PhD.
QIAGEN, Inc.
19300 Germantown Rd
Germantown, MD 20874
USA
E-mail: E.lader@qiagensciences.com

Laurence Lamarcq
BD Biosciences Clontech
1020 East Meadow Circle
Palo Alto, CA 94303
USA

Peter Lapan
Wyeth Research
35 Cambridge Park Drive
Cambridge, MA 02140
USA

Robert Larsen
BD Biosciences Clontech
1020 East Meadow Circle
Palo Alto, CA 94303
USA

Kathy Latham, PhD.
Ambion, Inc.
2130 Woodward Street
Austin, Texas 78744
USA

Charles E. Lawrence, PhD.
New York State Health Department
Wodsworth Center
Division of Molecular Medicine, Room C-660
Empire State Plaza
Albany, NY 12201-0509
USA

Joe D. Lewis
Anadys Pharmaceuticals Europe GmbH
Meyerhofstr.1
69117 Heidelberg
Germany

Patrick Y. Lu, Ph.D.
Intradigm Corporation
Rockville, Maryland
USA
E-mail: patricklu@intradigm.com

Michael P. Manns, M.D.
Department of Gastroenterology
Medical School of Hannover
Carl-Neuberg-Str. 1
30623 Hannover
Germany
E-mail: manns.michael@mh-hannover.de

Ying Mao
BD Biosciences Clontech
1020 East Meadow Circle
Palo Alto, CA 94303
USA

William S. Marshall, PhD.
Dharmacon, Inc.
2650 Crescent Dr, Suite #100
Lafayette, CO 80026
USA
E-mail: Marshall.b@dharmacon.com

Sahohime Matsumoto
Department of Chemistry and Biotechnology
School of Engineering
The University of Tokyo
Hongo, Tokyo 113-8656
Japan
and
Gene Function Research Center
National Institute of Advanced Industrial Science and Technology (AIST)
Central 4, 1-1-1 Higashi
Tsukuba Science City 305-8562
Japan
and
Department of Internal Medicine
Graduate School of Medicine
The University of Tokyo
Hongo, Tokyo 113-8655
Japan

Anton P. McCaffrey, PhD.
Stanford University School of Medicine
Departments of Pediatrics and Genetics
Program in Human Gene Therapy
Stanford, CA 94305
USA
E-mail: anton-mccaffrey@uiowa.edu

Chris Mello
BD Biosciences Clontech
1020 East Meadow Circle
Palo Alto, CA 94303
USA

Chris Miller
Wyeth Research
35 Cambridge Park Drive
Cambridge, MA 02140
USA

Makoto Miyagishi
Department of Chemistry and Biotechnology
School of Engineering
The University of Tokyo
Hongo, Tokyo 113-8656
Japan
and
Gene Function Research Center
National Institute of Advanced Industrial Science and Technology (AIST)
Central 4, 1-1-1 Higashi
Tsukuba Science City 305-8562
Japan

James C. Morris
Department of Genetics, Biochemistry and Life Science Studies
Clemson University
Clemson, SC 29634
USA

Brooke T. Mossman, MD.
Environmental Pathology Program, Department of Pathology
University of Vermont, College of Medicine
89 Beaumont Ave. HSRF 218
Burlington, VT 05405
USA
E-mail: brooke.mossman@uvm.edu

Shawn A. Motyka
Department of Biological Chemistry
Johns Hopkins Medical School
725 N. Wolfe St.
Baltimore, MD 21205
USA
E-mail: smotyka@jhmi.edu

Jason W. Myers, PhD.
Department of Molecular Pharmacology
Stanford University School of Medicine
269 Campus Drive, CCSR Rm 3160
Stanford, CA 94305-5174
USA
E-mail: jmyers@stanford.edu

Ryozo Nagai
Department of Internal Medicine
Graduate School of Medicine
The University of Tokyo
Hongo, Tokyo 113-8655
Japan

Nigel J. O'Neil
The Wellcome Trust Sanger Institute
Hinxton, Cambridge CB10 1SA
UK
E-mail: njo@sanger.ac.uk

Antje Ostareck-Lederer
Anadys Pharmaceuticals Europe GmbH
Meyerhofstr.1
69117 Heidelberg
Germany
E-mail: aostareck@biochemtech.uni-halle.de

Vince Pallotta
Ambion, Inc.
2130 Woodward Street
Austin, Texas 78744
USA

Amy E. Pasquinelli, Ph.D.
Molecular Biology Section
Division of Biology 0368
Bonner Hall, Room 2214
9500 Gilman Drive
University of California, San Diego
La Jolla, CA 92093-0368
USA
E-mail: apasquin@biomail.ucsd.edu

Maria Polycarpou-Schwarz
Anadys Pharmaceuticals Europe GmbH
Meyerhofstr.1
69117 Heidelberg
Germany

Thomas Quinn
BD Biosciences Clontech
1020 East Meadow Circle
Palo Alto, CA 94303
USA

Maria E. Ramos-Nino, PhD.
Environmental Pathology Program
Department of Pathology
University of Vermont, College of Medicine
89 Beaumont Ave. HSRF 218
Burlington, VT 05405
UK
E-mail: mramos@zoo.uvm.edu

Christopher J. A. Ring
Cellular Genomics
GlaxoSmithKline R&D
Stevenage, Herts
UK

John J. Rossi, PhD.
Division of Molecular Biology, Graduate School of Biological Sciences
Beckman Research Institute of the City of Hope
City of Hope, Duarte, CA 91010
USA
E-mail; jrossi@bricoh.edu

Rosa M. Ruiz-Vaźque, PhD.
Department of Genetics and Microbiology
Faculty of Biology
University of Murcia
Campus de Espinardo
30071 Murcia
Spain
E-mail: rmruiz@um.es

Rejina Sadhu
University of Zurich, Institute of Zoology
Winterthurerstrasse 190, CH-8057
Zurich
Switzerland

Dmitry Samarsky, PhD.
Invitrogen Corporation
14 Tech Circle
Natick, MA 01760
USA
E-mail: dsamarsky@oligo.com

Brad Scherer, PhD.
BD Biosciences Clontech
1020 East Meadow Circle
Palo Alto, CA 94303
USA

Lisa Scherer, PhD.
Division of Molecular Biology, Graduate School of Biological Sciences
Beckman Research Institute of the City of Hope
City of Hope, Duarte, CA 91010
USA

Oded Singer, PhD.
Laboratory of Genetics
The Salk Institute
10010 North Torrey Pines Road
La Jolla, CA 92037
USA
E-mail: singer@salk.edu

Mouldy Sioud, DEA Pharm, PhD.
Department of Immunology, Molecular Medicine Group
The Norwegian Radium Hospital
Montebello, 0310
Norway
E-mail: mouldy.sioud@biotek.uio.no

Muhammad Sohail, D. Phil.
MRC Research Associate
University of Oxford,
Department of Biochemistry
South Parks Road, Oxford OX1 3QU
UK
E-mail: muhammad.sohail@bioch.ox.ac.uk

Dag R. Sørensen, PhD.
Department of Immunology, Molecular Medicine Group
The Norwegian Radium Hospital
Montebello, 0310
Norway

Karen E. Stephens
The Wellcome Trust Sanger Institute
Hinxton, Cambridge CB10 1SA
UK
E-mail: kes@sanger.ac.uk

Esther T. Stoeckli
University of Zurich, Institute of Zoology
Winterthurerstrasse 190, CH-8057
Zurich
Switzerland
E-mail: esther.stoeckli@zool.unizh.ch

Yerramilli V. B. K. Subrahmanyam, PhD.
QIAGEN, Inc.
19300 Germantown Rd
Germantown, MD 20874
USA
E-mail: subu.yerramilli@qiagen.com

Asako Sugimoto, Ph.D.
Laboratory Head
Laboratory for Developmental Genomics
RIKEN Center for Developmental Biology
2-2-3 Minatojima-minamimachi, Chuo-ku
Kobe 650-0047
Japan
E-mail: sugimoto@cdb.riken.go.jp

Shizuyo Sutou
iGENE Therapeutics, Inc.
c/o AIST
Central 4, 1-1-1 Higashi
Tsukuba Science City 305-8562
Japan

Kazunari Taira, PhD.
Department of Chemistry and Biotechnology
School of Engineering, The University of Tokyo
Hongo, Tokyo 113-8656
Japan
E-mail: taira@chembio.t.u-tokyo.ac.jp
and
Gene Function Research Center
National Institute of Advanced Industrial Science and Technology (AIST)
Central 4, 1-1-1 Higashi
Tsukuba Science City 305-8562
Japan

Yasuomi Takagi
iGENE Therapeutics, Inc.
c/o AIST
Central 4, 1-1-1 Higashi
Tsukuba Science City 305-8562
Japan

Marcia Tan
BD Biosciences Clontech
1020 East Meadow Circle
Palo Alto, CA 94303
USA

Margaret Taylor, PhD.
Invitrogen Corporation
14 Tech Circle
Natick, MA 01760
USA

Rolf Thermann
Department of Biochemistry and Biotechnology
Institute of Biochemistry
Martin- Luther-University
Kurt-Mothes-Str. 3, 06120 Halle (Saale)
Germany
and
Anadys Pharmaceuticals Europe GmbH
and
European Molecular Biology Organization
Meyerhofstr. 1, 69117 Heidelberg
Germany
E-mail: sworland@anadyspharma.com

Gustavo Tiscornia, PhD.
Laboratory of Genetics
The Salk Institute
10010 North Torrey Pines Road
La Jolla, CA 92037
USA
E-mail: coyne@salk.edu

Li-Huei Tsai
Department of Neurology and Center for Neurologic Diseases
Brigham and Women's Hospital
Harvard Medical School
4 Blackfan Circle, HIM 760
Boston, MA 02115
USA

Ramachandran Vanitharani, PhD.
International Laboratory for Tropical Agricultural Biotechnology
Donald Danforth Plant Science Center
975 N. Warson Rd.
St Louis, MO 63132
USA
E-mail: iltab@danforthcenter.org; VRamachandran@danforthcenter.org

Inder M. Verma, PhD.
Laboratory of Genetics
The Salk Institute
10010 North Torrey Pines Road
La Jolla, CA 92037
USA
E-mail: Verma@salk.edu

Hans-Peter Vornlocher
Research and Development
Alnylam Europe AG
Fritz-Hornschuch-Strasse 9
95326 Kulmbach
Germany
E-mail: hpvornlocher@alnylam.de

Nancy N. Wang
Departments of Biochemistry and Dermatology
Stanford University School of Medicine
Stanford, CA 94305
USA

Zefeng Wang
Dept. of Mol. Microbiology, Rm. 9210
Washington University School of Medicine
Box 8230, 4940 Parkview Place
St. Louis, MO 63110
USA
and
Department of Biological Chemistry
Johns Hopkins Medical School
725 N. Wolfe St.
Baltimore, MD 21205
USA

Matthias Wilm
European Molecular Biology Organization
Meyerhofstr.1
69117 Heidelberg
Germany

Patty Wong
BD Biosciences Clontech
1020 East Meadow Circle
Palo Alto, CA 94303 USA

Martin C. Woodle, Ph.D.
Intradigm Corporation
Rockville, Maryland
USA
E-mail: mwoodle@intradigm.com

Paul Yaworsky
Wyeth Research,
35 Cambridge Park Drive
Cambridge, MA 02140, USA

Lars Zender, M.D.
Department of Gastroenterology
Medical School of Hannover
Carl-Neuberg-Str. 1
30623 Hannover
Germany
E-mail: Zender.Lars@mh-hannover.de

Olivier Zugasti
The Wellcome Trust Sanger Institute
Hinxton, Cambridge CB10, 1SA
UK
E-mail: omz@sanger.ac.uk

Introduction

Krishnarao Appasani

"If the RNAi technology can be made to work, there's a long list of diseases it can be applied to."
–Phillip A. Sharp, Nobel laureate, MIT From CBSNews.com July 18, 2003

Gene Expression (genomics) is a new discipline within molecular biology that narrates the functional organization of genes. Most of you are aware of the terms *Genome* (study of the expression of all the genes in an organism known as Genomics), *Proteome* (study of the expression of all the proteins in an organism known as Proteomics), and *Glycome* (study of the expression of all the glycoproteins in an organism known as Glycomics). Another scientific buzz word that is spreading fast in the research community these days is *RNome*, the RNA equivalent of the 'proteome,' 'genome' or 'glycome,' with the subject referred to as *RNomics*. *RNomics* is a newly emerging sub-discipline that categorically studies the structure, function and processes of noncoding RNAs and the mechanism of RNA interference in a cell.

RNA Interference Technology: From Basic Science to Drug Development is primarily intended for readers in the molecular cell biology and genomics fields but may be useful in more advanced graduate level courses. Much of the text should be of interest to those in applied sciences such as molecular medicine, genome science, and biotechnology. This book, which focuses on the concepts of basic RNAi biology and applications in drug development, consists of thirty four chapters, grouped into seven sections. Most of the chapters are written by the original discoverers or their associated scientists from academia, biotech and pharma, exclusively from the RNAi field. This is the first book of its kind that integrates the academic science with industry applications in drug validation and therapeutic development. This book will serve as a reference for graduate students, post-docs, and professors from academic research institutions who wish to initiate RNAi and siRNA research in their own laboratories. This book will also serve as a descriptive and in-depth analysis for executives, directors and scientists in

1

research and development from biotech and pharmaceutical companies. This will provide a valuable executive summary for investors and those responsible for business development in the life sciences, who need to keep abreast with the RNAi revolution.

In the post-genomic era, elucidation of the physiological function of genes has become a major rate-limiting step in the quest to develop gene-based-drugs. As we advance in the 'functional omics' arena, with the hope of discovering novel drug targets and therapies, their validation is a pivotal step before their use in clinical practice. Such an endeavor can be tested using small interference RNAs (siRNAs) or RNA-mediated genetic interference (RNAi). This elegant and revolutionary reverse genetic approach has tremendous commercial promise with regard to developing new drugs and therapeutics for human diseases. This book provides an in-depth summary of the field of RNAi by international experts and predicts some of the potential applications of gene suppression strategies in the pre-clinical drug discovery process within the biotech and pharmaceutical industries.

We are now beginning to exploit the information gleaned from the genome sequencing projects of humans and other organisms. However, this wealth of genetic information opens new challenges in deciphering the complete list of protein-encoding genes. In addition, transcriptional events such as RNA splicing and post-translational modifications make it difficult to predict the exact number of genes or proteins. With this degree of complexity, monitoring the entire proteome expression levels as a means of elucidating the functions of genes and proteins and developing them as potential drug targets are significant challenges for the biotech industry. Despite the "proteome" sequencing efforts, the "RNome" also has to be studied in depth to fully understand and tally the number of genes encoded by a genome and their regulation. The challenge for scientists in both academia and industry is to identify the whole complement of non-coding RNAs and elucidate their functions in gene expression and regulation. From a healthcare point of view, it is important to identify the disease relevant genes from these functional "OMES."

Six years ago, Mello and his colleagues discovered the phenomenon of RNAi. Since then, this technique has become widely used within the cell and molecular biology communities as a tool to aid understanding of gene expression and regulation. In addition, scientists have begun to take an *RNomics* approach to understanding the nature and function of microRNAs and siRNAs in order to utilize them as a mechanism of gene silencing. In broad terms, "post-transcriptional gene silencing," "co-suppression," "quelling" and "siRNA" are collectively included in the phenomenon of "RNA interference."

The RNAi mechanism and siRNAs have tremendous commercial potential in the bio-industry. This molecular tool will allow investigators to routinely implement "loss-of-function" screens and enable the development of rapid tests for genetic interactions in mammalian cells, which until now have been difficult to perform quickly. Many of the current and potential applications are described within the various themes of the book, which are summarized below.

Section 1: Basic biology of RNAi, siRNA, microRNAs and gene silencing mechanisms

The seminal work of Sydney Brenner, Robert Horovitz and John Sulston paved the way for the development of *Caenorhabditis elegans* as a model organism for studying development and behavior, for which they received the Nobel Prize for Medicine in 2002. *C. elegans* is the same organism in which the RNA interference was first demonstrated. The term "RNA interference" (hereafter RNAi) was coined by Andrew Fire and Craig Mello to describe a sequence-specific gene silencing phenomenon in *C. elegans*. In this book, Chapter 1 by Alla Grishok, one of the scientists involved in the discovery of the RNAi phenomenon in the laboratory of Craig Mello, describes the historical beginnings and genetics of RNAi, transgene silencing, and the systemic nature of the phenomenon in *C. elegans*. RNAi shares a remarkable degree of similarity with the gene silencing phenomenon observed in other organisms, and this supports the notion that they were derived from an ancient, conserved pathway used to regulate gene expression.

The discovery of RNAi in *C. elegans* initiated a flurry of biochemical and genetic experiments aimed at identifying the molecular components involved in this amazing phenomenon. The biochemistry of RNAi is well described here in Chapter 2 written by Myers and Ferrell. This chapter also includes the role and utility of Dicer enzyme both *in vivo* and *in vitro* and details its application in high throughput *in vitro* dicing, split-pool screening and small hairpin RNA expression library construction. Most importantly, this chapter highlights the advantages of double-stranded siRNAs (d-siRNAs) in the silencing of multiple targets.

The cell biology of RNAi is detailed by Dudley et al. in Chapter 3 in this section, giving the details of the two-step model for RNAi, and the genes required for RNAi (including initiators, effectors, RISC components and RNA-dependent RNA polymerases). This chapter also details the roles for chromatin-modifying proteins in RNAi. The continued identification of genes required for RNAi will further our understanding of the mechanisms of RNAi and its biology in general.

The emergence of RNAi has helped to clarify another enigma of non-coding temporal RNAs or microRNAs (miRNA) that were thought to antagonize gene expression by binding to the 3'-untranslated region. Genetic studies in *C. elegans* revealed the existence of miRNA genes which are now recognized to permeate the genomes of all multi-cellular organisms. These tiny RNA regulators are being implicated in diverse biological pathways, ranging from development to neuronal differentiation to fat metabolism. Chapter 4 written by Pasquinelli summarizes the historical aspects and the birth of tiny non-coding temporal RNAs or miRNAs from the laboratories of Victor Ambros and Gary Ruvkun that include *lin-4* and *let-7* respectively. Her chapter also details the connection between miRNAs and RNAi. In contrast to siRNAs, miRNAs are derived from the processing of endogenously encoded short hairpin RNAs. However, miRNAs are dependent on dicer for processing and associate with a complex that shares components present in RISC, suggesting that a mechanistic link between the RNAi and miRNA pathways exists.

Several hundred microRNAs have been cloned to date from a wide range of organisms. The discovery of predominantly brain-specific miRNAs suggests a role for RNAi-mediated mechanisms of gene regulation in neuronal development and functioning. In Chapter 5 Krichevsky et al. detail the recent findings concerning the discovery of miRNAs in mammalian brain development. This chapter also highlights the 'miRNA oligonucleotide array' method for the display of expression profiles that uniquely correspond to developmental epochs of brain development, and for the study of cell lineages during stem cell differentiation. The discovery of extensively regulated predominantly brain-specific miRNAs, sharing processing pathways with siRNAs, suggests a role of RNAi-related mechanisms of gene regulation in neuronal development.

Section 2: Design and synthesis of siRNAs

The fundamental challenge for successfully implementing RNAi in mammalian systems rests with designing specific and potent siRNAs. The rules that govern siRNA design and target silencing are largely undefined. Some targets are easy to silence while others are more difficult, requiring multiple screens of siRNAs to identify a single potent duplex. Therefore, various siRNA selection strategies have been developed and summarized in this section. Boese et al. describe the conventional and functional siRNA design methods and specificity studies in Chapter 6.

Subrahmanyam and Lader describe online design tools and rules based on observations and mechanisms. Their Chapter 7 also deals about the TOM-Amidite chemistry based synthesis of siRNAs, as well as high-throughput and automation strategies. Chemical modification of siRNAs will lead to improved activity, duration of effect, efficacy of delivery, and minimization of off-target effects. The current advantages of using chemically synthesized siRNAs as well as potential future improvements are likely to ensure the continued dominance of chemically synthesized siRNA as the reagent of choice for high-throughput target identification.

Ding and Lawrence summarize in Chapter 8 the Web-based software called *Sfold*, used for the prediction of target accessibility based on RNA secondary structure. This method bypasses the long-standing difficulty of selecting a single structure for accessibility evaluation. Integration of computational approaches for maximizing both potency and specificity will facilitate high-throughput applications for functional genomics, drug validation, and the development of RNAi-based human therapeutics.

Sohail and Doran from Edwin Southern's research group summarize enzymatic production of siRNAs by *in vitro* transcription in Chapter 9. Use of this method not only gives huge quantities of siRNAs, but also extends to the production of single or double-stranded RNAs of defined length and sequence. This is a method of choice for the production of long RNA sequences with defined ends for which chemical synthesis of siRNA molecules may be difficult. In addition, this chapter also describes the use of the RNaseIII-based method, for which the initial optimal target site selection procedure do not need to be considered. The combination of well-defined, sophisticated design strategies coupled with reliable and flexible methods for siRNA synthesis makes siRNA-mediated RNAi among the most promising reverse genetics tool available to date.

Section 3: Vector development and *in vivo, in vitro* and *in ovo* delivery methods

The choice of the siRNA production method will depend on a number of factors – such as the amenability of the method to long-term studies or to siRNA labeling. Latham et al. in this section describe six different methods commonly used to induce gene silencing in mammalian cells (see Chapter 10), and demonstrate that it is easier to produce cell lines that stably express shRNAs using RNA polymerase II promoters than pol III. The use of pol II promoters to express siRNAs may also lead to the development of tissue-specific siRNA expression constructs in the near future. Retroviruses offer several advantages as vector delivery systems, for example (i) they have relatively small and simple genomes, making it is easy to clone DNA of interest into them; (ii) since they are enveloped viruses, it is possible to change their target cell specificity simply by changing the envelope; (iii) they can integrate stably into the genome of the host cell providing long-term expression of any gene. In Chapter 11 Mao et al. describe the moloney murine leukemia virus-based retroviral vectors for efficient delivery of siRNAs into the mammalian cells.

Lentiviral-based siRNA vectors will have the ability to generate transgenic animals carrying an siRNA cassette in order to achieve silencing of endogenous genes. Singer et al. in Chapter 12 detail the development of transgenic rodents by *in vitro* transduction of fertilized eggs at different pre-implantation stages. They have also demonstrated the delivery of siRNA for green fluorescent protein into the embryos, and measured the lentivirus integration copy number using real-time PCR. Thus, lentiviral vectors will be ideal for generating large number of cell lines, tissues, and whole animals where the expression of targeted genes can be reduced substantially to influence biological function.

The benefits of stable silencing that are offered by lentiviral delivery systems are balanced by the concern that the integration of the virus in particular genomic locations could potentially lead to the activation of gene expression and the possibility of oncogene transformation. The ability to knock down specific genes could mean that these lentiviral vectors are used to generate transgenics not only in mice and rodents but also in other species.

The success of siRNA as therapeutic agents largely depends on the development of a delivery vehicle that can efficiently deliver them *in vivo*. Because of the simplicity and potent nature of cationic liposomes (positively charged lipid bilayers, which could form a complex with negatively charged siRNA duplexes), Sioud and Sørensen have adopted these molecules for the efficient *in vivo* delivery of siRNA molecules. In addition, they have evaluated the delivery of FITC-labeled siRNA against tumor necrosis factor-α by both intravenous and intraperitoneal injection into mice and noted significant uptake by some organs such as the spleen. Another important factor to be considered in terms of delivery is the effectiveness of siRNAs and their stability and bioavailability in the targeted tissues or organs. These issues are all discussed in Sioud and Sørensen's Chapter 13.

To gain stability and bioavailability, chemical modifications of siRNAs are generally needed. Such an approach is described by Hadwiger and Vornlocher in Chapter 14. Modifications in the sugar phosphate back bones, or 2'-O-methyl nucleotides in the siRNAs, or 2'deoxy-fluro-nucleotide containing siRNAs, or complete replacement of all phosphodiesters with phosophotriesters in the siRNAs, or terminal modifications of siRNAs (by an inverted deoxy abasic residue or an aminohexyl phosphodiester) could all knock down the gene expression efficiently. Overall, these modifications should improve resistance towards cellular degradation and uptake. Chemical modifications and conjugation strategies for various biomolecules will play a pivotal role in the transition of siRNA research from the lab bench to the development of therapeutics.

The information gained using classical gene knock-out experiments is often limited, especially when the gene of interest reveals an embryonic lethal phenotype. In Chapter 15 Buchholz et al. describe a fast, powerful, and economical platform to analyze gene function during mouse development by combing RNAi technology with mouse whole embryo culture technology. In this chapter they also detail the applications of endoribonuclease-prepared siRNAs and their *in vivo* delivery into post-implantation mouse embryos using an electroporation method. One of the potential side effects of this method is cell death and tissue damage, which can be minimized by optimizing electroporation conditions.

In a chick embryo model system, Bourikas et al. (Chapter 16) combine *in ovo* electroporation with an RNAi approach to identify axonal guidance clues of commissural neurons for neuro-glial cell adhesion molecule and axonin-1. This strategy allows them to silence the genes in a temporally and spatially controlled manner for the analysis of loss-of-function phenotypes during the development of nervous system. While this *in ovo* RNAi approach has been used here for axon guidance, it can be adapted to other biological processes, such as the analysis of cell migration, cell survival, and cell proliferation.

Section 4: Gene silencing in model organisms

Genome-sequencing projects and large-scale screens provide a tremendous amount of information about the genetic makeup of an organism. However, the long list of genes expressed in specific tissues or distinct phases of an organism's life provide little information about the function of the expressed proteins. Functional gene analysis remains a time-consuming and challenging step that requires analysis of changes in gene expression in the context of a living organism. Therefore, the availability of suitable model organisms is the key to our progress in understanding the role of genes in biological processes. For many studies, vertebrate model systems such as mice or rats are required. However, the high cost and the length of time required, along with technical constraints, restrict the usefulness of the mouse as a model system for functional genomics. In order to increase the pace of functional gene analysis, new model systems for large-scale screens are required. *Xenopus* and zebrafish have been established as alternative model systems to the mouse. The chicken embryo have been used as a developmental model system in experimental biology for more than a century. In contrast to mammals, chicken embryos are easily accessible for manipulations during development, both *in ovo* and in *ex ovo* cultures.

Model organisms have been instrumental in advancing our understanding of the mechanisms underlying different forms of gene regulation. *C. elegans* has established itself as an excellent model organism for biomedical research, in part because of its complete cell lineage, transparency, short generation time, and ease of genetic manipulation. The RNAi phenomenon was first observed in *C. elegans* and the availability of its whole genome sequence helped researchers to quickly develop whole genome RNAi screens. In Chapter 17, Stephens et al. summarize the practical applications of RNAi phenomenon in *C. elegans* by detailing its ease of use and the availability of the complete genome sequence. About 75% of known human disease genes have homologues in *C. elegans*, which makes this worm an elegant model organism in which to study developmental and disease biology. This chapter also highlights the RNAi feeding library method that has made it possible to identify genes affecting chromosome morphogenesis, lifespan, nucleotide excision repair, fat storage, and transposon-gene silencing. The power of RNAi combined with the genome sequence and biology of *C. elegans* provides a potent mix for exploring gene function and gaining insights into basic biology and human disease. This model organism approach has also been extended to several multi-cellular organisms, including parasites, yeast, filamentous fungi, and plants. This section summarizes the details of RNAi phenomenon in such model organisms.

The arsenal of genetic techniques available to the researcher working on the parasite Trypanosome is limited. However, the RNAi approach has raised arrays of hope for the development of a potentially powerful and convenient genetic approach. In Chapter 18 Drew et al. summarize the accomplishments made in the development of an inducible RNAi system in these parasites, which have helped them to achieve the goal of *bona fide* RNAi-based forward genetics in *Trypanosoma brucei*. Most importantly, the RNAi approach has served as an invaluable tool to study the parasite's kinetoplast DNA replication. In addition, the development of an RNAi library using a random fragmentation of purified genomic DNA method helped them to study the metabolic regulation and tubercidin toxicity in this parasite.

Arndt has describes the use of the fission yeast *Schizosacchromyces pombe*, a single-cell eukaryotic model system, for examining different types of RNA-mediated gene silencing. His Chapter 19 highlights the results supporting the existence of a common pathway for gene suppression by antisense and dsRNA, and the role of RNA in controlling the formation of silent heterochromatin.

RNA silencing was first described and molecularly characterized in *Neurospora crassa*, where it was referred to as "quelling." Post-transcriptional gene silencing was observed in this Ascomycete filamentous fungus, but was not detected in the Zygomycete *Mucor circinelloides*. Chapter 20 by Ruiz-Vázquez in this section details the use of this Zygomycete as a model organism to study gene silencing. Low silencing frequency, bi-directional transitive silencing has been discovered in this model, however, no dicer gene(s) have been yet identified. This is the only organism, to date, in which the differential accumulation of two size classes of antisense RNAs has been described. The study of RNA silencing in these pathogenic fungi is instructive for the future development of antifungal drugs.

Cassava mosaic disease is a complex one caused by eight different strains of Gemini virus. An understanding of these viruses at the molecular level would be very useful for the development of gene silencing based strategies to control these devastating viruses. In addition, the use of viral-vectors to silence host genes has advantages over making transgenic plants, an enormously time- and energy-dependent method for studying functional genomics. The first evidence for RNA silencing was observed in petunia plants transformed with an additional copy of chalcone synthase gene, which is required for floral pigment production and which resulted in a lack of pigmentation. This phenomenon was referred as "co-suppression." Vanitharani et al. in Chapter 21 summarize post-transcriptional gene silencing in both single-stranded DNA and single-stranded RNA plant viruses. In addition this chapter details the delivery of siRNA into protoplasts and discusses the observation that siRNAs can induce gene silencing in a transitive manner in cultured plant cells. This will serve as a valuable tool to understand the mechanism of gene silencing in plants and how the plant viral suppressors could act in this process.

Section 5: Drug target validation

Target validation is one of the biggest problems for the biopharmaceutical industry, and RNAi offers the prospect of reducing this bottleneck and speeding up the drug development process. Target validation determines whether a known candidate gene is responsible for a disease and whether altering expression of the gene is likely to result in a therapeutic effect. Functional genomics and target validation are critical stages for providing pharmaceutical and biotechnology companies with new gene targets that are involved in the disease processes.

The identification and validation of targets that map chemically tractable gene families (the 'druggable' genome or 'pharmone') is now recognized as a fundamental challenge to the pharmaceutical industry. RNAi and siRNA have taken center stage as effective, biologically relevant mechanisms for gene silencing. One of the key hurdles for drug target discovery or validation directly in animal disease models has been the lack of effectiveness of *in vivo* nucleic acid delivery methods. Recently cancer researchers have utilized an siRNA approach for the validation of several drug targets.

Lu and Woodle summarize unique target identification (Efficiency-First) approach in Chapter 22, which helped them to select a pool of gene targets that were over-expressed in growth factor–treated tumors. This chapter also highlights the target validation of genes for several growth factors and their counter receptors. To identify host cell factors potentially involved in the formation of Hepatitis C Virus internal ribosome entry site-dependent translation initiation process, Ostareck-Lederer et al. used RNAi as well as affinity purification assay methods, and identified protein p40 as a candidate small molecule drug (Chapter 23).

There is an intense competitive pressure to identify and validate novel therapeutic targets from the available genomic information. Pharmaceutical companies have embraced the full range of currently available functional genomic technologies, both in small scale and as 'platforms' in order to expedite this process. Haney et al. in Chapter 24 summarize the results of drug target validation of β Catenin, PTEN, CDK2, and molecules involved in various cellular pathways using RNAi, and cell culture approaches in conjunction with real time polymerase chain reaction. From a large pharmaceutical perspective, Clarke et al. (Chapter 25) validate several members in the G-Protein Coupled Receptor family including $GABA_B$ R1 and R2 receptors. In addition, this chapter also details the use of a small hairpin RNA library to identify therapeutically relevant targets.

Section 6: Therapeutic and drug development

Selection and validation of molecular targets is of great importance for drug development in the post-genomic era. Although phenotypes of many diseases are well known, the identification of the genes responsible for these phenotypes is a major challenge in the drug development process. The RNAi technology offers an alternative method to achieve this goal in a rapid and economical way. One can use a library of several hundred to thousands of chemical compounds and identify candidate target genes through transcriptional expression profiling in a chemical genomics approach. Subsequently, the function of the target genes identified for a specific chemical compound can be evaluated in a high-throughput manner using RNAi transfections directly into micro-titer plates, seeded with mammalian cells. In addition, RNAi could facilitate drug screening and development by identifying genes that can confer drug resistance, or genes whose mutant phenotypes are ameliorated by drug treatment. This approach will not only allow the modes of action for novel compounds to be determined, but could also help in the development of a new generation of antibiotics. RNAi methods could be extended to study gene expression of insect and parasite genomes and subsequently better gene-based insecticides or infection controlling drugs could be developed.

RNAi is a powerful technology with tremendous potential for functional genomic analysis, drug discovery and therapeutic applications. siRNA technologies are already widely used as a tool for reverse genetics in mammalian cells and their potential therapeutic applications are likely to come in the future. One exciting application has been the use of siRNA-based gene silencing technologies in the inhibition of viral replication and infection. The introduction of chemically synthesized siRNAs into mammalian cells has been shown to be an effective mechanism for the suppression of viral replication that confers a degree of protection to the host cells. In this section several clinically relevant examples are discussed as well as potential challenges in using RNAi-based technology against viral targets.

Chapter 26 by Dykxhoorn in this section documents the applications of RNAi and siRNA approaches towards the inhibition of replication of human papilloma virus, poliovirus, hepatitis virus, dengue virus, herpes simplex, influenza A, and immuno-deficiency viral genomes. His chapter details the widely used approach of silencing viruses by directly targeting viral protein expression and replication in host cells, and also by blocking their host cellular counterpart receptors. Samarsky et al. caution in Chapter 27 that if RNAi is to be effective, siRNA or shRNA expression vectors must be able to reach the appropriate sub-cellular locale within the organ of interest. Therefore, they inhibited the replication of hepatitis B virus by delivering siRNA molecules directly into the liver by hydrodynamic injection methods. This chapter also details the localized delivery of vascular endothelial growth factor in a mouse model.

Although many previous anti-viral approaches have been shown to have clinical benefits, their application in some cases has been hampered by the emergence of drug resistance and the accumulation of serial mutations. This same hurdle could be faced by RNAi researchers in the development of resistance to siRNAs. Just as viruses can undergo mutations which allow them to become resistant to antiviral drugs, it is assumed that viruses could also mutate the recognition site for the siRNA, resulting in the production of viruses which are resistant to particular siRNAs. This problem could be overcome by the use of a number of potentially effective siRNAs for any target, so that viruses can be targeted with a number of siRNAs in order to decrease the chances of 'escaping' through mutation.

The acute and chronic liver disease-related death of hepatocytes is predominantly mediated by apoptosis. Inhibition of apoptosis has not been investigated so far in clinical trials involving patients with liver disease. For successful molecular therapy of patients with acute liver failure, approaches have to be established that specifically target essential mediators of apoptosis in the liver, but not in extrahepatic tissues. Since the Fas/FasL-system plays a major role in the cell death of hepatocytes, Zender et al. in Chapter 28 describe how to target and inhibit this system using a siRNA strategy. They used a hydrodynamic injection method for the successful delivery of siRNAs into mouse hepatocytes, but caution that high-volume hydrodynamic injection is not a feasible option for clinical trials. In addition, the long-term effects of siRNA-mediated gene silencing have to be investigated in animal models before any potential use in clinical trials. In Chapter 29, Scherer and Rossi describe an RNAi delivery system in post-natal animals and other RNAi applications in living animal systems.

Section 7: High-throughput genome-wide RNAi analysis

The availability of entire genome sequences has stimulated the advancement of functional genome approaches, both to accelerate the comprehensive identification of components in biological systems and to understand their functional conservation across species. Gene expression profiling methods have brought a new revolution in the classification of tumors and have helped the development of new prognostic indicators for studying various forms of cancers. However, the detailed study of individual genes and proteins remains critical in terms of understanding the basic science and for the generation of new therapeutics. Gene suppression by siRNAs is a powerful tool to analyze the function of proteins *in-vitro*, particularly, for the rational design of drugs to block tumor-relevant genes.

C. elegans and *Drosophila* are the best-studied organisms in terms of their genetics and this knowledge has been key to the identification of conserved pathways between these species with important roles. This section of the book deals with genome-wide RNAi applications using high-throughput methods, such as microarrays and cell-based assays, particularly in *C. elegans* and *Drosophila*, *in vivo* in mice and in human cell systems. RNAi has the potential to combine forward and reverse genetics to provide the direct connection between gene sequencing and biological functions in different model systems. Large scale RNAi-based approaches show great promise in providing novel windows to identify essential components which were missed in the traditional forward genetic screens.

Sugimoto summarizes in Chapter 30 how the cDNA library approach is a much more efficient at producing more phenotypes in *C. elegans* than the PCR-based genomic library method. In addition, she also describes the RNAi-by-soaking method used to characterize genes that play multiple roles in different developmental (embryonic and post-embryonic) stages of the worm, as well as the large-scale RNAi results that provide important insights into the genome organization of *C. elegans*. In conjunction with other genome-wide data, such as expression profiles and protein–protein interaction maps, the RNAi phenotype data promises to provide a rich source of information for use in deciphering *C. elegans* biology.

Functional analysis by RNAi reveals previously unknown and evolutionarily conserved gene functions. This technique has the powerful ability to comprehensively and quantitatively determine the contribution of potentially every gene involved in a particular process. Large-scale loss-of-function studies will provide important clues about the disease-relevant cellular pathways and might allow the rapid identification of functionally confirmed targets for therapeutics. To identify gene functions using cell-based RNAi screens, Boutros and Hild (Chapter 31) systematically constructed genome-wide siRNA libraries for all the genes in *Drosophila* genome (approximately 21,306) and expressed in cultured cells. These systematic screens for cell types can thus be effective in characterizing functionally related genes on a genome-wide scale and they describe this work here.

The use of knock-out techniques can be a laborious process, which sometimes produces negative results. Establishing reliable methods to reduce or knock out gene expression has been the goal of molecular biologists for a number of years. Though many techniques have been developed, the RNAi method has emerged as the most robust and efficient method for gene suppression. To understand the processes involved in the initiation and development of malignant pleural mesothelioma, and most importantly to establish possible targets for therapy, Ramos-Nino and Mossman have combined the use of microarrays and RNAi methods in Chapter 32, and demonstrated the importance of the MAPK pathway as a possible target for therapy in human mesotheliomas. RNAi strategies will be required to assess gene function in the process of carcinogenesis and to exploit modifications in their functions as possible drug targets for cancer therapy. The arrival of RNAi technology provides hope in the battle to defeat and treat this often lethal disease.

RNAi analysis on a genome-wide scale is a versatile and powerful tool that holds great promise for the functional annotation of the human genome and in accelerating drug target discovery and validation. High-throughput methods such as microarrays can complement genetic studies in traditional model organisms in order to extend our understanding and probe the mechanisms of well-defined pathways. Chapter 33 by Chang et al. from Patrick Brown's laboratory, in this section reveals high-throughput RNAi in mammalian systems, and highlights the potential advances and pitfalls of these new technologies.

Mammalian cell culture systems and *in vivo* mouse models are particularly relevant for the study of human diseases. Several efforts have been initiated, both in academia and in industry, to develop mouse- and human-specific si/shRNA libraries for the systematic analysis of various signaling cascades, metabolic circuits and other biological pathways involved in human pathologies. Several groups have generated siRNA libraries directed against the entire human genome, and Miyagashi et al. discuss the building of such libraries in more economical and robust ways (Chapter 34).

Implementation of the emerging RNAi and high-throughput siRNA technology in drug discovery should open many more avenues in experimental biology and human therapeutics. The rapid developments in RNAi-based technologies and our increased understanding of the cell biology of gene silencing will help us to overcome the barriers which currently prevent the use of RNAi as effective therapeutic modalities in humans. Whether the technology and range of applications for RNAi meet our full expectations in the future, and prove to be robust enough approaches to help deliver and validate high quality therapeutic targets, remains to be seen. Within ten years, RNAi will undoubtedly emerge as a routine molecular tool for studying problems in bio-molecular medicine and we anticipate that all these approaches will help to develop new diagnostic reagents and novel molecular interventions for a range of human diseases.

A new age of predictive medicine lies before us. The RNAi gene suppression strategies and their applications in drug discovery described in this book will serve as examples of the possibilities for integrated genomic networks, functional genomics and personalized medicine. One day RNAi based technology platform derived drugs will be developed for a range of several diseases including cancer, vascular, neurological and infectious diseases and these methods could replace the classical drug development pipeline in the healthcare industry.

Many people have contributed to making my involvement in this project possible. I thank my teachers who have shown me the wonder of science and helped me become a scientist. Many thanks to my scientific mentors Debi P. Burma and Maharani Chakravorty-Burma (associates of Severo Ochoa and Bernard L. Horecker), who taught me the basics from enzymology to the molecular cloning of genes, and nurtured me as an young scientist during my Ph.D. studies at Banarus Hindu University, India. It is a privilege and honor to thank my postdoctoral mentors Edward B. Goldberg (associate of Nobel laureate A. D. Hershey), Robert E. Hausman (associate of Aaron Moscona), and Nobel laureate Har Gobind Khorana for their excellent guidance and brilliant ideas, all of which I have adopted in my daily life.

I am thankful to all of the contributors to this book. Without their commitment this book may not have emerged for the reader. Many people have had a hand in the launch of this book. Each chapter has been passed back and forth between the authors for criticism and revision, so that each chapter represents a joint composition. I am indebted to the staff of the Cambridge University Press, and in particular to Dr. Katrina Halliday and Khristine Queja for their kindness, generosity, and efficiency throughout, which have made the enterprise of editing a pleasure as well as an education for me.

June 2004

Basic RNAi, siRNA, microRNAs and gene-silencing mechanisms

1 RNAi beginnings, overview of the pathway in *C. elegans*

Alla Grishok

Introduction

The term "RNA interference" (RNAi) was coined by Andy Fire and Craig Mello to describe a sequence-specific gene silencing phenomenon of remarkable potency. They originally identified the main features of RNAi in *C. elegans*: initiation by double-stranded RNA (dsRNA) and ability to spread systemically (Fire et al., 1998). RNAi is recognized now as an ancient mechanism utilized by metazoans for silencing of foreign genetic elements and for the precise regulation of endogenous genes during development.

Initial studies of RNAi in *C. elegans* indicated that silencing was transient, did not affect the sequence of genomic DNA, and likely targeted mature mRNA (Fire et al., 1998; Montgomery et al., 1998). These observations identified RNAi as a sequence specific post-transcriptional gene silencing (PTGS) mechanism similar to those described in plants and fungi (Montgomery and Fire, 1998). The discovery of 21–25 nt long short interfering RNAs (siRNAs) as common intermediates in PTGS (Hamilton and Baulcombe, 1999) and RNAi (Zamore et al., 2000; Parrish et al., 2000; Tijsterman et al., 2002) further confirmed shared mechanistic features of sequence-specific silencing processes.

Steps of RNAi pathway in *C. elegans*

The remarkable response to introduction of just a few molecules of dsRNA per cell in *C. elegans* (Fire et al., 1998) suggested the existence of a pathway of genes responsible for the silencing process at the level of the whole organism. Screens for RNAi deficient (*rde*) mutants in the Mello lab (Tabara et al., 1999) identified two categories of mutants. The first category, represented by *rde-1* and *rde-4*, was required only for RNAi, lacking any other obvious phenotypes. The second category of mutants, which included *rde-2*, *rde-3* and *mut-7*, was required for RNAi only in the germline. Mutants of this category also displayed temperature-dependent sterility, a high incidence of males and mobilization of normally silenced transposons in the germline, which are the features of mutator mutants (Collins et al.,

1987; Ketting et al., 1999). Ketting et al., 1999, demonstrated that most of the mutator mutants were resistant to RNAi.

One of the important features of RNAi in *C. elegans* is its heritability. Injecting hermaphrodite *C. elegans* with dsRNAs targeting maternal genes required for the embryonic development caused production of dead embryos by the injected mothers. However, the RNAi effect was not instant; it took around 12 hours for the animals to convert and produce exclusively inviable progeny. Between 4 and 12 hours post-injection, the viable progeny worms produced were likely to be affected by RNAi themselves, as they lay dead eggs upon reaching adulthood. These animals were called "carriers" of the RNAi effect. RNAi can be carried in the inherited form for more than one generation of progeny in *C. elegans* and this inheritance is transmitted by extrachromosomal factor (Grishok et al., 2000).

Inheritance of RNAi allowed genetic analysis of the pathway (Grishok et al., 2000). It was found that genes of the RNAi pathway were not equally required in the animals exposed to dsRNA compared to those affected by the heritable form of RNAi. *rde-1* and *rde-4* genes were required in the injected animals for the initiation of RNAi and were dispensable in the progeny, while *rde-2* and *mut-7* genes were not necessary in the injected animals for the initiation of heritable RNAi but required in the progeny for the response to the inherited factor. These findings put *rde-1* and *rde-4* genes upstream in the RNAi pathway and implicated them in the production of a heritable factor distinct from the originally introduced dsRNA. siRNAs are the best candidates for the inherited RNAi factor in *C. elegans* (Sharp and Zamore, 2000), although the exact nature of this heritable complex containing siRNA has not been identified yet. Genetic analysis of RNAi in *C. elegans* predicted that *rde-2* and *mut-7* act in response to heritable factor and are likely to be part of the pathway common to RNAi and other silencing processes. Indeed, *rde-2* and *mut-7*, but not *rde-1* and *rde-4*, were implicated in co-suppression in *C. elegans* germline (Ketting and Plasterk, 2000; Dernburg et al., 2000).

The *rde-1* gene encodes a protein of the evolutionarily conserved PAZ-PIWI family (Tabara et al., 1999). Members of this family have been shown to play an important role in PTGS in different organisms (Fagard et al., 2000; Hammond et al., 2001, Aravin et al., 2001, Williams and Rubin, 2002; Martinez et al., 2002). *rde-4* encodes a dsRNA binding protein (Tabara et al., 2002). Rde-1 and Rde-4 have been shown to interact *in vivo* and this interaction was important for the accumulation of dsRNA initiating RNAi in *C. elegans* (Tabara et al., 2002).

Figure 1.1. Model of RNAi pathway in *C. elegans*. Transmembrane protein SID-1 allows dsRNA to enter the cell. In the cytoplasm, dsRNA gets processed by DCR-1, existing in a complex with RDE-4, RDE-1 and DRH-1. The resulting siRNA is shown bound to RDE-1. Double stranded siRNA is unwound by a helicase, allowing RISC complex to bind single strands of siRNA. RDE-2 and MUT-7 might help in bringing together target mRNA and antisense siRNA (white) bound to RISC. RISC, guided by siRNA (white), degrades mRNA. Antisense siRNA also serves as a primer for RdRp, which uses mRNA as a template and produces more antisense siRNAs (orange). Sense siRNAs (blue) do not accumulate in *C. elegans*. (See color section.)

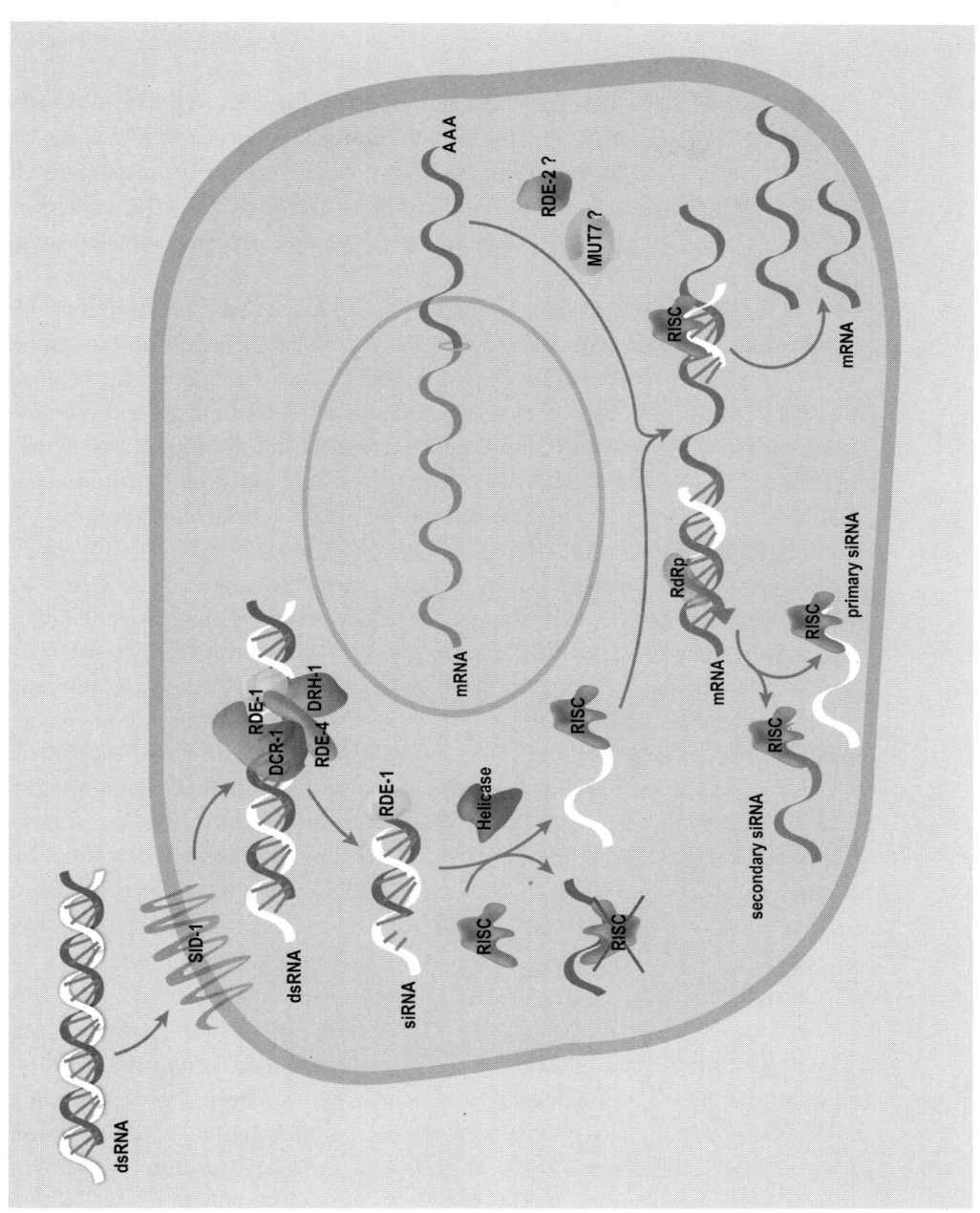

19

Production of siRNAs from dsRNA is an essential step in the RNAi process. This reaction is carried out by the multidomain RNase III Dicer (Bernstein et al., 2001; Ketting et al., 2001; Knight and Bass, 2001). Molecular analysis of the initiation step of RNAi in *C. elegans* revealed that Dicer (Dcr-1) protein exists in a complex with Rde-4, Rde-1 and DexH box dicer-related-helicase Drh-1 [(Tabara et al., 2002) (Figure 1.1)]. *rde-1* and *rde-4* were shown to be required for the accumulation of dsRNA in *C. elegans*, and approximately 60% of the dsRNA was estimated to be in the complex with Rde-4.

R2D2, *Drosophila* RNA binding protein homologous to Rde-4, has been recently shown to be required for RNAi (Liu et al., 2003). Similarly to Rde-4, it was shown to exist in a complex with Dicer. However, R2D2 was not required for the binding of the substrate dsRNA, but was important for binding siRNAs and efficient channeling of siRNAs to the downstream complex in the RNAi pathway. Stability of R2D2 *in vivo* was dependent on the presence of Dcr-2 and *vice versa* (Liu et al., 2003). Dcr-2 was shown to be the only one of two *Drosophila* Dicer proteins required for RNAi.

Although Rde-4 and R2D2 are homologous, both are required for RNAi and exist in a complex with Dicer, they might not function absolutely identically in RNAi pathway in two systems. For example, it is not very likely that the presence of Rde-4 is necessary for the stability of Dicer in *C. elegans*. *rde-4* mutants develop normally (Tabara et al., 1999), while *dcr-1* worms have multiple developmental phenotypes (Grishok et al., 2001). Since there is only one Dicer in *C. elegans*, dependence of its function on *rde-4* would suggest that *rde-4* mutants have developmental defects similar to *dcr-1*, but this is not the case. It is still remains possible that in *C. elegans* both Dcr-1 and Rde-4 contribute to the substrate dsRNA binding.

siRNA accumulation during RNAi in *C. elegans*

Production of siRNAs associated with PTGS *in vivo* was originally described in plants (Hamilton and Baulcombe, 1999). In this study, a strong signal corresponding to siRNAs was detected by Northern blot analysis. A gene encoding an RNA-dependent RNA polymerase has been implicated in PTGS in plants (Dalmay et al., 2000; Mourrain et al., 2000) and the large amount of the produced siRNAs was likely due to the amplification of either dsRNA or siRNAs.

Despite the potency of RNAi in *C. elegans*, only a very small amount of siRNAs was detected after introduction of radiolabeled dsRNA into the worm (Parrish et al., 2000). In this assay *rde-4* gene was required for the production of siRNAs, but *rde-1* was dispensable (Parrish and Fire, 2001). Later studies of siRNA production by Dicer in an *in vitro* system similar to that established in *Drosophila* embryo lysates (Zamore et al., 2000) confirmed that neither *rde-1* nor *mut-7* were required for this reaction (Tijsterman et al., 2002). While it is easy to imagine that *mut-7* function might be essential for the siRNA stability and not siRNA production (Tijsterman et al., 2002), the finding that dsRNA can be processed in the absence of *rde-1* seems to contradict genetic data on the upstream role of *rde-1* in the RNAi process and existence of Rde-1 in the Dicer complex. One possible interpretation of the Rde-1 role is that it is an important component present both in the Dicer complex and in more downstream complexes providing a link in the RNAi pathway. In this view, the role of *rde-1* seems most similar to that described for *Drosophila* R2D2 (Liu et al., 2003).

Interestingly, analysis of siRNA accumulation during RNAi in *C. elegans* by RNase protection and Northern blot analyses yielded results different from those obtained *in vitro* (Tijsterman et al., 2002; Grishok and Mello, unpublished data). siRNA accumulation was not detected in any of RNAi resistant mutants except *mut-14*, a gene encoding for a putative DEAD box RNA helicase (Tijsterman et al., 2002). Moreover, unlike other organisms, *C. elegans* accumulated only siRNAs antisense to the target mRNA (Tijsterman et al., 2002; Grishok and Mello, unpublished data) and this accumulation was strictly dependent on the presence of the target mRNA (Grishok and Mello, unpublished data). These findings indicated that the majority of the siRNAs present during RNAi in *C. elegans* are not likely to be the derivatives of initially injected dsRNA, but rather to be produced in an amplification process using target mRNA.

One gene encoding a putative RdRP in *C. elegans*, *ego-1*, was already shown to be required for the RNAi in the germline (Smardon et al., 2000). Furthermore, Sijen et al., 2001, described the process of transitive RNAi and showed that another RdRP gene, *rrf-1*, was required for the RNAi in somatic tissues. In the transitive RNAi assay, the fusion target mRNA containing portions of two endogenous genes was used and dsRNA specific to only the 3' gene of this fusion mRNA was introduced for the initiation of RNAi. Sijen et al. detected the secondary siRNAs (Figure 1.1) corresponding to the 5' gene of the fusion mRNA closest to the fusion point but not present in the initial trigger. Those secondary siRNAs were able to induce silencing of the corresponding endogenous gene. These experiments provide insight into the mechanism of target-dependent amplification of siRNAs in *C. elegans*, although it is hard to imagine why production of only one strand of siRNA results from the amplification.

Biochemical analysis of RNAi in *Drosophila* tissue culture and *in vitro* systems established the existence of the RNA induced silencing complex (RISC) responsible for the degradation of target mRNA in the final step of RNAi [(Hammond et al., 2000; Nykanen et al., 2001; Martinez et al., 2002) (Figure 1.1)]. This complex is believed to contain both endonuclease (Zamore at al., 2000) and exonuclease (Hammond et al., 2000) activities. Arguments for the possible association of a particular factor with the RISC complex are usually made when the factor is acting downstream of the production of siRNAs, e.g. when siRNAs can be detected in the mutant background (Catalanotto, 2002). Since *mut-14* is the only mutant where siRNA accumulation is described (Tijsterman et al., 2002), it is a likely *C. elegans* candidate for a RISC component. PAZ-PIWI protein Ago2 was the first *Drosophila* RISC component identified biochemically (Hammond et al., 2001) followed by VIG, FXR (Caudy et al., 2002) and, most recently, micrococcal nuclease homolog

Tudor-SN (Caudy et al., 2003). *C. elegans* homologs of VIG and Tudor-SN were shown to be components of RISC in worm extracts (Caudy et al., 2003). Thus, exonuclease component of the RISC has been identified but endonuclease remains elusive.

Recent analysis of the RISC activity in mammalian tissue culture determined that active RISC complex contains only one strand of the siRNAs (Martinez et al., 2002). Since only antisense siRNAs have been shown to accumulate in *C. elegans* it is possible that after target dependent siRNA amplification and unwinding, only catalytic complexes active in the mRNA degradation are retained. Thus, both amplification and catalytic mechanisms might serve the remarkable potency of the RNAi in *C. elegans*.

Systemic nature of RNAi in *C. elegans*

One of the features of RNAi in *C. elegans* is its systemic nature. Originally, dsRNA was introduced in the gonads of the hermaphrodites in order to get progeny affected by RNAi, similar to the production of the transgenic animals (Mello et al., 1991). It was accidentally found that injections into any part of the animal (body cavity or intestine) would lead to the spread of the effect and to the production of the affected progeny (Fire at al., 1998). Timmons and Fire described in 1998 a system for the RNAi induction by bacteria expressing dsRNA. This advance made possible genome wide application of the technology and creation of a database of RNAi phenotypes for virtually any gene in *C. elegans* (Gonczy et al., 2000; Kamath et al., 2003). Soaking of the worms in the dsRNA solution also proved effective in the initiation of RNAi (Tabara et al., 1999).

Screens aimed to identify genes specifically required for the systemic effect of RNAi have been conducted in Craig Hunter's lab. The first systemic RNA interference deficient *(sid)* mutant, *sid-1*, has been described (Winston at al., 2002). *sid-1* is required for the import of dsRNA cell-autonomously. It encodes a transmembrane protein containing 11 transmembrane domains that was shown to localize to cell periphery (Figure 1.1). It functions possibly as a channel or a receptor for RNA. Interestingly, there are human and mouse homologs of *sid-1* suggesting the existence of nucleic acids transport in higher organisms. Similarly to *rde-1* and *rde-4*, *sid-1* mutants have no obvious developmental phenotypes.

What is transported during RNAi in *C. elegans*, long dsRNA or siRNAs? There are several arguments against siRNAs being the systemic component of RNAi. First, injections of long dsRNA are much more efficient than short siRNAs (Parrish et al., 2000; Tijsterman et al., 2002). Also, *rde-4* does not prevent systemic transport of RNAi to the next generation (Tabara et al., 1999), but it is required for the production of siRNAs (Parrish et al., 2000), indicating that siRNAs are not the transported agent. Finally, introduction of 25-nt long antisense RNAs into both gonads of *C. elegans* resulted in 100% interference effect when the target gene was expressed in the germline. Only 50% effect was achieved when antisense RNA was introduced into one gonad arm, and it did not spread to the other one (Tijsterman et al., 2002). Short antisense RNAs described above are likely to act in the target-dependent amplification step of RNAi. Findings described above indicate that RNAi amplification is contained within the tissues expressing the target mRNA and systemic RNAi occurs before the amplification step.

RNAi and transgene silencing in *C. elegans*

The phenomenon of transgene silencing was first observed in plants. In attempts to create flowers producing extra pigment by introducing extra copies of the corresponding gene researchers have found that all the extra copies along with the endogenous gene were silenced (Napoli et al., 1990; Van der Krol et al., 1990). In *C. elegans*, transgenes are usually expressed in the somatic tissues, but are silenced in the germline (Kelly and Fire, 1997). Co-suppression, silencing of the endogenous gene along with the transgene, is also observed in the germline of *C. elegans* (Ketting and Plasterk, 2000; Dernburg et al., 2000). Kelly and Fire demonstrated that the repetitive nature of the DNA arrays introduced into the worms contribute to the efficient silencing and that it is possible to obtain transgenic animals expressing germline transgenes when the transgenes are introduced as "complex" arrays with genomic DNA added to the injected mixture (Kelly and Fire, 1998). Production of transgenic animals by particle bombardment rather than traditional injection methods allowed establishment of lines containing a single copy of the transgene with continuous expression (Praitis et al., 2001).

It is conceivable that dsRNA is produced from transgene repeats in opposite orientations and transgene silencing in *C. elegans*, like in other species, is mediated by siRNA containing complexes. Indeed, RNAi pathway genes *mut-7* and *rde-2* were implicated in transgene silencing at higher temperatures (Tabara et al., 1999). Surprisingly, *rde-1* and *rde-4* genes absolutely required for the initiation of RNAi were dispensable for transgene silencing (Tabara et al., 1999) or co-suppression in the germline (Ketting and Plasterk, 2000; Dernberg et al., 2000).

Recent report by Knight and Bass, 2002, describes transgene silencing in the somatic tissues of *C. elegans* in the absence of activities of two adenosine deaminase genes *adr-1* and *adr-2*. This silencing was initiated by the double-stranded RNA produced from the transgenes and required *rde-1* activity. Apparently, dsRNA produced from the somatic transgenes presented a target both for deamination and RNAi. In wild-type worms deamination prevailed and did not prevent the expression of transgenes, while in *adr-1*, *adr-2* mutant background transgenes were silenced by the RNAi pathway. Why there is a difference in genetic requirements for transgene silencing in the germline and in the somatic tissues is not yet clear.

Silencing of repetitive elements in *S. pombe* have been recently shown to be initiated by the RNAi machinery and followed by the establishment of the heterochromatin (Volpe et al., 2002; Hall et al., 2002; Reinhart and Bartel, 2002). Formation of the heterochromatin included methylation on lysine 9 of histone H3 by Clr4, methyltransferase containing SET domain, and recruitment of the Swi6, homolog of mammalian heterochromatin protein one (HP1). Parallel to this scenario can certainly be suggested for the silencing of transgenes in *C. elegans*. As was mentioned above, RNAi pathway genes were implicated in the initiation of the silencing process. It has been shown earlier that maintenance of the silenced repetitive transgenes in the germline of *C. elegans* required several mathernal effect sterile *(mes)* genes (Kelly and Fire, 1998), some of those genes contained SET domains and were homologous to Polycomb group genes in *Drosophila* (Holdeman et al., 1998; Korf et al., 1998). Methylation on lysine 9 of histone H3 has been detected on the silenced transgenic extrachromosomal arrays (Kelly et al., 2002). Finally, the *C. elegans* homolog of HP1, *hpl-2*, was shown to be required for transgene silencing in the germline and fertility (Couteau et al., 2002).

Possibilities for RNAi in regulation of endogenous genes expression

Mutations in most of the genes implicated in RNAi in *C. elegans* have no obvious developmental phenotypes. This can partially be explained by the design of the original screens and it can be expected that newly found genes implicated in RNAi would also play role in development.

Dicer *(dcr-1)*, critical for the RNAi function, also plays a role in the processing of microRNA precursors (Grishok et al., 2001; Ketting et al., 2001; Hütvagner et al., 2001). Thus, multiple developmental phenotypes of the *dcr-1* mutant (Grishok et al., 2001; Knight and Bass, 2001; Ketting et al., 2001) likely represent a cumulative effect of the downregulation of multiple developmental pathways where miRNAs play a role. Nevertheless, there are indications of the involvement of RNAi in development apart from the miRNA function. *ego-1* gene encoding RdRp plays a role in germline development, likely at the multiple steps (Smardon et al., 2000). Mutator genes are important for fertility (Ketting et al., 1999; Tabara et al., 1999). Also, mutations in the *rrf-3* gene encoding RdRp homolog cause enhanced susceptibility for RNAi (Simmer et al., 2002). This finding indicates the possibility of the competition between RdRp homologs for the RNAi machinery and the existence of the natural regulatory pathway utilizing *rrf-3* and natural siRNAs. Cloning of numerous putative endogenous siRNAs has been reported recently (Ambros et al., 2003). Most of the cloned siRNAs were perfectly complementary to the coding regions of the genes and were antisense to the target; siRNAs accumulating during RNAi in *C. elegans* had antisense polarity as well (Tijsterman et al., 2002; Grishok and Mello, unpublished results). Interestingly, a large number of identified siRNAs were complimentary to the genes encoding transcription factors, similar to the plant miRNAs with identified or predicted function (Llave et al., 2002; Rhoades et al., 2002). The significance of the cloned siRNAs and possible links to the phenotypes displayed by RNAi pathway mutants remain to be confirmed.

One of the most attractive possible natural roles for the RNAi pathway is the initiation of the heterochromatin formation at the loci regulated during development. RNAi pathway is closely connected to the lysine 9 histone H3 methylation in *S. pombe* (Volpe et al., 2002; Hall et al., 2002). Kelly and colleagues demonstrated that a number of germline specific genes on X chromosome of *C. elegans* are regulated by heterochromatin formation and associated with lysine 9 histone H3 methylation (Kelly et al., 2002). It will be interesting to connect natural siRNAs to this regulation. Also, a homolog of heterochromatin protein 1 in *C. elegans* was implicated both in fertility and vulval development (Couteau et al., 2002). As many protein complexes implicated in the negative regulation of vulval development in *C. elegans* act by chromatin remodeling (Lu and Horvitz, 1998; Ceol and Horvitz, 2001), participation of siRNAs in the initiation of such regulation is a tantalizing possibility.

REFERENCES

Ambros, V., Lee, R. C., Lavanway, A., Williams, P. T. and Jewell, D. (2003). MicroRNAs and other tiny endogenous RNAs in *C. elegans*. *Current Biology*, **13**, 807–818.

Aravin, A. A., Naumova, N. M., Tulin, A. V., Vagin, V. V., Rozovsky, Y. M. and Gvozdev, V. A. (2001). Double-stranded RNA-mediated silencing of genomic tandem repeats and transposable elements in the D. melanogaster germline. *Current Biology*, **11**, 1017–1027.

Bernstein, E., Caudy, A. A., Hammond, S. M. and Hannon, G. J. (2001). Role for a bidentate ribonuclease in the initiation step of RNA interference. *Nature*, **409**, 363–366.

Catalanotto, C., Azzalin, G., Macino, G. and Cogoni, C. (2002). Involvement of small RNAs and role of the qde genes in the gene silencing pathway in Neurospora. *Genes & Development*, **16**, 790–795.

Caudy, A. A., Ketting, R. F., Hammond, S. M., Denli, A. M., Bathoorn, A. M., Tops, B. B., Silva, J. M., Myers, M. M., Hannon, G. J. and Plasterk, R. H. (2003). A micrococcal nuclease homologue in RNAi effector complexes. *Nature*, **425**, 411–414.

Caudy, A. A., Myers, M., Hannon, G. J. and Hammond, S. M. (2002). Fragile X-related protein and VIG associate with the RNA interference machinery. *Genes & Development*, **16**, 2491–2496.

Ceol, C. J. and Horvitz, H. R. (2001). dpl-1 DP and efl-1 E2F act with lin-35 Rb to antagonize Ras signaling in *C. elegans* vulval development. *Molecular Cell*, **7**, 461–473.

Collins, J., Saari, B. and Anderson, P. (1987). Activation of a transposable element in the germ line but not the soma of *Caenorhabditis elegans*. *Nature*, **328**, 726–728.

Couteau, F., Guerry, F., Muller, F. and Palladino, F. (2002). A heterochromatin protein 1 homologue in *Caenorhabditis elegans* acts in germline and vulval development. *European Molecular Biology Organization Reports*, **3**, 235–241.

Dalmay, T., Hamilton, A., Rudd, S., Angell, S. and Baulcombe, D. C. (2000). An RNA-dependent RNA polymerase gene in Arabidopsis is required for posttranscriptional gene silencing mediated by a transgene but not by a virus. *Cell*, **101**, 543–553.

Dernburg, A. F., Zalevsky, J., Colaiacovo, M. P. and Villeneuve, A. M. (2000). Transgene-mediated cosuppression in the *C. elegans* germ line. *Genes & Development*, **14**, 1578–1583.

Fagard, M., Boutet, S., Morel, J. B., Bellini, C. and Vaucheret, H. (2000). AGO1, QDE-2, and RDE-1 are related proteins required for post-transcriptional gene silencing in plants, quelling in fungi, and RNA interference in animals. *Proceedings of the National Academy of Sciences USA*, **97**, 11650–11654.

Fire, A., Xu, S., Montgomery, M. K., Kostas, S. A., Driver, S. E. and Mello, C. C. (1998). Potent and specific genetic interference by double-stranded RNA in *Caenorhabditis elegans*. *Nature*, **391**, 806–811.

Gonczy, P., Echeverri, C., Oegema, K., Coulson, A., Jones, S. J., Copley, R. R., Duperon, J., Oegema, J., Brehm, M., Cassin, E., Hannak, E., Kirkham, M., Pichler, S., Flohrs, K., Goessen, A., Leidel, S., Alleaume, A. M., Martin, C., Ozlu, N., Bork, P. and Hyman, A. A. (2000). Functional genomic analysis of cell division in *C. elegans* using RNAi of genes on chromosome III. *Nature*, **408**, 331–336.

Grishok, A., Tabara, H. and Mello, C. C. (2000). Genetic requirements for inheritance of RNAi in *C. elegans*. *Science*, **287**, 2494–2497.

Grishok, A., Pasquinelli, A. E., Conte, D., Li, N., Parrish, S., Ha, I., Baillie, D. L., Fire, A., Ruvkun, G. and Mello, C. C. (2001). Genes and mechanisms related to RNA interference regulate expression of the small temporal RNAs that control *C. elegans* developmental timing. *Cell*, **106**, 23–34.

Hall, I. M., Shankaranarayana, G. D., Noma, K., Ayoub, N., Cohen, A. and Grewal, S. I. (2002). Establishment and maintenance of a heterochromatin domain. *Science*, **297**, 2232–2237.

Hamilton, A. J. and Baulcombe, D. C. (1999). A species of small antisense RNA in posttranscriptional gene silencing in plants. *Science*, **286**, 950–952.

Hammond, S. M., Bernstein, E., Beach, D. and Hannon, G. J. (2000). An RNA-directed nuclease mediates post-transcriptional gene silencing in *Drosophila* cells. *Nature*, **404**, 293–296.

Hammond, S. M., Boettcher, S., Caudy, A. A., Kobayashi, R. and Hannon, G. J. (2001). Argonaute2, a link between genetic and biochemical analyses of RNAi. *Science*, **293**, 1146–1150.

Holdeman, R., Nehrt, S. and Strome, S. (1998). MES-2, a maternal protein essential for viability of the germline in *Caenorhabditis elegans*, is homologous to a *Drosophila* Polycomb group protein. *Development*, **125**, 2457–2467.

Hütvagner, G., McLachlan, J., Pasquinelli, A. E., Balint, E., Tuschl, T. and Zamore, P. D. (2001). A cellular function for the RNA-interference enzyme Dicer in the maturation of the let-7 small temporal RNA. *Science*, **293**, 834–838.

Kamath, R. S., Fraser, A. G., Dong, Y., Poulin, G., Durbin, R., Gotta, M., Kanapin, A., Le Bot, N., Moreno, S., Sohrmann, M., Welchman, D. P., Zipperlen, P. and Ahringer, J. (2003). Systematic functional analysis of the *Caenorhabditis elegans* genome using RNAi. *Nature*, **421**, 231–237.

Kelly, W. G. and Fire, A. (1998). Chromatin silencing and the maintenance of a functional germline in *Caenorhabditis elegans*. *Development*, **125**, 2451–2456.

Kelly, W. G., Schaner, C. E., Dernburg, A. F., Lee, M. H., Kim, S. K., Villeneuve, A. M. and Reinke, V. (2002). X-chromosome silencing in the germline of *C. elegans*. *Development*, **129**, 479–492.

Kelly, W. G., Xu, S., Montgomery, M. K. and Fire, A. (1997). Distinct requirements for somatic and germline expression of a generally expressed Caernorhabditis elegans gene. *Genetics*, **146**, 227–238.

Ketting, R. F., Haverkamp, T. H., van Luenen, H. G. and Plasterk, R. H. (1999). Mut-7 of *C. elegans*, required for transposon silencing and RNA interference, is a homolog of Werner syndrome helicase and RNaseD. *Cell*, **99**, 133–141.

Ketting, R. F. and Plasterk, R. H. (2000). A genetic link between co-suppression and RNA interference in *C. elegans*. *Nature*, **404**, 296–298.

Ketting, R. F., Fischer, S. E., Bernstein, E., Sijen, T., Hannon, G. J. and Plasterk, R. H. (2001). Dicer functions in RNA interference and in synthesis of small RNA involved in developmental timing in *C. elegans*. *Genes & Development*, **15**, 2654–2659.

Knight, S. W. and Bass, B. L. (2001). A role for the RNase III enzyme DCR-1 in RNA interference and germ line development in *Caenorhabditis elegans*. *Science*, **293**, 2269–2271.

Korf, I., Fan, Y. and Strome, S. (1998). The Polycomb group in *Caenorhabditis elegans* and maternal control of germline development. *Development*, **125**, 2469–2478.

Liu, Q., Rand, T. A., Kalidas, S., Du, F., Kim, H. E., Smith, D. P. and Wang, X. (2003). R2D2, a bridge between the initiation and effector steps of the *Drosophila* RNAi pathway. *Science*, **301**, 1921–1925.

Llave, C., Xie, Z., Kasschau, K. D. and Carrington, J. C. (2002). Cleavage of Scarecrow-like mRNA targets directed by a class of Arabidopsis miRNA. *Science*, **297**, 2053–2056.

Lu, X. and Horvitz, H. R. (1998). lin-35 and lin-53, two genes that antagonize a *C. elegans* Ras pathway, encode proteins similar to Rb and its binding protein RbAp48. *Cell*, **95**, 981–991.

Martinez, J., Patkaniowska, A., Urlaub, H., Luhrmann, R. and Tuschl, T. (2002). Single-stranded antisense siRNAs guide target RNA cleavage in RNAi. *Cell*, **110**, 563–574.

Mello, C. C., Kramer, J. M., Stinchcomb, D. and Ambros, V. (1991). Efficient gene transfer in *C. elegans*: Extra chromosomal maintenance and integration of transforming sequences. *European Molecular Biology Organization Journal*, **10**, 3959–3970.

Montgomery, M. K. and Fire, A. (1998). Double-stranded RNA as a mediator in sequence-specific genetic silencing and co-suppression. *Trends in Genetics*, **14**, 255–258.

Montgomery, M. K., Xu, S. and Fire, A. (1998). RNA as a target of double-stranded RNA-mediated genetic interference in *Caenorhabditis elegans*. *Proceedings of the National Academy of Sciences USA*, **95**, 15502–15507.

Mourrain, P., Beclin, C., Elmayan, T., Feuerbach, F., Godon, C., Morel, J. B., Jouette, D., Lacombe, A. M., Nikic, S., Picault, N., Remoue, K., Sanial, M., Vo, T. A. and Vaucheret, H. (2000). Arabidopsis SGS2 and SGS3 genes are required for posttranscriptional gene silencing and natural virus resistance. *Cell*, **101**, 533–542.

Napoli, C., Lemieux, C. and Jorgensen, R. (1990). Introduction of a chalcone synthase gene into petunia results in reversible co-suppression of homologous genes in trans. *Plant Cell*, **2**, 279–289.

Nykanen, A., Haley, B. and Zamore, P. D. (2001). ATP requirements and small interfering RNA structure in the RNA interference pathway. *Cell*, **107**, 309–321.

Parrish, S. and Fire, A. (2001). Distinct roles for RDE-1 and RDE-4 during RNA interference in *Caenorhabditis elegans*. *RNA*, **7**, 1397–13402.

Parrish, S., Fleenor, J., Xu, S., Mello, C. and Fire, A. (2000). Functional anatomy of a dsRNA trigger: differential requirement for the two trigger strands in RNA interference. *Molecular Cell*, **6**, 1077–1087.

Praitis, V., Casey, E., Collar, D. and Austin, J. (2001). Creation of low-copy integrated trans-genic lines in *Caenorhabditis elegans*. *Genetics*, **157**, 1217–1226.

Reinhart, B. J. and Bartel, D. P. (2002). Small RNAs correspond to centromere heterochromatic repeats. *Science*, **297**, 1831.

Rhoades, M. W., Reinhart, B. J., Lim, L. P., Burge, C. B., Bartel, B. and Bartel, D. P. (2002). Prediction of plant microRNA targets. *Cell*, **110**, 513–520.

Sharp, P. A. and Zamore, P. D. (2000). Molecular biology of RNA interference. *Science*, **287**, 2431–2433.

Sijen, T., Fleenor, J., Simmer, F., Thijssen, K. L., Parrish, S., Timmons, L., Plasterk, R. H. and Fire, A. (2001). On the role of RNA amplification in dsRNA-triggered gene silencing. *Cell*, **107**, 465–476.

Simmer, F., Tijsterman, M., Parrish, S., Koushika, S. P., Nonet, M. L., Fire, A., Ahringer, J. and Plasterk, R. H. (2002). Loss of the putative RNA-directed RNA polymerase RRF-3 makes *C. elegans* hypersensitive to RNAi. *Current Biology*, **12**, 1317–1319.

Smardon, A., Spoerke, J. M., Stacey, S. C., Klein, M. E., Mackin, N. and Maine, E. M. (2000). EGO-1 is related to RNA-directed RNA polymerase and functions in germ-line development and RNA interference in *C. elegans*. *Current Biology*, **10**, 169–178.

Tabara, H., Sarkissian, M., Kelly, W. G., Fleenor, J., Grishok, A., Timmons, L., Fire, A. and Mello, C. C. (1999). The rde-1 gene, RNA interference, and transposon silencing in *C. elegans*. *Cell*, **99**, 123–132.

Tabara, H., Yigit, E., Siomi, H. and Mello, C. C. (2002). The dsRNA binding protein RDE-4 interacts with RDE-1, DCR-1, and a DExH-box helicase to direct RNAi in *C. elegans*. *Cell*, **109**, 861–871.

Tijsterman, M., Ketting, R. F., Okihara, K. L., Sijen, T. and Plasterk, R. H. (2002). RNA helicase MUT-14-dependent gene silencing triggered in *C. elegans* by short antisense RNAs. *Science*, **295**, 694–697.

Timmons, L. and Fire, A. (1998). Specific interference by ingested dsRNA. *Nature*, **395**, 854.

Van der Krol, A. R., Mur, L. A., Beld, M., Mol, J. N. and Stuitje, A. R. (1990). Flavoid genes in Petunia: Addition of a limited number of gene copies may lead to a suppression of gene expression. *Plant Cell*, **2**, 291–299.

Volpe, T. A., Kidner, C., Hall, I. M., Teng, G., Grewal, S. I. and Martienssen, R. A. (2002). Regulation of heterochromatic silencing and histone H3 lysine-9 methylation by RNAi. *Science*, **297**, 1833–1837.

Williams, R. W. and Rubin, G. M. (2002). ARGONAUTE1 is required for efficient RNA interference in *Drosophila* embryos. *Proceedings of the National Academy of Sciences USA*, **99**, 6889–6894.

Winston, W. M., Molodowitch, C. and Hunter, C. P. (2002). Systemic RNAi in *C. elegans* requires the putative transmembrane protein SID-1. *Science*, **295**, 2456–2459.

Zamore, P. D., Tuschl, T., Sharp, P. A. and Bartel, D. P. (2000). RNAi: double-stranded RNA directs the ATP-dependent cleavage of mRNA at 21 to 23 nucleotide intervals. *Cell*, **101**, 25–33.

2 Dicer in RNAi: Its roles *in vivo* and utility *in vitro*

Jason W. Myers and James E. Ferrell, Jr.

Introduction

In its infancy, RNA interference (RNAi) was simply an intriguing curiosity; now it has leapt to the forefront of science (Couzin, 2002). The growth spurt of RNAi-related studies has been partly pragmatic – RNAi clearly had the potential to evolve into an incredibly powerful technology for manipulating gene expression in diverse cell types, and to a great extent this potential has been realized. Moreover, as a physiological phenomenon, RNAi represents a fascinating and previously unrecognized level of cellular regulation. RNAi is essential for silencing of heterochromatin (Reinhart and Bartel, 2002; Volpe et al., 2002; Schramke and Allshire, 2003; Volpe et al., 2003) and gene expression via methylation (Grant, 1999; Jones et al., 1999), for antiviral defense, at least in plants (Baulcombe, 1999; Grant, 1999; Ratcliff et al., 1999), for controlling the expression of transposable elements and repetitive sequences (Ketting et al., 1999; Tabara et al., 1999; Ambros et al., 2003a&b; Sijen and Plasterk, 2003), and for proper embryonic development (Grishok et al., 2001; Hütvagner et al., 2001; Ketting et al., 2001; Knight and Bass, 2001; Bernstein et al., 2003; Houbaviy et al., 2003).

This chapter focuses on Dicer, an RNase III family enzyme essential for sequence-specific gene suppression. We begin with an overview of the discovery of Dicer and its implication in RNAi, followed by a discussion of its hypothesized roles *in vivo*. In addition, we discuss the use of recombinant Dicer *in vitro* as a way of generating pools of small interfering RNAs (siRNAs) for gene silencing, and the advantages and disadvantages of this approach. For excellent reviews of other aspects of RNAi, see (Fire, 1999; Hunter, 2000; Carthew, 2001; Hammond et al., 2001; Sharp, 2001; Grishok and Mello, 2002; Hannon, 2002; Hütvagner and Zamore, 2002a&b; Pasquinelli and Ruvkun, 2002; Tijsterman et al., 2002; Zamore, 2002).

RNAi, cosuppression, and quelling

RNA-induced gene suppression in the nematode *Caenorhabditis elegans* was first accomplished with antisense RNA (Fire et al., 1991), but it was also noticed that sense RNA was, curiously, just as effective (Fire et al., 1991; Guo and Kemphues, 1995). This result raised the possibility that each RNA preparation contained some contaminant responsible for inhibiting gene expression, the most likely candidate being dsRNA, which may be present in single-stranded RNA preparations. Fire, Mello and colleagues hypothesized that a synergy between sense and antisense RNAs might result in more potent gene silencing, and to their surprise they were correct: the dsRNA was at least two orders of magnitude more effective than either strand alone (Fire et al., 1998). dsRNA-triggered gene silencing was evident in both injected worms and their progeny. Moreover, only a few molecules of dsRNA molecules were required to attain gene suppression and, gene after gene, RNAi seemed to recapitulate the null phenotype. It was also established that dsRNA instigated degradation of the complementary mRNA (Fire et al., 1998).

This seminal discovery was important for several reasons. First, RNAi added to the repertoire of tools available for studying gene function in *C. elegans* and other invertebrates. It is simple to generate and inject dsRNA (Fire et al., 1998) or to feed worms bacteria expressing dsRNA (Timmons and Fire, 1998). Consequently, numerous systematic, large-scale functional genomic analyses have now been carried out in worms (Gonczy et al., 2000; Ashrafi et al., 2003; Kamath and Ahringer, 2003; Kamath et al., 2003; Lee et al., 2003; Pothof et al., 2003; Simmer et al., 2003; Vastenhouw et al., 2003). Likewise, Carthew and colleagues took advantage of RNAi to demonstrate the necessity of two proteins for proper embryonic development in *Drosophila melanogaster* embryos (Kennerdell and Carthew, 1998), and since then RNAi has been frequently used for reverse genetic analysis in *Drosophila* (Kiger et al., 2003; Lum et al., 2003).

Second, these studies paved the way for the development of RNAi methods for vertebrate cells, where, in contrast to *Drosophila* and *C. elegans*, the existing tools for loss of function experiments had been minimal. Over the past few years RNAi has greatly facilitated somatic cell genetics. A large number of genes have now been successfully silenced in a variety of mammalian model systems (Paddison and Hannon, 2002; Dykxhoorn et al., 2003).

Third, RNAi yielded insights into other homology-dependent gene-silencing phenomena, including transcriptional gene silencing (TGS), post-transcriptional gene silencing (PTGS) and cosuppression in plants (Baulcombe, 1999), and quelling in fungi (Cogoni et al., 1996). Cosuppression, defined as specific gene silencing induced by transgene expression, was first noticed in plants when attempts to overexpress a transgene caused both the endogenous gene and transgene to be suppressed (Jorgensen, 1990). Since the discovery of RNAi, all homology-dependent gene-silencing phenomena have become better understood, and the link between these processes suggests that the core machinery of RNAi has been well conserved throughout evolution.

Fourth, the emergence of RNAi has helped clarify another enigma. Nearly a decade before the discovery of RNAi, small non-coding RNAs were discovered as temporal regulators of embryonic development in *C. elegans* (Arasu et al., 1991; Lee et al., 1993). These RNAs have since been referred to as small temporal RNAs (stRNAs) or micro RNAs [(miRNAs) (Ambros et al., 2003)], and were thought to antagonize gene expression by binding to the 3′-untranslated region [(3′-UTR)

(Arasu et al., 1991; Lee et al., 1993)]. However, most of the mechanistic details of miRNA action remained unclear until the discovery of RNAi. As described below, miRNAs are similar in many ways to siRNAs, and it has been confirmed that the RNAi machinery is essential for the biogenesis and regulation of these miRNAs (Grishok et al., 2001; Hütvagner et al., 2001). Before the discovery of RNAi only a few miRNAs had been identified; now a large number of these and other small regulatory RNAs have been identified in a wide variety of organisms. The discovery of RNAi has therefore aided in the understanding of an important mechanism of gene regulation that had gone virtually undetected for quite some time.

Mechanisms of RNAi

The discovery of RNAi in *C. elegans* initiated a flurry of biochemical and genetic experiments aimed at identifying the molecular components involved in this amazing phenomenon. Because biochemical experiments are difficult in worms and plants, investigators turned to *Drosophila* embryo extracts and cultured S2 cells. dsRNA, but not sense or antisense RNA, triggered specific gene silencing in the cell free extract; firefly and Renilla luciferase mRNAs were silenced only by their cognate dsRNAs (Tuschl et al., 1999), similar to RNAi in worms. Silencing was accompanied by specific mRNA degradation (Tuschl et al., 1999), again as had been found in *C. elegans*. Silencing was potentiated by preincubation of dsRNA in the lysate, implying that the dsRNA must be modified in order to trigger silencing (Tuschl et al., 1999). Irrelevant dsRNA, unable to target luciferase mRNA, acted as a competitive inhibitor (Tuschl et al., 1999), indicating that silencing was saturable and mediated by a defined amount and set of proteins or enzymes.

The specificity of mRNA degradation and gene suppression suggests that complementary base pairing between the antisense strand of the dsRNA and the (sense) mRNA is involved, since nucleic acid interactions are very accurate. What remained unclear was how the antisense strand was generated. A clue came from investigating PTGS in plants when Baulcombe discovered that antisense RNA strands about 25 nucleotides in length were detected only in plants undergoing PTGS (Hamilton and Baulcombe, 1999). It was noticed that the corresponding 25-nt sense strand was also present, suggesting that a small dsRNA was generated during nucleotide-mediated gene silencing and that the 25-nt antisense RNA acted as a specificity determinant. Subsequently the fate of dsRNA in *Drosophila* extract was determined; radiolabeled dsRNA was cleaved at 21–23 nucleotide intervals in an ATP-dependent fashion when mixed with *Drosophila* embryo extract and mRNA was cleaved at similar intervals, suggesting the small RNAs were indeed guiding mRNA cleavage (Zamore et al., 2000). Erickson and colleagues also found that dsRNA was cleaved into 21–23 nucleotide fragments in *Drosophila* embryos (Yang et al., 2000).

Hannon and colleagues also carried out biochemical studies of the mechanism of RNAi in *Drosophila*. Cell-free extract from cultured S2 cells transfected with dsRNA was shown to contain a nuclease activity, which they termed the RNA-induced silencing complex (RISC), that degraded the mRNA corresponding to the transfected dsRNA, but did not degrade other mRNAs (Hammond et al., 2000).

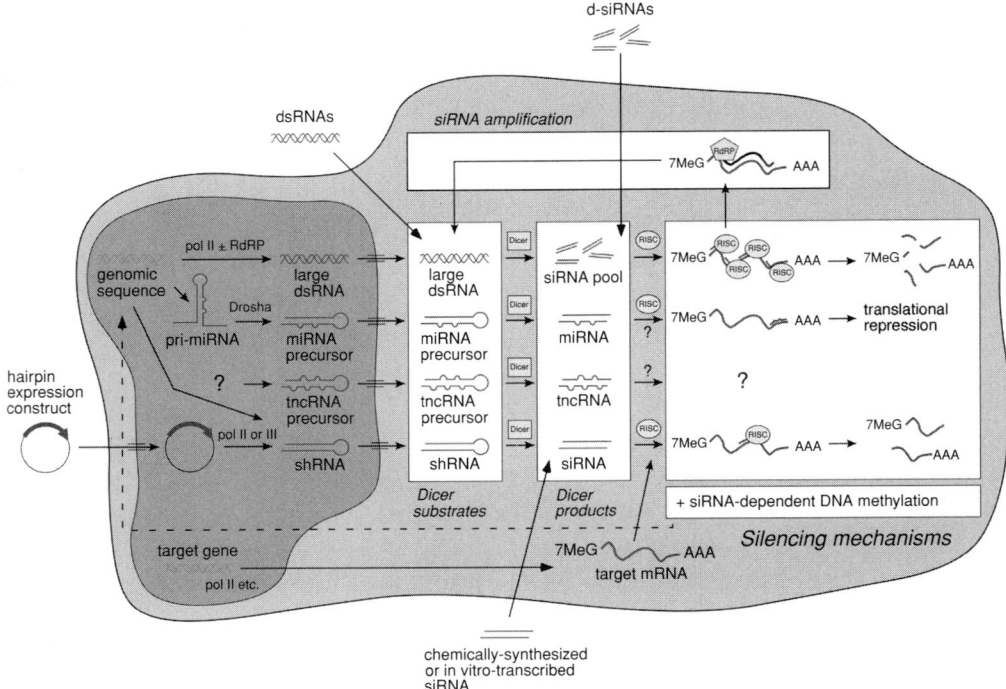

Figure 2.1. Sequence dependent regulation of gene expression. Expression of particular genomic sequences produces various dsRNAs. Dicer is responsible for processing the dsRNAs into small RNAs. The small RNA is then incorporated into RISC, guiding the protein complex to a specific mRNA. RISC enforces specific gene suppression either by cleavage of the mRNA or inhibition of translation. The degree of complementarity between the small RNA and the mRNA determines the fate of the mRNA; perfect or near perfect complementarity generally results in cleavage of the mRNA, whereas, a few mismatches results in suppression of translation. In a sequence specific manner, small RNAs also regulate chromatin structure and hence gene expression. Some of the processes depicted here may not be present in all organisms or cell types. From a practical standpoint small RNAs can be used to investigate gene function; several types of small dsRNAs can either be introduced into the cell or produced inside the cell. In all cases the small RNA is funneled into the RNAi pathway and triggers specific gene suppression. (See color section.)

RISC contained a ~25-nt RNA species that co-purified with the sequence-specific nuclease (Hammond et al., 2000). The ~25-nt RNA appeared to be required for cleavage of the corresponding mRNA, since nuclease treatment abolished mRNA degradation (Hammond et al., 2000).

Taken together, these results amounted to what has become the standard model for dsRNA-triggered mRNA degradation (Figure 2.1): a large dsRNA is cleaved into smaller (21–26 nt) dsRNAs, which are then used to guide a nuclease to the complementary mRNA, enforcing sequence specific silencing through cleavage of the small RNA-mRNA complex. The next important challenge was to identify the enzymes responsible for these steps.

Identification of Dicer

Hannon and colleagues knew that a nuclease activity capable of converting large dsRNAs to small ones was present in both *Drosophila* embryo and S2 cell-free extracts, and the activity was separable from the multicomponent nuclease RISC (Bernstein et al., 2001). Therefore, the *Drosophila* genome was scanned for RNase III family enzymes, nucleases that show specificity for dsRNA. The search resulted in several types of RNase III enzymes. One in particular was shown to be capable of cleaving dsRNA, but not ssRNA, into ~22 nucleotide fragments; this enzyme was appropriately named Dicer (Bernstein et al., 2001). At about the same time, Tuschl and colleagues demonstrated that 21–23 nucleotide guide dsRNAs, termed small interfering RNAs or siRNAs, mediated gene-specific silencing by directing cleavage of the mRNA at the site where the siRNA interacted with the mRNA (Elbashir et al., 2001a,b,&c). Thus, Dicer appeared to play a key role in the production of siRNAs, which in turn appeared to be critical for RNAi. Subsequently, other components of the pathway were identified through genetic and biochemical approaches (Denli and Hannon, 2003).

In vivo Dicing: Production of miRNAs, siRNAs, and tncRNAs

All organisms in which dsRNA-mediated gene silencing has been observed possess a Dicer homolog, and in some cases, more than one, with *Drosophila* having two (Bernstein et al., 2001) and *Arabadopsis thaliana* having four [(DCL1–4) (Schauer et al., 2002)]. The canonical Dicer protein is composed of six recognizable domains (Figure 2.2). At or near the N-terminus is a ~550 aa DExH-box RNA helicase domain, which is immediately followed by a conserved ~100 aa domain of unknown function (DUF283). Just C-terminal to the DUF283 domain is the PAZ (for Piwi/Argonaute/Zwille) domain. The function of the PAZ domain is now known and is not important for protein-protein interaction. The PAZ domain in Argonaute proteins has been shown to bind RNA (Lingel et al., 2003; Song et al., 2003; Yan et al., 2003; Lingel et al., 2004; Ma et al., 2004) and presumably the PAZ domain in Dicer could also bind RNA to position the catalytic domains for cleavage (Zhang et al., 2004). The C-terminus of the Dicer protein is composed of two RNase III catalytic domains and a putative dsRNA-binding domain.

The issue of whether Dicer might be responsible for the cleavage of dsRNAs *in vivo* was addressed in an elegant "RNAi of RNAi" experiment (Bernstein et al., 2001). In S2 cells, a mixture of dsRNAs complementary to the two *Drosophila* Dicer genes, but not an irrelevant dsRNA, was shown to reduce the *in vitro* activity of Dicer 6–7 fold, and similarly reduced Dicer mRNA levels (Bernstein et al., 2001). Under these same depletion conditions, the ability to silence exogenous green fluorescent protein (GFP) expression with GFP dsRNA was substantially compromised, but not completely inhibited, suggesting an *in vivo* role for Dicer in the RNAi pathway. The lack of complete inhibition of silencing may have been due to residual Dicer, although it is also possible that Dicer-independent silencing mechanisms exist (Bernstein et al., 2001).

miRNAs

In *C. elegans*, the *lin-4* and *let-7* miRNAs were demonstrated to regulate stage-specific development (Arasu et al., 1991; Lee et al., 1993; Moss, 2000). These

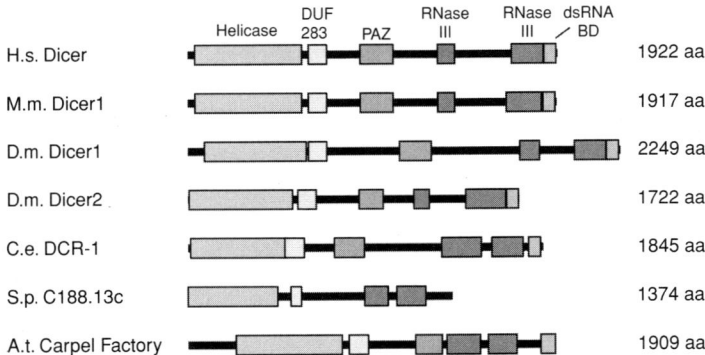

Figure 2.2. Alignment of Dicer homologs. Most Dicer proteins contain six conserved domains: a DExH helicase domain; a domain of unknown function (DUF283); a PAZ domain; two RNase III catalytic domains; and finally a double stranded RNA binding domain. Spacing between the domains is different for the homologs, which may explain the different size classes of small RNAs found in some species. (See color section.)

miRNAs are single-stranded 21-nucleotide oligomers that bind the 3′ untranslated region (UTR) of specific mRNAs, thereby inhibiting their translation (Olsen and Ambros, 1999; Moss, 2000; Pasquinelli and Ruvkun, 2002). miRNAs are derived from a hairpin precursor through the agency of Dicer, as inferred from "RNAi of RNAi" studies in HeLa cells (Hütvagner et al., 2001). Consistent with these findings, Mello and colleagues found that Dicer was required for generating mature miRNAs in *C. elegans* (Grishok et al., 2001). Moreover, the phenotype of Dicer-deficient worms was similar to those of *lin-4* and *let-7* mutants (Grishok et al., 2001; Ketting et al., 2001; Knight and Bass, 2001), suggesting that the production of miRNAs is one of the main functions of Dicer in *C. elegans* development. To date, several hundred miRNAs have been predicted or cloned from *C. elegans*, *Drosophila*, plants, fish, and mammals (Ambros et al., 2003a&b; Lim et al., 2003) and some have been shown to play a role in embryonic development (Houbaviy et al., 2003).

Recent work has added a number of important details to our understanding of the biogenesis of miRNAs. miRNAs are transcribed as long primary transcripts, primiRNAs, which are cleaved into pre-miRNAs in the nucleus by Drosha, an RNase III family enzyme (Lee et al., 2003). The ~70 nucleotide pre-miRNA is then transported into the cytosol (Lee et al., 2002) where it is cleaved into a mature miRNA by Dicer (Grishok et al., 2001; Hütvagner et al., 2001; Ketting et al., 2001). At this point the miRNA is double stranded, like an siRNA, but subsequent unwinding by an unidentified helicase (with Dicer itself being one plausible candidate) promotes asymmetrical incorporation of the appropriate strand into RISC, although in some cases both strands are incorporated (Schwarz et al., 2003). The activated RISC complex somehow finds the complementary mRNA, triggering cleavage or translational repression (Figure 2.1). Cleavage occurs only if base pairing between the miRNA and mRNA is exact; otherwise, the mismatches cause translational repression (Hütvagner and Zamore, 2002a&b; Llave et al., 2002a&b; Doench et al., 2003).

A role for Dicer in the production of siRNAs *in vivo*

Dicer is, under some conditions, required for dsRNA-mediated gene silencing in *C. elegans*. Dicer-null worms are defective for RNAi in the germ line, but surprisingly not in somatic cells (Ketting et al., 2001; Knight and Bass, 2001). Inconsistencies were also obtained depending on the method used to deliver the dsRNA, Dicer was required when the dsRNA was derived from a heat-inducible transgene, but not when dsRNA was directly injected (Knight and Bass, 2001). It is possible that gene silencing by dsRNA can occur through multiple pathways, or that maternal Dicer can compensate for some but not all RNAi (Grishok et al., 2001; Knight and Bass, 2001). Recently about 750 endogenous siRNAs derived from protein coding sequence have been identified in *C. elegans* (Ambros et al., 2003a&b). The siRNAs appear to target about 550 genes throughout the genome (Ambros et al., 2003) suggesting that at least some of the siRNAs may be involved in gene regulation important for embryonic development. Thus, the Dicer null phenotype may be due in part to an absence of endogenous siRNAs.

tncRNAs

Tiny non-coding RNAs (tncRNAs) are similar to miRNAs in that they are from genomic locations outside of protein coding sequence, are detectable as ~20–22 nucleotide species on northern blots, and display distinct temporal expression patterns throughout development (Ambros et al., 2003a&b). However, tncRNAs are not classified as miRNAs because they do not meet key secondary-structure criteria (Ambros et al., 2003a&b). tncRNAs are predicted not to form the same type of stem-loop foldback structure characteristic of miRNAs: tncRNAs exhibit excessive numbers of bulged nucleotides within the stem or have fewer than 16 base pairs involving the small RNA (Ambros et al., 2003a&b). Also, tncRNAs are not well conserved outside of *C. elegans*, further distinguishing them from miRNAs (Ambros et al., 2003). tncRNA precursors probably do fold into some sort of duplex, as indicated by the fact that 13 of the 16 tncRNAs tested do not accumulate in the absence of Dicer (Ambros et al., 2003a&b). Sequence analysis suggests that 5 of the 33 identified tncRNAs may in fact be siRNAs (Ambros et al., 2003). Although the function of tncRNAs is unclear, the fact that Dicer is required for their production again highlights the importance of Dicer in the biogenesis of small regulatory RNAs.

RNAi in the silencing of transposons, retrotransposons, and heterochromatin

In several *C. elegans* mutants that fail to carry out RNAi, there is a marked derepression of transposon mobilization (Ketting et al., 1999; Tabara et al., 1999; Sijen and Plasterk, 2003), suggesting that one function of RNAi (and, presumably, of Dicer) is to defend the genome against transposons. In support of this idea, siRNAs specific for endogenous transposons have been identified in *C. elegans* (Ambros et al., 2003). Silencing of genomic tandem repeats and transposable elements in *Drosophila* is also mediated by dsRNA (Aravin et al., 2001). Likewise, RNAi components, including Dicer, are required for the maintenance of heterochromatin (Volpe et al., 2002; Volpe et al., 2003) and for retrotransposon silencing in

fission yeast (Schramke and Allshire, 2003). A role for Dicer in genome maintenance is further supported by the fact that 12 siRNAs with sequence matching the
centromeric repeats were cloned from fission yeast (Reinhart and Bartel, 2002).
Finally, in fission yeast, Dicer is required for chromosome segregation and gene
silencing (Provost et al., 2002).

Dicer in gene suppression in plants

In transformed plants, PTGS is targeted against the transgene and homologous
endogenous genes so that the corresponding gene products fail to accumulate
(Kumagai et al., 1995; Angell and Baulcombe, 1997; Ruiz et al., 1998; reviewed in
Vaucheret et al., 1998). Similarly, viral genes are suppressed via PTGS (Dougherty
et al., 1994; Baulcombe, 1999). It is speculated that siRNAs, or at least some small
RNAs other than miRNAs, function in heterochromatic gene silencing in plants
(Hamilton et al., 2002; Bartel and Bartel, 2003). This gene suppression results
from sequence-specific degradation of the mRNA and methylation of genomic
sequence (Grant, 1999; Jones et al., 1999; Mette et al., 2000). In addition, small
dsRNAs or siRNAs are present only in plants undergoing PTGS (Hamilton and
Baulcombe, 1999). The presence of siRNAs suggests that a plant homolog of Dicer
is involved, assuming that the precursor was a large dsRNA. Mutations in the
RNA-dependent RNA polymerase SDE-1 makes plants defective for PTGS (Dalmay
et al., 2000), suggesting that SDE-1 produces the dsRNAs that are processed into
siRNAs.

However, mutants in DCL1 (Dicer-like 1; also called Carpel Factory or CAF)
are defective for normal development (Errampalli et al., 1991; Robinson-Beers
et al., 1992; Castle et al., 1993; Lang et al., 1994; Ray et al., 1996; Ray et al., 1996;
Jacobsen et al., 1999; McElver et al., 2001; Golden et al., 2002; Schauer et al.,
2002) and are incapable of efficient miRNA production (Llave et al., 2002; Park
et al., 2002; Reinhart et al., 2002), but are still capable of PTGS (Finnegan et al.,
2003). This finding may be explained by the fact that there are four Dicer-like
genes in plants [(DCL1–4) (Schauer et al., 2002)] and each may have a distinct
role (Tang et al., 2003), possibly dependent on subcellular localization (Papp et al.,
2003). For example, DCL2 and DCL3, which, unlike DCL1 and DCL4, lack nuclear
localization sequences (Schauer et al., 2002; Papp et al., 2003), could be responsible for cleaving larger dsRNAs into siRNAs in the cytoplasm. Thus DCL1 might
be required for nuclear miRNA and siRNA biogenesis but not cytoplasmic siRNA
generation and the cytoplasmic degradation of mRNA associated with PTGS.

Furthermore, two populations of small dsRNAs are produced in plants, 21–
22 nucleotides and 24–26 nucleotides. The shorter RNAs are required for mRNA
degradation but not DNA methylation or systemic gene suppression and vice
versa for the larger RNAs (Hamilton et al., 2002). It is possible that DCL1 or DCL4
cleaves large, nuclear dsRNA into 24–26 nucleotide fragments which trigger DNA
methylation and systemic gene suppression, whereas DCL2 or DCL3 cleave cytoplasmic dsRNA into 21–22 nucleotide fragments necessary for mRNA degradation.
Because Dicer is presented with various forms of dsRNAs, some organisms, such
as plants, may process distinct dsRNAs with different Dicer isoforms.

The presence of miRNAs and the diversity of siRNA function in plants further suggests that PTGS, or some form of homology-dependent silencing, is essential for heterochromatin silencing, embryonic development, and defense against viral infection and retrotransposon and transposon movement. In each case, the trigger is some form of dsRNA, implicating Dicer as being critical for plant development and defense.

Other organisms

In mammalian systems Dicer is required for gene silencing in cell culture when exogenous siRNAs are introduced experimentally (Doi et al., 2003). In addition, Dicer is essential for maintenance of the stem cell population in mice (Bernstein et al., 2003). This may be due to a requirement for miRNA production; along these lines, Sharp and colleagues have identified miRNAs possibly essential for maintenance of the pluripotent state and early embryonic development (Houbaviy et al., 2003). Thus Dicer appears to be important for miRNA production during embryogenesis in diverse organisms.

RNAi as a new tool for loss of function investigations

In genetically tractable organisms like *C. elegans* and *Drosophila*, RNAi was an important addition to an already powerful arsenal of genetic and reverse genetic tools. In vertebrate systems, where the existing tools were much more limited, RNAi has been a godsend. However, the approaches required in most mammalian cells are somewhat different from those used in *C. elegans* and *Drosophila*. For example, long (~500 bp) dsRNAs bring about specific gene silencing in *C. elegans* and *Drosophila*, but in most mammalian cell lines dsRNAs of this size lead to induction of the interferon response (Stark et al., 1998). dsRNAs greater than 30 base pairs in length activate the protein kinase PKR, causing a non-specific inhibition of translation through phosphorylation of eIF2α (Kostura and Mathews, 1989; Manche et al., 1992) and a general destruction of mRNA through RNase L activation (Minks et al., 1979; Samuel, 2001). These responses can be avoided in some cell lines; for example, cultured embryonic cells, like P19 or embryonic stem cells (ES), are devoid of the interferon response, and 500 bp dsRNAs cause specific gene silencing in these cells (Paddison et al., 2002). To extend RNAi to more commonly studied cell lines, Tuschl and colleagues devised a clever way of avoiding the non-specific interferon response. The 21 nucleotide intermediates in the RNAi pathway, siRNAs, proved to be small enough to avoid the interferon response, but large enough to provide specific and efficacious gene silencing [(Figure 2.1) (Elbashir et al., 2001)].

Many aspects of the mechanism of RNAi in *C. elegans* and *Drosophila* are also relevant to RNAi in mammalian cells [(Figure 2.1) (Elbashir et al., 2001)]. There are some differences, however. In mammalian cells, silencing is not spread to the progeny, probably because lack of or inactivation of a Sid-1 ortholog, a transmembrane protein responsible for transport of large (greater than 100 base pairs), but not small dsRNAs across cellular membranes (Winston et al., 2002; Feinberg and Hunter, 2003). Also, mammalian cells do not regenerate siRNAs, presumably because they lack the RNA-dependent RNA polymerase (RdRp) capable of amplifying an mRNA when primed by the antisense strand of the siRNA (Schwarz et al., 2002).

siRNAs make a number of genetic approaches feasible in frequently used mammalian model systems. For example, epi-allelic, complementation, and mutational analysis experiments can now be completed in cultured somatic cells (Hemann et al., 2003). One important concern in an RNAi experiment is that there might be off-target silencing of specific genes with similar nucleotide identity, such as a gene family member or a gene that adventitiously has a similar stretch of sequence. Another concern is non-specific gene silencing, since high concentrations of even siRNAs result in elevated levels of eIF2α phosphorylation and thereby general inhibition of translation (Myers et al., 2003). However, initial experiments suggested that siRNAs are, in fact, very specific in silencing their cognate mRNAs. For example, the GL2 and GL3 forms of firefly luciferase differ by only a handful of point mutations, yet GL3 can be specifically silenced by synthetic siRNAs that match its sequence perfectly (21/21 nt) and match the GL2 sequence imperfectly [(17/21 nt) (Elbashir et al., 2001)].

This issue has now been addressed more thoroughly through microarray experiments. Two groups reassuringly found minimal off-target and non-specific effects in siRNA-treated cells, based on analysis of the expression of tens of thousands of genes (Chi et al., 2003; Semizarov et al., 2003). However, a third group reported a disconcertingly large degree of off-target and non-specific effects (Jackson et al., 2003). At present it is unclear why RNAi seems so specific in some contexts and so problematic in others; see Chapters 6 and 33 (Boese et al.; Chang et al.) for further discussion of these issues.

A good test for specificity of an RNAi-induced effect is through an add-back or reconstitution experiment. One way to achieve this is to alter the exogenous cDNA's codon usage without changing the amino acid sequence it encodes. Several experiments suggest that three-point mutations within the 21-nucleotide target region should be sufficient to make the cDNA resistant to the siRNA (Parrish et al., 2000; Elbashir et al., 2001; Elbashir et al., 2001; Lassus et al., 2002; Lassus et al., 2002). A second approach is to target an untranslated region of the endogenous mRNA and express only the coding region of the cDNA. Finally, the co-expressed cDNA could be derived from an alternative species, since a majority of genes have a 21-nucelotide region that lacks a high degree of nucleotide homology when compared to an ortholog from another species, even when the amino acid sequence is highly conserved. In any of these instances, there may be a concern with overexpression; some care should be taken to ensure that the siRNA-resistant cDNA is expressed at an appropriate level.

A similar strategy can be used to carry out a "knock-in" experiment, which allows structure-function studies to be carried out in a clean genetic background – that is, in the absence of the wild-type version of the protein whose structure has been altered. These types of analyses have been carried out to great advantage in yeast, where the prevalence of homologous recombination makes knock-in experiments relatively easy, and a number of important insights have come from the analysis of knock-in mice as well. RNAi certainly promises to give new life to genetic investigations in mammalian cultured cells.

siRNA sequence selection

Originally siRNAs were exclusively produced by chemical synthesis (Elbashir et al., 2001; Elbashir et al., 2001). There are two main limitations to this approach: synthesis is very expensive, and several different oligos may need to be tried before a particular gene is successfully silenced. Thus, alternative, inexpensive methods to produce siRNAs have been pursued (Figure 2.3 and Table 2.1). One option is to synthesize less expensive DNA oligonucleotide templates, and then transcribe them *in vitro* using promoters for both strands [(Figure 2.3B) (Yu et al., 2002)]. Alternatively, DNA oligonucleotides can be cloned into an expression vector employing RNA polymerase III promoters, from which the sense and antisense strands can be transcribed *in vivo* [(Figure 2.3D) (Lee et al., 2002; Miyagishi and Taira, 2002)]. Similarly, a DNA oligo encoding a short hairpin RNA (sh-RNA) can be constructed and expressed [(Figure 2.3C, E) (Brummelkamp et al., 2002; Castanotto et al., 2002; Paddison et al., 2002; Paul et al., 2002)]. The resulting sh-RNA is trimmed into a siRNA by Dicer *in vivo*, and then feeds into the RNAi pathway.

All of these approaches are substantially less expensive than chemically synthesizing siRNAs. However, there is still the problem of designing and identifying effective siRNAs. In general the perception is that about one in four randomly chosen siRNAs will prove to be effective (Khvorova et al., 2003; Schwarz et al., 2003). Thus, if one synthesizes three siRNAs, the chances are about 60% ($1-0.75^3$) at least one will be effective, and if one synthesizes four siRNAs the chances are about 70% ($1-0.75^4$). This still leaves open the possibility that some critical target will require the synthesis of tens of different siRNAs before an effective one is found. Thus it will be important to try to work out reliable rules for the synthesis of effective siRNAs. Two recent reports indicate that if the internal stability within the first 4 base pairs at the 5'-end of the antisense strand is lower than at the 5'-end of the sense strand, then the antisense strand is preferentially incorporated into RISC, thereby enhancing the potency (Khvorova et al., 2003; Schwarz et al., 2003). Still, the predictive power of this approach is not yet well-established; it is not clear that siRNAs that satisfy these criteria are much more likely to be effective than randomly selected siRNAs are.

Individual siRNAs may lack silencing efficacy because they are incapable of effectively binding to the target mRNA, due to secondary or tertiary structure within the mRNA, or because they cannot effectively compete with mRNA binding proteins (Dykxhoorn et al., 2003; Vickers et al., 2003). Because mRNAs are quite long and siRNAs are merely 21 nucleotides in length, there are hundreds of possible siRNA sequences to choose from. Narrowing in on the ideal target sequence is facilitated by incorporating certain characteristics in the siRNA, and some qualities appear to make some siRNAs better than others: low GC content,

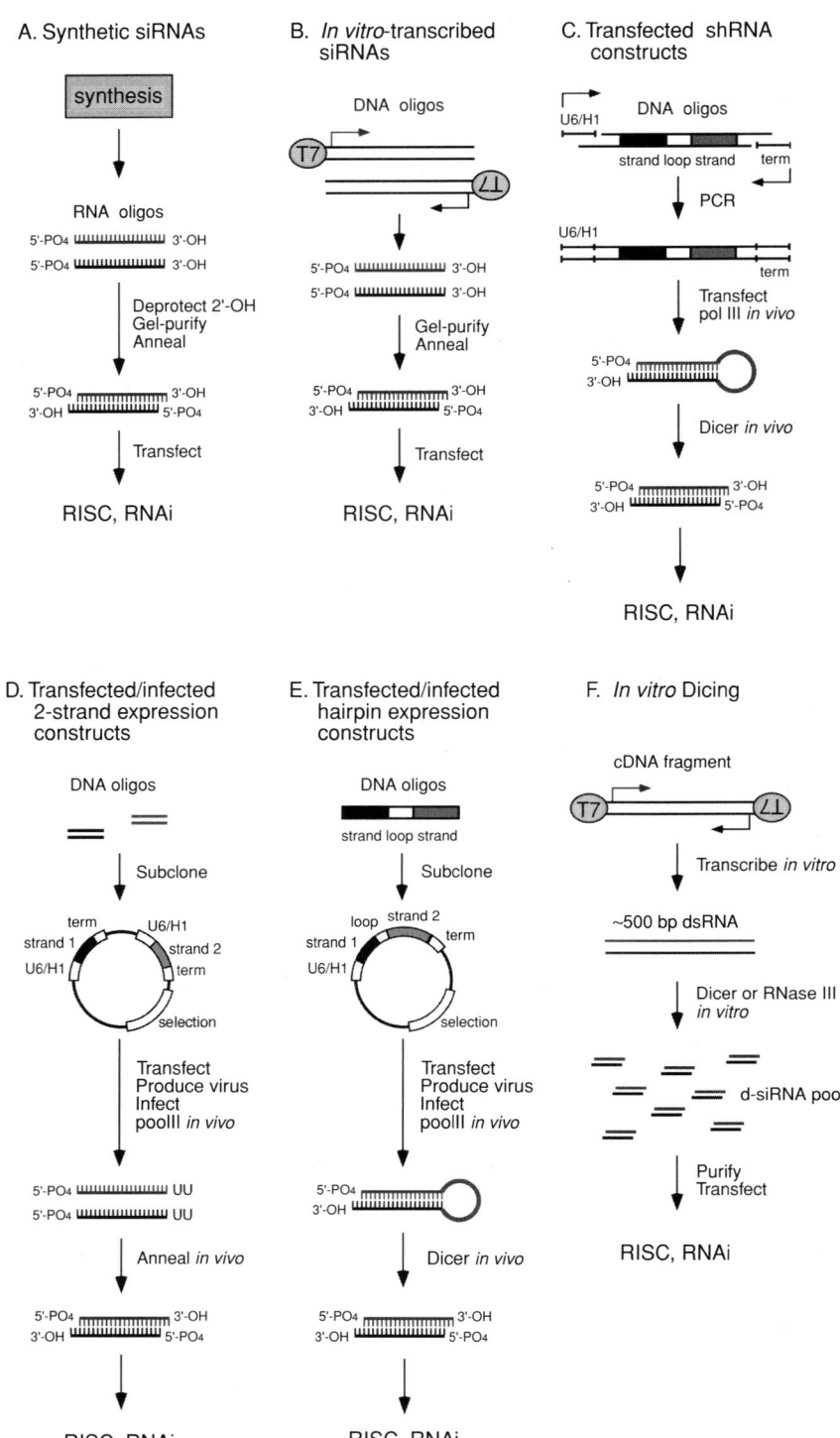

Figure 2.3. Strategies for siRNA production. siRNAs can either be generated *in vitro* and then introduced into the cell (A, B or F), or produced *in vivo* from transcription of a DNA template (C, D, or E). In some cases the siRNA is derived from a hairpin precursor (C and E) and Dicer is responsible for cleaving the hairpin into the siRNA. Most methods use an individual siRNA or a few siRNAs pooled together (A–E), whereas one method employs a complex pool of siRNAs (F). Each method of siRNA production has advantages and disadvantages concerning cost, efficacy, and longevity of gene suppression. *In vitro* Dicing compares favorably to other methods because d-siRNAs are inexpensive to produce, easy to design, and are very potent for gene suppression, but in some cases, when long-term gene suppression is required, the hairpin expression constructs are preferred.

Table 2.1. Advantages and disadvantages of various siRNA approaches

Method	Cost	Transient/ sustained silencing	Certainty of effective silencing	Amenability to "reconstitution" experiments
Synthetic siRNAs	High	Transient	~25–40%	Good
In vitro transcribed siRNAs	Low	Transient	~25–40%	Good
Transfected shRNA constructs	Low	Transient	~25–40%	Good
Transfected/infected 2-strand expression constructs	Low	Sustained	~25–40%	Good
Transfected/infected hairpin expression constructs (e.g. pSUPER)	Low	Sustained	~25–40%	Good
In vitro Dicing	Low	Transient	~90%	More difficult

AA dinucleotides at the 5′-end, two-nucleotide 3′-overhangs, 5′-phosphorylations (Elbashir et al., 2002), low internal strand stability at the 5′-end of the antisense strand (Khvorova et al., 2003; Schwarz et al., 2003) and at positions 9–14 (Khvorova et al., 2003), as well as a lack of inverted repeats and base preference in the sense strand at positons 3, 10, 13, and 19 (Reynolds et al., 2004). One approach that has been helpful in designing effective antisense DNA oligonucleotides has been to select those that hybridize most tightly to the intended target. The idea is that binding affinity will correlate with potency, and indeed this appears to be the case for insulin-like growth factor receptor 1 siRNAs (Bohula et al., 2003). Determining the binding affinities of several DNA or RNA oligonucleotides to a single mRNA target is not too difficult but does require additional effort and expense.

Individual siRNAs suffer from other shortcomings as well. None of the methods is easily scaled up for screens, since for each member of a library, one or more oligonucleotides must be individually designed and synthesized. Also, since individual siRNAs cause cleavage of target mRNAs at a single site (Elbashir et al., 2001), it is possible that the remaining 3′-mRNA fragment will be translated, with the resulting N-terminal truncated protein potentially functioning as a dominant negative or positive rather than a protein-null (Thoma et al., 2001). Individual siRNAs may also be selective for particular splice variants, which could be either an advantage or a shortcoming. Recently it has been discovered that particular individual chemically synthesized siRNAs can trigger the interferon response (Sledz et al., 2003). Likewise, out of two shRNAs targeting the same gene, one initiated an interferon response and the other did not (Bridge et al., 2003), adding yet another complication to siRNA selection. Finally, an enormous number of single nucleotide polymorphisms are spread throughout the genome (Oefner, 2002) complicating gene suppression, since discrepancies between the siRNA and mRNA can render silencing ineffective (Elbashir et al., 2001). These problems motivated the development of *in vitro* dicing (Figure 2.3F), using either recombinant Dicer (Provost et al., 2002; Zhang et al., 2002; Kawasaki et al., 2003; Myers et al., 2003) or recombinant bacterial RNAse III (Yang et al., 2002) to cleave long, *in vitro* transcribed dsRNAs into a complex pool of diced-siRNAs (d-siRNAs).

How does *in vitro* dicing measure up?

Large quantities of human recombinant Dicer can easily be produced in and purified from insect cells (Provost et al., 2002; Zhang et al., 2002; Myers et al., 2003). One group has reported successful expression of active Dicer in bacteria as well (Kawasaki et al., 2003). The partially purified Dicer enzyme can dice large amounts of *in vitro* transcribed dsRNA into a pool of diced siRNAs (d-siRNAs), with yields up to 70% (Figure 2.4A). The resulting pool of siRNA products is highly complex, containing hundreds of distinct siRNAs (D. Gong, personal communication). The d-siRNAs produced are monophosphorylated at the 5′-terminus, have a 2-nucleotide overhang at the 3′-terminus, and are appropriate in size (20–21 nt) to be optimally effective for gene silencing (Myers et al., 2003).

Indeed, d-siRNAs compare favorably to individual siRNAs in terms of silencing potency (Figure 2.4B and Myers et al., 2003). The mechanistic basis for this high potency is not completely understood. It is possible that d-siRNA pools contain a few super-high potency siRNAs that contribute most of the gene silencing. Alternatively, many siRNAs might contribute to the silencing, either individually or cooperatively. In any case, d-siRNA pools largely obviate the need to develop predictive rules for designing siRNAs. Frequently a large amount of gene suppression, usually 90%, is attained on the first attempt. To date many gene products have been effectively silenced in several cell types (Figure 2.4C and Table 2.2), and the number of genes suppressed and diversity of cell types is continually increasing.

Bacterial RNase III can also be used to cleave *in vitro* transcribed dsRNA into smaller dsRNA fragments (Yang et al., 2002). The bacterial RNase III is easier to produce and more active than recombinant Dicer, but the resulting siRNAs are not optimal in size. Exhaustive cleavage of dsRNA with bacterial RNase III generates duplex fragments averaging 12–15 nucleotides in length (Robertson and Hunter, 1975; Dunn, 1982), and some commercially available RNase III systems suggest the use of the 12–15 nucleotide siRNAs. However, under appropriate reaction conditions larger fragments of 21–23, 24–26, and 27–30 nucleotides can be produced, and each respective size is reportedly capable of specific gene silencing (Yang et al., 2002). Less work is required to prepare bacterial RNase III and cleave dsRNAs but more work is required to obtain efficacious siRNAs.

One important potential problem with using a pool of siRNAs, like d-siRNAs or siRNAs generated by bacterial RNase III digestion, is the possibility of increased off-target effects. A complex mixture of siRNAs will certainly contain more off-target matches than a single siRNA would, particularly if the pool consists of short siRNAs like those produced by RNase III digestion. On the other hand, the concentration of any individual siRNA in a complex d-siRNA pool is likely to be quite low – perhaps ~1/300th of the total siRNA concentration – and so unless many different members of the pool share the same off-target effects, the effects might be insignificant. In support of this view, the gene silencing achieved in *C. elegans* and *D. melanogaster* by the complex pools of d-siRNAs produced from large dsRNAs *in vivo* has proven to be quite specific. Likewise, the low concentrations of individual siRNAs in a d-siRNA pool may decrease the chance of

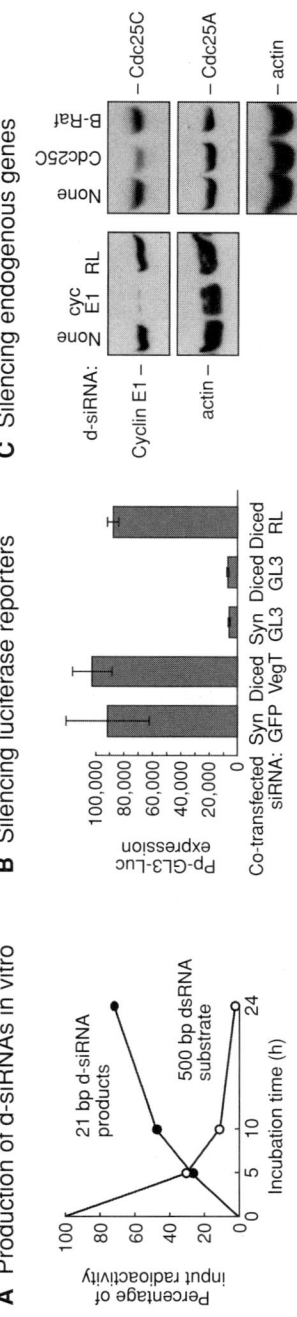

A Production of d-siRNAs in vitro

B Silencing luciferase reporters

C Silencing endogenous genes

Figure 2.4. *In vitro* Dicing and efficacy of d-siRNAs. Recombinant Dicer efficiently cleaves large dsRNAs (≥500 bp) into a complex pool of 20–21 nt siRNAs, also termed d-siRNAs. (A) Dicing is efficient, with yields up to 70%. (B) d-siRNAs are as effective as an individual chemically synthesized siRNA for specific silencing of an exogenous reporter. (C) d-siRNAs can be used to specifically suppress endogenous gene expression without affecting the expression of closely related gene products. Adapted from Myers et al., 2003.

Table 2.2. Gene silencing with d-siRNA pools

Target	Cell type	Silencing	References
ATR	HEK 293, HeLa	~85%	Byun, Cimprich, Hendrickson, Myers and Ferrell, unpublished
CaMKII-α	Rat hippocampal neurons	~90%	(Fink et al., 2003)
CaMKII-β	Rat hippocampal neurons	~90%	(Fink et al., 2003)
Cdc25C	HEK 293, HeLa	~90%	(Myers et al., 2003)
Cyclin A2	HeLa	~95%	Myers and Ferrell, unpublished
Cyclin B1	HEK 293, HeLa	~95%	Gong, Myers, Po and Ferrell, unpublished
Cyclin B2	HEK 293, HeLa	~90%	Gong, Myers, Po and Ferrell, unpublished
Cyclin E1	HEK 293, HeLa	~90%	(Myers et al. 2003)
Mad2	RBL	~80%	Jones, Myers, Ferrell and Meyer, Nature Biotechnology, in review
Mos	HeLa, Xenopus oocytes	negligible	Yue, Xiong, Myers, Hendrickson, Rios-Cardona, Ferrell, unpublished
Mre11	HEK 293, HeLa	~85%	Byun, Cimprich, Hendrickson, Myers and Ferrell, unpublished
PKC-δ	NIH 3T3	~60%	Myers, Chen, Mochly-Rosen and Ferrell, unpublished
PKC-ε	NIH 3T3	~60%	Myers, Chen, Mochly-Rosen and Ferrell, unpublished
B-Raf	HEK 293, HeLa	~90%	Myers and Ferrell, unpublished

triggering the interferon response (Bridge et al., 2003; Myers et al., 2003; Sledz et al., 2003).

The experience to date suggests that d-siRNAs are, in fact, effective and highly specific for gene silencing in mammalian cells. For example, Cdc25C was suppressed without measurable decreases in Cdc25A levels (Figure 2.4C), despite the fact that the two cDNAs share stretches of up to 15 nucleotides in common (Myers et al., 2003). Similar results have been reported for two other pairs of closely related genes, cyclin B1 and B2 (Gong and Myers, unpublished data) and CaMKIIα and CaMKIIβ (Fink et al., 2003); when one is silenced the other is not. It would be of interest to test the specificity further, at least for a few targets, through global microarray analysis, as has been reported recently for individual siRNAs (Chi et al., 2003; Jackson et al., 2003; Semizarov et al., 2003).

d-siRNAs provide only transient gene silencing, which is problematic for some applications. Stable gene silencing is probably best accomplished by expression of an shRNA from a plasmid with a selectable marker (Figure 2.3, Table 2.1). d-siRNAs are amenable to complementation approaches if the 3'-UTR can be used for silencing; individual siRNAs (stably or transiently expressed) offer the additional possibility of introducing a codon-usage variant to check specificity (Table 2.1).

Delivery, detection, and controls

Many forms of siRNA production suffer from delivery problems, limiting their utility to highly transfectable cell lines (Table 2.1). One way around this is to use viral delivery of a vector encoding an siRNA or sh-siRNA, including retroviral (Barton and Medzhitov, 2002; Brummelkamp et al., 2002), lentiviral (Abbas-Terki et al., 2002; An et al., 2003; Dirac and Bernards, 2003; Matta et al., 2003; Rubinson et al., 2003; Scherr et al., 2003), and adenoviral [(Shen et al., 2003; Tomar et al., 2003) (see also Chapter 11)].

Another problem is the validation of siRNA-induced protein reduction. Suitable antibodies are unavailable for some proteins. Measuring the remaining mRNA levels, by northern blotting, quantitative PCR, or RNase protection, may give an idea of the amount of gene silencing, but a reduction in mRNA levels and protein levels are not necessarily equivalent. For example, if a protein has a long half-life, its mRNA might disappear long before the protein does. On the other hand, if an siRNA causes more translational silencing than mRNA degradation, the protein might disappear without a decrease in its mRNA level.

Another problem with gene silencing experiments is deciding on an appropriate control for non-specific effects. Often the control is an siRNA lacking homology to any known gene in the genome. However, some control siRNAs may have more severe off-target (Jackson et al., 2003) or non-specific (Bridge et al., 2003; Sledz et al., 2003) effects than others. In addition, siRNAs that lack a target may persist in the cell longer than siRNAs whose targets are present, and so they may have greater toxic and off-target effects. Therefore, the best control may be an siRNA that targets a non-essential gene (perhaps an exogenously expressed non-essential gene, to avoid the possibility that the "control" gene is unexpectedly linked to the intended target gene).

Strategies for systematic and large-scale screens

1. *In vitro* dicing

In vitro dicing seems well-suited to experiments aimed at silencing scores or hundreds of genes in parallel. The *in vitro* dicing protocol can easily be adapted to a 96-well format, and even without robotics, several 96-well plates containing ready-to-use d-siRNAs can be generated in a single day. The first step is to obtain *in vitro* transcription templates. Any region of the gene, or the entire gene with or without untranslated regions, can be PCR amplified from one or several cDNA collections. Alternatively, the dsRNA can be transcribed from a vector, if a cDNA library is already constructed, and then arrayed in 96-well format. In either case, target selection is not an issue, eliminating the tedious and time-consuming bioinformatics involved in selecting an efficacious individual siRNA. Furthermore, only one template has to be designed for each gene, rather than three or four that are required with an individual siRNA to ensure successful gene suppression. The subsequent steps in generating the d-siRNAs, *in vitro* transcription, *in vitro* dicing, and purification of the d-siRNAs are also very simple. Thus, screening is limited only by the number of reliable assays available and the number of genes that are reasonable to obtain. In fact, the entire human or mouse genome could be diced.

2. Split-pool screens

In addition, a genome could be screened in a split-pool format (Smith and Harland, 1992; Lustig et al., 1997). This approach has facilitated understanding of gene function, especially in the realm of overexpression, gain of function approaches. Pools of genes, 10, 50, 100, or even more genes can be diced and screened simultaneously. Once a "hit" is obtained, the pool can be systematically split until the culprit is identified. This approach is appealing if the screen is intended to identify regulators of a specific cellular function from a large number of genes. Really, any type of siRNA would work in split-pool format, but it is very appealing to apply *in vitro* dicing to this format because the screen can be done on a random set of genes, whereas each siRNA would have to be rationally designed and arrayed before beginning.

3. shRNA expression libraries

The alternative approach is to build collections of individual siRNAs or shRNA constructs. Bernards and coworkers have carried out this sort of approach, using a collection of shRNAs to investigate the functions of de-ubiquitinating enzymes. They identified the de-ubiquitating enzyme CYLD as a novel tumor suppressor affecting the NF-kB pathway (Brummelkamp et al., 2003).

siRNAs and d-siRNAs as therapeutic agents

As discussed in other chapters in this book, there is a great deal of excitement over the possibility of using RNAi in the treatment of ailments that result from over-expression or aberrant expression of endogenous genes, for example cancer, and for the treatment of viral and parasitic infections. The key advantage of RNAi is specificity, the hallmark of RNAi since its discovery. Nucleic acid interactions are amazingly specific, and even more intriguing is the fact that base pairing needs to be exact or near exact in order to get mRNA destruction or inhibition of translation, respectively. It is very difficult to rationally design a drug that will affect a single protein, simply because domains within proteins have similar binding pockets allowing the drug to affect several pathways. In some cases the unintended affects are not problematic; however, it is often the case that unwanted side effects arise from non-specific drug action. Therefore, RNAi could potentially eliminate the undesired side effects of drugs by specifically silencing the target.

Individual siRNAs provide a means for a therapeutic agent to distinguish between an aberrant gene and a wild-type gene (Brummelkamp et al., 2002). However, there are instances where this specificity is not necessary, and in fact a complex pool of many distinct siRNAs would be preferred. One such situation might be the treatment of viral infections. Here the problem with individual siRNAs as therapeutic agents is two-fold: certain retroviruses mutate rapidly like HIV and polio and thus may be able to escape the effect of an individual siRNA (Gitlin et al., 2002; Andino, 2003), and loss of only one viral component may only cripple the virus. However, the probability of generating enough mutations to render the virus insensitive to a complex pool of d-siRNAs should be vanishingly small. And even if a highly mutated virus did arise that was capable of reproduction in the presence of the d-siRNAs, the virus could be obtained from the patient, and a new set of d-siRNAs could be derived from the current genome of the virus, via simple molecular biology. This process could be repeated until the viral titer is lowered enough for the immune system to catch up and eliminate the infection altogether. Moreover, since d-siRNAs do not have to be screened for efficacy and potency, it is very simple to silence multiple targets.

Finally, if an endogenous human gene is the target, polymorphisms within certain populations could render individual siRNAs ineffective. The complexity of the d-siRNAs silences this issue.

Acknowledgements

We thank Delquin Gong, Dave Hendrickson, Tom Wehrman, and members of the Ferrell lab for numerous invaluable discussions. Our work in this area is supported by NIH Grant GM46383.

REFERENCES

Abbas-Terki, T., Blanco-Bose, W., Deglon, N., Pralong, W. and Aebischer, P. (2002). Lentiviral-mediated RNA interference. *Human Gene Therapy*, **13**, 2197–2201.

Ambros, V., Bartel, B., Bartel, D. P., Burge, C. B., Carrington, J. C., Chen, X., Dreyfuss, G., Eddy, S. R., Griffiths-Jones, S., Marshall, M., Matzke, M., Ruvkun, G. and Tuschl, T. (2003a). A uniform system for microRNA annotation. *RNA*, **9**, 277–279.

Ambros, V., Lee, R. C., Lavanway, A., Williams, P. T. and Jewell, D. (2003b). MicroRNAs and other tiny endogenous RNAs in C. elegans. *Current Biology*, **13**, 807–818.

An, D. S., Xie, Y., Mao, S. H., Morizono, K., Kung, S. K. and Chen, I. S. (2003). Efficient lentiviral vectors for short hairpin RNA delivery into human cells. *Human Gene Therapy*, **14**, 1207–1212.

Andino, R. (2003). RNAi puts a lid on virus replication. *Nature Biotechnology*, **21**, 629–630.

Angell, S. M. and Baulcombe, D. C. (1997). Consistent gene silencing in transgenic plants expressing a replicating potato virus X RNA. *European Molecular Biology Organization Journal*, **16**, 3675–3684.

Arasu, P., Wightman, B. and Ruvkun, G. (1991). Temporal regulation of lin-14 by the antagonistic action of two other heterochronic genes, lin-4 and lin-28. *Genes & Development*, **5**, 1825–1833.

Aravin, A. A., Naumova, N. M., Tulin, A. V., Vagin, V. V., Rozovsky, Y. M. and Gvozdev, V. A. (2001). Double-stranded RNA-mediated silencing of genomic tandem repeats and transposable elements in the D. melanogaster germline. *Current Biology*, **11**, 1017–1027.

Ashrafi, K., Chang, F. Y., Watts, J. L., Fraser, A. G., Kamath, R. S., Ahringer, J. and Ruvkun, G. (2003). Genome-wide RNAi analysis of *Caenorhabditis elegans* fat regulatory genes. *Nature*, **421**, 268–272.

Bartel, B. and Bartel, D. P. (2003). MicroRNAs: at the root of plant development? *Plant Physiology*, **132**, 709–717.

Barton, G. M. and Medzhitov, R. (2002). Retroviral delivery of small interfering RNA into primary cells. *Proceedings of the National Academy of Sciences USA*, **99**, 14943–14945.

Baulcombe, D. C. (1999). Fast forward genetics based on virus-induced gene silencing. *Current Opinion in Plant Biology*, **2**, 109–113.

Bernstein, E., Caudy, A. A., Hammond, S. M. and Hannon, G. J. (2001). Role for a bidentate ribonuclease in the initiation step of RNA interference. *Nature*, **409**, 363–366.

Bernstein, E., Kim, S. Y., Carmell, M. A., Murchison, E. P., Alcorn, H., Li, M. Z., Mills, A. A., Elledge, S. J., Anderson, K. V. and Hannon, G. J. (2003). Dicer is essential for mouse development. *Nature Genetics*, **35**, 215–217.

Bohula, E. A., Salisbury, A. J., Sohail, M., Playford, M. P., Riedemann, J., Southern, E. M. and Macaulay, V. M. (2003). The efficacy of small interfering RNAs targeted to the type 1 insulin-like growth factor receptor (IGF1R) is influenced by secondary structure in the IGF1R transcript. *Journal of Biological Chemistry*, **278**, 15991–15997.

Bridge, A. J., Pebernard, S., Ducraux, A., Nicoulaz, A. L. and Iggo, R. (2003). Induction of an interferon response by RNAi vectors in mammalian cells. *Nature Genetics*, **34**, 263–264.

Brummelkamp, T. R., Bernards, R. and Agami, R. (2002a). Stable suppression of tumorigenicity by virus-mediated RNA interference. *Cancer Cell*, **2**, 243–247.

Brummelkamp, T. R., Bernards, R. and Agami, R. (2002b). A system for stable expression of short interfering RNAs in mammalian cells. *Science*, **296**, 550–553.

Brummelkamp, T. R., Nijman, S. M., Dirac, A. M. and Bernards, R. (2003). Loss of the cylindromatosis tumour suppressor inhibits apoptosis by activating NF-kappaB. *Nature*, **424**, 797–801.

Carthew, R. W. (2001). Gene silencing by double-stranded RNA. *Current Opinion in Cell Biology*, **13**, 244–248.

Castanotto, D., Li, H. and Rossi, J. J. (2002). Functional siRNA expression from transfected PCR products. *RNA*, **8**, 1454–1460.

Castle, L. A., Errampalli, D., Atherton, T. L., Franzmann, L. H., Yoon, E. S. and Meinke, D. W. (1993). Genetic and molecular characterization of embryonic mutants identified following seed transformation in Arabidopsis. *Molecular and General Genetics*, **241**, 504–514.

Chi, J. T., Chang, H. Y., Wang, N. N., Chang, D. S., Dunphy, N. and Brown, P. O. (2003). Genomewide view of gene silencing by small interfering RNAs. *Proceedings of the National Academy of Sciences USA*, **100**, 6343–6346.

Cogoni, C., Irelan, J. T., Schumacher, M., Schmidhauser, T. J., Selker, E. U. and Macino, G. (1996). Transgene silencing of the al-1 gene in vegetative cells of Neurospora is mediated by a cytoplasmic effector and does not depend on DNA-DNA interactions or DNA methylation. *European Molecular Biology Organization Journal*, **15**, 3153–3163.

Couzin, J. (2002). Breakthrough of the year. Small RNAs make big splash. *Science*, **298**, 2296–2297.

Dalmay, T., Hamilton, A., Rudd, S., Angell, S. and Baulcombe, D. C. (2000). An RNA-dependent RNA polymerase gene in Arabidopsis is required for posttranscriptional gene silencing mediated by a transgene but not by a virus. *Cell*, **101**, 543–553.

Denli, A. M. and Hannon, G. J. (2003). RNAi: An ever-growing puzzle. *Trends in Biochemical Sciences*, **28**, 196–201.

Dirac, A. M. and Bernards, R. (2003). Reversal of senescence in mouse fibroblasts through lentiviral suppression of p53. *Journal of Biological Chemistry*, **278**, 11731–11734.

Doench, J. G., Petersen, C. P. and Sharp, P. A. (2003). siRNAs can function as miRNAs. *Genes & Development*, **17**, 438–442.

Doi, N., Zenno, S., Ueda, R., Ohki-Hamazaki, H., Ui-Tei, K. and Saigo, K. (2003). Short-interfering-RNA-mediated gene silencing in mammalian cells requires Dicer and eIF2C translation initiation factors. *Current Biology*, **13**, 41–46.

Dougherty, W. G., Lindbo, J. A., Smith, H. A., Parks, T. D., Swaney, S. and Proebsting, W. M. (1994). RNA-mediated virus resistance in transgenic plants: Exploitation of a cellular pathway possibly involved in RNA degradation. *Mol Plant Microbe Interact*, **7**, 544–552.

Dunn, J. J. (1982). Ribonuclease III. *In The Enzymes*, P. D. Boyer, ed. (New York, Academic Press), pp. 485–499.

Dykxhoorn, D. M., Novina, C. D. and Sharp, P. A. (2003). Killing the messenger: Short RNAs that silence gene expression. *Nature Reviews Molecular Cell Biology*, **4**, 457–467.

Elbashir, S. M., Harborth, J., Lendeckel, W., Yalcin, A., Weber, K. and Tuschl, T. (2001a). Duplexes of 21-nucleotide RNAs mediate RNA interference in cultured mammalian cells. *Nature*, **411**, 494–498.

Elbashir, S. M., Lendeckel, W. and Tuschl, T. (2001b). RNA interference is mediated by 21- and 22-nucleotide RNAs. *Genes & Development*, **15**, 188–200.

Elbashir, S. M., Martinez, J., Patkaniowska, A., Lendeckel, W. and Tuschl, T. (2001c). Functional anatomy of siRNAs for mediating efficient RNAi in *Drosophila melanogaster* embryo lysate. *European Molecular Biology Organization Journal*, **20**, 6877–6888.

Elbashir, S. M., Harborth, J., Weber, K. and Tuschl, T. (2002). Analysis of gene function in somatic mammalian cells using small interfering RNAs. *Methods*, **26**, 199–213.

Errampalli, D., Patton, D., Castle, L., Mickelson, L., Hansen, K., Schnall, J., Feldmann, K. and Meinke, D. (1991). Embryonic lethals and T-DNA insertional mutagenesis in Arabidopsis. *Plant Cell*, **3**, 149–157.

Feinberg, E. H. and Hunter, C. P. (2003). Transport of dsRNA into cells by the transmembrane protein SID-1. *Science*, **301**, 1545–1547.

Fink, C. C., Bayer, K. U., Myers, J. W., Ferrell, J. E., Jr., Schulman, H. and Meyer, T. (2003). Selective regulation of neurite extension and synapse formation by the beta but not the alpha isoform of CaMKII. *Neuron*, **39**, 283–297.

Finnegan, E. J., Margis, R. and Waterhouse, P. M. (2003). Posttranscriptional gene silencing is not compromised in the Arabidopsis CARPEL FACTORY (DICER-LIKE1) mutant, a homolog of Dicer-1 from *Drosophila*. *Current Biology*, **13**, 236–240.

Fire, A. (1999). RNA-triggered gene silencing. *Trends in Genetics*, **15**, 358–363.

Fire, A., Albertson, D., Harrison, S. W. and Moerman, D. G. (1991). Production of antisense RNA leads to effective and specific inhibition of gene expression in C. elegans muscle. *Development*, **113**, 503–514.

Fire, A., Xu, S., Montgomery, M. K., Kostas, S. A., Driver, S. E. and Mello, C. C. (1998). Potent and specific genetic interference by double-stranded RNA in Caenorhabditis elegans. *Nature*, **391**, 806–811.

Gitlin, L., Karelsky, S. and Andino, R. (2002). Short interfering RNA confers intracellular antiviral immunity in human cells. *Nature*, **418**, 430–434.

Golden, T. A., Schauer, S. E., Lang, J. D., Pien, S., Mushegian, A. R., Grossniklaus, U., Meinke, D. W. and Ray, A. (2002). SHORT INTEGUMENTS1/SUSPENSOR1/CARPEL FACTORY, a Dicer homolog, is a maternal effect gene required for embryo development in Arabidopsis. *Plant Physiology*, **130**, 808–822.

Gonczy, P., Echeverri, C., Oegema, K., Coulson, A., Jones, S. J., Copley, R. R., Duperon, J., Oegema, J., Brehm, M., Cassin, E., Hannak, E., Kirkham, M., Pichler, S., Flohrs, K., Goessen, A., Leidel, S., Alleaume, A. M., Martin, C., Ozlu, N., Bork, P. and Hyman, A. A. (2000). Functional genomic analysis of cell division in C. elegans using RNAi of genes on chromosome III. *Nature*, **408**, 331–336.

Grant, S. R. (1999). Dissecting the mechanisms of posttranscriptional gene silencing: Divide and conquer. *Cell*, **96**, 303–306.

Grishok, A. and Mello, C. C. (2002). RNAi (Nematodes: *Caenorhabditis elegans*). *Advances in Genetics*, **46**, 339–360.

Grishok, A., Pasquinelli, A. E., Conte, D., Li, N., Parrish, S., Ha, I., Baillie, D. L., Fire, A., Ruvkun, G. and Mello, C. C. (2001). Genes and mechanisms related to RNA interference regulate expression of the small temporal RNAs that control C. elegans developmental timing. *Cell*, **106**, 23–34.

Guo, S. and Kemphues, K. J. (1995). par-1, a gene required for establishing polarity in C. elegans embryos, encodes a putative Ser/Thr kinase that is asymmetrically distributed. *Cell*, **81**, 611–620.

Hamilton, A., Voinnet, O., Chappell, L. and Baulcombe, D. (2002). Two classes of short interfering RNA in RNA silencing. *European Molecular Biology Organization Journal*, **21**, 4671–4679.

Hamilton, A. J. and Baulcombe, D. C. (1999). A species of small antisense RNA in posttranscriptional gene silencing in plants. *Science*, **286**, 950–952.

Hammond, S. M., Bernstein, E., Beach, D. and Hannon, G. J. (2000). An RNA-directed nuclease mediates post-transcriptional gene silencing in *Drosophila* cells. *Nature*, **404**, 293–296.

Hammond, S. M., Caudy, A. A. and Hannon, G. J. (2001). Post-transcriptional gene silencing by double-stranded RNA. *Nature Reviews Genetics*, **2**, 110–119.

Hannon, G. J. (2002). RNA interference. *Nature*, **418**, 244–251.

Hemann, M. T., Fridman, J. S., Zilfou, J. T., Hernando, E., Paddison, P. J., Cordon-Cardo, C., Hannon, G. J. and Lowe, S. W. (2003). An epi-allelic series of p53 hypomorphs created by stable RNAi produces distinct tumor phenotypes *in vivo*. *Nature Genetics*, **33**, 396–400.

Houbaviy, H. B., Murray, M. F. and Sharp, P. A. (2003). Embryonic stem cell-specific Micro-RNAs. *Developmental Cell*, **5**, 351–358.

Hunter, C. P. (2000). Gene silencing: shrinking the black box of RNAi. *Current Biology*, **10**, R137–140.

Hütvagner, G., McLachlan, J., Pasquinelli, A. E., Balint, E., Tuschl, T. and Zamore, P. D. (2001). A cellular function for the RNA-interference enzyme Dicer in the maturation of the let-7 small temporal RNA. *Science*, **293**, 834–838.

Hütvagner, G. and Zamore, P. D. (2002a). A microRNA in a multiple-turnover RNAi enzyme complex. *Science*, **297**, 2056–2060.

Hütvagner, G. and Zamore, P. D. (2002b). RNAi: Nature abhors a double-strand. *Current Opinion in Genetics and Development*, **12**, 225–232.

Jackson, A. L., Bartz, S. R., Schelter, J., Kobayashi, S. V., Burchard, J., Mao, M., Li, B., Cavet, G. and Linsley, P. S. (2003). Expression profiling reveals off-target gene regulation by RNAi. *Nature Biotechnology*, **21**, 635–637.

Jacobsen, S. E., Running, M. P. and Meyerowitz, E. M. (1999). Disruption of an RNA helicase/RNAse III gene in Arabidopsis causes unregulated cell division in floral meristems. *Development*, **126**, 5231–5243.

Jones, L., Hamilton, A. J., Voinnet, O., Thomas, C. L., Maule, A. J. and Baulcombe, D. C. (1999). RNA-DNA interactions and DNA methylation in post-transcriptional gene silencing. *Plant Cell*, **11**, 2291–2301.

Jorgensen, R. (1990). Altered gene expression in plants due to trans-interactions between homologous genes. *Trends in Biotechnology*, **8**, 340–344.

Kamath, R. S. and Ahringer, J. (2003). Genome-wide RNAi screening in Caenorhabditis elegans. *Methods*, **30**, 313–321.

Kamath, R. S., Fraser, A. G., Dong, Y., Poulin, G., Durbin, R., Gotta, M., Kanapin, A., Le Bot, N., Moreno, S., Sohrmann, M., Welchman, D. P., Zipperlen, P. and Ahringer, J. (2003). Systematic functional analysis of the *Caenorhabditis elegans* genome using RNAi. *Nature*, **421**, 231–237.

Kawasaki, H., Suyama, E., Iyo, M. and Taira, K. (2003). siRNAs generated by recombinant human Dicer induce specific and significant but target site-independent gene silencing in human cells. *Nucleic Acids Research*, **31**, 981–987.

Kennerdell, J. R. and Carthew, R. W. (1998). Use of dsRNA-mediated genetic interference to demonstrate that frizzled and frizzled 2 act in the wingless pathway. *Cell*, **95**, 1017–1026.

Ketting, R. F., Fischer, S. E., Bernstein, E., Sijen, T., Hannon, G. J. and Plasterk, R. H. (2001). Dicer functions in RNA interference and in synthesis of small RNA involved in developmental timing in C. elegans. *Genes & Development*, **15**, 2654–2659.

Ketting, R. F., Haverkamp, T. H., van Luenen, H. G. and Plasterk, R. H. (1999). Mut-7 of C. elegans, required for transposon silencing and RNA interference, is a homolog of Werner syndrome helicase and RNaseD. *Cell*, **99**, 133–141.

Khvorova, A., Reynolds, A. and Jayasena, S. D. (2003). Functional siRNAs and miRNAs exhibit strand bias. *Cell*, **115**, 209–216.

Kiger, A., Baum, B., Jones, S., Jones, M., Coulson, A., Echeverri, C. and Perrimon, N. (2003). A functional genomic analysis of cell morphology using RNA interference. *Journal of Biology*, **2**, 27.

Knight, S. W. and Bass, B. L. (2001). A role for the RNase III enzyme DCR-1 in RNA interference and germ line development in Caenorhabditis elegans. *Science*, **293**, 2269–2271.

Kostura, M. and Mathews, M. B. (1989). Purification and activation of the double-stranded RNA-dependent eIF-2 kinase DAI. *Molecular and Cellular Biology*, **9**, 1576–1586.

Kumagai, M. H., Donson, J., della-Cioppa, G., Harvey, D., Hanley, K. and Grill, L. K. (1995). Cytoplasmic inhibition of carotenoid biosynthesis with virus-derived RNA. *Proceedings of the National Academy of Sciences USA*, **92**, 1679–1683.

Lang, J. D., Ray, S. and Ray, A. (1994). sin 1, a mutation affecting female fertility in Arabidopsis, interacts with mod 1, its recessive modifier. *Genetics*, **137**, 1101–1110.

Lassus, P., Opitz-Araya, X. and Lazebnik, Y. (2002). Requirement for caspase-2 in stress-induced apoptosis before mitochondrial permeabilization. *Science*, **297**, 1352–1354.

Lassus, P., Rodriguez, J. and Lazebnik, Y. (2002). Confirming specificity of RNAi in mammalian cells. *Science's STKE: Signal Transduction Knowledge Environment*, **2002**, PL13.

Lee, N. S., Dohjima, T., Bauer, G., Li, H., Li, M. J., Ehsani, A., Salvaterra, P. and Rossi, J. (2002a). Expression of small interfering RNAs targeted against HIV-1 rev transcripts in human cells. *Nature Biotechnology*, **20**, 500–505.

Lee, Y., Jeon, K., Lee, J. T., Kim, S. and Kim, V. N. (2002b). MicroRNA maturation: Stepwise processing and subcellular localization. *European Molecular Biology Organization Journal*, **21**, 4663–4670.

Lee, R. C., Feinbaum, R. L. and Ambros, V. (1993). The *C. elegans* heterochronic gene lin-4 encodes small RNAs with antisense complementarity to lin-14. *Cell*, **75**, 843–854.

Lee, S. S., Lee, R. Y., Fraser, A. G., Kamath, R. S., Ahringer, J. and Ruvkun, G. (2003a). A systematic RNAi screen identifies a critical role for mitochondria in *C. elegans* longevity. *Nature Genetics*, **33**, 40–48.

Lee, Y., Ahn, C., Han, J., Choi, H., Kim, J., Yim, J., Lee, J., Provost, P., Radmark, O., Kim, S. and Kim, V. N. (2003b). The nuclear RNase III Drosha initiates microRNA processing. *Nature*, **425**, 415–419.

Lim, L. P., Glasner, M. E., Yekta, S., Burge, C. B. and Bartel, D. P. (2003). Vertebrate microRNA genes. *Science*, **299**, 1540.

Lingel, A., Simon, B., Izaurralde, E. and Sattler, M. (2003). Structure and nucleic-acid binding of the *Drosophila* Argonaute 2 PAZ domain. *Nature*, **426**, 465–469.

Lingel, A., Simon, B., Izaurralde, E. and Sattler, M. (2004). Nucleic acid 3′-end recognition by the Argonaute2 PAZ domain. *Nature, Structural and Molecular Biology*, **11**, 576–577.

Llave, C., Kasschau, K. D., Rector, M. A. and Carrington, J. C. (2002a). Endogenous and silencing-associated small RNAs in plants. *Plant Cell*, **14**, 1605–1619.

Llave, C., Xie, Z., Kasschau, K. D. and Carrington, J. C. (2002b). Cleavage of scarecrow-like mRNA targets directed by a class of Arabidopsis miRNA. *Science*, **297**, 2053–2056.

Lum, L., Yao, S., Mozer, B., Rovescalli, A., Von Kessler, D., Nirenberg, M. and Beachy, P. A. (2003). Identification of hedgehog pathway components by RNAi in *Drosophila* cultured cells. *Science*, **299**, 2039–2045.

Lustig, K. D., Stukenberg, P. T., McGarry, T. J., King, R. W., Cryns, V. L., Mead, P. E., Zon, L. I., Yuan, J. and Kirschner, M. W. (1997). Small pool expression screening: Identification of genes involved in cell cycle control, apoptosis, and early development. *Methods in Enzymolgy*, **283**, 83–99.

Ma, J. B., Ye, K. and Patel, D. J. (2004). Structural basis for overhang-specific small interfering RNA recognition by the PAZ domain. *Nature*, **429**, 318–322.

Manche, L., Green, S. R., Schmedt, C. and Mathews, M. B. (1992). Interactions between double-stranded RNA regulators and the protein kinase DAI. *Molecular and Cellular Biology*, **12**, 5238–5248.

Matta, H., Hozayev, B., Tomar, R., Chugh, P. and Chaudhary, P. M. (2003). Use of lentiviral vectors for delivery of small interfering RNA. *Cancer Biology and Therapy*, **2**, 206–210.

McElver, J., Tzafrir, I., Aux, G., Rogers, R., Ashby, C., Smith, K., Thomas, C., Schetter, A., Zhou, Q., Cushman, M. A., Tossberg, J., Nickle, T., Levin, J. Z., Law, M., Meinke, D. and Patton, D. (2001). Insertional mutagenesis of genes required for seed development in Arabidopsis thaliana. *Genetics*, **159**, 1751–1763.

Mette, M. F., Aufsatz, W., van der Winden, J., Matzke, M. A. and Matzke, A. J. (2000). Transcriptional silencing and promoter methylation triggered by double-stranded RNA. *European Molecular Biology Organization Journal*, **19**, 5194–5201.

Minks, M. A., West, D. K., Benvin, S. and Baglioni, C. (1979). Structural requirements of double-stranded RNA for the activation of 2′,5′-oligo(A) polymerase and protein kinase of interferon-treated HeLa cells. *Journal of Biological Chemistry*, **254**, 10180–10183.

Miyagishi, M. and Taira, K. (2002). U6 promoter-driven siRNAs with four uridine 3′ overhangs efficiently suppress targeted gene expression in mammalian cells. *Nature Biotechnology*, **20**, 497–500.

Moss, E. G. (2000). Non-coding RNAs: lightning strikes twice. *Current Biology*, **10**, R436–439.

Myers, J. W., Jones, J. T., Meyer, T. and Ferrell, J. E., Jr. (2003). Recombinant Dicer efficiently converts large dsRNAs into siRNAs suitable for gene silencing. *Nature Biotechnology*, **21**, 324–328.

Oefner, P. J. (2002). Sequence variation and the biological function of genes: Methodological and biological considerations. *Journal of Chromatography. B, Analytical Technologies in the Biomedical and Life Sciences*, **782**, 3–25.

Olsen, P. H. and Ambros, V. (1999). The lin-4 regulatory RNA controls developmental timing in *Caenorhabditis elegans* by blocking LIN-14 protein synthesis after the initiation of translation. *Developmental Biology*, **216**, 671–680.

Paddison, P. J., Caudy, A. A., Bernstein, E., Hannon, G. J. and Conklin, D. S. (2002a). Short hairpin RNAs (shRNAs) induce sequence-specific silencing in mammalian cells. *Genes & Development*, **16**, 948–958.

Paddison, P. J., Caudy, A. A. and Hannon, G. J. (2002b). Stable suppression of gene expression by RNAi in mammalian cells. *Proceedings of the National Academy of Sciences USA*, **99**, 1443–1448.

Paddison, P. J. and Hannon, G. J. (2002). RNA interference: The new somatic cell genetics? *Cancer Cell*, **2**, 17–23.

Papp, I., Mette, M. F., Aufsatz, W., Daxinger, L., Schauer, S. E., Ray, A., van der Winden, J., Matzke, M. and Matzke, A. J. (2003). Evidence for nuclear processing of plant micro RNA and short interfering RNA precursors. *Plant Physiology*, **132**, 1382–1390.

Park, W., Li, J., Song, R., Messing, J. and Chen, X. (2002). CARPEL FACTORY, a Dicer homolog, and HEN1, a novel protein, act in microRNA metabolism in Arabidopsis thaliana. *Current Biology*, **12**, 1484–1495.

Parrish, S., Fleenor, J., Xu, S., Mello, C. and Fire, A. (2000). Functional anatomy of a dsRNA trigger. Differential requirement for the two trigger strands in RNA interference. *Molecular Cell*, **6**, 1077–1087.

Pasquinelli, A. E. and Ruvkun, G. (2002). Control of developmental timing by micrornas and their targets. *Annual Review of Cell and Developmental Biology*, **18**, 495–513.

Paul, C. P., Good, P. D., Winer, I. and Engelke, D. R. (2002). Effective expression of small interfering RNA in human cells. *Nature Biotechnology*, **20**, 505–508.

Pothof, J., van Haaften, G., Thijssen, K., Kamath, R. S., Fraser, A. G., Ahringer, J., Plasterk, R. H. and Tijsterman, M. (2003). Identification of genes that protect the C. elegans genome against mutations by genome-wide RNAi. *Genes & Development*, **17**, 443–448.

Provost, P., Dishart, D., Doucet, J., Frendewey, D., Samuelsson, B. and Radmark, O. (2002a). Ribonuclease activity and RNA binding of recombinant human Dicer. *European Molecular Biology Organization Journal*, **21**, 5864–5874.

Provost, P., Silverstein, R. A., Dishart, D., Walfridsson, J., Djupedal, I., Kniola, B., Wright, A., Samuelsson, B., Radmark, O. and Ekwall, K. (2002b). Dicer is required for chromosome segregation and gene silencing in fission yeast cells. *Proceedings of the National Academy of Sciences USA*, **99**, 16648–16653.

Ratcliff, F. G., MacFarlane, S. A. and Baulcombe, D. C. (1999). Gene silencing without DNA. rna-mediated cross-protection between viruses. *Plant Cell*, **11**, 1207–1216.

Ray, A., Lang, J. D., Golden, T. and Ray, S. (1996a). SHORT INTEGUMENT (SIN1), a gene required for ovule development in Arabidopsis, also controls flowering time. *Development*, **122**, 2631–2638.

Ray, S., Golden, T. and Ray, A. (1996b). Maternal effects of the short integument mutation on embryo development in Arabidopsis. *Developmental Biology*, **180**, 365–369.

Reinhart, B. J. and Bartel, D. P. (2002a). Small RNAs correspond to centromere heterochromatic repeats. *Science*, **297**, 1831.

Reinhart, B. J., Weinstein, E. G., Rhoades, M. W., Bartel, B. and Bartel, D. P. (2002b). MicroRNAs in plants. *Genes & Development*, **16**, 1616–1626.

Reynolds, A., Leake, D., Boese, Q., Scaringe, S., Marshall, W. S. and Khvorova, A. (2004). Rational siRNA design for RNA interference. *Nature Biotechnology*, **22**, 326–330.

Robertson, H. D. and Hunter, T. (1975). Sensitive methods for the detection and characterization of double helical ribonucleic acid. *Journal of Biological Chemistry*, **250**, 418–425.

Robinson-Beers, K., Pruitt, R. E. and Gasser, C. S. (1992). Ovule development in wild-type Arabidopsis and two female-sterile mutants. *Plant Cell*, **4**, 1237–1249.

Rubinson, D. A., Dillon, C. P., Kwiatkowski, A. V., Sievers, C., Yang, L., Kopinja, J., Rooney, D. L., Ihrig, M. M., McManus, M. T., Gertler, F. B., Scott, M. L. and Van Parijs, L. (2003). A lentivirus-based system to functionally silence genes in primary mammalian cells, stem cells and transgenic mice by RNA interference. *Nature Genetics*, **33**, 401–406.

Ruiz, M. T., Voinnet, O. and Baulcombe, D. C. (1998). Initiation and maintenance of virus-induced gene silencing. *Plant Cell*, **10**, 937–946.

Samuel, C. E. (2001). Antiviral actions of interferons. *Clinical Microbiology Reviews*, **14**, 778–809.

Schauer, S. E., Jacobsen, S. E., Meinke, D. W. and Ray, A. (2002). DICER-LIKE1: Blind men and elephants in Arabidopsis development. *Trends in Plant Sciences*, **7**, 487–491.

Scherr, M., Battmer, K., Ganser, A. and Eder, M. (2003). Modulation of gene expression by lentiviral-mediated delivery of small interfering RNA. *Cell Cycle*, **2**, 251–257.

Schramke, V. and Allshire, R. (2003). Hairpin RNAs and retrotransposon LTRs effect RNAi and chromatin-based gene silencing. *Science*, **301**, 1069–1074.

Schwarz, D. S., Hütvagner, G., Du, T., Xu, Z., Aronin, N. and Zamore, P. D. (2003). Asymmetry in the assembly of the RNAi enzyme complex. *Cell*, **115**, 199–208.

Schwarz, D. S., Hütvagner, G., Haley, B. and Zamore, P. D. (2002). Evidence that siRNAs function as guides, not primers, in the *Drosophila* and human RNAi pathways. *Molecular Cell*, **10**, 537–548.

Semizarov, D., Frost, L., Sarthy, A., Kroeger, P., Halbert, D. N. and Fesik, S. W. (2003). Specificity of short interfering RNA determined through gene expression signatures. *Proceedings of the National Academy of Sciences USA*, **100**, 6347–6352.

Sharp, P. A. (2001). RNA interference – 2001. *Genes & Development*, **15**, 485–490.

Shen, C., Buck, A. K., Liu, X., Winkler, M. and Reske, S. N. (2003). Gene silencing by adenovirus-delivered siRNA. *Federation of European Biochemical Society Letters*, **539**, 111–114.

Sijen, T. and Plasterk, R. H. (2003). Transposon silencing in the Caenorhabditis elegans germ line by natural RNAi. *Nature*, **426**, 310–314.

Simmer, F., Moorman, C., Van Der Linden, A. M., Kuijk, E., Van Den Berghe, P. V., Kamath, R., Fraser, A. G., Ahringer, J. and Plasterk, R. H. (2003). Genome-wide RNAi of C. elegans using the hypersensitive rrf-3 strain reveals novel Gene Functions. *Public Library of Science Biology*, **1**, E12.

Sledz, C. A., Holko, M., de Veer, M. J., Silverman, R. H. and Williams, B. R. (2003). Activation of the interferon system by short-interfering RNAs. *Nature Cell Biology*, **5**, 834–839.

Smith, W. C. and Harland, R. M. (1992). Expression cloning of noggin, a new dorsalizing factor localized to the Spemann organizer in Xenopus embryos. *Cell*, **70**, 829–840.

Song, J. J., Liu, J., Tolia, N. H., Schneiderman, J., Smith, S. K., Martienssen, R. A., Hannon, G. J. and Joshua-Tor, L. (2003). The crystal structure of the Argonaute2 PAZ domain reveals an RNA binding motif in RNAi effector complexes. *Nature Structural Biology*, **10**, 1026–1032.

Stark, G. R., Kerr, I. M., Williams, B. R., Silverman, R. H. and Schreiber, R. D. (1998). How cells respond to interferons. *Annual Review of Biochemistry*, **67**, 227–264.

Tabara, H., Sarkissian, M., Kelly, W. G., Fleenor, J., Grishok, A., Timmons, L., Fire, A. and Mello, C. C. (1999). The rde-1 gene, RNA interference, and transposon silencing in C. elegans. *Cell*, **99**, 123–132.

Tang, G., Reinhart, B. J., Bartel, D. P. and Zamore, P. D. (2003). A biochemical framework for RNA silencing in plants. *Genes & Development*, **17**, 49–63.

Thoma, C., Hasselblatt, P., Kock, J., Chang, S. F., Hockenjos, B., Will, H., Hentze, M. W., Blum, H. E., von Weizsacker, F. and Offensperger, W. B. (2001). Generation of stable mRNA fragments and translation of N-truncated proteins induced by antisense oligodeoxynucleotides. *Molecular Cell*, **8**, 865–872.

Tijsterman, M., Ketting, R. F. and Plasterk, R. H. (2002). The genetics of RNA silencing. *Annual Review of Genetics*, **36**, 489–519.

Timmons, L. and Fire, A. (1998). Specific interference by ingested dsRNA. *Nature*, **395**, 854.

Tomar, R. S., Matta, H. and Chaudhary, P. M. (2003). Use of adeno-associated viral vector for delivery of small interfering RNA. *Oncogene*, **22**, 5712–5715.

Tuschl, T., Zamore, P. D., Lehmann, R., Bartel, D. P. and Sharp, P. A. (1999). Targeted mRNA degradation by double-stranded RNA *in vitro*. *Genes & Development*, **13**, 3191–3197.

Vastenhouw, N. L., Fischer, S. E., Robert, V. J., Thijssen, K. L., Fraser, A. G., Kamath, R. S., Ahringer, J. and Plasterk, R. H. (2003). A genome-wide screen identifies 27 genes involved in transposon silencing in C. elegans. *Current Biology*, **13**, 1311–1316.

Vaucheret, H., Beclin, C., Elmayan, T., Feuerbach, F., Godon, C., Morel, J. B., Mourrain, P., Palauqui, J. C. and Vernhettes, S. (1998). Transgene-induced gene silencing in plants. *Plant Journal*, **16**, 651–659.

Vickers, T. A., Koo, S., Bennett, C. F., Crooke, S. T., Dean, N. M. and Baker, B. F. (2003). Efficient reduction of target RNAs by small interfering RNA and RNase H-dependent antisense agents. A comparative analysis. *Journal of Biological Chemistry*, **278**, 7108–7118.

Volpe, T., Schramke, V., Hamilton, G. L., White, S. A., Teng, G., Martienssen, R. A. and Allshire, R. C. (2003). RNA interference is required for normal centromere function in fission yeast. *Chromosome Research*, **11**, 137–146.

Volpe, T. A., Kidner, C., Hall, I. M., Teng, G., Grewal, S. I. and Martienssen, R. A. (2002). Regulation of heterochromatic silencing and histone H3 lysine-9 methylation by RNAi. *Science*, **297**, 1833–1837.

Winston, W. M., Molodowitch, C. and Hunter, C. P. (2002). Systemic RNAi in C. elegans requires the putative transmembrane protein SID-1. *Science*, **295**, 2456–2459.

Yan, K. S., Yan, S., Farooq, A., Han, A., Zeng, L. and Zhou, M. M. (2003). Structure and conserved RNA binding of the PAZ domain. *Nature*, **426**, 468–474.

Yang, D., Buchholz, F., Huang, Z., Goga, A., Chen, C. Y., Brodsky, F. M. and Bishop, J. M. (2002). Short RNA duplexes produced by hydrolysis with Escherichia coli RNase III mediate effective RNA interference in mammalian cells. *Proceedings of the National Academy of Sciences USA*, **99**, 9942–9947.

Yang, D., Lu, H. and Erickson, J. W. (2000). Evidence that processed small dsRNAs may mediate sequence-specific mRNA degradation during RNAi in *Drosophila* embryos. *Current Biology*, **10**, 1191–1200.

Yu, J. Y., DeRuiter, S. L. and Turner, D. L. (2002). RNA interference by expression of short-interfering RNAs and hairpin RNAs in mammalian cells. *Proceedings of the National Academy of Sciences USA*, **99**, 6047–6052.

Zamore, P. D. (2002). Ancient pathways programmed by small RNAs. *Science*, **296**, 1265–1269.

Zamore, P. D., Tuschl, T., Sharp, P. A. and Bartel, D. P. (2000). RNAi: Double-stranded RNA directs the ATP-dependent cleavage of mRNA at 21 to 23 nucleotide intervals. *Cell*, **101**, 25–33.

Zhang, H., Kolb, F. A., Brondani, V., Billy, E. and Filipowicz, W. (2002). Human Dicer preferentially cleaves dsRNAs at their termini without a requirement for ATP. *European Molecular Biology Organization Journal*, **21**, 5875–5885.

Zhang, H., Kolb, F. A., Jaskiewicz, L., Westhof, E. and Filipowicz, W. (2004). Single processing center models for human Dicer and bacterial RNase III. *Cell*, **118**, 57–68.

3 Genes required for RNA interference

Nathaniel R. Dudley, Ahmad Z. Amin, and Bob Goldstein

Introduction

RNA interference (RNAi) is a recently discovered phenomenon in which double-stranded RNA (dsRNA) silences endogenous gene expression in a sequence-specific manner (Fire et al., 1998). Since its discovery, the use of RNAi has become widely employed in many organisms to specifically knock down gene function. RNAi shares a remarkable degree of similarity with silencing phenomena in other organisms (Cogoni et al., 1999a; Sharp, 1999). For instance, RNAi, post-transcriptional gene silencing in plants and cosuppression in fungi can all be activated by the presence of aberrant RNAs (Maine, 2000; Tijsterman et al., 2002a). Additionally, plant, worm, and fly cells or extracts undergoing RNA-mediated interference all contain small dsRNAs, around 25 nucleotides in length, identical to the sequences present in the silenced gene (Baulcombe, 1996; Hammond et al., 2000; Zamore et al., 2000; Catalanotto et al., 2000).

The high degree of similarity between these RNA-mediated silencing phenomena supports the notion that they were derived from an ancient and conserved pathway used to regulate gene expression, presumably to eliminate defective RNAs and to defend against viral infections and transposons (Zamore, 2002). Components of RNAi have also been implicated in developmental processes, suggesting that RNAi may play a broader role in regulating gene expression (Smardon et al., 2000; Knight et al., 2001; et al., Ketting et al., 2001).

Although we have learned much about the general mechanisms underlying RNAi, a detailed understanding of how RNAi works remains to be elucidated. In this chapter we will discuss first the biology of RNAi, then the genes required for its function, and we will end with a discussion on recent findings that have implicated chromatin silencing in the mechanism of RNAi.

The biology of RNAi

Both genetic and biochemical analyses have significantly increased our understanding of how RNAi works. The RNAi mechanism involves an early step, in

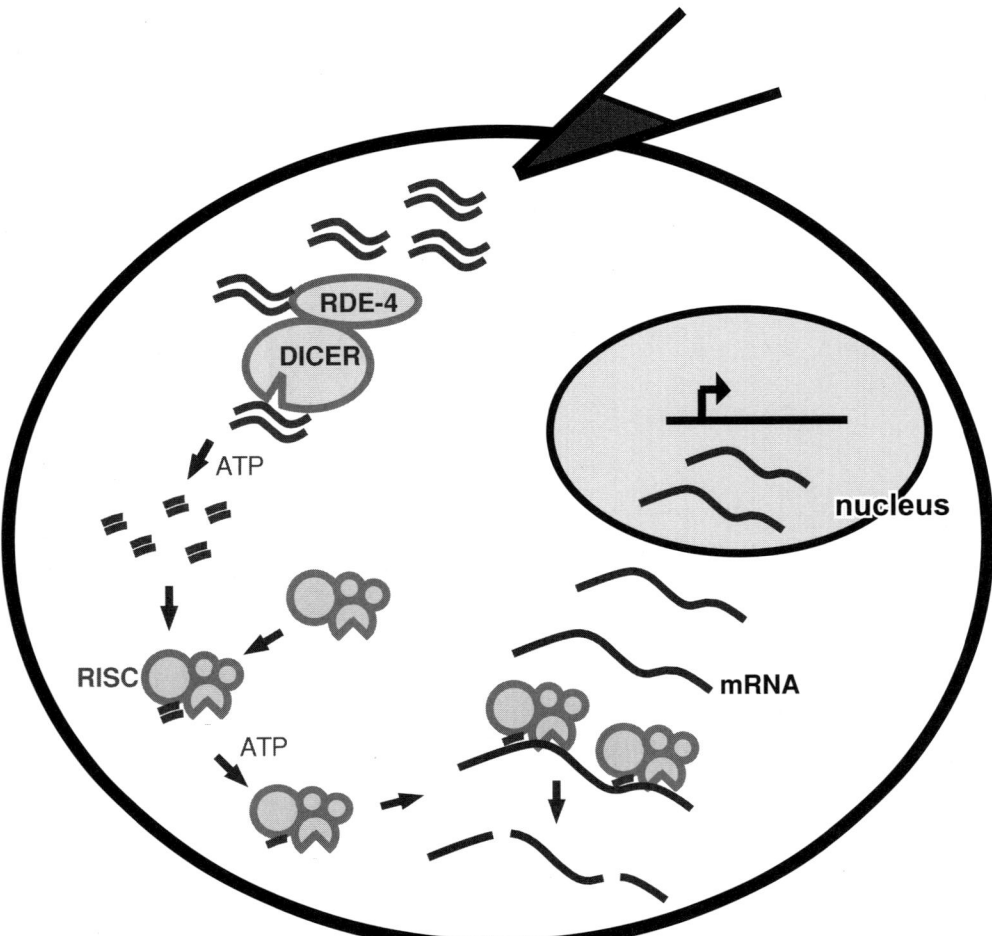

Figure 3.1. Model for mRNA degradation in the cytoplasm by RNAi. Introduced dsRNAs are recognized by RDE-4/R2D2, a dsRNA binding protein. These dsRNAs are then processed by Dicer into 21-23nt duplexes that can associate with an enzyme complex called RISC. After unwinding of the siRNAs, RISC becomes competent to target homologous mRNA transcripts for degradation. (See color section.)

which the dsRNA is recognized and is targeted for RNase-dependent digestion, and a late step, which comprises the downstream events that lead to the silencing of the target gene.

Two-step model for RNAi

Figure 3.1 illustrates a current model by which introduction of dsRNAs into a cell can result in degradation of targeted mRNAs. Introduced dsRNAs are recognized by a dsRNA-binding protein, RDE-4/R2D2 (Tabara et al., 2002; Liu et al., 2003), that facilitates the subsequent dicing of these RNAs into small interfering RNAs (siRNAs) 21–25 nucleotides in length, with 2-nucleotide overhangs at both 3′-ends. Dicing is catalyzed by an RNase III enzyme named Dicer (Hammond et al., 2000; Bernstein et al., 2001; Hütvagner and Zamore, 2002; Elbashir et al., 2001a). These siRNAs then act as guides in association with a protein complex to target homologous transcripts for degradation. This siRNA/protein complex, termed RISC, for RNA-Induced Silencing Complex (Hammond et al., 2000), becomes competent to target degradation of homologous mRNAs upon the ATP-dependent unwinding of the siRNAs (Nykanen et al., 2001).

siRNAs

siRNAs are complementary to both the sense and antisense strands of the targeted mRNA and have a distinct chemical polarity that is essential for their function (Zamore et al., 2000; Elbashir et al., 2001a; Nykanen et al., 2001; Schwarz et al., 2002; Schwarz et al., 2003). For instance, efficient siRNA mediated interference requires that the siRNAs contain 2 bp overhangs at their 3′ ends as well as a 5′ phosphate and, at least in *Drosophila*, a 3′ hydroxyl group. Additionally, the base composition at the 5′ end can influence which siRNA strand can initiate RNAi. It has also been shown that the sequence composition of the antisense strand is more important than that of the sense strand, as modifications on the antisense strand of the dsRNA trigger preferentially blocks RNAi (Elbashir et al., 2001b).

Long dsRNAs cannot be used for gene silencing in some organisms, including mammals, because of the presence of dsRNA defense mechanisms. In these organisms, the introduction of dsRNA leads to the activation of the protein kinase PKR and 2′,5′-oligoadenylate synthetase. The activation of these two proteins leads to non-sequence-specific effects including the inhibition of translation and the degradation of mRNA (Sen et al., 1976; Stark et al., 1998). However, siRNAs alone can elicit a potent and effective RNAi response. It has been shown that chemically synthesized siRNAs, introduced into a variety of mammalian cell lines, could also specifically inhibit endogenous gene expression (Elbashir et al., 2001c). Moreover, RNAi has been shown to work in model organisms such as mice, *Drosophila*, *C. elegans*, and zebrafish, thus expanding the potential for its use in understanding gene function in a variety of model systems (Fire et al., 1998; Kennerdell and Carthew, 1998; Wargelius et al. 1999; Hunter, 1999; Wianny and Zernicka-Goetz, 2000).

Systemic nature of RNAi

One interesting aspect of RNAi, which was originally observed in plants, is that the effect can spread between tissues, in that gene silencing induced in one tissue can result in gene silencing in other tissues (Palauqui et al., 1997; Fire et al., 1998; Grishok et al., 2000). This suggests the existence of a mechanism to uptake and transport a silencing signal between cells.

Amplification of the RNAi response and transitive RNAi

Another intriguing aspect of RNAi is that the effect is surprisingly robust. Concentrations as low as a few molecules of dsRNA per cell can elicit a strong and persistent response (Fire et al., 1998), suggesting that an amplification step may occur within the RNAi pathway.

In some organisms, it is likely that the amplification mechanism involves RISC acting catalytically (Sijen et al., 2001) as amplification could be effected by

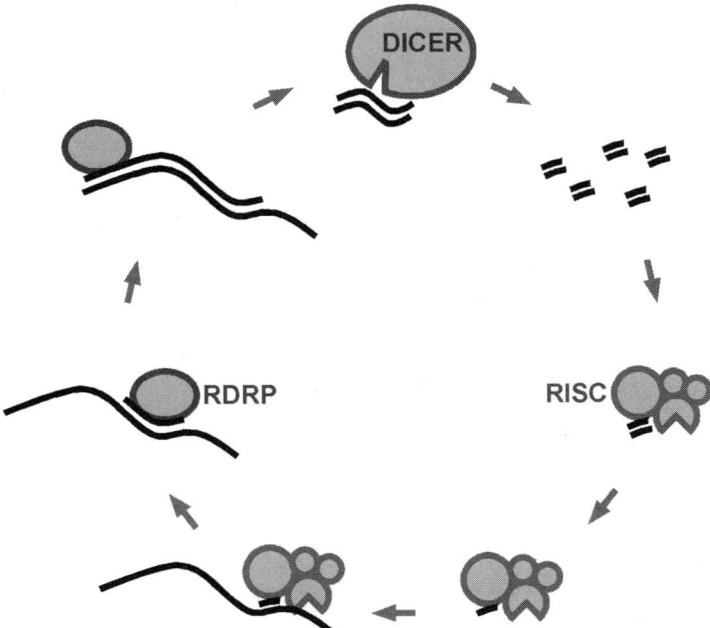

Figure 3.2. Amplification of dsRNA by an RNA-dependent RNA polymerase (RDRP). In certain organisms, new dsRNAs can be generated by RDRPs, primed by siRNAs on mRNA targets. The new dsRNAs can be used subsequently by Dicer to create more siRNAs, which can lead to additional rounds of amplification. (See color section.)

multiple turnover of RISC (Hammond et al., 2001; Sharp, 2001). In other organisms, the activity of RNA-dependent RNA polymerases (RDRPs) has been implicated in the amplification mechanism (Smardon et al., 2000; Maine, 2000; Sijen et al., 2001; Simmer et al., 2002). The presence of siRNAs with sequences 5′ to the initial trigger have been demonstrated in some animals undergoing RNAi (Figure 3.2). These siRNAs, called secondary siRNAs, correspond to sequences just upstream of targeted sequence on the same transcript and are produced by de novo RNA synthesis by an RDRP (Sijen et al., 2001). However, transitive RNAi does not appear to occur in cultured *Drosophila* cells (Celotto and Graveley, 2002; Schwarz et al., 2002; Roignant et al., 2003).

These observations suggest that in certain organisms, mRNAs are not only targets of the RNAi machinery but can also be used to amplify the original signal. One consideration for researchers using RNAi as a gene silencing tool is that transitive RNAi may also lead to inactivation of mRNA species that were not originally targeted by the initial trigger sequence; for example, RNAi of Green Fluorescent Protein (GFP) in a strain bearing a GFP fusion transgene can in certain cases result in silencing of the gene to which GFP is fused (Sijen et al., 2001; Alder et al., 2003).

A link between RNAi and microRNAs
Although not a product of dsRNA-mediated interference, microRNAs (miRNAs) have also been shown to be key components of RNA-based gene regulation in

Table 3.1. Proteins implicated in RNAi and related RNA silencing phenomena

	C. elegans	Drosophila	Humans	Plants	Fungi
Dicer RNase:	DCR-1	Dicer	Dicer	CAF/Sin-1	Dicer
RNA-dependent RNA polymerases:	EGO-1, RRF-1, RRF-3			SGS2/S DE1	QDE1, RDRP, RrpA
Proteins with PAZ/piwi domains:	RDE-1	AGO2, Piwi, Aubergine	eIF2C1/2	AGO1, AGO4	QDE2, Ago1, Ago2
Nucleases:	MUT-7	Tudor-SN		WEX-1	
Helicases:	MUT-14, DRH-1/2	p68, Spindle-E		MUT6, SDE3	QDE3
Chromatin modifiers:	MES-3, -4, -6			DDM1, MET1	
dsRNA-binding protein:	RDE-4	R2D2			
Nonsense-mediated decay	SMG2, SMG-5, SMG-6				
Other proteins:	SID-1	FMRP, dFXR, VIG		SGS3, HEN1	

For references, see text and Cogoni and Macino, 1999a,b; Dalmay et al., 2000; Domeier et al., 2000; Mourrain et al., 2000; Wu-Scharf et al., 2000; Dalmay et al., 2001; Vaucheret et al., 2001; Kennerdell et al., 2002; Martens et al., 2002; Schauer et al., 2002; Bateman, 2002; Boutet et al., 2003; Doi et al., 2003; Glazov et al., 2003; Zilberman et al., 2003.

many organisms. In contrast to siRNAs, miRNAs are derived from the processing of endogenously encoded short hairpin RNAs. However, miRNAs are dependent on Dicer for processing (Carrington and Ambros, 2003) and associate with a complex that shares components present in RISC (Hütvagner and Zamore, 2002), suggesting a mechanistic link between the RNAi and miRNA pathways. Recent studies suggest that miRNAs play important roles in controlling development in both plants and animals (Carrington and Ambros, 2003). Anti-sense binding of miRNAs to target mRNA sequences can silence expression by either inhibiting translation (in animals) or directing the target mRNA for degradation (in plants) (Carrington and Ambros, 2003).

Genes required for RNAi

Genes required for RNAi (Table 3.1) have been identified by genetic screens aimed at isolating mutants that are defective in RNA-mediated silencing, by using RNAi knockdown of components of the RNAi machinery, and by biochemical methods. Determining how these proteins function in RNAi is an ongoing challenge. Here, we discuss the roles of genes with defined functions in RNAi.

Initiators

The *C. elegans* genes *rde-1* and *rde-4* (rde stands for "RNAi deficient") are involved in the early step of RNAi. RDE-4 is a double-stranded RNA binding protein required for efficient recognition of the dsRNA trigger and, at least in *Drosophila*, for helping siRNAs transit from Dicer to RISC (Tabara et al., 2002; Liu et al., 2003). Consistent with RDE-4 acting early, *rde-4* mutant animals do not produce siRNAs, and the introduction of siRNAs can bypass the requirement for this gene (Parrish and Fire,

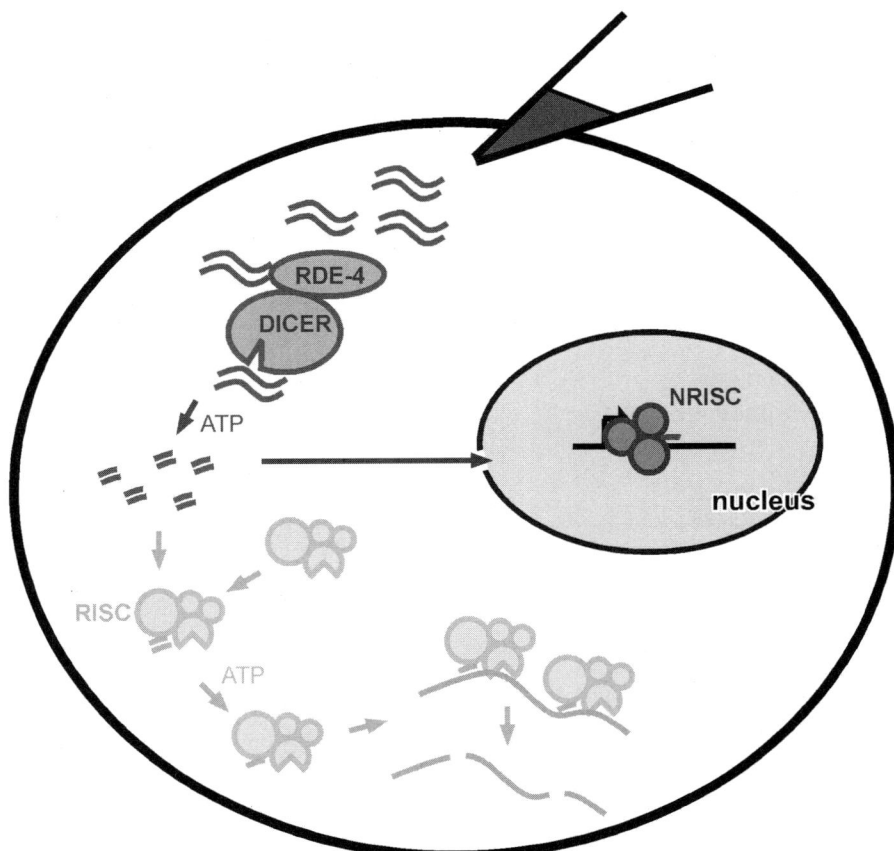

Figure 3.3. Model for gene silencing in the nucleus by RNAi. RNAi can also silence the transcription of targeted genes in certain organisms. In this model a signal can direct a putative nuclear RNAi silencing complex (NRISC), composed of chromatin modifying proteins, to the targeted locus, silencing gene expression at the level of transcription. (See color section.)

2001). RDE-4 has also been shown to physically associate with Dicer, RDE-1, and a Dicer-related helicase called DRH-1/2 [(Tabara et al., 2002) (Figure 3.3)].

The *rde-1* gene is a member of a large family (composed of over 20 genes in *C. elegans*), and has homologs in *Drosophila*, plants, fungi, and mammals (Tabara et al., 1999; Fagard et al., 2000). RDE-1 is a member of the PPD (PAZ and Piwi domain) family of proteins and can physically associate with RDE-4 (Tabara et al., 2002). However, unlike *rde-4*, siRNAs cannot bypass the requirement for *rde-1*, suggesting that *rde-1* is required downstream of siRNA production. Interestingly, secondary siRNAs are not produced in *rde-1* mutants (Parrish and Fire, 2001; Tijsterman et al., 2002b).

Effectors

Genes required in the late step of RNAi, after formation of siRNAs, include the *C. elegans* genes *rde-2* and *mut-7*. These genes were initially identified in screens for mutant animals unable to respond to RNAi and from screens designed to identify components that silence DNA transposition in the germline (Ketting et al., 1999).

The fact that transposon silencing depends on factors required for RNAi suggests that silencing of transposons and RNAi may share related mechanisms. The *mut-7* gene encodes a putative 3′-to-5′ exonuclease sharing homology to the nuclease domains of RNase D and a helicase implicated in Werner Syndrome, a human premature aging disorder (Ketting et al. 1999; Ketting and Plasterk, 2000). The *rde-2* gene product has not yet been identified.

Current research supports a scheme in which RDE-1 and RDE-4 are required to generate/stabilize siRNAs, whereas RDE-2 and MUT-7 are required to respond to this signal to silence the targeted gene.

RISC components

RISC is an siRNA/protein complex that directs the cleavage of targeted transcripts. There have been several RISC factors thus far identified. *Drosophila Argonaute-2* (AGO2, a homolog of *C. elegans* RDE-1, *Neurospora* QDE-2, and *Arabidopsis* AGO1) has been shown to associate with RISC during processing of miRNAs (Hammond et al., 2001; Williams and Rubin, 2002).

Other components found to be associated with RISC are two RNA-binding proteins: VIG (Vasa intronic gene), and FMR1 [(Fragile X mental retardation syndrome protein) (Caudy et al., 2002)]. VIG is encoded within an intron of the *Vasa* gene in *Drosophila* and contains a protein motif able to bind RNA (Heaton et al. 2001). FMR1 is the *Drosophila* homolog of the mammalian Fragile X mental retardation protein and encodes an RNA-binding protein that associates with translating ribosomes and is presumed to act as a negative regulator of translation (Ishizuka et al., 2002).

Drosophila Tudor-SN, a nuclease containing domains in common with Tudor and staphylococcal nuclease, also associates with RISC (Caudy et al., 2003), and Dicer has been shown to associate with RISC, although Dicer is not required for RISC activity (Bernstein et al., 2001).

RNA-dependent RNA polymerases

RDRPs have been implicated in RNAi in *C. elegans*. The *ego-1* gene encodes an RDRP required for RNAi specifically in the germline (Smardon et al., 2000; Sijen et al., 2001). The *C. elegans* genome contains three additional RDRP loci – *rrf-1*, *rrf-2*, and *rrf-3*. Mutations in *rrf-1* result in an RNAi-deficient phenotype, however, only somatic cells are RNAi-deficient. RRF-2 appears to have no role in RNAi, whereas mutations in *rrf-3* make *C. elegans* hypersensitive to RNAi suggesting that it may act as an antagonist to the RNAi pathway (Simmer et al., 2002; Sijen et al., 2001).

Certainly, the existence of an RDRP might explain the remarkable efficiency of dsRNA-induced silencing if it amplified either the dsRNA prior to cleavage or the siRNAs directly. RDRP activity has been reported in *Drosophila* embryo lysates. However, no homolog of an RDRP has been found in the *Drosophila* and human genomes (Lipardi et al., 2001). Additionally, siRNAs that have been modified at their 3′ ends, which can no longer associate with an RDRP, can produce efficient RNAi in human tissue culture systems (Holen et al., 2002).

Novel component required for systemic RNAi

In *C. elegans*, a genetic screen was recently developed to identify components required for systemic RNAi (Winston et al., 2002; Feinberg et al., 2003). Over 200 mutants were identified and placed into five complementation groups: *sid-1*, *sid-2*, *sid-3*, *sid-4*, and *sid-5* (for systemic RNA interference deficient). To date, *sid-1* has been cloned. SID-1 localizes to the cell periphery and encodes a novel multi-pass transmembrane protein that is required cell autonomously and enables the passive uptake of dsRNA into cells. Interestingly, dsRNAs of approximately 100bp in length function as preferred substrates for *sid-1*-dependent RNAi over shorter dsRNAs (Winston et al., 2002; Feinberg et al., 2003).

Roles for chromatin-modifying proteins in RNAi

Until recently, most models predicted RNAi to act exclusively at the post-transcriptional level to target mRNA stability. However, recent work may have revealed an additional role for RNAi, a role in regulating transcription. Work in fission yeast has revealed that RNAi silences transcription of genes via chromatin modifications. Chromatin modification proteins have also been implicated in RNAi in *C. elegans*, and RNAi machinery components have been implicated in transcriptional silencing in *Drosophila*, suggesting that the mechanisms of gene silencing found in fission yeast might be operating in animal cells as well.

C. elegans: Using RNAi to identify genes required for RNAi

As discussed above, the molecular components of the RNAi machinery have been identified by screening for mutants that are RNAi-deficient, and by isolating proteins that interact with the known RNAi machinery. Recently, an additional approach to identify RNAi components using RNAi itself was fortuitously developed out of a screen originally intended to identify genes required for early development. The screen, conducted in *C. elegans*, involved pooling together eight dsRNAs at a time, co-injecting these into hermaphrodites, and identifying pools containing genes essential for development by scoring progeny for embryonic lethality. For pools that resulted in lethality, sub-pooling was performed until the dsRNA responsible for the effect was isolated.

One such pool contained a dsRNA corresponding to the essential gene, *glp-1* (germ line proliferation 1). GLP-1 is a Notch-like molecule and loss of *glp-1* function results in a maternal-effect lethal phenotype. Surprisingly, no lethality was observed in the pool that included *glp-1* dsRNA. However, injecting *glp-1* dsRNA alone resulted in a high degree of lethality. In an attempt to identify the dsRNA responsible for suppressing *glp-1* (RNAi), each dsRNA in this pool was co-injected with *glp-1* dsRNA. It was found that only one of the dsRNAs was able to suppress *glp-1* (RNAi), and removing this dsRNA from the pool restored *glp-1* dsRNA-mediated lethality.

Interestingly, suppression was not limited to *glp-1* (RNAi), as the suppressing dsRNA was also able to suppress the lethality associated with other dsRNAs, suggesting that the suppressing dsRNA might work by affecting RNAi in general, and not just RNAi of *glp-1*. Consistent with this, known components of the RNAi machinery have been found to behave similarly (Hammond et al., 2000; Bernstein et al., 2001; Grishok et al., 2001; Dudley et al., 2002). These results suggested that co-injection of multiple dsRNAs could be used as an effective method to identify new genes required for RNAi (Dudley et al., 2002).

Surprisingly, several genes identified using this method in *C. elegans* encode proteins predicted to associate with chromatin, raising the possibility that RNAi in animal cells may also work at the level of chromatin, as it does in plants (Baulcombe, 1996; Morel et al., 2000). Some of the proteins identified form a complex that includes a protein homologous to a *Drosophila* polycomb group protein, which in flies functions to repress gene expression through the modification of chromatin (Cao et al., 2002; Pirrotta, 2002), raising the possibility that repression of transcription by chromatin modifying proteins might play a role in RNAi-dependent gene silencing.

Drosophila: A piece of the RNAi machinery required for transgene-mediated transcriptional silencing

Recent experiments have shown that polycomb-dependent transcriptional gene silencing is disrupted in an RNAi-defective background in *Drosophila*. (Pal-Bhadra et al., 2002). In mutants defective for *piwi*, a homolog of the *C. elegans* RNAi component RDE-1, transcriptional gene silencing was impaired alongside with posttranscriptional gene silencing. Transcription from an inserted transgene was monitored in both normal and *piwi* mutant flies. In normal flies, transcriptional gene silencing was induced by high copy numbers of the inserted transgene, while in the *piwi* mutants transcription was unaffected (Pal-Bhadra et al., 2002). These results show that an RNAi component is required for at least one aspect of transcriptional silencing.

Fission yeast: Silencing of centromeric DNA and the mat locus

Recent work in the fission yeast *Schizosaccharomyces pombe* implicates RNAi in both the initiation and the maintenance of centromeric silencing (Hall et al., 2002). Centromeric DNA is flanked on either side by repetitive sequences. It has recently been shown that a DNA fragment normally present within the centromere can be inserted into a transcriptionally active region of chromatin elsewhere and induce chromatin remodeling at the new site. Modifications such as the methylation of lysine 9 of histone H3 subsequently leads to gene silencing. Interestingly, methylation of lysine 9 on Histone H3 and silencing is impaired in yeast lacking functional RNAi components such as Dicer, RDRP and Argonaute (Volpe et al., 2002; Hall et al., 2003). Additionally, transcripts complementary to silenced centromeric sequences tend to accumulate in RNAi mutant backgrounds, suggesting that RNAi components respond to endogenous dsRNAs produced at centromeres, silencing centromeric sequences (Volpe et al., 2002). Indeed, it has been shown that non-coding overlapping centromeric sequences can produce transcripts that hybridize to form dsRNAs that are processed by the RNAi machinery. Additionally, ectopically expressed short hairpin RNAs can also induce heterochromatin formation at homologous loci. This process requires components of the RNAi machinery and a histone methyltransferase (Schramke and Allshire, 2003). These data suggest that aberrant transcription within highly repetitive sequences can lead to the formation of hairpin RNAs, which are processed by the RNAi pathway, leading to heterochromatin formation and thus silencing of gene expression.

It was discovered recently that RNAi components are also required to initiate silencing at the yeast *mat* locus. Fission yeast can exist in a haploid state, or in a diploid state – which is formed when two haploid cells fuse. This fusion can only occur between two haploid cells that are of different mating types. The mating type of the haploid cells is determined by a single locus, termed the mating type locus (*mat*). The *mat* region consists of three loci, two of which are always silent and are required for the switching of the transcriptionally active locus. Repression of transcription from the *mat* locus is essential for mating and is accompanied by changes in chromatin structure (Kayne et al., 1988; Grewal et al., 1996).

RNAi components function to initiate heterochromatin formation at the *mat* locus. Mutants in yeast RNAi machinery failed to initiate heterochromatin assembly after removal of epigenetic markers by the use of the drug trichostatin A. Additionally, the levels of H3 lysine 9 methylation as well as Swi6, a heterochromatin protein, were greatly reduced at the *mat* locus in RNAi mutants (Hall et al., 2002).

Conclusion

RNA-mediated gene silencing has become an important tool to analyze gene function in many organisms as well as in mammalian tissue culture systems. As genome sequences become readily available, RNAi makes the ability to analyze the function of each and every gene an attainable goal. Moreover, the ability to readily silence specific genes holds promise in designing effective therapies to combat a range of illnesses. The continued identification of genes required for RNAi should further our understanding of the mechanisms of RNAi.

Acknowledgements

Our work using RNAi is supported by grants NSF MCB-0235654 and NIH R01GM68966.

REFERENCES

Alder, M. N., Dames, S., Gaudet, J. and Mango, S. E. (2003). Gene silencing in *Caenorhabditis elegans* by transitive RNA interference. *RNA*, **9**, 25–32.

Bateman, A. (2002). The SGS3 protein involved in PTGS finds a family. *BMC Bioinformatics*, **3**, 21.

Baulcombe, D. C. (1996). RNA as a target and an initiator of post-transcriptional gene silencing in transgenic plants. *Plant Molecular Biology*, **32**, 79–88.

Boutet, S., Vazquez, F., Liu, J., Beclin, C., Fagard, M., Gratias, A., Morel, J. B., Crete, P., Chen, X. and Vaucheret, H. (2003). Arabidopsis HEN1. A Genetic link between endogenous miRNA controlling development and siRNA controlling transgene silencing and virus resistance. *Current Biology*, **13**, 843–848.

Bernstein, E., Caudy, A. A., Hammond, S. M. and Hannon, G. J. (2001). Role for a bidentate ribonuclease in the initiation step of RNA interference. *Nature*, **409**, 363–366.

Cao, R., Wang, L., Wang, H., Xia, L., Erdjument-Bromage, H., Tempst, P., Jones, R. S. and Zhang, Y. (2002). Role of histone H3 lysine 27 methylation in polycomb-group silencing. *Science*, **298**, 1039–1043.

Carrington, J. C. and Ambros, V. (2003). Role of microRNAs in plant and animal development. *Science*, **301**, 336–338.

Catalanotto, C., Azzalin, G., Macino, G. and Cogoni, C. (2000). Gene silencing in worms and fungi. *Nature*, **404**, 245.

Caudy, A. A., Myers, M., Hannon, G. J. and Hammond, S. M. (2002). Fragile X-related protein and VIG associate with the RNA interference machinery. *Genes & Development*,**16**, 2491–2496.

Caudy, A. A., Ketting, R. F., Hammond, S. M., Denli, A. M., Bathoorn, A. M., Tops, B. B., Silva, J. M., Myers, M. M., Hannon, G. J. and Plasterk, R. H. (2003). A micrococcal nuclease homologue in RNAi effector complexes. *Nature*, **425**, 411–414.

Celotto, A. M. and Graveley, B. R. (2002). Exon-specific RNAi: a tool for dissecting the functional relevance of alternative splicing. *RNA*, **8**, 718–724.

Cogoni, C. and Macino, G. (1999a). Homology-dependent gene silencing in plants and fungi: A number of variations on the same theme. *Current Opinion in Microbiology*, **2**, 657–662.

Cogoni, C. and Macino, G. (1999b). Posttranscriptional gene silencing in Neurospora by a RecQ DNA helicase. *Science*, **286**, 2342–2344.

Dalmay, T., Hamilton, A., Rudd, S., Angell, S. and Baulcombe, D. C. (2000). An RNA-dependent RNA polymerase gene in Arabidopsis is required for posttranscriptional gene silencing mediated by a transgene but not by a virus. *Cell*, **101**, 543–553.

Dalmay, T., Horsefield, R., Braunstein, T. H. and Baulcombe, D. C. (2001). SDE3 encodes an RNA helicase required for post-transcriptional gene silencing in Arabidopsis. *European Molecular Biology Organization Journal*, **20**, 2069–2078.

Doi, N., Zenno, S., Ueda, R., Ohki-Hamazaki, H., Ui-Tei, K. and Saigo, K. (2003). Short-interfering-RNA-mediated gene silencing in mammalian cells requires Dicer and eIF2C translation initiation factors. *Current Biology*, **13**, 41–46.

Domeier, M. E., Morse, D. P., Knight, S. W., Portereiko, M., Bass, B. L. and Mango, S. E. (2000). A link between RNA interference and nonsense-mediated decay in *Caenorhabditis elegans*. *Science*, **289**, 1928–1931.

Dudley, N. R., Labbe, J. C. and Goldstein, B. (2002). Using RNA interference to identify genes required for RNA interference. *Proceedings of the National Academy of Sciences USA*, **99**, 4191–4196.

Elbashir, S. M., Lendeckel, W. and Tuschl, T. (2001a). RNA interference is mediated by 21- and 22-nucleotide RNAs. *Genes & Development*, **15**, 188–200.

Elbashir, S. M., Martinez, J., Patkaniowska, A., Lendeckel, W. and Tuschl, T. (2001b). Functional anatomy of siRNAs for mediating efficient RNAi in *Drosophila melanogaster* embryo lysate. *European Molecular Biology Organization Journal*, **20**, 6877–6888.

Elbashir, S. M., Harborth, J., Lendeckel, W., Yalcin, A., Weber, K. and Tuschl, T. (2001c). Duplexes of 21-nucleotide RNAs mediate RNA interference in cultured mammalian cells. *Nature*, **411**, 494–498.

Fagard, M., Boutet, S., Morel, J. B., Bellini, C. and Vaucheret, H. (2000). AGO1, QDE-2, and RDE-1 are related proteins required for post-transcriptional gene silencing in plants, quelling in fungi, and RNA interference in animals. *Proceedings of the National Academy of Sciences USA*, **97**, 11650–11654.

Feinberg, E. H. and Hunter, C. P. (2003). Transport of dsRNA into cells by the transmembrane protein SID-1. *Science*, **301**, 1545–1547.

Fire, A., Xu, S., Montgomery, M. K., Kostas, S. A., Driver, S. E. and Mello, C. C. (1998). Potent and specific genetic interference by double-stranded RNA in *Caenorhabditis elegans*. *Nature*, **391**, 806–811.

Glazov, E., Phillips, K., Budziszewski, G. J., Meins, F. and Levin, J. Z. (2003). A gene encoding an RNase D exonuclease-like protein is required for post-transcriptional silencing in Arabidopsis. *Plant Journal*, **35**, 342–349.

Grewal, S. I. and Klar, A. J. (1996). Chromosomal inheritance of epigenetic states in fission yeast during mitosis and meiosis. *Cell*, **86**, 95–101.

Grishok, A., Tabara, H. and Mello, C. C. (2000). Genetic requirements for inheritance of RNAi in *C. elegans*. *Science*, **287**, 2494–2497.

Grishok, A., Pasquinelli, A. E., Conte, D., Li, N., Parrish, S., Ha, I., Baillie, D. L., Fire, A., Ruvkun, G. and Mello, C. C. (2001). Genes and mechanisms related to RNA interference regulate expression of the small temporal RNAs that control *C. elegans* developmental timing. *Cell*, **106**, 23–34.

Hall, I. M., Shankaranarayana, G. D., Noma, K., Ayoub, N., Cohen, A. and Grewal, S. I. (2002). Establishment and maintenance of a heterochromatin domain. *Science*, **297**, 2232–2237.

Hall, I. M., Noma, K. and Grewal, S. I. (2003). RNA interference machinery regulates chromosome dynamics during mitosis and meiosis in fission yeast. *Proceedings of the National Academy of Sciences USA*, **100**, 193–198.

Hammond, S. M., Bernstein, E., Beach, D. and Hannon, G. J. (2000). An RNA-directed nuclease mediates post-transcriptional gene silencing in *Drosophila* cells. *Nature*, **404**, 293–296.

Hammond, S. M., Boettcher, S., Caudy, A. A., Kobayashi, R. and Hannon, G. J. (2001). Argonaute2, a link between genetic and biochemical analyses of RNAi. *Science*, **293**, 1146–1150.

Heaton, J. H., Dlakic, W. M., Dlakic, M. and Gelehrter, T. D. (2001). Identification and cDNA cloning of a novel RNA-binding protein that interacts with the cyclic nucleotide-responsive sequence in the Type-1 plasminogen activator inhibitor mRNA. *Journal of Biological Chemistry*, **276**, 3341–3347.

Holen, T., Amarzguioui, M., Wiiger, M. T., Babaie, E. and Prydz, H. (2002). Positional effects of short interfering RNAs targeting the human coagulation trigger Tissue Factor. *Nucleic Acids Research*, **30**, 1757–1766.

Hunter, C. P. (1999). Genetics: A touch of elegance with RNAi. *Current Biology*, **9**, R440-R4422.

Hütvagner, G. and Zamore, P. D. (2002). A microRNA in a multiple-turnover RNAi enzyme complex. *Science*, **297**, 2056–2060.

Ishizuka, A., Siomi, M. C. and Siomi, H. (2002). A *Drosophila* Fragile X protein interacts with components of RNAi and ribosomal proteins. *Genes & Development*, **16**, 2497–2508.

Kayne, P. S., Kim, U. J., Han, M., Mullen, J. R., Yoshizaki, F. and Grunstein, M. (1988). Extremely conserved histone H4 N terminus is dispensable for growth but essential for repressing the silent mating loci in yeast. *Cell*, **55**, 27–39.

Kennerdell, J. R. and Carthew, R. W. (1998). Use of dsRNA-mediated genetic interference to demonstrate that frizzled and frizzled 2 act in the wingless pathway. *Cell*, **95**, 1017–1026.

Kennerdell, J. R., Yamaguchi, S. and Carthew, R. W. (2002). RNAi is activated during *Drosophila* oocyte maturation in a manner dependent on aubergine and spindle-E. *Genes & Development*, **16**, 1884–1889.

Ketting, R. F., Haverkamp, T. H., van Luenen, H. G. and Plasterk, R. H. (1999). Mut-7 of *C. elegans*, required for transposon silencing and RNA interference, is a homolog of Werner syndrome helicase and RNaseD. *Cell*, **99**, 133–141.

Ketting, R. F. and Plasterk, R. H. (2000). A genetic link between co-suppression and RNA interference in *C. elegans*. *Nature*, **404**, 296–298.

Ketting, R. F., Fischer, S. E., Bernstein, E., Sijen, T., Hannon, G. J. and Plasterk, R. H. (2001). Dicer functions in RNA interference and in synthesis of small RNA involved in developmental timing in *C. elegans*. *Genes & Development*, **15**, 2654–2659.

Knight, S. W. and Bass, B. L. (2001). A role for the RNase III enzyme DCR-1 in RNA interference and germ line development in *Caenorhabditis elegans*. *Science*, **293**, 2269–2271.

Lipardi, C., Wei, Q. and Paterson, B. M. (2001). RNAi as random degradative PCR: siRNA primers convert mRNA into dsRNAs that are degraded to generate new siRNAs. *Cell*, **107**, 297–307.

Liu, Q., Rand, T. A., Kalidas, S., Du, F., Kim, H. E., Smith, D. P. and Wang, X. (2003). R2D2, a bridge between the initiation and effector steps of the *Drosophila* RNAi pathway. *Science*, **301**, 1921–1925.

Maine, E. M. (2000). A conserved mechanism for post-transcriptional gene silencing? *Genome Biology*, **1**, REVIEWS1018.

Martens, H., Novotny, J., Oberstrass, J., Steck, T. L., Postlethwait, P. and Nellen, W. (2002). RNAi in Dictyostelium: The role of RNA-directed RNA polymerases and double-stranded RNase. *Molecular Biology Cell*, **13**, 445–453.

Morel, J. B., Mourrain, P., Beclin, C. and Vaucheret, H. (2000). DNA methylation and chromatin structure affect transcriptional and post-transcriptional transgene silencing in Arabidopsis. *Current Biology*, **10**, 1591–1594.

Mourrain, P., Beclin, C., Elmayan, T., Feuerbach, F., Godon, C., Morel, J. B., Jouette, D., Lacombe, A. M., Nikic, S., Picault, N., Remoue, K., Sanial, M., Vo, T. A. and Vaucheret, H. (2000). Arabidopsis SGS2 and SGS3 genes are required for posttranscriptional gene silencing and natural virus resistance. *Cell*, **101**, 533–542.

Nykanen, A., Haley, B. and Zamore, P. D. (2001). ATP requirements and small interfering RNA structure in the RNA interference pathway. *Cell*, **107**, 309–321.

Pal-Bhadra, M., Bhadra, U. and Birchler, J. A. (2002). RNAi related mechanisms affect both transcriptional and posttranscriptional transgene silencing in *Drosophila*. *Molecular Cell*, **9**, 315–327.

Palauqui, J. C., Elmayan, T., Pollien, J. M. and Vaucheret, H. (1997). Systemic acquired silencing: Transgene-specific post-transcriptional silencing is transmitted by grafting from silenced stocks to non-silenced scions. *European Molecular Biology Organization Journal*, **16**, 4738–4745.

Parrish, S., and Fire, A. (2001). Distinct roles for RDE-1 and RDE-4 during RNA interference in *Caenorhabditis elegans*. *RNA*, **7**,1397–1402.

Pirrotta, V. (2002). Silence in the germ. *Cell*, **110**, 661–664.

Roignant, J. Y., Carre, C., Mugat, B., Szymczak, D., Lepesant, J. A. and Antoniewski, C. (2003). Absence of transitive and systemic pathways allows cell-specific and isoform-specific RNAi in *Drosophila*. *RNA*, **9**, 299–308.

Schauer, S. E., Jacobsen, S. E., Meinke, D. W. and Ray, A. (2002). DICER-LIKE1: Blind men and elephants in Arabidopsis development. *Trends in Plant Sciences*, **7**, 487–491.

Schwarz, D. S., Hütvagner, G., Haley, B. and Zamore, P. D. (2002). Evidence that siRNAs function as guides, not primers, in the *Drosophila* and human RNAi pathways. *Molecular Cell*, **10**, 537–548.

Schwarz, D. S., Hütvagner, G., Du, T., Xu, Z., Aronin, N. and Zamore, P. D. (2003). Asymmetry in the assembly of the RNAi enzyme complex. *Cell*, **115**, 199–208.

Sen, G. C., Lebleu, B., Brown, G. E., Kawakita, M., Slattery, E. and Lengyel, P. (1976). Interferon, double-stranded RNA and mRNA degradation. *Nature*, **264**, 370–373.

Sharp, P. A. (1999). RNAi and double-strand RNA. *Genes & Development*, **13**, 139–141.

Sharp, P. A. (2001). RNA interference – 2001. *Genes & Development*, **15**, 485–490.

Shramke, V. and Allshire, R. (2003) Hairpin RNAs and retrotransposon LTRs effect RNAi and chromatin-based gene silencing. *Science*, **301**, 1069–1074.

Sijen, T., Fleenor, J., Simmer, F., Thijssen, K. L., Parrish, S., Timmons, L., Plasterk, R. H. and Fire, A. (2001). On the role of RNA amplification in dsRNA-triggered gene silencing. *Cell*, **107**, 465–476.

Simmer, F., Tijsterman, M., Parrish, S., Koushika, S. P., Nonet, M. L., Fire, A., Ahringer, J. and Plasterk, R. H. (2002). Loss of the putative RNA-directed RNA polymerase RRF-3 makes *C. elegans* hypersensitive to RNAi. *Current Biology*, **12**, 1317–1319.

Smardon, A., Spoerke, J. M., Stacey, S. C., Klein, M. E., Mackin, N., and Maine, E. M. (2000). EGO-1 is related to RNA-directed RNA polymerase and functions in germ-line development and RNA interference in *C. elegans*. *Current Biology*, **10**, 169–178.

Stark, G. R., Kerr, I. M., Williams, B. R., Silverman, R. H. and Schreiber, R. D. (1998). How cells respond to interferons. *Annual Reviews of Biochemistry*, **67**, 227–264.

Tabara, H., Sarkissian, M., Kelly, W. G., Fleenor, J., Grishok, A., Timmons, L., Fire, A. and Mello, C. C. (1999). The rde-1 gene, RNA interference, and transposon silencing in *C. elegans*. *Cell*, **99**, 123–132.

Tabara, H., Yigit, E., Siomi, H. and Mello, C. C. (2002). The dsRNA binding protein RDE-4 interacts with RDE-1, DCR-1, and a DExH-box helicase to direct RNAi in *C. elegans*. *Cell*, **109**, 861–871.

Tijsterman, M., Ketting, R. F. and Plasterk, R. H. (2002a). The genetics of RNA silencing. *Annual Reviews of Genetics*, **36**, 489–519.

Tijsterman, M., Ketting, R. F., Okihara, K. L., Sijen, T. and Plasterk, R. H. (2002b). RNA helicase MUT-14-dependent gene silencing triggered in *C. elegans* by short antisense RNAs. *Science*, **295**, 694–697.

Vaucheret, H., Beclin, C. and Fagard, M. (2001). Post-transcriptional gene silencing in plants. *Journal of Cell Sience*, **114**, 3083–3091.

Volpe, T. A., Kidner, C., Hall, I. M., Teng, G., Grewal, S. I. and Martienssen, R. A. (2002). Regulation of heterochromatic silencing and histone H3 lysine-9 methylation by RNAi. *Science*, **297**, 1833–1837.

Wargelius, A., Ellingsen, S. and Fjose, A. (1999). Double-stranded RNA induces specific developmental defects in zebrafish embryos. *Biochemical Biophysical Research Communications*, **263**, 156–1561.

Wianny, F. and Zernicka-Goetz, M. (2000). Specific interference with gene function by double-stranded RNA in early mouse development. *Nature Cell Biology*, **2**, 70–75.

Williams, R. W. and Rubin, G. M. (2002). ARGONAUTE1 is required for efficient RNA interference in *Drosophila* embryos. *Proceedings of the National Academy of Sciences USA*, **99**, 6889–6894.

Winston, W. M., Molodowitch, C. and Hunter, C. P. (2002). Systemic RNAi in *C. elegans* requires the putative transmembrane protein SID-1. *Science*, **295**, 2456–2459.

Wu-Scharf, D., Jeong, B., Zhang, C. and Cerutti, H. (2000). Transgene and transposon silencing in Chlamydomonas reinhardtii by a DEAH-box RNA helicase. *Science*, **290**, 1159–1162.

Zamore, P. D., Tuschl, T., Sharp, P. A. and Bartel, D. P. (2000). RNAi: Double-stranded RNA directs the ATP-dependent cleavage of mRNA at 21 to 23 nucleotide intervals. *Cell*, **101**, 25–33.

Zamore, P. D. (2002). Ancient pathways programmed by small RNAs. *Science*, **296**, 1265–1269.

Zilberman D, Cao X. and Jacobsen S. E. (2003). ARGONAUTE4 control of locus-specific siRNA accumulation and DNA and histone methylation. *Science*, **299**, 716–719.

4 MicroRNAs: A small contribution from worms

Amy E. Pasquinelli

Introduction

The 2002 Nobel Prize for Medicine was awarded to Sydney Brenner, Robert Horvitz and John Sulston for their seminal work in establishing the nematode *Caenorhabditis elegans* as a model genetic organism for studying development and behavior. One of the most sensational discoveries to emerge from *C. elegans* is the identification of tiny, non-coding RNA genes that regulate development. The original report of a 22 nucleotide (nt) RNA essential for controlling temporal patterning in the worm was unprecedented. Seven years passed before another 22-nt RNA gene was found, once again through genetic studies of developmental timing in *C. elegans*. This second tiny RNA gene turned out not to be a worm oddity, but instead it was shown to be expressed in most animal species. Thus, the general existence of 22-nt RNA genes was established and the hunt for additional RNAs of this type intensified. Hundreds of ∼22-nt RNA genes have now been uncovered in plants and animals. It is not a coincidence that the tiny size of these endogenous RNAs, called microRNAs (miRNAs), is similar to that of the small interfering RNAs (siRNAs) that direct RNA interference (RNAi); common cellular factors and mechanisms participate in the expression and function of these ∼22-nt RNAs. This chapter focuses on how the discovery of 22-nt RNA genes in *C. elegans* revealed the broad existence of miRNAs and how the regulation of gene expression by miRNAs compares to RNAi.

The discovery of tiny RNA genes in *C. elegans*

The fathers of the *C. elegans* system accomplished the remarkable feat of identifying the timing and fate of all cell divisions that produce the ∼960-cell adult worm [(Figure 4.1) (Sulston and Horvitz, 1977)]. This achievement set the stage for genetic screens to identify genes that regulated development of the embryo to four larval stages (L1-L4) and finally adulthood (Sulston and Horvitz, 1981). An interesting category of mutants emerged from such screens as potential regulators of temporal patterning (Chalfie et al., 1981; Ambros and Horvitz, 1984). These

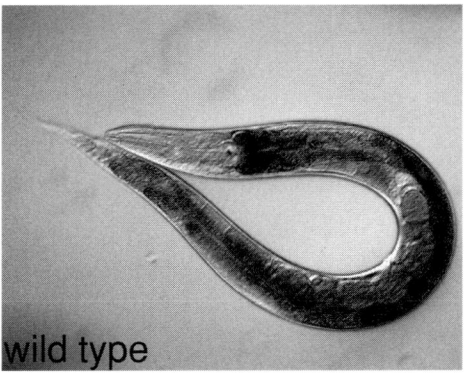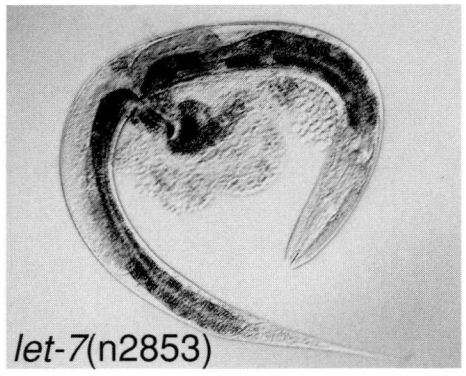

Figure 4.1. The nematode *Caenorhabditis elegans*. The left panel shows a wild-type adult *C. elegans* worm. The right panel shows an adult *let-7*(n2853) mutant worm, which has burst through its midsection because of abnormal temporal patterning of cell divisions.

mutants were classified as heterochronic because they caused cell divisions of a type that was inappropriate for a particular time in development. For example, the hypodermal seam cells of *C. elegans* undergo precise patterns of division at each of the four larval stages, but mutations in the *lin-4* (*lin*=lineage) gene produce reiterations of the first larval type of divisions throughout the lifetime of the worm [(Figure 4.2) (Chalfie et al., 1981; Ambros and Horvitz, 1984)]. Amazingly, many other tissues in *lin-4* mutants maintain the normal pattern of cell divisions and fates, thus producing an adult worm that is a mosaic of larval and adult cellular identities.

In contrast to the phenotype produced by *lin-4* mutants, worms that contain loss of function mutations in the *lin-14* gene skip the L1-specific seam cell division patterns and instead precociously express the L2 fates while the worm is still at the first larval stage [(Figure 4.2) (Ambros and Horvitz, 1984)]. Curiously, gain of function mutations in *lin-14* result in the opposite phenotype – a reiteration of L1 seam cell fates, just as in *lin-4* mutants [(Figure 4.2) (Ambros and Horvitz, 1984)]. Identification of the molecular nature of these heterochronic genes and the mutations was critical for understanding their roles in developmental timing. The *lin-14* gene encodes a nuclear-localized protein that is expressed in many cell types of first larval stage worms and then disappears in the next larval stages (Ruvkun and Giusto, 1989). Although the presence of LIN-14 protein is critical for the expression of L1 type cell fates, the mechanism by which it defines this stage is yet to be determined.

Genetic studies indicated that *lin-4* was a negative regulator of *lin-14* (Ambros and Horvitz, 1987; Ambros, 1989), but nobody could have predicted the extraordinary nature of the *lin-4* gene product. Through tedious mapping and rescue experiments, the Ambros lab found that a segment of genomic DNA less than 700 nucleotides long could produce functional *lin-4* gene activity. And, they detected a gene product of merely 22 nt that first appeared during the L1 stage (Figure 4.3), exactly when the LIN-14 protein levels began to plummet (Lee et al., 1993). The pieces of the puzzle of how *lin-4* might negatively regulate *lin-14* came together with identification of the *lin-4* 22-nt RNA sequence and mapping

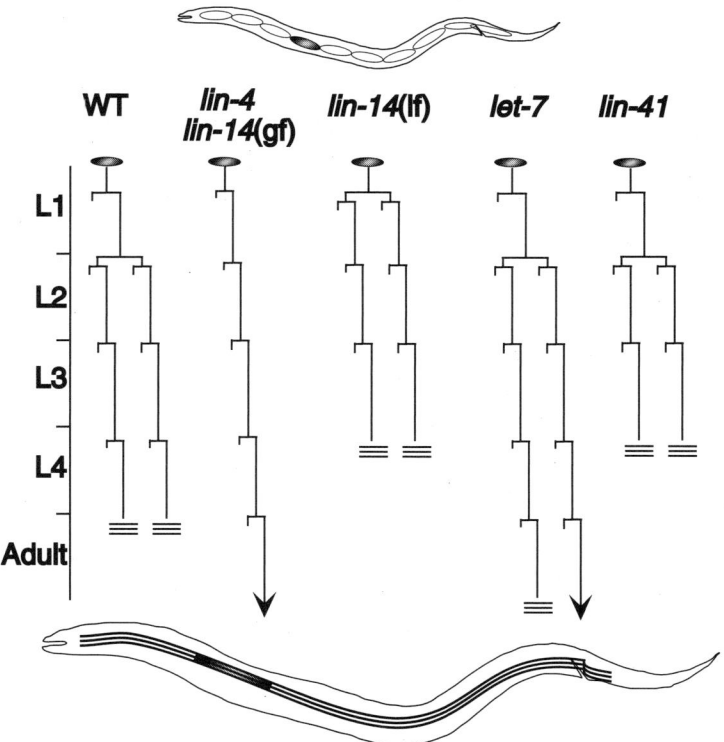

Figure 4.2. Seam cell division patterns in wild-type (WT) and heterochronic mutants of *C. elegans*. The V1 hypodermal seam cell (shaded) divides at each larval stage (L1-L4) until the transition to adulthood when the V1 descendents terminally differentiate and fuse with the other seam cells, as indicated by three horizontal stripes. The *lin-4*, *lin-14*(gf) (gain of function) and *let-7* mutants display reiterations of seam cell divisions characteristic of particular larval stages that often continue throughout the lifetime of the animal, as indicated by an arrowhead (Ambros and Horvitz, 1984; Reinhart et al., 2000). The *lin-14*(lf) (loss of function) mutants skip L1 type divisions and *lin-41* mutants skip the last larval seam cell divisions and precociously express adult fates (Ambros and Horvitz, 1984; Slack et al., 2000).

of the *lin-14* gain of function mutations that result in the loss of regulation by *lin-4*. The *lin-4* 22-nt RNA is complementary to multiple sites in the 3′ untranslated region (UTR) of the *lin-14* mRNA, which are deleted from the *lin-14* gain of function mutants (Lee et al., 1993; Wightman et al., 1993). The Ambros and Ruvkun labs proposed that the *lin-4* RNA is expressed midway through the first larval stage to direct down regulation of LIN-14 protein expression by recognizing antisense sequences in the *lin-14* 3′UTR (Lee et al., 1993; Wightman et al., 1993). This work uncovered an entirely new system of gene regulation, but it would be almost a decade before this phenomenon would be established in organisms beyond worms.

At the turn of the millennium, a second 22-nt RNA gene was found to regulate developmental timing in *C. elegans*, this time at the transition from the last larval stage to adulthood. Mutations in the *let-7* (*let*=lethal) gene cause reiterations of larval cell fates at the adult stage (Figure 4.2), which may contribute to the premature lethality observed in such mutants [(Figure 4.1) (Reinhart et al.,

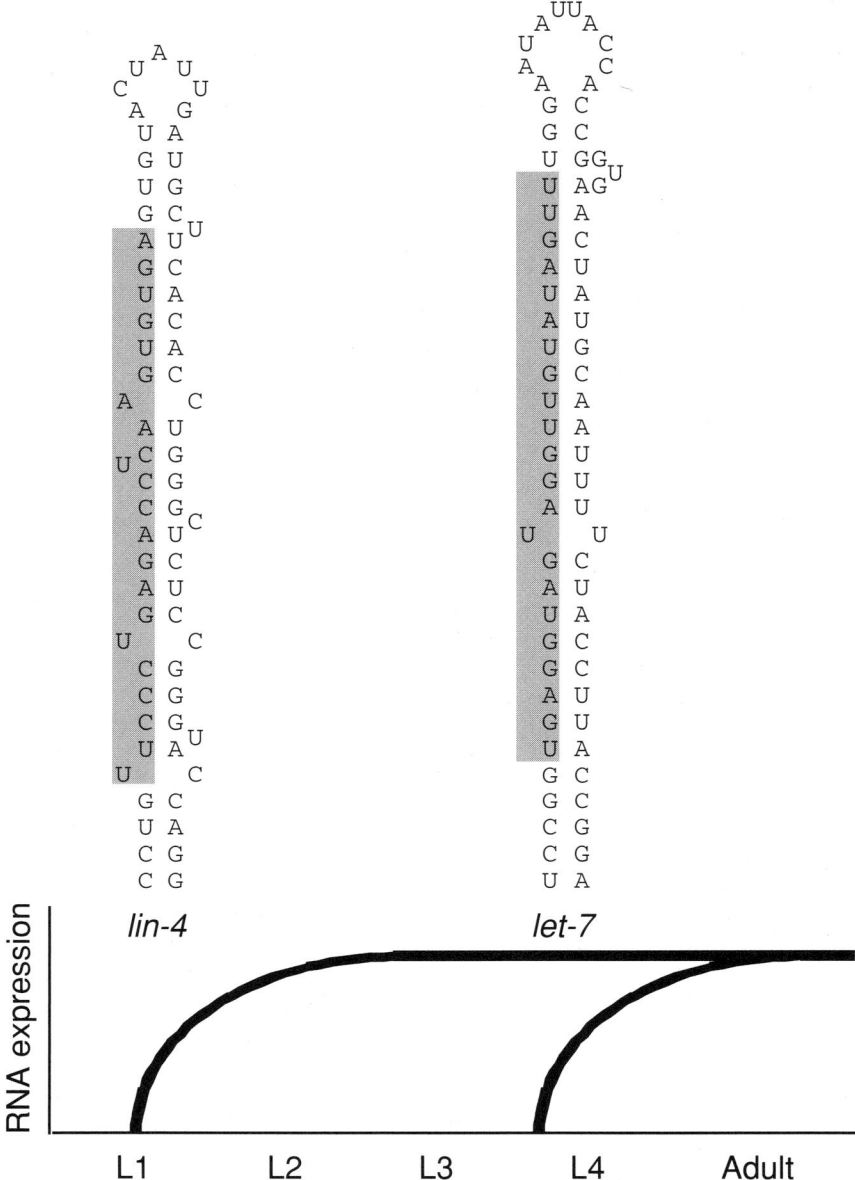

Figure 4.3. The *lin-4* and *let-7* RNAs are generated from stem loop precursors expressed at precise times in development. The 22nt *lin-4* and *let-7* RNAs (shaded) derive from partially base-paired precursor RNAs that initially appear in the first and last larval stages, respectively (Lee et al., 1993; Reinhart et al., 2000).

2000)]. The *let-7* gene was found to encode a 22-nt noncoding RNA and genetic suppressor screens revealed *lin-41* as a target of negative regulation mediated by *let-7* complementary sites in the 3′UTR of its mRNA [(Figure 4.4) (Reinhart et al., 2000; Slack et al., 2000)]. Analogous to the *lin-4/lin-14* relationship, the *let-7* RNA is expressed in the last larval stages (Figure 4.3), at which point it inhibits LIN-41 protein expression in certain cell types to direct the transition to adult fates (Reinhart et al., 2000; Slack et al., 2000). Although the molecular function

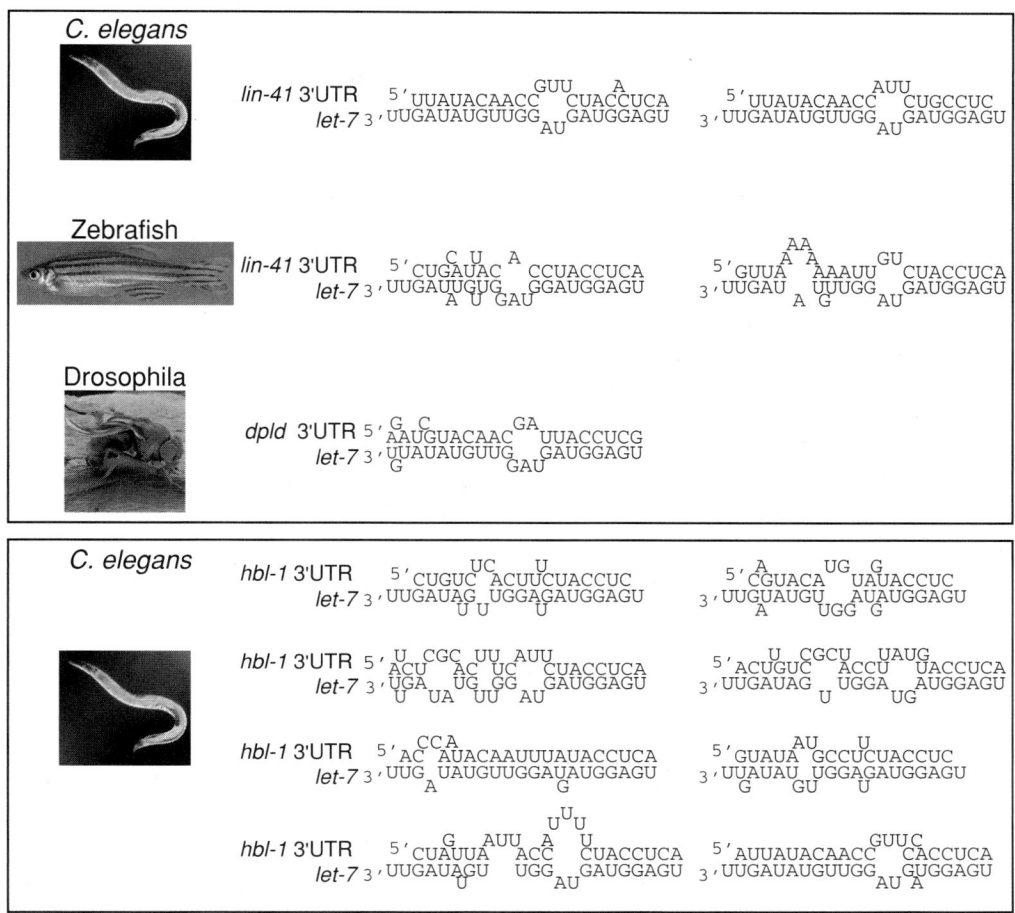

Figure 4.4. The *let-7* RNA recognizes sites of partial complementarity in the 3' untranslated regions (UTRs) of its target mRNAs. The top panel shows the predicted duplexes formed between *let-7* RNA and sites in the 3'UTR of the *lin-41* mRNA target in *C. elegans* and putative homologs in Zebrafish (cDNA AI794385) and *Drosophila* (*dpld* = dappled) (Pasquinelli et al., 2000; Reinhart et al., 2000; Slack et al., 2000). The bottom panel shows multiple duplexes predicted to form between *let-7* RNA and sites in the 3'UTR of the *hbl-1* (*hbl* = hunchback like) gene, which is the *C. elegans* homolog of *Drosophila hunchback* (Abrahante et al., 2003; Lin et al., 2003).

of *lin-41* is still unknown, the protein shares several motifs, such as a RING finger, B box, coiled coil and several copies of the NHL (Ncl, HT2A and LIN-41, the first proteins recognized to contain this amino acid motif (Slack and Ruvkun, 1998)) domain with proteins conserved throughout animal phylogeny, hinting that the temporal patterning pathway regulated by *lin-41* in worms could extend to other species (Slack et al., 2000).

let-7 heralds the existence of 22-nt RNA genes beyond worms

The discovery of two distinct tiny RNA genes in *C. elegans* prompted the question of whether they were unique to nematodes or whether they exemplified a new family of regulatory RNAs common to other species as well. Analyses of the newly

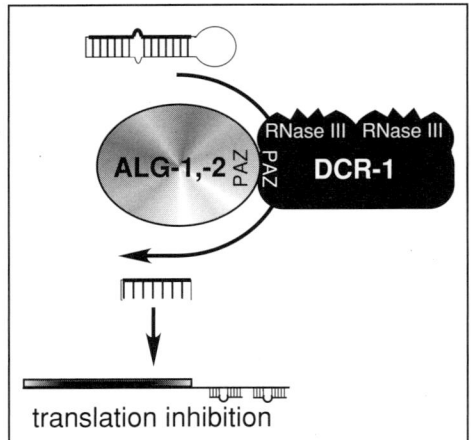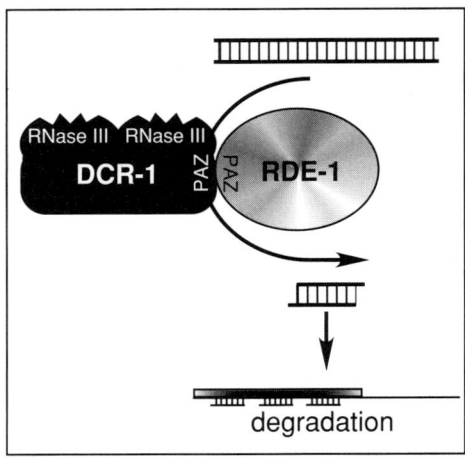

Figure 4.5. Comparison of the miRNA and RNAi pathways in *C. elegans*. The left panel depicts processing of partially base paired stem loop precursor miRNAs to the single-stranded 22nt forms by DCR-1 RNase (Grishok et al., 2001; Hütvagner et al., 2001; Ketting et al., 2001). The ALG-1,-2 proteins may participate in this process by interacting with DCR-1 through shared PAZ domains (Grishok et al., 2001). MiRNAs target mRNAs for down-regulated protein expression via imperfect antisense sites in their 3′UTRs (Olsen and Ambros, 1999; Seggerson et al., 2002). The right panel shows cleavage of long double stranded RNAs to duplex ∼22-nt forms, called small interfering RNAs (siRNAs) by the action of DCR-1 and possibly the PAZ domain containing factor RDE-1 (Tabara et al., 1999; Bernstein et al., 2001; Elbashir et al., 2001b). The siRNAs target mRNAs with sites of perfect homology for degradation (Elbashir et al., 2001b).

sequenced *Drosophila* and human genomes revealed exact sequence matches to the 22-nt *let-7* RNA. The Ruvkun lab demonstrated that these organisms indeed express *let-7* RNA and, as in *C. elegans*, appearance of the 22-nt RNA is temporally regulated (Pasquinelli et al., 2000). Interestingly, *let-7* RNA is undetectable in animals considered most basal as well as non-metazoans (Pasquinelli et al., 2000 and 2003). Furthermore, *let-7* complementary sites are also found in the 3′UTRs of potential *lin-41* orthologs, supporting the possibility that the targets and mechanisms of *let-7* regulation may be broadly conserved in the most recently diverging metazoans [(Figure 4.4) (Pasquinelli et al., 2000)].

The RNAi connection

Coincident with the discovery of the *let-7* 22-nt RNA gene was the realization that RNAi was mediated by ∼22-nt RNAs (Figure 4.5). Fire and Mello demonstrated that long double-stranded RNAs (dsRNA) could cause inactivation of target mRNAs that shared homologous sequences in *C. elegans* (Fire et al., 1998). This amazing technological advance, termed RNAi, immediately became the subject of intense research aimed at understanding the mechanism. An important clue towards recognizing the fate of the dsRNA came from studies of post-transcriptional gene silencing (PTGS) in plants. The PTGS phenomenon is similar to RNAi in that foreign sequences can cause inactivation of endogenous genes that share stretches of sequence identity. The Baulcombe lab observed that in many cases of PTGS there is a significant accumulation of 20–25 nt sense and antisense RNAs that match sequences of the gene being inactivated (Hamilton and Baulcombe, 1999). Shortly thereafter it was shown that the input dsRNA in RNAi is processed to ~22-nt duplexes, called small interfering RNAs [(siRNAs)(Elbashir et al., 2001b)], that function to target homologous mRNA sequences for degradation [(Figure 4.5) (Hammond et al., 2000; Parrish et al., 2000; Yang et al., 2000; Zamore et al., 2000).

The double-stranded nature of the *lin-4* and *let-7* miRNA precursors and the ~22-nt size of the mature forms are strikingly reminiscent of the RNAs that function in RNAi (Figure 4.5). This observation led to the discovery that the Dicer RNase (called DCR-1 in *C. elegans*), which is responsible for generating the ~22-nt siRNAs that function in RNAi (Bernstein et al., 2001), also processes *lin-4* and *let-7* precursors to the mature ~22-nt forms [(Figure 4.5) (Grishok et al., 2001; Hütvagner et al., 2001; Ketting et al., 2001; Knight and Bass, 2001)]. Another gene essential for RNAi in *C. elegans* is *rde-1* [(RNAi defective) (Tabara et al., 1999)], which is a member of the extensive and highly conserved PAZ (Piwi, Argonaute, Zwille – the first proteins recognized to share this amino acid motif) Piwi Domain (PPD) protein family (Cerutti et al., 2000; Schwarz and Zamore, 2002). The PAZ motif is also present in the Dicer protein and may serve as a platform for protein–protein interactions. Although loss of *rde-1* function in *C. elegans* does not seem to affect the health of the worms, mutations in the related *argonaute* gene in Arabidopsis result in abnormal plant development as well as diminished ability to perform PTGS (Bohmert et al., 1998; Fagard et al., 2000). Inactivation of the two genes in the worm PPD family that show the greatest degree of homology to the plant *argonaute* gene, appropriately called *alg-1* and *alg-2* (argonaute like gene), produces severe developmental abnormalities (Grishok et al., 2001). Some of the phenotypes caused by inactivation of *alg-1/-2* are linked to the inability to generate mature *lin-4* and *let-7* miRNAs [(Figure 4.5) (Grishok et al., 2001)].

Despite shared factors and mechanisms between the miRNA and siRNA pathways of gene regulation, there are also striking differences (Figure 4.5). The stems of the *lin-4* and *let-7* miRNA precursors exhibit only partial complementarity and the mature miRNA is generated from one side of the hairpin stem (Lee et al., 1993; Pasquinelli et al., 2000; Grishok et al., 2001; Hütvagner et al., 2001; Ketting et al., 2001). In contrast, the dsRNAs used to initiate RNAi work best if they are perfect duplexes and both sense and antisense RNAs are generated by the processing reaction (Fire et al., 1998; Tuschl et al., 1999; Caplen et al., 2000; Hammond et al., 2000; Parrish et al., 2000; Yang et al., 2000; Zamore et al., 2000; Bernstein et al., 2001; Elbashir et al., 2001a; Elbashir et al., 2001b; Elbashir et al., 2001c). How does Dicer recognize the imperfect stem structures of miRNAs and the complete duplexes of RNAi inputs and why are the 22-nt product forms different? Perhaps specific members of the PPD family aid Dicer in recognizing the different RNA substrates and/or participate in the generation or stabilization of single-stranded miRNA or duplex siRNA products. For example, *alg-1,-2* are essential for the production of *lin-4* and *let-7* mature RNAs, but not for RNAi function (Grishok et al., 2001). In contrast, *rde-1* is apparently dispensable for worm development, but not for RNAi (Tabara et al., 1999). The molecular differences that underlie the divergent functions of the PPD family members are yet to be uncovered.

Another conspicuous difference between the miRNA and siRNA pathways is the fate of the mRNA targeted for regulation (Figure 4.5). The siRNAs employ RISC (RNA Induced Silencing Complex) to degrade homologous mRNAs (Hammond et al., 2000) but the *lin-4* and *let-7* miRNAs silence their targets without appreciably affecting mRNA stability (Wightman et al., 1993; Olsen and Ambros, 1999; Seggerson et al., 2002). In fact, mRNAs regulated by *lin-4* remain associated with polysomes (Olsen and Ambros, 1999; Seggerson et al., 2002), but the mechanism by which *lin-4* inhibits translation is unknown.

Similar to the differences in the 22-nt products formed by Dicer processing of miRNA and siRNA precursors, the degree of complementarity also seems to dictate the final outcome of gene silencing by 22-nt RNAs. All of the natural targets of *lin-4* and *let-7* regulation contain sites of complementarity to the miRNAs in their 3'UTRs, but these sequences fail to support perfect base-pairing with the 22-nt RNAs (Lee et al., 1993; Wightman et al., 1993; Ha et al., 1996; Moss et al., 1997; Reinhart et al., 2000; Slack et al., 2000; Abrahante et al., 2003; Lin et al., 2003). However, the Zamore lab demonstrated that if *let-7* is faced with a target containing an exact antisense sequence, it can direct RNA degradation (Hütvagner and Zamore, 2002). Thus, the sequence of a 22-nt RNA not only guides the regulatory machinery to its homologous target but also dictates the mechanism by which the gene will be silenced (Hütvagner and Zamore, 2002; Zeng et al., 2002; Doench et al., 2003; Zeng et al., 2003).

A new class of RNA genes is born

The revelation that the 22-nt *let-7* RNA gene was present from worms to humans set the stage for seeking other tiny RNA genes. The Tuschl, Bartel and Ambros labs succeeded in using biochemical techniques to identify numerous ~22-nt RNAs in humans, flies and worms (Lagos-Quintana et al., 2001; Lau et al., 2001; Lee and Ambros, 2001). The new class of tiny RNA genes was christened microRNA (miRNA) and was soon found to extend into the plant world (Llave et al., 2002; Park et al., 2002; Reinhart et al., 2002). The general guidelines for defining a miRNA require that it be processed by Dicer from the stem of a hairpin precursor to a ~22-nt RNA (Ambros et al., 2003a). The founding miRNAs, *lin-4* and *let-7*, retain their original names but all new members of this RNA family are presented with the "miR" prefix, assigned a number (i.e. miR-1) and are curated at the Rfam database (Ambros et al., 2003a).

How many miRNA genes exist? Continued cloning efforts and bioinformatic analyses produce estimates of up to 300 in *C. elegans* (Ambros et al., 2003b; Grad et al., 2003; Lim et al., 2003b), 110 in *Drosophila* (Lai et al., 2003) and approximately 250 in vertebrates (Lim et al., 2003a). Many of the miRNA genes implicated by computer searches could not be verified by traditional experimental approaches, perhaps because the RNA levels are below detection limits or expression of the miRNA is restricted to certain conditions (Lee and Ambros, 2001;

Grad et al., 2003; Lai et al., 2003; Lim et al., 2003b). However, new techniques have been devised already to detect low-abundance miRNAs (Grad et al., 2003; Lim et al., 2003b). Another potential problem is that the bioinformatic scans of genomic sequences search for particular stem-loop precursor structures that might not be common to all miRNAs (Lee and Ambros, 2001; Grad et al., 2003; Lai et al., 2003; Lim et al., 2003b). Thus, we may yet be surprised by the final tabulation of miRNA genes.

Generation and function of miRNAs

Although hundreds of miRNAs have now been isolated from diverse species, the actual genomic codes that signal their production are yet to be defined. In *C. elegans*, ~700 nt or ~2500 nt of genomic DNA sequences containing *lin-4* or *let-7*, respectively, are sufficient to fully rescue null mutations in these miRNA genes (Lee et al., 1993; Reinhart et al., 2000). What are the regulatory elements contained in these sequences that direct expression at the appropriate times in development? The Slack lab demonstrated that the *let-7* rescuing fragment harbors a 116 base pair (bp) region located ~1200 nt upstream of the *let-7* RNA coding sequence that is essential for full *let-7* activity (Johnson et al., 2003). Interestingly, this region, called the temporal regulatory element (TRE), can also restrict expression of a protein-coding gene to the same time in development as the appearance of *let-7* RNA (Johnson et al., 2003). The ability of the presumed *let-7* promoter sequences to direct expression of an mRNA may indicate that Polymerase II is responsible for miRNA synthesis, although Polymerase III type promoters have been employed to generate miRNA precursors in mammalian systems (McManus et al., 2002). Considering the varied temporal and spatial expression patterns of miRNAs, specialized transcriptional complexes may recognize specific groups of miRNA genes.

MiRNAs derive from longer stem loop precursors, but what is the nature of the initial transcripts? Some miRNA genes are clustered in discrete genomic locations and exhibit overlapping expression patterns (Lagos-Quintana et al., 2001; Lau et al., 2001; Mourelatos et al., 2002; Bashirullah et al., 2003; Sempere et al., 2003). At least some of these miRNAs are transcribed as part of a common primary transcript, a "pri-miRNA" [(Figure 4.6) (Lee et al., 2002)]. A nuclear activity excises the individual ~70-nt stem loop precursors, which are then exported to the cytoplasm to serve as substrates for Dicer processing to the ~22-nt forms (Lee et al., 2002). The other protein factors responsible for the multistep process of miRNA biogenesis are yet to be uncovered.

Virtually all of the newly identified miRNAs are yet to be partnered with their targets. No *bona fide* animal miRNA has been found that exhibits exact antisense complementarity to a protein coding gene, thus complicating our ability to scan genomes for miRNA targets. Moreover, a pattern to the structure of the imperfect duplexes formed between miRNA and target remains elusive. The *lin-14* 3′UTR contains seven sites of complementarity to the *lin-4* RNA that are conserved between *C. elegans* and its cousin nematode *C. briggsae* (Wightman et al., 1993).

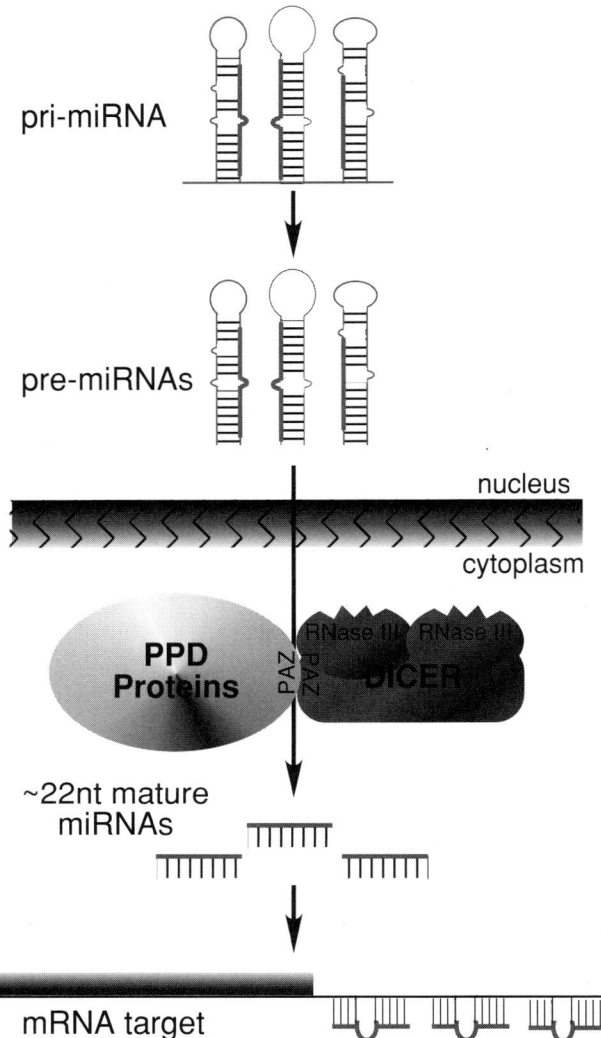

pri-miRNA

pre-miRNAs

nucleus

cytoplasm

PPD Proteins

PAZ PAZ

RNase III RNase III

DICER

~22nt mature miRNAs

mRNA target

Figure 4.6. A general model for animal miRNA biogenesis. Some miRNAs are expressed as longer primary transcripts (pri-miRNAs) that may contain multiple miRNAs (Lagos-Quintana et al., 2001; Lau et al., 2001; Mourelatos et al., 2002; Bashirullah et al., 2003; Sempere et al., 2003). A nuclear activity excises the individual stem-loop precursor miRNAs (pre-miRNAs), which are then exported to the cytoplasm where they undergo further processing by Dicer and accessory PPD (PAZ Piwi Domain) proteins (Grishok et al., 2001; Hütvagner et al., 2001; Ketting et al., 2001; Lee et al., 2002). The mature ~22-nt miRNAs recognize target mRNAs that contain sites of imperfect complementarity in their 3′UTRs.

The predicted duplexes between *lin-4* RNA and the antisense sites in the *lin-14* 3′UTR vary in structure, and this factor seems to affect the efficiency of inhibition of protein expression (Wightman et al., 1993; Ha et al., 1996). Additionally, the number of miRNA recognition sites may contribute to the potency of regulation (Doench et al., 2003). The *lin-28* protein-coding gene is also controlled by *lin-4* but the reduction in LIN-28 protein level is delayed compared to that of LIN-14, perhaps because *lin-28* only contains one *lin-4* complementary site in its 3′UTR (Moss et al., 1997). Thus, a single miRNA might regulate multiple target mRNAs at distinct times in development depending on the type and number of duplexes that can form.

Sequences in 3′UTRs that are known to direct regulation are rich sources to match potential miRNA partners. For example, the *Drosophila K* (cUGUGAUa) and Brd (AGCUUUA) box motifs in the 3′UTRs of several Notch pathway target genes dictate post-transcriptional gene regulation by an unknown mechanism (Lai et al., 1998; Lai and Posakony, 1997). These short motifs share perfect anti-sense complementarity with the 5′ half of several fly miRNAs, suggesting that regulation mediated by K and Brd boxes could be guided by miRNAs (Lai, 2002). Another clue for identifying miRNA/ target pairs is the observation that the miRNA and target recognition sites may be conserved. For example, putative *lin-41* homologs in *Drosophila* and Zebrafish contain sites in their 3′UTRS that could base pair with the *let-7* RNA present in these species [(Figure 4.4) (Pasquinelli et al., 2000)]. Likewise, the *C. elegans lin-28* gene, a target of *lin-4* regulation, is homologous to vertebrate genes and the 3′UTRs of these genes contain sites complementary to miR-125a (Moss and Tang, 2003), which is closely related to the *lin-4* RNA sequence (Lagos-Quintana et al., 2002).

However, in some cases matching conserved protein coding genes with miRNAs may be more complicated. The *Drosophila* homolog of *lin-28* apparently lacks sites complementary to miR-125, which is the closest sequence match to *lin-4* RNA in this organism (Moss and Tang, 2003). Similarly, the *C. elegans* homolog of *Drosophila hunchback* is regulated by *let-7* and contains eight sites of imperfect complementarity to this miRNA in its 3′UTR [(Figure 4.4) (Abrahante et al., 2003; Lin et al., 2003)]. These sites are not conserved in the *Drosophila hunchback* 3′UTR, but this sequence does contain multiple regions that could be recognized by other *Drosophila* miRNAs (Abrahante et al., 2003; Lin et al., 2003). At this point it is unclear if miRNA recognition sites are missing from otherwise highly conserved genes, or if we are just unable to recognize them. Clearly, predicted miRNA complementary sites require the scrutiny of functional testing before patterns can be derived for identifying genuine sites of regulation.

Outlook

Genetic studies in *C. elegans* revealed the existence of miRNA genes, which are now recognized to permeate the genomes of all multicellular organisms (reviewed in Ambros, 2001; Carthew, 2001; Ruvkun, 2001; Banerjee and Slack, 2002; Dennis, 2002; Grosshans and Slack, 2002; Moss and Poethig, 2002; Pasquinelli, 2002; Pasquinelli and Ruvkun, 2002; Zamore, 2002; Ambros, 2003; Hunter and Poethig, 2003; Hüttenhofer et al., 2003). These tiny RNA regulators are being implicated in diverse biological pathways, ranging from development to neuronal differentiation to fat metabolism (reviewed in Ambros, 2003). The factors and mechanisms employed to generate functional miRNAs are just beginning to be uncovered by combined biochemical and genetic approaches in diverse organisms. Although the identification by bioinformatics of genes regulated by miRNAs is confounded by the lack of perfect antisense complementarity between animal miRNAs and their targets, candidate pairings have been proposed and their functional significance awaits testing. So far, genetic suppressor screens have been the most reliable method for identifying genes regulated by miRNAs. But, this approach first requires a genetic mutation in a particular miRNA that produces an obvious phenotype, and it assumes that expression of targets will be negatively regulated. Inhibition of translation by antisense pairing to 3′UTR sequences may be just one of many modes of negative or positive regulation employed by miRNAs. The unanticipated existence of miRNA genes prompts the question of what other elusive regulators of gene expression lurk in our genomes. Perhaps a genetic screen in a worm or other humble organism has already identified one.

REFERENCES

Abrahante, J. E., Daul, A. L., Li, M., Volk, M. L., Tennessen, J. M., Miller, E. A. and Rougvie, A. E. (2003). The *Caenorhabditis elegans* hunchback-like gene lin-57/hbl-1 controls developmental time and is regulated by microRNAs. *Developmental Cell*, **4**, 625–637.

Ambros, V. and Horvitz, H. R. (1984). Heterochronic mutants of the nematode *Caenorhabditis elegans*. *Science*, **226**, 409–416.

Ambros, V. and Horvitz, H. R. (1987). The lin-14 locus of *Caenorhabditis elegans* controls the time of expression of specific postembryonic developmental events. *Genes & Development*, **1**, 398–414.

Ambros, V. (1989). A hierarchy of regulatory genes controls a larva-to-adult developmental switch in *C. elegans*. *Cell*, **57**, 49–57.

Ambros, V. (2001). microRNAs: Tiny regulators with great potential. *Cell*, **107**, 823–826.

Ambros, V. (2003). MicroRNA pathways in flies and worms: Growth, death, fat, stress, and timing. *Cell*, **113**, 673–676.

Ambros, V., Bartel, B., Bartel, D. P., Burge, C. B., Carrington, J. C., Chen, X., Dreyfuss, G., Eddy, S. R., Griffiths-Jones, S., Marshall, M., Matzke, M., Ruvkun, G. and Tuschl, T. (2003a). A uniform system for microRNA annotation. *RNA*, **9**, 277–279.

Ambros, V., Lee, R. C., Lavanway, A., Williams, P. T. and Jewell, D. (2003b). MicroRNAs and other tiny endogenous RNAs in *C. elegans*. *Current Biology*, **13**, 807–818.

Banerjee, D. and Slack, F. (2002). Control of developmental timing by small temporal RNAs: A paradigm for RNA-mediated regulation of gene expression. *Bioessays*, **24**, 119–129.

Bashirullah, A., Pasquinelli, A. E., Kiger, A. A., Perrimon, N., Ruvkun, G. and Thummel, C. S. (2003). Coordinate regulation of small temporal RNAs at the onset of *Drosophila* metamorphosis. *Developmental Biology*, **259**, 1–8.

Bernstein, E., Caudy, A. A., Hammond, S. M. and Hannon, G. J. (2001). Role for a bidentate ribonuclease in the initiation step of RNA interference. *Nature*, **409**, 363–366.

Bohmert, K., Camus, I., Bellini, C., Bouchez, D., Caboche, M. and Benning, C. (1998). AGO1 defines a novel locus of Arabidopsis controlling leaf development. *European Molecular Biology Organization Journal*, **17**, 170–180.

Caplen, N. J., Fleenor, J., Fire, A. and Morgan, R. A. (2000). dsRNA-mediated gene silencing in cultured *Drosophila* cells: A tissue culture model for the analysis of RNA interference. *Gene*, **252**, 95–105.

Carthew, R. W. (2001). Gene silencing by double-stranded RNA. *Current Opinions in Cell Biology*, **13**, 244–248.

Cerutti, L., Mian, N. and Bateman, A. (2000). Domains in gene silencing and cell differentiation proteins: The novel PAZ domain and redefinition of the Piwi domain. *Trends in Biochemical Sciences*, **25**, 481–482.

Chalfie, M., Horvitz, H. R. and Sulston, J. E. (1981). Mutations that lead to reiterations in the cell lineages of *C. elegans*. *Cell*, **24**, 59–69.

Dennis, C. (2002). The brave new world of RNA. *Nature*, **418**, 122–124.

Doench, J. G., Petersen, C. P. and Sharp, P. A. (2003). siRNAs can function as miRNAs. *Genes & Development*, **17**, 438–442.

Elbashir, S. M., Harborth, J., Lendeckel, W., Yalcin, A., Weber, K. and Tuschl, T. (2001a). Duplexes of 21-nucleotide RNAs mediate RNA interference in cultured mammalian cells. *Nature*, **411**, 494–498.

Elbashir, S. M., Lendeckel, W. and Tuschl, T. (2001b). RNA interference is mediated by 21- and 22-nucleotide RNAs. *Genes & Development*, **15**, 188–200.

Elbashir, S. M., Martinez, J., Patkaniowska, A., Lendeckel, W. and Tuschl, T. (2001c). Functional anatomy of siRNAs for mediating efficient RNAi in *Drosophila melanogaster* embryo lysate. *European Molecular Biology Organization Journal*, **20**, 6877–6888.

Fagard, M., Boutet, S., Morel, J. B., Bellini, C. and Vaucheret, H. (2000). AGO1, QDE-2, and RDE-1 are related proteins required for post-transcriptional gene silencing in plants, quelling in fungi, and RNA interference in animals. *Proceedings of the National Academy of Sciences USA*, **97**, 11650–11654.

Fire, A., Xu, S., Montgomery, M. K., Kostas, S. A., Driver, S. E. and Mello, C. C. (1998). Potent and specific genetic interference by double-stranded RNA in *Caenorhabditis elegans*. *Nature*, **391**, 806–811.

Grad, Y., Aach, J., Hayes, G. D., Reinhart, B. J., Church, G. M., Ruvkun, G. and Kim, J. (2003). Computational and experimental identification of *C. elegans* microRNAs. *Molecular Cell*, **11**, 1253–1263.

Grishok, A., Pasquinelli, A. E., Conte, D., Li, N., Parrish, S., Ha, I., Baillie, D. L., Fire, A., Ruvkun, G. and Mello, C. C. (2001). Genes and mechanisms related to RNA interference regulate expression of the small temporal RNAs that control *C. elegans* developmental timing. *Cell*, **106**, 23–34.

Grosshans, H. and Slack, F. J. (2002). Micro-RNAs: Small is plentiful. *Journal of Cell Biology*, **156**, 17–21.

Ha, I., Wightman, B. and Ruvkun, G. (1996). A bulged lin-4/lin-14 RNA duplex is sufficient for *Caenorhabditis elegans* lin-14 temporal gradient formation. *Genes & Development*, **10**, 3041–3050.

Hamilton, A. J. and Baulcombe, D. C. (1999). A species of small antisense RNA in posttranscriptional gene silencing in plants. *Science*, **286**, 950–952.

Hammond, S. M., Bernstein, E., Beach, D. and Hannon, G. J. (2000). An RNA-directed nuclease mediates post-transcriptional gene silencing in *Drosophila* cells. *Nature*, **404**, 293–296.

Hunter, C. and Poethig, R. S. (2003). miSSING LINKS: miRNAs and plant development. *Current Opinion in Genetics and Development*, **13**, 372–378.

Hüttenhofer, A., Brosius, J., Bachellerie, J.-P. (2003). RNomics: identification and function of small, non-messenger RNAs. *Current Opinion in Chemical Biology*, **6**, 835–843.

Hütvagner, G., McLachlan, J., Pasquinelli, A. E., Balint, E., Tuschl, T. and Zamore, P. D. (2001). A cellular function for the RNA-interference enzyme Dicer in the maturation of the let-7 small temporal RNA. *Science*, **293**, 834–838.

Hütvagner, G. and Zamore, P. D. (2002). A microRNA in a multiple-turnover RNAi enzyme complex. *Science*, **297**, 2056–2060.

Johnson, S. M., Lin, S. Y. and Slack, F. J. (2003). The time of appearance of the *C. elegans* let-7 microRNA is transcriptionally controlled utilizing a temporal regulatory element in its promoter. *Developmental Biology*, **259**, 364–379.

Ketting, R. F., Fischer, S. E., Bernstein, E., Sijen, T., Hannon, G. J. and Plasterk, R. H. (2001). Dicer functions in RNA interference and in synthesis of small RNA involved in developmental timing in *C. elegans*. *Genes & Development*, **15**, 2654–2659.

Knight, S. W. and Bass, B. L. (2001). A role for the RNase III enzyme DCR-1 in RNA interference and germ line development in *Caenorhabditis elegans*. *Science*, **293**, 2269–2271.

Lagos-Quintana, M., Rauhut, R., Lendeckel, W. and Tuschl, T. (2001). Identification of novel genes coding for small expressed RNAs. *Science*, **294**, 853–858.

Lagos-Quintana, M., Rauhut, R., Yalcin, A., Meyer, J., Lendeckel, W. and Tuschl, T. (2002). Identification of tissue-specific microRNAs from mouse. *Current Biology*, **12**, 735–739.

Lai, E. C. and Posakony, J. W. (1997). The Bearded box, a novel 3′ UTR sequence motif, mediates negative post-transcriptional regulation of Bearded and Enhancer of split Complex gene expression. *Development*, **124**, 4847–4856.

Lai, E. C., Burks, C. and Posakony, J.W. (1998). The K box, a conserved 3′ UTR sequence motif, negatively regulates accumulation of enhancer of split complex transcripts. *Development*, **125**, 4077–4088.

Lai, E. C. (2002). Micro RNAs are complementary to 3′ UTR sequence motifs that mediate negative post-transcriptional regulation. *Nature Genetics*, **30**, 363–364.

Lai, E. C., Tomancak, P., Williams, R. W. and Rubin, G. M. (2003). Computational identification of *Drosophila* microRNA genes. *Genome Biology*, **4**, R42.

Lau, N. C., Lim le, E. P., Weinstein, E. G. and Bartel, D. P. (2001). An abundant class of tiny RNAs with probable regulatory roles in *Caenorhabditis elegans*. *Science*, **294**, 858–862.

Lee, R. C. and Ambros, V. (2001). An extensive class of small RNAs in *Caenorhabditis elegans*. *Science*, **294**, 862–864.

Lee, R. C., Feinbaum, R. L. and Ambros, V. (1993). The *C. elegans* heterochronic gene lin-4 encodes small RNAs with antisense complementarity to lin-14. *Cell*, **75**, 843–854.

Lee, Y., Jeon, K., Lee, J. T., Kim, S. and Kim, V. N. (2002). MicroRNA maturation: stepwise processing and subcellular localization. *European Molecular Biology Organization Journal*, **21**, 4663–4670.

Lim, L. P., Glasner, M. E., Yekta, S., Burge, C. B. and Bartel, D. P. (2003a). Vertebrate microRNA genes. *Science*, **299**, 1540.

Lim, L. P., Lau, N. C., Weinstein, E. G., Abdelhakim, A., Yekta, S., Rhoades, M. W., Burge, C. B. and Bartel, D. P. (2003b). The microRNAs of *Caenorhabditis elegans*. *Genes & Development*, **17**, 991–1008.

Lin, S. Y., Johnson, S. M., Abraham, M., Vella, M. C., Pasquinelli, A., Gamberi, C., Gottlieb, E. and Slack, F. J. (2003). The *C. elegans* hunchback homolog, hbl-1, controls temporal patterning and is a probable microRNA target. *Developmental Cell*, **4**, 639–650.

Llave, C., Xie, Z., Kasschau, K. D. and Carrington, J. C. (2002). Cleavage of Scarecrow-like mRNA targets directed by a class of Arabidopsis miRNA. *Science*, **297**, 2053–2056.

McManus, M.T., Petersen, C. P., Haines, B. B., Chen, J. and Sharp, P. A. (2002). Gene silencing using micro-RNA designed hairpins. *RNA*, **8**, 842–850.

Moss, E. G., Lee, R. C. and Ambros, V. (1997). The cold shock domain protein LIN-28 controls developmental timing in *C. elegans* and is regulated by the lin-4 RNA. *Cell*, **88**, 637–646.

Moss, E. G. and Poethig, R. S. (2002). MicroRNAs: Something new under the sun. *Current Biology*, **12**, R688–690.

Moss, E. G. and Tang, L. (2003). Conservation of the heterochronic regulator Lin-28, its developmental expression and microRNA complementary sites. *Developmental Biology*, **258**, 432–442.

Mourelatos, Z., Dostie, J., Paushkin, S., Sharma, A., Charroux, B., Abel, L., Rappsilber, J., Mann, M. and Dreyfuss, G. (2002). miRNPsA novel class of ribonucleoproteins containing numerous microRNAs. *Genes & Development*, **16**, 720–728.

Olsen, P. H. and Ambros, V. (1999). The lin-4 regulatory RNA controls developmental timing in *Caenorhabditis elegans* by blocking LIN-14 protein synthesis after the initiation of translation. *Developmental Biology*, **216**, 671–680.

Park, W., Li, J., Song, R., Messing, J. and Chen, X. (2002). CARPEL FACTORY, a Dicer homolog, and HEN1, a novel protein, act in microRNA metabolism in Arabidopsis thaliana. *Current Biology*, **12**, 1484–1495.

Parrish, S., Fleenor, J., Xu, S., Mello, C. and Fire, A. (2000). Functional anatomy of a dsRNA trigger. Differential requirement for the two trigger strands in RNA interference. *Molecular Cell*, **6**, 1077–1087.

Pasquinelli, A. E., Reinhart, B. J., Slack, F., Martindale, M. Q., Kuroda, M. I., Maller, B., Hayward, D. C., Ball, E. E., Degnan, B., Muller, P., Spring, J., Srinivasan, A., Fishman, M., Finnerty, J., Corbo, J., Levine, M., Leahy, P., Davidson, E. and Ruvkun, G. (2000). Conservation of the sequence and temporal expression of let-7 heterochronic regulatory RNA. *Nature*, **408**, 86–89.

Pasquinelli, A. E. (2002). MicroRNAs: Deviants no longer. *Trends in Genetics*, **18**, 171–173.

Pasquinelli, A. E. and Ruvkun, G. (2002). Control of developmental timing by micrornas and their targets. *Annual Reviews of Cell and Developmental Biology*, **18**, 495–513.

Pasquinelli, A. E., McCoy, A., Jimenez, E., Salo, E., Ruvkun, G., Martindale, M. Q. and Baguna, J. (2003). Expression of the 22 nucleotide let-7 heterochronic RNA throughout the Metazoa: A role in life history evolution? *Evolution & Development*, **5**, 372–378.

Reinhart, B. J., Slack, F. J., Basson, M., Pasquinelli, A. E., Bettinger, J. C., Rougvie, A. E., Horvitz, H. R. and Ruvkun, G. (2000). The 21-nucleotide let-7 RNA regulates developmental timing in *Caenorhabditis elegans*. *Nature*, **403**, 901–906.

Reinhart, B. J., Weinstein, E. G., Rhoades, M. W., Bartel, B. and Bartel, D. P. (2002). MicroRNAs in plants. *Genes & Development*, **16**, 1616–1626.

Ruvkun, G. and Giusto, J. (1989). The *Caenorhabditis elegans* heterochronic gene lin-14 encodes a nuclear protein that forms a temporal developmental switch. *Nature*, **338**, 313–319.

Ruvkun, G. (2001). Molecular biology. Glimpses of a tiny RNA world. *Science*, **294**, 797–799.

Schwarz, D. S. and Zamore, P. D. (2002). Why do miRNAs live in the miRNP? *Genes & Development*, **16**, 1025–1031.

Seggerson, K., Tang, L. and Moss, E. G. (2002). Two genetic circuits repress the *Caenorhabditis elegans* heterochronic gene lin-28 after translation initiation. *Developmental Biology*, **243**, 215–225.

Sempere, L. F., Sokol, N. S., Dubrovsky, E. B., Berger, E. M. and Ambros, V. (2003). Temporal regulation of microRNA expression in *Drosophila melanogaster* mediated by hormonal signals and broad-Complex gene activity. *Developmental Biology*, **259**, 9–18.

Slack, F. J. and Ruvkun, G. (1998). A novel repeat domain that is often associated with RING finger and B- box motifs. *Trends Biochem Sci*, **23**, 474–475.

Slack, F. J., Basson, M., Liu, Z., Ambros, V., Horvitz, H. R. and Ruvkun, G. (2000). The lin-41 RBCC gene acts in the *C. elegans* heterochronic pathway between the let-7 regulatory RNA and the LIN-29 transcription factor. *Molecular Cell*, **5**, 659–669.

Sulston, J. E. and Horvitz, H. R. (1977). Post-embryonic cell lineages of the nematode, *Caenorhabditis elegans*. *Developmental Biology*, **56**, 110–156.

Sulston, J. E. and Horvitz, H. R. (1981). Abnormal cell lineages in mutants of the nematode *Caenorhabditis elegans*. *Developmental Biology*, **82**, 41–55.

Tabara, H., Sarkissian, M., Kelly, W. G., Fleenor, J., Grishok, A., Timmons, L., Fire, A. and Mello, C. C. (1999). The rde-1 gene, RNA interference, and transposon silencing in *C. elegans*. *Cell*, **99**, 123–132.

Tuschl, T., Zamore, P. D., Lehmann, R., Bartel, D. P. and Sharp, P. A. (1999). Targeted mRNA degradation by double-stranded RNA *in vitro*. *Genes & Development*, **13**, 3191–3197.

Wightman, B., Ha, I. and Ruvkun, G. (1993). Posttranscriptional regulation of the heterochronic gene lin-14 by lin- 4 mediates temporal pattern formation in *C. elegans*. *Cell*, **75**, 855–862.

Yang, D., Lu, H. and Erickson, J. W. (2000). Evidence that processed small dsRNAs may mediate sequence-specific mRNA degradation during RNAi in *Drosophila* embryos. *Current Biology*, **10**, 1191–1200.

Zamore, P. D., Tuschl, T., Sharp, P. A. and Bartel, D. P. (2000). RNAi: double-stranded RNA directs the ATP-dependent cleavage of mRNA at 21 to 23 nucleotide intervals. *Cell*, **101**, 25–33.

Zamore, P. D. (2002). Ancient pathways programmed by small RNAs. *Science*, **296**, 1265–1269.

Zeng, Y., Wagner, E. J. and Cullen, B. R. (2002). Both natural and designed micro RNAs can inhibit the expression of cognate mRNAs when expressed in human cells. *Molecular Cell*, **9**, 1327–1333.

Zeng, Y., Yi, R. and Cullen, B. R. (2003). MicroRNAs and small interfering RNAs can inhibit mRNA expression by similar mechanisms. *Proceedings of the National Academy of Sciences U S A*, **100**, 9779–9784.

5 miRNAs in the brain and the application of RNAi to neurons

Anna M. Krichevsky, Shih-Chu Kao, Li-Huei Tsai, and Kenneth S. Kosik

I. miRNAs in the brain

Several hundred microRNAs (miRNAs) have been cloned from a wide range of organisms across phylogeny. miRNAs are 19–23 nucleotide transcripts with characteristic 3′ hydroxyl and 5′ phosphate termini cleaved from a ~70-nt hairpin precursor by Dicer Ribonuclease III (Hütvagner et al., 2001; Ketting et al., 2001). Many miRNAs, often with highly conserved sequences, have been mapped in the genomes of *C. elegans, Drosophila*, rodents and humans (Lagos-Quintana et al., 2001; Lau et al., 2001; Lee and Ambros, 2001; Lagos-Quintana et al., 2002; Mourelatos et al., 2002; Dostie et al., 2003). Based on the length, hairpin structure and conservation the total number of miRNAs in the *Drosophila* genome has been estimated to be 110 (Lai et al., 2003) and in the human genome to be 255 (Lim et al., 2003). Some miRNAs are organized in the genome as clusters which can be separated by intervals as short as a few nucleotides (Lagos-Quintana et al., 2001; Lau et al., 2001). Despite the high degree of conservation of miRNAs, their functions in general, and in mammals particularly, have not been well defined. The first two miRNAs discovered, *lin-4* and *let-7*, were found in *C. elegans* where they control developmental timing. These short transcripts form imperfect base pairing with elements within the 3′ UTR of target mRNAs and attenuate their translation (Lee et al., 1993; Wightman et al., 1993; Olsen and Ambros, 1999; Reinhart et al., 2000; Slack et al., 2000). Several studies conducted in plants, worms, flies and fission yeast suggest a very broad variety of molecular functions for this novel class of tiny RNA molecules. These functions include the regulation of mRNA translation and stability, heterochromatin formation, genome rearrangement and DNA excision (reviewed in Baulcombe, 2002; Carrington and Ambros, 2003).

Most apparent is the important role miRNAs play in development. Inactivation of the miRNA-producing enzyme Dicer causes heterochronic phenotypes similar to *let-7* and *lin-4* mutations in *C. elegans* (Grishok et al., 2001). In vertebrates (Zebrafish) Dicer inactivation results in developmental arrest (Wienholds et al., 2003). In *Drosophila*, bantam miRNA stimulates cell proliferation and tissue growth (Brennecke et al., 2003). In Arabidopsis, suppression of RNA silencing, which shares a common pathway with miRNA processing, results in developmental defects (Kasschau et al., 2003). However, no function has been demonstrated for miRNA in mammals yet. Recently, a role of mir-181 in controling of mouse hematopoiesis was demonstrated by Chen et al. (2004).

Many miRNAs have been isolated from mammalian embryonic neurons and mature brain and some of them are neuronal-specific (Dostie et al., 2003; Lagos-Quintana M, 2002; Kim J. et al., 2004). The extensive use of post-transcriptional regulation of gene expression in neuronal development, synaptic plasticity and possibly long-term memory storage (Okabe et al., 2001; Miller et al., 2002) could potentially require miRNAs. In fact, the idea of "RNA-mediated memory transfer" was enthusiastically investigated as early as in the 1960s in flatworms and in the 1970s in rats (McConnell, 1966; Golub et al., 1970). *lin-4* miRNA regulates synaptic remodelling in *C. elegans* by controlling *lin-14* expression (Hallam and Jin, 1998). Interestingly, numerous miRNAs are found in ribonucleoprotein complexes with Fragile X mental retardation protein [(FMRP) (O'Donnell and Warren, 2002)] and with Survival of Motor Neuron (SMN) protein (Mourelatos et al., 2002). Mutations of these two proteins cause major neurological disorders, Fragile X syndrome and spinal muscular atrophy, correspondingly. These data strongly suggest a role for miRNAs in neuronal function.

miRNAs in mammalian brain development

To study the involvement of miRNAs in mammalian brain development and physiology, we have designed an oligonucleotide array capable of determining expression patterns of many neuronal miRNAs (Krichevsky et al., 2003). The first generation array used probes for 43 miRNAs that were cloned from mouse brain (Lagos-Quintana et al., 2002) and from cultured rat primary neurons (Kim J. et al., 2004). Sequences complementary to these mature miRNAs were spotted onto a nylon membrane. Specific probes for the loop region of a few highly abundant pre-miRNAs were also included to distinguish between a specific pre-miRNA and mature miRNA. First, miRNA expression profiling was performed at stages of cortical neurogenesis when cell proliferation, migration, regional specification and the establishment of circuitry occur. For the purpose of detecting the expression of mature miRNA, we used a labeled RNA sample highly enriched in low molecular weight (LMW) RNA molecules below 70 nt. These arrays revealed many miRNAs that are differentially expressed in brain during the developmental period from rat embryonic day 12 to adult, and at least nine of them changed more than 2-fold.

Northern blots validated the differential expression of all nine miRNAs (Figure 5.1) and the fold differences detected by Northern blots were in most cases within the range predicted by the oligoarrays. The expression patterns fell into four categories: 1) postnatal expression, e.g. miR-128; 2) prenatal expression, e.g. miR-19b; 3) peak expression at E21, e.g. miR-9, miR-125b, miR-131, miR-181a; 4) increasing during the embryonic period followed by stable expression in the

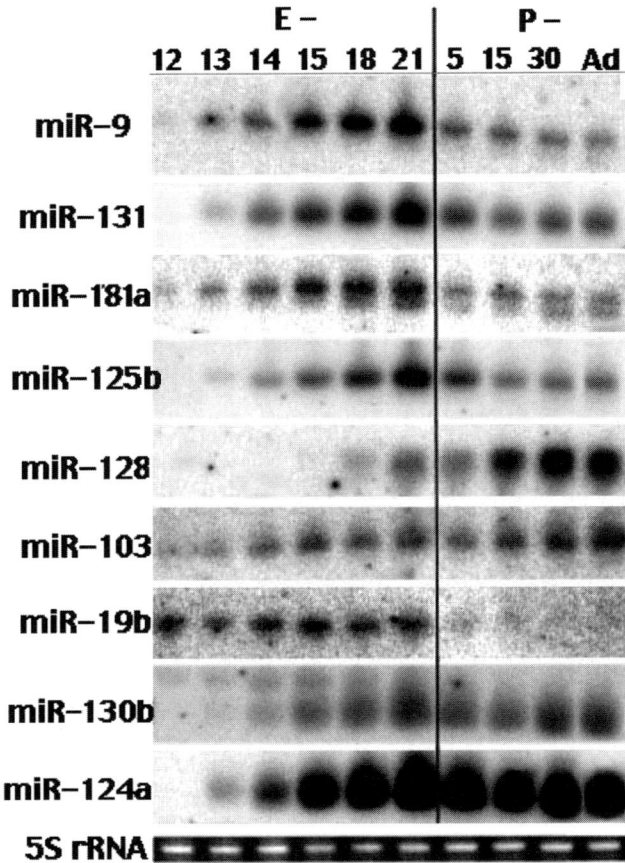

Figure 5.1. Regulated expression of miRNAs during brain development was confirmed by Northern blot hybridizations (E-prenatal and P-postnatal days). Expression patterns of these miRNAs in corresponding stages of corticogenesis have been predicted by oligonucleotide arrays. The figure is reprinted from: Krichevsky et al., A microRNA array reveals extensive regulation of microRNAs during brain development. *RNA*. 2003 Oct; 9(10):1274–81.

post-natal period, e.g. miR-124a and miR-181a; 5) gradually increasing expression, e.g. miR-103. Because several factors, including transcription, maturation and degradation, contribute to expression, these patterns suggest finely tuned regulatory mechanisms resulting in up or down regulation of specific miRNA or groups of miRNAs at precise developmental time points. The tuned temporal expressions of specific groups of miRNAs suggest that these groups may play distinct roles in brain development and physiology. Interestingly, miR-9, miR-125b, miR-131 and miR-181a showed a similar pattern of up-regulation from E12 to E21 followed by down-regulation after birth and a steady-state level after P5. The similar expression pattern of these miRNAs suggests that they share common regulatory elements. Two of these miRNAs, miR-9 and miR-131, are brain-enriched transcripts processed from both arms of the stem of a ~65-nt common precursor. Usually just one mature miRNA can be detected as a stable product cleaved from a hairpin-shaped pre-miRNA. Therefore, similar expression patterns of miR-9 and miR-131 most likely reflect their co-transcription and co-maturation from the same precursor. Two other co-expressed miRNAs, mir-125b and mir-181a, are

Figure 5.2. miRNAs are differentially associated with translating polyribosomes. A. Cytoplasmic extracts of primary cortical neurons were fractionated on 15%–45% sucrose gradients after sedimentation of nuclei and the majority of mitochondria. Fractions were collected with continuous monitoring of RNA at 254 nm. B. Representative oligo arrays showing hybridization with ^{33}P-labeled LMW RNA that was isolated from mRNP and polysomal fractions of cultured primary neurons. Signals corresponding to miR-9 and miR-131 hybridizations are indicated.

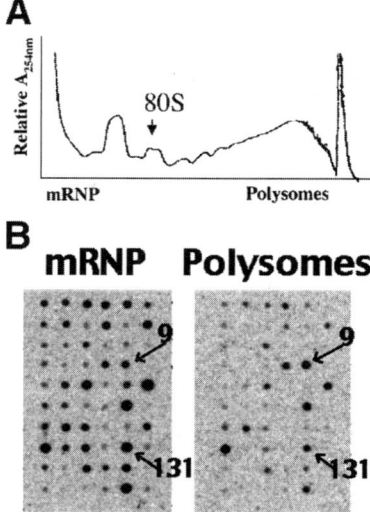

encoded on different chromosomes. Since transcription of miRNA genes is a completely unstudied process, the analysis of genomic sequences upstream of miR-9/131, miR-125b and miR-181a may reveal common regulatory sites. Creating clusters of miRNAs based on their expression may be helpful for discovering possible molecular functions such as regulation of the same mRNA target or a target family that acts in a specific biological process.

Neuronal miRNAs are differentially associated with polyribosomes

One of the proposed roles for miRNAs is the regulation of translation by forming imperfectly base-paired duplexes with mRNAs. In *C. elegans* both *lin-4* miRNA and its target *lin-14* mRNA associate with polyribosomes, the sites of translation, and *lin-4* inhibits the translation of *lin-14* mRNA after initiation of its translation (Ambros, 2000). Therefore we studied the association of miRNAs with translating polyribosomes in primary neurons. Cytosolic lysates from cerebro-cortical cells were fractionated in sucrose gradients and RNA was isolated separately from fractions of mRNP, free (untranslating) ribosomes, polyribosomes and RNA granules as described by (Krichevsky and Kosik, 2001; Figure 5.2a). RNA granules are macromolecular structures observed in neurons, where they localize mRNA before its activity-dependent translation. LMW RNA samples from each of the four principal fractions were hybridized to the oligonucleotide arrays. Strong hybridization signals were detected with probes corresponding to mRNP- and polyribosomes-associated RNA (Figure 5.2b). Co-sedimentation of most (if not all) miRNAs with polyribosomes strongly suggests their role in the regulation of neuronal translation. Overall a stronger signal was observed in the mRNPs, however, the proportion of specific miRNAs in polysomes versus mRNPs differed widely: some miRNAs were over-represented in polysomes, whereas others are under-represented. This data suggest that miRNA association with translation machinery, and consequently miRNA effect on mRNA target translation, could be regulated.

miRNAs may be rapidly generated and dynamically localized when required (compare to proteins and other RNA molecules) and therefore may facilitate precise temporal regulation during developmental transitions or in the context of neuronal activity. Neuronal activity causes changes in the translation of somatodendritic, plasticity-related mRNAs (Aakalu et al., 2001; Eberwine et al., 2001; Krichevsky and Kosik, 2001; Steward and Schuman, 2001). Regulation can occur locally in a region of synaptic activity, and affect only specific synapses. miR-NAs would fit well into this model, if their expression, maturation or localization were modulated by synaptic activity. Indeed, processing of the miRNA precursor to the mature miRNA via Dicer is known to occur in the cytoplasm (Lee et al., 2003). While depolarization of primary neurons with KCl neither alter the expression nor the association of several miRNAs tested with polyribosome fractions (Kim J. et al., 2004), other miRNAs may have a role in an activity-dependent translation in neurons. miRNA arrays will be most valuable tool to discover whether this class of RNAs can serve as regulators of local dendritic translation.

miRNA expression in Presenilin-1 knockout mice

Because miRNAs demonstrate precisely regulated patterns of expression during brain development, we sought evidence for their dysregulation in a brain developmental disorder. Presenilin-1 (PS1) is part of a multi-protein complex that mediates intra-membranous proteolysis and release of receptor endodomain for cell signaling. Because the Notch receptors are key substrates of the complex, PS1-deficient mice show lethal developmental defects in the CNS similar to those observed in mice deficient in Notch 1 (Shen et al., 1997). Although PS1(–/–) mice die shortly after birth and exhibit severe CNS defects and axial skeleton malformations, a 50% reduction of PS1 activity in PS1(+/–) mice does not lead to any detectable developmental abnormalities. Premature neuronal differentiation in the PS1(–/–) brain is associated with aberrant neuronal migration and disorganization of the laminar architecture of the developing cerebral hemisphere (Handler et al., 2000). We applied the array to the study of miRNA expression in PS1 knockout mice. This experiment demonstrated that expression of the developmentally regulated pair, miR-9 and miR-131, was reduced in the forebrains of PS1-deficient mice compared to their PS1(+/–) littermates. While expression of the two miR-NAs is strongly up-regulated during brain development in all normal embryos (see Figure 5.1), in PS1(–/–) brains miR-9 and miR-131 levels did not increase during the observed period from E13.5 to E17.5. The failure of miR-9 and miR-131 to up regulate in PS1(–/–) embryonic brain development may be due to impaired Notch signaling. Whether lowered mir-9/mir-131 expression contributes to developmental defects in PS1-deficient mice or just represents a consequence of impaired development remains to be determined. No differences in the expression of other miRNAs in PS1(–/–) mice have been detected, suggesting that Notch signaling affected mir-9/mir-131 expression in a specific way.

The functions of mammalian miRNAs have been extrapolated from other species. Heterochronic RNAs in *C. elegans* control cell cycle progression and a particular sequence of cell divisions that is necessary for the generation of a

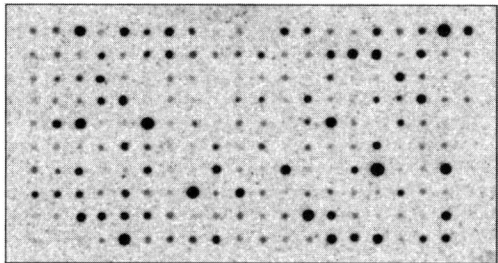

Figure 5.3. Representative oligonucleotide array on which ^{33}P-labeled LMW RNA from rat forebrain was hybridized.

specific cell type (Ambros, 2000). Stem cell lineages appear to be a fundamental principle of development and miRNAs are likely to play a role in modifying these lineages. miRNA arrays can reveal specific miRNAs that modify lineages as stem cells undergo staged differentiation in many organs including the brain. The dysregulation of the miR-9/miR-131 pair in the PS(−/−) mouse supports this idea. Here the lack of PS1 leads to premature differentiation of neural progenitor cells, while neural proliferation and apoptotic cell death during neurogenesis are unaltered (Handler et al., 2000). This shift in the developmental timing resembles the heterochronic phenotypes seen in *C. elegans*.

These oligonucleotide arrays that detect miRNA expression provide an informative tool for assigning biological significance to this novel class of molecules. miRNAs, which are likely to share with other genes related regulatory control elements, display expression profiles that uniquely correspond to developmental epochs of brain development. A major challenge in understanding miRNA function is target identification. The expression pattern of a miRNA may narrow the search for and contribute to the validation of targets. The miRNA and its putative target must be expressed in the same tissue and in the same cells. Putative targets presumably undergo post-transcriptional regulation that is coordinated with the expression of a specific miRNA. Creating clusters of co-expressed miRNAs, such as miR-9/131, miR-181a and miR-125b, will contribute to finding mRNA targets as well as to bioinformatic analysis of upstream regulatory motifs and, in general, to understanding their functions. The estimated number of the total miRNAs in vertebrates is ~250 molecules (Lim et al., 2003). This prediction is based on the observation that miRNAs derive from evolutionary conserved hairpin precursor RNAs. However, mammalian brain may develop unique regulatory molecules, and express many more miRNAs. In addition, many tiny non-coding RNA molecules that are not processed from a miRNA-like hairpin precursor and are not phylogenetically conserved have been recently cloned from *C. elegans* and are likely to be expressed in vertebrates as well (Ambros et al., 2003). Theoretically, there is an enormous number of these different molecules, and they may represent another uncovered layer for gene regulation. In any case, annotating the genes for small non-coding RNAs will involve establishing their expression profiles and the use of oligonucleotide arrays is ideally suited to this task. We are currently applying a new version of arrays to profile 180 miRNAs that have been cloned from vertebrates (Figure 5.3).

II. RNAi in neurons

In a wide range of organisms endogenously expressed or exogenously supplied double-stranded RNA (dsRNA) triggers post-transcriptional gene silencing or RNA interference (RNAi). dsRNAs are cleaved by RNase III Dicer, the miRNA-producing enzyme, into 21–22 nucleotide RNA duplexes or small interfering RNAs (siRNAs). siRNAs trigger the degradation of the cognate mRNA (described in the other chapters of this book), however, they can also function similarly to miRNAs and inhibit translation of mRNA bearing partially complementary sequences without inducing detectable RNA cleavage (Doench et al., 2003). On the other hand, endogenous miRNA is able to mediate degradation of mRNA bearing fully complementary target sites (Zeng et al., 2003). Therefore, miRNAs and siRNAs are produced in a similar pathway and can use similar mechanisms to inhibit gene expression.

siRNAs, the 21-nt dsRNA intermediates of this natural pathway, have became a powerful tool to knock-down specific gene expression in mammalian cell lines and are widely used for generation of loss-of-function phenotypes. In mammalian primary neuronal cultures, where genetic manipulations are especially difficult and inefficient, RNAi might be developed into a highly efficacious tool to study the roles of specific genes in neuron development and functioning. However, for unclear reasons neurons appeared more resistant to RNAi than other cell types, perhaps due to differences related to the RNA transport across the cell membrane or the RNAi pathway in these cells. Although RNAi is systemic in *C. elegans* when fed or injected to whole animals and even silences genes into the next generation, worm neurons seemed resistant to RNAi (Timmons et al., 2001). However, the discovery of predominantly brain specific miRNAs suggested a role of RNAi-related mechanisms of gene regulation in neuronal development and functioning.

Here we discuss the applications of RNAi to post-mitotic primary hippocampal and cortical neuronal cultures. Importantly, synthetic siRNA can be readily introduced into neurons and effectively inhibit the expression of endogenous and transfected genes (Krichevsky and Kosik, 2002).

To assess the effectiveness of RNAi on cultured hippocampal and cortical neurons, the 21-nt sense and antisense ssRNAs targeted against EGFP-C2 and DsRed2 reporter vectors were selected from the coding regions and designed as recommended (Elbashir et al., 2001). Two strands of synthetic ssRNAs were annealed to create the duplex with the characteristics of siRNAs. Primary neurons were transfected with these two plasmids and one cognate siRNA 5–10 days after plating. 24–48 h after transfections cells were fixed, and green and red fluorescence were monitored. Fluorescent microscopy showed that EGFP or DsRed2 expressions were significantly and specifically inhibited by their corresponding siRNAs, but not by unrelated siRNAs, in nearly all transfected neurons. siRNA cognate to GFP significantly reduced EGFP expression without affecting dsRed expression and, therefore, the cells appeared red. The quantitative analysis demonstrated that GFP expression was inhibited by the cognate dsRNA duplex as detected by almost a two-fold reduction in the green fluorescence intensity, whereas the red fluorescence remained unchanged. Moreover, the accurate analysis of fluorescence distribution functions of control versus targeted cells clearly indicates that nearly all transfected neurons were affected by siGFP even in its lowest concentration (6 ng /ml). The suppression was specific for siRNA duplex: both sense and antisense ssRNAs did not inhibit expression of the reporter gene in the range of concentrations used.

We next attempted to inhibit endogenous mRNA expression by RNAi. In the course of these experiments, we noticed that chemically synthesized siRNA could be readily introduced into neurons as compared to plasmid DNA: over 90% of the neurons appeared fluorescent when transfected with fluorescein-labeled siRNA. Therefore, there was no need to introduce the reporter vector as a marker of cells transfected with siRNA. The fact that siRNAs can be introduced into nearly all primary neurons when formulated with lipophilic reagents is very important, because transfections of post-mitotic primary neurons with DNA are very inefficient (rarely more than 1–8% of cells are transfected) and often toxic. Therefore, the approach to gene suppression by intracellular expression of siRNAs from transfected plasmids would not be useful for neurons *en masse*. We first tested the effect of siRNAs on endogenous mRNA for microtubule-associated protein 2 (MAP2). MAP2 was targeted by transfections of cognate 21-nucleotide siRNAs with non-lyposomal cationic TransMessenger Transfection Reagent. Double immunofluorescence labeling of MAP2 and three unrelated proteins (cortactin, actin, YB-1) was performed 48–68 h after transfection. The expression of MAP2 was specifically reduced by the cognate siRNA duplex in 70–80% of cells, whereas expression of unrelated proteins was unaffected (Figure 5.4). Among these cells the decrease in MAP2 fluorescent intensity was ~4-fold. The fluorescent intensity over the total cell population varied greatly due to variation in transfection efficiency and possibly differential sensitivity of the neurons to siRNA. Nevertheless, even when the fluorescent intensity was measured over the total population of cells, a decrease in MAP2 fluorescence was readily detectable (Figure 5.4). Phenotypic effects of MAP2 suppression on cultured neurons can be observed early on after plating as neurons consolidate their lamellae and elaborate filopodia. It was shown previously that MAP2 is required to overcome early transitions in neurite development and to begin processes elongation (Caceres et al., 1992; Gonzalez-Billault et al., 2002). We targeted primary cultures with the cognate siRNAs 3 h after plating and analyzed the effect over 28–40 h. Within this shorter time frame, fewer neurons suppressed MAP2 expression. However, the degree of MAP2 suppression correlated with the recognized defect in neurite elaboration.

In addition to synthetic siRNA, double-stranded 21-mers generated by recombinant Dicer protein from long double-stranded RNA template may induce RNAi in neuronal cultures (Fink et al., 2003). Another approach for RNAi in neurons employed a Pol-III promoter-driven transcription of hairpin RNA-from a plasmid (Gaudilliere et al., 2002). Although, as indicated above, in this case just a small population of neurons expresses siRNA, this approach might be useful for a single-cell analysis.

The finding that siRNA can be effectively delivered to neurons and specifically suppress target gene expression opens new perspectives in molecular therapeutics.

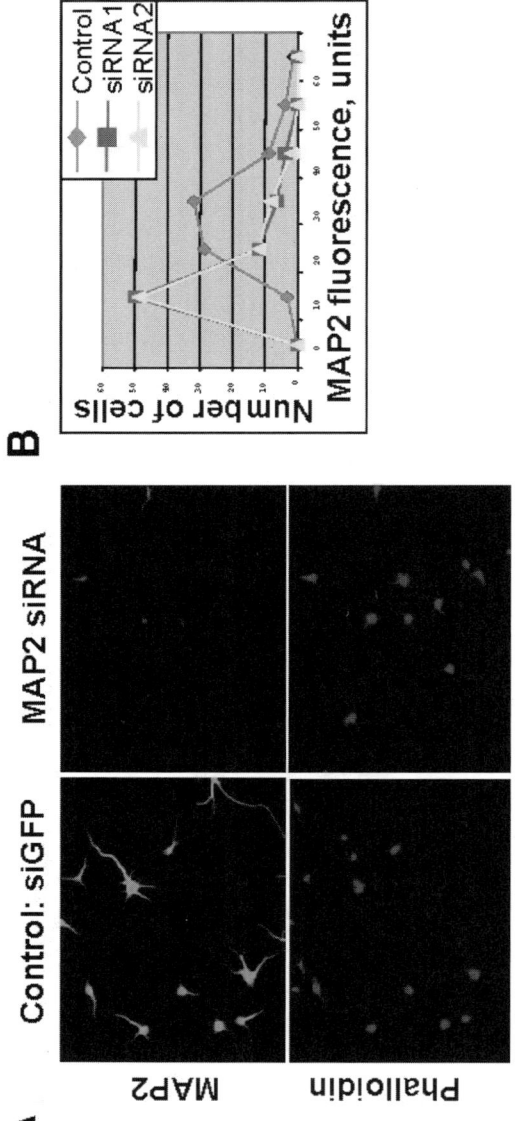

Figure 5.4. Microtubule Associated Protein 2 (MAP2) suppression in primary cortical neurons by cognate 21nt-siRNAs. A. Double fluorescence staining of neurons transfected with non-specific siRNA or with MAP2-siRNA. Upper panels-staining with MAP2 monoclonal antibody (green); lower panels-staining with actin-bound toxin phalloidin (red). B. Distribution of MAP2 expression levels in control and targeted cells, two different siRNA (siRNA1 and siRNA2) show a very similar effect. In each experiment, at least 70 random neurons per experimental condition were analyzed and gene expression was quantified in both control and targeted cells. The figure is reprinted from: Krichevsky, A. M. and Kosik, K. S. "RNAi functions in cultured mammalian neurons." *Proc Natl Acad Sci U S A.*, **99**(18):11926–9 (2002). (See color section.)

Among several genes we have successfully suppressed in primary neurons is β-secretase (BACE 1), one of the key therapeutic targets for treating Alzheimer's disease [(AD) (Kao et al., 2003)]. Some physiological outcomes of this suppression are discussed below.

III. BACE1 targeting by RNAi

The common pathology of all AD patients includes severe brain atrophy, neuronal loss, neurofibrillary tangles and senile plaques composed of aggregated β-amyloid ($A\beta$) peptides (Whitehouse et al., 1985). Since all known mutations that are associated with early onset AD enhance $A\beta$ peptide production (Price et al., 1998), it is likely that the peptide plays a critical role in the pathogenesis of AD (Hardy and Higgins, 1992; Price et al., 1995; Selkoe, 2000). $A\beta$ peptides accumulate from proteolysis of the amyloid precursor proteins (APP) by β- and γ-secretases. β-secretase (BACE 1) and its homologue BACE2 are integral membrane glycoproteins that are essential for the first ectodomain cleavage of APP generating 99 or 89 amino acid membrane-bound fragments (βCTFs) containing the $A\beta$ peptides. The next step requires the enzyme complex termed γ-secretase, which cleaves the transmembrane region to generate $A\beta$ peptides and the release of an APP intracellular fragment (AICD).

In brain samples of AD patients the protein level and activity of BACE1 is upregulated (Fukumoto et al., 2002; Holsinger et al., 2002; Yang et al., 2003). The essential role of BACE1 in the generation of $A\beta$ peptides is demonstrated by the finding that no $A\beta$ peptides can be detected in mice with a homozygous deletion of BACE1 (Luo et al., 2001). Importantly, BACE1 null mice do not exhibit any developmental abnormality or outward behavioral phenotype. All these make BACE1 a strong target for anti-amyloid therapy. However, the structure of the enzyme poses challenges for the development of small molecule inhibitors; we therefore sought to apply RNAi for BACE1 suppression.

We have used RNAi technology to inhibit BACE1 expression in neurons derived from both wild type and β-amyloid over-expressing cells (Kao et al., 2003). BACE1 siRNAs efficiently inhibited the generation of APP βCTFs and secretion of $A\beta$ peptides without affecting subcellular distribution of APP. Furthermore, BACE siRNA also reduced $A\beta$ production from neurons derived from transgenic mice harboring the Swedish APP mutant (APPsw). Importantly, pretreatment with BACE1 siRNA reduced the neurotoxicity of H_2O_2-induced oxidative stress. These results indicate that BACE1 siRNA specifically impacted on β-cleavage of APP and may be a potential therapeutic approach for combating Alzheimer's disease.

We chose 19-nt mRNA sequences that are conserved between human, rat and mouse BACE1, but not BACE2 as target sequences and designed four specific siRNAs against these target sequences. Both BACE1 and BACE2 cleave within the $A\beta$ sequence; however, the BACE1 cleavage product remains amyloidogenic, but the BACE2 product is not (Farzan et al., 2000; Yan et al., 2001). Therefore, inhibiting BACE2 is not desirable and RNAi provides the specificity to suppress selectively only one of these isoforms. To test the efficacy and specificity of these siRNAs, we first co-transfected either human embryonic kidney (HEK) 293 cells or CAD neuroblastoma cells with BACE1 siRNAs, human BACE1 or BACE2 cDNA and human APP cDNA. A non-silencing fluorescent RNA duplex (fluorescein siRNA) that does not overlap with any known mammalian genome sequence was used as a negative control. In those cells, 4 different siRNAs decreased BACE1 level in a reduction range of 50 to 99%. We observed the strongest knockdown of BACE1 expression by siRNA1 and siRNA3 whereas expression of BACE2 was not affected by any siRNA sequences tested. siRNA1 was most efficient in knocking down BACE1 expression with an efficiency of 99% after 24 hours of treatment. siRNA3 also reduced BACE1 protein level by about 90%, whereas siRNA2 and siRNA4 only affected BACE1 expression with about 70% and 50% reduction, respectively. Correspondingly, the levels of C89/C99, the two major BACE1 cleavage products of APP were also decreased.

We next sought to investigate whether BACE1 siRNAs also affect the expression of endogenous BACE1 in mouse primary cortical neurons. Dissociated mouse cortical neurons were treated with the 4 siRNAs for 1.5 hours. BACE1 levels were evaluated 72 hours after siRNA treatment by Western analysis. We observed a ~70% reduction of BACE1 in siRNA treated neurons, but not in control fluorescein RNA treated neurons (Figure 5.5). In contrast to CAD and HEK 293 cells, in primary cortical neurons all 4 siRNAs were comparably efficient in inducing RNAi. The effect of siRNA on BACE1 expression could be also demonstrated by immunocytochemistry. In primary cortical cultures co-transfected with BACE1 siRNA and fluorescein RNA duplex as a transfection indicator over 90% of the neurons were fluorescein positive, indicating that siRNAs were efficiently introduced into primary cortical neurons. Most importantly, the expression of BACE1 was significantly reduced by the cognate BACE1 siRNA in most the neurons.

Since BACE1 cleavage of APP is the initial step for Aβ generation, we investigated the consequence of knocking-down endogenous BACE1 expression on APP processing. Mouse primary cortical neurons were transfected with cognate BACE1 siRNA for 1.5 hours. In some experiments, neurons were infected with herpes simplex virus expressing human APP (HSV-APP) to facilitate detection of Aβ generation. 72 hours after siRNA treatment, we observed significant reduction in protein levels of BACE1 cleavage products of APP, C99/C89 in both noninfected and HSV-APP infected neurons (Figure 5.5). To determine if the reduced C99 levels influenced secretion of Aβ, we measured Aβ1–40 and 1–42 in culture media of neurons infected with HSV-APP. Those experiments demonstrated that lowering BACE1 level by siRNAs effectively reduced Aβ secretion. There was a 40–60% significant reduction of Aβ1–40 and Aβ1–42 secretion compared to control samples.

We have also tested the effects of siRNA on BACE1 expression and APP processing in primary cortical neurons from transgenic mice that express the Swedish double mutation (KM670/671NL) of the human APP (APPsw). Similarly,

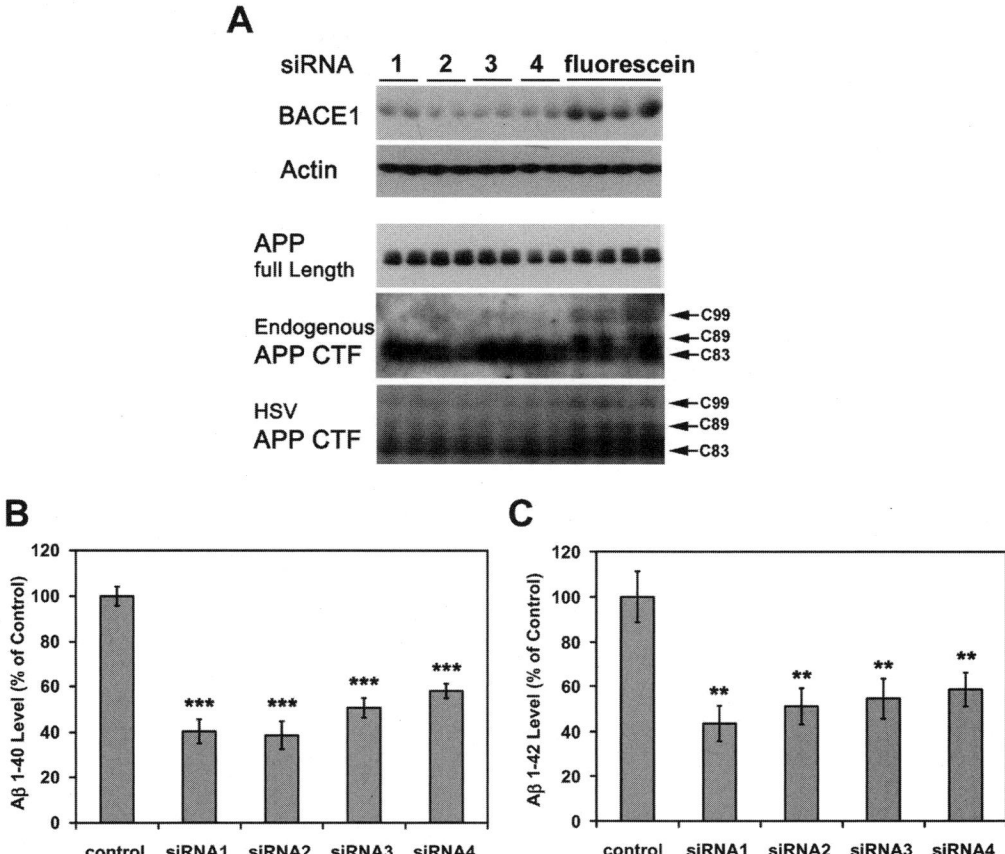

Figure 5.5. BACE1 suppression by siRNAs results in decreased βCTF generation and Aβ secretion. Mouse primary cortical neurons were transfected with BACE1 siRNA or fluorescein RNA duplex as a control for 1.5 hrs. The cells were later infected with herpes simplex virus expressing human APP (HSV-APP). A. After additional 72 hrs incubation, endogenous BACE1 levels as well as protein levels of APP full length and C-terminal fragments were detected by Western blot analyses. B, C. Conditioned media were collected and secreted Aβ measured by sandwich ELISA. The conditioned media of neurons without HSV-APP infection were used as background and were subtracted from the ELISA reading for all Aβ measurements. The figure is reprinted from: Kao SC, Krichevsky AM, Kosik KS, Tsai LH. "BACE1 suppression by RNA interference in primary cortical neurons." *J. Biol Chem.* 2004 Jan 16;**279**(3):1942−9.

introducing BACE1 siRNA into APPsw cortical neurons led to a decrease in BACE1 protein expression, as well as in C99/89 generation and Aβ secretion. Basi et al. published a similar study demonstrating that selective inactivation of endogenous BACE1 by RNAi resulted in decreased APP and Aβ secretion from HEK293 cells stably expressing recombinant wild-type APP or APPsw (Basi et al., 2003). Together, these results support the principle of reducing Aβ peptide generation using the BACE1 siRNA approach.

Oxidative stress has been implicated in pathogenesis and progression of AD. Many oxidation products including hydroxyl radicals (•OH) are found accumulating in AD brains. Hydrogen peroxide (H_2O_2) as an experimental system to

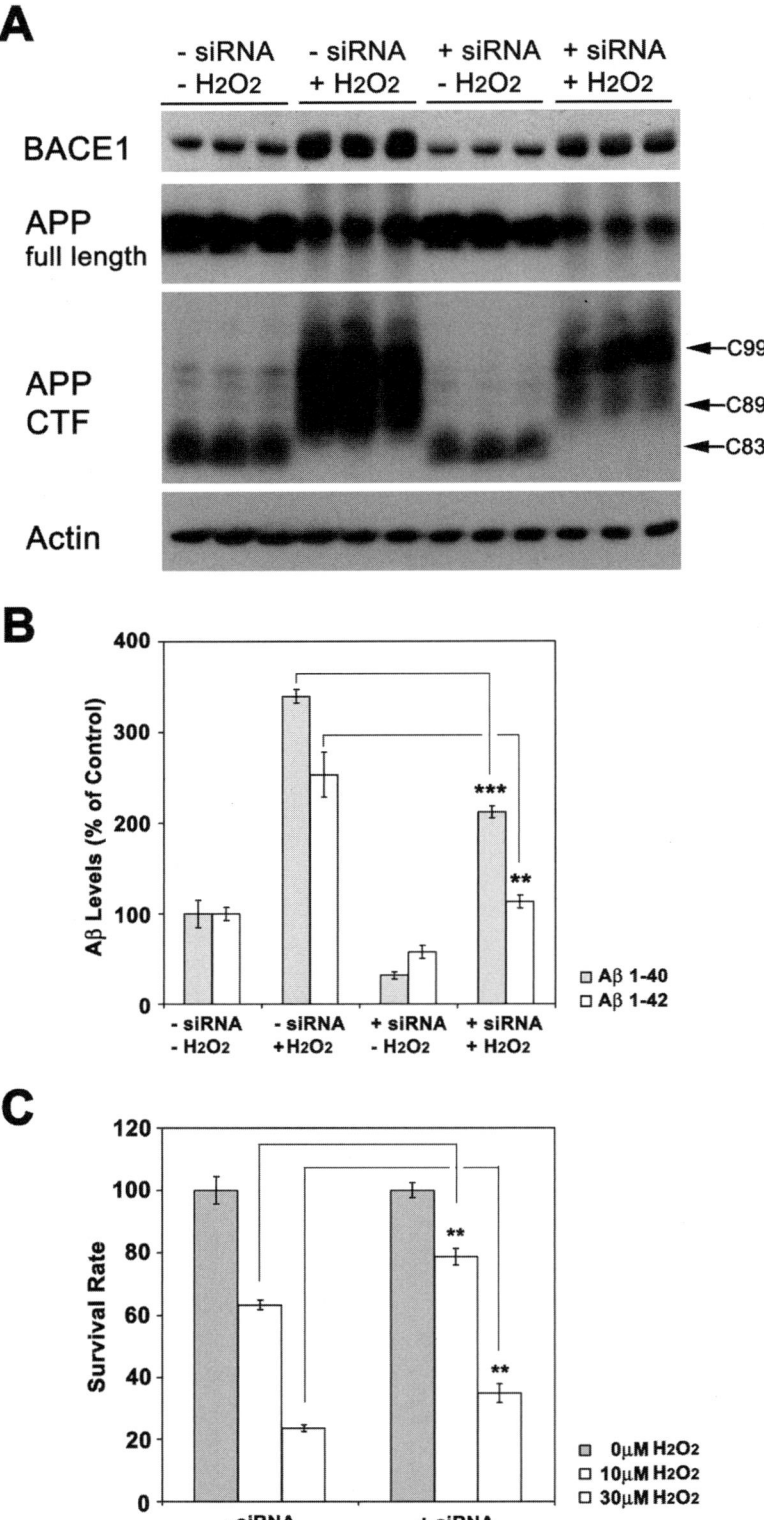

induce oxidative stress has been widely applied to the study of oxidative stress induced neurodegenerative responses. H_2O_2 induces intracellular accumulation of Aβ in neuroblastoma cells (Misonou et al., 2000). It is currently unclear how H_2O_2 induces Aβ generation. In primary cortical neurons treated with 10 μM H_2O_2 for 6 hours both BACE1 protein and its products, the βCTFs, are significantly increased (Figure 5.6A and also see (Tamagno et al., 2002)). Correspondingly, Aβ1–40 and 1–42 levels in conditioned media of neurons infected with HSV-APP also increase significantly after H_2O_2 treatment (Figure 5.6B). H_2O_2 treatment also induces cell death (Figure 5.6C). However, transfections of BACE1 siRNA lowered the BACE1, βCTF and Aβ induction levels following H_2O_2 treatment (Figure 5.6A and B). Furthermore, pretreatment with BACE1 siRNA prevented neurons from H_2O_2 induced cell death to a significant level (Figure 5.6C). Together, these data show that suppression of BACE1 expression protects neurons from oxidative stress induced cell death. This finding suggests that increased BACE1 expression by oxidative stress may trigger further downstream responses, particularly Aβ accumulation, which can be prevented by reducing BACE1 with RNAi.

Previous experiments have shown that no apparent adverse effects are observed associated with BACE1 deficient mice (Luo et al., 2001; Roberds et al., 2001). Similarly, we found that siRNA suppresses BACE1 expression specifically, and that loss of BACE1 using siRNA elicits no profound cellular defects. Together, these data support BACE1 as an excellent therapeutic target for the treatment of AD. At this juncture, improved delivery methods for siRNAs need to be developed to bring this approach to fruition.

Conclusion

The discovery of extensively regulated predominantly brain specific microRNAs, a novel class of noncoding RNAs sharing processing pathways with siRNAs, suggests a role of RNAi-related mechanisms of gene regulation in neuronal development and functioning. While the potential of these natural regulators is huge, where and how they operate *in vivo* remains to be uncovered. Our ability to manipulate this system for regulating expression of specific genes will have broad applications in general, and in neurons, where genetic manipulations have proven difficult, specifically.

Figure 5.6. Pretreatment of BACE1 siRNA protected neurons from H_2O_2 induced cell death. Wild-type primary cortical neurons were transfected with BACE1 siRNA1 or with fluorescein RNA duplex as a control. A. 72 hr post-transfection neurons were treated with 10μM H_2O_2 for an additional 6 hours. The protein levels of BACE1, full length APP and APP C-terminal fragments were detected by Western blot analysis. The blots were then reprobed with anti-actin antibody to verify protein loading. C. Cell viability was determined by MTS assay. Reduced cell death was detected in neurons transfected with BACE1 siRNA1. B. 48 hr post-transfection, neurons were infected with HSV-APP for 24 hrs and then treated with 100 nM H_2O_2 for an additional 24 hours. Secreted Aβ in conditioned media was measured by sandwich ELISA. The figure is reprinted from: Kao S. C., Krichevsky A. M., Kosik K. S., Tsai L. H. "BACE1 suppression by RNA interference in primary cortical neurons." *J. Biol Chem.* 2004 Jan 16;**279**(3):1942–9.

REFERENCES

Aakalu, G., Smith, W. B., Nguyen, N., Jiang, C. and Schuman, E. M. (2001). Dynamic visualization of local protein synthesis in hippocampal neurons. *Neuron*, **30**, 489–502.

Ambros, V. (2000). Control of developmental timing in *Caenorhabditis elegans*. *Current Opinion in Genetics and Development*, **10**, 428–433.

Ambros, V., Lee, R. C., Lavanway, A., Williams, P. T. and Jewell, D. (2003). MicroRNAs and other tiny endogenous RNAs in *C. elegans*. *Current Biology*, **13**, 807–818.

Basi, G., Frigon, N., Barbour, R., Doan, T., Gordon, G., McConlogue, L., Sinha, S. and Zeller, M. (2003). Antagonistic effects of beta-site amyloid precursor protein-cleaving enzymes 1 and 2 on beta-amyloid peptide production in cells. *Journal of Biological Chemistry*, **278**, 31512–31520.

Baulcombe, D. (2002). DNA events. An RNA microcosm. *Science*, **297**, 2002–2003.

Brennecke, J., Hipfner, D. R., Stark, A., Russell, R. B. and Cohen, S. M. (2003). Bantam encodes a developmentally regulated microRNA that controls cell proliferation and regulates the proapoptotic gene hid in *Drosophila*. *Cell*, **113**, 25–36.

Caceres, A., Mautino, J. and Kosik, K. S. (1992). Suppression of MAP-2 in cultured cerebellar macroneurons inhibits minor neurite formation. *Neuron*, **9**, 607–618.

Carrington, J. C. and Ambros, V. (2003). Role of microRNAs in plant and animal development. *Science*, **301**, 336–338.

Chen, C. Z., Li, L., Lodish, H. F. and Bartel, D. P. (2004). MicroRNAs modulate hematopoietic lineage differentiation. *Science*, 2;**303**(5654):83–6.

Doench, J. G., Petersen, C. P. and Sharp, P. A. (2003). siRNAs can function as miRNAs. *Genes & Development*, **17**, 438–442.

Dostie, J., Mourelatos, Z., Yang, M., Sharma, A. and Dreyfuss, G. (2003). Numerous microRNPs in neuronal cells containing novel microRNAs. *RNA*, **9**, 180–186.

Eberwine, J., Miyashiro, K., Kacharmina, J. E. and Job, C. (2001). Local translation of classes of mRNAs that are targeted to neuronal dendrites. *Proceedings of the National Academy of Sciences USA*, **98**, 7080–7085.

Elbashir, S. M., Martinez, J., Patkaniowska, A., Lendeckel, W. and Tuschl, T. (2001). Functional anatomy of siRNAs for mediating efficient RNAi in *Drosophila melanogaster* embryo lysate. *European Molecular Biology Organization Journal*, **20**, 6877–6888.

Farzan, M., Schnitzler, C. E., Vasilieva, N., Leung, D. and Choe, H. (2000). BACE2, a beta – secretase homolog, cleaves at the beta site and within the amyloid-beta region of the amyloid-beta precursor protein. *Proceedings of the National Academy of Sciences USA*, **97**, 9712–9717.

Fink, C. C., Bayer, K. U., Myers, J. W., Ferrell, J. E., Jr., Schulman, H. and Meyer, T. (2003). Selective regulation of neurite extension and synapse formation by the beta but not the alpha isoform of CaMKII. *Neuron*, **39**, 283–297.

Fukumoto, H., Cheung, B. S., Hyman, B. T. and Irizarry, M. C. (2002). Beta-secretase protein and activity are increased in the neocortex in Alzheimer disease. *Archives of Neurology*, **59**, 1381–1389.

Gaudilliere, B., Shi, Y. and Bonni, A. (2002). RNA interference reveals a requirement for myocyte enhancer factor 2A in activity-dependent neuronal survival. *Journal of Biological Chemistry*, **277**, 46442–46446.

Golub, A. M., Masiarz, F. R., Villars, T. and McConnell, J. V. (1970). Incubation effects in behavior induction in rats. *Science*, **168**, 392–395.

Gonzalez-Billault, C., Engelke, M., Jimenez-Mateos, E. M., Wandosell, F., Caceres, A. and Avila, J. (2002). Participation of structural microtubule-associated proteins (MAPs) in the development of neuronal polarity. *Journal of Neuroscience Research*, **67**, 713–719.

Grishok, A., Pasquinelli, A. E., Conte, D., Li, N., Parrish, S., Ha, I., Baillie, D. L., Fire, A., Ruvkun, G. and Mello, C. C. (2001). Genes and mechanisms related to RNA interference regulate expression of the small temporal RNAs that control *C. elegans* developmental timing. *Cell*, **106**, 23–34.

Hallam, S. J. and Jin, Y. (1998). lin-14 regulates the timing of synaptic remodelling in *Caenorhabditis elegans*. *Nature*, **395**, 78–82.

Handler, M., Yang, X. and Shen, J. (2000). Presenilin-1 regulates neuronal differentiation during neurogenesis. *Development*, **127**, 2593–2606.

Hardy, J. A. and Higgins, G. A. (1992). Alzheimer's disease: The amyloid cascade hypothesis. *Science*, **256**, 184–185.

Holsinger, R. M., McLean, C. A., Beyreuther, K., Masters, C. L. and Evin, G. (2002). Increased expression of the amyloid precursor beta-secretase in Alzheimer's disease. *Annals of Neurology*, **51**, 783–786.

Hütvagner, G., McLachlan, J., Pasquinelli, A. E., Balint, E., Tuschl, T. and Zamore, P. D. (2001). A cellular function for the RNA-interference enzyme Dicer in the maturation of the let-7 small temporal RNA. *Science*, **293**, 834–838.

Kao, S. C., Krichevsky, A. M., Kosik, K. S. and Tsai, L. H. (2003). BACE1 suppression by RNA interference in primary cortical neurons. *Journal of Biological Chemistry*, **279**, 1942–1949.

Kasschau, K. D., Xie, Z., Allen, E., Llave, C., Chapman, E. J., Krizan, K. A. and Carrington, J. C. (2003). P1/HC-Pro, a viral suppressor of RNA silencing, interferes with Arabidopsis development and miRNA unction. *Developmental Cell*, **4**, 205–217.

Ketting, R. F., Fischer, S. E., Bernstein, E., Sijen, T., Hannon, G. J. and Plasterk, R. H. (2001). Dicer functions in RNA interference and in synthesis of small RNA involved in developmental timing in *C. elegans*. *Genes & Development*, **15**, 2654–2659.

Krichevsky, A. M. and Kosik, K. S. (2001). Neuronal RNA granules: A link between RNA localization and stimulation-dependent translation. *Neuron*, **32**, 683–696.

Krichevsky, A. M. and Kosik, K. S. (2002). RNAi functions in cultured mammalian neurons. *Proceedings of the National Academy of Sciences USA*, **99**, 11926–11929.

Krichevsky, A. M., King, K. S., Donahue, C. P., Khrapko, K. and Kosik, K. S. (2003). A microRNA array reveals extensive regulation of microRNAs during brain development. *RNA*, **9**, 1274–1281.

Lagos-Quintana, M., Rauhut, R., Lendeckel, W. and Tuschl, T. (2001). Identification of novel genes coding for small expressed RNAs. *Science*, **294**, 853–858.

Lagos-Quintana M, R. R., Yalcin A, Meyer J, Lendeckel W. and Tuschl T. (2002). Identification of tissue-specific MicroRNA's from mouse. *Current Biology*, **12**, 735–739.

Lai, E. C., Tomancak, P., Williams, R. W. and Rubin, G. M. (2003). Computational identification of *Drosophila* microRNA genes. *Genome Biology*, **4**, R42.

Lau, N. C., Lim, L. P., Weinstein, E. G. and Bartel, D. P. (2001). An abundant class of tiny RNAs with probable regulatory roles in *Caenorhabditis elegans*. *Science*, **294**, 858–862.

Lee, R. C. and Ambros, V. (2001). An extensive class of small RNAs in *Caenorhabditis elegans*. *Science*, **294**, 862–864.

Lee, R. C., Feinbaum, R. L. and Ambros, V. (1993). The *C. elegans* heterochronic gene lin-4 encodes small RNAs with antisense complementarity to lin-14. *Cell*, **75**, 843–854.

Lee, Y., Ahn, C., Han, J., Choi, H., Kim, J., Yim, J., Lee, J., Provost, P., Radmark, O., Kim, S. and Kim, V. N. (2003). The nuclear RNase III Drosha initiates microRNA processing. *Nature*, **425**, 415–419.

Lim, L. P., Glasner, M. E., Yekta, S., Burge, C. B. and Bartel, D. P. (2003). Vertebrate microRNA genes. *Science*, **299**, 1540.

Luo, Y., Bolon, B., Kahn, S., Bennett, B. D., Babu-Khan, S., Denis, P., Fan, W., Kha, H., Zhang, J., Gong, Y., et al. (2001). Mice deficient in BACE1, the Alzheimer's beta-secretase, have normal phenotype and abolished beta-amyloid generation. *Nature Neuroscience*, **4**, 231–232.

McConnell, J. V. (1966). Comparative physiology: learning in invertebrates. *Annual Reviews of Physiology*, **28**, 107–136.

Miller, S., Yasuda, M., Coats, J. K., Jones, Y., Martone, M. E. and Mayford, M. (2002). Disruption of dendritic translation of CaMKIIalpha impairs stabilization of synaptic plasticity and memory consolidation. *Neuron*, **36**, 507–519.

Misonou, H., Morishima-Kawashima, M. and Ihara, Y. (2000). Oxidative stress induces intracellular accumulation of amyloid beta-protein (Abeta) in human neuroblastoma cells. *Biochemistry*, **39**, 6951–6959.

Mourelatos, Z., Dostie, J., Paushkin, S., Sharma, A., Charroux, B., Abel, L., Rappsilber, J., Mann, M. and Dreyfuss, G. (2002). miRNPs: a novel class of ribonucleoproteins containing numerous microRNAs. *Genes & Development*, **16**, 720–728.

O'Donnell, W. T. and Warren, S. T. (2002). A decade of molecular studies of Fragile X syndrome. *Annual Reviews of Neuroscience*, **25**, 315–338.

Okabe, M., Imai, T., Kurusu, M., Hiromi, Y. and Okano, H. (2001). Translational repression determines a neuronal potential in *Drosophila* asymmetric cell division. *Nature*, **411**, 94–98.

Olsen, P. H. and Ambros, V. (1999). The lin-4 regulatory RNA controls developmental timing in *Caenorhabditis elegans* by blocking LIN-14 protein synthesis after the initiation of translation. *Developmental Biology*, **216**, 671–680.

Price, D. L., Sisodia, S. S. and Gandy, S. E. (1995). Amyloid beta amyloidosis in Alzheimer's disease. *Current Opinion in Neurology*, **8**, 268–274.

Price, D. L., Tanzi, R. E., Borchelt, D. R. and Sisodia, S. S. (1998). Alzheimer's disease: Genetic studies and transgenic models. *Annual Reviews of Genetics*, **32**, 461–493.

Reinhart, B. J., Slack, F. J., Basson, M., Pasquinelli, A. E., Bettinger, J. C., Rougvie, A. E., Horvitz, H. R. and Ruvkun, G. (2000). The 21-nucleotide let-7 RNA regulates developmental timing in *Caenorhabditis elegans*. *Nature*, **403**, 901–906.

Roberds, S. L., Anderson, J., Basi, G., Bienkowski, M. J., Branstetter, D. G., Chen, K. S., Freedman, S. B., Frigon, N. L., Games, D., Hu, K., et al. (2001). BACE knockout mice are healthy despite lacking the primary beta-secretase activity in brain: Implications for Alzheimer's disease therapeutics. *Human Molecular Genetics*, **10**, 1317–1324.

Selkoe, D. J. (2000). Toward a comprehensive theory for Alzheimer's disease. Hypothesis: Alzheimer's disease is caused by the cerebral accumulation and cytotoxicity of amyloid beta-protein. *Annals of New York Academy of Sciences*, **924**, 17–25.

Shen, J., Bronson, T., Chjen, D. F., Xia, W., Selkoe, D. J. and Tonegawa, S. (1997). Skeletal and CNS defects in *presenilin-1*-deficient mice. *Cell*, **89**, 629–639.

Slack, F. J., Basson, M., Liu, Z., Ambros, V., Horvitz, H. R. and Ruvkun, G. (2000). The lin-41 RBCC gene acts in the *C. elegans* heterochronic pathway between the let-7 regulatory RNA and the LIN-29 transcription factor. *Molecular Cell*, **5**, 659–669.

Steward, O. and Schuman, E. M. (2001). Protein synthesis at synaptic sites on dendrites. *Annual Reviews of Neuroscience*, **24**, 299–325.

Tamagno, E., Bardini, P., Obbili, A., Vitali, A., Borghi, R., Zaccheo, D., Pronzato, M. A., Danni, O., Smith, M. A., Perry, G. and Tabaton, M. (2002). Oxidative stress increases expression and activity of BACE in NT2 neurons. *Neurobiology Disorders*, **10**, 279–288.

Timmons, L., Court, D. L. and Fire, A. (2001). Ingestion of bacterially expressed dsRNAs can produce specific and potent genetic interference in *Caenorhabditis elegans*. *Gene*, **263**, 103–112.

Whitehouse, P. J., Struble, R. G., Hedreen, J. C., Clark, A. W. and Price, D. L. (1985). Alzheimer's disease and related dementias: Selective involvement of specific neuronal systems. *Critical Reviews of Clinical Neurobiology*, **1**, 319–339.

Wienholds, E., Koudijs, M. J., van Eeden, F. J., Cuppen, E. and Plasterk, R. H. (2003). The microRNA-producing enzyme Dicer1 is essential for zebrafish development. *Nature Genetics*, **35**, 217–218.

Wightman, B., Ha, I. and Ruvkun, G. (1993). Posttranscriptional regulation of the heterochronic gene lin-14 by lin-4 mediates temporal pattern formation in *C. elegans*. *Cell*, **75**, 855–862.

Yan, R., Munzner, J. B., Shuck, M. E. and Bienkowski, M. J. (2001). BACE2 functions as an alternative alpha-secretase in cells. *Journal of Biological Chemistry*, **276**, 34019–34027.

Yang, L. B., Lindholm, K., Yan, R., Citron, M., Xia, W., Yang, X. L., Beach, T., Sue, L., Wong, P., Price, D., et al. (2003). Elevated beta-secretase expression and enzymatic activity detected in sporadic Alzheimer disease. *Nature Medicine*, **9**, 3–4.

Zeng, Y., Yi, R. and Cullen, B. R. (2003). MicroRNAs and small interfering RNAs can inhibit mRNA expression by similar mechanisms. *Proceedings of the National Academy of Sciences USA*, **100**, 9779–9784.

Design, synthesis
of siRNAs

6 Design and synthesis of small interfering RNA (siRNA)

Queta Boese, William S. Marshall, and Anastasia Khvorova

Efficient RNA interference (RNAi) depends on siRNA design and synthesis

RNAi is a powerful technology with tremendous utility for functional genomic analysis, drug discovery strategies and therapeutic applications (Appasani 2003). While this pathway for post-transcriptional gene regulation is ubiquitous among eukaryotes, species-specific variations in the mechanism impact the utility of this pathway. These species-specific distinctions have strong implications with regard to the design, production, and delivery of the functional silencing intermediates. For example, in *Caenorhabditis elegans* (*C. elegans*), simple exposure by soaking (Tabara et al. 1998; Timmons and Fire 1998), feeding (Fraser et al. 2000; Timmons et al. 2001), or injecting (Fire et al. 1998) the nematode with long dsRNA is sufficient to induce prolonged and potent gene knockdown. Silencing efficiency appears to be due to siRNA-primed amplification of additional dsRNA from the mRNA target resulting in a secondary pool of Dicer processed duplexes (Sijen et al. 2001; Tijsterman et al. 2002). This mechanism is characteristic of post-transcriptional gene silencing in nematodes and other lower eukaryotes and is mediated by an RNA-dependent RNA polymerase [(RdRP) (Sijen et al. 2001; Martens et al. 2002)]. Invariably, several of the newly generated siRNAs will be capable of proficient gene-specific knockdown thereby eliminating the need to carefully design and synthesize a single siRNA silencing intermediate.

In mammalian cell culture models, preliminary attempts to induce RNAi using long dsRNA met with limited success (Tuschl et al. 1999; Caplen et al. 2000; Zhao et al. 2001). This is due in part to the dsRNA-dependent activation of the interferon (IFN) response, which results in a global inhibition of protein synthesis and cell death (Baglioni 1979; Minks et al. 1979). However, in subsequent landmark studies, researchers demonstrated the ability to bypass the cytotoxic IFN response by introducing chemically synthesized siRNAs that lead to targeted gene knockdown (Caplen et al. 2001; Elbashir et al. 2001a). These synthetic siRNAs function catalytically at nanomolar concentrations and were capable of inducing cleavage of up to 95% of the target mRNA.

RNAi mediated silencing is clearly dependent on the functionality and specificity of the siRNA silencing intermediates. In lower eukaryotes, target site selection is unnecessary given that Dicer processing of the long dsRNA molecules results in a heterogeneous population of siRNAs capable of efficient gene silencing (Bernstein et al. 2001; Billy et al. 2001; Ketting et al. 2001; Knight and Bass 2001). In contrast, in higher eukaryotes, artificially induced RNAi relies on the functionality and specificity of unique, discrete duplexes. Thus, the fundamental challenge for successfully implementing RNAi in mammalian systems rests on designing specific and potent siRNAs.

The rules that govern siRNA design and target silencing are largely undefined. Several reports document that different siRNAs directed against the same target exhibit widely variable silencing efficiencies (Holen et al. 2002; Krichevsky and Kosik 2002; Spankuch-Schmitt et al. 2002; Laposa et al. 2003). Furthermore, some targets are easy to silence while others are more difficult, requiring multiple screens of siRNAs to identify a single potent duplex. For these reasons, siRNA selection strategies represent a critical first step that can have significant downstream implications in terms of the time and labor commitment one invests to achieve efficient siRNA-mediated RNAi.

Conventional methods of siRNA design

siRNA design strategies should include those attributes that contribute to duplex functionality and specificity. The most widely used strategy for selecting 21-base mRNA target sequences is based on early *in vitro* biochemical studies that assessed functional and structural requirements for siRNA duplexes using *Drosophila* and mammalian cell lysates (Elbashir et al. 2001b; Elbashir et al. 2002; Martinez et al. 2002). The conventional method for selection begins by searching the coding sequence for regions that are most likely to be free of translational or regulatory proteins (e.g. 100 bases downstream of the start codon). Sequence motifs characterized by an AA (or NA) dinucleotide preceding a 19-base sequence with 30–70% G/C content are then selected. The dinucleotide leader defines the sequence composition of the antisense 3′ overhangs so that 19-base duplexes targeting $AA(N_{19})$ would have 3′ termini of UU or dTdT. The subsequent antisense strand would be completely complementary to the target. In general, these motifs can also be found in the untranslated regions (UTRs) of the transcript but are selected with care as the UTRs tend to be more polymorphic or to include regulatory sequences shared among family members.

On average, 60–70% of duplexes identified by conventional design methods silence their target by at least 50–70% but manifest significant variability in silencing efficiencies (Bernstein et al. 2001; Aza-Blanc et al. 2003). This is best illustrated for a functional assay of conventionally designed siRNAs targeting two different targets, human glyceraldehyde-3-phosphate dehydrogenase (GAPDH) and diazepam binding inhibitor (DBI) (see Figure 6.1). In many cases, the low level of silencing (less than 70%) induced by the majority of the conventionally designed duplexes may not be biologically significant, thus requiring additional screens to identify highly potent siRNA.

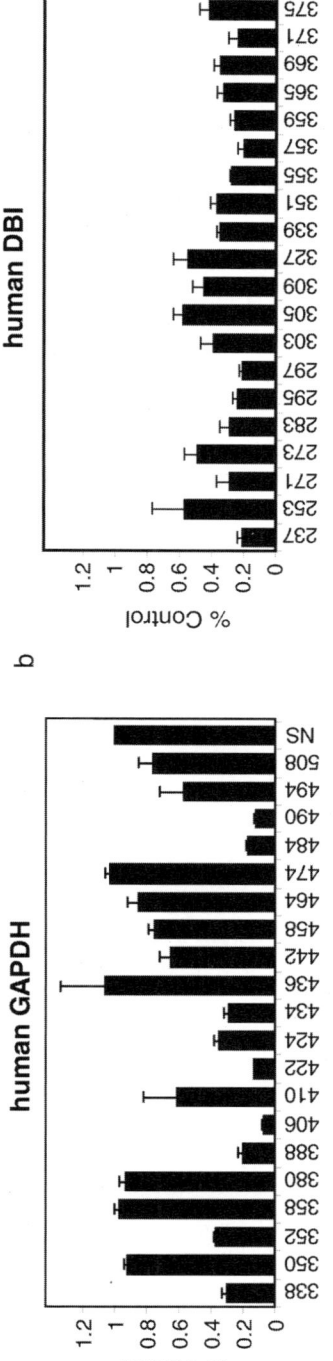

Figure 6.1. siRNAs targeting human glyceraldehyde-3-phosphate dehydrogenase (GAPDH) and diazepam binding inhibitor (DBI) using conventional siRNA design methods exhibit a wide range of silencing efficiencies. siRNAs are designated by their position relative to the start codon. HEK-293 cells were transfected at 75% confluency with 100 nM siRNA complexed with Lipofectamine™ 2000 (Invitrogen Life Technologies). Gene silencing was assessed by measuring mRNA 24 hrs post-transfection using the QuantiGene® branched DNA assay (Genospectra, Inc.).

Other variations of conventional design assess target site accessibility, which is often used as a predictor for antisense oligonucleotide (ASO) silencing efficiency. While a correlation has been observed between the functionality of ASOs and siRNAs targeting identical sites (Vickers et al. 2003), target sites that are otherwise refractory to silencing by ASOs are readily silenced by the comparable siRNAs. One possible explanation for this disparity is that the silencing machinery may eliminate secondary structures through the activity of putative helicase(s) or other proteins associated with the RNA induced silencing complex (RISC). This notion is supported by the silencing data for the panel of siRNA target sites illustrated in Figure 6.1. If secondary structure or relative target positions were primary determinants of functionality, the reduction in potency between siRNAs surrounding a good target would not be expected to be so disparate.

Conventional design methods are based on a subset of features attributed to functional siRNAs. The inherent uncertainty of these selection strategies and unpredictable range of silencing efficiencies achieved with the resulting siRNAs add a costly and time-consuming screening step to RNAi. The identification of sophisticated *in silico* procedures that eliminate unnecessary screening is the key to advancing RNAi into new research and therapeutic venues.

Primary determinants of siRNA functionality

To identify the primary determinants of siRNA functionality, it is necessary to consider the siRNA-mediated silencing pathway and accompanying nucleic acid-protein interactions. After Dicer processing of the dsRNA trigger, the resulting siRNAs are incorporated into the RISC. One or more helicases within the RISC then unwind the siRNA duplex, enabling the complementary antisense strand to guide target recognition (Nykanen et al. 2001; Schwarz et al. 2002). Finally, one or more endonucleases within the RISC cleave the mRNA to induce silencing (Elbashir et al. 2001; Caudy et al. 2003). This process illustrates that siRNA produced naturally or introduced by transfection is involved in multiple protein interactions including (1) initial duplex recognition by the pre-RISC complex, (2) ATP-dependent RISC activation and siRNA unwinding, (3) target recognition, and (4) target cleavage and product release (see Figure 6.2). Due to the nature of these RNA-protein interactions, thermodynamic and sequence-specific characteristics of functional siRNA duplexes are likely to bias strand selection during siRNA-RISC assembly and activation, and contribute to the overall efficiency of RNAi.

In recent work, it was determined that naturally occurring microRNA precursor duplexes (pre-miRNA) exhibit a bias towards low internal thermodynamic stability at the 5′-antisense (AS) end (Reynolds et al. 2004). Prior to being processed to their regulatory, single stranded form, pre-miRNAs can be considered endogenously expressed counterparts of siRNAs. Because these molecules share a duplex state during processing, the observation of a biased thermodynamic profile was extended to siRNAs for which functionality had been assessed in tissue culture (Khvorova et al. 2003). The resulting data supports concurrent work by Dr. Phillip

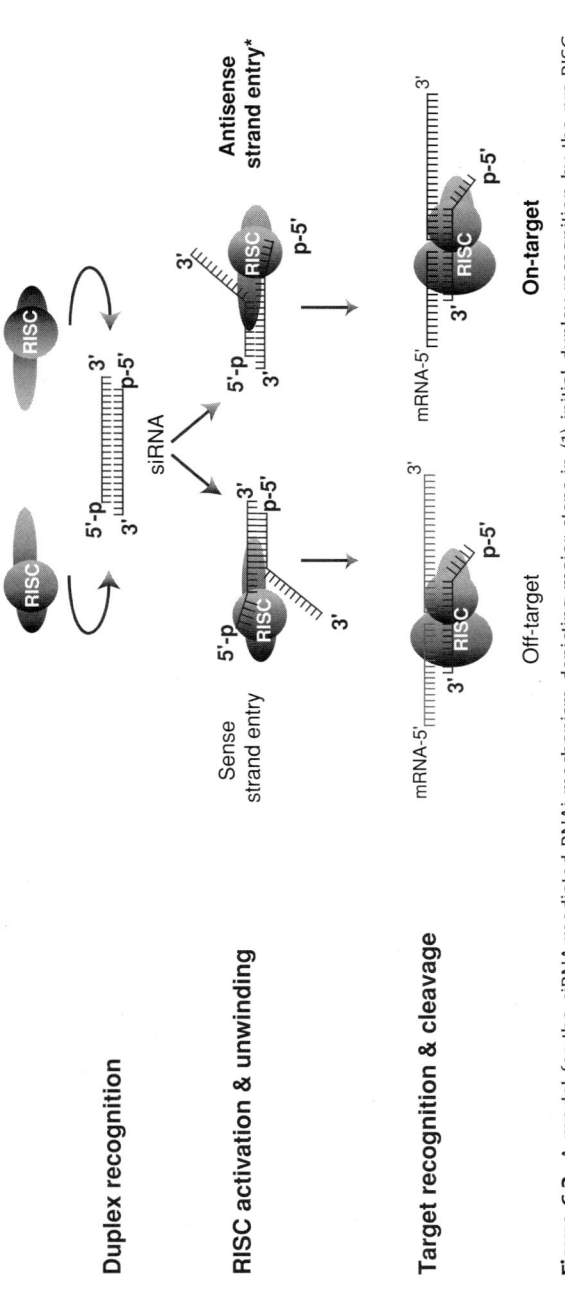

Figure 6.2. A model for the siRNA-mediated RNAi mechanism depicting major steps in (1) initial duplex recognition by the pre-RISC complex, (2) ATP-dependent RISC activation and siRNA unwinding, (3) target recognition, (4) target cleavage and product release. The desired outcome is directed by antisense strand entry* into the RISC to produce "on-target" silencing effects. Undesirable off-target effects are thought to be directed by sense strand identity to unrelated sequences but can be minimized using rational design coupled with comprehensive sequence analyses. (See color section.)

107

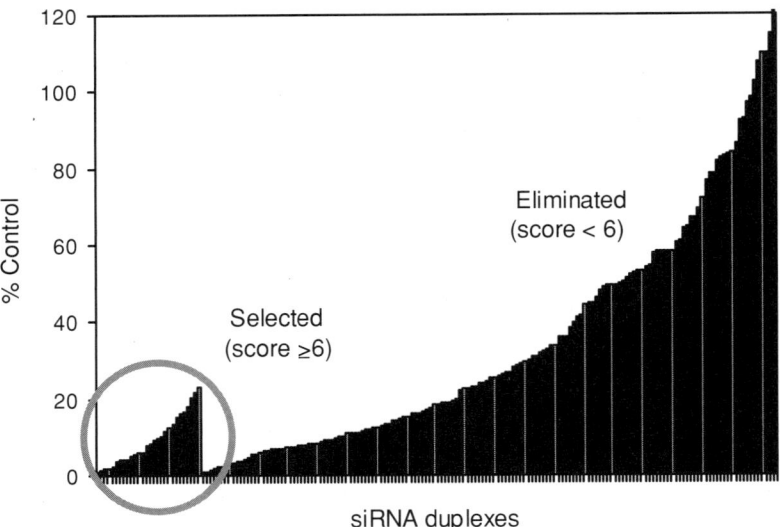

Figure 6.3. Application of all eight criteria to a test panel of siRNAs targeting firefly luciferase and human cyclophilin. The algorithm assigns a score between 1 and 10. Duplexes with scores of greater than 6 were "Selected" while duplexes with scores less than 6 were "Eliminated". Although many of the eliminated, low-scoring siRNAs were functional, all of the duplexes in the selected group were functional, reducing target expression by at least 75% or better (Reynolds et al. 2003).

Zamore and colleagues in *Drosophila* cell lysates where asymmetry or differential stability of miRNA/siRNA duplex precursors determined strand selection and entrance into the RISC (Schwarz et al. 2003).

While these studies demonstrate that low internal stability of the 5'AS end of the siRNA duplex is necessary for functionality, this attribute alone is clearly not sufficient to induce high levels of gene silencing. Analysis of multiple siRNAs with low internal stability at the 5' AS revealed that not all of these molecules were functional in the mRNA silencing reaction (Khvorova et al. 2003). Therefore, to identify additional siRNA sequence features that promote efficient silencing, and to quantify the importance of certain currently accepted conventional factors (e.g. G/C content), the silencing potential of 180 siRNAs targeting firefly luciferase and human cyclophilin was assessed. The results of these studies identified three thermodynamic and five sequence-related criteria that correlate with siRNA functionality (Reynolds et al. 2003). These eight determinants are hypothesized to affect one of the major steps associated with the siRNA processing. Specifically, the propensity to form internal hairpins is thought to affect the efficiency of initial duplex recognition by RISC, while additional attributes – G/C content, low internal stability of the 5'AS end, and base preference for an adenosine and not cytosine in position 19 of the sense strand – may be involved at the RISC duplex unwinding and activation step. Lastly, the remaining three attributes – base preferences at position 10 (uridine), position 3 (adenosine), and position 13 (any base but guanosine) – are thought to play a role in catalytic cleavage of the target (Reynolds et al. 2003).

When applied individually, no single attribute is sufficient to ensure silencing. However, when all eight factors are integrated into a single algorithm and used to evaluate a test panel of 180 duplexes targeting firefly luciferase and human cyclophilin, these eight criteria successfully eliminate all non-functional duplexes (Figure 6.3). The power of this refined method for siRNA design is striking when one compares two recent studies that used either conventional methods or a more sophisticated algorithm that incorporates approximately 30 criteria. When the conventional approach was applied to 22 genes of the insulin-signaling pathway, only 20% of the 120 siRNAs designed induced more than 80% gene silencing (Hsieh et al. 2004). This is in stark contrast to the second study involving an analysis of 12 genes involved in clathrin dependent endocytosis. Of the siRNAs designed using the improved version of the design strategy, all forty-eight induced greater than 80% gene silencing (Huang et al. 2004). Thus, based on these analyses, siRNA functionality can be effectively predicted by evaluating a complex set of biophysical and sequence-related characteristics. These studies demonstrate the complexity of siRNA design and emphasize that *in silico* prescreening of potential target sequences with algorithms based on thermodynamic and sequence specific attributes can significantly improve silencing efficiency.

Benefits of functional siRNA design

An unexpected benefit of this improved selection method was discovered when the effects of pooling rationally designed siRNA were investigated. Previous studies by a number of laboratories (Martinez et al. 2002; McManus et al. 2002; Holen et al. 2003) had shown that addition of non- or semi-functional siRNA to functional duplexes led to attenuation in silencing. This outcome was believed to be the result of saturating the silencing machinery (e.g. RISC) and suggested that a functional limitation existed for the concentration or number of sequences that could be directed against a single or multiple targets. As certain therapeutic and functional genomic applications of RNAi may require knockdown of two or more targets, this phenomenon represents an undesirable shortcoming in the technique.

The unanticipated improvement presented by rationally designed siRNA is that on average, the potency (and longevity) is greater than that of conventionally designed counterparts. These qualities enable significant levels of silencing to be achieved with low or sub-nanomolar concentrations. Subsequent studies with pools of highly functional siRNA directed against a single target confirm that rational design eliminates the ceiling previously set when using multiple conventionally designed molecules. Pools of four of these highly potent siRNAs directed against a single target typically perform as well or better than any single duplex (see Figure 6.4). In addition, our preliminary data with individual (or pools) of siRNA directed against multiple genes appear to successfully knock down expression of multiple targets simultaneously. Thus while conventional design methodologies frequently develop reagents with variable functionality, rational design reinforces the broad utility of this new technology.

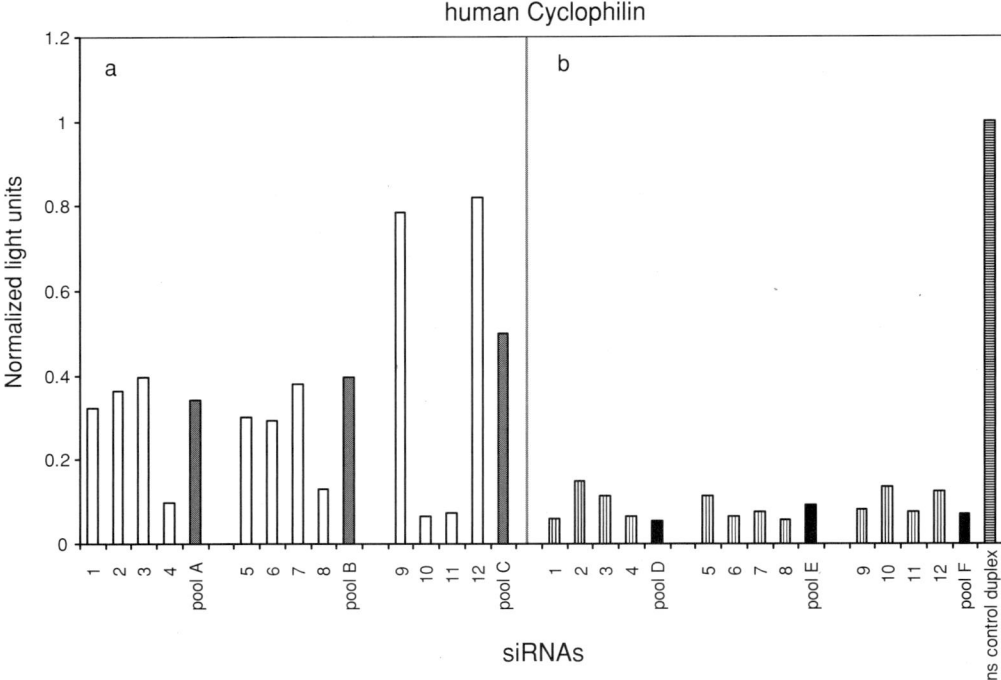

Figure 6.4. Rationally selected duplexes targeting human cyclophilin perform better as individuals or as pools than duplexes selected at random and used individually or as pools. (a) siRNAs were selected at random and tested as individuals (clear) and as pools of four each (grey). (b) siRNAs were selected by rational design and tested as individuals (stripe) and as pools of four (black). siRNAs targeting human cyclophilin were transfected into HEK-293 cells at 75% confluency using a final concentration of 100 nM complexed with Lipofectamine 2000. Gene silencing was assessed by measuring mRNA 24 hrs post-transfection using the QuantiGene® branched DNA assay.

Functional siRNA design and specificity

The importance of rational design as it relates to off-target effects became clear in recent publications that employed microarray analysis to monitor the genome-wide effects of siRNA-induced silencing. In one instance, studies were performed using conventionally designed siRNA (Jackson et al. 2003) while in another case siRNAs were selected based on conventional design coupled with secondary measures that took into consideration sequence and biophysical properties (Semizarov et al. 2003). The distinctly different results obtained by these two groups drew attention to the importance of rational design. The group led by Semizarov observed few off-target effects that were readily eliminated by optimizing transfection conditions (e.g. adjusting siRNA:lipid ratios, reducing siRNA concentrations). In contrast, using conventionally designed siRNA directed against MAPK14 and IGF1R, Jackson and colleagues observed down-regulation of multiple unrelated transcripts, some of which exhibited as little as 11 bases of identity with the sense or antisense strand.

Upon closer inspection, the level of off-target down regulation was 2-3 fold below that of normal gene expression and, thus, may not be functionally significant.

Moreover, among the modulated genes, the target gene remained suppressed while the off-target genes recovered expression over the course of the study. Therefore, these results further support the application of *in silico* methods that serve to minimize factors contributing to off-target effects (e.g. partial sequence identity, secondary structure, or binding affinities).

One striking result observed by Jackson and colleagues was unanticipated off-target silencing induced by the sense strand. Given current models for siRNA-RISC processing, one plausible approach for reducing off-target effects and potentially enhancing siRNA specificity is to alter strand affinities for the RISC. By incorporating chemical modifications, base pair mismatches, and/or biased internal strand stability, it may be possible to drive RISC affinity to the strand complementary to the target, thus eliminating off-target effects associated with the sense strand. For example, we recently completed a modification-interference analysis using modified nucleosides in select positions of the duplex to demonstrate that addition of certain groups in selected positions of the antisense strand completely eliminated duplex functionality. Because equivalent modifications to the sense strand had no effect on the molecule's silencing capacity, this technique provided a means for functionally distinguishing sense and antisense strands, and suggested a strategy for lessening the off-target effects induced by the sense strand. The viability of this approach was confirmed in subsequent expression-profiling studies, where sense strand modifications were observed to eliminate off-target effects (data not shown).

Methods of siRNA synthesis

There are currently several methods of synthesizing siRNA including *in vitro* transcription (Capodici et al. 2002), Dicer digestion of dsRNA (Calegari et al. 2002; Yang et al. 2002), PCR amplification of DNA expression cassettes (Castanotto et al. 2002), and plasmid or viral expression of siRNA or short hairpin RNA (shRNA) (Brummelkamp et al. 2002; Lee et al. 2002; Paul et al. 2002; Sui et al. 2002). These techniques are discussed at length in the adjoining chapters and have various cost-saving attributes and limitations with regard to reliability, purity, and flexibility in design and scalability. In general, production methods should yield high quality RNA quickly, reliably, and economically and at the same time be amenable to high-throughput formats and bulk syntheses. Furthermore, the synthetic platform of choice should be flexible enough to introduce chemical modifications that have been shown to enhance silencing specificity, potency, and longevity.

Chemically synthesized versions of the 21-23 base siRNA duplexes were the first entities used to demonstrate RNAi activity in mammalian cells (Caplen et al. 2001; Elbashir et al. 2001a). By directly introducing these functional intermediates, researchers were able to effectively bypass the dsRNA-induced IFN response that leads to widespread changes in gene expression and ultimately cell death. Among the available methods for siRNA production, chemical synthesis remains the predominant and most reliable method for generating siRNAs because it

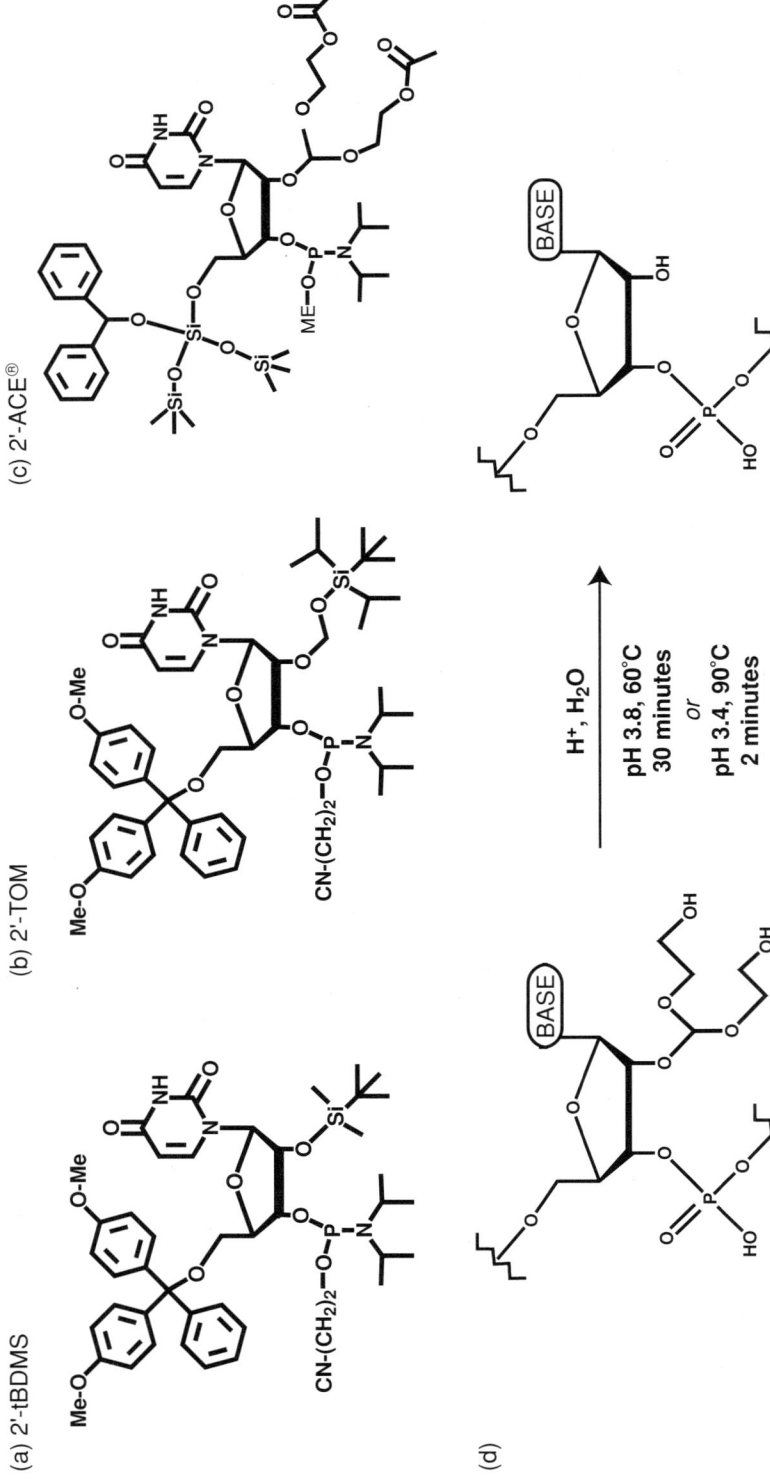

(a) 2'-tBDMS

(b) 2'-TOM

(c) 2'-ACE®

(d)

H⁺, H₂O

pH 3.8, 60°C
30 minutes

or

pH 3.4, 90°C
2 minutes

Figure 6.5. Chemical RNA synthetic strategies. Structure of protected phosphoramidite monomers used in (a) 2′-tBDMS, (b) 2′-TOM and (c) 2′-ACE® RNA synthesis, respectively. (d) The 2′-deprotection reaction for 2′-ACE protected RNA oligos is achieved rapidly under mildly acidic conditions. The reaction is efficient, quantitative, and irreversible.

provides the fastest production capability, the highest purity, and the easiest scalability for high-throughput strategies.

Chemical methods are available through commercial suppliers who employ solid phase synthesis strategies using automated DNA/RNA synthesizers to generate RNA oligonucleotides. Synthesis proceeds in the 3′ to 5′ direction by addition of modified, protected nucleoside phosphoramidites corresponding to the desired sequence. The 3′ most terminal nucleoside is typically attached to a solid support serving as an anchor for complex syntheses so all reagents and reaction intermediates are removed by extensive washes between each nucleoside addition cycle. RNA oligonucleotide yields are determined by UV spectrophotometry (absorbance at 260 nm); the sequence fidelity and purity is readily assessed via mass spectrometry, strong anion exchange HPLC, or PAGE.

Currently there are three chemical RNA synthesis platforms that are commercially available. These platforms are known by the protecting group that modifies the 2′ position of the ribose sugar and distinguishes the phosphoramidites used in each synthetic strategy: 2′-tBDMS (2′-*O*-t-butyldimethylsilyl; Ambion, Proligo, etc) (Caruthers 1985; Usman 1987), 2′-TOM (2′-*O*-triisopropylsilyloxymethyl; Qiagen) (Wu and Pitsch 1998), and 2′-ACE® (2′-acetoxyethoxy; Dharmacon) (Scaringe 2000; Scaringe 2001) (see Figure 6.4). The end products derived from any one of these platforms result in siRNA duplexes of sufficient quality for most *in vitro* or *in vivo* work. As an example, the typical yield from a standard 0.02-μmole scale synthesis of a 21mer duplex (average molecular weight of 13300 g/mole) is 15 nmoles (approximately 200 ng). As a general guide, under optimized transfection conditions, when using a final concentration of 100 nM siRNA complexed with a lipid-reagent, one nmole is sufficient for transfecting cells plated in 20-24 wells in a 24-well plate format.

Of these RNA synthetic methods, the 2′-ACE RNA chemistry offers the greatest flexibility because it is readily scalable and amenable to high-throughput production – both critical attributes for gene functional analyses in animal models or large-scale synthesis for therapeutic applications. Another key attribute of 2′-ACE chemistry is the ability to synthesize long RNA sequences (>80 bases in length) including sequences with a high propensity for secondary structure (e.g. hairpins) that would otherwise interfere with synthesis and purification yields. The 2′-ACE protecting group present throughout synthesis and purification serves to reduce secondary structure that may interfere with the efficiency and yields. The 2′-protected RNA is water-soluble, resistant to nuclease degradation, and stable during long-term storage. The protecting group is easily removed using mildly acidic conditions and can be carried out at any point after receipt of protected oligos (see Figure 6.5). This facet of RNA oligo production makes 2′-ACE chemistry unique among all others as no other chemical platform permits delivery of the final product in a protected form.

Conclusion

The promise of RNAi as an enabling technology for functional genomics, discovery biology, and therapeutic applications is characterized by key attributes of the functional silencing intermediate. siRNAs can be (1) highly sequence specific, (2) functional at low concentrations, (3) comparatively rapid and cost effective to produce, and (4) amenable to large-scale screens and high-throughput platforms. The combination of well-defined, sophisticated design strategies coupled with reliable and flexible methods for siRNA synthesis makes siRNA-mediated RNAi among the most promising reverse genetics technologies available to date. There are clearly numerous opportunities to improve this technology. Studies are ongoing to characterize additional features that can be applied to further improve functionality and specificity.

Acknowledgements

We would like to thank the following individuals for their valuable contributions: Angela Reynolds for sharing data; Julia Kendall for help preparing figures and graphics; and Jon Karpilow for helpful discussions and careful review of the manuscript.

REFERENCES

Appasani, K. (2003). RNA interference: A technology platform for target validation, drug discovery and therapeutic development. *Drug Discovery World*, Summer Issue, 61–66.

Aza-Blanc, P., Cooper, C., Wagner, K., Batalov, S., Deveraux, Q. and Cooke, M. (2003). Identification of modulators of TRAIL-induced apoptosis via RNAi-based phenotypic screening. *Molecular Cell*, **12**, 627–637.

Baglioni, C. (1979). Interferon-induced enzymatic activities and their role in the antiviral state. *Cell*, **17**, 255.

Bernstein, E., Caudy, A. A., Hammond, S. M. and Hannon, G. J. (2001). Role for a bidentate ribonuclease in the initiation step of RNA interference. *Nature*, **409**, 363–366.

Billy, E., Brondani, V., Zhang, H., Muller, U. and Filipowicz, W. (2001). Specific interference with gene expression induced by long, double-stranded RNA in mouse embryonal teratocarcinoma cell lines. *Proceedings of the National Academy of Sciences U S A*, **98**, 14428–14433.

Brummelkamp, T. R., Bernards, R. and Agami, R. (2002). A system for stable expression of short interfering RNAs in mammalian cells. *Science*, **296**, 550–553.

Calegari, F., Haubensak, W., Yang, D., Huttner, W. B. and Buchholz, F. (2002). Tissue-specific RNA interference in postimplantation mouse embryos with endoribonuclease-prepared short interfering RNA. *Proceedings of the National Academy of Sciences U S A*, **99**, 14236–14240.

Caplen, N., Fleenor, J., Fire, A. and Morgan, R. (2000). dsRNA-mediated gene silencing in cultured *Drosophila* cells: A tissue culture model for the analysis of RNA interference. *Gene*, **252**, 95–105.

Caplen, N. J., Parrish, S., Imani, F., Fire, A. and Morgan, R. A. (2001). Specific inhibition of gene expression by small double-stranded RNAs in invertebrate and vertebrate systems. *Proceedings of the National Academy of Sciences U S A*, **98**, 9742–9747.

Capodici, J., Kariko, K. and Weissman, D. (2002). Inhibition of HIV-1 infection by small interfering RNA-mediated RNA interference. *Journal of Immunology*, **169**, 5196–5201.

Caruthers, M. H. (1985). Gene synthesis machines: DNA chemistry and its uses. *Science*, **230**, 281–285.

Castanotto, D., Li, H. and Rossi, J. J. (2002). Functional siRNA expression from transfected PCR products. *RNA*, **8**, 1454–1460.

Caudy, A., Ketting, R., Hammond, S., Denli, A., Bathoorn, A., Tops, B., Silva, J., Myers, M., Hannon, G. and Plasterk, R. (2003). A micrococcal nuclease homologue in RNAi effector complexes. *Nature*, **425**, 411–414.

Elbashir, S. M., Harborth, J., Lendeckel, W., Yalcin, A., Weber, K. and Tuschl, T. (2001a). Duplexes of 21-nucleotide RNAs mediate RNA interference in cultured mammalian cells. *Nature*, **411**, 494–498.

Elbashir, S. M., Lendeckel, W. and Tuschl, T. (2001b). RNA interference is mediated by 21- and 22-nucleotide RNAs. *Genes & Development*, **15**, 188–200.

Elbashir, S. M., Harborth, J., Weber, K. and Tuschl, T. (2002). Analysis of gene function in somatic mammalian cells using small interfering RNAs. *Methods*, **26**, 199–213.

Fire, A., Xu, S., Montgomery, M. K., Kostas, S. A., Driver, S. E. and Mello, C. C. (1998). Potent and specific genetic interference by double-stranded RNA in *Caenorhabditis elegans*. *Nature*, **391**, 806–811.

Fraser, A. G., Kamath, R. S., Zipperlen, P., Martinez-Campos, M., Sohrmann, M. and Ahringer, J. (2000). Functional genomic analysis of *C. elegans* chromosome I by systematic RNA interference. *Nature*, **408**, 325–330.

Holen, T., Amarzguioui, M., Babaie, E. and Prydz, H. (2003). Similar behaviour of single-strand and double-strand siRNAs suggests they act through a common RNAi pathway. *Nucleic Acids Research*, **31**, 2401–2407.

Holen, T., Amarzguioui, M., Wiiger, M. T., Babaie, E. and Prydz, H. (2002). Positional effects of short interfering RNAs targeting the human coagulation trigger Tissue Factor. *Nucleic Acids Research*, **30**, 1757–1766.

Hsieh, A., Bo, R., Manola, J., Bare, O., Khvorova, A., Scaringe, S. and Sellers, W. (2004). A library of siRNA duplexes targeting the phosphoinositide 3-kinase pathway: Determinants of gene silencing for use in cell-based screens. *Nucleic Acids Research*, **32**, 893–901.

Huang, F., Khvorova, A., Marshall, W. S. and Sorkin, A. (2004). Analysis of clathrin-mediated endocytosis of epidermal growth factor receptor by RNA interference. *Journal of Biological Chemistry*, **279**, 16657–16661.

Jackson, A. L., Bartz, S. R., Schelter, J., Kobayashi, S. V., Burchard, J., Mao, M., Li, B., Cavet, G. and Linsley, P. S. (2003). Expression profiling reveals off-target gene regulation by RNAi. *Nature Biotechnology*, **21**, 635–637.

Ketting, R. F., Fischer, S. E., Bernstein, E., Sijen, T., Hannon, G. J. and Plasterk, R. H. (2001). Dicer functions in RNA interference and in synthesis of small RNA involved in developmental timing in *C. elegans*. *Genes & Development*, **15**, 2654–2659.

Khvorova, A., Reynolds, A. and Jayasena, S. (2003). Functional siRNAs and miRNAs exhibit strand bias. *Cell*, **115**, 209–216.

Knight, S. W. and Bass, B. L. (2001). A role for the RNase III enzyme DCR-1 in RNA interference and germ line development in *Caenorhabditis elegans*. *Science*, **293**, 2269–2271.

Krichevsky, A. M. and Kosik, K. S. (2002). RNAi functions in cultured mammalian neurons. *Proceedings of the National Academy of Sciences U S A*, **99**, 11926–11929.

Laposa, R. R., Feeney, L. and Cleaver, J. E. (2003). Recapitulation of the cellular xeroderma pigmentosum-variant phenotypes using short interfering RNA for DNA Polymerase H. *Cancer Research*, **63**, 3909–3912.

Lee, N. S., Dohjima, T., Bauer, G., Li, H., Li, M. J., Ehsani, A., Salvaterra, P. and Rossi, J. (2002). Expression of small interfering RNAs targeted against HIV-1 rev transcripts in human cells. *Nature Biotechnology*, **20**, 500–505.

Martens, H., Novotny, J., Oberstrass, J., Steck, T. L., Postlethwait, P. and Nellen, W. (2002). RNAi in Dictyostelium: The role of RNA-directed RNA polymerases and double-stranded RNase. *Molecular Biology of the Cell*, **13**, 445–453.

Martinez, J., Patkaniowska, A., Urlaub, H., Luhrmann, R. and Tuschl, T. (2002). Single-stranded antisense siRNAs guide target RNA cleavage in RNAi. *Cell*, **110**, 563–574.

McManus, M. T., Haines, B. B., Dillon, C. P., Whitehurst, C. E., van Parijs, L., Chen, J. and Sharp, P. A. (2002). Small interfering RNA-mediated gene silencing in T lymphocytes. *Journal of Immunology*, **169**, 5754–5760.

Minks, M. A., West, D. K., Benvin, S. and Baglioni, C. (1979). Structural requirements of double-stranded RNA for the activation of 2′,5′-oligo(A) polymerase and protein kinase of interferon-treated HeLa cells. *Journal of Biological Chemistry*, **254**, 10180–10183.

Nykanen, A., Haley, B. and Zamore, P. D. (2001). ATP requirements and small interfering RNA structure in the RNA interference pathway. *Cell*, **107**, 309–321.

Paul, C. P., Good, P. D., Winer, I. and Engelke, D. R. (2002). Effective expression of small interfering RNA in human cells. *Nature Biotechnology*, **20**, 505–508.

Reynolds, A., Leake, D., Scaringe, S., Marshall, W. S., Boese, Q. and Khvorova, A. (2004). Rational siRNA design for RNA interference. *Nature Biotechnology*, **22**, 326-330.

Scaringe, S. A. (2000). Advanced 5′-silyl-2′-orthoester approach to RNA oligonucleotide synthesis. *Methods in Enzymology*, **317**, 3–18.

Scaringe, S. A. (2001). RNA oligonucleotide synthesis via 5′-silyl-2′-orthoester chemistry. *Methods*, **23**, 206–217.

Schwarz, D. S., Hütvagner, G., Du, T., Xu, Z., Aronin, N. and Zamore, P. D. (2003). Unexpected asymmetry in the assembly of the RNAi Enzyme complex. *Cell*, **115**, 199–208.

Schwarz, D. S., Hütvagner, G., Haley, B. and Zamore, P. D. (2002). Evidence that siRNAs function as guides, not primers, in the *Drosophila* and human RNAi pathways. *Molecular Cell*, **10**, 537–548.

Semizarov, D., Frost, L., Sarthy, A., Kroeger, P., Halbert, D. N. and Fesik, S. W. (2003). Specificity of short interfering RNA determined through gene expression signatures. *Proceedings of the National Academy of Sciences U S A*, **100**, 6347–6352.

Sijen, T., Fleenor, J., Simmer, F., Thijssen, K. L., Parrish, S., Timmons, L., Plasterk, R. H. and Fire, A. (2001). On the role of RNA amplification in dsRNA-triggered gene silencing. *Cell*, **107**, 465–476.

Spankuch-Schmitt, B., Bereiter-Hahn, J., Kaufmann, M. and Strebhardt, K. (2002). Effect of RNA silencing of polo-like kinase-1 (PLK1) on apoptosis and spindle formation in human cancer cells. *Journal of the National Cancer Institute*, **94**, 1863–1877.

Sui, G., Soohoo, C., Affar el, B., Gay, F., Shi, Y. and Forrester, W. C. (2002). A DNA vector-based RNAi technology to suppress gene expression in mammalian cells. *Proceedings of the National Academy of Sciences U S A*, **99**, 5515–5520.

Tabara, H., Grishok, A. and Mello, C. C. (1998). RNAi in *C. elegans*: Soaking in the genome sequence. *Science*, **282**, 430–431.

Tijsterman, M., Ketting, R. F., Okihara, K. L., Sijen, T. and Plasterk, R. H. (2002). RNA helicase MUT-14-dependent gene silencing triggered in *C. elegans* by short antisense RNAs. *Science*, **295**, 694–697.

Timmons, L., Court, D. L. and Fire, A. (2001). Ingestion of bacterially expressed dsRNAs can produce specific and potent genetic interference in *Caenorhabditis elegans*. *Gene*, **263**, 103–112.

Timmons, L. and Fire, A. (1998). Specific interference by ingested dsRNA. *Nature*, **395**, 854.

Tuschl, T., Zamore, P. D., Lehmann, R., Bartel, D. P. and Sharp, P. A. (1999). Targeted mRNA degradation by double-stranded RNA *in vitro*. *Genes & Development*, **13**, 3191–3197.

Usman, N. O., KK. Jiang, MY. Cedergren, RJ. (1987). The automated chemical synthesis of long oligoribuncleotides using 2′-O-silylated ribonucleoside 3′-O-phosphoramidites on a controlled-pore glass support: synthesis of a 43-nucleotide sequence similar to the 3′-half molecule of an Escherichia coli formylmethionine tRNA. *Journal of the American Chemical Society*, **109**, 7845.

Vickers, T. A., Koo, S., Bennett, C. F., Crooke, S. T., Dean, N. M. and Baker, B. F. (2003). Efficient reduction of target RNA's by small interfering RNA and RNase H dependent antisense agents: A comparative analysis. *Journal of Biological Chemistry*, **278**, 7108–7118.

Wu, X. and Pitsch, S. (1998). Synthesis and pairing properties of oligoribonucleotide analogues containing a metal-binding site attached to beta-D-allofuranosyl cytosine. *Nucleic Acids Research*, **26**, 4315–4323.

Yang, D., Buchholz, F., Huang, Z., Goga, A., Chen, C.-Y., Brodsky, F. M. and Bishop, J. M. (2002). Short RNA duplexes produced by hydrolysis with Escherichia coli RNase III mediate effective RNA interference in mammalian cells. *Proceedings of the National Academy of Sciences U S A*, **99**, 9942–9947.

Zhao, Z., Cao, Y., Li, M. and Meng, A. (2001). Double-stranded RNA injection produces nonspecific defects in zebrafish. *Developmental Biology*, **229**, 215–223.

7 Automated design and high throughput chemical synthesis of siRNA

Yerramilli V. B. K. Subrahmanyam and Eric Lader

Introduction

The application of RNA interference (RNAi) to mammalian cells (Caplen et al., 2001; Elbashir et al., 2001a) has the potential to revolutionize the field of functional genomics and target validation in drug discovery research (Harborth et al., 2001; Tuschl and Borkhardt, 2002). In addition RNAi is also viewed as potential means for nucleic acid-based therapeutic applications. The ability to simply, effectively, and specifically down-regulate the expression of genes in mammalian cells holds enormous scientific, commercial, and therapeutic potential. To ensure success in gene silencing experiments, the design and synthesis of the targeting molecule (the siRNA) must be carefully optimized. The present chapter describes the design criteria for successful gene silencing, followed by a description of recent improvements in high-throughput chemical synthesis of siRNAs, and transcriptome wide gene silencing studies.

Advantages of using RNAi over other technologies for functional genomics

A variety of methods and technologies are available for studying gene function such as gene knockouts and transgenic animal models (Capecchi, 1989; Feil et al., 1996; Jaenisch, 1998; Le and Sauer, 2001; Perkins, 2002), the majority of which are time-consuming and expensive. To facilitate functional genomics in mammalian cells at a transcriptome-wide level, an efficient approach that enables rapid and high throughput analysis is required. Use of synthetic siRNA for RNA interference offers a rapid and cost-effective alternative to previous methods, and can be used to silence expression of several genes simultaneously (Harborth et al., 2001; Kamath et al., 2003; Kamath and Ahringer, 2003). A potent siRNA is far easier to design, in comparison to the effort needed in developing a specific and potent antisense oligonucleotide. Recent work demonstrating the use of siRNA to attenuate the expression of ICAM-1 mRNA has shown that the biological activity of siRNA is approximately 100-fold higher (e.g. siRNA was effective at

0.01× the concentration of an effective antisense RNA) in comparison with antisense oligonucleotides (Kretschmer-Kazemi Far and Sczakiel, 2003). These results have been confirmed by other studies using a luciferase reporter system (Miyagishi et al., 2003).

siRNA Design

Although any stretch of 21 nt on a message sequence could theoretically be sufficient for targeting by RNAi, the experience of several groups indicate that only one in four or five sequences selected in this manner is functionally active in RNAi (Holen et al., 2002; Kumar et al., 2003). Such observations strongly indicate that target site selection and siRNA design have profound influence on the level and specificity of gene silencing. Recent studies addressing the mechanistic aspects of RNAi indicate that siRNAs that are designed based on our understanding of the mechanism of RNAi have a higher likelihood of being potent and also specific to the target (Khvorova et al., 2003; Schwarz et al., 2003). A 'good' siRNA should meet certain criteria: It must be potent (ability to achieve silencing at lower concentrations) to facilitate efficient gene silencing. It should also be highly specific, leading to the post-transcriptional gene silencing (PTGS) of the intended target gene, while avoiding cross-reactivity to unintended mRNA targets. There is an obvious need for a design algorithm that meets the above requirements with high efficiency. A thorough informatics analysis in conjunction with such a design tool should enhance our ability to successfully identify optimal siRNA targets for transcriptome-wide gene silencing experiments. In the following sections we will discuss these design criteria in more detail.

Choosing a target sequence

Any region of an mRNA could be targeted in principle, although regions with a potential for interference from the binding of mRNA-regulatory proteins are typically avoided (such as sequences around translational start sites, and exon/exon boundaries). It is also desirable to avoid regions containing repetitive sequences, due to the potential for off-target effects. Although targeting the open reading frame (ORF) is preferred (Elbashir et al., 2002), the 3'-UTRs have also been successfully targeted in RNAi experiments (Dykxhoorn et al., 2003; Elbashir et al., 2001a). Targeting the 3' UTR is efficacious if silencing of endogenous expression is complemented by the expression of a mutated or tagged form of the gene on a vector ('knock-in' experiment) and also to prove specificity of the phenotype by rescue with the normal transcript. Another reason for targeting the ORF is that the sequence information available in the public databases is generally more reliable for the protein coding regions. In addition, selecting the coding region could afford the option, if needed, to target a specific exon (unique to a particular splice variant), or a common exon (that targets all mRNA variants) of a given gene.

Table 7.1. Performance comparison of siRNAs designed by standard and advanced design tools: siRNAs designed by standard and the advanced design criteria for a total of 100 genes were tested for their knock down performance. The relative success rates were evaluated and presented in this table

Silencing activity (percentage)	Percentage of total siRNAs tested	
	Standard design	Advanced design
Functional siRNAs (with >50% activity)	63%	85%
siRNAs with >70% activity	57%	75%
siRNAs with >91% activity	17%	46%

Design considerations

While general guidelines for designing siRNAs (Elbashir et al., 2002) are available, the exact requirements for effective silencing are not clear and hence the process of designing siRNAs is essentially empirical (Dykxhoorn et al., 2003). Short 21-nt double-stranded RNAs with a 2-nt overhang on the 3′-ends resembling the naturally occurring products of dicer are shown to be efficient mediators of RNAi (Elbashir et al., 2001a, 2001b, 2001c; Elbashir et al., 2002). Based on this general rule, targeting sequences with the pattern AA (N19) TT, (essentially random) gives adequate results. Additional criteria such as GC content (35–65%) and avoiding stretches of Gs or Cs longer than 3 bases have been recommended to enhance the silencing activity/ability of the siRNA. There is a general preference to select the target regions starting with AA, so that the 2-nt overhangs on the 3′-end would contain Uridine residues. These Uridine overhangs can be replaced with 2′-deoxy-thymidine without any loss of silencing activity. This may enhance the nuclease resistance of the siRNA and also is cost effective in chemical synthesis.

Until recently, with these general rules a researcher must typically try several siRNAs for a target gene to find at least one functional siRNA. However, recent observations (Khvorova et al., 2003; Schwarz et al., 2003) provide some clues towards successful siRNA design. It appears that functional siRNAs have a lower internal stability at the 5′-antisense end and that this end is less stable due to the presence of bases that confer such lower stability at positions 9–14 (from the 5′-antisense end) compared to the 5′-sense terminus, suggesting that the helicase associated with RISC may be biased towards the unstable end (Khvorova et al., 2003). Schwarz et al. (2003) showed that the siRNA duplexes can be functionally asymmetric, and that small changes in siRNA sequence have profound and predictable effects on the extent to which the individual strands of an siRNA duplex enter the RNAi pathway. Once the siRNA duplex is unwound only one strand enters RISC and the other strand is rapidly degraded. Their data suggest that an enzyme that governs RISC assembly selects that strand of a siRNA whose 5′-end is less tightly paired to its complement. Use of such mechanism-based rules to design siRNAs clearly increases the likelihood of designing potent and specific RNAi reagents. SiRNA design involving such functional asymmetry may also minimize the sense strand participation in RNAi and related off-target effects (Jackson et al., 2003).

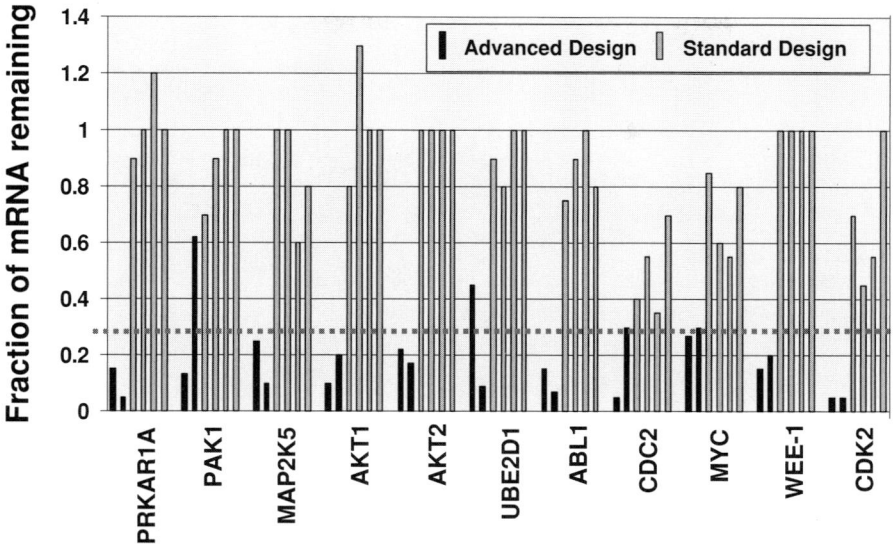

Figure 7.1. Performance comparison of siRNAs designed using standard (gray bars) and advanced design (solid bars): siRNAs designed using standard and advanced tools against a panel of 11 genes, were transfected into Hela S3 cells (in triplicate) at a concentration of 100 nM using Qiagen's RNAifect transfection reagent. Total RNA was isolated 24 hours post transfection and Q-RT-PCR analysis was performed using SYBRgreen assay for the target genes. Either glyceraldehyde-3-phosphate dehydrogenase (GAPDH) or Cyclophilin were used as internal reference genes. The following genes were used in the study: PRKAR1A (NM_002734), PAK1 (NM_002576), MAP2K5 (MEK5, NM_002757), AKT1 (NM_005163), AKT2 (NM_001626), UBCH5 (UBE2D1, NM_003338), ABL1 (NM_007313), CDC-2 (CDK1, NM_001786), MYC (NM_002467), WEE1 (NM_003390), CDK2 (NM_001798).

A detailed functional analysis of siRNAs designed against more than 100 genes by Qiagen and our collaborators confirms the advantages of siRNAs designed using such newer recommendations. Implementation of advanced design rules decreases failure rates [(Table 7.1) (defined as siRNAs which reduce mRNA levels by less than 50%)] substantially, from 37% to 15%. Thus, more than 85% of siRNAs designed using these enhanced rules are active. In addition, the number of highly functional siRNAs (defined as siRNAs which reduce mRNA levels by more than 90%) increases from 17% to 46%. Perhaps most indicative of the benefits of this advanced design is the ability to design potent and functional siRNAs even against the targets for which the 'standard' design failed to result in functional siRNAs. Data on mRNA knockdown for 11 genes are shown in Figure 7.1. In each case, 4 siRNAs designed with standard guidelines failed to deliver a potent silencing effect. However, in all cases, siRNAs designed using the advanced rules resulted in at least one, and usually two, highly active duplexes, out of two duplexes tested.

Specificity of imperfect base and mismatch pairing sequences in siRNA strands

Aptamers and antisense oligonucleotides have been reported to show undesirable interactions with cellular proteins (Shi et al., 1999; Cho et al., 2001 Fisher et al., 2002) indicating that specificity of gene silencing is critical for accurate analysis.

Target gene: NM_000123 (ERCC5)

Candidate siRNA(Sense strand): AACAGACTTTGGTGAAGAGAA

Target specific alignment:

Query Accession	Hit Accession	Gene Name	DB XRef	Species
NM_000123	D16305	ERCC5	Homo sapiens	GI:303607

Score	Align
21	1 aacagactttggtgaagagaa 21 ********************* 254 aacagactttggtgaagagaa 274

Related Alignment:

Smith-Waterman Homology search places this related sequence within the top 10 alignments.

BLAST search has not ranked this even in the top 100 alignments

Query Accession	Hit Accession	Gene Name	DB XRef	Species
NM_000123	NM_005863	Hs.25155	Homo sapiens	

Score	Align
17	1 aacagactttggtgaagagaa 21 ** *********** ****** 1759 aagagactttggtgtagagaa 1779

Figure 7.2. Homology alignment using Smith-Waterman and BLAST: We selected NM_000123 (ERCC5 mRNA) as a test gene. We selected a candidate siRNA for homology search in GenBank by either Smith-Waterman or BLAST analysis. Smith-Waterman analysis identified matches of 19/21 bases to an unrelated sequence (NM_005863, NET1 mRNA, Hs.25155) that BLAST search failed to identify (in the top 100 alignments).

Recently, specificity of siRNAs in human cells was addressed by global gene expression profiling analysis using cDNA microarrays (Chi et al., 2003; Semizarov et al., 2003). Their results suggest that the expression profiles for different siRNAs against the same target gene correlate closely when siRNA design and transfection conditions are optimized. This demonstrates that siRNA-mediated gene knockdown can be a highly specific and reliable technique.

In addition, recent observations indicated that siRNAs with mismatches up to 3–4 base pairs can act as micro RNAs (miRNAs) causing translational repression (Doench et al., 2003; Saxena et al., 2003; Zeng and Cullen, 2003). This implies that siRNAs showing imperfect base pairing to any non-intended targets could lead to non-specific effects by a miRNA mechanism. Some mismatches also affect the ability of RISC to target an mRNA more than other mismatches (Hamada et al., 2002). The Smith-Waterman algorithm is an ideal tool to analyze candidate siRNAs for potential cross-reactivity to other sequences (Smith and Waterman, 1981). While BLAST can be much quicker than Smith-Waterman, it has the potential to miss extremely important 'hits' in the database. BLAST requires 7 contiguous bases of match to score as a hit. If a siRNA has as little as two strategically placed mismatches to a related sequence, BLAST may rank the homology very low (Figure 7.2).

Requirement of 5′-Hydroxyl (OH) group to participate in RNAi

Synthetic siRNAs containing 5-OH termini can successfully induce RNAi effects in *Drosophila* embryo lysates (Nykanen et al., 2001) and in cultured mammalian cells (Elbashir et al., 2001b). A potent kinase activity has been implicated in *Drosophila* embryo extracts that phosphorylates the 5′-OH termini of the synthetic siRNAs. (Nykanen et al., 2001). Recent evidence suggests that blocking the 5-OH group of the antisense strand by modification abolishes silencing completely (Chiu and Rana, 2002). Such a modification renders the 5′- end hydroxyl group inaccessible for phosphorylation. However similar modification blocking the 5′-OH on the sense strand has no effect on gene silencing activity. Thus, this strand is appropriate to incorporate fluorogenic moieties during chemical synthesis. These results indicate that a free 5-OH on the antisense strand of the siRNA duplex is important for RNAi activity in human cells. Blocking the 3′-OH of either sense or antisense strand of siRNA duplex has little effect on RNAi activity. These results imply that it is possible to avoid undesirable off-target effects resulting from the sense strand by blocking its 5′-OH group.

The TOM-Amidite chemistry of synthetic siRNA and purification

The chemical synthesis of RNA is intrinsically more demanding than the synthesis of DNA, because the additional 2′-OH group, present in RNA-nucleotides, must be protected during assembly. As a consequence, additional chemical steps are required, both for the introduction of the 2′-O-protecting groups into monomers (phosphoramidites), and for their removal from the assembled RNA sequences. Reliable protecting groups have to meet the following requirements: Their removal at the end of the synthesis must occur quantitatively without concomitant destruction of the product; they have to be completely stable under all reaction conditions; they should not interfere with the coupling reactions; and they should be easy to introduce into monomers. RNA synthesis at Qiagen is carried out by the recently introduced, patented TOM-chemistry (Pitsch et al., 1999a; Pitsch et al.,1999b; Pitsch et al., 2001; Pitsch and Weiss, 2001), which meets all these requirements and is at the same time as closely as possible adapted to the established DNA-chemistry. With 2′-O-TOM protected monomers (TOM = [(triisopropylsilyl)oxy]methyl), excellent coupling yields of >99.5% are achieved routinely under DNA-assembly conditions (2 minutes coupling time). The final deprotection under very mild conditions is fast, easy and efficient, and is not interfering with the integrity of the product RNA sequence. Furthermore, the chemical structure of the TOM-group prevents partial formation of 2′-5′-linked

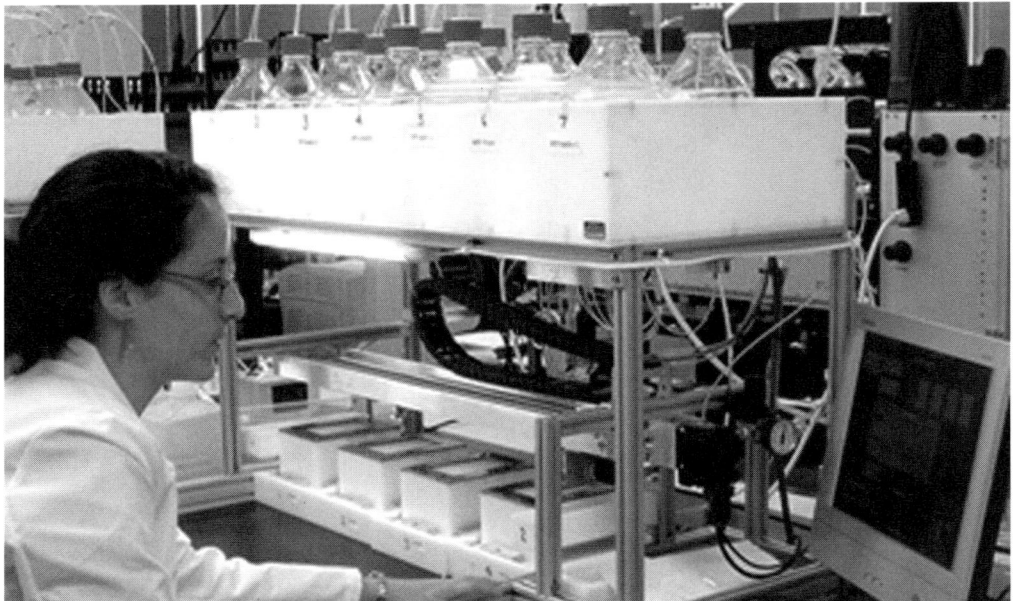

Figure 7.3. High throughput purification on Qiagen HPP robot: siRNAs are purified >90% purity using a proprietary solid phase, high throughput purification system. One HPP robot can purify 384 single stranded RNAs in one purification cycle.

phosphodiester bonds, a problem frequently encountered with traditional RNA protecting groups. As a consequence of the excellent coupling efficiency and the clean and reliable deprotection achieved with TOM-protected RNA building blocks, up to 100mer sequences can be prepared with the same quality in the same amounts as the analogous DNA sequences. TOM-chemistry is not only fully compatible with DNA synthesis and the traditional TBDMS (*tert*-butyldimethylsilyl) chemistry, but also with all known modified building blocks, such as the various fluorescent labels, sequence and terminus modifiers, backbone modifications, and unnatural nucleobases. The close relation to DNA and the reliable chemistry of TOM-protected phosphoramidites allowed Qiagen the development of the straightforward HPP-process (HPP = High Performance Purification), which combines high-throughput RNA synthesis with high-throughput purification. The resulting siRNA sequences are obtained in excellent yields and with a reproducible purity of >90%. This makes the HPP process a robust and economical approach to meet the high throughput demand delivering highest purity siRNAs.

High throughput and automation strategy

It is now possible that every mRNA in a cell can be screened by knockout in a high throughput phenotypic or functional screen. However, its ability to screen such a genome-wide library of siRNAs depend upon robust transfection and automated screening technologies. Such tools already exist in most if not all large biotech and pharma research laboratories. The challenge is in the design and production of such genome-wide libraries, which requires a combination of good design and

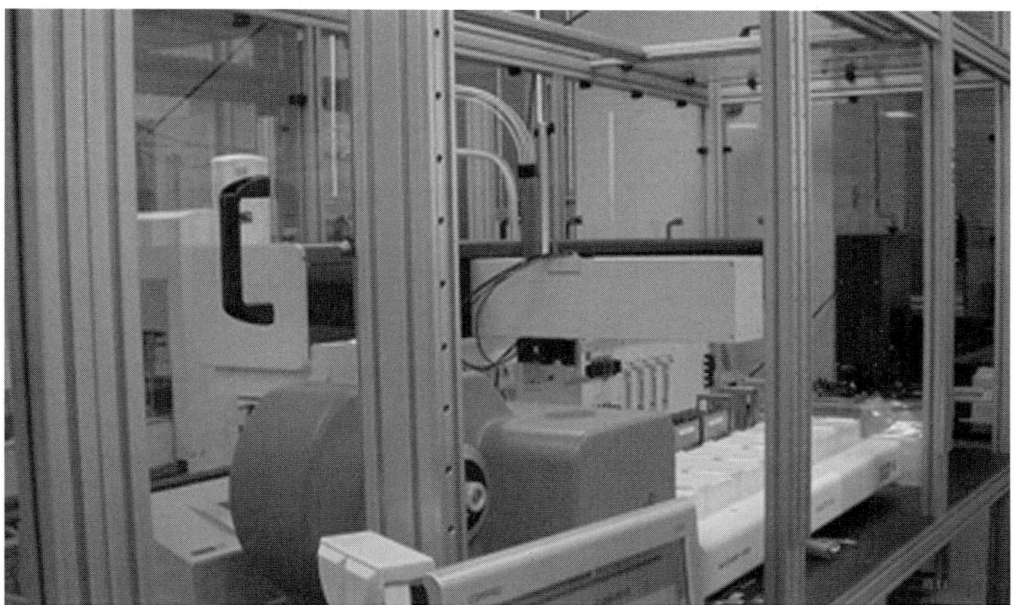

Figure 7.4. BioRobot 3000 system for QC sampling: Single strand siRNAs are sampled by the BioRobot 3000 for MALDITOF-MS analysis and quantification by ultraviolet (UV) absorbance at 260 nm. Plates are sealed by an integrated ALPS plate sealer and held until QC data is available.

informatics tools to develop and deliver about 100,000 siRNA duplexes to a single customer in a realistic timeframe. To address these logistical issues and meet the demands we at Qiagen developed a high-throughput synthesis laboratory in combination with a custom Laboratory Information Management System (LIMS). Using this tracking system from the data entry of an mRNA target sequence to chemical synthesis in 96-well plates, quality control (QC) and final processing could be measurable and manageable at each and every step.

We at Qiagen synthesize siRNA on dual 96-well plate synthesizers, which enable real-time monitoring of coupling efficiency by trityl monitoring. Newly synthesized RNA is subject to gas-phase deprotection, and is then purified using the automated, high throughput HPP process (Figure 7.3). Post HPP steps, which are performed in an aqueous environment, are carried out on Qiagen BioRobot automated workstations. These workstations are fully enclosed in HEPA-filtered cabinets to avoid any possible contaminations (Figure 7.4). The BioRobot 3000 system samples each single-stranded RNA preparation for MALDI-TOF Mass-Spectrometry analysis, and for quantification by absorbance at 260 nm (carried out by an integrated 96 well plate based spectrophotometer). Once sampling is done, these plates are sealed and held until all the QC data are available. Sequence information of any sample that fails QC requirements is automatically entered back into the system for resynthesis. Once replacements are synthesized the system automatically requests the stored master plates and those plates containing the replacement RNAs. These plates are loaded onto a BioRobot 8000 system (Figure 7.5) that re-arrays all RNAs that passed QC, to generate plates containing sense and antisense strands. The BioRobot 8000 in a contained environment

Figure 7.5. BioRobot 8000 system for rearraying, normalization, and annealing: Once QC data is available, the BioRobot 8000 system normalizes all samples for molarity and combines the corresponding sense and antisense strands for annealing. The annealed duplexes are either dried directly in the same plates by speed vac system, or dried after transferring the samples into individual tubes.

carries out molar normalization of all the strands, and anneals sense and antisense strands into duplex in a new master plate. The annealed siRNAs in these plates are then subjected to vacuum drying by speed vac [(Thermo Electron) (or aliquoted into individual tubes and subjected to drying)] and the samples are then ready for shipment or storage.

Conclusions

As reviewed in this chapter, the technology that supports RNA interference as a research tool is far from mature. The past two years have seen major advances in design, synthesis, and delivery of chemically synthesized siRNA. The future holds even further refinements of siRNA design to improve activity and specificity. In addition, chemical modification of siRNA will lead to improved activity, duration of effect, efficacy of delivery, and minimization of off-target effects. Taken together, the current advantages of using chemically synthesized siRNA and future improvements will insure the continued dominance of chemically synthesized siRNA as the reagent of choice for high throughput target identification.

Acknowledgements

We thank Prof. Stefan Pitsch for helpful suggestions and inputs.

REFERENCES

Capecchi, M. R. (1989). Altering the Genome by homologous recombination. *Science*, **244**, 1288–1292.

Caplen, N. J., Parrish, S., Imani, F., Fire, A. and Morgan, R. A. (2001). Specific inhibition of gene expression by small double-stranded RNAs in invertebrate and vertebrate systems. *Proceedings of the National Academy of Sciences USA*, **98**, 9742–9747.

Chi, J-T., Chang, H. Y., Wang, N. N., Chang, D. S., Dunphy, N. and Brown, P. O. (2003). Genomewide view of gene silencing by small interfering RNAs. *Proceedings of the National Academy of Sciences USA*, **100**, 6343–6346.

Chiu, Y. L. and Rana, T. M. (2002). RNAi in human cells: Basic structural and functional features of small interfering RNA. *Molecular Cell*, **10**, 549–561.

Cho, Y. S., Kim, M-K., Cheadle, C., Neary, C, Becker, K. G. and Cho-Chung, Y. S. (2001). Antisense DNAs as multisite genomic modulators identified by DNA microarray. *Proceedings of the National Academy of Sciences USA*, **98**, 9819–9823.

Doench, J. G., Petersen, C. P. and Sharp, P. A. (2003). SiRNAs can function as miRNAs. *Genes & Development*, **17**, 438–442.

Dykxhoorn, D. M., Novina, C. D. and Sharp, P. A. (2003). Killing the messenger: Short RNAs that silence the gene expression. *Nature Reviews Molecular Cell Biology*, **4**, 457–467.

Elbashir, S. M., Harborth, J., Lendeckel, W., Yalcin, A., Weber, K. and Tuschl, T. (2001a). Duplexes of 21-nucleotide RNAs mediate RNA interference in cultured mammalian cells. *Nature*, **411**, 494–498.

Elbashir, S. M., Lendeckel, W. and Tuschl, T. (2001b). RNA interference is mediated by 21 and 22 nt RNAs. *Genes & Development*, **15**, 188–200.

Elbashir, S. M., Martinez, J., Patkaniowska, A., Lendeckel, W. and Tuschl, T. (2001c). Functional anatomy of siRNAs for mediating efficient RNAi in *Drosophila melanogaster* embryo lysate. *European Molecular Biology Organization Journal*, **20**, 6877–6888.

Elbashir, S. M., Harborth, J., Weber, K. and Tushl, T. (2002). Analysis of gene function in somatic mammalian cells using small interfering RNAs. *Methods*, **26**, 199–213.

Feil, R., Brocard, J, Mascrez, B, LeMeur, M, Metzger, D. and Chambon, P. (1996). Ligand activated site specific recombination in mice. *Proceedings of the National Academy of Sciences USA*, **93**, 10887–90.

Fisher, A. A., Ye, D., Sergueev, D. S., Fisher, M. H., Shaw, B. R. and Juliano, R. L. (2002). Evaluating the specificity of Antisense oligonucleotide conjugates: A DNA array analysis. *Journal of Biological Chemistry*, **277**, 22980–22984.

Hamada, M., Ohtsuka, T., Kawaida, R., Koizumi, M., Morita, K., Furukawa, H., Imanishi, T., Miyagishi, M. and Taira, K. (2002). Effects on RNA interference in gene expression (RNAi) in cultured mammalian cells of mismatches and the introduction of chemical modifications at the 3′-ends of siRNAs. *Antisense and Nucleic Acid Drug Development*, **12**, 301–309.

Harborth, J., Elbashir, S. M., Bechert, K., Tuschl, T. and Weber, K. (2001). Identification of essential genes in cultured mammalian cells using small interfering RNAs. *Journal of Cell Science*, **114**, 4557–4565.

Holen, T., Amarzguioui, M., Wiiger, M. T., Babaie, E. and Prydz, H. (2002). Positional effects of short interfering RNAs targeting the human coagulation trigger tissue factor. *Nucleic Acids Research*, **30**, 1757–1766.

Jackson, A. L., Bartz, S. R., Schelter, J., Kobayashi, S. V., Burchard, J., Mao, M., Li, B., Cavet, G. and Linsley, P. S. (2003). Expression profiling reveals off-target gene regulation by RNAi. *Nature Biotechnology*, **21**, 635–637.

Jaenisch, R. (1988). Transgenic animals. *Science*, **240**, 1468–1474.

Kamath, R. S., Fraser, A. G., Dong, Y., Poulin, G., Durbin, R., Gotta, M., Kanapin, A., LeBot, N., Moreno, S., Sohrmann, M., Welchman, D. P., Zipperlen, P. and Ahringer, J. (2003). Systematic functional analysis of the *Caenorhabditis elegans* genome using RNAi. *Nature*, **421**, 231–237.

Kamath, R. S. and Ahringer, J. (2003). Genome-wide RNAi screening in *Caenorhabditis elegans*. *Methods*, **4**, 313–321.

Khvorova, A., Reynolds, A. and Jayasena, S. D. (2003). Functional siRNAs and miRNAs exhibit strand bias. *Cell*, **115**, 209–216.

Kretschmer-Kazemi Far, R. and Sczakiel, G. (2003). The activity of siRNA in mammalian cells is related to structural target accessibility: A comparison with antisense oligonucleotides, *Nucleic Acids Research*, **31**, 4417–4424.

Kumar, R., Conklin, D. S. and Mittal, V. (2003). High-throughput selection of effective RNAi probes for gene silencing. *Genome Research*, **13**, 2333–2340.

Le, Y. and Sauer, B. (2001). Conditional gene knockout using Cre Recombinase. *Molecular Biotechnology*, **3**, 269–275.

Miyagishi, M., Hayashi, M. and Tiara, K. (2003). Comparison of the suppressive effects of Antisense oligonulceotides and siRNAs directed against the same targets in mammalian cells. *Antisense and Nucleic Acid Drug Development*, **13**, 1–7.

Nykanen, A., Haley, B. and Zamore, P. D. (2001). ATP requirements and small interfering RNA structure in the RNA interference pathway. *Cell*, **107**, 309–321.

Perkins, A. S. (2002). Functional Genomics in mouse. *Functional & Integrative Genomics*, **3**, 81–91.

Pitsch, S., Weiss, P. A., Jenny, L., Stutz, A., Wu, X. (2001). Reliable chemical synthesis of oligoribonucleotides (RNA) with 2'-*O*-[(Triisopropylsilyl)oxy]methyl (2'-O-tom) protected phosphoramidites. *Helvetica Chimica Acta*, **84**, 3773–3795.

Pitsch, S., Weiss, P. A., Jenny, L. (1999a). US Patent 5, 986,084.

Pitsch, S., Weiss, P. A., Wu, X., Ackermann, D. and Honegger, T. (1999b). Fast and reliable automated synthesis of RNA and partially 2'-*O*-Protected precursors ("caged RNA") based on two novel, orthogonal 2'-*O*-protecting groups. *Helvetica Chimica Acta*, **82**, 1753–1761.

Pitsch, S. and Weiss, P. A. (2001). Chemical Synthesis of RNA-Sequences with 2'-*O*-[(Triisopropylsilyl)oxy]methyl-protected Ribonucleoside Phosphoramidites (=TOM-Phosphoramidites) (Unit; 3.9). In: *Current Protocols of Nucleic Acid Chemistry* Ed. S. Beaucage, John Wiley & Sons, Inc., New York, NY.

Saxena, S., Jonsson, Z. O. and Dutta, A. (2003). Small RNAs with imperfect match to endogenous mRNA repress translation: Implications for off-target activity of small inhibitory RNA in mammalian cells. *Journal of Biological Chemistry*, **278**, 44312–44319.

Schwarz, D. S., Hütvagner, G., Du, T., Xu, Z., Aronin, N. and Zamore, P. D. (2003). Asymmetry in the assembly of the RNAi Enzyme Complex. *Cell*, **115**, 199–208.

Semizarov, D., Frost, L., Sarthy, A., Kroeger, P., Halbert, D. N. and Fesik, S. W. (2003). Specificity of short interfering RNA determined through gene expression signatures. *Proceedings of the National Academy of Sciences USA*, **100**, 6347–6352.

Shi, H., Hoffman, B. E. and Lis, J. T. (1999). RNA Aptamers as effective protein antagonists in a multicellular organism. *Proceedings of the National Academy of Sciences USA*, **96**, 10033–10038.

Smith, T. F. and Waterman, M. S. (1981). Identification of common molecular subsequences. *Journal of Molecular Biology*, **147**:195–197.

Tuschl, T. and Borkhardt, A. (2002). Small interfering RNAs: A revolutionary tool for the analysis of gene function and gene therapy. *Molecular Interventions*, **2**, 158–167.

Zeng, Y., Yi, R. and Cullen, B. R. (2003). MicroRNAs and small interfering RNAs can inhibit mRNA expression by similar mechanisms. *Proceedings of the National Academy of Sciences USA*, **100**, 9779–84.

8 Rational design of siRNAs with the *S*fold software

Ye Ding and Charles E. Lawrence

Introduction

In eukaryotic organisms, RNA interference (RNAi) is the sequence-specific gene silencing that is induced by double-stranded RNA (dsRNA) homologous to the silenced gene. In the cytoplasm of mammalian cells, long dsRNAs (>30 nt) can activate the potent interferon and a protein kinase-mediated pathway, which lead to non-sequence-specific effects that can include apoptosis (Kumar and Carmichael, 1998). Elbashir and coworkers (2001a) made the important discovery that small interfering RNAs (siRNAs) of about 21 nt specifically inhibit gene expression, because siRNAs are too short to activate the interferon or protein kinase pathway. The silencing by synthetic siRNAs is transient. This limitation can be overcome by stably expressed short hairpin RNAs (shRNAs), which are processed by Dicer into siRNAs (Paddison et al., 2002; Brummelkamp et al., 2002). However, it was recently reported that shRNA vectors can induce an interferon response (Bridge et al., 2003).

Because target recognition presumably depends on Watson-Crick base pairing, the RNAi machinery is widely believed to be exquisitely specific. As a reverse genetic tool, RNAi has set the standard in high-throughput functional genomics (Barstead, 2001; Tuschl, 2003). RNAi has also become an important tool in the identification and validation of drug targets in preclinical therapeutic development (Thompson, 2002; Appasani, 2003). Furthermore, RNAi-based human therapeutics are under development.

Initial empirical rules have been established by the Tuschl lab for the design of siRNAs. However, large variation in the potency of siRNAs is commonly observed, and often only a small proportion of the tested siRNAs are effective. Increasingly, emerging experimental evidence suggests that secondary structure and accessibility of target RNA are important factors in determining the potency of siRNAs (Lee et al., 2002; Vickers et al., 2003; Bohula et al., 2003; Far and Sczakiel, 2003).

In this chapter, we first review empirical rules for the design of siRNAs. A novel method is described for improving siRNA design, through combination of empirical rules with prediction of target accessibility. We report a new RNA folding software package, named *Sfold*, that offers rational siRNA design tools based on this novel methodology. *Sfold* is available through Web servers at http://sfold.wadsworth.org/ and http://www.bioinfo.rpi.edu/applications/sfold/. The issue of specificity and its relationship to potency are discussed. Finally, we propose a computational strategy for maximizing both potency and specificity in high throughput siRNA applications.

Empirical siRNA design rules

Tuschl rules and expansions

The first set of empirical rules for siRNA design was compiled by Tuschl's group (http://www.rockefeller.edu/labheads/tuschl/sirna.html; Elbashir et al., 2001b). The Tuschl rules can be summarized as follows:

(1) siRNA duplexes should be composed of 21-nt sense and 21-nt antisense strands, paired so as to each have a 2-nt 3′ dTdT overhang;

(2) The targeted region is selected from a given cDNA sequence beginning 50 to 100 nt downstream of the start codon (3′ UTRs also have been successfully targeted);

(3) The target motif is selected in the following order of preferences: i) NAR(N17)YNN, where N is any nucleotide, R is purine (A or G) and Y is pyrimidine (C or U); ii) AA(N19)TT; iii) NA(N21);

(4) Nucleotides 1–19 of the sense siRNA strand correspond to positions 3–21 of the 23-nt target motif;

(5) The target sequence is selected to have around 50% GC content; and

(6) Selected siRNA sequences should be aligned against EST libraries to ensure that only one gene will be targeted.

The AA(N19)TT motif has a low frequency of occurrence. AA(N19) is a popular motif advocated by two reagent companies, Ambion and Qiagen. It has also been suggested to avoid more than three Gs or three Cs in a row, because polyG and polyC sequences can hyperstack to form agglomerates that may interfere in the siRNA silencing mechanism (http://www.qiagen.com/catalog/auto/cget.asp?p=RNAi_support). Many siRNAs with 60% GC content are effective, but siRNAs with ≥70% GC are not as effective. From the perspective of target accessibility, a possible explanation is that a high GC motif is more likely to occur in a highly structured region, because G•C pairing is energetically stronger than A-U pairing. Examples are provided below to illustrate the rule-based design.

Examples

Example 1. AA(N19) motif.

Targeted gene	Rabbit β-globin
	(GenBank Accession no. V00879; coding region nt 54–nt 497)
Target position and GC%	418–438; 52.4%
Target sequence	5′-AAUUCACUCCUCAGGUGCAGG-3′

| Sense siRNA | 5'-UUCACUCCUCAGGUGCAGGTT-3' |
| Antisense siRNA | 5'-CCUGCACCUGAGGAGUGAATT-3' |

Example 2. AA(N19)TT motif.

Targeted gene	Enhanced green fluorescent protein (GFP)
	(GenBank Accession no. U55762; nt 583–nt 1602)
Target position and GC%	571–593; 47.6%
Target sequence	5'-AAGAACGGCAUCAAGGUGAACUU-3'
Sense siRNA	5'-GAACGGCAUCAAGGUGAACTT-3'
Antisense siRNA	5'-GUUCACCUUGAUGCCGUUCTT-3'

Exceptions to the rules

The Tuschl design rules were largely based on studies with *Drosophila melanogaster* embryo lysate (Elbashir et al., 2001b). Exceptions to the rules have been reported. Active siRNAs with a dinucleotide leader other than AA or NA can be found in the literature. siRNAs without specific nucleotide overhangs can be highly efficient (Czauderna et al., 2003). It has been shown that the antisense strand of the siRNA duplex can be almost as effective as the siRNA duplex, and the antisense strand and the double-strand siRNA appear to share the RNAi pathway (Amarzguioui et al., 2003; Holen et al., 2003). Hybrid DNA:RNA molecules were reported to be more effective that RNA:RNA duplexes in both duration and degree of silencing (Lamberton and Christian, 2003).

Target secondary structure and accessibility

RNAs form stable secondary structures through Watson-Crick and wobble G-U base pairing. Structural elements in the secondary structure include both helices and single-stranded loop regions. The single-stranded regions are likely to be accessible for RNA-targeting nucleic acids through base pairing interactions. Target accessibility has long been established as an important factor for the potency of antisense oligonucleotides and *trans*-cleaving ribozymes. For siRNAs, the empirical rules do not take into account the target secondary structure and accessibility. This lack of consideration is perhaps because siRNAs are more potent than antisense oligonucleotides. However, based on the large variation in the potency of siRNAs synthesized against different sites on the same target mRNA, it has been speculated that the low activity of the majority of siRNAs may be the result of limited accessibility of the target sequence due to secondary structure of target mRNA (Holen et al., 2002). Recently, increasing experimental evidence based on a number of experimental approaches has emerged to support the importance of target secondary structure and accessibility. Lee and coworkers (2002) demonstrated that the potency of siRNAs is determined by the target accessibility, with accessibility assessed by an oligo library. Bohula and coworkers (2003), using an oligo array method, and Vickers and coworkers (2003) both reported that the target secondary structure has an important effect on the potency of siRNAs. More recently, based on computational

Probability Profile of Target RNA (from position 1 to 589)

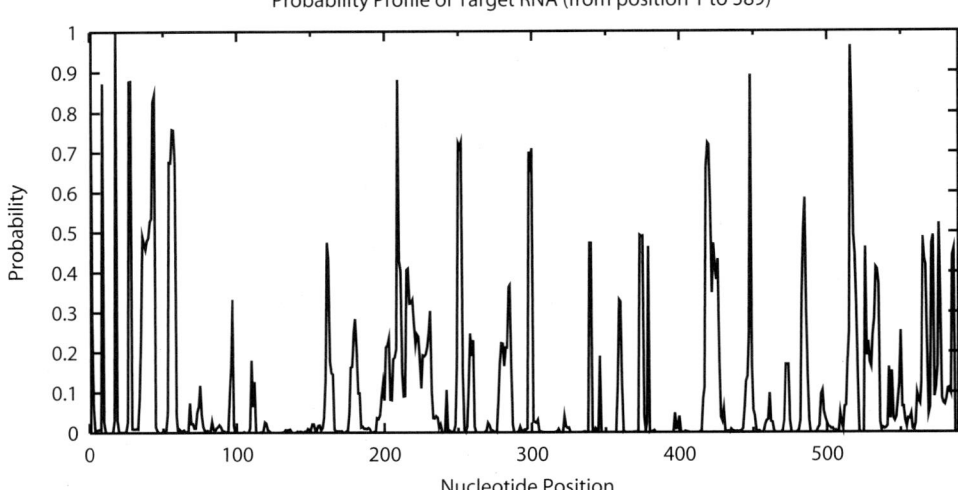

Figure 8.1. Probability Profile of Target RNA: The profile is shown for the region from position 1 to 589 nt that is targeted by antisense oligonucleotides. The vertical axis shows the probability for the profile with inhibition corresponding to probability multiplied by 100%.

predictions and validation by antisense oligonucleotides for accessibility, Far and Sczakiel (2003) demonstrated that target accessibility determined by local RNA structure is important for the potency of both antisense oligonucleotides and siRNAs.

Sfold software for rational siRNA design

It has been widely assessed that usually one out of four or five siRNAs designed by empirical rules is effective. For example, in a study that screened 356 siRNAs against 64 genes, the average number of siRNAs screened, in order to find one that induced >80% knockdown on the mRNA level, was 4.2 (Liszewski 2003). Because synthetic siRNA duplexes are expensive, methods to improve siRNA design are clearly needed.

We have developed a novel method for prediction of target accessibility, based on a probability profiling approach for predicting single-stranded regions in RNA secondary structure (Ding and Lawrence, 2001). It is believed that an mRNA can exist in a population of structures (Christoffersen et al., 1994). High probability regions in the profile reveal target sites that are *commonly accessible* for a large number of statistically representative structures for the target RNA (see Figure 8.1). Through assignment of statistical confidence in predictions, this novel approach bypasses the long-standing difficulty of selecting a single structure for accessibility evaluation.

We have employed the algorithm to develop a software package, Sfold, for the rational design of RNA-targeting nucleic acids. An application module, Sirna, is included in the software. Sirna aims to provide tools for improved siRNA design, by combining secondary structure and accessibility prediction with the established empirical design rules. *S*fold is available through Web servers at http://sfold.wadsworth.org/ and http://www.bioinfo.rpi.edu/applications/sfold/. Documentation, frequently asked questions (FAQs), and a summary of promising preliminary validation data are available on the Web sites. The basic steps in using *Sirna* for siRNA design are 1) selection of accessible sites from the probability profile of the target RNA and 2) for such selected accessible sites, choice of siRNAs that meet the requirements of the empirical rules, as well as the requirement of favorable binding energy between the antisense siRNA strand and its target sequence. Stronger binding is indicated by lower binding energy (stacking energies are negative valued). For example, an antisense siRNA with a binding energy of –20 kcal/mol is more effective than an antisense siRNA with a binding energy of –10 kcal/mol. The module provides both graphical files and text files to facilitate the design process. *Sirna* provides output files for both the popular AA(N19) target motif and other Tuschl motifs. Based on the empirical rules, both filtered output and complete output without filtering are available. Portions of two output files for rabbit β-globin are shown below.

Sample output

File *aan19–f.out* gives filtered output for siRNAs targeting AA(N19) motifs:

Line 1:	ID of AA(N19) motif	target position	target sequence
	GC content	antisense siRNA binding energy (kcal/mol)
Line 2:	sense siRNA (r′ → 3′)	antisense siRNA (5′ → 3′)

Filter criteria:

A. $40\% \leq GC\% \leq 60\%$;

B. Antisense siRNA binding energy ≤ -15 kcal/mol;

C. Exclusion of target sequence with at least one of AAAA, CCCC, GGGG, or UUUU.

4	43–63	AAACAGACAGAAUGGUGCAUC	42.9%	–16.9
		ACAGACAGAAUGGUGCAUCTT	GAUGCACCAUUCUGUCUGUTT	
5	44–64	AACAGACAGAAUGGUGCAUCU	42.9%	–16.0
		CAGACAGAAUGGUGCAUCUTT	AGAUGCACCAUUCUGUCUGTT	
6	53–73	AAUGGUGCAUCUGUCCAGUGA	47.6%	–16.2
		UGGUGCAUCUGUCCAGUGATT	UCACUGGACAGAUGCACCATT	
18	249–269	AAGAAGGUGCUGGCUGCCUUC	57.1%	–15.8
		GAAGGUGCUGGCUGCCUUCTT	GAAGGCAGCCAGCACCUUCTT	

File *sirna–f.out* gives filtered output of siRNAs targeting Tuschl motifs:

Column 1:	target position (starting – ending)
Column 2:	sense siRNA (5′ → 3′)
Column 3:	antisense siRNA (5′ → 3′)
Column 4:	GC content of target sequence
Column 5:	antisense siRNA binding energy (kcal/mol)
Column 6:	Pattern code for Tuschl target motifs

Filter criteria:

A. 40% $\leq$ GC% $\leq$ 60%;

B. Antisense siRNA binding energy $\leq$ −15 kcal/mol;

C. Exclusion of target sequence with at least one of AAAA, CCCC, GGGG, or UUUU.

43–63	ACAGACAGAAUGGUGCAUCTT	GAUGCACCAUUCUGUCUGUTT
		42.9% −16.9 BCD
44–64	CAGACAGAAUGGUGCAUCUTT	AGAUGCACCAUUCUGUCUGTT
		42.9% −16.0 B D
53–73	UGGUGCAUCUGUCCAGUGATT	UCACUGGACAGAUGCACCATT
		47.6% −16.2 B
195–215	CCUGUCCUCUGCAAAUGCUTT	AGCAUUUGCAGAGGACAGGTT
		50.0% −16.7 B D
243–263	UGGCAAGAAGGUGCUGGCUTT	AGCCAGCACCUUCUUGCCATT
		55.0% −16.8 B D

Statistical modeling of experimental data

Computational modeling of the RNAi pathway is not possible, because the molecular mechanism for gene silencing is as yet only partially understood. However, experimental data from RNAi experiments have been emerging. The melting temperature T_m of the siRNA duplex can be calculated with RNA thermodynamic parameters (Xia et al., 1998). T_m, the antisense siRNA binding energy calculated by *S*fold, secondary structural features at target site, GC content, and other primary sequence composition features can be included as variables for building statistical models, such as logistic regression for predicting the potency of siRNAs. In one study, T_m was not associated with the potency of siRNAs (Hohjoh, 2002). This suggests that thermodynamics of the siRNA duplex is not an important factor for potency. The lack of correlation is perhaps because duplex unwinding is a ATP-dependent process performed by an unknown helicase (Elbashir et al., 2001a; Nykänen et al., 2001). On the other hand, it has been speculated that low GC content at the 5′ end of the duplex facilitates unwinding (G. Hütvagner, personal communication, 2003). Should experimental evidence emerge to support this speculation, inclusion of a rule in modeling and design could improve target prediction. Clearly, there is considerable room for improvement in siRNA design. We expect that statistical modeling of experimental findings will provide an efficient means to assess the contribution to silencing potency by each of these factors.

Specificity of siRNAs

Specificity of siRNAs is important for interpretation of experimental results and for the reliability of newly discovered phenotypes from large-scale RNAi screening. For *Drosophila melanogaster* embryo lysate, a single base mismatch in the center of the siRNA duplex prevented target RNA cleavage (Elbashir et al., 2001b). The exquisite specificity is also supported by two studies using gene expression profiling (Chi et al., 2003; Semizarov et al., 2003). However, there is evidence that challenges the requirement of perfect identity between siRNA and its target mRNA. Holen and coworkers (2002) demonstrated that one or two central mutations in the siRNA targeting position 167 in human tissue factor (TF) did not abolish the siRNA's ability to reduce mRNA levels. Also, for human TF it was found that siRNAs generally tolerated mutations at the 5' end of the target, whereas they exhibited low tolerance for mutations at the 3' end (Amarzgouioui et al. 2003). More strikingly, an expression profiling study revealed off-target effects on non-targeted genes that possessed as few as 11 contiguous nucleotides of identity to the siRNA (Jackson et al. 2003). A separate concern about specificity is that siRNAs can also function as microRNAs (miRNAs; Doench et al., 2003; Zeng Yi, and Cullen, 2003). These authors report reduction in protein levels even through the use of siRNAs having only partial complementarity to the 3'UTRs of target mRNAs. miRNAs can trigger a translational repression mechanism, which is distinct from the catalytic mechanism for RNAi and is not yet well understood at the molecular level. However, this does not explain the off-target effects reported by Jackson and coworkers (2003), because translational repression does not trigger target mRNA cleavage, and it is thus unlikely to alter array profiling. For plants, a majority of predicted targets for miRNAs are transcription factors (Rhoades et al., 2002). Regulation of transcription factors can alter the levels of mRNAs. Thus, the non-specific effects observed by Jackson and coworkers (2003) may result from secondary effects of siRNAs acting as miRNAs that induce regulation of transcription factors. Recently, Zeng and coworkers (2003) reported that miRNAs can act as (antisense) siRNAs. Furthermore, in a plant embryo extract, a miRNA that lacks perfect complementarity to its target acts as a siRNA, by guiding an endonuclease for efficient target cleavage (Tang et al., 2003).

Despite the mixed reports on specificity, it is advisable to BLAST siRNAs against NCBI's EST or UniGene database, to ensure that a selected siRNA does not have strong homology to non-targeted genes. It is understood that the RNA-induced silencing complex (RISC) can take either strand of the siRNA duplex after duplex unwinding, because RISC does not know the identity of its substrate before the contained siRNA strand identifies the target, presumably through complementary base pairing (G. Hütvagner, personal communication, 2003). Thus, both the sense siRNA strand and the antisense strand may be available for binding to non-targeted genes, before they are digested by nucleases. It is therefore a good practice to perform alignment for both strands. The UniGene database contains transcript sequences that are either of known genes or are longest region of high-quality sequence data derived from ESTs. An Encyclopedia of DNA Elements (ENCODE) project was recently launched by the National Human Genome Research Institute (http://www.genome.gov/Pages/Research/ENCODE/). The goal of this project is to comprehensively identify functional elements in the human genome sequence. Such elements include alternative splicing sites, transcriptional start sites, translational start sites, polyadenylation sites, and protein-coding and non-protein-coding regions. The reliability of the BLAST search will improve with more accurate genome annotation. BLAST capability on the

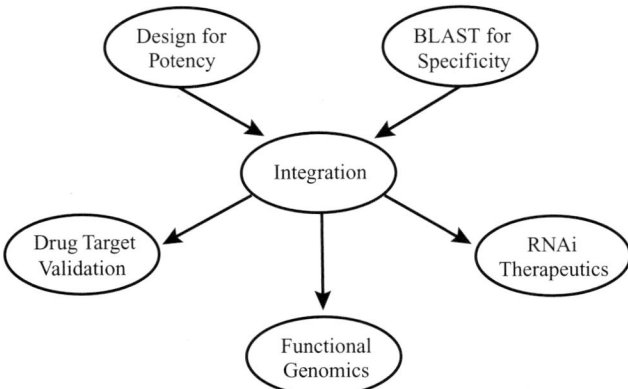

Figure 8.2. A potential high-throughput antisense framework for functional genomics and drug target validation: Integration of computational antisense design with greater specificity and higher potency approaches will provide better RNA therapeutics.

*S*fold server is currently under development and will be made available in the future.

Enhancing specificity through potency

Potency and specificity are related issues. In RNAi screening applications based on siRNA libraries, a design method with high likelihood for potent silencing can improve specificity, because off-target effects will be minimized when as few siRNAs as possible are needed to knock down their targets. Off-target effects are more likely to occur for higher concentration of siRNAs (Semizarov et al., 2003). The strong binding by a potent siRNA to its target may discourage cross-hybridization with genes of partial complementarity, because cross-hybridization is less energetically favorable. On the other hand, an ineffective siRNA can bind only weakly to its target, and is thus more readily available for cross-hybridization with non-targeted genes.

In the construction of a siRNA library for high-throughput RNAi screening, the issues of potency and specificity must be addressed together. For each target gene, numerous accessible target sites are usually revealed by *S*fold that satisfy the empirical criteria. Thus, how to select one or two siRNAs for each of hundreds or even thousands of genes is a design challenge. Proper computational optimization is necessary to maximize both potency and specificity. This can be achieved through integration of target selection based on accessibility prediction and empirical rules and sequence homology analysis.

Conclusions

RNAi has become a powerful reverse genetics tool that promises to revolutionize molecular biology. Currently, there is no golden rule for the design of siRNAs. Improved design may be achieved through a combination of empirical rules, target accessibility evaluation, and statistical modeling. *S*fold aims to provide such tools to the scientific community through Web service. Specificity is an important issue for RNAi applications, and more experimental work is needed to reconcile conflicting results. Specificity can be enhanced through improvement in potency of silencing. Integration of computational approaches for maximizing both potency and specificity will facilitate high-throughput applications to functional genomics, drug target validation, and development of RNAi-based human therapeutics (Figure 8.2).

REFERENCES

Amarzguioui, M., Holen, T., Babaie, E. and Prydz, H. (2003). Tolerance for mutations and chemical modifications in a siRNA. *Nucleic Acids Research*, **31**, 589–595.

Appasani, K. (2003). RNAi interference: A technology platform for target validation, drug discovery and therapeutic development. *Drug Discovery World*, Summer Issue, 61–68.

Barstead, R. (2001). Genome-wide RNAi. *Current Opinion in Chemical Biology*, **5**, 63–66.

Bohula, E. A., Salisbury, A. J., Sohail, M., Playford, M. P., Riedemann, J., Southern, E. M. and Macaulay, V. M. (2003). The efficacy of small interfering RNAs targeted to the type 1 Insulin-like growth factor receptor (IGF1R) is influenced by secondary structure in the IGF1R transcript. *Journal of Biological Chemistry*, **278**, 15991–15997.

Bridge, A., Pebernard, S., Duxraux, A., Nicaulaz, A. and Iggo, R. (2003). Induction of an interferon response by RNAi vectors in mammalian cells. *Nature Genetics*, **34**, 263–264.

Brummelkamp, T. R., Bernards, R. and Agami, R. (2002). A system for stable expression of short interfering RNAs in mammalian cells. *Science*, **296**, 550–553.

Chi, J. T., Chang, H. Y., Wang, N. N., Chang, D. S., Dunphy, N. and Brown, P. O. (2003). Genome wide view of gene silencing by small interfering RNAs. *Proceedings of the National Academy of Sciences USA*, **100**, 6343–6346.

Christoffersen, R. E., McSwiggen, J. A. and Konings, D. (1994). Application of computational technologies to ribozyme biotechnology products. *Journal of Molecular Structure (Theochem)*, **311**, 273–284.

Czauderna. F., Fechtner, M., Dames, S., Aygun, H., Klippel, A., Pronk, G. J., Giese, K. and Kaufmann, J. (2003). Structural variations and stabilising modifications of synthetic siRNAs in mammalian cells. *Nucleic Acids Research*, **31**, 2705–2716.

Ding, Y. and Lawrence, C. E. (2001). Statistical prediction of single-stranded regions in RNA secondary structure and application to predicting effective antisense target sites and beyond. *Nucleic Acids Research*, **29**, 1034–1046.

Doench, J. G., Petersen, C. P. and Sharp, P. A. (2003). siRNAs can function as miRNAs. *Genes & Development*, **17**, 438–442.

Elbashir, S. M., Harborth, J., Lendeckel, W., Yalcin, A., Weber, K. and Tuschl, T. (2001a). Duplexes of 21-nucleotide RNAs mediate RNA interference in cultured mammalian cells. *Nature*, **411**, 494–498.

Elbashir, S. M., Martinez J., Patkaniowska, A., Lendeckel, W. and Tuschl, T. (2001b). Functional anatomy of siRNAs for mediating efficient RNAi in *Drosophila melanogaster* embryo lysate. *European Molecular Biology Organization Journal*, **20**, 6877–6888.

Far, R. K. and Sczakiel, G. (2003). The activity of siRNA in mammalian cells is related to structural target accessibility: A comparison with antisense oligonucleotides. *Nucleic Acids Research*, **31**, 4417–4424.

Hohjoh, H. (2002). RNA interference (RNA(i)) induction with various types of syntheticoligonucleotide duplexes in cultured human cells. *Federation of European Biological Society Letters*, **521**, 195–199.

Holen, T., Amarzguioui, M., Babaie, E. and Prydz, H. (2003). Similar behavior of single-strand and double-strand siRNAs suggests they act through a common RNAi pathway. *Nucleic Acids Research*, **31**, 2401–2407.

Holen, T., Amarzguioui, M., Wiiger, M. T., Babaie, E. and Prydz, H. (2002). Positional effects of short interfering RNAs targeting the human coagulation trigger Tissue Factor. *Nucleic Acids Research*, **30**, 1757–1766.

Jackson, A. L., Bartz, S. R., Schelter, J., Kobayashi, S. V., Burchard, J., Mao, M., Li, B., Cavet, G. and Linsley, P. S. (2003). Expression profiling reveals off-target gene regulation by RNAi. *Nature Biotechnology*, **21**, 635-637.

Kumar, M. and Carmichael, G. G. (1998). Antisense RNA: Function and fate of duplex RNA in cells of higher eukaryotes. *Microbiological and Molecular Biological Reviews*, **62**, 1415–1434.

Lamberton, J. and Christian, A. (2003). Varying the nucleic acid composition of siRNA molecules dramatically varies the duration and degree of gene silencing. *Molecular Biotechnology*, **24**, 111–120.

Lee, N. S., Dohjima, T., Bauer, G., Li, H., Li, M. J., Ehsani, A., Salvaterra, P. and Rossi, J. (2002). Expression of small interfering RNAs targeted against HIV-1 rev transcripts in human cells. *Nature Biotechnology*, **20**, 500–505.

Liszewski, K. (2003). Progress in RNA interference. *Genetic Engineering News*, **23**, 1, 11–12, 14–15, 59.

Nykänen, A., Haley, B. and Zamore, P. D. (2001). ATP requirements and small interfering RNA structure in the RNA interference pathway. *Cell*, **107**, 309–321.

Paddison, P. J., Caudy, A. A., Bernstein, E., Hannon, G. J. and Conklin, D. S. (2002). Short hairpin RNAs (shRNAs) induce sequence-specific silencing in mammalian cells. *Genes & Development*, **16**, 948–958.

Rhoades, M. W., Reinhart, B. J., Lim, L. P., Burge, C. B., Bartel, B. and Bartel, D. P. (2002). Prediction of plant microRNA targets. *Cell*, **110**, 513–520.

Semizarov, D., Frost, L., Sarthy, A., Kroeger, P., Halbert, D. N. and Fesik, S.W. (2003). Specificity of short interfering RNA determined through gene expression signatures. *Proceedings of the National Academy of Sciences of the USA*, **100**, 6347–6352.

Sfold. WebURLs: http://sfold.wadsworth.org/; http://www.bioinfo.rpi.edu/applications/sfold/.

Tang, G., Reinhart, B. J., Bartel, D. P. and Zamore, P. D. (2003). A biochemical framework for RNA silencing in plants. *Genes & Development*, **17**, 49–63.

Thompson, J. D. (2002). Applications of antisense and siRNAs during preclinical drug development. *Drug Discovery Today*, **7**, 912–917.

Tuschl, T. (2003). Functional genomics: RNA sets the standard. *Nature*, **421**, 220–221.

Vickers,T. A., Koo, S., Bennett, C. F., Crooke, S. T., Dean, N. M. and Baker, B. (2003). Efficient reduction of target RNAs by small interfering RNA and RNase H-dependent antisense agents. A comparative analysis. *Journal of Biological Chemistry*, **278**, 7108–7118.

Xia, T., SantaLucia, J. Jr, Burkard, M. E., Kierzek, R., Schroeder, S. J., Jiao, X., Cox, C. and Turner, D. H. (1998). Thermodynamic parameters for an expanded nearest-neighbor model for formation of RNA duplexes with Watson-Crick base pairs. *Biochemistry*, **37**, 14719–14735.

Zeng, Y., Yi, R. and Cullen, B. R. (2003). MicroRNAs and small interfering RNAs can inhibit mRNA expression by similar mechanism. *Proceedings of the National Academy of Sciences USA*, **100**, 9779–9784.

9 Enzymatic production of small interfering RNAs

Muhammad Sohail and Graeme Doran

Introduction

RNA interference is a powerful natural phenomenon of post-transcriptional gene silencing that has been found in several biological systems. Small interfering RNAs (siRNAs) are important reagents in the RNA interference pathways that determine gene-specificity of the pathway. RNA interference can be induced in a system by the application of artificial siRNAs. Small interfering RNAs are vital intermediates in natural RNA interference that determine gene specificity of the phenomenon. RNA interference can be engineered for a gene for which this process does not occur in nature by applying artificial siRNAs, to study its function or for therapeutic purpose (Dykxhoorn et al., 2003). The original way of producing siRNAs was by chemical synthesis, which is expensive and usually there is no guarantee that reagents would produce desired gene silencing effects. A number of laboratories have developed alternative methods of producing siRNAs involving plasmid- or virus-mediated intra cellular expression and *in vitro* with the use of bacterial or viral enzymes. This chapter provides an overview of methods of producing siRNAs by enzymatic means.

Production of siRNAs by *in vitro* transcription

Donze and Picard (2002) reported a method based on the use of T7 RNA polymerase and short synthetic oligonucleotides as template (Milligan et al., 1987) in producing siRNAs and showed that siRNAs produced with this *in vitro* transcription method were capable of silencing GFP expression in human HeLa and that of protein kinase PKR in HEK293T cells.

The two siRNA strands are produced in separate *in vitro* transcription reactions and are annealed to produce siRNA duplex. A gene-specific oligonucleotide carrying a T7 promoter sequence is designed and is annealed with a 18mer oligonucleotide complementary to the T7 sequence. This partially double-stranded (ds) template is used in *in vitro* transcription reactions. Since the last G residue of the T7 promoter appears as the first ribonucleotide of *in vitro* transcripts and, therefore, all siRNA transcripts designed by this method start with a G residue. Thus the method requires that the sequence starts with a G residue and has a C residue at position 19 to allow annealing with the complementary strand (that also has to begin with a G). Therefore, to obtain dsRNA, only two deoxyribo-oligonucleotides corresponding to the sense and the antisense strands of the target site (with added T7 promoter sequence) are needed since the 18mer T7 promoter oligonucleotide is common to all targets. Since deoxyribo-oligonucleotides are considerably cheaper to synthesise than RNA, this strategy substantially reduces the cost of siRNA synthesis, even taking into account the reagents for *in vitro* transcription and purification steps, without detectably compromising the efficacy of the reagents.

Application of deoxyribozymes in siRNA production

The useful method of Donze and Picard (2002) is limited by inherent sequence requirements (5′ G-N_{17}-C rule) by the viral DNA-dependent RNA polymerase. Sohail et al. (2003) reported an approach to overcome this limitation. This method combines *in vitro* transcription with an additional deoxyribozyme digestion step of the siRNA transcripts that enables the use of any sequence in siRNA design. Deoxyribozymes are catalytic nucleic acids that consist of a catalytic domain flanked by two substrate-specific guide arms. The target RNA is cleaved at a specific di-nucleotide sequence (Breaker, 1997; Santoro and Joyce, 1997; Feldman and Sen, 2001). The requirement for this di-nucleotide sequence varies with different deoxyribozyme groups: Deoxyribozymes with the '10–23' catalytic domain cleave RNA at 5′AT, 5′AC, 5′GC and 5′GT while '8–17' type deoxyribozymes cleave at 5′AG di-nucleotide; the Bipartite II type deoxyribozymes cleave 5′AA. Thus in principle siRNA sequences starting and ending with any nucleotide can be chosen.

The two strands of siRNA are produced in separate reactions. For each strand a gene-specific synthetic duplex carrying a viral promoter/leader sequence is used as a template in *in vitro* transcription. A partial duplex (Donze and Picard 2002) that is double-stranded in the viral RNA polymerase promoter/leader sequence region can also be used as template. The leader sequence appears in the transcripts and so one arm of the deoxyribozyme (the 3′ binding arm) can be designed to bind with this sequence. Generally, two or three extra nucleotides are added downstream of the leader sequence in order to increase the length of the binding sequence and also to provide one nucleotide of the deoxyribozyme-specific di-nucleotide sequence at the catalytic site (the second nucleotide is provided by the first nucleotide of the gene-specific sequence). The second arm (5′ binding arm) is made complementary to the gene-specific sequence in the transcripts. Therefore, the sequences of the 3′ binding arm and the catalytic domain remain constant for one type of deoxyribozyme and the only variable is the sequence of the gene-specific 5′ binding arm.

siRNAs produced with this method were tested against type I insulin-like growth factor receptor (IGF1R) in MDA-MB231 breast cancer cells and their activity was also compared with chemically synthesised siRNAs with the same sequence. Western blot analysis using an anti-IGFR antibody showed that inhibition of IGF1R expression by siRNAs produced with this method was comparable to that by chemically synthesised siRNAs. The technical advantages provided by this method are not limited to the production of siRNAs, but extend to the production of single or double-stranded RNAs of defined length and sequence. This method is particularly useful in the production of long RNA sequences with defined ends that may be difficult to synthesise chemically or where the cost of synthesis is high.

siRNAs production by hydrolysis of long dsRNA with *Escherichia coli* RNase III

It is becoming increasingly obvious that not all siRNAs are effective in gene silencing and very often several may have to be screened to find a potent reagent (Holen et al., 2002; Vickers et al., 2003; Bohula et al., 2003). This difficulty could be partly overcome by the use of a library of siRNAs targeting several sites in the target. Yang et al. (2002) developed a method of generating a heterogeneous population of siRNAs that could potentially target multiple sites in the target. *E. coli* RNase III can efficiently digest dsRNA into 12–15 bp long fragments, carrying termini similar to those of siRNAs (Amarasinghe et al., 2001; Elbashir et al., 2001; Yang et al., 2002). However, these short dsRNA fragments are unable to trigger RNA interference. Yang et al. (2002) performed limited digestion of long dsRNA with RNase III to produce fragments that were 21–25 bp long and were efficient in producing RNA interference response.

DsRNA for digestion with RNase III is produced by *in vitro* transcription using PCR products (generally of the order of half a kilobase) that has added bacteriophage promoters at each end, representing the mRNA target. After partial digestion with RNase III, the products are purified by column chromatography or by gel electrophoresis. Purified products (called esiRNAs) are transfected into appropriate cell cultures and analysed for gene silencing effects. The obvious advantage of the RNase III-based method is that one does not need to worry about initial optimal target site selection procedures that may be needed when using single siRNAs.

Yang et al. (2002) tested esiRNAs in a number of different cells lines, including C33A human cervical carcinoma cells, HeLa human cervical epithiloid carcinoma cells, 293 transformed human embryonic kidney cells and *Drosophila* S2 cells. They targeted F-luc and R-luc genes (plus nearly 20 other genes) with esiRNAs and found these potent gene silencers and also found that in *Drosophila* S2 cells unprocessed dsRNAs were 2–10 fold less potent than esiRNAs. Calegari et al. (2002) later used esiRNAs to perform gene expression knockdown during the development of mammalian post-implantation embryos using the developing CNS system of day 10 mouse embryos as a model tissue (see Chapter 15 by Buchholz et al.). The esiRNAs were delivered by injecting into the lumen of the neural tube at specific regions and also by directed electroporation into neuro-epithelial cells that resulted in efficient gene silencing.

How specific esiRNAs are in targeting intended genes remains to be established. It was recently reported that single siRNAs could have noticeable non-specific effects on unintended targets as determined by global gene expression analysis (Jackson et al., 2003; Chi et al., 2003; Chang et al., Chapter 33). These non-specific effects are difficult to detect by standard molecular techniques used in most studies. Therefore, it is intriguing to discover what effect the population of esiRNAs will have on other genes using global gene expression analysis approaches. Since digestion of long dsRNAs is also carried out naturally by RNase III-related Dicer enzymes that produces a library of siRNAs related to the target gene, such global gene expression studies would therefore also shed light on target gene-specificity of natural RNA interference.

Acknowledgements

We would like to thank Ed Southern for useful comments on the manuscript. Research in the authors' laboratories is largely funded by the Medical Research Council, UK.

REFERENCES

Amarasinghe, A. K., Calin-Jageman, I., Harmouch, A., Sun, W. and Nicholson, A. W. (2001). Escherichia coli ribonuclease III: Affinity purification of hexahistidine-tagged enzyme and assays for substrate binding and cleavage. *Methods in Enzymology*, **342**, 143–158.

Bohula, E. A., Salisbury, A. J., Sohail, M., Playford, M. P., Riedemann, J., Southern, E. M. and Macaulay, V. M. (2003). The efficacy of small interfering RNAs targeted to the type 1 insulin-like growth factor receptor (IGF1R) is influenced by secondary structure in the IGF1R transcript. *Journal of Biological Chemistry*, **278**, 15991–15997.

Breaker, R. R. (1997). DNA enzymes. *Nature Biotechnology*, **15**, 427–431.

Calegari, F., Haubensak, W., Yang, D., Huttner, W. B. and Buchholz, F. (2002). Tissue-specific RNA interference in postimplantation mouse embryos with endoribonuclease-prepared short interfering RNA. *Proceedings of the National Academy of Sciences USA*, **99**, 14236–14240.

Chi, J. T, Chang, H. Y., Wang, N. N., Chang, D. S., Dunphy, N. and Brown, P. O. (2003). Genomewide view of gene silencing by small interfering RNAs. *Proceedings of the National Academy of Sciences USA*, **100**, 6343–6346.

Donze, O. and Picard, D. (2002). RNA interference in mammalian cells using siRNAs synthesized with T7 RNA polymerase. *Nucleic Acids Research*, **30**, e46.

Dykxhoorn, D. M., Novina, C. D. and Sharp, P. A. (2003). Killing the messenger: short RNAs that silence gene expression. *Nature Reviews Cell Biology*, **4**, 457–467.

Elbashir, S.M., Martinez, J., Patkaniowska, A., Lendeckel, W. and Tuschl, T. (2001). Functional anatomy of siRNAs for mediating efficient RNAi in *Drosophila melanogaster* embryo lysate. *European Molecular Biology Organization Journal*, **20**, 6877–6888.

Feldman, A. R. and Sen, D. (2001). A new and efficient DNA enzyme for the sequence-specific cleavage of RNA. *Journal of Molecular Biology*, **313**, 283–294.

Holen, T., Amarzguioui, M., Wiiger, M. T., Babaie, E. and Prydz, H. (2002). Positional effects of short interfering RNAs targeting the human coagulation trigger Tissue Factor. *Nucleic Acids Research*, **30**, 1757–1766.

Jackson, A. L., Bartz, S. R., Schelter, J., Kobayashi, S. V., Burchard, J., Mao, M., Li, B., Cavet, G. and Linsley, P. S. (2003). Expression profiling reveals off-target gene regulation by RNAi. *Nature Biotechnology*, **21**, 635–637.

Milligan, J. F., Groebe, D. R., Witherell, G. W. and Uhlenbeck, O. C. (1987). Oligoribonu-cleotide synthesis using T7 RNA polymerase and synthetic DNA templates. *Nucleic Acids Research*, **15**, 8783–8798.

Santoro, S. W. and Joyce, G. F. (1997). A general purpose RNA-cleaving DNA enzyme. *Proceedings of the National Academy of Sciences USA*, **94**, 4262–4266.

Sohail, M., Doran, G., Riedemann, J., Macaulay, V. and Southern, E. M. (2003). A simple and cost-effective method for producing small interfering RNAs with high efficacy. *Nucleic Acids Research*, **31**, e38.

Vickers, T. A., Koo, S., Bennett, C. F., Crooke, S. T., Dean, N. M. and Baker, B. F. (2003). Efficient reduction of target RNAs by small interfering RNA and RNase H-dependent antisense agents: A comparative analysis. *Journal of Biological Chemistry*, **278**, 7108–7118.

Yang, D., Buchholz, F., Huang, Z., Goga, A., Chen, C., Brodsky, F. M. and Bishop, M. J. (2002). Short RNA duplexes produced by hydrolysis with *Escherichia coli* RNase III mediate effective RNA interference in mammalian cells. *Proceedings of the National Academy of Sciences USA*, 99, **15**, 9942–9947.

Vector development and *in vivo, in vitro* and *in ovo* delivery methods

10 Six methods of inducing RNAi in mammalian cells

Kathy Latham, Vince Pallotta, Lance Ford, Mike Byrom, Mehdi Banan, Po-Tsan Ku, and David Brown

Using siRNAs to silence gene expression

The capacity to utilize a cell's naturally occurring RNA interference (RNAi) pathway to silence target gene expression has precipitated a new era in functional genomic research (McManus and Sharp, 2003; Dillin, 2003). RNAi can be induced in mammalian cells by small interfering RNAs (siRNAs) that target complementary mRNAs for degradation. Because of its specificity, reproducibility, and ease of use, RNAi is greatly accelerating the functional characterization of disease-relevant genes for drug discovery, target validation, and basic research efforts.

There are six basic methods for generating siRNAs. siRNAs can be prepared *in vitro* by chemical synthesis, *in vitro* transcription, or RNase III/Dicer digestion of long double-stranded RNAs (dsRNAs). The *in vitro* prepared siRNAs can then be delivered to mammalian cells by a variety of methods including lipofection and electroporation. Alternatively, siRNAs can be expressed in mammalian cells from DNA plasmids, viral vectors, or PCR products bearing an siRNA template adjacent to a compatible promoter. As with the *in vitro* prepared siRNAs, DNA plasmids and PCR products can be delivered to mammalian cells by a variety of methods – methods such as transfection agents and electroporation. Viral vectors, on the other hand, are first packaged into viral particles that are then used to infect cells.

Various methods for inducing RNAi in mammalian cells are described in the following sections. The relative advantages and disadvantages for each method are discussed. In addition, design criteria for both siRNAs and siRNA expression templates are described.

Designing siRNAs

Typically the first step in an RNAi experiment in a mammalian system is to design the siRNA sequence. In early studies, Tuschl and colleagues determined that the most effective siRNAs are 21 nt in length and have 2-nt 3′ overhangs. Moreover, their results suggested that the most effective siRNA overhangs have a UU or TT sequence (Elbashir et al., 2001). Since then, most of the reported siRNA sequences have been 21 nt in length, although there have been a few reports of effective 23-nt siRNAs (Shang and Brown, 2002). Almost without exception, the *in vitro* prepared siRNAs in use today include 2-nt 3′ overhangs. UU and TT overhangs are still the most common, although other overhang sequences are also being used with success.

The potency of siRNAs does not seem to correlate with their location of binding along the transcript. Initially, it was recommended that the siRNA target site be located 80–100 nt downstream of the start codon. Later, this was extended to include the 5′ or 3′ untranslated regions (UTRs). At present, many researchers design siRNAs such that they target $AA(N_{19})$ sequences and have a ~50% GC content (http://www.rockefeller.edu/labheads/tuschl/sirna.html). Upon selection, a BLAST search is done to ensure target specificity. This simple approach will lead to the identification of many potential siRNA sequences, about 1/4 of which lead to effective (>70%) gene silencing.

The activity of siRNAs may in part be due to the accessibility of targeted mRNA regions. Rossi and colleagues, for example, have been able to correlate effective siRNAs with accessible regions of target transcripts – determined by hybridizing DNA oligonucleotides to different regions of the target mRNA and assaying for RNase H dependent digestion [(RNase H digests DNA-RNA hybrids) (Lee et al., 2002)]. RNA folding programs, however, have been inadequate at predicting accessible RNA regions and thus have not been used for siRNA design.

To develop better siRNA design criteria, several research groups and companies have performed experiments to evaluate characteristics of effective siRNAs. The characteristics of potent and specific siRNAs have then been incorporated into intelligent siRNA design algorithms. Using this approach, Cenix BioScience GmbH, in partnership with Ambion Inc., has developed an algorithm that can predict potent siRNA sequences with a high likelihood (Figure 10.1). To develop their algorithm, Cenix tested multiple siRNAs targeting hundreds of different human genes that were expressed at detectable levels in several different cell lines. More than 900 siRNAs have been designed and tested to date. The effective and ineffective siRNAs were used to judge the impact of a number of physical characteristics on siRNA potency and specificity – including T_m, nucleotide content of the 3′ overhangs, siRNA length, nucleotide distribution over the length of the siRNA, and presence and location of mismatches. A key step in the algorithm is a stringent analysis of each siRNA sequence in comparison to all the known genes in an organism's genome to maximize target specificity.

Figure 10.1A shows data from 79 siRNAs designed using the algorithm developed by Cenix. In this experiment, 94% (74 of 79) of the siRNAs reduced target mRNA levels by at least 70% when assayed 48 hours after transfection. Recent studies have shown that minimizing the working concentration of siRNAs in an experiment maximizes the specificity of the RNAi effect (Semizarov et al., 2003). As a result, we tested the potency of several of the above siRNAs at different concentrations. Figure 10.1B shows the potency of six different siRNAs designed using the algorithm. Significantly, all were as effective at reducing target mRNA levels at 10 nM and 30 nM, as at 100 nM.

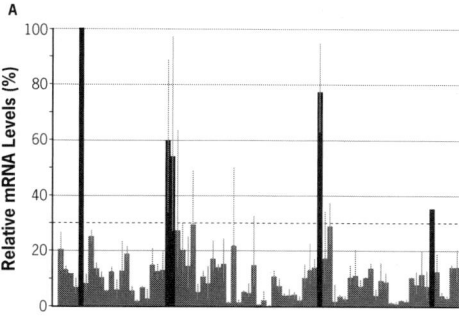 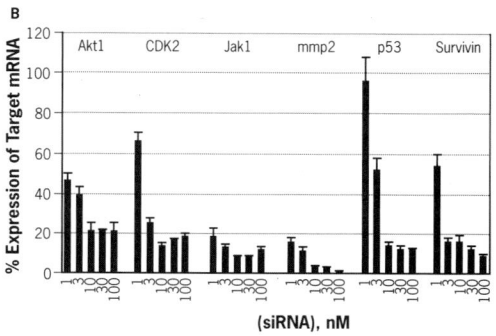

Figure 10.1. A. The Effectiveness of Rationally Designed siRNAs. siRNAs targeting 79 human genes were designed using the algorithm developed by Cenix. Individual siRNAs for each target were prepared and transfected at a 100 nM final concentration. Forty eight hours post-transfection, target gene expression was quantified by real-time RT-PCR. Relative reduction in mRNA expression was measured against cells transfected with a negative control siRNA. Sample size was normalized by measuring 18S rRNA in the various samples using a real-time PCR. Each bar represents the result from an individual siRNA. **B. Effectiveness of Rationally Designed siRNAs at Various Concentrations.** The indicated siRNAs were complexed with siPORT™ Lipid and the resulting complexes were added to HeLa cells in twenty-four well plates at the final concentration of siRNA shown. Forty-eight hours after transfection, RNA from the treated cells was recovered and reverse transcribed. Target cDNA levels were measured by real-time PCR. Expression of target genes in the transfected cells was compared to cells transfected with an equal concentration of *Silencer*™ Negative Control #1. Input cDNA in the different samples was normalized using real-time data for 18S rRNA. The bar graphs represent an average of three data points.

In general, improved siRNA design criteria are paramount for reducing the costs associated with most RNAi experiments. In addition, they are useful for the creation of genomewide siRNA libraries, which will have great utility for drug discovery, target validation, and functional genomics efforts. The utility of genomewide dsRNA libraries has already been demonstrated in screening experiments using *Drosophila* and *Caenorhabditis elegans* dsRNA libraries. For instance, RNAi libraries targeting more than 10,000 genes have been used in *C. elegans* to identify genes that regulate fat metabolism (Ashrafi et al., 2003), life expectancy (Lee et al., 2003) and mutation control (Pothof et al., 2003). A similar library for *Drosophila* has been used to identify the genes responsible for regulating the phosphorylation of Down Syndrome cell-adhesion molecule (Muda et al., 2002). In each of these screening applications, the keys to the experiments have been robust phenotypic assays and high quality RNAi libraries.

Chemically synthesized siRNA

Once an siRNA sequence is chosen, the easiest way to prepare it is to have it chemically synthesized and purified. Chemically synthesized siRNAs are available from several manufacturers. Most manufacturers offer a variety of synthesis scales and purification options. Many provide the siRNAs pre-annealed and ready to use. The turnaround time for obtaining chemically synthesized RNA is typically four days to two weeks depending on the synthesis scale and purification desired – an important point to consider when planning experiments. While synthesis chemistry differs among manufacturers, all the RNA end products are chemically identical. Because chemically synthesized siRNAs are available with HPLC or PAGE purification, researchers can obtain large amounts of high purity siRNA with this preparation method.

The relative ease of using chemically synthesized siRNAs for RNAi experiments comes at a price. Chemical synthesis is the most expensive siRNA preparation method. Because not all siRNAs are equally active, most scientists find it beneficial to test several siRNAs per gene to identify the most active one. To reduce the cost of RNAi research, many researchers screen siRNAs using a less expensive preparation method – such as *in vitro* transcription – and then have the most effective sequence(s) synthesized chemically. Another limitation of chemically synthesized siRNA is that it is not amenable to long-term studies.

In vitro transcribed siRNAs

siRNAs can be readily prepared in the lab by *in vitro* transcription (Donze and Picard, 2002). siRNAs produced by this method are considerably less expensive than their chemically synthesized counterparts. However, they function just as well as chemically synthesized siRNAs to reduce target gene expression. This makes *in vitro* transcribed siRNAs ideal for screening. In addition, *in vitro* transcription can be used to prepare siRNAs in less than a day, making it possible to conceive of an experiment, produce an siRNA, and transfect cells in the same week.

To prepare siRNAs by *in vitro* transcription, DNA templates are prepared with the desired siRNA sequence behind a phage promoter (e.g., T7). Templates encoding the sense and antisense strands of the desired siRNA are then transcribed in a high yield *in vitro* transcription reaction. During or after transcription, the sense and antisense strand synthesis reactions are combined to permit annealing of the two siRNA strands. The siRNA preparation is then treated with DNase to remove the DNA template. Prior to final purification, some protocols include an RNase digestion step to remove the unannealed siRNA strands. Using a modified protocol, *in vitro* transcription can also be used to generate hairpin RNAs rather than traditional siRNAs (see below for more information about hairpin RNAs).

Disadvantages of this method include the limited scale-up potential and the hands-on time, which requires more work from the researcher than simply purchasing the siRNAs. Like chemically synthesized siRNA, *in vitro* transcribed siRNA is not suitable for long-term studies.

siRNA cocktails generated by cleavage of long dsRNA

A major downside of most siRNA production methods is the need to screen siRNAs for identification of effective ones. Using siRNA cocktails overcomes this limitation (Calegari et al., 2002; Yang et al., 2002; Trotta et al., 2003). At least one, and often many, of the siRNA sequences included in the cocktail are typically effective at inducing RNAi. siRNA cocktails can be conveniently prepared

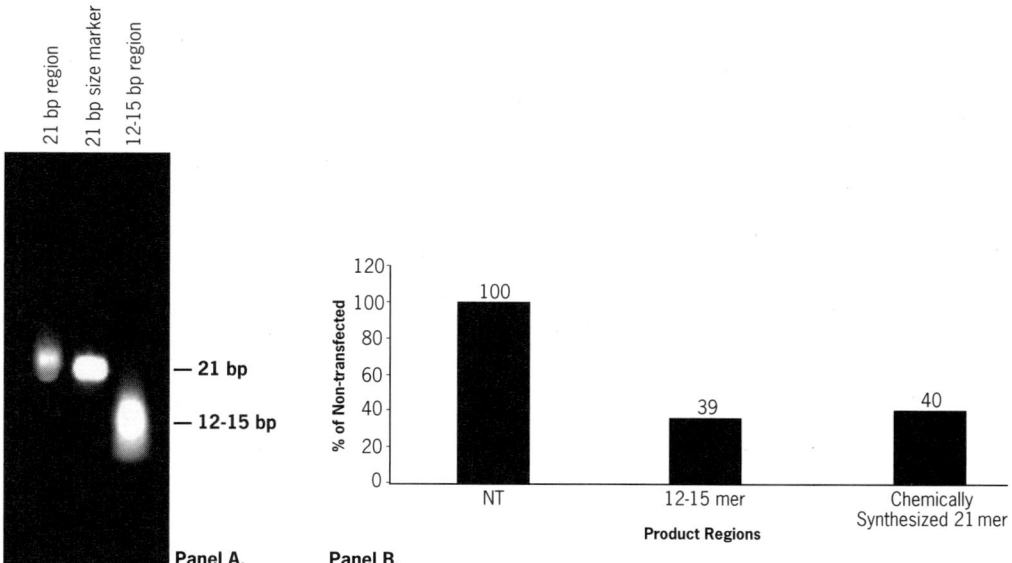

Panel A. Panel B.

Figure 10.2. 12–15 bp RNase III Digestion Products Elicit Silencing. A 200 bp GAPDH dsRNA (30 µg) was digested with RNase III (30 U) for 1 hour at RT. Digestion products were run on a 15% non-denaturing acrylamide gel and the 12–15 bp products were excised, eluted, and ethanol precipitated. A sample was run on a 15% non-denaturing acrylamide gel for visualization (3A). HeLa cells were transfected with 100 nM of the 12–15 bp RNase III generated GAPDH siRNAs or a 21 bp chemically synthesized GAPDH siRNA. GAPDH protein levels were monitored by immunofluorescence 48 hours after transfection and the resulting images were quantitated (3B).

by first generating long dsRNA by *in vitro* transcription using a template that typically encodes a 200–1000 nt region of the target mRNA. dsRNA preparation can be accomplished by transcription from either a single DNA template with opposing phage polymerase promoters (e.g., T7) or two separate DNA templates with promoters on opposite ends of the region to be transcribed. The resulting dsRNA is then digested to siRNAs with one of the double-strand specific RNases – RNase III or Dicer. After a purification step to remove any remaining undigested dsRNAs, the siRNA cocktail can be transfected into cells in a manner similar to an individual siRNA.

Both recombinant human Dicer and *Escherichia coli* RNase III can be used to prepare siRNA cocktails. Dicer is the enzyme responsible for cleavage of long dsRNAs to siRNA in the endogenous RNAi pathway. The siRNAs produced by Dicer are ~21 nt in length and contain 3′-dinucleotide overhangs with 5′-phosphate and 3′-hydroxyl termini (Bass, 2000; Sharp, 2001). *E. coli* RNase III cleavage products are ~15 nt in length with termini identical to those produced by Dicer (Miyagashi and Taira, 2002). Although RNase III produces shorter cleavage products than Dicer, these shorter products are active mediators of gene silencing when transfected into cells (Figure 10.2). Indeed, there appears to be little difference in the silencing efficacy between RNase III- and Dicer-generated siRNA cocktails. Cytotoxic effects are low for both methods. Because Dicer is more difficult to over-express and purify than RNase III, its price is usually much higher. In addition, purified Dicer is a relatively inefficient enzyme, requiring long digestion times (12–16 hour). In contrast, RNase III digestions take only about an hour. For these reasons, many researchers choose RNase III over Dicer to generate siRNA cocktails.

The siRNA cocktail approach is perhaps the easiest method for inducing RNAi because it eliminates the need to design and test individual sequences. A downside of the method, however, is the potential for nonspecific silencing effects, particularly for closely related genes.

Expression of siRNA *in vivo*

All of the methods described so far rely on the *in vitro* preparation of siRNAs. The use of siRNA expression vectors and PCR-derived siRNA expression cassettes, however, relies on the expression of siRNAs *in vivo*. Undoubtedly, the primary advantage of siRNA expression vectors is their amenability to long-term studies.

Plasmids expressing siRNAs have been used extensively to evaluate gene function in mammalian cells. They have also been used to test the feasibility of siRNA-based therapeutics and to affect gene expression in animals (Brummelkamp et al., 2002; Lee et al., 2002; Miyagashi and Taira, 2002; Paddison et al., 2002; Paul et al., 2002; Sui et al., 2002; Yu et al., 2002). Moreover, siRNA expression vectors containing antibiotic resistance markers (neo^R, puro^R, or hygro^R) have been used to create cell lines that stably express siRNAs (Brummelkamp et al., 2002; Matsuguchi et al., 2003), making it possible to evaluate the long-term effects of reducing target-gene expression. Transient selection of cells transfected with selectable marker-containing plasmids also permits for the enrichment of transfected cells. This, in turn, can help compensate for the low transfection efficiencies in difficult-to-transfect cells. Viral vectors have been used successfully to express siRNAs in mammalian cells (Devroe and Silver, 2002; Xia et al., 2002; Hemann et al., 2003; Miller et al., 2003; Qin et al., 2003; Robinson et al., 2003).

Plasmid constructs and viral vectors that permit stable integration of siRNA expression cassettes provide a means for creating cell lines and transgenic animals with reduced levels of a target gene(s). Inducible promoters for siRNA expression enable studies of essential genes as well as experiments requiring reduction in gene expression for a defined period of time. Finally, viral vectors facilitate delivery of an siRNA expression construct into cell types refractory to transfection, permitting RNAi studies in primary cells and animals.

Designing shRNA sequences for siRNA expression vectors and cassettes

Most siRNA expression vectors and cassettes described to date have been engineered to express short hairpin RNAs (shRNAs), which are then processed *in vivo* into siRNAs. Initially, shRNAs were expressed from pol III expression units. RNA Pol III is primarily involved in expressing small RNAs such as small nuclear RNAs. Most shRNA expression plasmids and PCR-derived shRNA expression cassettes described to date have featured the RNA Pol III promoters, U6 and H1. This is primarily because these RNA Pol III transcription units have defined transcription initiation and termination sites and their transcripts lack poly(A) tails. The RNA Pol III termination site consists of a stretch of 5–6 thymidines (Ts) and results in the addition of 1–4 uridine residues prior to transcript termination. Pol III termination sites are therefore ideal for creating the U overhangs that characterize the desired ends of many siRNAs.

In the initial reports, the design criteria of shRNA templates varied. Most of the template designs featured two inverted repeats, which were separated by a short spacer sequence, and ended with a string of 5–6 thymidines for transcription termination. The shRNA templates, however, varied with respect to their target sequence, the length of the inverted repeats (stem), the order of the inverted repeats (sense-loop-antisense or antisense-loop-sense), the length and composition of the spacer sequence (loop), and the presence or absence of 5′-overhangs (Brummelkamp et al., 2002; Lee et al., 2002; Miyagishi and Taira, 2002; Paddison et al., 2002; Paul et al., 2002; Sui et al., 2002; Yu et al., 2002).

Order of the inverted repeat

An shRNA expression vector or cassette that features an RNA pol III promoter is usually constructed to encode the sense strand of the siRNA, followed by a short spacer, the antisense strand of the siRNA, and 5–6 T's as transcription terminator, in that order. One group of researchers has found that reversing the order of sense and antisense strands within the siRNA expression constructs does not reduce the effectiveness of the shRNA (Yu et al., 2002). In contrast, another group of researchers has found that a similar reversal of order (using a different siRNA expression cassette) caused a partial reduction in gene silencing (Paul et al., 2002). The reason for this difference is not clear. At present, it is still advisable to construct the siRNA expression cassette in the following order: sense strand, short spacer, antisense strand, and transcription terminator.

Length of the siRNA stem

There appears to be some degree of variation in the length of nucleotide sequence coding for the stem of shRNAs. Most research groups have used templates coding for a 19-nt stem (Brummelkamp et al., 2002; Miyagishi and Taira, 2002; Paul et al., 2002; Yu et al., 2002). In contrast, other research groups have used templates coding for shRNA stems ranging in length from 21 nt (Lee et al., 2002; Sui et al., 2002) to 29 nt (Paddison et al., 2002). shRNAs with these various stem lengths have all been found to function well in gene silencing studies.

Length and sequence of the loop linking sense and antisense strands of hairpin siRNA

Various research groups have reported successful gene silencing results using shRNAs with loop sizes ranging from 3 nt to 23 nt in length (Brummelkamp

Table 10.1.

Loop size (# of nucleotides)	Specific loop sequence	Reference
3	AUG	(Yu et al., 2002)
3	CCC	(Jacque et al., 2002)
4	UUCG	(Paul et al., 2002)
5	CCACC	(Jacque et al., 2002)
6	CTCGAG	(Yu et al., 2002)
6	AAGCUU	(Sui et al., 2002)
7	CCACACC	(Jacque et al., 2002)
9	UUCAAGAGA	(Brummelkamp et al., 2002)
9	UUUGUGUAG	(Lee et al., 2002)
23	Not reported	(Paddison et al., 2002)

et al., 2002; Paddison et al., 2002; Paul et al., 2002; Sui et al., 2002; Yu et al., 2002). Table 10.1 shows a summary of loop size and specific loop sequences used by various research groups:

siRNA expression from Pol II promoters

More recently, functional shRNAs have been successfully expressed from a pol II transcription unit consisting of a cytomegalovirus (CMV) promoter and minimal SV40 polyadenylation signal (Xia et al., 2002). Although the CMV/SV40 transcription unit results in addition of a short 30–50 bp poly(A) tail to the shRNA, the resulting shRNA can elicit potent gene silencing. It is hypothesized that the poly(A) tail is cleaved *in vivo* to create functional siRNAs. An adenoviral vector has been developed using the CMV/SV40 transcription unit and has been used to demonstrate allele specific silencing of disease genes in a model system (Miller et al., 2003).

The advent of siRNA expression vectors with Pol II promoters may solve one of the issues with pol III promoters – creation of stable cell lines that constitutively express siRNA. While it is possible to use vectors with pol III promoters to create stable cell lines, there are few published accounts of doing so (Brummelkamp et al., 2002; Matsuguchi et al., 2003). At this point, it is unclear whether pol III promoters are more readily switched off upon integration into the host genome, or if there is perhaps interference between the pol III promoter driving siRNA expression and the pol II promoter responsible for transcription of the antibiotic resistance gene. In our experiments, we have found that it is easier to produce cell lines that stably express shRNAs using RNA Pol II promoters than RNA Pol III promoters (Figure 10.3). Perhaps more exciting, however, is that use of Pol II promoters to express siRNAs may also lead to the development of tissue-specific siRNA expression constructs in the future.

Inducible promoters

To be able to stably silence genes that are involved in cell growth or survival, it is desirable to express siRNAs from an inducible promoter. Several groups have

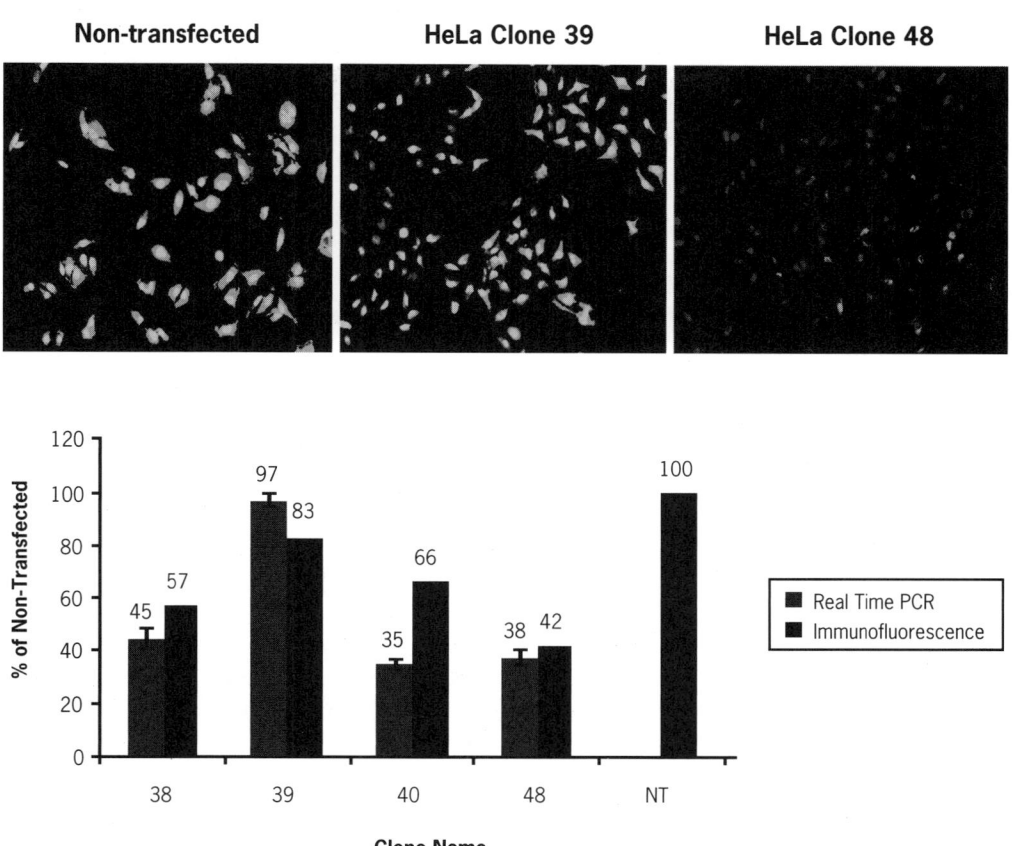

Figure 10.3. Long Term Silencing of GAPDH with CMV Puro Plasmid. HeLa cells were transfected with a CMV puro plasmid expressing GAPDH-specific siRNAs. The cells were cloned, and clonal populations were selected in 2.5 μg/ml puromycin. Three weeks after selection, GAPDH expression was analyzed by (A) RT-PCR or (B) immunofluorescence. Expression levels of several cell clones are shown. Green: GAPDH. Blue: DAPI stained nuclei. (See color section.)

already developed tetracycline-inducible siRNA expression systems based on the Tet operator/repressor and RNA pol III promoters (Matsukura et al., 2003; van de Wetering et al., 2003). In one system, an inducible H1 promoter was created by replacing 19 bp of the promoter close to transcription start site (positions −23 to −5) with that of the Tet operator region (van de Wetering et al., 2003). When introduced into a cell line that stably expresses the Tet repressor, binding of the Tet repressor to the Tet operator prevented transcription of the shRNA. shRNA expression was induced upon addition of the tetracycline derivative, doxycycline – a molecule that binds to the Tet repressor, changes its conformation, and prevents it from binding to the Tet operator region.

PCR-derived siRNA expression cassettes

For testing expression of siRNAs from RNA pol III promoters, siRNA expression cassettes (SECs) provide the simplest and fastest method. SECs are PCR-derived shRNA expression templates that can be introduced into cells directly (without being first cloned into a vector). First described by Castanotto and colleagues (Castanotto et al., 2002), SECs include an RNA pol III promoter upstream of a sequence encoding an shRNA followed by an RNA pol III termination site. SECs featuring other promoters will likely be developed in the near future. Preparing SECs is a simple, three-step process: (1) one or two oligonucleotides encoding the siRNA sequence are designed and obtained, (2) the oligonucleotide(s) are used as primers in one or more PCRs with an RNA Pol III promoter-containing DNA template, and (3) the resulting PCR product is purified. The SECs are then transfected into cells.

SECs circumvent some of the shortcomings of siRNA expression vectors: (1) In contrast to siRNA expression vectors, which require cloning and sequencing prior to use and thus take 1–2 weeks to prepare, SECs can be generated by PCR in less than a day. SEC production only requires 2–3 PCR reactions with target-specific oligonucleotides. (2) Because they are small (160–360 bp, depending on the length of the RNA polymerase promoter), SECs can be transfected into a broad range of mammalian cells with high efficiency.

SECs are a perfect complement to siRNA expression vectors. SECs can be used to screen for the optimal promoter and siRNA sequences – those giving the highest level of knockdown in a given cell line. Based on evaluations of several different RNA Pol III promoters, the activities of the different promoters are likely to vary from cell type to cell type. The localization of expressed RNA is also likely to vary with the cell type and RNA Pol III promoter (Ilves et al., 1996). To optimize siRNA expression, it may be beneficial to assess knockdown using SECs with at least two different RNA Pol III promoters. The most effective SEC can then be readily cloned into a plasmid or viral vector for subsequent studies.

SECs, however, have several disadvantages. First, they are not as easily transfected into cells as siRNAs. As new transfection agents and protocols are developed, this difference in transfection efficiency is expected to decrease. In addition, SECs are not amenable to scale-up without being first cloned into plasmids.

siRNA Expression Plasmids

There are now several commercially available siRNA expression plasmids. Some also include antibiotic resistance genes (neo^R, $puro^R$, or $hygro^R$) to permit selection of cells that have taken up the construct. To produce shRNA expression plasmids, two oligodeoxynucleotides encoding the desired shRNA sequence are ordered, annealed, and cloned into the vector downstream of the shRNA promoter. Alternatively, SECs (which already contain a pol III promoter) are cloned into neo^R, $puro^R$, or $hygro^R$ plasmids that do not contain the shRNA promoter. Because cloning is involved, the procedure takes several days, and sequencing the region of the insert is required. However, this inconvenience is balanced by the ability to produce large quantities of vector once the vector (or the incorporated SEC) is shown to work well in gene silencing experiments.

Viral vectors

A disadvantage of using plasmids and PCR cassettes for siRNA expression is that they are difficult to introduce into certain cell types (e.g. primary cells) and into animals. Viral vectors have been created to overcome the hurdles associated with siRNA delivery. To date, both adenoviral (Xia et al., 2002; Miller et al., 2003) and retroviral siRNA-expression vectors (Devroe and Silver, 2002; Hemann et al., 2003; Qin et al., 2003; Robinson et al., 2003) have been developed. In these viral vectors, siRNA expression has been driven by either a Pol III or a CMV transcription unit.

Adenoviral vectors are capable of infecting dividing cells, non-dividing cells, and tissues via a receptor-mediated infection – recognizing the widely expressed coxsackievirus and adenovirus receptor (CAR) as well as the integrin $\alpha v \beta 1$ molecule (Kay et al., 2001). Upon infection, these viruses can provide transient expression of transduced gene(s). Adenoviral vectors have been successfully used to transiently express siRNAs in both cells and tissues (e.g. mouse brain and liver).

Retroviral vectors, including ones derived from the murine stem cell virus (MSCV) (Devroe and Silver, 2002; Hemann et al., 2003) and the lentiviral sub-class (Qin et al., 2003; Robinson et al., 2003) have been developed to express shRNAs in cells. Retroviruses, like adenoviruses, can infect a range of dividing and non-dividing cells – a range that depends on glycoproteins expressed on the viral envelope. Unlike adenoviral vectors, however, retroviral vectors can easily integrate into the chromatin of target cells. As a result, retroviral vectors are especially suitable for creation of stable cell lines. So far, lentiviral vectors have been used to stably express siRNAs in primary T cells, hematopoietic stem cells, and mouse embryonic stem (ES) cells (Qin et al., 2003; Robinson et al., 2003). More-over, siRNA-expressing hematopoietic stem cells have been used to reconstitute immuno-compromised animals (Robinson et al., 2003). A similar approach may be used for siRNA-mediated therapeutics in humans.

Viral vectors can potentially be used to deliver siRNAs in a tissue-specific manner. This would depend on 1) expression of cell-specific recognition elements on the viral surface or 2) driving siRNA expression by tissue-specific promoters. To date, there has been limited success in using viral vectors for targeted gene-delivery (Vigna and Naldini, 2000; Kay et al., 2001). Whether viral vectors will successfully be used to deliver and express siRNAs in a tissue-specific manner remains to be determined.

Creating 'knockdown' animals

Recently, shRNA expression constructs with Pol III transcription units have been used to create transgenic animals. Generating these so-called 'knockdown' animals, which have some residual target gene expression, has a number of advantages over creating 'knockouts.' First, the process of generating transgenic animals is less laborious than that of gene targeting. Furthermore, one can use this system to correlate gene knockdown levels with the severity of loss-of-function phenotypes since different siRNAs can be used to achieve different levels of gene silencing. For example, this approach was effectively used to generate *Rasa-1* "knockdown" embryos with a range of phenotypic defects – the severity of which correlated with the *Rasa-1* knockdown levels [(i.e. those with low levels of *Rasa-1* silencing resembled wild-type animals, while those with high levels of gene silencing appeared like *Rasa-1* knockout animals) (Kunath et al., 2003.)] With high levels of gene silencing, it may be possible to obtain 'knockdown' mice with phenotypes resembling that of their 'knockout' counterparts.

Choosing amongst the alternatives

The choice of the siRNA production method will depend on a number of factors – such as the amenability of the method to long-term studies or to siRNA labeling. Nevertheless, it should be noted that many of these six methods are complementary and that collectively, they have greatly expanded the scope of RNAi experiments in mammalian cells. Using these methods, for example, it is now possible to transiently or stably express siRNAs to knockdown target gene expression – or to silence target genes in primary cells by producing siRNAs through viral vectors. Significantly, it may become possible to express siRNAs from tissue-specific pol II promoters in the near future – a feat that would expand the utility of siRNAs in transgenic studies.

REFERENCES

Ashrafi, K., Chang, F. Y., Watts, J. L., Fraser, A. G., Kamath, R. S., Ahringer, J. and Ruvkun, G. (2003). Genome-wide RNAi analysis of *Caenorhabditis elegans* fat regulatory genes. *Nature*, **421**, 268–272.

Bass, B. L. (2000). Double-stranded RNA as a template for gene silencing. *Cell*, **101**, 235–238.

Brummelkamp, T. R., Bernards, R. and Agami, R. (2002). A system for stable expression of short interfering RNAs in mammalian cells. *Science*, **296**, 550–553.

Calegari, F., Haubensak, W., Yang, D., Huttner, W. B. and Buchholz, F. (2002). Tissue-specific RNA interference in postimplantation mouse embryos with endoribonuclease-prepared short interfering RNA. *Proceedings of the National Academy of Sciences USA*, **99**, 14236–14240.

Castanotto, D., Li, H. and Rossi, J. J. (2002). Functional siRNA expression from transfected PCR products. *RNA*, **8**, 1454–1460.

Devroe, E. and Silver, P.A. (2002). Retrovirus-delivered siRNA. *BMC Biotechnology*, **2(1)**, 15.

Dillin, A. (2003). The specifics of small interfering RNA specificity. *Proceedings of the National Academy of Sciences USA*, **100**, 6289–6291.

Donzé, O. and Picard, D. (2002). RNA interference in mammalian cells using siRNAs synthesized with T7 RNA polymerase. *Nucleic Acids Research*, **30(10)**, e46.

Elbashir, S. M., Martinez, J., Patkaniowska, A., Lendeckel, W. and Tuschl, T. (2001). Functional anatomy of siRNAs for mediating efficient RNAi in *Drosophila melanogaster* embryo lysate. *European Molecular Biology Organization Journal*, **20**, 6877–6888.

Ilves, H., Barske., C., Junker., U., Bohnlein, E. and Veres, G. (1996). Retroviral vectors designed for targeted expression of RNA polymerase III-driven transcripts: A comparative study. *Gene*, **171**, 203–208.

Hemann, M. T., Fridman, J. S., Zilfou, J. T., Hernando, E., Paddison, P. J., Cordon-Cardo, C., Hannon, G. J. and Lowe, S. W. (2003). An epi-allelic series of p53 hypomorphs created by stable RNAi produces distinct tumor phenotypes *in vivo*. *Nature Genetics*, **33**, 396–400.

Jacque, J.-M., Triques, K. and Stevenson, M. (2002). Modulation of HIV-1 replication by RNA interference. *Nature*, **418**, 435–438.

Kay, M. A., Glorioso, J. C. and Naldini, L. (2001). Viral vectors for gene therapy: The art of turning infectious agents into vehicles of therapeutics. *Nature Medicine*, **7(1)**, 33–40.

Kunath, T., Gish, G., Lickert, H., Jones, N., Pawson, T. and Rossant, J. (2003). Transgenic RNA interference in ES cell-derived embryos recapitulates a genetic null phenotype. *Nature Biotechnology*, **21**, 559–561.

Lee, N. S., Dohjima, T., Bauer, G., Li, H., Li, M. J., Ehsani, A., Salvaterra., P. and Rossi, J. (2002). Expression of small interfering RNAs targeted against HIV-1 rev transcripts in human cells. *Nature Biotechnology*, **19**,500–505.

Lee, S. S., Lee, R. Y., Fraser, A. G., Kamath, R. S., Ahringer, J. and Ruvkun, G. A. (2003). Systematic RNAi screen identifies a critical role for mitochondria in *C. elegans* longevity. *Nature Genetics*, **33**, 40–48.

Matsuguchi, T., Masuda, A., Sugimoto, K., Nagai, Y. and Yoshikai, Y. (2003). JNK-interacting protein 3 associates with Toll-like receptor 4 and is involved in LPS-mediated JNK activation. *European Molecular Biology Organization Journal*, **22**, 4455–4464.

Matsukura, S., Jones, P. A. and Takai, D. (2003) Establishment of conditional vectors for hairpin siRNA knockdowns. *Nucleic Acids Research*, **31(15)**, e77.

McManus, M. T. and Sharp, P. A. (2003). Gene silencing in mammals by small interfering RNAs. *Nature Medicine*, **3**, 737–747.

Miller, V. M., Xia, H., Marrs, G. L., Gouvion, C. M., Lee, G., Davidson, B. L. and Paulson, H. L. (2003). Allele-specific silencing of dominant disease genes. *Proceedings of the National Academy of Sciences USA*, **100**, 7195–7200.

Miyagishi, M. and Taira, K. (2002). U6-promoter-driven siRNAs with four uridine 3′ over-hangs efficiently suppress targeted gene expression in mammalian cells. *Nature Biotechnology*, **19**, 497–500.

Muda, M., Worby, C. A., Simonson-Leff, N., Clemens, J. C. and Dixon, J. E. (2002). Use of double-stranded RNA-mediated interference to determine the substrates of protein tyrosine kinases and phosphatases. *Biochemical Journal*, **366**, 73–77.

Paddison, P. J., Caudy, A. A., Bernstein, E., Hannon, G. J. and Conklin, D. S. (2002). Short hairpin RNAs (shRNAs) induce sequence-specific silencing in mammalian cells. *Genes & Development*, **16**, 948–958.

Paul, C. P., Good, P. D., Winer, I. and Engelke, D. R. (2002). Effective expression of small interfering RNA in human cells. *Nature Biotechnology*, **19**, 505–508.

Pothof, J., van Haaften, G., Thijssen, K., Kamath, R. S., Fraser, A. G., Ahringer, J., Plasterk, R. H. A. and Tijsterman, M. (2003). Identification of genes that protect the *C. elegans* genome against mutations by genome-wide RNAi. *Genes & Development*, **17**, 443–448.

Qin, X-F., An, D. S., Chen, I. S. Y. and Baltimore, D. (2003). Inhibiting HIV-1 infection in human T cells by lentiviral-mediated delivery of small interfering RNA against CCR5. *Proceedings of the National Academy of Sciences USA*, **100**, 183–188.

Robinson, D. A., Dillon, C. P., Kwiatkowski, A. V., Sievers, C., Yang, L., Kopinja, J., Zhang, M., McManus, M. T., Gertler, F.B., Scott, M.L. and Parijs, L.V. (2003). A lentivirus-based system to functionally silence genes in primary mammalian cells, stem cells and transgenic mice by RNA interference. *Nature Genetics*, **33**, 401–406.

Semizarov, D., Frost, L., Sarthy, A., Kroeger, P., Halbert, D. N. and Fesik, S. W. (2003). Specificity of short interfering RNA determined through gene expression signatures. *Proceedings of the National Academy of Sciences USA*, **100**, 6347–6352.

Shang, Y. and Brown, M. (2002). Molecular Determinants for the Tissue Specificity of SERMs. *Science* **295**, 2465–2468.

Sharp, P. A. (2001). RNA interference – 2001. *Genes & Development*, **15**, 485–490.

Sui, G., Soohoo, C., Affar, E.-B., Gay, F., Shi, Y., Forrester, W. C. and Shi, Y. (2002). A DNA vector-based RNAi technology to suppress gene expression in mammalian cells. *Proceedings of the National Academy of Sciences USA*, **99**, 5515–5520.

Trotta, R., Vignudelli, T., Candini, O., Intine, R. V., Pecorari, L., Guerzoni, C., Santilli, G., Byrom, M. W., Goldoni, S., Ford, L. P., Caligiuri, M. A., Maraia, R. J., Perrotti, D. and

Calabretta, B. (2003). BCR/ABL activates mdm2 mRNA translation via the La antigen. *Cancer Cell*, **3**, 145–160.

van de Wetering, M., Oving, I., Muncan, V., Fong, M. T. P., Brantjes, H., van Leenen, D., Holstege, F. C. P., Brummelkamp, T. R., Agami, R. and Clevers, H. (2003). Specific inhibition of gene expression using a stably integrated, inducible small-interfering-RNA vector. *European Molecular Biology Organization Reports*, **4**, 609–615.

Vigna, E. and Naldini, L. (2000). Lentiviral vectors: Excellent tools for experimental gene transfer and promising candidates for gene therapy. *The Journal of Gene Medicine*, **2**, 308–316.

Xia, H., Paulson, H. L. and Davidson, B. L. (2002). siRNA-mediated gene silencing *in vitro* and *in vivo*. *Nature Biotechnology*, **20**, 1006–1010.

Yang, D., Buchholz, F., Huang, Z., Goga, A., Chen, C-Y., Brodsky, F. M. and Bishop, M. J. (2002). Short RNA duplexes produced by hydrolysis with *Escherichia coli* RNase III mediate effective RNA interference in mammalian cells. *Proceedings of the National Academy of Sciences USA*, **99**, 9942–9947.

Yu, J.-Y., DeRuiter, S. L. and Turner, D. L. (2002). RNA interference by expression of short-interfering RNAs and hairpin RNAs in mammalian cells. *Proceedings of the National Academy of Sciences USA*, **99**, 6047–6052.

11 Viral delivery of shRNA

Ying Mao, Chris Mello, Laurence Lamarcq, Brad Scherer, Thomas Quinn, Patty Wong, and Andrew Farmer

A. Introduction

The completion of the human genome has made available the sequences of thousands of genes (Baltimore, 2001), allowing researchers to switch focus from identifying genes to understanding their function. In broad terms, gene function studies can be classified into two categories: those where the gene of interest is introduced into a system in which it is not expressed, and those in which the gene is disrupted or removed. While over-expression studies are fairly straightforward, methods for gene inactivation have been hampered in higher eukaryotes by the difficulty in manipulating their genetic material. Thus, although it is possible to generate mice lacking genes of interest by homologous recombination (Capecchi, 1989; van der Weyden et al., 2002), such studies remain technically challenging and expensive. Moreover, in some cases, deletion of a gene may be lethal, preventing its analysis (*e.g.*, Lui et al., 1996). Alternatively, the phenotype produced may differ from that expected in humans (Harlow, 1992; Lee et al., 1992). A simple method for effective genetic inactivation in somatic cells *in vitro* is greatly needed, but has remained elusive (Sedivy and Dutriaux, 1999). Not surprisingly, recent years have seen considerable interest in a novel method for inactivating gene function in somatic cells that exploits the phenomenon of RNA interference (RNAi), first described by Fire et al. (1998). In their seminal study, they showed that double-stranded (ds)RNA homologous to a gene of interest could inhibit its expression. The dsRNA is digested into 21–23 nucleotide small interfering RNAs (siRNAs). These bind a nuclease complex (RNA-induced silencing complex; RISC), which then targets the endogenous mRNA by base-pairing and cleaves it, so preventing expression of the encoded protein (Hammond et al., 2001; Sharp, 2001; Hütvagner and Zamore, 2002). Thus, the gene is inactivated at the mRNA level, rather than by deletion of the DNA encoding it. In contrast to traditional knock-out methods, RNAi is proving much simpler to induce in higher eukaryotes.

Initial experiments with siRNAs relied on chemical synthesis of RNA oligos that were transiently transfected into the cells (Elbashir et al., 2001). It has also been possible to generate siRNAs using T7-based *in vitro* transcription systems

(Donze and Picard, 2002). This is cheaper than oligo synthesis, but still relies upon transient transfection. A problem with such methods is that the effect is not permanent – lasting only a few days at most. Moreover, it may be difficult to deliver the siRNA to all the cells in a population, especially in the case of primary cells. Any potential effect of the siRNAs will then be masked against the background of untransfected cells.

In early 2002, several groups developed expression vectors that used Pol III promoters to drive sustained expression of shRNAs within cells (Brummelkamp et al., 2002a; Miyagishi and Taira, 2002; Paul et al., 2002;, Yu et al., 2002, Lee et al., 2002; Paddison et al., 2002). Commonly used Pol III promoters include the U6 promoter (Miyagishi and Taira, 2002); the H1 promoter (Brummelkamp et al., 2002a); and tRNA promoters (Kawasaki and Taira, 2003). In some of these vectors, one promoter is used to express a single transcript containing both sense and antisense strands, connected by a short linker sequence. (Brummelkamp et al., 2002a; Paddison et al., 2002; Paul et al., 2002; and Yu et al., 2002). The transcript generated, being self-complementary, folds into a stem-loop structure known as a short hairpin RNA (shRNA). The ends of the shRNAs are processed *in vivo*, converting them into ~21 base-pair siRNA-like molecules, which in turn initiate RNAi (Brummelkamp et al., 2002a). The development of these vectors represents a definite improvement over chemical and *in vitro* synthesis methods of inducing RNAi, since they permit stable expression of shRNAs in cells by virtue of the fact that they can integrate stably into the genome of the cell.

For RNAi to be truly useful as a research tool and, ultimately, as a therapy, it is critical that the siRNAs be delivered effectively to the bulk, if not all, of the target cells of interest. Moreover, this delivery must be effective in animals and, most preferably, stable. While chemical- and plasmid-based delivery methods can be effective in certain situations, they are unable to fully address the issue of delivery. In contrast, viruses are exquisitely efficient at delivering genetic material to cells – both *in vitro* and in whole animals. Moreover, they are capable of infecting both dividing and non-dividing cells. For this reason, considerable effort has been expended over the last decade in the development of viral vectors for gene delivery – both as research tools and for gene therapy. In particular, there has been extensive development of retroviral and adenoviral vectors, and these viral systems are now commonly used to express genes in cells. Here we describe the development of retroviral and adenoviral systems for the efficient delivery of shRNAs to cells.

B. Retroviral delivery

Retroviruses offer several advantages as delivery systems. First, they have relatively small and simple genomes, making it easy to clone DNA of interest into them. Virus production is also rapid and can be amplified. Second, because they are enveloped viruses, it is possible to change their target cell specificity (tropism) simply by changing the envelope. (This is called pseudo-typing.) Finally, they can integrate stably into the genome of the host cell, so providing long-term expression of any introduced genetic sequence.

To generate a retroviral shRNA expression vector, the complete human U6 promoter (see Genbank record: M14486.1) was amplified by PCR from a human genomic library (human Caucasian placental library: HL1067j; BD Biosciences Clontech). The PCR product was then cloned downstream of the viral packaging signal in the Moloney murine leukemia virus (MMLV)-based retroviral vector pQCXIX (BD Biosciences Clontech; Julius et al., 2000). This backbone was chosen for two reasons. First, pQCXIX uses a hybrid 5′LTR containing CMV enhancer elements in place of the U3 region. This hybrid promoter is more strongly expressed than the wild-type MMLV LTR in 293-based packaging lines, resulting in greater production of viral genomic transcripts and, thus, higher viral titers (Julius et al., 2000). Second, the vector carries a deletion in the U3 region of the 3′LTR. During reverse transcription of the wild-type virus, the U3 region of the 3′LTR is copied to the 5′LTR to regenerate the promoter sequences of the 5′LTR absent in the RNA form of the virus. Thus, the deletion in the 3′LTR leads to disruption of the 5′LTR in the integrated provirus, rendering both LTRs incapable of driving transcription. Such self-inactivating retroviral vectors have several advantages. First, self-inactivation minimizes the risk of generating replication competent retrovirus, since the 5′LTR is no longer capable of expressing the full-length viral transcript required for virus production (Yu et al., 1986). Second, it reduces the likelihood for aberrant expression from the LTRs of cellular coding sequences located adjacent to the integration site. This represents a significant development for gene therapy research, where a particular concern is preventing incidental activation of endogenous oncogenes (Yu et al., 1986; Hacein-Bey-Abina et al., 2003). Finally, potential transcriptional interference between the LTR and the internal promoter driving the transgene is prevented (Emerman and Temin, 1984). It has been suggested that expression of internal Pol III promoters is strongly reduced in vectors with wild-type LTRs (Ilves et al., 1996; Armentano et al., 1987). We therefore chose to use a self-inactivating vector to avoid any potential for interference. Several others who have developed retroviral vectors for RNAi have also used self-inactivating vectors (Barton and Medzhitov, 2002; Brummelkamp et al., 2002b; Tiscornia et al., 2003, Qin et al., 2003; Rubinson et al., 2003). Nonetheless, others have successfully demonstrated RNAi using non-self-inactivating viruses (Devroe and Silver, 2002; Hemann et al., 2003). A careful comparison between such systems remains to be done to determine what impact, if any, the activity of the LTRs have on shRNA expression.

To simplify the selection of virus-infected cells, a puromycin resistance marker, driven by the PGK (phosphoglycerate kinase) promoter, was amplified from pMSCVpuro (BD Biosciences Clontech) and cloned downstream of the U6 cassette in place of the CMV promoter and internal ribosome entry site (IRES) sequences of pQCXIX. By placing the resistance marker downstream of the U6 cassette, we aimed to lessen any potential interference from the PGK promoter on the expression of the U6 promoter. Puromycin resistance was chosen for its fast action, permitting the rapid selection of stable populations of virus-infected cells. The PGK promoter has been shown to be expressed well in a variety of cell types and may be less susceptible to down-regulation than some viral promoters (Correll

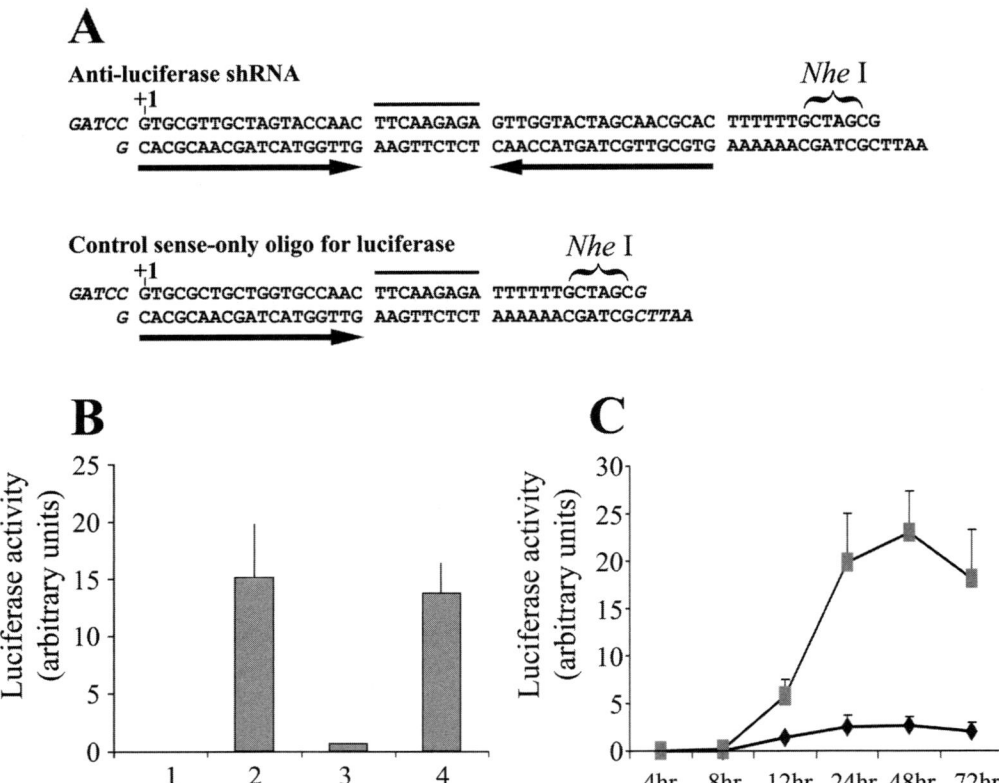

Figure 11.1. (A) Stable knockdown of luciferase expression by retroviral delivery of shRNA. Oligos used to generate pSIREN-RetroQ constructs for expression of an shRNA against luciferase and a sense-only control. Arrows mark the sense and antisense sequences against luciferase. The bar marks the loop sequence. +1 defines the transcription start point. The sequence of 6 'T' residues is used as a terminator and the NheI site is added to simplify screening for insertion of the oligos into the vector. (B) Knockdown of luciferase expression in stably infected 293 cells. Pools of clones stably infected with pSIREN-RetroQ expressing either an shRNA against luciferase or a negative control (sense strand only) were tested for their ability to repress expression of lucifierase from a transiently transfected plasmid. 1). parental HEK293 cells transiently. 2). parental HEK 293 cells transfected with a luciferase expression plasmid. 3). 293-anti-luc cells, stably expressing an shRNA against luciferase. 4). 293-control cells, stably expressing a sense only control sequence. (C) Stably infected 293 cells continue to show knockdown even after 2 months in culture. After 2 months in culture, the 293-anti-luc cells ([*squares*]). and 293-control cells ([*diamonds*]). were retested for ability to knockdown luciferase expression at various time points following transient transfection.

et al., 1994, Hawley et al., 1989). The final vector was named pSIREN-RetroQ. A similar construct has been described by others (Barton and Medzhitov, 2002).

To quantify the effectiveness of this construct in inducing RNAi, we cloned an shRNA sequence against firefly luciferase (Genbank record: M15077) downstream of the U6 promoter. As a control, a similar construct was generated except that in place of the shRNA we cloned a sense-only strand of luciferase. The oligo pairs used were designed with an on-line design tool (URL: http://bioinfo.clontech.com/rnaidesigner) and are shown in Figure 11.1A.

To generate virus, each construct was co-transfected together with an expression plasmid for the VSV-G envelope into the packaging line GP2 293

(BD Biosciences, Clontech) according to the instructions in the User Manual (URL: http://www.clontech.com/techinfo/manuals/PDF/PT3132–1.pdf). Virus–containing supernatant from the transfected cells was then used to infect HEK 293 cells, and stable pools of clones were selected in 1.1 μg/mL puromycin for 10 days. In this way, two virus-infected cell populations were isolated – 293-anti-luc (expressing the shRNA against luciferase) and 293-control (expressing a sense-only control). By studying a pool of transduced cells rather than individual clones, we aimed to mitigate effects of clonal variation on expression of the shRNAs. Retroviral systems are particularly suited to the study of cell populations, due to their high infectivity and the fact that the viral cassette integrates in its entirety. This is in contrast to plasmid integration, which occurs at low frequency and can result in disruption of the expression cassette.

It is perhaps surprising that virus is produced given that the viral RNA genome must contain the same sequence as the shRNA that it is designed to express. It should therefore be subject to targeted down-regulation. It is possible, since this target is itself a hairpin, that it is unavailable to the RISC complex. Alternatively, the expression level of the viral transcript may not be rate-limiting for virus production. Whatever the reason, it is clear from this and other reports that virus is generated in these systems (Barton and Medzhitov, 2002; Brummelkamp et al., 2002b; Devroe and Silver, 2002).

To test the effectiveness of the integrated shRNA, we then challenged each cell population by transient transfection with a plasmid expressing luciferase under control of the CMV promoter. As shown in Figure 11.1B, 293-anti-luc cells, expressing the shRNA against luciferase, showed over 90% reduction in luciferase activity compared to either the 293-control cells or the original parental 293 cells. Given that expression from the CMV promoter is high in 293 cells (Julius et al., 2000), the ability of the integrated shRNA cassette to knock down expression of the transiently expressed luciferase to such an extent suggests that expression levels of the shRNA produced from the U6 promoter in this context are more than sufficient even for highly expressed genes. Replication-incompetent retroviruses, as used here, typically generate one to a few integrants per cell, depending on the multiplicity of infection. Since the cells tested here are a population of virus-infected cells, the number of integrants per cell cannot be determined directly. It remains to be seen if a single integrant is sufficient for full knockdown of the target gene or whether increasing copy number will further enhance knockdown. In addition to super-infection, one simple way to increase copy number using retroviral systems would be to place the expression-cassette within the U3 region of the 3'LTR. It will then be duplicated along with the U3 region during viral replication and integration. We are currently testing the efficacy of such systems in comparison to the single expression-cassette described here.

We have heard reports, not to our knowledge published at this time, that integrated shRNA cassettes may be down-regulated over time in culture. Such a possibility for down-regulation is not unexpected, since it is commonly known that exogenous Pol II-based expression constructs can also be down-regulated with time in culture. To date, however, we have observed the 293-anti-luc population over a period of 2 months in culture and have seen no loss in knockdown activity (Figure 11.1C). It is possible that retroviral integrants are less susceptible to long-term down-regulation, due to their preference for integration into actively transcribed regions (Wu et al., 2003; Schroder et al., 2002). Indeed, others have used similar vectors to make transgenic animals showing stable knockdown of a target gene – arguing strongly that shRNA expression can be long lived (Rubinson et al., 2003; Tiscornia et al., 2003; Hemann et al., 2003).

Before turning to discuss adenoviral systems, one additional advantage of retroviral systems as expression vehicles for RNAi should be raised. That is their utility for the generation and screening of libraries. Several factors account for this, including: the high titers achievable, broad host range, simple construction, and high infectivity. All these factors make retroviral vectors an ideal choice for building shRNA libraries. These could enable whole genome screening for shRNAs that cause a loss of phenotype; offering a novel method for identifying critical regulators of physiological and pathological processes. These libraries would be an excellent complement to the use of retroviruses to generate gene-knockouts by insertional mutagenesis (Robertson et al., 1986).

C. Adenoviral Delivery

MMLV-based retroviral vectors, such as pSIREN-RetroQ described above, can be pseudo-typed to have very broad tropisms. Additionally, they are able to stably integrate into the genome of the target cell. However, they are limited to the infection of dividing cells. Thus, they are not suitable in all applications. For infection of many non-dividing cell types, adenoviral vectors have proven to be very effective. Unlike retroviruses, adenoviruses are non-enveloped DNA viruses with relatively large and complex genomes (>30kb), which are maintained episomally within the infected cells. This has, until recently, made standard cloning into adenoviral vectors impractical due to the lack of convenient restriction sites. Previously cloning into adenovirus was done by the rather inefficient and time-consuming process of homologous recombination in human cells. This method requires plaque purification of the virus to ensure the removal of contaminating non-recombinant virus (Graham and Prevec, 1991).

To address the need for simpler and faster methods for adenovirus construction, we have developed a ligation cloning kit based on a technology from Mizuguchi and Kay (1998). In this system, genes of interest are cloned by standard methods into a shuttle plasmid that bears two rare restriction sites (I-*Ceu*I and PI-*Sce*I) flanking the insertion site for the gene of interest. These rare sites are similarly engineered into the E1a region of an adenoviral backbone carried on a low-copy plasmid. Using these rare sites, it is possible to transfer inserts from the shuttle vector directly into the adenoviral backbone by standard ligation. This is not only simpler and faster than traditional methods of homologous recombination, but also has the advantage that all cloning is done *in vitro* in *E. coli*. Thus, a clonal virus is generated without the need to plaque purify. This alone represents considerable savings in time and effort. More recently, we have further simplified

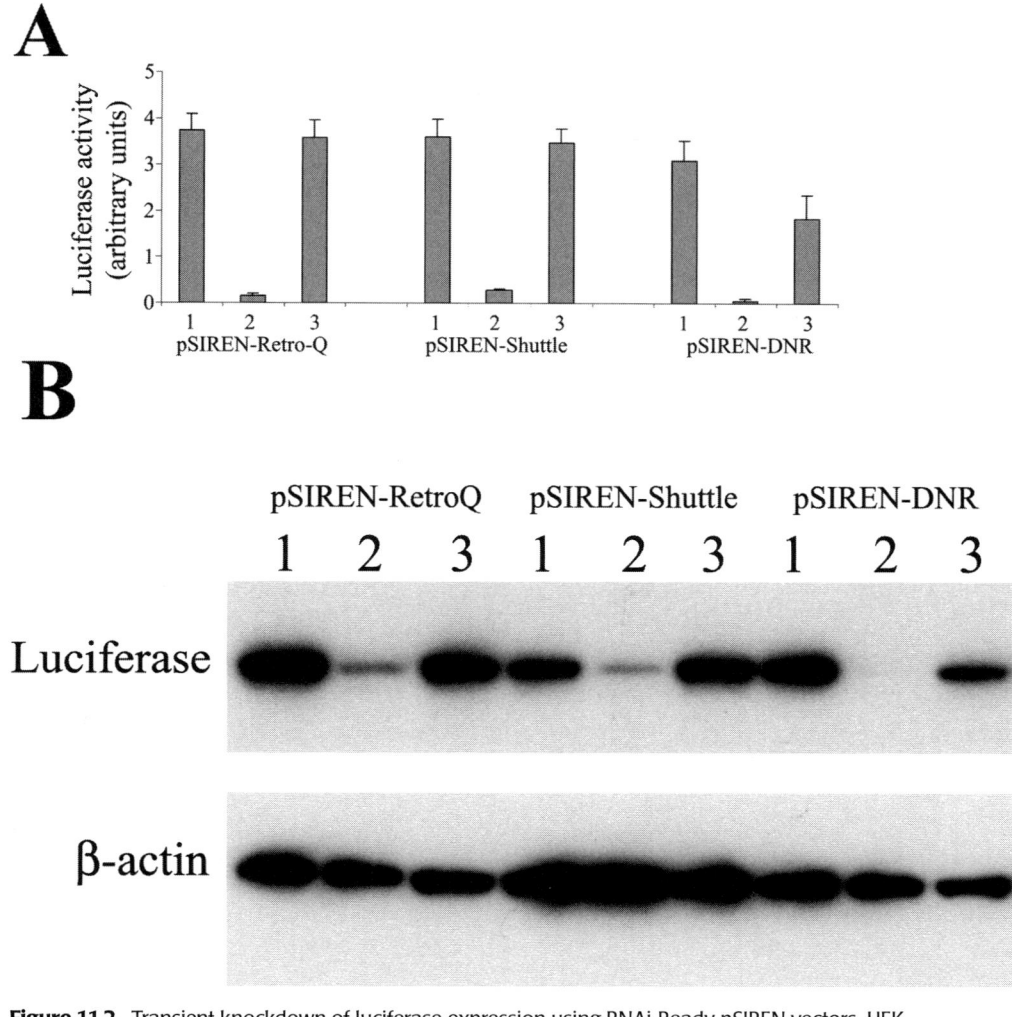

Figure 11.2. Transient knockdown of luciferase expression using RNAi-Ready pSIREN vectors. HEK 293 cells were transiently transfected with a luciferase expression vector either alone (Lane 1) or together with a pSIREN vector expressing either an shRNA against luciferase (Lane 2) or a sense-only control (Lane 3). Twenty-four hours post transfection, the cells were lysed and assayed for luciferase activity (Panel A). Additionally, lysates were western blotted and probed using an anti-luciferase antibody (Panel B). To control for loading, the same blot was additionally probed with an anti-beta-actin antibody.

this process by adapting the vectors to a Cre/*lox*P-based recombination system (BD Creator™ system). In this case, genes of interest are first cloned into a specialized recombination-adapted plasmid (donor vector). In this vector, the gene of interest is flanked by two recombination sites (*lox*P sites), recognized by the Cre enzyme (Abremski and Hoess, 1984). The gene of interest can then be efficiently transferred into a second (acceptor) vector by combining the two plasmids together with Cre recombinase *in vitro*. There is no need for either restriction digestion or ligation.

We have adapted both adenoviral systems for expression of shRNAs. To do this, we cloned the human U6 promoter cassette from pSIREN-RetroQ into the

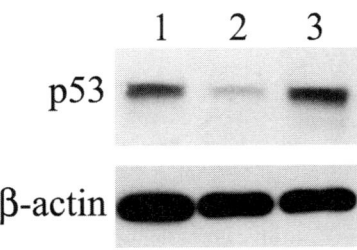

Figure 11.3. Knockdown of p53 expression in MCF7 cells. MCF7 cells were either mock infected (Lane 1) or infected with an adenovirus expressing either an shRNA against p53 (Lane 2) or a sense-only control (Lane 3). Forty-eight hours after infection, the cells were lysed and probed by western blot for p53 expression (upper panel). To control for loading, the lysates were also probed with an anti-beta-actin antibody (lower panel).

respective adenoviral plasmids. The resulting constructs are pSIREN-Shuttle (ligation-based system) and pSIREN-DNR (BD Creator™ system). Maps for these vectors and pSIREN-RetroQ can be found at the following URL: (http://www.clontech.com/techinfo/vectors/catrnai.shtml).

To confirm the functionality of the U6 promoter in these vectors, we cloned the shRNA against luciferase (Figure 11.1A), as well as the sense-only control, into both vectors. These were then tested in a transient co-transfection assay for ability to down-regulate luciferase. As shown in Figure 11.2, both vectors behaved equivalently to pSIREN-RetroQ; reducing luciferase expression over 90% when carrying an shRNA against luciferase. This result further demonstrates that these vectors, designed to facilite the production of recombinant viruses, can also be used effectively as simple plasmids in transient transfections.

As discussed above, one advantage of adenoviral delivery is the ability to efficiently infect non-dividing or hard-to-transfect cell types that may be difficult to transfect by standard methods. In such cases, action of the shRNA may be masked against the background of untransfected cells. The breast cancer cell line MCF7 (HTB-22: ATCC), which over-expresses p53, is typically hard to transfect (Agami and Bernards, 2000; Shen et al., 2003). It thus represents a good model system to test the effectiveness of adenoviral infection for delivery of shRNA. Although it has been possible to knockdown p53 expression in MCF7 with shRNAs expressed from plasmids (Brummelkamp et al., 2002a), electroporation was required to achieve efficient transfection. Indeed, others have reported that when lipid-based transfection protocols are used, knockdown of p53 expression in MCF7 could not be detected (Shen et al., 2003). To test the ability of our adenoviral system to knock down p53 in MCF7, we cloned an shRNA against p53 (Brummelkamp et al., 2002a) into the adenoviral backbone using our ligation-based cloning system. Virus was then generated by transfecting the linearized adenoviral DNA into 293 cells, as described in the User Manual (see URL: http://www.clontech.com/techinfo/manuals/PDF/PT3414–1.pdf). Virus expressing either the shRNA against p53 or a sense-only control was then used to infect MCF7 cells. As shown in Figure 11.3, cells infected with adenovirus expressing the shRNA against p53 showed significantly reduced expression of p53, compared to cells infected with an adenovirus expressing a sense-only control. Consistent with this result, others have shown knockdown of p53 expression in MCF7 cells infected with an adenovirus expressing an shRNA against p53 under control of the H1 promoter (Shen et al., 2003).

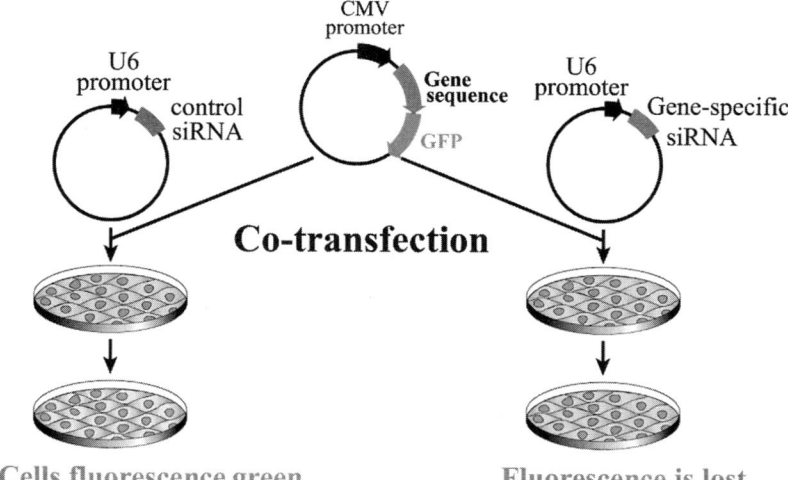

Figure 11.4. Method for identifying effective shRNA sequences. To screen for shRNAs that are effective against a gene of interest, the gene to be targeted is cloned into an expression vector as a translational fusion to a fluorescent protein. This construct is then co-transfected with test and control siRNA sequences against the gene. If the shRNA sequence is effective, then expression of the fusion protein will be reduced, resulting in a loss of fluorescence. (See color section.)

Other examples of the use of adenoviral vectors for RNAi, both *in vitro* and *in vivo*, have been presented by Davidson and colleagues (Xia et al., 2002; Miller et al., 2003). Interestingly, these authors used a modified CMV promoter and SV40 poly A sequence rather than a Pol III promoter to express shRNAs. It will be interesting to see if this expression system can be used effectively in other vectors, since the use of the Pol II promoters would simplify the development of inducible and tissue-specific expression systems for siRNA.

D. Rapid Screening for shRNAs

In addition to the question of delivery of shRNA to cells, we have focused on developing methods for rapid functional validation of shRNA sequences. Although Dharmacon recently published (Reynolds et al., 2004), extensive rules for siRNA design, information in the public domain has been limited. Thus, most investigators are using the fairly broad and simple rules described by Tuschl and collegues (see URL: http://www.rockefeller.edu/labheads/tuschl/sirna.html). While these rules are able to identify sequences that work as shRNAs, they are broad in scope and typically identify many potential sequences per gene. On average 1 in 3 these sequences works well (*i.e.*, gives greater than 70% knockdown). A simple and rapid screen to identify those shRNAs that function well is, therefore, highly desirable; especially if one wishes to develop shRNAs against many genes. The approach we have chosen is outlined in Figure 11.4. In this assay, shRNAs against a gene of interest are co-transfected with a reporter construct in which the gene of interest is cloned in-frame with a fluorescent protein. Hence, any shRNA that targets the mRNA for the gene will also target the mRNA of the fusion construct, resulting

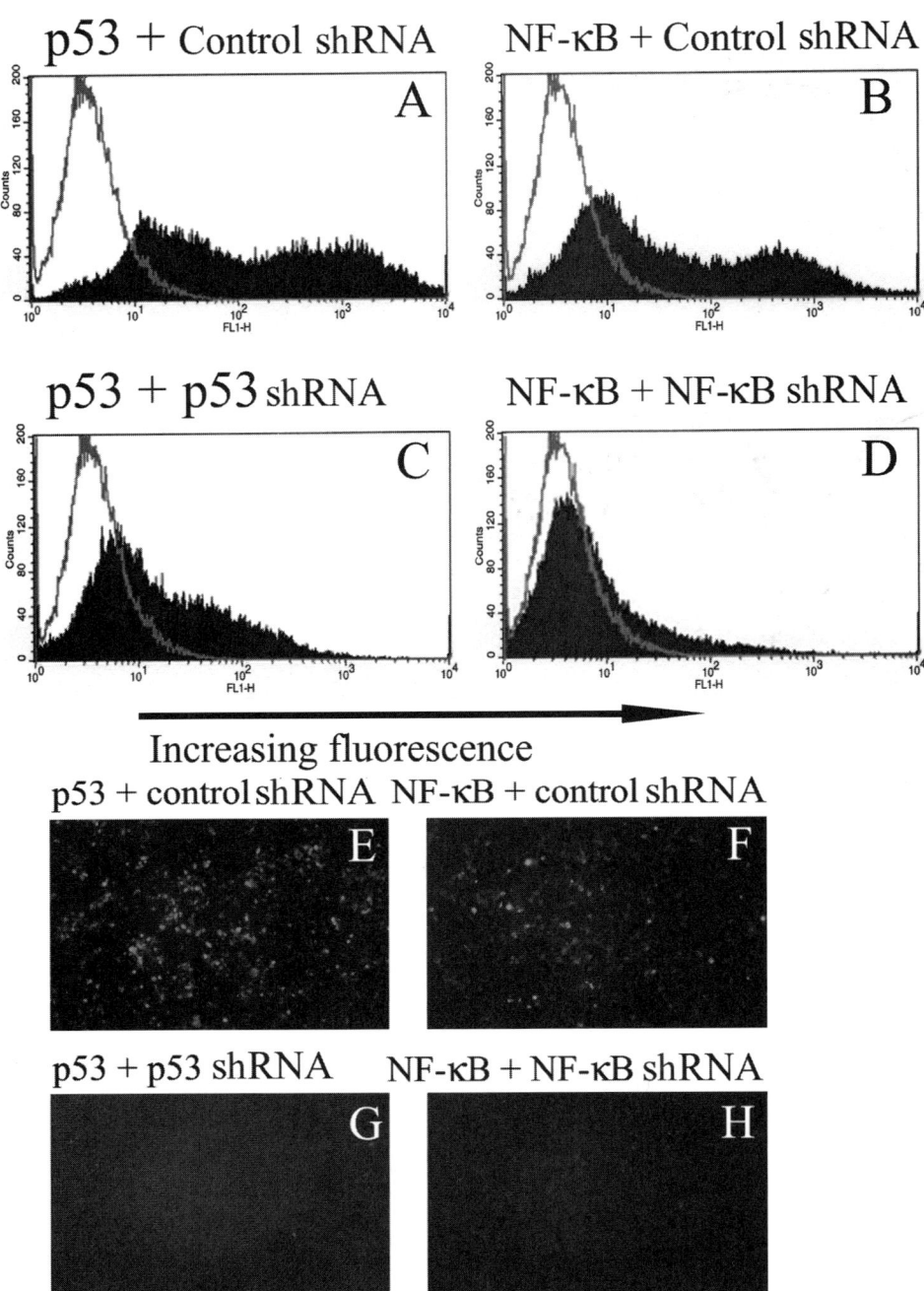

Figure 11.5. Screening for shRNAs against p53 and NF-κB using the fluorescent co-transfection assay. HEK 293 cells were co-transfected with a reporter plasmid expressing either p53 or NF-κB fused to EGFP at their C-terminus, and an shRNA expression-construct. After 48 hours, the co-transfected cells were analyzed for EGFP expression by fluorescence microscopy and flow cytometry. Fluorescence micrographs of cells cotransfected with either an shRNA (C, D), or a sense only control (A, B). Fluorescence profiles generated by flow cytometry. In each plot, the line represents the fluorescence profile of untransfected cells and the shaded region gives the profile of co-transfected cells. Panels G, H: cells co-transfected with an shRNA; Panels E, F: cells co-transfected with a sense only control (E, F). Experiments on p53: panels A, C, E, G. Experiments on NF-κB: panels B, D, F, H.

in down-regulation of the fluorescence signal. As proof of principle, we have used this assay to identify shRNAs against p53 and NF-κB (Figure 11.5). There are two distinct of advantages to this method. First, multiple genes can be screened all with the same readout, i.e. fluorescence. Second, the loss of fluorescence occurs irrespective of where in the gene the shRNA targets. This is in contrast to Q-PCR methods, where it is necessary to design PCR primers flanking each shRNA target site.

E. Summary

RNAi appears to be a very promising technology for simplifying gene inactivation in mammalian cells, greatly enhancing our ability to study gene function. Critical to the application of RNAi are the effective design of siRNA or shRNA sequences and their efficient delivery to the cell. We and others have shown that it is possible to use viral vectors for efficient delivery of shRNA expression-cassettes to cells (e.g., Barton and Medzhitov, 2002; Brummelkamp et al., 2002b; Xia et al., 2002; Shen et al., 2003).

We have also developed a simple screening method to rapidly identify functional shRNAs for multiple genes. We are currently testing this method on a large set of genes. At time of press, we have screed for shRNAs against over 70 genes using the method described here. Our goal is to use these tools to generate a collection of shRNAs against human genes that can be effectively delivered to a wide variety of cell types by viral infection.

REFERENCES

Abremski, K. and Hoess, R. (1984). Bacteriophage P1 site-specific recombination. Purification and properties of the Cre recombinase protein. *Journal of Biological Chemistry*, **259**, 1509–1514.

Agami, R. and Bernards, R. (2000). Distinct initiation and maintenance mechanisms cooperate to induce G1 cell cycle arrest in response to DNA damage. *Cell*, **102**, 55–66.

Armentano, D., Yu S-F., Kantoff, P. W., von Ruden, T., Anderson, W. F., and Gilboa, E. (1987). Effect of internal viral sequences on the utility of retroviral vectors. *Journal of Virology*, **61**, 1647–1650.

Baltimore, D. (2001). Our genome unveiled. *Nature*, **409**, 814–816.

Barton, G. M., and Medzhitov, R. (2002). Retroviral delivery of small interfering RNA into primary cells. *Proceedings of the National Academy of Sciences USA*, **99**, 14943–14945.

Brummelkamp, T. R., Bernards, R., and Agami, R. (2002a). A system for stable expression of short interfering RNAs in mammalian cells. *Science*, **296**, 550–553.

Brummelkamp, T. R., Bernards, R., and Agami, R. (2002b). Stable suppression of tumorigenicity by virus-mediated RNA interference. *Cancer Cell*, **2**, 243–247.

Capecchi, M. R. (1989). Altering the genome by homologous recombination. *Science*, **244**, 1288–1292.

Correll, P. H., Colilla, S., and Karlsson, S. (1994). Retroviral vector design for long-term expression in murine hematopoietic cells *in vivo*. *Blood*, **84**, 1812–1822.

Devroe, E., and Silver, P. A. (2002). Retrovirus-delivered siRNA. *BMC Biotechnology*, **2**, 15.

Donze, O., and Picard, D. (2002). RNA interference in mammalian cells using siRNAs synthesized with T7 RNA polymerase. *Nucleic Acids Research*, **30**, e46.

Elbashir, S. M., Harborth, J., Lendeckel, W., Yalcin, A., Weber, K., and Tuschl, T. (2001). Duplexes of 21-nucleotide RNAs mediate RNA interference in cultured mammalian cells. *Nature*, **411**, 494–498.

Emerman, M., and Temin, H. M. (1984). Genes with promoters in retrovirus vectors can be independently suppressed by an epigenetic mechanism. *Cell*, **39**, 449–467.

Fire, A., Xu, S., Montgomery, M. K., Kostas, S. A., Driver, S. E., and Mello C. C. (1998). Potent and specific genetic interference by double-stranded RNA in *Caenorhabditis elegans*. *Nature*, **391**, 806–811.

Graham, F. L., and Prevec L. (1991). Manipulation of adenovirus vectors. *Methods in Molecular Biology*, **7**, 109–128.

Hacein-Bey-Abina, S., von Kalle, C., Schmidt, M., Le Deist, F., Wulffraat, N., McIntyre, E., Radford, I., Villeval, J. L., Fraser, C. C., Cavazzana-Calvo, M., and Fischer A. (2003). A serious adverse event after successful gene therapy for X-linked severe combined immunodeficiency. *New England Journal of Medicine*, **348**, 255–256.

Hammond, S. M., Caudy, A. A., and Hannon, G. J. (2001). Post-transcriptional gene silencing by double-stranded RNA. *Nature Reviews Genetics*, **2**, 110–119.

Harlow, E. (1992). Retinoblastoma. For our eyes only. *Nature*, **359**, 270–271.

Hawley, T. S., Sabourin, L. A., and Hawley, R. G. (1989). Comparative analysis of retroviral vector expression in mouse embryonal carcinoma cells. *Plasmid*, **22**, 120–131.

Hemann, M. T., Fridman, J. S., Zilfou, J.T., Hernando, E., Paddison, P. J., Cordon-Cardo, C., Hannon, G. J., and Lowe, S. W. (2003). An epi-allelic series of p53 hypomorphs created by stable RNAi produces distinct tumor phenotypes *in vivo*. *Nature Genetics*, **33**, 396–400.

Hütvagner, G., and Zamore, P. D. (2002). RNAi: Nature abhors a double-strand. *Current Opinion in Genetics and Development*, **12**, 225–232.

Ilves, H., Barske, C., Junker, U., Bohnlein, E., and Veres, G. (1996). Retroviral vectors designed for targeted expression of RNA polymerase III-driven transcripts: A comparative study. *Gene*, **171**, 203–208.

Julius, M. A., Yan, Q., Zheng, Z., and Kitajewski, J. (2000). Q Vectors, Bicistronic retroviral vectors for gene transfer. *BioTechniques*, **28**, 702–707.

Kawasaki, H., and Taira, K. (2003). Short hairpin type of dsRNAs that are controlled by tRNA(Val). promoter significantly induce RNAi-mediated gene silencing in the cytoplasm of human cells. *Nucleic Acids Research*, **31**, 700–707.

Lee, E. Y., Chang, C. Y., Hu, N., Wang, Y. C., Lai, C. C., Herrup, K., Lee, W. H., and Bradley, A. (1992). Mice deficient for Rb are nonviable and show defects in neurogenesis and haematopoiesis. *Nature*, **359**, 288–294.

Lee, N. S., Dohjima, T., Bauer, G., Li, H., Li, M-J., Ehsani, A., Salvaterra, P., and Rossi, J. (2002). Expression of small interfering RNAs targeted against HIV-1 rev transcripts in human cells. *Nature Biotechnology*, **20**, 500–505.

Liu, C. Y., Flesken-Nikitin, A., Li, S., Zeng, Y., and Lee, W.H. (1996). Inactivation of the mouse Brca1 gene leads to failure in the morphogenesis of the egg cylinder in early postimplantation development. *Genes & Development*, **10**, 1835–1843.

Miller, V. M., Xia, H., Marrs, G. L., Gouvion, C. M., Lee, G., Davidson, B. L., and Paulson, H. L. (2003). Allele-specific silencing of dominant disease genes. *Proceedings of the National Academy of Sciences USA*, **100**, 195–7200.

Miyagishi, M., and Taira, K. (2002). U6-promoter-driven siRNAs with four uridine 3′ overhangs efficiently suppress targeted gene expression in mammalian cells. *Nature Biotechnology*, **20**, 497–500.

Mizuguchi, H., and Kay, M.A. (1998). Efficient construction of a recombinant adenovirus vector by an improved *in vitro* ligation method. *Human Gene Therapy*, **9**, 2577–2583.

Nykanen, A., Haley, B., and Zamore, P. D. (2001). ATP requirements and small interfering RNA structure in the RNA interference pathway. *Cell*, **107**, 309–321.

Paddison, P. J., Caudy, A. A., Bernstein, E., Hannon, G. J., and Conklin, D. S. (2002). Short hairpin RNAs (shRNAs) induce sequence-specific silencing in mammalian cells. *Genes & Development*, **16**, 948–958.

Paul, C. P., Good, P. D., Winer, I., and Engelke, D. R. (2002). Effective expression of small interfering RNA in human cells. *Nature Biotechnology*, **20**, 505–508.

Qin, X. F., An, D. S., Chen, I. S., and Baltimore, D. (2003). Inhibiting HIV-1 infection in human T cells by lentiviral-mediated delivery of small interfering RNA against CCR5. *Proceedings of the National Academy of Sciences USA*, **100**, 183–188.

Reynolds, A., Leake, D., Boese, Q., Scaringe, S., Marshall, W. S. and Khvorova, A. (2004). Rational siRNA design for RNA interference. *Nature Biotechnology*, **22**, 326–330.

Robertson, E., Bradley, A., Kuehn, M., and Evans, M. (1986). Germ-line transmission of genes introduced into cultured pluripotential cells by retroviral vector. *Nature*, **323**, 445–448.

Rubinson, D. A., Dillon, C. P., Kwiatkowski, A. V., Sievers, C., Yang, L., Kopinja, J., Zhang, M., McManus, M. T., Gertler, F. B., Scott, M. L., and Van Parijs. L. (2003). A lentivirus-based system to functionally silence genes in primary mammalian cells, stem cells and transgenic mice by RNA interference. *Nature Genetics*, **33**, 401–406.

Schroder, A. R., Shinn, P., Chen, H., Berry, C., Ecker, J. R., and Bushman, F. (2002). HIV-1 integration in the human genome favors active genes and local hotspots. *Cell*, **110**, 521–529.

Sedivy, J. M., and Dutriaux, A. (1999). Gene targeting and somatic cell genetics – a rebirth or a coming of age? *Trends in Genetics*, **15**, 88–90.

Sharp, P. A. (2001). RNA Interference – 2001. *Genes & Development*, **15**, 485–490.

Shen, C., Buck, A.K., Liu, X., Winkler, M., and Reske, S. N. (2003). Gene silencing by adenovirus-delivered siRNA. *Federation of European Biochemical Society Letters*, **539**, 111–114.

Tiscornia, G., Singer, O., Ikawa, M., and Verma, I. M. (2003). A general method for gene knockdown in mice by using lentiviral vectors expressing small interfering RNA. *Proceedings of the National Academy of Sciences USA*, **100**, 1844–1848.

Yu, J-Y., DeRuiter, S. L., and Turner, D. L. (2002). RNA interference by expression of short-interfering RNAs and hairpin RNAs in mammalian cells. *Proceedings of the National Academy of Sciences USA*, **99**, 6047–6052.

Yu, S. F., von Ruden, T., Kantoff, P. W., Garber, C., Seiberg, M., Ruther, U., Anderson, W. F., Wagner, E. F., and Gilboa, E. (1986). Self-inactivating retroviral vectors designed for transfer of whole genes into mammalian cells. *Proceedings of the National Academy of Sciences USA*, **83**, 3194–3198.

Van der Weyden, L., Adams, D.J., and Bradley, A. (2002). Tools for targeted manipulation of the mouse genome. *Physiological Genomics*, **11**, 133–164.

Wu, X., Li, Y., Crise, B., and Burgess, S.M. (2003). Transcription start regions in the human genome are favored targets for MLV integration. *Science*, **300**, 1749–1751.

Xia, H., Mao, Q., Paulson, H.L., and Davidson, B.L. (2002). siRNA-mediated gene silencing *in vitro* and *in vivo*. *Nature Biotechnology*, **20**, 1006–1010.

12 siRNA delivery by lentiviral vectors: Design and applications

Oded Singer, Gustavo Tiscornia, and Inder Verma

Introduction

A major challenge in the post genomic era of biology is to decipher the molecular function of over 30,000 genes. The gene knock-out by homologous recombination has proven to be very useful but is laborious and expensive. RNA interference has recently emerged as a novel pathway that allows modulation of gene expression. The basic biology of RNAi is described in the next section. Briefly, long dsRNA molecules are processed by the endonuclease Dicer into short 21–23 nucleotide small interfering RNAs (siRNAs), which are then incorporated into RISC (RNA-induced silencing complex), a multi-component nuclease complex that selects and degrades mRNAs that are homolgous to the dsRNA initially delivered (Fjose et al., 2001; Hannon, 2002). In mammalian systems, synthetic siRNA's can be delivered exogenously (Elbashir et al., 2001) or can be expressed endogenously from RNA Pol III promoters, resulting in a powerful tool for achieving specific downregulation of target mRNA's (Miyagashi and Taira, 2002; Paul et al., 2002; Oliviera and Goodell, 2003). The delivery of synthetic siRNAs to cells in culture is hampered by limitations in transfection efficiency for many cell types and the transient nature of the silencing effect. *In vivo*, delivering siRNAs to target cells is difficult due to lack of stability of siRNA and low uptake efficiency in the absence of transfection agents (Isacson et al., 2003). Thus in order to apply this potent technique to both basic biological questions and therapeutic strategies, efficient siRNA delivery methods must be developed. In this chapter, we describe the design of lentiviral vectors expressing siRNA and their use both *in vitro* and *in vivo*.

Overview of lentiviral vectors

During the past decade gene delivery vehicles based on HIV-1, the best characterized of the lentiviruses, have been developed. Lentiviral vectors derived from HIV-1 are capable of infecting a wide variety of dividing and non-dividing cells, and integrate stably into the host genome, resulting in long-term expression of the transgene. The HIV-1 genome contains nine open reading frames encoding

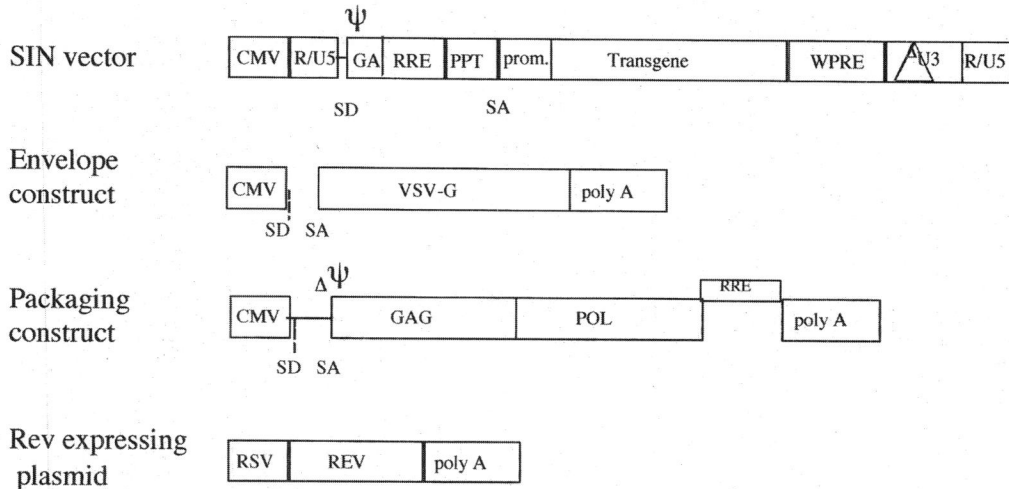

Figure 12.1. Generation of infectious lentiviral vectors. The vector plasmid containing the transgene and the DNA sequences needed for packaging is transfected together with three plasmids expressing the packaging proteins (VSV-G, GagPol and Rev). Concentration of lentiviral vector particles is measured using an ELISA kit for p24 capsid protein.

at least 15 distinct proteins involved in the infectious cycle, including structural and regulatory proteins. In addition, there are a number of *cis*-acting elements required at various stages of the viral life cycle. These include the long terminal repeats (LTRs), the TAT activation region (TAR), primer binding site (PBS), splice donor and acceptor sites, packaging and dimerization signal (Ψ), Rev-responsive element (RRE) and the central and terminal polypurine tracts (PPT). The general strategy used to produce vector particles that are replication defective has been to eliminate all dispensable genes from the HIV-1 genome and separate the *cis*-acting sequences from those *trans*-acting factors that are absolutely required for viral particle production, infection and integration. The widely used third generation of lentiviral vectors consists of four plasmids: The transfer vector contains the transgene to be delivered in a lentiviral backbone containing all the *cis*-acting sequences required for genomic RNA production and packaging. The packaging system involves 3 additional plasmids, which provide the required *trans*-acting factors, namely Gag-pol, Rev and VSVG respectively. Gag-pol codes for structural proteins, integrase and reverse transcriptase. While the structural proteins are required for particle production, integrase and reverse transcriptase molecules are packaged in the particle and are required upon subsequent infection. Rev interacts with the Rev-responsive element (RRE), a sequence contained in the transfer vector, enhancing the nuclear export of unspliced viral genomic RNA. Vesicular Stomatitis Virus envelope protein G (VSVG) is incorporated into the viral membrane and increases the viral tropism, allowing the transduction of a wide arrays of cells, tissues, organs and organisms. In addition, a deletion in the 3′LTR causes self-inactivation, as during reverse transcription the proviral 5′LTR is copied from the 3′LTR; the deleted 5′LTR is transcriptionally inactive, blocking viral genomic RNA production from the integrated provirus. When these four plasmids are transfected into 293T human embryonic kidney cells, viral particles accumulate in the supernantant and high titer viral preparations can be prepared by ultracentrifugation. A scheme of the four plasmids is presented in Figure 12.1. For further details on lentiviral vector design and production, see (Trono, 2002).

Design of lentivectors expressing siRNAs

A crucial breakthrough occurred with the report that siRNAs could be expressed as hairpins from Pol III promoters cloned into plasmids (Brummelkamp et al., 2002; Miyagashi and Taira, 2002b). The two Pol III promoters most commonly used are H1 and U6 (both human and mouse). Pol III promoters are characterized by their compact size [(less than 400 bp) (Myslinski et al., 2001)], and by the fact that all sequences required for promoter function are upstream of the +1 transcriptional start site. Pol III promoters have ubiquitous expression and efficiently express short RNAs (Paule and White, 2000; Miyagashi and Taira, 2002a). Thus, they are ideally suited to express short hairpins consisting of a 21–23 nucleotide sense sequence that is identical to the target sequence in the mRNA to be downregulated, followed by a 9 bp loop and an antisense 21–23 nucleotide sequence. A stretch of 5 T's provides a Pol III transcriptional termination signal. Thus, when this construct is expressed, a short 21–23 base pair hairpin is formed; the loop is digested by Dicer and the resulting siRNA oligonucleotide triggers degradation of the mRNA target (Brummelkamp et al., 2002). Once these plasmid-based silencing cassettes were shown to be effective both *in vitro* and *in vivo*, we and other groups have sought to use viral vectors to deliver them *in vivo*, effectively attempting to couple a potent downregulation strategy with an effective delivery system (Barton and Medzhitov, 2002; An et al., 2003; Matta et al., 2003; Rubinson et al., 2003; Tiscornia et al., 2003).

Ideally, a silencing lentiviral vector would contain both a marker gene such as GFP or an antibiotic resistance gene and the siRNA silencing cassette. Previous experience in our lab indicates that the precise combination of elements in a lentiviral construct can have unforeseen effects both on the viral titer and the efficiency of expression of both the marker and the silencing cassette. Initially we designed a lentiviral vector carrying GFP as a marker and an H1 promoter–driven siRNA cloned into the lentiviral 3'LTR (Tiscornia et al., 2003). Upon infection of target cells by the lentivector, reverse transcriptase generates a viral cDNA, which then stably integrates into the host DNA. During this process, the 5'LTR is generated from the 3'LTR. Thus, cloning the H1-driven siRNA into the 3'LTR results in duplication of the silencing cassette, presumably resulting in more potent silencing (Figure 12.2a). When a lentiviral construct containing only GFP as a marker is transfected along with the three packaging plasmids into 293T cells, two GFP-coding RNAs are produced. One is the viral genomic mRNA (transcribed from the 5'RSV promoter), which will eventually be packaged, and the second is a GFP mRNA (transcribed from the internal promoter driving GFP); both RNAs extend into the 3'LTR. Typically, strong GFP expression is seen and high viral

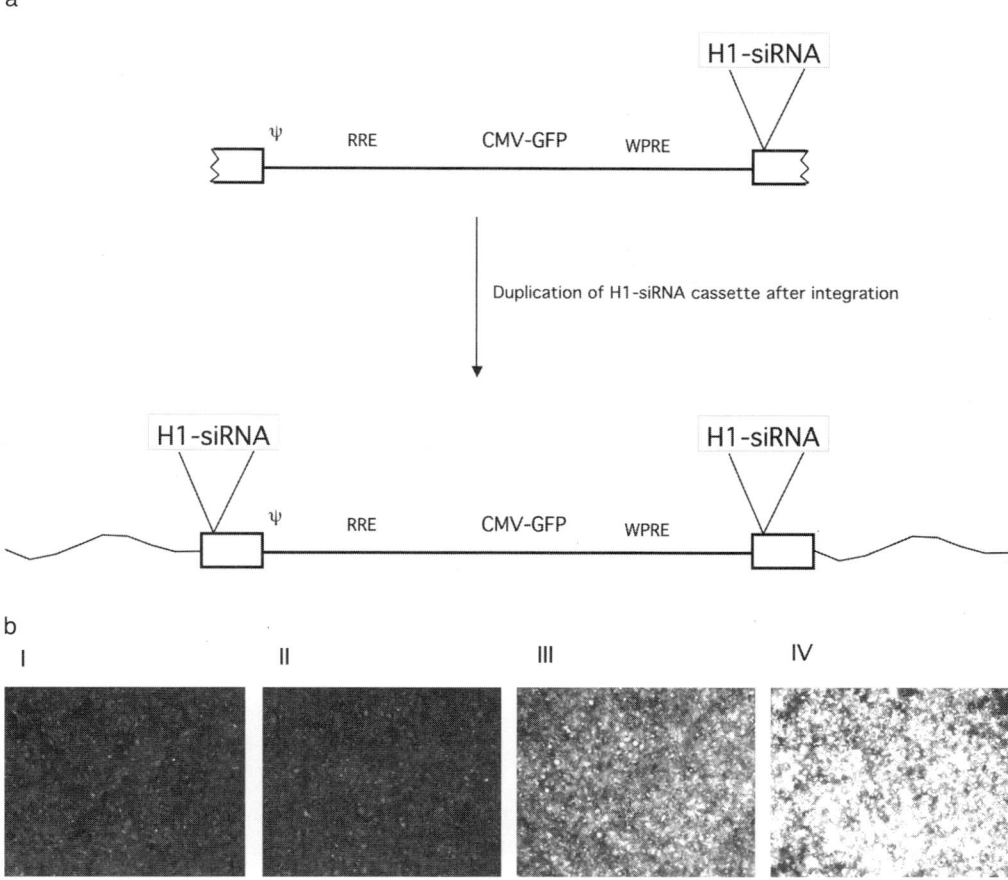

Figure 12.2. Design of lentivirus vectors expressing siRNAs. a. An expression cassette containing the H1 promoter, siRNA hairpin sequence and Pol III termination signal is inserted in the 3'LTR portion of the lentiviral vector plasmid. This vector will express both a marker gene (GFP) and siRNA. Following integration of the viral vector, duplication of the 3'LTR will results in duplication of the H1-siRNA cassette. b. Expression of the marker gene (GFP) in the packaging cells following transfection. Expression of the siRNA molecules is also noticed since it results in silencing of the GFP when siGFP is used, some reduction in GFP expression is noticed when other siRNAs are used. The orientation of the silencing cassette (S or AS) did not show a difference (I. LV-CMV-GFP-siGFP-S, II. LV-CMV-GFP-siGFP-AS, III. LV-CMV-GFP-siRNA, IV. LV-CMV-GFP).

titers are obtained. When a lentiviral construct containing both a GFP marker and a silencing cassette against GFP cloned in the 3'LTR was designed, an initial concern was the fact that expression of the siRNA hairpin in the lentiviral producing cells might target both the viral genomic RNA and the GFP mRNA, as the siRNA target sequence (ie, the 21–23 sense nucleotide strand) is contained in both RNAs, possibly resulting in lack of viral production. When this construct was transfected along with the packaging vectors into 293T cells, GFP expression (albeit noticeably weaker than in the case of the GFP only lentiviral construct) was observed in the producer cells (Figure 12.2b). While the weaker GFP expression suggests that the siRNA against GFP is active and that the genomic viral RNA and/or the GFP mRNA are being targeted in the producer cells, viral particles

c

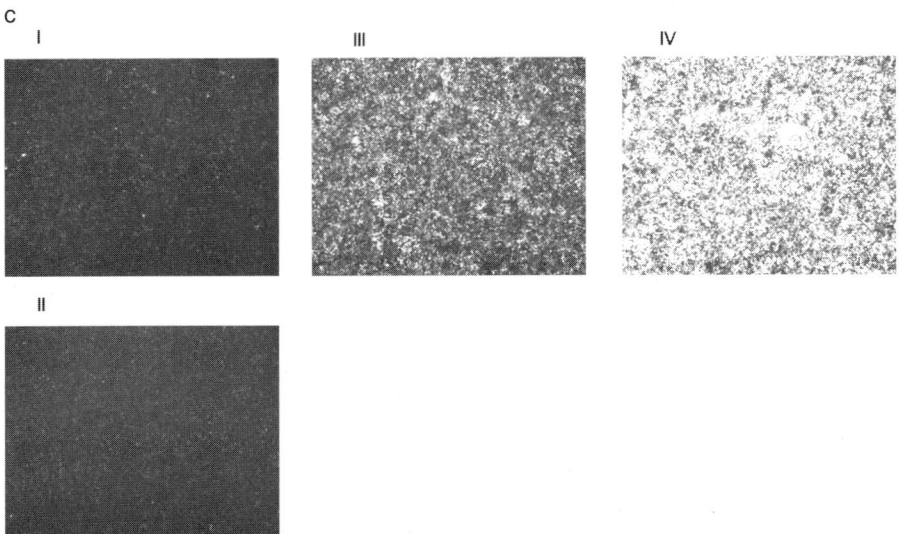

Figure 12.2. *(cont.)* c. Expression of marker gene (GFP) in cells transduced by virus collected from packaging cells. Percent of cells expressing GFP was similar in all panels (data not shown) however strong reduction in GFP was noticed when siGFP was used and some reduction when other siRNAs used (I. LV-CMV-GFP-siGFP-S, II. LV-CMV-GFP-siGFP-AS, III. LV-CMV-GFP-siRNA, IV. LV-CMV-GFP).

capable of conferring weak GFP expression in infected cells were obtained, suggesting that the kinetics of viral production overwhelm those of siRNA expression (Figure 12.2c). Thus, despite the fact that expression of siRNA in producer cells can to a certain extent target viral genomic RNA, the lentiviral delivery system remains a viable option for siRNA delivery. Subsequent experience in our laboratory with a number of siRNA-expressing lentiviral vectors shows that viral titers, while somewhat lower, are not significantly affected.

Lentiviral mediated gene silencing *in vitro*

We then sought to test the silencing capabilities of the construct further. In order to do so, the CMV-GFP marker cassette was deleted from the construct, leaving only the H1-siRNA against GFP in the 3′LTR. A high titer viral preparation was prepared and used to infect a 293T cells line or different primary cells stably expressing GFP, with increasing multiplicity of infection. A similar construct containing an H1-siRNA silencing cassette against p53 in the lentiviral 3′LTR was used as a control. The lentiviral vector expressing siRNA against GFP achieved GFP downregulation in a dosage specific manner, while the lentiviral construct expressing siRNA against p53 did not (see Figure 12.3). Two observations are worth noting. First, significant downregulation of GFP levels required relatively high multiplicity of infection. While the multiplicity of infection required to achieve silencing of a given target will depend on several factors, such as the levels of expression of the gene in question and the effectiveness of the siRNA used, this

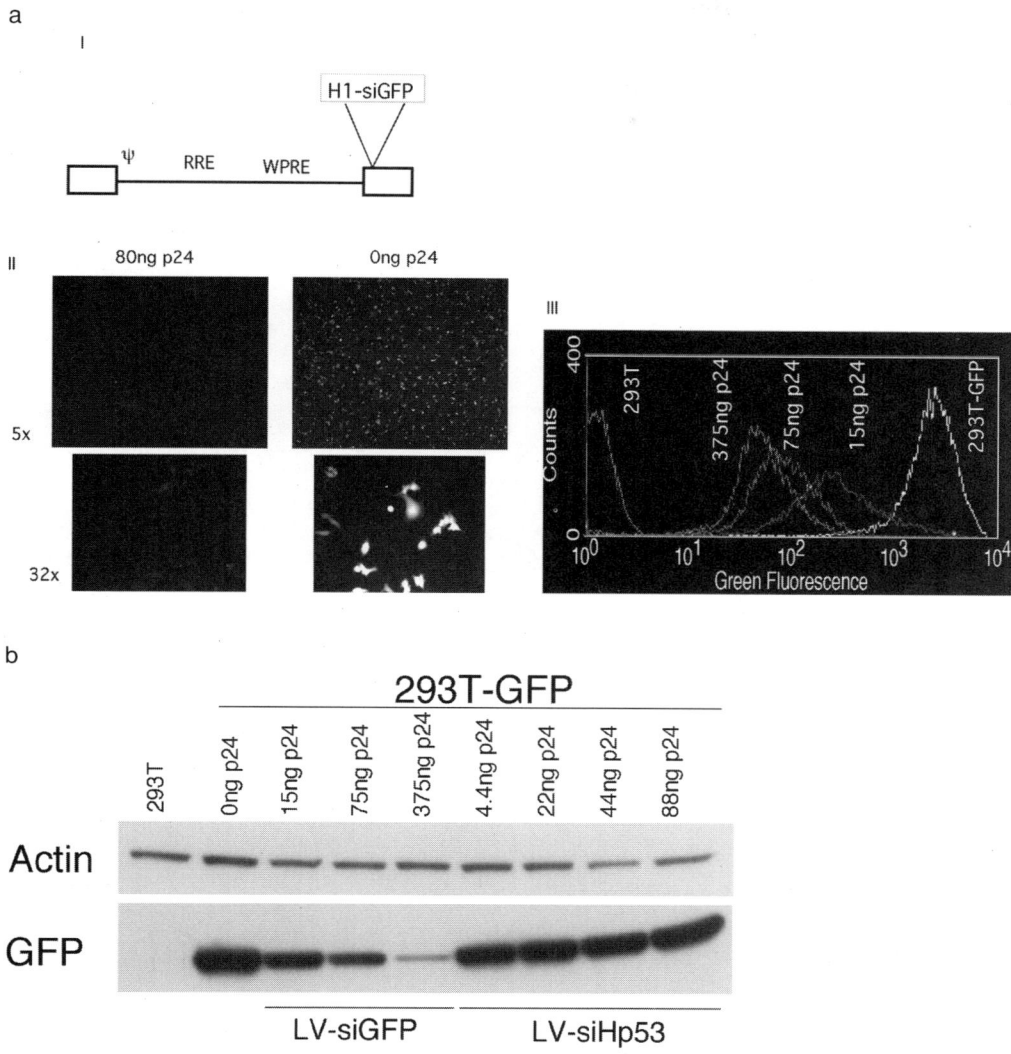

Figure 12.3. Inhibition of GFP in cells stably expressing GFP. a. CMV-GFP expressing cassette was deleted to generate LV-siGFP (I). Transduction of GFP positive 293T cells (293T-GFP) with increasing amounts of LV-siGFP resulted in decreasing levels of fluorescence detected by fluorescence microscopy (II) and FACS analysis (III). b. Western analysis showed that reduction of GFP protein was correlated to the increasing amounts of virus p24 used for transduction. Transduction with a virus expressing irrelevant siRNA (siHp53) did not have any effect on GFP proteins. TaqMan analysis showed that the amounts of integrated proviral DNA was comparable in LV-siGFP and LV-siHp53 transduced cells (data not shown).

result suggests that in most cases a high moi will be necessary. This diminishes the perceived advantage of cloning the silencing cassette into the lentiviral 3′LTR, as increasing the level of silencing by a factor of two is not likely to impact silencing significantly, while the position of the cassette in the 3′LTR would weaken the expression of any marker cloned upstream. Therefore, we are currently developing lentiviral vectors in which the silencing cassette is cloned upstream of a marker gene. Second, the silencing effect was stable for up to three months of continuous culture. Despite the limitations of the current LV-siRNA vectors, our laboratory

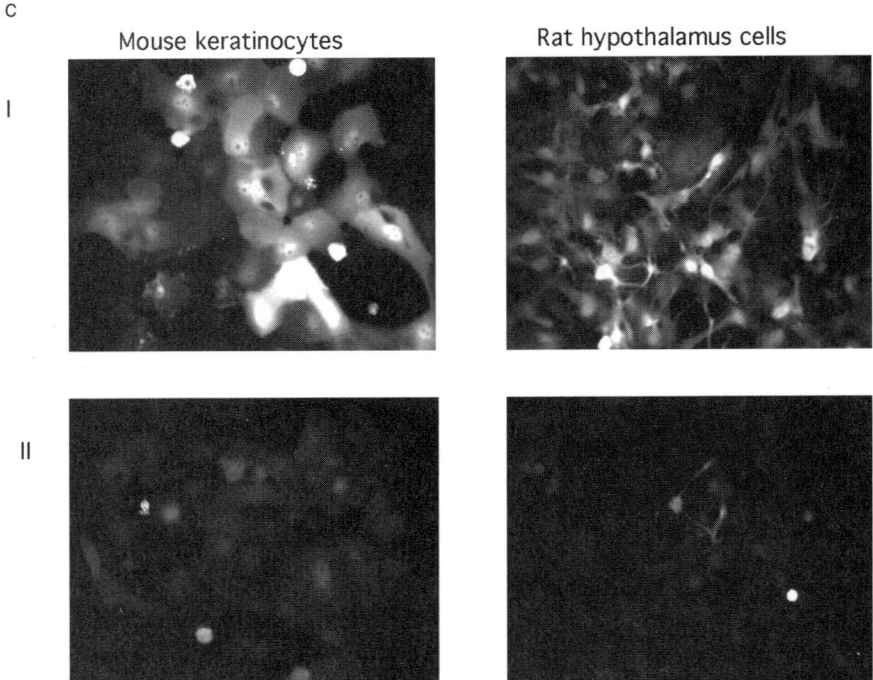

Figure 12.3. *(cont.)* c. Primary cells transduced with LV-CMV-GFP alone (I) or in combination with LV-siGFP (II).

in the last year has generated over 30 LV-siRNA vectors capable of silencing the intended gene products. Table 12.1 lists the name of genes and the appropriately effective siRNA oligonucleotide.

Lentiviral mediated gene silencing *in vivo*

We were interested in testing the effectiveness of lentiviral silencing vectors (LV-siRNA vectors) *in vivo*. A powerful application of LV-siRNA vectors would be the ability to generate transgenic animals carrying a siRNA cassette in order to achieve silencing of endogenous genes. Our lab and other groups have recently developed methods to generate transgenic rodents by *in vitro* transduction of fertilized eggs at different pre-implantation stages. This approach involves delivering high titer lentiviral preparations to early stage embryos (two-to-four-cell stage). Two alternative delivery methods are available. One option is to digest the zona pellucida of the fertilized oocytes with acidic tyrode or pronase treatment, followed by infection with a recombinant lentiviral vector (Pfeifer et al., 2002). The second possibility is the direct injection of lentiviral particles under the zona pellucida using micromanipulator equipment (Lois et al., 2002). In both cases, the transduced oocytes are transferred to pseudopregnant females and carried to term. We tested whether the lentiviral virus carrying an H1-siRNA silencing cassette against GFP cloned in the 3′LTR (LV-siGFP) was able to silence GFP expression in GFP transgenic mice (TgGFP); which are homozygous for an array of multiple copies of

Table 12.1. siRNAs selected for different genes. Usually, effective siRNAs were selected by designing of 5–7 siRNA's followed by co-transfection of siRNA expressing cassette with exogenous expression of the target cDNA. Best effective siRNA was selected by western blot analysis.

Gene name	Target sequence	Comments
GFP	AAGCAAGCTGACCCTGAAGTTC	(Tiscornia et al., 2003)
hp53	AAGACTCCAGTGGTAATCTAC	(Brummelkamp et al., 2002)
Vif	AAGCAGGACATAACAAGGTAG	
mTyr	AACTGGGGATGAGAACTTCAC	
mGlut-4	AAGGTGATTGAACAGAGCTAC	
mIGF1R	AAGCTCACAGTCATCACCGAG	
hBACE-1	AATGGACTGCAAGGAGTACAA	
hELKS	AAATGCTGCCTATGCCACCTC	
mIKBβ	AAGAGATGCCTCAGATACCTA	
mIKBε	AAGTCCTGGACATTCAGAACA	
mp100	AAGACCTATCCTACTGTCAAG	
mp105	AACACTGGAAGCACGGATGA	
hXPC	AAGAAAGTGGCCAAGGTGACT	
hXPA	AAGTTGTTCATCAACCAGGAC	
hXPD	AAGAACTGTGCCAGAGATTGA	
mGlut-1	AAGAGTGTGCTGAAGAAGCTT	
hBRCA-1	AACTTGATGCTCAGTATTTGCAG	
hBRCA-2	AACTGTCAATCCAGACTCTGA	
mIKK-1	AACGTCAGTCTGTACCAGCAC	
hGadd45α	AACGTCGACCCCGATAACGTG	
hGadd45β	AAGTTGATGAATGTGGACCCA	
hGadd45γ	AACGAGGACGCCTGGAAGGAT	
hPDEF	AAGTGCTCAAGGACATCGAGA	
hESE-1	AACTACTTCAGTGCGATGTAC	
mRDC-1	GCTATAAGAAGAAGATGGTAC	
NMDAR-1	AAGATACAGCTCAACGCCAC	
hHD	GATAAATTTGTGTTGAGAGATG	

GFP. We reasoned that by generating a TgGFP mouse that is also transgenic for the GFP silencing cassette, we should be able to show whole body knockdown of GFP expression. Fertilized eggs were collected from females that had been mated with homozygous Tg-GFP males and transduced with LV-siGFP virus. Reduction of GFP fluorescence in virus treated blastocysts was observed 48 hs after transduction (see Figure 12.4a). The transduced blastocysts were transplanted into pseudopregnant females. Live pups were analyzed for levels of fluorescence and the presence of integrated virus was confirmed by PCR using primers specific for siGFP cassette (Figure 12.4b). Two F_0 females (F_0-2 and F_0-4, Figure 12.4b) carrying the siGFP cassette were mated to wild-type males to obtain F_1 progeny. One pup (F_1-36) from female F_0-2 was found to be positive for the siGFP cassette (Figure 12.4c). Interestingly, this pup also showed remarkably lowered overall GFP fluorescence (Figure 12.4c). The second siGFP positive F_0 female (F_0-4) was also crossed to a wild-type male and E12 embryos were harvested and examined. Of nine F_1 embryos analyzed, six were genotyped to be positive for the GFP array (#1 through #6, Figure 12.4d) while three (#7 through #9, Figure 12.4d) were not. Only embryos #6 and #8

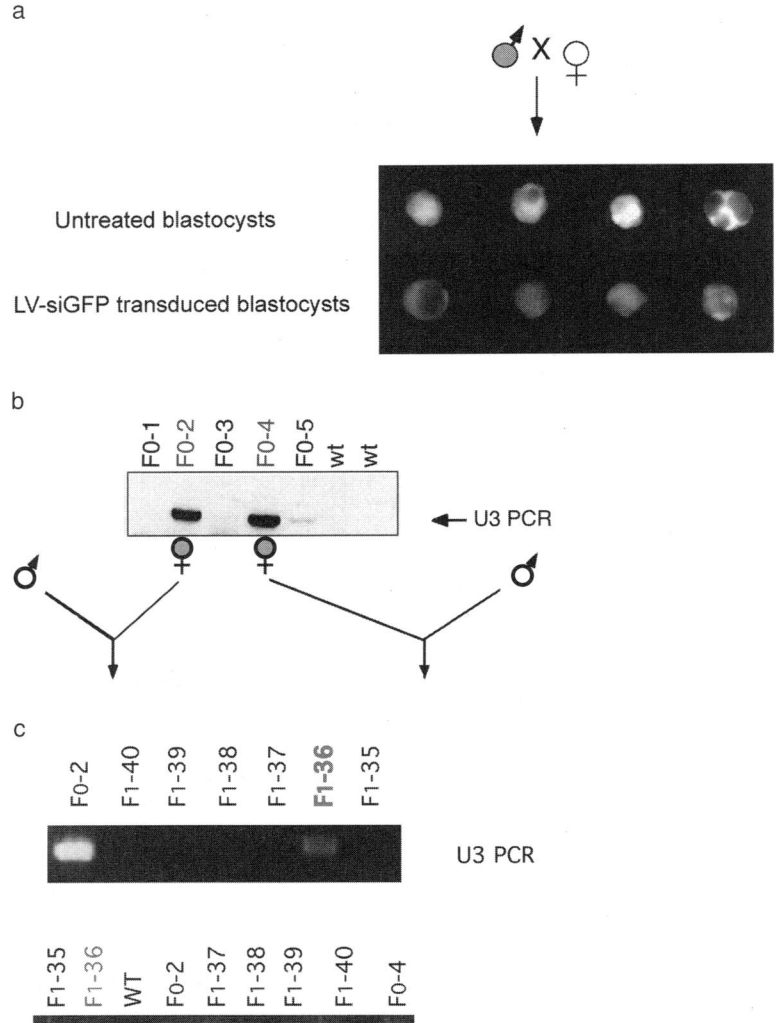

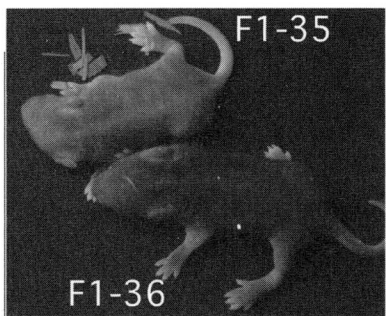

Figure 12.4. Inhibition of GFP in transgenic mice expressing GFP. a. TgGFP males were mated with WT females. Embryos (2−4 cells stage) were harvested and incubated with LV-siGFP. Shown are blastocysts following two days incubation with LV-siGFP. b. Five F_0 mice were genotype for presence of H1-siGFP cassette by PCR using U3 specific primers. Two positive female (F_0-2 and F_0-4) were mated with WT males to obtain F_1 progeny. c. Six F_1 mice (progeny of F_0-2) were genotype for presence of H1-siGFP cassette by PCR using U3 specific primers and for the presence of GFP using GFP specific primers. One positive male (F_1-36) showed overall reduced fluorescence.

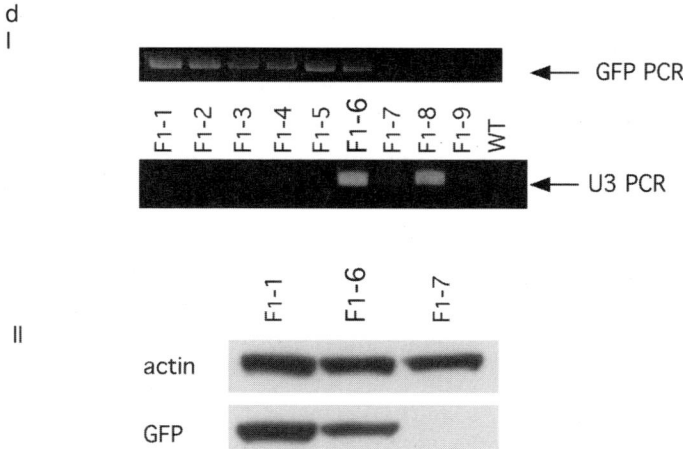

Figure 12.4. d. *(cont.)* d. Nine F_1 embryos (taken from F_0-4) were genotype for presence of H_1-siGFP cassette by PCR using U3 specific primers and for the presence of GFP using GFP specific primers (I). Proteins extracted from embryos 1, 6 and 7 were probed for GFP and actin (II).

were positive for the siGFP cassette. Embryo #1 (GFP +, siGFP–) and embryo #6 (GFP+, siGFP+) were compared by Western blot using GFP specific antibodies; interestingly, embryo #6 showed levels of GFP reduced to approximately 50% of the levels found in embryo #1 (Figure 12.4d). Using TaqMan PCR, we have estimated the lentivirus integration copy number in the F1-6 embryo as 10 (Tiscornia et al., 2003).

Perspectives

We have described a methodology by which lentiviral vectors expressing small interfering RNAs can be used to knock down target genes *in vitro*. Furthermore, we have successfully generated transgenic mice in which expression of specific genes can be down-regulated substantially. The technology builds on the use of lentiviral vectors for transgenesis combined with the use of RNAi to silence gene expression. Additionally, the ability of lentiviral vectors to transduce nondividing cells (Naldini et al., 1996b) would allow direct injection into a tissue or organ to knock down the expression of specific genes (Naldini et al., 1996a; Blomer et al., 1997). Because lentiviral vectors integrate in the chromosome, the succeeding progeny are likely to inherit the provirus and express the appropriate genes. Although the work reported here deals with mice, lentiviral vectors have been shown previously to generate transgenic rats (Lois et al., 2002), and we foresee that they can be used to generate transgenics in other species, including nonhuman primates.

Although the combination of lentiviral vectors with siRNA gene-silencing approaches shows great promise, a number of aspects of the technology require further development. At present there is no well-defined set of rules for designing siRNA oligonucleotides. Because some candidate siRNAs work well and others do not, we regularly test a panel of five to seven siRNAs empirically. Furthermore, it is not clear what molar ratio of siRNA molecules to target gene mRNA is required for effective knockdown or whether this parameter will be target-specific. It will be useful to know how much siRNA is required to knock down expression from a one-copy gene, which most likely will be the case for most functional genomics studies. A number of variations can be envisioned. Lentiviral vectors can be modified where multiple siRNAs are expressed from a single provirus, achieving knockdown of several targets simultaneously. Several reports have been recently published describing tetracycline regulatable Pol III promoters (Matsukura et al., 2003; Wiznerowicz and Trono, 2003). We are currently developing vectors in which the CRE recombinase loxP system will be used to achieve silencing in a tissue-specific manner. It is also not clear whether expression from integrated lentiviral vectors will be silenced in certain progeny or tissues. Transcriptional shut-off from integrated retroviral DNA has been observed (Svoboda et al., 2000), whereas the problem seems to be less acute with lentiviral proviruses. Despite these caveats and limitations, we believe the use of lentiviral vectors capable of expressing siRNA will prove to be a very useful tool to study gene function. The technology will allow the creation of novel murine disease models that recapitulate the human disease. In particular, the ability to knock down specific genes expands gene therapy approaches to gain of function diseases such as Huntington's disease, Spinal and Cerebellar Ataxias and other disorders arising from triplet repeat expansions.

In conclusion we believe that use of LV-siRNA vectors can play a significant role in generating a large number of cell lines, tissues, and whole animals where the expression of targeted genes can be reduced substantially to influence biological function.

REFERENCES

An, D. S., Xie, Y., Mao, S. H., Morizono, K., Kung, S. K. and Chen, I. S. (2003). Efficient lentiviral vectors for short hairpin RNA delivery into human cells. *Human Gene Therapy*, **14**, 1207–1212.

Barton, G. M. and Medzhitov, R. (2002). Retroviral delivery of small interfering RNA into primary cells. *Proceedings of the National Academy of Sciences USA*, **99**,14943–14945.

Blomer, U., Naldini, L., Kafri, T., Trono, D., Verma, I. M. and Gage, F. H. (1997). Highly efficient and sustained gene transfer in adult neurons with a lentivirus vector. *Journal of Virology*, **71**, 6641–6649.

Brummelkamp, T. R., Bernards, R. and Agami, R. (2002). A system for stable expression of short interfering RNAs in mammalian cells. *Science*, **296**, 550–553.

Elbashir, S. M., Harborth, J., Lendeckel, W., Yalcin, A., Weber, K. and Tuschl, T. (2001). Duplexes of 21-nucleotide RNAs mediate RNA interference in cultured mammalian cells. *Nature*, **411**, 494–498.

Fjose, A., Ellingsen, S., Wargelius, A. and Seo, H. C. (2001). RNA interference: mechanisms and applications. *Biotechnology Annual Review*, **7**, 31–57.

Hannon, G. J. (2002). RNA interference. *Nature*, **418**, 244–251.

Isacson, R., Kull, B., Salmi, P. and Wahlestedt, C. (2003). Lack of efficacy of 'naked' small interfering RNA applied directly to rat brain. *Acta Physiology Scandinavia*, **179**, 173–177.

Lois, C., Hong, E. J., Pease, S., Brown, E. J. and Baltimore, D. (2002). Germline transmission and tissue-specific expression of transgenes delivered by lentiviral vectors. *Science*, **295**, 868–872.

Matsukura, S., Jones, P. A. and Takai, D. (2003). Establishment of conditional vectors for hairpin siRNA knockdowns. *Nucleic Acids Research*, **31**(15), e77.

Matta, H., Hozayev, B., Tomar, R., Chugh, P. and Chaudhary, P. M. (2003). Use of lentiviral vectors for delivery of small interfering RNA. *Cancer Biololgy and Therapeutics*, **2**, 206–210.

Miyagishi, M. and Taira, K. (2002a). Development and application of siRNA expression vector. *Nucleic Acids Research*, Suppl (2), 113–114.

Miyagishi, M. and Taira, K. (2002b). U6 promoter-driven siRNAs with four uridine 3′ over-hangs efficiently suppress targeted gene expression in mammalian cells. *Nature Biotechnology*, **20**, 497–500.

Myslinski, E., Ame, J. C., Krol, A. and Carbon, P. (2001). An unusually compact external promoter for RNA polymerase III transcription of the human H1RNA gene. *Nucleic Acids Research*, **29**, 2502–2509.

Naldini, L., Blomer, U., Gage, F. H., Trono, D. and Verma, I. M. (1996a). Efficient trans-fer, integration, and sustained long-term expression of the transgene in adult rat brains injected with a lentiviral vector. *Proceedings of the National Academy of Sciences USA*, **93**, 11382–11388.

Naldini, L., Blomer, U., Gallay, P., Ory, D., Mulligan, R., Gage, F. H., Verma, I. M. and Trono, D. (1996b). *In vivo* gene delivery and stable transduction of nondividing cells by a lentiviral vector. *Science*, **272**, 263–267.

Oliveira, D. M. and Goodell, M. A. (2003). Transient RNA interference in hematopoietic progenitors with functional consequences. *Genesis*, **36**, 203–208.

Paul, C. P., Good, P. D., Winer, I. and Engelke, D. R. (2002). Effective expression of small interfering RNA in human cells. *Nature Biotechnology*, **20**, 505–508.

Paule, M. R. and White, R. J. (2000). Survey and summary: Transcription by RNA polymerases I and III. *Nucleic Acids Research*, **28**, 1283–1298.

Pfeifer, A., Ikawa, M., Dayn, Y. and Verma, I. M. (2002). Transgenesis by lentiviral vectors: Lack of gene silencing in mammalian embryonic stem cells and preimplantation embryos. *Proceedings of the National Academy of Sciences USA*, **99**, 2140–2145.

Rubinson, D. A., Dillon, C. P., Kwiatkowski, A. V., Sievers, C., Yang, L., Kopinja, J., Rooney, D. L., Ihrig, M. M., McManus, M. T., Gertler, F. B., Scott, M. L. and Van Parijs, L. (2003). A lentivirus-based system to functionally silence genes in primary mammalian cells, stem cells and transgenic mice by RNA interference. *Nature Genetics*, **33**, 401–406.

Svoboda, J., Hejnar, J., Geryk, J., Elleder, D. and Vernerova, Z. (2000). Retroviruses in foreign species and the problem of provirus silencing. *Gene*, **261**, 181–188.

Tiscornia, G., Singer, O., Ikawa, M. and Verma, I. M. (2003). A general method for gene knockdown in mice by using lentiviral vectors expressing small interfering RNA. *Proceedings of the National Academy of Sciences USA*, **100**, 1844–1848.

Trono, D., ed. *Lentiviral Vectors* (2002). Springer-Verlag, Berlin-Heidelberg.

Wiznerowicz, M. and Trono, D. (2003). Conditional suppression of cellular genes: lentivirus vector-mediated drug-inducible RNA interference. *Journal of Virology*, **77**, 8957–8961.

13 Liposomal delivery of siRNAs in mice

Mouldy Sioud and Dag R. Sørensen

Introduction

Novel tools for evaluating gene function *in vivo* such as ribozymes and RNA interference (RNAi) are emerging as the most highly effective strategies (Sioud, 2001; Sioud, 2004; Hannon, 2002). RNAi is sequence-specific posttranscriptional gene silencing, which is triggered by double stranded RNA (dsRNA). This evolutionarily conserved gene-silencing pathway triggered by dsRNAs was first described in the nematode worm *Caenorhabditis elegans* (Fire et al., 1998). This process has been linked to many previously described phenomena such as post-transcriptional gene silencing (PTGS) in plants (Jorgensen, 1990). The difficulty of using RNAi in somatic mammalian cells was overcome when Tuschl and colleagues discovered that siRNAs (21 nt), normally generated from long dsRNA during RNAi, could be used to inhibit specific gene targets (Elbashir et al., 2001). Currently, there is a great interest in the use of small interfering RNA as a research tool to study gene function and drug target validation (Dykxhoorn et al., 2003; Sørensen et al., 2003).

The therapeutic application of siRNAs, however, is largely dependent on the development of a delivery vehicle that can efficiently deliver the siRNAs to target cells. In addition, such delivery vehicles should be administered efficiently, safely and repeatedly. Cationic liposomes represent one of the few examples that can meet these requirements (Templeton, 2002). These agents are composed of positively charged lipid bilayers, and can be complexed to negatively charged siRNA duplexes. The routes of delivery include direct injection (e.g. intratumoral), intravenous, intraperitoneal, intraarterial, intracranial, and others. Although the first generation of liposomal delivery reagents were hampered by low levels of *in vivo* delivery and subsequent low gene expression, optimization strategies directed at the lipid formulation have indicated that liposomal delivery remains an appropriate approach for somatic gene therapy (Ulrich, 2002). In this respect, cationic liposome-mediated intravenous gene delivery has been shown to produce significant levels of reporter gene expression in all tissues examined (Zhu et al., 1993). Furthermore, Thierry and colleagues (1995) demonstrated that the expression of a transgene through systemic gene transfer via liposomal delivery was efficient and depends on both liposome preparation and the design of the plasmid expression vector.

Recently, the demonstration that small interfering RNAs (siRNAs) can be used to inhibit gene expression in mammals has opened a new avenue for analysis of gene function and drug target validation. The success of siRNAs as therapeutics, however, is largely dependent on the development of a delivery vehicle that can efficiently deliver them *in vivo*. For drug delivery, liposomes have been widely investigated as versatile carriers and shown to facilitate gene transfer *in vivo*. We used a fluorescein-labelled siRNA to examine the efficacy of cationic liposome-mediated intravenous and intraperitoneal delivery of siRNAs in adult mice. Because of its simplicity and potency, this approach is useful for *in vivo* testing of siRNAs.

1. Cationic liposome-mediated intravenous nucleic acid delivery

Similar to antisense oligonucleotides and ribozymes, synthetic siRNAs are generally delivered to cells via liposome-based transfection reagents (Elbashir et al., 2001). These reagents offer the possibility of developing pharmaceutical siRNAs for local or systemic delivery. Notably, most patients who die from cancer have subclinical metastatic disease present at the time of diagnosis (Timme et al., 2003). Thus, it is essential that molecular medicine-based therapies to treat cancer be adjusted to a systemic administration. In addition, when catalytic RNAs such as siRNAs are made *in vitro*, a variety of 2′ and backbone chemical modifications can be introduced into the molecules to increase their half-life in biological fluids (Harborth et al., 2003). In the past several groups have shown that lipid mediated intravenous delivery of gene can produce high level, systemic expression of biologically and therapeutically relevant genes (Liu et al., 1995). In addition, DNA has been successfully complexed with cationic, anionic and neutral liposomes, as well as various mixtures thereof.

Being unstable in biological fluids, the *in vivo* stability and uptake of antisense oligonucleotides and ribozymes have been improved by the use of cationic and anionic liposomes after intravenous administration. In this respect, intravenous pre-treatment of mice with cationic lipids and antisense against intercellular adhesion molecule-1 (ICAM) significantly decreased its expression in the lung following LPS challenge (Ma et al., 2002). Systemic administration of cationic liposomes and a plasmid encoding for a ribozyme against the transcription factor NF-κB suppressed NF-κB expression in metastatic melanoma cells and significantly reduced metastatic spread (Kashani-Sabet et al., 2002). Recently, we have found that intravenous injection of liposome-formulated siRNAs against GFP inhibited GFP gene expression in various organs such as the liver and spleen (Sørensen et al., 2003). In the liver and spleen, most of the uptake was in endothelial, Kupffer cells and macrophages (unpublished results). Taken together these studies demonstrated the utility of *in vivo* gene targeting via intravenous delivery of

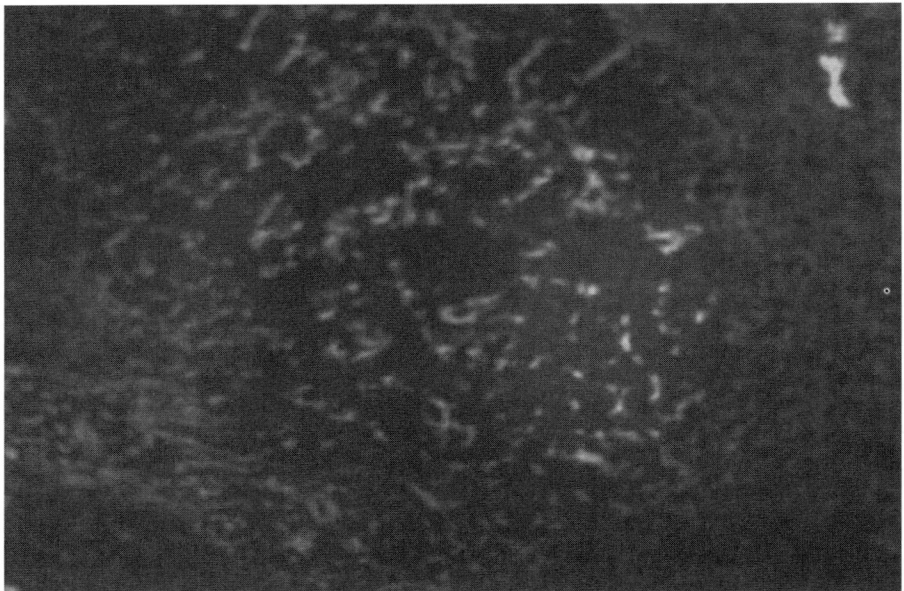

Figure 13.1. *Distribution of a FITC-labeled siRNA in mouse spleen cells 6 hours following i.v. injection of FITC-labeled siRNA liposome complexes.* Six hours after vein tail injection, mice were killed and organs were taken and then frozen in OCT with liquid nitrogen. Ten-micrometer cryosections were cut, fixed in ethanol, washed twice with PBS and then analyzed by an epifluorescence microscope. FITC-labeled siRNAs were chemically synthesized and purified by Eurogentec. The siRNAs were annealed in transfection buffer 20 mM HEPES, 150 mM NaCl pH 7.4. (For experimental conditions see Sioud and Sørensen, 2003.)

nucleic acids such as antisense oligonucleotides, ribozymes, and siRNAs. However, still lipid-based systems have important drawbacks, including the lack of specific targeting and variation arising during fabrication. Setting the optimal liposome formulations for intravenous delivery of siRNAs would be an important step to identify gene function in whole animals and to develop therapeutic siRNAs.

To evaluate the ability of liposomal carriers such as DOTAP, we have assessed the intravenous delivery of a FITC-labeled siRNA directed against TNF-α. Tissues from a group of mice receiving siRNA showed a significant uptake by some organs such as the spleen (Figure 13.1). A substantial majority of the siRNA molecules were found to be localizing around the vessels 6 hours after intravenous injection via the tail vein. Endothelial cells are more likely to be the most targeted by the siRNA complexes.

2. Cationic liposome-mediated intraperitoneal (ip) nucleic acid delivery

The appropriate delivery route to assess the efficacy of siRNAs can facilitate the development of therapeutic siRNAs. Notably, malignant ascites is a major cause of morbidity in patients with intra-abdominal dissemination of various neoplasms such as ovarian, colorectal and breast cancer (Parsons et al., 1996). Treatment of patients with this troubling clinical condition would involve systemic or

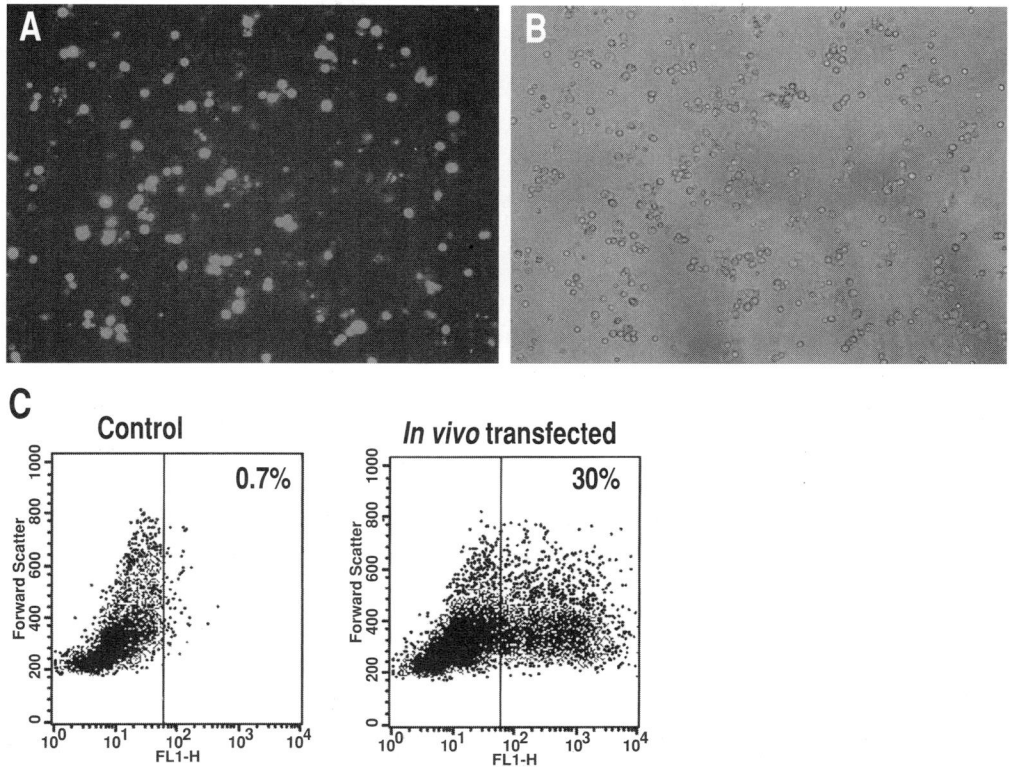

Figure 13.2. *Analysis of mouse peritoneal cells by an epifluorescence microscope after i.p. delivery of 100 μg FITC-labeled siRNA.* 20 hours after delivery mice were killed, peritoneal cells were prepared washed with RPMI medium and then analysed by an epifluouorescence microscope (A). Light image of the same field is shown (B). Dot plots of flow cytometric analysis of the same cells are shown in C. Nearly 30% of the cells show a significant siRNA uptake. (For experimental conditions see Sioud and Sørensen, 2003).

intraabdominal chemotherapy, generally not successful due to drug resistance of such tumours. Notably, the peritoneal cavity represents the largest cavity of the human body and its anatomic structures, along with residual leukocyte populations, play an important role in the defence against invading microorganisms, in particular those breaching the gut integrity. Thus, the development of efficient intraperitoneal delivery agents would represent a new potential to explore the efficacy of new therapeutics *in vivo*. In this respect, we have found that cationic liposomes can enhance the uptake of anti-TNF-α ribozymes and siRNAs by peritoneal macrophages and were able to downregulate the expression of TNF-α *in vivo* (Sioud, 1996; Sørensen et al., 2003). Similarly, Kisick et al. (1999) demonstrated that cationic liposomes could deliver ribozymes to peritoneal cells. Figure 13.2A, B and C documents the uptake of a FITC-labeled siRNA by peritoneal cells 20 hours after intraperitoneal delivery of 100 μg siRNAs. A substantial number of cells have taken up the siRNA-liposome complexes. The uptake could be under estimated, since FITC may get quenched at the low pH in the endosomes. Regarding siRNA labelling, Cy3 and Cy5 are expected to be more stable than FITC.

Adherent peritoneal cells are more susceptible to *in vivo* transfection than their non-adherent counterparts

The *in vivo* uptake of siRNAs can differ dramatically with the cell types as well as with the status of cell differentiation. Different cell types may require significantly different transfection conditions to yield optimal uptake of siRNAs. For biological activity, liposomes must deliver their contents into the cytoplasm, which can be cell-dependent. We have analysed the *in vivo* transfection efficiency of adherent and non-adherent peritoneal cells. As shown in Figure 13.3, a large proportion of adherent peritoneal cells were *in vivo* transfected with a FITC-labeled siRNA. In contrast, most of the non-adherent peritoneal cells were not transfected. Thus, there might be some mechanism in adherent macrophages for effectively taking up the liposome formulated siRNA molecules. These macrophages could be the major source of TNF-α production, since intraperitoneal delivery of siRNAs against mouse TNF-α reduced the expression of TNF-α (Sørensen et al., 2003). Mice pre-treated with anti-TNF-α siRNA prior LPS challenge showed less severe clinical symptoms than those treated with an inactive siRNA. These data would suggest that liposomes-mediated delivery of anti-TNF-α siRNAs might be valuable in a number of conditions such as rheumatoid arthritis in which TNF-α is a pathogenic mediator (Beutler, 1999). In contrast to mouse cells, however, we have found that certain synthetic siRNAs activated the production of inflammatory cytokines (eg. TNF-alpha) in human freshly isolated monocytes via the activation of NF-κB signalling pathway (Sioud and Sørensen, 2003; Sioud, unpublished results). More recently, Sledz and colleagues (2003) found that transfection of siRNAs induced the activation of the Jak-Stat pathway and global upregulation of interferon-stimulated genes. In our case, the activation of TNF-α production by siRNAs seems to be cell and sequence dependent that need further investigation.

Concluding remarks

The study of nucleic acids has revealed remarkable properties of RNA molecules that could make them attractive therapeutic agents, independent of their well-known ability to encode biologically active proteins. This application is now mainly driven by small interfering RNAs that serve as guide sequences to induce target specific mRNA cleavage via the activation of a highly conserved regulatory mechanism, called RNA interference or posttranscriptional gene silencing (Elbashir et al., 2001). However, crucial for the success of siRNAs as a pharmaceutical or a basic research tool is *in vivo* transfection efficiency. A significant degree of *in vivo* cell uptake of siRNAs via liposomal delivery was observed in our studies. Similarly, hydrodynamic force (rapid injection via tail vein) has been applied to deliver naked DNA and siRNAs to hepatocytes, with significant delivery efficiencies (Liu et al., 1995; Lewis et al., 2002; McCaffrey et al., 2002). Although, delivery of siRNAs via such method can induce gene silencing *in vivo*, the technique is not feasible for clinical application in humans. From the perspective of human cancer therapy, however, cationic liposomes have been shown to be safe and effective for *in vivo* gene delivery, and are currently being used in many approved clinical

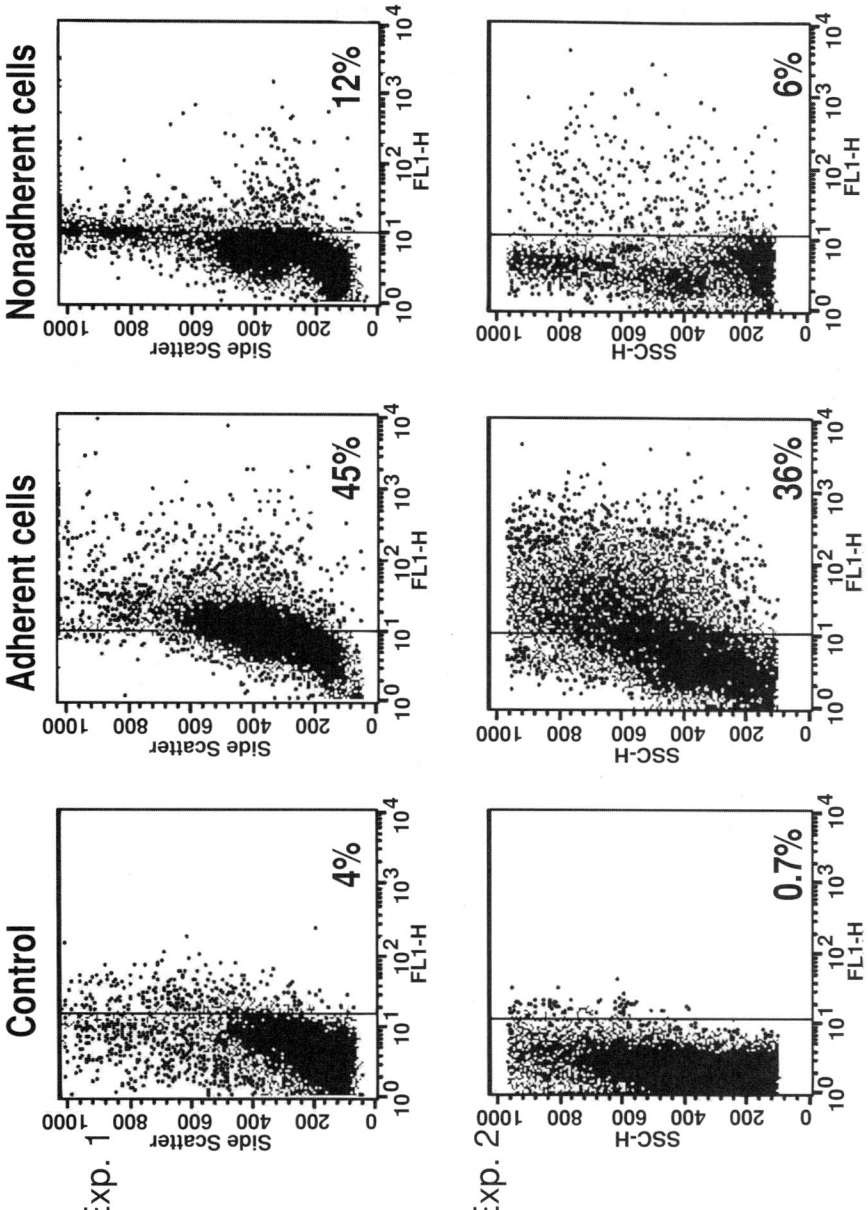

Figure 13.3. *Analysis of adherent and nonadherent peritoneal cells by flow cytometry.* 18 hours after i.v. delivery of a FITC-labeled siRNA, mice were killed and peritoneal cells were prepared, washed, resuspended in RPMI medium supplemented with 10% FCS and then plated in 75 cm³ tissue culture dishes. After 2 hours, the cultures were gently washed with RPMI to remove nonadherent cells. Adherent cells were harvested by gentle scrapping. Both cell populations were analysed by flow cytometry. (For experimental conditions see Sioud and Sørensen, 2003.)

trials (Cattel et al., 2003). Notably, chemically synthesized siRNAs are expected to be useful in pathological situations where immediate and/or short-term effects are required.

In addition to delivery, the effectiveness of siRNAs as therapeutic agents may depend on their stability and bioavailability into the targeted tissues or organs. Backbone and 2' modifications can also contribute to changes in the tissue distribution and pharmacokinetics of siRNAs. The analysis of these parameters (e.g. stability in blood circulation, extravasations into tissues) in whole animals should provide important information for liposome optimization. Despite being the most commonly used methods of nucleic acids delivery, however, liposome based delivery still have important drawbacks, including the lack of specific targeting and potential side effects. Under our experimental conditions, mice appeared normal and no signs of toxicity were recorded.

In contrast to early reports (Caplan et al., 2001; Elbashir et al., 2001), siRNAs were found to activate the interferon pathway. While the activation of this pathway must be controlled when siRNAs are used for the treatment of inflammatory diseases, it can be, however, beneficial for the treatment of certain diseases such as cancer. In experimental models for human gliomas we found that the use of GFP double-stranded RNA as adjuvant enhanced vaccine efficacy via the activation of innate immunity response genes (unpublished data). Thus, despite the potential induction of interferon and the activation of NF-κB signalling pathway, siRNAs will almost certainly be an effective means in cancer therapy.

Acknowledgment

This work was supported by the Norwegian Cancer Society to M. Sioud. We thank Dr. Anne Dybwad for critical reading of the manuscript and Khu Ky Cuong for technical assistance.

REFERENCES

Beutler, B.A. (1999). The role of tumour necrosis factor in health and diseases. *Journal of Rheumatology*, **57**, 16–21.

Caplan, L., Ceruti, M. and Dosio, F. (2003). From conventional to stealth liposomes: a new frontier in cancer chemotherapy. *Tumori*, **89**, 237–249.

Caplan, N.J., Parrish, S., Imani, F., Fire, A. and Morgen, R.A. (2001). Specific inhibition of gene expression by small double-stranded RNAs in invertebrate and vertebrate systems. *Proceedings of National Academy of Sciences USA*, **98**, 9742–9747.

Dykxhoorn, D.M., Novina, C.D. and Sharp, P.A. (2003). Killing the messenger: short RNAs that silence gene expression. *Nature Reviews Molecular Cell Biology*, **4**, 457–467.

Elbashir, S.M., Harborth, J., Lendeckel, W., Yalcin, A., Weber, K., and Tuschl, T. (2001). Duplexes of 21-nucleotide RNAs mediate RNA interference in cultured mammalian cells. *Nature*, **411**, 494–498.

Fire, A., Xu, S., Montgomery, M.K., Kostas, S.A., Driver, S.E. and Mello, C.C. (1998). Potent and specific genetic interference by double-stranded RNA in *Caenorhabditis elegans*. *Nature*, **391**, 806–811.

Hannon, G.J. (2002). RNA interference. *Nature*, **418**, 244–251.

Harborth, J., Elbashir, S.M., Vandenburgh, K. Manninga, H., Scaringe, S.A., Weber, K. and Tuschl, T. (2003). Sequence, chemical, and structural variation of small interfering RNAs and short hairpin RNAs and the effect on mammalian gene silencing. *Antisense and Nucleic Acid Drug Development*, **13**, 83–105.

Jorgensen, R. (1990). Altered gene expression in plants due to trans-interaction between homologous genes. *Trends in Biotechnology*, **8**, 340–344.

Kashani-Sabet, M., Liu, Y., Fong, S., Desprez, P.-Y., Liu, S., Tu, G., Nosrati, M., Handumrongkul, C., Liggitt, D., Thor, A.D. and Debs, R.J. (2002). Identification of gene function and functional pathways by systemic plasmid-based ribozyme targeting in adult mice. *Proceedings of National Academy of Sciences USA*, **99**, 3878–3883.

Kisich, K.O., Malone, R.W., Feldstein, P.A. and Erickson, K.L. (1999). Specific inhibition of macrophage TNF-α expression by *in vivo* ribozyme treatment. *Journal of Immunology*, **163**, 2008–2016.

Lewis, D.L., Hagstom, G., Haley, B. and Zamore, P.D. (2002). Efficient delivery of siRNA for inhibition of gene expression in postnatal mice. *Nature Genetics*, **32**, 107–108.

Liu, Y., Liggitt, D., Zhong, W., Tu, G., Gaensler, K. and Debs, R. (1995). Cationic liposome-mediated intravenous gene delivery. *Journal of Biological Chemistry*, **270**, 24864–24870.

Ma, Z., Zhang, J., Alder, S., Dileo, J., Negishi, Y., Stolz, D., Watkins, S., Huang, L., Pitt, B. and Li, S. (2002). Lipid-mediated delivery of oligonucleotide to pulmonary endothelium. *American Journal of Respiratory Cell Molecular Biology*, **27**, 151–159.

McCaffrey, A.P., Meuse, L., Pham, T.T., Conklin, D.S., Hannon, G.J. and Kay, M.A. (2002). RNA interference in adult mice. *Nature*, **418**, 38–39.

Parsons, P., Lang, M. and Steele, R. (1996). Malignant ascites: A 2-year review from a technical hospital. *Journal of Surgery and Oncology*, **22**, 237–239.

Sioud, M. (1996). Ribozyme modulation of lipopolysaccharide-induced tumor necrosis factor-aplha production by peritoneal cells *in vitro* and *in vivo*. *European Journal of Immunology*, **26**, 1026–1031.

Sioud, M. (2001). Nucleic acid enzymes as a novel generation of anti-gene agents. *Current Molecular Medicine*, **1**, 575–588.

Sioud, M. (2004). Therapeutic siRNAs. *Trends in Pharmacological Sciences*, **25**, 22–28.

Sioud, M. and Sørensen, D.R. (2003). Cationic liposome-mediated delivery of siRNAs in adult mice. *Biochemical and Biophysical Research Communications*, **312**, 1220–1225.

Sledz, C.A., Holko, M., de Veer, M.J., Silverman, R.H., Williams, B.R.G. (2003). Activation of the interferon system by short-interfering RNAs. *Nature Cell Biology*, **5**, 834–839.

Sørensen, D.R., Leirdal, M. and Sioud, M. (2003). Gene silencing by systemic delivery of synthetic siRNAs in adult mice. *Journal of Molecular Biology*, **327**, 761–766.

Templeton, S.N. (2002). Liposomal delivery of nucleic acids *in vivo*. *DNA Cellular Biology*, **21**, 859–867.

Thierry, A.R., Lunardi-Iskandar, Y., Bryant, J.L., Rabinovich, P., Gallo, R.C. and Mahan, L.C. (1995). Systemic gene therapy: Biodistribution and long-term expression of a transgene in mice. *Proceedings of National Academy of Sciences USA*, **92**, 9742–9746.

Timme, T.L., Satoh, T., Tahir, S.A., Wang, H., The, B.S., Butler, E.B., Miles, B.J., Amato, R.J., Kadmon, D. and Thompson, T.C. (2003). Therapeutic targets for metastatic prostate cancer. *Current Drug Targets*, **4**, 251–261.

Ulrich, A.S. (2002). Biophysical aspects of using liposomes as delivery vehicles. *Bioscience Reports*, **22**, 129–150.

Zhu, N., Liggitt, D., Liu, Y. and Debs, R. (1993). Systemic gene expression after intravenous DNA delivery in adult mice. *Science*, **261**, 209–211.

14 Chemical modifications to achieve increased stability and sensitive detection of siRNA

Philipp Hadwiger and Hans-Peter Vornlocher

Introduction

RNA interference (RNAi) was discovered in the worm *C. elegans* as an endogenous, double-stranded RNA (dsRNA) driven mechanism resulting in a specific inhibition of gene expression on a posttranscriptional level (Fire et al., 1998). Since then, gene suppression by RNA interference has been widely applied to study gene function in a variety of organisms (Hannon, 2002). Double-stranded small interfering RNA (siRNA) of 21 to 23 nucleotides was identified as mediator of this silencing signal in a *Drosophila* embryo lysate system (Elbashir et al., 2001b). In a seminal study it was demonstrated that exogenously delivered chemically synthesized siRNA can function as trigger for this specific silencing mechanism in mammalian cells (Elbashir et al., 2001a). Initial experiments indicated that, in contrast to long dsRNA, siRNA does not stimulate an unspecific inhibition of protein synthesis mediated by activation of protein kinase R (PKR) in mammals (Caplen et al., 2001). Because of its specificity and high efficiency as well as simple practicability, siRNA triggered RNAi has become rapidly accepted as the method of choice for studying gene function in cell culture systems. Moreover, recent reports demonstrated the successful siRNA-mediated down-regulation of reporter genes (McCaffrey et al., 2002; Lewis et al., 2002) as well as endogenous target genes in mice (Xia et al., 2002; Song et al., 2003; Rubinson et al., 2003). Now, researchers in academia and industry are attempting to utilize RNAi as a platform for the development of therapeutics. Whereas there is no need to alter the all-*ribo* nature of siRNA for *in vitro* knock-down experiments (besides the possibility that modified siRNA might be helpful in elucidating the still not fully understood underlying mechanism of RNAi) this picture may change when considering the development of drugs based on siRNA.

Pre-requisites for a clinical development of siRNA based drugs are, besides the *in vivo* reconfirmation of target down-regulation found in cell culture, a sufficient stability towards nucleases present in a biological environment and a favorable pharmacokinetic profile. Therefore, in order to be successful in the transformation of siRNA from a powerful and well established tool for manipulating gene

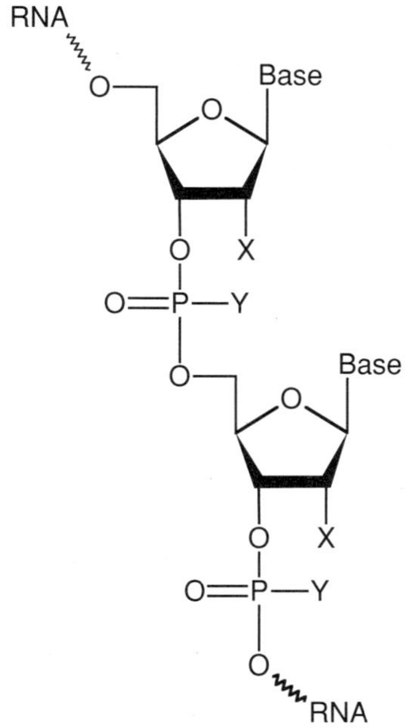

Figure 14.1. Chemical structure of the sugar phosphate backbone of natural single stranded RNA (X=OH, Y=O⁻) and modified derivatives (X=H: 2'-deoxy, X=OCH₃: 2'-O-methyl, X=OCH₂CHCH₂: 2'-O-allyl, X=F: 2'-deoxy-2'-fluoro, X=NH₂: 2'-amino-2'-deoxy, Y=S⁻ phosphorothioate.

expression in mammalian cells into a functional therapeutic molecule, it might be necessary to utilize chemical modifications of the siRNA. Ideally, these modifications should improve both resistance towards cellular degradation processes as well as cellular up-take (Cook, 1999). Although it is well known that double-stranded RNA is more stable towards nucleases than their single-stranded counterparts, it is likely that an all-*ribo* siRNA carrying no further chemical modification would not match the criteria for drug development mentioned above.

Chemical alterations have substantially contributed to the advancement of antisense oligonucleotide (ASON) (Manoharan, 1999; Crooke, 2001; Opalinska and Gewirtz, 2002) and ribozyme based therapeutics (James and Gibson, 1998; Sun et al., 2000). These modifications were vital as they prolonged the presence of the active compounds in a biologic milieu significantly. For example, the half-life of an unmodified all-*ribo* RNA hammerhead ribozyme in human serum was determined to be less than six seconds (Jarvis et al., 1996). By the aid of suitable chemically modified nucleotides in the ribozyme the half-life could be extended to as much as 8 hours without significantly compromising the enzymatic activity (Heidenreich et al., 1994; Beigelman et al., 1995). Among the great variety of different chemical modifications, the two most frequently used are the 2'-O-methyl nucleotides and the phosphorothioate linkage, in which a non-bridging oxygen of the phosphodiester between adjacent nucleotides in the oligomer chain is replaced by a sulphur atom (Eckstein, 2002). These two modifications were also successfully employed in the gapmer approach of second generation ASONs (for a review see Kurreck, 2003). Moreover, especially in the ribozyme field, further 2'-sugar modifications (see Figure 14.1) such as 2'-O-allyl, 2'-amino-2-deoxy and 2'-deoxy-2'-fluororibofuranose nucleotides were exploited (Usman and Blatt, 2000).

Generally, chemical modifications in oligoribonucleotides are synthesized using the well-established phosphoramidite technology (Beaucage and Radhakrishnan, 1992; Wincott et al., 1995). This solid-phase chemistry adds suitably protected nucleotides stepwise to the growing ribonucleotide chain immobilized on a solid support. By means of this approach, it is possible to synthesize small quantities of oligoribonucleotides in a high-throughput fashion as well as to manufacture kilogram amounts of material for clinical trials.

Up to now, published reports assess the consequences of modifications in siRNA on activity *in vitro* and stability in biological media such as serum. Most reports focus on sugar phosphate backbone modifications, whereas studies on alterations of the nucleobases in siRNA are rare (Chiu and Rana, 2003).

Probing RNAi's tolerance towards chemical modifications

In an early study on the function of long dsRNA, chemical modifications were introduced on both RNA strands (Parrish et al., 2000). Alteration of the sense strand proved to be generally better tolerated than of the antisense strand. Furthermore, modified nucleotides which promote the A-form helical structure of a natural RNA-RNA duplex, such as 2'-deoxyfluoro nucleotides, did not adversely affect RNAi activity, whereas modified nucleotides with a tendency to destabilize the A form type duplex, such as 2'-deoxy or 2'-aminodeoxy nucleotides, resulted in a reduced inhibitory activity of the dsRNA. Several reports carefully address the influence of sugar-phosphate backbone modifications on siRNA function.

2'-deoxy-modified siRNA

Analysis of siRNA function in a *Drosophila* embryo lysate revealed that a change from the all-*ribo* design to all-2'-deoxy oligonucleotides in one or both siRNA strands abolished RNAi (Elbashir et al., 2001c). Interestingly, up to four consecutive 2'-deoxyribonucleotide substituents at both 3'-ends produced appreciably efficient siRNA which was marginally less active than the all-*ribo* siRNA variants in an *in vitro* mRNA cleavage assay. Nevertheless, one should be careful in adding stretches of 2'-deoxy residues to siRNA since the resulting heteroduplex could be prone to RNase H mediated cleavage of the RNA-containing strand. This may help explain why substitution of either RNA strand in a siRNA by 2'-deoxy oligonucleotides results in the described reduction of RNAi activity in the aforementioned *Drosophila* embryo lysate (Elbashir et al., 2001c), in a *Drosophila in vivo* assay (Boutla et al., 2001), and in HeLa cells (Chiu and Rana, 2003).

Effect of 2'-*O*-methyl nucleotides in siRNA

Full modification of either one or both siRNA strands with 2'-O-methyl nucleotides almost abolished siRNA activity in different systems (Elbashir et al., 2001c) and siRNA with only an all-2'-O-methyl sense strand retained significant residual activity (Chiu and Rana, 2003; Czauderna et al., 2003b). For 19 and 21 basepair (bp) blunt-ended siRNA, the sense and antisense strands were synthesized in an alternating fashion of 2'-*O*-methyl-modified and unmodified nucleotides that furnished an overall duplex structure in which the 2'-*O*-methyl nucleotides were either facing each other or a modified nucleotide was vis-á-vis a regular one. Some of these siRNA molecules mediated a protein knock-down in HeLa cells comparable to an all-*ribo* blunt end siRNA, but positional effects of 2'-*O*-methyl modifications were observed. Similarly, a siRNA design with alternating blocks of five 2'-*O*-methyl modified nucleotides in a row followed by five regular ribose nucleotides proved to be as efficient as the parent all-*ribo*/dTdT-overhang siRNA (Czauderna et al., 2003b). In another study, six 2'-*O*-methyl nucleotides were introduced into each siRNA strand (four at the 3'-ends and two at the 5'-ends) without losing significant activity. Increasing the number of 2'-*O*-methyl modification in increments of two up to 28 out of 42 nucleotides in the siRNA duplex resulted in a gradual attenuation of inhibitory activity (Amarzguioui et al., 2003).

2'-deoxyfluoro nucleotide – containing siRNA

Several groups studied the effects of 2'-deoxyfluoro (2'-F) modifications on siRNA function. Introducing up to four 2'-F-uridines in the flanks of the duplex did not affect siRNA activity (Braasch et al., 2003). Replacing all pyrimidines in a siRNA with the corresponding 2'-F nucleotides in either the sense strand (ten 2'-F residues), the antisense strand (nine 2'-F residues) or the whole siRNA had no effect on siRNA ability to inhibit gene expression in a HeLa cell system. Combining the 2'-F modifications with partially or full replacement of purine residues by the corresponding 2'-deoxy nucleotide led to a maximum of 50 % reduction of siRNA activity. Restricting the deoxy-residues in the 2'-F background to the 3'- or the 5'-half of the antisense strand demonstrates that the 5'-region of the siRNA (defined by the antisense strand) reacts more sensitively towards these modifications (Chiu and Rana, 2003). Similarly, replacing all the 19 pyrimidines of a given siRNA with 2'-F nucleotides and four 2'-deoxythymidines on the overhangs did not affect activity in a HeLa cell system. Additional substitution of three phosphodiesters with phosphorothioate linkages (PTL) at the 3'-end of each strand also did not influence the RNAi machinery significantly (Harborth et al., 2003).

Other siRNA modifications

Complete replacement of all phosphodiesters with PTLs on either strand or in the entire siRNA reduced silencing activity by roughly 50% (Braasch et al., 2003; Chiu and Rana, 2003). Again, the antisense strand was more sensitive towards modification than the sense strand (Chiu and Rana, 2003). When up to twelve PTLs were incorporated at the flanks of the siRNA molecule, activity was marginally lower compared to unmodified siRNA (Amarzguioui et al., 2003; Braasch et al., 2003). At the same time it was demonstrated that long stretches of consecutive PTLs caused toxicity in HaCaT cells (Amarzguioui et al., 2003). Also siRNA with alternating phosphodiesters and phosphorothioates still displayed full silencing efficiency. The observed reduction in cell growth and viability was attributed to cytotoxic effects of the PTL modifications (Harborth et al., 2003).

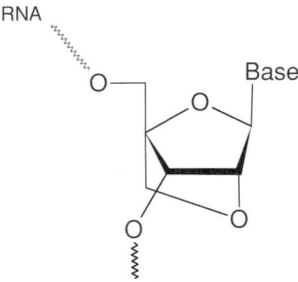

Figure 14.2. General chemical structure of locked nucleic acid (LNA).

The effect of locked nucleic acid (LNA, Figure 14.2)/RNA chimeras with the LNA placed at terminal and internal positions on siRNA efficacy was analyzed in a HeLa cell system. Some siRNA with up to 4 LNA residues was still fully active in mediating RNAi. Nevertheless, incorporation of LNA displayed a subtle position dependency. It seems that the antisense strand does not tolerate LNA in its core region, but may tolerate a limited number of LNAs on flanking positions (Braasch et al., 2003).

Analyzing the role of siRNA termini

The siRNA user guide published by the Tuschl group recommends the use of 2′-deoxythymidine (dT) for the two overhanging nucleotides on both termini of the 19-bp siRNA duplex, although it was shown in the *Drosophila* system that mRNA-sequence complementarity of the *ribo*-overhang on the antisense strand was most efficient (Elbashir et al., 2001c). As pointed out in this study, the 3′-dTdT-design does not significantly affect RNAi but helps to reduce costs of RNA synthesis and may improve resistance against RNase degradation. In human HaCaT cells RNAi activity was shown to be dependent neither on the *ribo*- or deoxy-nature of the overhangs, nor on the base-pairing of the overhangs with the target-mRNA (Holen et al., 2002). These data were contrasted by a study in HeLa and NTera2D1 cells, demonstrating the superior efficiency of ribose-containing overhangs (Hohjoh, 2002).

The Zamore laboratory employed end-modified siRNA to confirm the conservation of the RNAi mechanism between flies and mammals. After blocking the 5′-OH of the antisense strand by methylation, the endogenous kinase, which usually phosphorylates the 5′-OH of siRNA, was not able to add a 5′-phosphate to the siRNA and consequently RNAi activity was largely abolished, both in *Drosophila* extracts as well as in cultured human HeLa cells (Schwarz et al., 2002). However, a 5′-phosphodiester on the antisense strand, *i.e.* 6-aminohexyl phosphodiester, rendered the resulting duplex active in target mRNA silencing, although activity was reduced compared to a 5′-OH-bearing duplex. In contrast, modifying both 3′-termini with 2′,3′-dideoxy ribose or a 3′-aminopropyl phosphodiester resulted in siRNA behaving like standard siRNA with respect to efficiency and specificity (Schwarz et al., 2002). In addition, a siRNA with a 3′-fluorescein-labeled antisense strand had an activity identical to the unmodified parent molecule in a human cell assay (Holen et al., 2002). These data support the view that in flies and mammals siRNA-mediated target mRNA degradation does not involve a RNA dependent RNA polymerase driven degradative PCR mechanism as originally suggested for *Drosophila* and *C. elegans* (Sijen et al., 2001; Lipardi et al., 2001).

The contribution of the 3'- and 5'-termini to siRNA activity were investigated by employing siRNA with end-modified sense or antisense strands in reporter-cotransfection experiments in HeLa cells (Chiu and Rana, 2002). Whereas 5'-termini of both strands where conjugated to an aminopropyl phosphodiester, puromycin, and biotin were attached to both siRNA 3'-ends. Modification of the 3'-ends was well tolerated, irrespectively whether the sense or the antisense strand was altered. This picture changed for the 5'-modification, where the modifying entity was tolerated exclusively on the sense strand. In contradiction to the results of the Zamore laboratory (see above) 5'-phosphodiester-modification of the antisense strand abolished RNAi activity (Chiu and Rana, 2002). The importance of a functional 5'-end of the antisense strand was also demonstrated by allylation of the siRNA termini. Complete 2'-*O*-allyl modification of the siRNA termini led to diminished activity in down-regulation of endogenous tissue factor expression. In contrast, exclusive allylation of both 3'-ends had no effect on activity. It was speculated that the sterically more demanding allyl substituent on the 5'-terminal nucleotide could interfere with the *in vivo* phosphorylation of the 5'-OH shown to be necessary for siRNA activity (Amarzguioui et al., 2003).

Alteration of all four siRNA termini with either an inverted deoxy abasic residue (iB) or an aminohexyl phosphodiester resulted in a dramatic reduction of RNAi activity. End modification of the sense strand either on the 5'- or on the 3'-terminus or on both ends had no effect on the silencing activity. On the other hand, antisense strand modification of the 3'-end was well tolerated whereas a modification on the 5'-end diminished target down-regulation. Even with all termini except the 5'-end of the antisense strand simultaneously altered in the duplex, no reduction in RNAi activity could be observed (Czauderna et al., 2003b). These data indicate that the 5'-end of the antisense strand is most sensitive towards modifications and that this end has to be either 5'-OH or 5'-phosphate. A 5'-phosphodiester structure seems to be tolerated as well, although information on potential intracellular processing of such a terminus is not available yet.

These results were in part contrasted in a study in which transfection of 3'-end modified siRNA into NIH3T3 cells resulted in a strongly reduced knock-down activity (Hamada et al. 2002). Here, researchers used 2-hydroxyethylphosphate (hp) modified thymidine or 2'-*O*,4'-*O*-ethylene thymidine (eT), two rather unusual residues, as modifying nucleotides (Figure 14.3). Strikingly, two consecutive eT residues on the 3'-end, irrespective of whether placed in the sense or antisense strand or on both strands of the duplex, rendered the resulting siRNA drastically less active. On the other hand, one hp residue on the 3'-end was well tolerated when presented on the sense strand, yielding a siRNA as active as the parent compound. Incorporation of this modification on the antisense strand 3'-end or on both 3'-termini of the duplex abolished RNAi. Therefore, it seems that not

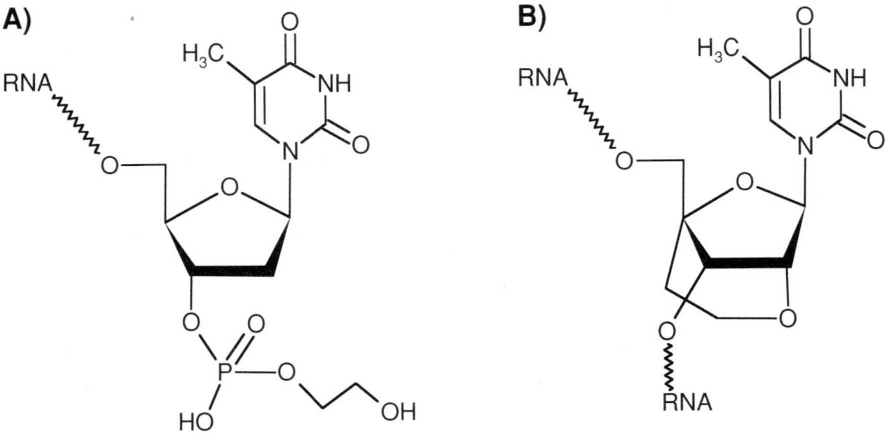

Figure 14.3. A) Chemical structure of 2-hydroxyethylphosphate thymidine (hp) and B) chemical structure of 2′-O,4′-O-ethylene thymidine (eT).

only the site of incorporation of modifications but also the chemical structure of the modifier is crucial.

In another study siRNA carrying fluorescein on the four different termini was synthesized. In agreement with the literature the fluorophore conjugated to the 5′- or 3′-end of the sense strand did not perturb efficient protein down-regulation in HeLa SS6 cells. However, RNAi activity was abolished when the reporter molecule was attached to the 3′-end of the antisense strand, but was fully active with the re-porter placed on the 5′-end (Harborth et al., 2003). These results disagree with pre-viously published data demonstrating the crucial role of a 5′-OH or 5′-phosphate as well as 5′-phosphodiester (Chiu and Rana, 2002; Schwarz et al., 2002). The authors discussed this result in terms of differences in the linker structures and in terms of siRNA sequence dependent effects.

Improving siRNA's resistance towards nucleolytic degradation

siRNA with sugar phosphate backbone modifications was tested for improved stability towards nucleolytic degradation and for prolonged inhibitory potency. Comparing the results of different research groups is difficult, as diverse assay systems were utilized to address siRNA stability. Complete 2′-O-methylation ren-dered the siRNA serum nuclease-resistant, although gene silencing activity was almost completely abolished. Five nucleotide-blockwise modification of siRNA with 2′-O-methyl nucleotides preserved activity and at the same time resulted in an increased stability. siRNA with every other nucleotide in both strands modified by a 2′-O-methyl nucleotide were notably stabilized, but only the version with a modified nucleotide in one strand facing a non-modified nucleotide in the other strand retained activity. Stabilization of siRNA by either blockwise or alternating 2′-O-methyl modification results in a drastically prolonged target knock-down, with inhibition of target expression extended from 48 hours to 120 hours. After that time the parent unmodified siRNA had completely lost its ability to reduce protein expression (Czauderna et al., 2003b). The enhanced serum stability of the modified siRNA may be paralleled by an improved intracellular stability resulting in prolonged gene silencing. 2'-*O*-methyl modifications could also change association kinetics of siRNA with the RNA induced silencing complex (RISC) or directly influence the enzymatic activity of a siRNA-loaded RISC. 2'-*O*-methylation of terminal siRNA residues also resulted in a prolonged inhibitory potency of the corresponding siRNA (Amarzguioui et al., 2003). Although not demonstrated, an increased intracellular stability might be responsible for this observation.

The serum stability of single- and double-stranded phosphodiester RNA and the corresponding PTL containing counterparts was determined (Braasch et al., 2003). As expected, after a 30-second incubation in 5% fetal bovine serum, single-stranded RNA was no longer detectable. Strikingly, single-stranded phosphorothioate-RNA was also readily degraded in this assay system. Even more surprisingly, an unmodified siRNA was stable for up to 72 hours and this stability could not be further improved by a complete phosphorothiolation of both siRNA strands. On the other hand, upon incubation in a HeLa cell extract, the beneficial effect of phosphorothioates on the half-life of siRNA and single-stranded RNA was demonstrated (Chiu and Rana, 2003).

Introduction of an inverted deoxy abasic modification (iB) at the termini of the siRNA, which is known to protect nucleic acids against degradation by exonucleases, did not result in a protection towards degradation by serum nucleases. Furthermore, siRNA carrying aminohexyl phosphodiester modifications on all termini was tested for stability in serum. Because of the rapid degradation of this "end-capped," two 2'-deoxythymidine 3'-overhang carrying siRNA in serum, it was reasoned that siRNA is predominantly degraded by serum endonucleases (Czauderna et al., 2003b). Very interestingly, serum incubation of this siRNA revealed that the unmodified blunt-end version was significantly more stable after 120 minutes than the "end-capped" siRNA carrying the dTdT– overhangs, which were initially thought to protect against degradation by nucleases (Elbashir et al., 2001c). Unfortunately, the authors did not comment on the stability of unmodified blunt end siRNA versus their end-modified counterparts.

Monomolecular siRNA

Alternative to a linker nucleotide sequence, both siRNA strands can be covalently connected via non-nucleotide linkers, e.g., by ethylene glycol derivatives of different length (Thomson et al. 1993). Independent work by two groups demonstrates that in synthetically derived hairpin RNA, chemical structure and size of the linker sequence interlocking the complementary strands had no significant influence on RNAi activity (Harborth et al., 2003; Czauderna et al., 2003a). In the later study, the relative position of the linker, i.e. clamping via the 3' or 5' end of the antisense strand, does not govern silencing efficiencies, at least for hairpins with 19-bp stems. For stems consisting of more than 19 bp this linker orientation effect seems to be negligible. The other study demonstrated a bias towards more active siRNA with the linker located on the 5'-terminus of the antisense strand (Czauderna et al., 2003a). Most importantly, these constructs show a target down-regulation efficiency which is similar and in some cases even better than that of bimolecular standard siRNA (Harborth et al., 2003). On the other hand, Brummelkamp demonstrated that the RNAi activity of plasmid-vector derived, intracellularly transcribed short hairpin RNA (shRNA) is dependent on the number of nucleotides in the loop (Brummelkamp et al., 2002). Whether a processing of monomolecular siRNA is required for activity is not completely clear yet. Whereas for transcribed shRNA a processing step was demonstrated to be essential (Brummelkamp et al., 2002), no processing products of expressed shRNA or synthesized hairpin siRNA could be detected by Northern blot after strongly denaturing gel electrophoresis (Czauderna et al., 2003a).

Different non-nucleotide linkers did not inhibit activity of the corresponding monomolecular siRNA in HeLa cells and therefore can be used to replace the more expensive nucleotide loops (Harborth et al., 2003; Czauderna et al., 2003a). Non-nucleotide linker containing siRNA is of specific interest since these molecules are characterized by additional advantageous features. First, a non-nucleotide linker should help to enhance nuclease resistance. Second, these constructs help to reduce synthesis costs as only one equivalent of solid support is needed. Third, the subsequent post-synthesis handling (deprotection and purification) is reduced. Especially when considering the manufacture of an active pharmaceutical ingredient only one synthesis column and half the amount of solid support is needed. This is not only a major cost factor in large synthesis, but also an economically attractive characteristic of likker-containing siRNA. However, as these molecules are twice as long as standard siRNA the RNA synthesis technology must be capable of allowing stepwise coupling yields greater than 99% in order to achieve useful final yields. In addition a non-nucleotide linker might serve as an anchor for the attachment of a tissue or organ targeting enabling entity.

Fluorescent labeling of siRNA

Despite the change in the overall physico-chemical properties of siRNA caused by the attachment of bulky hydrophobic fluorescence dyes, these conjugates are widely used to address the following issues: 1) Analysis of cellular distribution of siRNA; 2) Assessment of transfection efficiency; 3) Expression attenuation; 4) Correlation of siRNA up-take with down-regulation of target protein (in combination with labeled antibodies). The 5'-end of the sense strand is ideally suited to attach the fluorescent reporter group as the dye can be linked to the oligoribonucleotide during solid-phase synthesis yielding a full-length product which differs in the chromatographic properties from the (n-1) product, thus facilitating purification by means of reversed phase or ion exchange high-performance liquid chromatography. Fluorescent siRNA is commonly used to assess transfection efficiencies and to test for the optimal transfection reagent for a given cell type. Fluorescein and the Cy-dyes™ are preferentially selected for that purpose.

5'-Cy3-labeled siRNA was delivered to Kasumi-1 cells to analyze electroporation efficacy as well as intracellular distribution. As judged by fluorescence-activated cell sorting (FACS) analysis almost all cells harbored Cy3 fluorescence sixteen hours after transfection. Fluorescence microscopy revealed a cytoplasmic distribution of the fluorescent label, whereas the nuclear regions were only weakly stained (Heidenreich et al., 2003). The sub-cellular distribution of siRNA was monitored by confocal microscopy after lipofection of HeLa cells with 3′-end Alexa 488 labeled siRNA (Harborth et al., 2003). Fluorescence was localized to discrete foci on the cytoplasmatic side of the nuclear membrane. Nevertheless, it was suspected that cellular uptake of fluorescently labeled siRNA correlates with gene silencing, as formation of small foci of concentrated fluorescent dyes in endosomal compartments were observed independent of a knock-down phenotype. It was also pointed out that cell sorting could be misleading when utilizing the transfected fluorophore on the siRNA in order to enrich the silenced-cell population (Harborth et al., 2003). siRNA fluorescently labeled on the 5′-terminus of the sense strand was utilized in studies with non-adherent cell lines (Walters and Jelinek, 2002). With the aid of fluorescein-tagged siRNA, transfection efficiencies of different commercially available transfection agents were investigated employing FACS analysis. For some cells successful internalization of siRNA was not sufficient to achieve target silencing. A siRNA introduced into myeloma cells by means of lipofection did not result in gene silencing whereas the same siRNA delivered by electroporation triggered RNAi, indicating that these two different transfection techniques deliver the active agents to different cellular compartments.

Researchers from Ambion, Inc. used fluorescently labeled siRNA to track the molecules inside HeLa S3 cells (Ambion Technotes Newsletters, Volume 9, Number 3). They demonstrated that the labeled and unlabeled siRNA performed equivalently in terms of target down-regulation and specificity. These properties were not affected by the label being placed either on the sense and/or the antisense strand. Furthermore, fluorescence resonance energy transfer experiments were employed to confirm strand separation of siRNA as part of the RNAi mechanism in transfected mammalian cells. These data nicely complement results where a correlation between target silencing and siRNA strand separation was observed (Nykanen et al., 2001). Generally, fluorescence reporter groups are preferentially attached to at least one of the termini to follow the fate of siRNA inside cells employing fluorescence microscopy. As for radioactive phosphate end-labeled oligonucleotides, one should keep in mind that the fluorophore might potentially get cleaved by a cellular nuclease. As a consequence, analysis of the reporter may not yield the desired information on the parental molecule.

Conclusions

Numerous studies indicate that a substantial amount of chemical modification in both siRNA strands is tolerated by the RNAi machinery. These modifications have the potency to improve siRNA stability as well as pharmacokinetic properties of the molecule. Overall, we are optimistic that chemical modification in addition to conjugation strategies for various biomolecules will play a pivotal role in the transition of siRNA from a laboratory knock-down tool into a valuable platform for the development of therapeutics.

Acknowledgement

The authors thank Fritz Eckstein for valuable suggestions and Anke Geick, Birgit Bramlage and Matthias John for critical reading of the manuscript.

REFERENCES

Amarzguioui, M., Holen, T., Babaie, E., and Prydz, H. (2003). Tolerance for mutations and chemical modifications in a siRNA. *Nucleic Acids Research*, **31**, 589–595.

Beaucage, S. L. and Radhakrishnan, P. I. (1992). Advances in the synthesis of oligonucleotides by the phosphoramidite approach. *Tetrahedron*, **48**, 2223–2311.

Beigelman, L., McSwiggen, J. A., Draper, K. G., Gonzalez, C., Jensen, K., Karpeisky, A. M., Modak, A. S., Matulic-Adamic, J., DiRenzo, A. B., Haeberli, P., Sweedler, D., Tracz, D., Grimm, S., Wincott, F., Thackray, V.G. and Usman, N. (1995). Chemical modification of hammerhead ribozymes. Catalytic activity and nuclease resistance. *Journal of Biological Chemistry*, **270**, 25702–25708.

Boutla, A., Delidakis, C., Livadaras, I., Tsagris, M., and Tabler, M. (2001). Short 5′-phosphorylated double-stranded RNAs induce RNA interference in *Drosophila*. *Current Biology*, **11**, 1776–1780.

Braasch, D. A., Jensen, S., Liu, Y., Kaur, K., Arar, K., White, M. A., and Corey, D. R. (2003). RNA interference in mammalian cells by chemically modified RNA. *Biochemistry*, **42**, 7967–7975.

Brummelkamp, T. R., Bernards, R., and Agami, R. (2002). A system for stable expression of short interfering RNAs in mammalian cells. *Science*, **296**, 550–553.

Caplen, N. J., Parrish, S., Imani, F., Fire, A., and Morgan, R. A. (2001). Specific inhibition of gene expression by small double-stranded RNAs in invertebrate and vertebrate systems. *Proceedings of the National Academy of Science USA*, **98**, 9742–9747.

Chiu, Y. L., and Rana, T. M. (2002). RNAi in human cells: Basic structural and functional features of small interfering RNA. *Molecular Cell*, **10**, 549–561.

Chiu, Y. L., and Rana, T. M. (2003). siRNA function in RNAi: A chemical modification analysis. *RNA*, **9**, 1034–1048.

Cook, P. D. (1999). Making drugs out of oligonucleotides: A brief review and perspective. *Nucleosides & Nucleotides*, **18**, 1141–1162.

Crooke, S.T. (2001). *Antisense drug technology*, New York, NY and Basel, Switzerland; Marcel Dekker, Inc.

Czauderna, F., Fechtner, M., Aygun, H., Arnold, W., Klippel, A., Giese, K., and Kaufmann, J. (2003a). Functional studies of the PI(3)-kinase signalling pathway employing synthetic and expressed siRNA. *Nucleic Acids Research*, **31**, 670–682.

Czauderna, F., Fechtner, M., Dames, S., Aygun, H., Klippel, A., Pronk, G. J., Giese, K., and Kaufmann, J. (2003b). Structural variations and stabilising modifications of synthetic siRNAs in mammalian cells. *Nucleic Acids Research*, **31**, 2705–2716.

Eckstein, F. (2002). Developments in RNA chemistry, a personal view. *Biochimie*, **84**, 841–848.

Elbashir, S. M., Harborth, J., Lendeckel, W., Yalcin, A., Weber, K., and Tuschl, T. (2001a). Duplexes of 21-nucleotide RNAs mediate RNA interference in cultured mammalian cells. *Nature*, **411**, 494–498.

Elbashir, S. M., Lendeckel, W., and Tuschl, T. (2001b). RNA interference is mediated by 21- and 22-nucleotide RNAs. *Genes & Development*, **15**, 188–200.

Elbashir, S. M., Martinez, J., Patkaniowska, A., Lendeckel, W., and Tuschl, T. (2001c). Functional anatomy of siRNAs for mediating efficient RNAi in *Drosophila melanogaster* embryo lysate. *European Molecular Biology Organization Journal*, **20**, 6877–6888.

Fire, A., Xu, S., Montgomery, M. K., Kostas, S. A., Driver, S. E., and Mello, C. C. (1998). Potent and specific genetic interference by double-stranded RNA in *Caenorhabditis elegans*. *Nature*, **391**, 806–811.

Hamada, M., Ohtsuka, T., Kawaida, R., Koizumi, M., Morita, K., Furukawa, H., Imanishi, T., Miyagishi, M., and Taira, K. (2002). Effects on RNA interference in gene expression (RNAi) in cultured mammalian cells of mismatches and the introduction of chemical modifications at the 3'-ends of siRNAs. *Antisense and Nucleic Acid Drug Development*, **12**, 301–309.

Hannon, G. J. (2002). RNA interference. *Nature*, **418**, 244–251.

Harborth, J., Elbashir, S. M., Vandenburgh, K., Manninga, H., Scaringe, S. A., Weber, K., and Tuschl, T. (2003). Sequence, chemical, and structural variation of small interfering RNAs and short hairpin RNAs and the effect on mammalian gene silencing. *Antisense and Nucleic Acid Drug Development*, **13**, 83–105.

Heidenreich, O., Benseler, F., Fahrenholz, A., and Eckstein, F. (1994). High activity and stability of hammerhead ribozymes containing 2'-modified pyrimidine nucleosides and phosphorothioates. *Journal of Biological Chemistry*, **269**, 2131–2138.

Heidenreich, O., Krauter, J., Riehle, H., Hadwiger, P., John, M., Heil, G., Vornlocher, H.-P., and Nordheim, A. (2003). AML1/MTG8 oncogene suppression by small interfering RNAs supports myeloid differentation of t(8;21)-positive leukemic cells. *Blood*, **101**, 3157–3163.

Hohjoh, H. (2002). RNA interference (RNA(i)) induction with various types of synthetic oligonucleotide duplexes in cultured human cells. *Federation of the European Biochemical Society Letters*, **521**, 195–199.

Holen, T., Amarzguioui, M., Wiiger, M. T., Babaie, E., and Prydz, H. (2002). Positional effects of short interfering RNAs targeting the human coagulation trigger Tissue Factor. *Nucleic Acids Research*, **30**, 1757–1766.

James, H. A. and Gibson, I. (1998). The therapeutic potential of ribozymes. *Blood*, **91**, 371–382.

Jarvis, T. C., Wincott, F. E., Alby, L. J., McSwiggen, J. A., Beigelman, L., Gustofson, J., DiRenzo, A., Levy, K., Arthur, M., Matulic-Adamic, J., Karpeisky, A., Gonzalez, C., Woolf, T. M., Usman, N., and Stinchcomb, D. T. (1996). Optimizing the cell efficacy of synthetic ribozymes. Site selection and chemical modifications of ribozymes targeting the proto-oncogene c-myb. *Journal of Biological Chemistry*, **271**, 29107–29112.

Kurreck, J. (2003). Antisense technologies. Improvement through novel chemical modifications. *European Journal of Biochemistry*, **270**, 1628–1644.

Lewis, D. L., Hagstrom, J. E., Loomis, A. G., Wolff, J. A., and Herweijer, H. (2002). Efficient delivery of siRNA for inhibition of gene expression in postnatal mice. *Nature Genetics*, **32**, 107–108.

Lipardi, C., Wei, Q., and Paterson, B. M. (2001). RNAi as random degradative PCR: siRNA primers convert mRNA into dsRNAs that are degraded to generate new siRNAs. *Cell*, **107**, 297–307.

Manoharan, M. (1999). 2'-carbohydrate modifications in antisense oligonucleotide therapy: importance of conformation, configuration and conjugation. *Biochimica et Biophysica Acta*, **1489**, 117–130.

McCaffrey, A. P., Meuse, L., Pham, T. T., Conklin, D. S., Hannon, G. J., and Kay, M. A. (2002). RNA interference in adult mice. *Nature*, **418**, 38–39.

Nykanen, A., Haley, B., and Zamore, P. D. (2001). ATP requirements and small interfering RNA structure in the RNA interference pathway. *Cell*, **107**, 309–321.

Opalinska, J. B., and Gewirtz, A. M. (2002). Nucleic-acid therapeutics: Basic principles and recent applications. *Nature Reviews in Drug Discovery*, **1**, 503–514.

Parrish, S., Fleenor, J., Xu, S., Mello, C., and Fire, A. (2000). Functional anatomy of a dsRNA trigger: Differential requirement for the two trigger strands in RNA interference. *Molecular Cell*, **6**, 1077–1087.

Rubinson, D. A., Dillon, C. P., Kwiatkowski, A. V., Sievers, C., Yang, L., Kopinja, J., Rooney, D. L., Ihrig, M. M., McManus, M. T., Gertler, F. B., Scott, M. L., and van Parijs, L. (2003). A lentivirus-based system to functionally silence genes in primary mammalian cells, stem cells and transgenic mice by RNA interference. *Nature Genetics*, **33**, 401–406.

Schwarz, D. S., Hütvagner, G., Haley, B., and Zamore, P. D. (2002). Evidence that siRNAs function as guides, not primers, in the *Drosophila* and human RNAi pathways. *Molecular Cell*, **10**, 537–548.

Sijen, T., Fleenor, J., Simmer, F., Thijssen, K. L., Parrish, S., Timmons, L., Plasterk, R. H., and Fire, A. (2001). On the role of RNA amplification in dsRNA-triggered gene silencing. *Cell*, **107**, 465–476.

Song, E., Lee, S. K., Wang, J., Ince, N., Ouyang, N., Min, J., Chen, J., Shankar, P., and Lieberman, J. (2003). RNA interference targeting Fas protects mice from fulminant hepatitis. *Nature Medicine*, **9**, 347–351.

Sun, L. Q., Cairns, M. J., Saravolac, E. G., Baker, A., and Gerlach, W. L. (2000). Catalytic nucleic acids: From lab to applications. *Pharmacolocical Reviews*, **52**, 325–347.

Thomson, J. B., Tuschl, T., and Eckstein, F. (1993). Activity of hammerhead ribozymes containing non-nucleotidic linkers. *Nucleic Acids Research*, **21**, 5600–5603.

Usman, N. and Blatt, L. M. (2000). Nuclease-resistant synthetic ribozymes: Developing a new class of therapeutics. *Journal of Clinical Investigations*, **106**, 1197–1202.

Walters, D. K. and Jelinek, D. F. (2002). The effectiveness of double-stranded short inhibitory RNAs (siRNAs) may depend on the method of transfection. *Antisense and Nucleic Acid Drug Development*, **12**, 411–418.

Wincott, F., DiRenzo, A., Shaffer, C., Grimm, S., Tracz, D., Workman, C., Sweedler, D., Gonzalez, C., Scaringe, S., and Usman, N. (1995). Synthesis, deprotection, analysis and purification of RNA and ribozymes. *Nucleic Acids Research*, **23**, 2677–2684.

Xia, H., Mao, Q., Paulson, H. L., and Davidson, B. L. (2002). siRNA-mediated gene silencing *in vitro* and *in vivo*. *Nature Biotechnology*, **20**, 1006–1010.

15 RNA interference in postimplantation mouse embryos

Frank Buchholz, Federico Calegari, Ralf Kittler, and Wieland B. Huttner

Introduction

Sequencing of whole genomes has provided new perspectives into the blueprints of diverse organisms, including the genome of the mouse (Waterston et al., 2002). Although the complete sequence is now available, the estimation of total gene number encoded by the mouse genome is ranging approximately from 25,000 to 50,000 (Okazaki et al., 2002). This uncertainty about the functional units within the genome highlights the importance of a detailed analysis of the encoded genes.

A significant step toward a better understanding of the genome has been the development of large-scale gene expression analysis tools utilizing DNA microarrays (Bono et al., 2003). This technology allows the generation of gene expression profiles that can give important clues for the interpretation of biological processes. However, the obtained data do not directly address the function of individual genes. Rather, they present a snapshot of global gene expression changes. While this is a very useful parameter for understanding the genome, it is not very useful for studying detailed phenotypic changes after gene ablation.

About 15 years ago gene function analysis became available in the mouse through the development of gene knock-out technology (Capecchi, 1989). In this approach genes are targeted in embryonic stem (ES) cells through homologous recombination. The manipulated ES cells are subsequently injected into blastocysts, and chimeric offspring are checked for germline transmission. Successful germline transmission allows the production of animals deficient in the gene of interest. Careful phenotypic analyses of these animals can then disclose the function(s) of the knocked-out gene. This approach is very time consuming and cost intensive, and it would be a difficult undertaking to study all genes of the mouse genome using this technology. In addition, the amount of information gained by classical gene knock-outs is often limited, especially when the gene of interest reveals an embryonic lethal phenotype. Furthermore, gene redundancy and compensatory mechanisms often prevent the appearance of observable phenotypes, hence requiring the production of double or triple knock-out organisms.

Conditional knock-out technologies have been developed to circumvent early lethal phenotypes (reviewed in Lewandoski, 2001; Kuhn and Schwenk, 2003). In particular, the Cre/loxP system has been developed to allow the detailed temporal and spatial analyses of gene function in the mouse. In such studies, standard gene-targeting techniques are used to produce a mouse in which a region of a gene of interest is flanked by recognition target sites (loxP) of the Cre recombinase. By crossing such a mouse line to a mouse line that expresses Cre recombinase in a tissue-specific manner, progeny are produced in which the conditional allele is inactivated only in those cells that express Cre. While the power of this approach is evident, it has the disadvantage of being even more expensive and labour intensive than classical knock-outs.

The realization that RNA interference (RNAi) is a useful tool for mammalian gene function studies has had a big impact on molecular biology. The rapid acceptance of this technology in the research community resulted primarily from the ease of use of RNAi technology and the strong need for a reliable method to down-regulate individual genes to elucidate their function. RNAi is not limited to mammalian cells grown in tissue culture, but can also be used for gene knock-down studies in whole organisms, including gene silencing in the mouse. Specifically, RNA interference (RNAi) has become available to mammalian cells in general (Elbashir et al., 2001a), and to mouse genetics in particular (McCaffrey et al., 2002; Lewis et al., 2002; Calegari et al., 2002; Mellitzer et al., 2002; Hasuwa et al., 2002). RNA interference (RNAi) is a cellular mechanism that can be used to down-regulate the expression of genes through the destruction of the cognate mRNA. Because RNAi-mediated gene regulation occurs after the mRNA has been generated from the DNA template, this mechanism is also referred to as post-transcriptional gene silencing. Compared to the knock-out technology, loss of gene function analyses via RNAi are more cost effective and less labour intensive. Furthermore, simultaneous silencing of gene expression can be achieved by using a mixture of different siRNAs targeting different genes in order to circumvent compensatory mechanisms due to the existence of redundancy. It should be noted, however, that RNAi experiments produce gene knock-downs, not knock-outs.

To make RNAi useful to the study of mouse development, systems are required that allow embryo manipulation without affecting developmental processes that follow the manipulation. Such a system is provided by the technology of mouse whole embryo culture (Cockroft, 1990). This technique allows the normal development of mouse postimplantation embryos for up to two days *in vitro*. Importantly, several kind of manipulations of the mouse embryo have been developed, including topical injection followed by directional electroporation (Akamatsu et al., 1999).

Here we describe the production of endoribonuclease prepared short interfering RNA (esiRNA) as a cost-effective way to generate a cocktail of different siRNAs and its use for gene knock-down studies in the postimplantation mouse embryo. We combine (i) the manipulation of mouse embryos by topical injection and directional electroporation of esiRNA with (ii) whole embryo culture, to rapidly

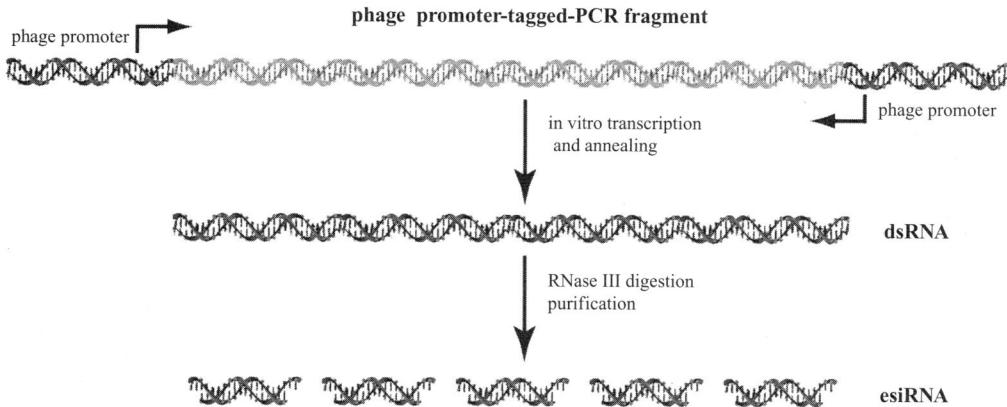

Figure 15.1. Schematic presentation of esiRNA production. Steps carried out to obtain a pool of functional siRNAs are illustrated.

analyse gene function in a tissue- and time-specific fashion in the developing mouse.

The combination provides a fast, powerful, and economical platform to analyse gene function during mouse development.

Generation of endoribonuclease-prepared siRNAs (esiRNAs)

The origin and mechanisms of RNA interference, design and synthesis have been described in depth in other chapters of this book. The present chapter highlights the generation of esiRNAs, a method that we employed to knock down genes important for mouse embryonic development. Initial RNAi experiments with siRNAs where carried out in mammalian tissue culture cells (Elbashir et al., 2001a). Soon after these reports several laboratories published the result that siRNAs could also silence gene expression in the adult (McCaffrey et al., 2002; Lewis et al., 2002; Hasuwa et al., 2002) and in the postimplantation mouse embryo (Calegari et al., 2002). As in the initial experiments in tissue culture cells most groups used reporter genes like luciferase and GFP to demonstrate the functionality of RNAi in the mouse. However, a recent report indicates that phenotypes observed via RNAi resemble phenotypes produced by classical gene knock-out through gene targeting in ES cells (Kunath et al., 2003). This study demonstrated that RNAi can be an efficient technology to study gene function in the whole animal.

Different methods have been developed for the generation of siRNA (reviewed in Kittler and Buchholz, 2003 and by Myers et al., Chapter 2 of this book). A different way to generate siRNA molecules is the enzymatic digestion of *in vitro* transcribed and annealed long double-stranded RNA (dsRNA) by an endoribonuclease like Dicer (Kawasaki et al., 2003; Myers et al., 2003), or *E.coli* RNase III (Yang et al., 2002). These enzymes bind to dsRNA and cleave them to produce endoribonuclease-prepared short interfering RNAs, or esiRNAs (Figure 15.1). The advantages of this method are that it is fast and cost effective, and that through this process a pool of different siRNAs is produced that usually contains effective silencing molecules. Therefore, screening for potent siRNAs is not necessary.

A potential disadvantage of this method that has been of concern is that the pool of different siRNAs produced may contain molecules that could induce cross-silencing (for a recent review see Check, 2003). However, published data (Yang et al., 2000) and unpublished data from our own laboratories (F. Buchholz, unpublished observations) suggest that cross-silencing of homologous genes is typically not observed. In fact, *in silico* experiments with a large number of esiRNAs directed against different genes showed that less than 10% of those contain any siRNA that matches a different gene (Henschel et al., 2004). Furthermore, we have not seen cross-silencing of highly homologous genes utilizing quantitative RT-PCR experiments (F. Buchholz, unpublished observations) and Western blot analyses (Yang et al., 2002).

Manipulation of whole postimplantation mouse embryos

Whole embryo culture

To understand the mechanisms controlling prenatal mammalian development, systems are required that allow observation and manipulation of the whole organism during its embryonic life. Because mammalian embryos develop inside the uterus, direct observation is not feasible and only a very limited kind of manipulation can be done. For this reason, for over a century great efforts have been made towards establishing reproducible techniques that allow manipulation and development *in vitro* indistinguishable at the morphological, cellular and molecular level from that occurring *in utero* (Heape, 1890; Nicholas and Rudnick, 1938; New, 1978; Cockroft, 1990).

A few days after fertilisation mammalian embryos descend the oviduct and adhere to the uterine wall (E4.5 in the mouse) in a process called implantation. Due to the reduced embryonic size and to a minor dependency on the uterine environment, preimplantation mouse embryos can be efficiently kept in culture for several days and through all stages of preimplantation development (for reviews see Biggers, 1998; Diaz-Cueto and Gerton, 2001). In contrast, the culture of postimplantation rodent embryos, which require more complex culture conditions, can be achieved only until the organogenetic stage and support normal development *in vitro* for up to two days (New, 1978; Cockroft, 1990). However, because many of the major processes of development (such as neurulation, somitogenesis and development of the cardiovascular, digestive and locomotor system) occur after implantation, the culture of postimplantation, rather than preimplantation, embryos is an attractive and powerful system towards a better understanding of the cellular and molecular processes of mammalian development.

The culture of postimplantation whole mouse embryos can be carried out from egg cylinder (E5.5) to about the 60th somite stage (E14), with the highest efficiency and reliability from early-somite (E7.5) to organogenetic stage (E11.5). Dissection and preparation of rodent embryos for culture are now very-well-established procedures, described in detail in several laboratory manuals (Cockroft, 1990; Nagy et al., 2003). According to standard protocols, embryos are freed from uterine walls and decidual tissue and then cultured with their yolk sac and ectoplacental cone. After dissection, the culture is carried out in medium containing 50 to 100% of immediately centrifuged, heat-inactivated rodent serum, and continuous rotation and oxygenation of this medium are carried out during the whole period of culture. These conditions are known to support normal embryonic development *in vitro* for up to two days both at the morphological (New, 1978; Cockroft, 1990) and, as shown for the onset and progression of the gradients of neurogenesis, at the cellular and molecular level (Calegari and Huttner, 2003 and unpublished data).

For several decades, different kinds of manipulation were combined with whole embryo culture. In the late 1970s, the use of whole embryo culture as a system of choice for toxicological and teratogenic studies reflected the tragic episode of thalidomide, sold as a sleeping drug but inducing severe abnormalities in human foetal development (for a review see Webster et al., 1997). For a long time, the application of potentially toxic compounds remained the almost exclusive kind of manipulation of mouse embryo developing *in vitro*. However, in recent years many other techniques have been successfully applied to whole embryo culture in order to study the mechanisms of mammalian development, such as cell or tissue transplantation, antibody interference, cell lineage tracing, infection with viral vectors and electroporation of nucleic acid (Drake and Little, 1991; Tam, 1998; Inoue et al., 2000; Oback et al., 2000; Osumi and Inoue, 2001).

The combination of these techniques with whole embryo culture offers a powerful system for understanding the mechanisms of mammalian prenatal development. In particular, the possibility of mediating gene transfer into target tissues of a developing organism offers a unique possibility for studies of gene function.

Gene transfer in postimplantation mouse embryos

Two methods of gene transfer have been applied in postimplantation mouse embryos viral vector-mediated gene transfer and electroporation.

One of the first means of obtaining gene transfer into postimplantation mouse embryos in culture (Figure 15.2B) involved viral vectors (Stuhlmann, 1984; Oback et al., 2000). Viral vector-mediated gene transfer as a means of delivering genes into target cells is based on the preparation of transgenic viral particles used to infect the target tissue. However, this method presents several disadvantages, such as the relatively time-consuming preparation of the viral particles, potentially toxic side effects of the viral infection, and, often, biosafety issues.

An alternative approach to mediating region-specific gene transfer into whole organisms is *in vivo* electroporation (Figure 15.2A). Electroporation, as an highly efficient means of delivering nucleic acids into target cells, is based on the capability of electric pulses to create transitory pores into the plasma membrane through which nuclei acids and other molecules such as drugs, peptides, proteins and polysaccharides can pass (Banga and Prausnitz, 1998). The electric field, in addition to "cell-poration," also induces the cathode-to-anode movement of the negatively charged nucleic acids, therefore mediating their migration into the permeabilized cells. When electroporation is applied upon injection of nucleic acids into the cavity of an organ, this allows the transfection to occur in only

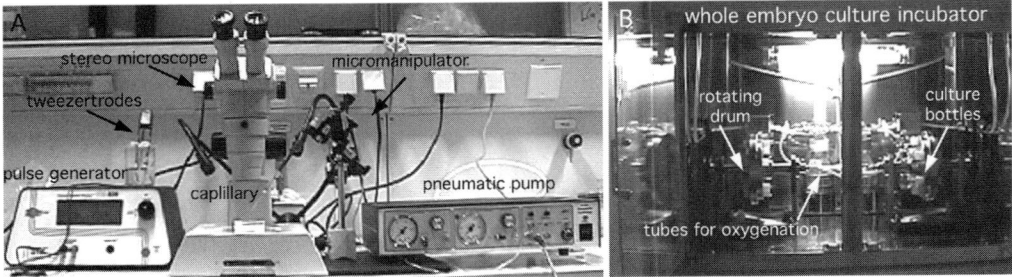

Figure 15.2. Equipment for *in vivo* electroporation and whole embryo culture. A. Picture of instruments used for *in vivo* electroporation. B. Picture of a whole embryo culture incubator (Ikemoto, Japan).

one half of the target organ, with the other, non-transfected, half serving as an internal negative control. Moreover, by using *in vivo* co-electroporation, multiple genes can easily be delivered into target cells simultaneously, with a high degree of efficiency (80–90%) (Saito and Nakatsuji, 2001; Calegari et al., 2002). Hence, electroporation of nucleic acids is emerging as a method of choice for gene function analyses (Swartz et al., 2001) and may constitute a fast and suitable approach for genome-wide functional screens. Finally, one additional advantage of electroporation is the extent of transfection of the target tissue achieved by using electrodes of the appropriate size and shape, which allows transfection to occur in entire regions, organs, tissues or even single cells of the organism (Figure 15.3). In addition, *in vivo* electroporation has also been carried out *in utero*, circumventing the limitation of whole embryo culture (Saito and Nakatsuji, 2001; Tabata and Nakajima, 2001; Takahashi et al., 2002), and can easily be applied to many other organisms and also to adult animals.

Possible disadvantages of *in vivo* electroporation are cell death and tissue damage caused by the electric fields. However, these side effects can be minimized by optimizing electroporation conditions such as voltage, size of the electrodes, etc. In addition to these classical methods of gene delivery, a recent report describes an alternative possibility of gene transfer into mouse embryos by tail injection of nucleic acids into pregnant mice (Gratsch et al., 2003). Although this novel approach does not allow tissue specificity, it opens the possibility for a very fast practical approach of gene transfer in developing embryos.

RNAi in postimplantation mouse embryos

The possibility of triggering silencing of gene expression by siRNAs in mammalian cells in culture (Elbashir et al., 2001a), together with the possibility of transfering nucleic acids by topical injection and directional electroporation *in vivo*, leads us to investigate RNAi in postimplantation mouse embryos (Calegari et al., 2002; Gratsch et al., 2003). In order to address this issue different experimental approaches were designed: first, silencing of exogenous gene expression in tissues facing cavities; second, silencing of exogenous gene expression in tissues lacking cavities; third, silencing of gene expression of an endogenously expressed

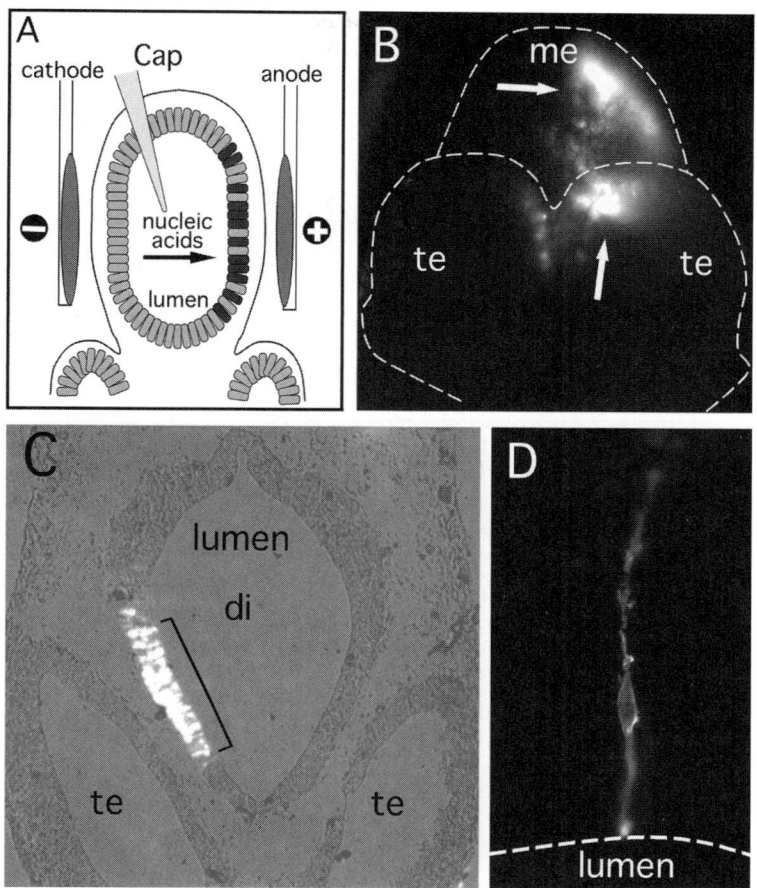

Figure 15.3. Gene transfer in postimplantation mouse embryos using directional electroporation. A. Cartoon illustrating the topical injection (Cap, capillary) of nucleic acids into the lumen of the neural tube of mouse embryo, with the electrodes in the lateral orientation, cathode-right/anode-left, and the uptake of nucleic acids into the left side of the neuroepithelium upon electroporation (darker cells). B. Frontal fluorescent view of the head of an unfixed mouse embryo (outlined by dashed lines) showing GFP expression in the left side of the dorso-lateral mesencephalon and caudal telencephalon (white region; arrows) upon injection of GFP reporter plasmid into the lumen of the anterior neural tube at E10, followed by electroporation and subsequent whole-embryo culture for 24 hours; te, telencephalic vesicles, me, mesencephalon. C. Superimposed phase contrast (grey) and fluorescence (white) micrograph of a transverse cryosection through the diencephalic (di) and telencephalinc (te) neural tube of a E10 mouse embryo upon injection into the lumen of the anterior neural tube and electroporation using a cathode-left/anode-right orientation, and subsequent 24 hours whole embryo culture. GFP expression in neuroepithelial cells of the right diencephalon (white cells) is indicated (black line). D. High magnification fluorescent micrograph of the mouse neuroepithelium (dashed line: apical, lumen facing side of the neuroepithelium) of a E10 mouse embryo showing plasma membrane localised GFP in a single neuroepithelial cell upon injection of GFP-GAP43 reporter plasmid into the lumen of the neural tube, electroporation and subsequent whole embryo culture for 24 hours.

transgene; and fourth, silencing of gene expression of an endogenously expressed gene.

First. Silencing of exogenous gene expression in tissues containing cavities, such as the neural tube and heart, was investigated as follows: Plasmid vectors

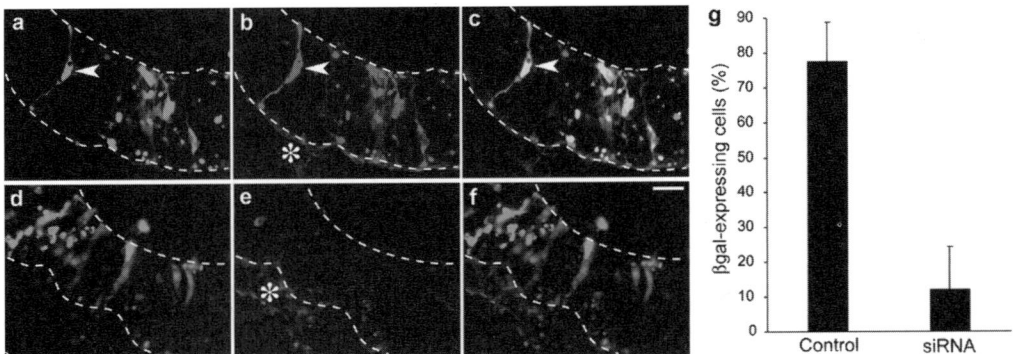

Figure 15.4. RNAi in the neuroepithelium of E10 mouse embryos. E10 mouse embryos were injected, into the lumen of the telencephalic neural tube, with the two reporter plasmids pEGFP-N2 (for GFP) and pSVpaXD (for βgal), either without (**a–c** and **g, Control**) or with (**d-f** and **g, siRNA**) βgal-directed esiRNAs, followed by directional electroporation and whole embryo culture for 24 hours. (**a-f**) Horizontal cryosections through the targeted region of the telencenphalon were analysed by double fluorescence for expression of GFP (green; **a** and **d**) and βgal immunoreactivity (red; **b** and **e**). Co-expression of GFP and βgal in neuroepithelial cells appears yellow in the merge (**c** and **f, arrowheads**). Note the lack of βgal expression in neuroepithelial cells in the presence of βgal-directed esiRNAs. Upper and lower dashed lines indicate the lumenal (apical) surface and basal border of the neuroepithelium, respectively. Asterisks in (**b** and **e**) indicate signal due to the cross-reaction of the secondary antibody used to detect βgal with the basal lamina and underlying mesenchymal cells. Scale bar in (**f**), 20 μm. (**g**) Quantitation of the percentage of GFP-expressing neuroepithelial cells that also express βgal without (**Control**) or with (**siRNA**) application of βgal-directed esiRNAs. Data are the mean of three embryos analyzed as in (**a-f**); bars indicate S.D. (Reprinted figure with permission from PNAS). (See color section.)

driving the constitutive expression of reporter genes were injected into the respective cavity, with or without reporter-directed esiRNAs, followed by directional electroporation and whole embryo culture for 24 hours.

When a mixture of two reporter genes, GFP and βgal, were co-injected and co-electroporated alone, i.e. without esiRNAs, into the E10 neuroepithelium of the anterior neural tube, almost all of the transfected neuroepithelial cells expressed, with an efficiency of up to 90%, both reporter genes after 24 hours of whole embryo culture. In contrast, when the two reporter genes were co-injected and co-electroporated together with a mixture of βgal-directed esiRNAs βgal expression was barely detectable [(Figure 15.4) (Calegari et al., 2002)]. Similarly, injection and electroporation into the E10 beating heart of a GFP reporter plasmid alone, i.e. without esiRNAs, showed, in the vast majority of cases, GFP expression after 24 hours of whole embryo culture. In contrast, when GFP reporter plasmids were co-injected and co-electroporated together with GFP-directed esiRNAs, no GFP expression could be seen. These experiments revealed a very high efficiency of co-transfection *in vivo* of two reporter genes and an almost complete, specific esiRNA-mediated silencing of exogenous gene expression in cavity-facing tissues. In addition, it is worth noting that in both cases analysed, the whole procedure of injection, electroporation and whole embryo culture did not affect, to any observable extent, the normal functionality and development in culture of the target organs.

Second. Silencing of exogenous gene expression in tissues lacking cavities, such as the ectoderm, was investigated as follows: Plasmid vectors driving the constitutive expression of reporter genes were released in proximity of the tissue, with or without reporter-directed esiRNAs, followed by directional electroporation and whole embryo culture for 24 hours.

When a mixture of two reporter genes, GFP and DsRed, together with unspecific esiRNAs as control, was co-administered into the culture medium in proximity of the limb bud ectoderm and co-electroporated, almost all transfected ectodermal cells co-expressed both reporter genes after 24 hours of whole embryo culture. In contrast, when the two reporter genes were co-injected and co-electroporated together with a mixture of DsRed-directed esiRNAs, DsRed expression was barely detectable (Calegari et al., submitted). These experiments confirmed the previously shown high efficiency of co-transfection achieved by *in vivo* electroporation (Saito and Nakatsuji, 2001), and the high efficiency and specificity of esiRNA-mediated gene silencing.

Third. Silencing of gene expression for the endogenously expressed transgene GFP was investigated by injecting, into the cavity of the neural tube, a mixture of GFP-directed esiRNAs, followed by directional electroporation and whole embryo culture for 24 hours.

RNAi specifically prevents the translation of complementary mRNA and has no effect on the pre-existing protein. For this reason, the effect of esiRNAs-mediated gene silencing should be particularly evident for genes whose transcription is succeeding the delivery of the siRNA. Knowing that during neurogenesis the gene TIS21 is specifically expressed in neuron-generating neuroepithelial cells, with an onset of expression in the mouse telencephalon at E10 (Iacopetti et al., 1999), we decided to investigate the possibility of silencing endogenous gene expression in the E10 neuroepithelium of mice obtained from a knock-in line expressing GFP from the TIS21 locus (Tis21[+/tm2(Gfp)Wbh], Haubensak et al. 2004).

Upon injection of GFP-directed esiRNAs into the anterior neural tube, electroporation using a cathode right/anode left orientation, and 24-hour whole embryo culture, we observed a dramatic reduction, by about 90%, of green fluorescence in the anode-facing left side of the targeted neural tube as compared to the controlateral, cathode-facing right side (Figure 15.5). Moreover, quantification of GFP expressing cells on fixed cryosections obtained from these regions of the neuroepithelium indicated a reduction of GFP-positive cells by about 75% as compared to the controlateral side of the neural tube (Calegari et al., 2002).

Fourth. We finally investigated the possibility to induce gene silencing for a truly endogenous, non-transgenic gene by injecting and electroporating TIS21-directed esiRNAs into the anterior neural tube of wild-type E10 mouse embryos, i.e. at the onset of TIS21 expression, followed by whole-embryo culture for 24 hours.

In situ hybridization on fixed slices obtained from these embryos showed that esiRNA-triggered RNAi may be achieved not only for an endogenously expressed transgene (GFP in the Tis21[+/tm2(Gfp)Wbh] mouse line), but also for TIS21 itself (Calegari et al., 2004). It is therefore conceivable that esiRNA-mediated gene

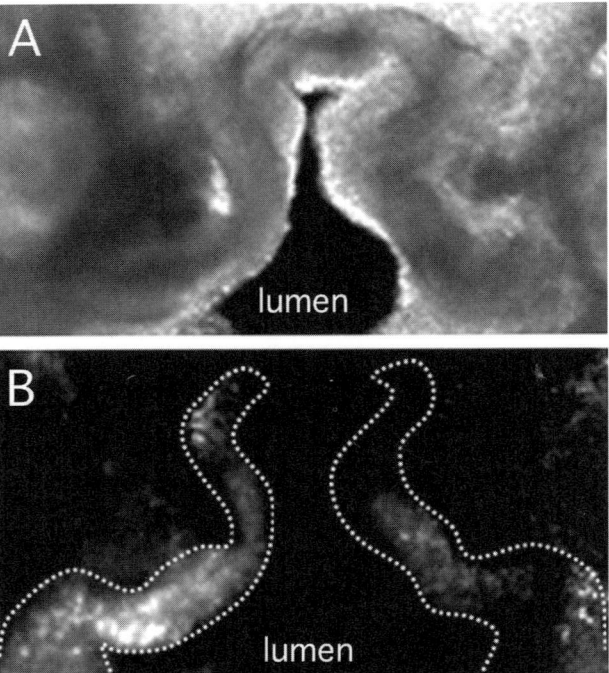

Figure 15.5. esiRNA-mediated RNAi of a gene endogenously expressed during the development of postimplantation mouse embryos. Heterozygous E10 *Tis21*[+/tm2(Gfp)Wbh] mouse embryos were injected with GFP-directed esiRNAs into the lumen of neural tube, followed by directed electroporation (lateral, cathode-right/anode-left orientation) and whole-embryo culture for 24 hours. A. Low-power dark-field micrographs of horizontal vibratome sections through the diencephalon. The left half of the brain is on the right side of the panels. B. Note the silencing of GFP expression in the selected neuroepithelium (dashed lines) in the left, anode-facing half of the embryo.

silencing may be achieved for any endogenously expressed gene by using topical injection and directional electroporation of esiRNAs.

These findings, together with reports from other laboratories showing specific knock-down of gene expression in postimplantation mouse embryos using either long dsRNAs (Mellitzer et al., 2002) or shRNAs (Gratsch et al., 2003), extend previous work showing siRNA-triggered RNAi in adult mice (Lewis et al., 2002; McCaffrey et al., 2002).

Conclusions

The use of topical injection and directional electroporation of esiRNAs to induce region- and tissue-specific silencing of gene expression during the development in whole embryo culture of postimplantation mouse embryos offers several advantages as compared to other classical techniques of gene knock-down.

The first advantage of this technique is the possibility to obtain gene knock-down without the labour-intensive and time-consuming generation of genetically modified animals. In addition, the combination of topical injection and directional electroporation (using electrodes of the appropriate size and shape) allows easy restriction of the silencing effect only to a selected portion of organ, tissue or even single cells. Another major advantage of this approach is the possibility of using a mixture of siRNAs directed against mRNAs of different genes as an easy and rapid way of preventing the possible appearance of compensatory effects due to gene redundancy. In fact, the knock-down of multiple genes using mixtures of siRNAs may constitute a feasible approach for genome-wide functional screens in order to systematically study gene function during mammalian development.

The present approach is mainly limited by the temporal restrictions of whole embryo culture, which can be efficiently carried out only for up to two days from E7 until E12. However, it should be noted that this limitation could be overcome both by *in utero* electroporation (Takahashi et al., 2002) and by tail injection into pregnant mice (Gratsch et al., 2003). Moreover, the use of DNA templates expressing shRNAs, as alternative to esiRNAs, would allow the extension of this technology to long-term studies of mouse development.

REFERENCES

Akamatsu, W., Okano, H. J., Osumi, N., Inoue, T., Nakamura, S., Sakakibara, S., Miura, M., Matsuo, N., Darnell, R. B. and Okano, H. (1999). Mammalian ELAV-like neuronal RNA-binding proteins HuB and HuC promote neuronal development in both the central and the peripheral nervous systems. *Proceedings of the National Academy of Sciences USA*, **96**, 9885–9890.

Banga, A. K. and Prausnitz, M. R. (1998). Assessing the potential of skin electroporation for the delivery of protein- and gene-based drugs. *Trends in Biotechnology*, **16**, 408–412.

Biggers, J. D. (1998). Reflections on the culture of the preimplantation embryo. *International Journal of Developmental Biology*, **42**, 879–884.

Bono, H., Yagi, K., Kasukawa, T., Nikaido, I., Tominaga, N., Miki, R., Mizuno, Y., Tomaru, Y., Goto, H., Nitanda, H. et al. (2003). Systematic expression profiling of the mouse transcriptome using RIKEN cDNA microarrays. *Genome Research*, **13**, 1318–1323.

Calegari, F., Haubensak, W., Yang, D., Huttner, W. B. and Buchholz, F. (2002). Tissue-specific RNA interference in postimplantation mouse embryos with endoribonuclease-prepared short interfering RNA. *Proceedings of the National Academy of Sciences USA*, **99**, 14236–14240.

Calegari, F. and Huttner, W. B. (2003). An inhibition of cyclin-dependent kinases that lengthens, but does not arrest, neuroepithelial cell cycle induces premature neurogenesis. *Journal of Cell Science*, **116**, 4947–4955.

Calegari, F., Marzesco, A. M., Kittler, R., Buchholz, F. and Huttner, W. B. (2004). Tissue-specific RNA interference in post-implantation mouse embryos using directional electroporation and whole embryo culture. *Differentiation*, **72**, 92–102.

Capecchi, M. R. (1989). Altering the genome by homologous recombination. *Science*, **244**, 1288–1292.

Check, E. (2003). Gene regulation: RNA to the rescue? *Nature*, **425**, 10–12.

Cockroft, D. L. (1990). Dissection and culture of postimplantation embryos. In *Postimplantatio mammalian embryos. A practical approach* (ed. Rickwood, D. and Hames, B. D.), pp. 15–40. Oxford: Oxford University Press.

Diaz-Cueto, L. and Gerton, G. L. (2001). The influence of growth factors on the development of preimplantation mammalian embryos. Archives of Medical Research, **32**, 619–626.

Drake, C. J. and Little, C. D. (1991). Integrins play an essential role in somite adhesion to the embryonic axis. *Developmental Biology*, **143**, 418–421.

Elbashir, S. M., Harborth, J., Lendeckel, W., Yalcin, A., Weber, K. and Tuschl, T. (2001a). Duplexes of 21-nucleotide RNAs mediate RNA interference in cultured mammalian cells. *Nature*, **411**, 494–498.

Elbashir, S. M., Lendeckel, W. and Tuschl, T. (2001b). RNA interference is mediated by 21- and 22-nucleotide RNAs. *Genes & Development*, **15**, 188–200.

Gratsch, T. E., De Boer, L. S. and O'Shea, K. S. (2003). RNA inhibition of BMP-4 gene expression in postimplantation mouse embryos. *Genetics*, **37**, 12–17.

Hasuwa, H., Kaseda, K., Einarsdottir, T. and Okabe, M. (2002). Small interfering RNA and gene silencing in transgenic mice and rats. *FEBS Letters*, **532**, 227–230.

Haubensak, W., Attardo, A., Denk, W. and Huttner, W. B. (2004). Neurons arise in the basal neuroepithelium of the early mammalian telencephalon: a major site of neurogenesis. *Proceedings of the National Academy of Science USA*, **101**, 3196–3201.

Heape, W. (1890). Preliminary note on the transplantation and growth of mammalian ova within a uterine foster-mother. *Proceedings of the Royal Society of London, B. Biological Sciences*, **48**, 457–459.

Henschel, A., Buchholz, F. and Habermann, B. (2004). DEQOR: a web-based tool for the design and quality control of siRNAs. *Nucleic Acids Res*. 2004 Jul 1;32 (Web Server issue):W113–20.

Iacopetti, P., Michelini, M., Stuckmann, I., Oback, B., Aaku-Saraste, E. and Huttner, W. B. (1999). Expression of the antiproliferative gene TIS21 at the onset of neurogenesis identifies single neuroepithelial cells that switch from proliferative to neuron-generating division. *Proceedings of the National Academy of Sciences* USA, **96**, 4639–4644.

Inoue, T., Nakamura, S. and Osumi, N. (2000). Fate mapping of the mouse prosencephalic neural plate. *Developmental Biology*, **219**, 373–383.

Kawasaki, H., Suyama, E., Iyo, M. and Taira, K. (2003). siRNAs generated by recombinant human Dicer induce specific and significant but target site-independent gene silencing in human cells. *Nucleic Acids Research*, **31**, 981–987.

Kittler, R. and Buchholz, F. (2003). RNA interference: Gene silencing in the fast lane. *Seminars in Cancer Biology*, **13**, 259–265.

Kuhn, R. and Schwenk, F. (2003). Conditional knockout mice. *Methods in Molecular Biology*, **209**, 159–185.

Kunath, T., Gish, G., Lickert, H., Jones, N., Pawson, T. and Rossant, J. (2003). Transgenic RNA interference in ES cell-derived embryos recapitulates a genetic null phenotype. *Nature Biotechnology*, **21**, 559–561.

Lewandoski, M. (2001). Conditional control of gene expression in the mouse. *Nature Reviews Genetics*, **2**, 743–755.

Lewis, D. L., Hagstrom, J. E., Loomis, A. G., Wolff, J. A. and Herweijer, H. (2002). Efficient delivery of siRNA for inhibition of gene expression in postnatal mice. *Nature Genetics*, **32**, 107–108.

McCaffrey, A. P., Meuse, L., Pham, T. T., Conklin, D. S., Hannon, G. J. and Kay, M. A. (2002). RNA interference in adult mice. *Nature*, **418**, 38–39.

Mellitzer, G., Hallonet, M., Chen, L. and Ang, S. L. (2002). Spatial and temporal 'knock down' of gene expression by electroporation of double-stranded RNA and morpholinos into early postimplantation mouse embryos. *Mechanisms of Development*, **118**, 57–63.

Myers, J. W., Jones, J. T., Meyer, T. and Ferrell, J. E., Jr. (2003). Recombinant Dicer efficiently converts large dsRNAs into siRNAs suitable for gene silencing. *Nature Biotechnology*, **21**, 324–328.

Nagy, A., Gertsenstein, M., Vintersten, K. and Behringer, R. R. (2003). In *Manipulating the mouse embryo. A laboratory manual* (eds. J. Inglis and J. Cuddihy), pp. 209–250. New York: Cold Spring Harbor Laboratory Press.

New, D. A. (1978). Whole-embryo culture and the study of mammalian embryos during embryogenesis. *Biological Reviews*, **53**, 81–122.

Nicholas, J. S. and Rudnick, D. (1938). The development of rat embryos in tissue culture. *Proceedings of the National Academy of Sciences USA*, 656–658.

Oback, B., Cid-Arregui, A. and Huttner, W. B. (2000). Gene transfer into cultured postimplantation mouse embryos using herpes simplex amplicons. In *Viral Vectors: Basic Science and Gene Therapy*, (ed. A. C.-A. A. García-Carrancá), pp. 277–293. Natick, MA: Eaton Publishing.

Okazaki, Y. Furuno, M. Kasukawa, T. Adachi, J. Bono, H. Kondo, S. Nikaido, I. Osato, N. Saito, R. Suzuki, H. et al. (2002). Analysis of the mouse transcriptome based on functional annotation of 60770 full-length cDNAs. *Nature*, **420**, 563–573.

Osumi, N. and Inoue, T. (2001). Gene transfer into cultured mammalian embryos by electroporation. *Methods*, **24**, 35–42.

Saito, T. and Nakatsuji, N. (2001). Efficient gene transfer into the embryonic mouse brain using *in vivo* electroporation. *Developmental Biology*, **240**, 237–246.

Stuhlmann, H., Cone, R., Mulligan, R. C. and Jaenisch, R. (1984). Introduction of a selectable gene into different animal tissue by a retrovirus recombinant vector. *Proceedings of the National Academy of Sciences USA*, **81**, 7151–7155.

Swartz, M., Eberhart, J., Mastick, G. S. and Krull, C. E. (2001). Sparking new frontiers: using *in vivo* electroporation for genetic manipulations. *Developmental Biology*, **233**, 13–21.

Tabata, H. and Nakajima, K. (2001). Efficient in utero gene transfer system to the developing mouse brain using electroporation: Visualization of neuronal migration in the developing cortex. *Neuroscience*, **103**, 865–872.

Takahashi, M., Sato, K., Nomura, T. and Osumi, N. (2002). Manipulating gene expressions by electroporation in the developing brain of mammalian embryos. *Differentiation*, **70**, 155–162.

Tam, P. P. (1998). Postimplantation mouse development: Whole embryo culture and micromanipulation. *International Journal of Developmental Biology*, **42**, 895–902.

Waterston, R. H. Lindblad-Toh, K. Birney, E. Rogers, J. Abril, J. F. Agarwal, P. Agarwala, R. Ainscough, R. Alexandersson, M. An, P. et al. (2002). Initial sequencing and comparative analysis of the mouse genome. *Nature*, **420**, 520–562.

Webster, W. S., Brown-Woodman, P. D. and Ritchie, H. E. (1997). A review of the contribution of whole embryo culture to the determination of hazard and risk in teratogenicity testing. *International Journal of Developmental Biology*, **41**, 329–335.

Yang, D., Buchholz, F., Huang, Z., Goga, A., Chen, C. Y., Brodsky, F. M. and Bishop, J. M. (2002). Short RNA duplexes produced by hydrolysis with Escherichia coli RNase III mediate effective RNA interference in mammalian cells. *Proceedings of the National Academy of Sciences USA*, **99**, 9942–9947.

16 *In ovo* RNAi opens new possibilities for functional genomics in vertebrates

Dimitris Bourikas, Thomas Baeriswyl, Rejina Sadhu, and Esther T. Stoeckli

Introduction

The development of high-throughput methods has changed the way genes are analyzed. In the past, a spontaneous or targeted mutation was a prerequisite for the identification of a gene and its function. Today, genome-sequencing projects and large-scale screens provide a tremendous amount of information about the genetic make-up of an organism. Unfortunately, the long lists of genes expressed in specific tissues or distinct phases of an organism's life provide little or no information about the function of the expressed proteins. Functional gene analysis remains a time-consuming and challenging step that requires changes in gene expression in the context of a living organism. Therefore, the availability of suitable model systems is key to our progress in understanding the role of genes in biological processes. Model systems need to be easily accessible and efficient in producing functional read-outs of gene manipulation. Due to constraints in time and money these criteria were met only by invertebrate animal models, such as *Drosophila* and *C. elegans* (Adams and Sekelsky, 2002; St. Johnston, 2002; Lee et al., 2003; Simmer et al., 2003). For many studies, however, vertebrate model systems are required. The mouse has been the most widely used vertebrate model system, because technologies for genetic manipulations based on homologous recombination in embryonic stem cells are available and allow for selected inactivation of target genes (Porter, 1998; Müller, 1999; Jackson, 2001). However, the high cost, the length of time required, and technical constraints restrict the usefulness of the mouse as a model system for functional genomics. In order to increase the pace of functional gene analysis new model systems for large-scale screens are required.

Choosing a model organism developmental studies

Depending on the specific question asked, further restraints in the selection of suitable model systems apply. As developmental neuroscientists, we depend on model systems that are accessible during development. Loss of gene function should be controllable both temporally and spatially in order to avoid early embryonic lethality that prevents the analysis of a gene's function during development of the nervous system. For different reasons, *Xenopus* and zebrafish have been established as alternative model systems to the mouse (Driever, 1995; Karlstrom et al., 1997; Chang and Hemmati-Brivanlou, 1998; Neuhauss, 2003). They are more easily accessible for developmental studies than the mouse, but the availability of genetic tools for gene manipulation is limited and often restricted to very early stages of development. Frogs were traditionally used by embryologists to study developmental processes (Chang and Hemmati-Brivanlou, 1998). The lack of genetic tools and their special life cycle limit their usefulness as model organisms. Zebrafish became very popular because their embryos are translucent and easy to handle. These properties made them very useful for modern imaging techniques (Dynes and Ngai, 1998; Jontes et al., 2000; Finley et al., 2001; O'Malley et al., 2003). In addition, zebrafish are small animals that can be easily bred under lab conditions.

To make up for the lack of genetic tools as compared to invertebrates or the mouse, the use of antisense oligonucleotides was established in both *Xenopus* and zebrafish. Morpholinos, modified antisense oligonucleotides, were developed to efficiently silence target genes when applied to very early stages of embryonic zebrafish development (Heasman, 2002). Morpholinos are designed to be more stable than conventional antisense oligonucleotides but they still suffer from some of the same disadvantages, namely the requirement of the 5′ UTR or the translation initiation site for their synthesis (Summerton and Weller, 1997; Morales and de Pablo, 1998; Lebedeva and Stein, 2001). Because they have to be injected early in development they can elicit unspecific effects and are limited in their efficiency for gene silencing at later stages (Nasevicius and Ekker, 2001).

In addition to *Xenopus* and zebrafish the chicken embryo was used as a developmental model system in experimental biology for more than a century. In contrast to mammals, chicken embryos are easily accessible for manipulation during development both *in ovo* as well as in *ex ovo* cultures (Figure 16.1). The use of chick/quail chimeras and microsurgical manipulations of chicken embryos have led to significant discoveries of developmental processes (Le Douarin, 1973; Le Douarin and Teillet, 1973; Le Lievre and Le Douarin, 1975; Selleck, 1996).

Unfortunately, the lack of genetic tools has hampered the usefulness of the chicken embryo as a model system in functional gene analyses (Bourikas and Stoeckli, 2003). We have used loss-of-function studies at the protein level to analyze axonal pathfinding in the chicken spinal cord (Stoeckli and Landmesser, 1995; Stoeckli et al., 1997; Burstyn-Cohen et al., 1999; Fitzli et al., 2000). For this purpose, we used function-blocking antibodies injected into the central canal of the spinal cord to interfere with guidance cues directing commissural axons toward and across the ventral midline of the spinal cord (reviewed in Stoeckli and Landmesser, 1998). In addition, we developed a similar approach to study subpopulation-specific pathfinding of sensory afferents in the gray matter of the spinal cord (Perrin et al., 2001).

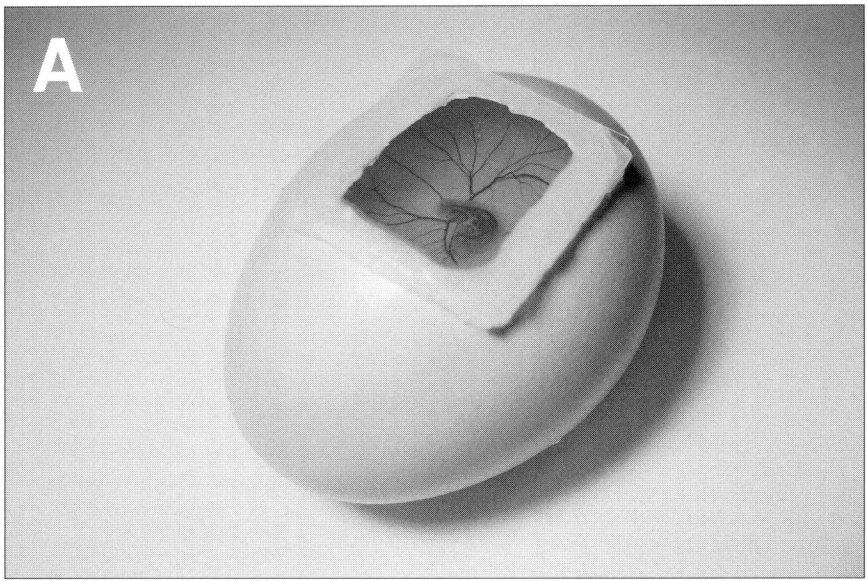

Figure 16.1. Chicken embryos are a good model system for developmental studies due to their accessibility. Chicken embryos can be accessed in ovo (A) through a window in the eggshell that can be resealed after manipulations with a coverslip and melted paraffin. As an alternative approach, chicken embryos can be used as ex ovo cultures (B). With both methods embryos can be kept alive throughout embryonic development. (See color section.)

The disadvantage of such studies is their limitation due to the requirement for specific and function-blocking antibodies. In order to extend the usefulness of the chicken embryo and to overcome these limitations, several methods were developed (reviewed in Bourikas and Stoeckli, 2003). For technical reasons gain-of-function studies were more readily available. Using RCAS vectors (Replication-Competent ASLV long terminal repeat with a Splice acceptor), a family of retroviral vectors derived from the Rous sarcoma virus (Bell and Brickell, 1997;

Petropoulos and Hughes, 1991), genes of interest could be expressed in different parts of the chicken embryo (Cepko, 1988; Fekete and Cepko, 1993; Morgan and Fekete, 1996; Logan and Tabin, 1998). The disadvantage of the RCAS vector is its limitation to 2.5 kb of insert size and the fact that it infects only dividing cells successfully. The same problems were encountered with other viral vectors, such as adenoviral or lentiviral vectors, although they are less restrictive with respect to insert size.

In 1997, *in ovo* electroporation was established as an efficient tool for gene transfer (Muramatsu et al., 1997). The efficiency of electroporation exceeded viral vectors and was soon applied also for loss-of-function studies using constructs encoding dominant negative mutants of proteins. Still, unless dominant negative mutants of genes could be expressed, loss-of-function approaches were not possible.

Because loss-of-function phenotypes are often more informative (Hudson et al., 2002) and because gain-of-function studies require the expression of the entire coding sequence, we were searching for a new tool that would allow us to study the function of candidate genes in axonal pathfinding. Because we had identified our candidates in a screen based on subtractive hybridization, we did not have full-length sequences of our cDNAs. In order to preselect those candidate genes for further analysis that were functionally relevant, we turned to RNAi. Combining the advantages of *in ovo* electroporation for efficient transfer of nucleic acids across cell membranes with the gene silencing effect of dsRNA allowed us to use the chicken embryo as an efficient model system for functional genomics.

Using *in ovo* RNAi to screen for axon guidance cues

We had identified guidance cues directing axons of commissural neurons from their dorsal origin toward the floor plate, the ventral midline of the spinal cord, with function-blocking antibodies (Stoeckli and Landmesser, 1995; Stoeckli et al., 1997; Fitzli et al., 2000). In these *in vivo* studies, we demonstrated an effect of the cell adhesion molecules axonin-1, NrCAM, and NgCAM in commissural axon growth and guidance (reviewed in Stoeckli and Landmesser, 1998). While axonin-1 interacting with NgCAM was required for the fasciculation of commissural axons, the interaction of axonin-1, expressed on commissural axons, with NrCAM, expressed on the floor plate surface, was required for commissural axons to cross the midline (see Figure 16.2). In the absence of axonin-1/NrCAM interactions some commissural axons failed to enter the floor plate and turned into the longitudinal axis prematurely (Figure 16.2B and C). The perturbation of NgCAM interactions resulted in a defasciculation of commissural axons both before (Figure 16.2J) and after midline crossing (Figure 16.2D). A defasciculation of commissural axons was also seen after perturbation of axonin-1 interactions (Figure 16.2B) but not after perturbation of NrCAM interactions (Figure 16.2C).

Using recombinant protein and function-blocking antibodies against different domains of the protein, we demonstrated the role of F-spondin in commissural axons' turn into the longitudinal axis (Burstyn-Cohen et al., 1999). Although their turning angle was no longer restricted to the stereotypic 90° found in control

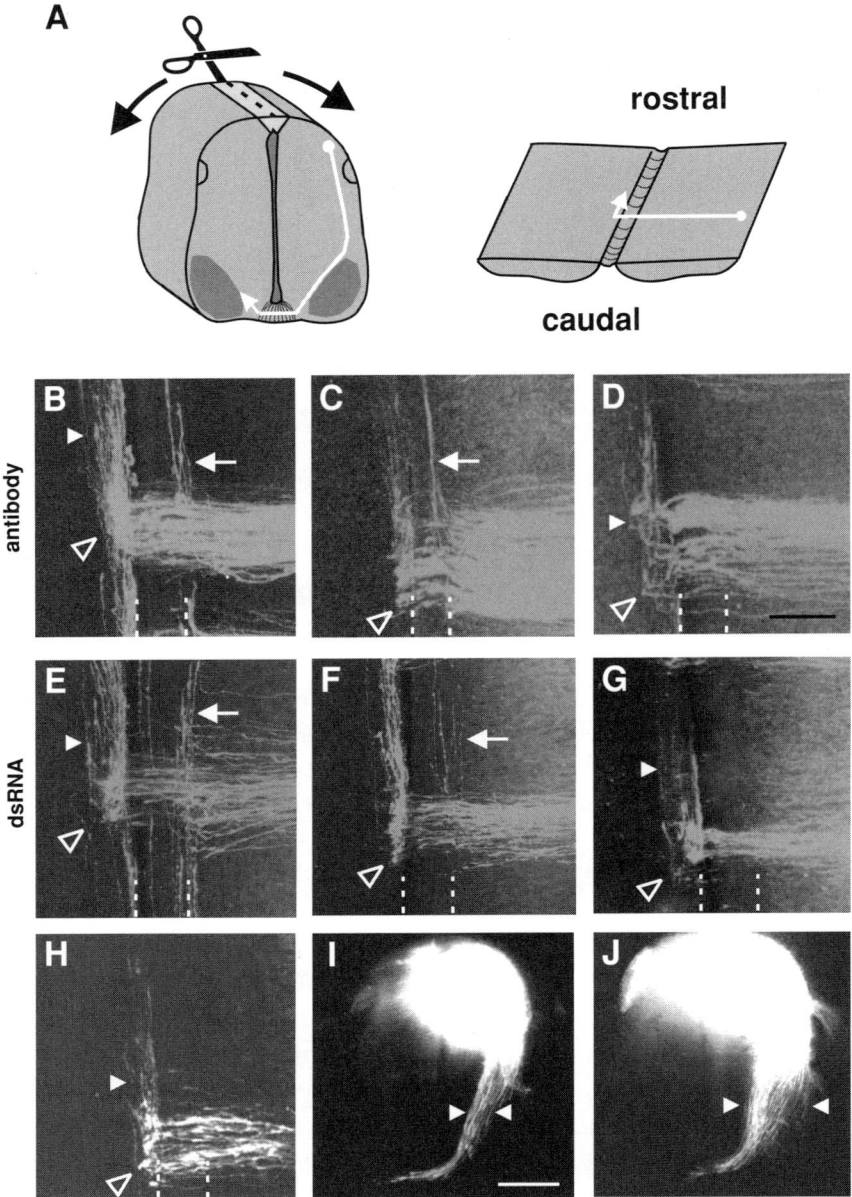

Figure 16.2. *In ovo* RNAi produces specific and reproducible phenotypes. In order to test for the specificity and reproducibility of loss-of-function phenotypes induced by *in ovo* RNAi we used dsRNA derived from axonin-1, NrCAM, and NgCAM. These cell adhesion molecules had been identified previously as short-range guidance cues for commissural axons in the embryonic chicken spinal cord (Stoeckli and Landmesser, 1995). Commissural neurons are located in the dorsal spinal cord (A). They extend their axons toward the floor plate, the ventral midline area of the spinal cord. Axons enter the floor plate to cross the midline. Reaching the contralateral border of the floor plate, axons make a 90° turn into the longitudinal axis. They always turn rostrally (A). For the analysis of commissural axon pathfinding transverse sections of the spinal cord (A, left side; I, J) or open-book preparations (A, right side; B – H) were used. Open-book preparations are obtained by dissecting the entire spinal cord out of the embryo and by cutting it open along the dorsal midline as indicated in the schematic drawing (A). Axons are visualized by the application of a lipophilic dye, FastDil, to the cell bodies of commissural neurons. Because the loss-of-function phenotypes of axonin-1, NrCAM, and NgCAM in commissural axon pathfinding were known from our

embryos, axons in embryos injected with anti-F-spondin antibodies still turned predominantly rostrally. In order to identify guidance cues directing commissural axons along the longitudinal axis of the spinal cord, we set up a screen based on subtractive hybridization (Pekarik et al., 2003). For this purpose, we isolated floor-plate cells of two different stages of development. One pool of cells was obtained from a stage just after commissural axons had turned into the longitudinal axis. The second pool was harvested from much younger embryos, at a stage when the floor plate had not yet been contacted by commissural axons. The differentially expressed genes were analyzed with respect to their expression pattern. For this purpose, the cDNAs obtained from our subtractive hybridization screen were cloned into a vector that was suitable for the generation of Dig-labeled *in situ* probes.

The challenge was, however, to select the functionally relevant clones from the pool of candidates. Therefore, we had to develop a novel functional *in vivo* assay that would allow us to study gene function based on the availability of a cDNA fragment. RNA interference (RNAi) seemed like the perfect approach. At that time, RNAi had just been used in invertebrates (Fire et al., 1998; Kennerdell and Carthew, 1998) and very early stages of mouse embryos (Svoboda et al., 2000; Wianny and Zernicka-Goetz, 2000). Results from RNAi studies in zebrafish were very controversial (Wargelius et al., 1999; Li et al., 2000; *contra* Oates et al., 2000; Zhao et al., 2001). Because conditions of gene silencing in our studies, where we wanted to analyze much older embryos than used for RNAi in mouse or zebrafish, were very different, we had to find alternative approaches for dsRNA transfer into cells. For this purpose, we took advantage of the fact that *in ovo* electroporation was a very efficient method for gene transfer in chicken embryos (Itasaki et al., 1999; Swartz et al., 2001). We combined *in ovo* electroporation with RNAi to develop a powerful method for specific gene silencing (Pekarik et al., 2003; Stoeckli 2003; Figure 16.3).

A particular asset of *in ovo* RNAi is the fact that it does not require additional cloning. After selecting the candidate clones from our screen that had the

---

Figure 16.2. (*cont.*) earlier *in vivo* perturbation studies at the protein level (see text for details), these short-range guidance cues were chosen to test the specificity of *in ovo* RNAi (Pekarik et al., 2003). The phenotypes obtained after perturbation of short-range guidance cues at the protein level (B − D) and after *in ovo* RNAi (E − G) were identical. After perturbation of axonin-1 interactions (B and E) some commissural axons failed to cross the midline, and turned into the longitudinal axis along the ipsilateral floor-plate border (arrows). In addition, commissural axons showed a defasciculated growth pattern indicated by a wider trajectory along the floor-plate border compared to control embryos (arrowheads). The perturbation of NrCAM interactions (C and F) had only little effect on the fasciculation of commissural axons but induced pathfinding errors (arrows) similar to the ones seen after axonin-1 perturbation. The interference with NgCAM interactions (D and G) resulted in a defasciculated growth pattern of commissural axons (arrowheads) but did not cause any pathfinding errors. The defasciculation after injection of NgCAM dsRNA (J) was even more obvious in transverse sections of the spinal cord (compare with control embryo in I). The trajectories of commissural axons in control embryos are shown in H and I. Pathfinding errors are indicated by arrows. Changes in the fasciculation of commissural axons are indicated by arrowheads. The floor plate is indicated by dashed lines in panels B − H. Bar: 50 μm in B − H; 100 μm in I, J. (Adapted from Pekarik et al., 2003).

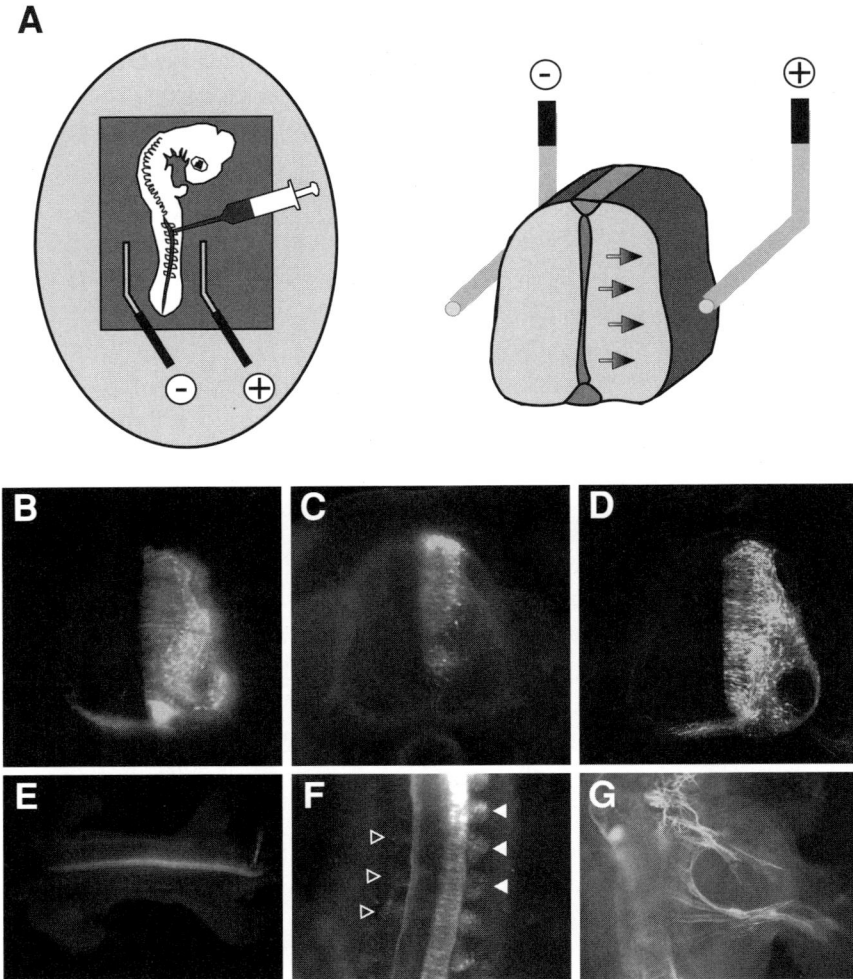

Figure 16.3. Different cell populations can be targeted with *in ovo* RNAi. In order to silence genes during the development of the embryonic chicken spinal cord, we access the embryo through a window in the eggshell (A). The dsRNA derived from the gene of interest is injected into the central canal of the spinal cord (for details see Perrin and Stoeckli, 2000, and Pekarik et al., 2003). Depending on the position of the electrodes and the developmental stage of the embryo at the time of electroporation, different cell populations take up the dsRNA. In order to follow the distribution of the injected dsRNA we mix it with a plasmid encoding YFP. Electroporation between embryonic stages 18 and 19 result in YFP expression in all cell types (B). If the electrodes are kept dorsally, ventral cell types are not transfected (C). The same electrode positions as used in (B) but for embryos at later stages of development prevent dsRNA uptake by lateral motoneurons (D). The distribution of the injected dsRNA along the anterior-posterior axis can be controlled similarly by the type and the position of the electrodes used (E). Electroporations between stages 14 and 17 will result in transfection of neural crest derived cells, including the dorsal root ganglia (F). Note that under these conditions the left side cannot be used as a control as neural crest cells can migrate to either side after electroporation at early stages. Therefore YFP can also be seen on the left side (open arrowheads) although at lower levels compared to the right side (arrowheads). Whole-mount staining allows the analysis of affected sensory and motor axons in the intact embryo (G).

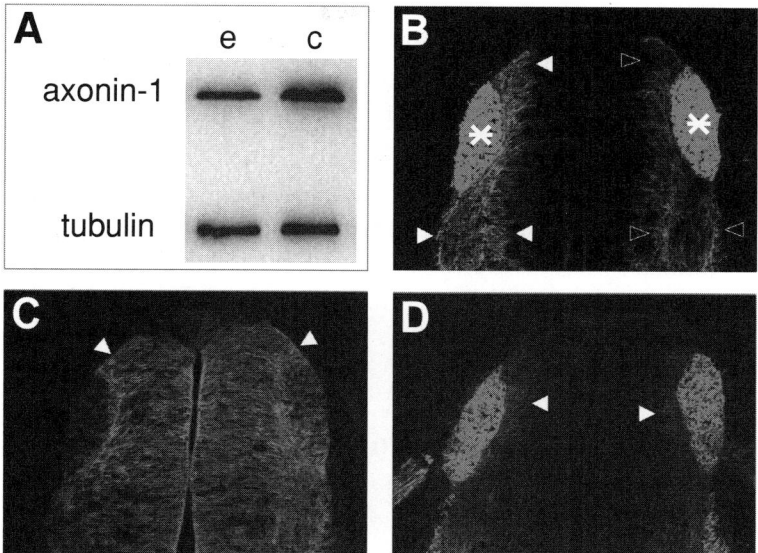

Figure 16.4. *In ovo* RNAi results in the specific decrease of the targeted gene product. A decrease of the protein expression level was measured with different techniques. Western blots of proteins isolated from the control (c) and the electroporated side (e) of an embryo treated with axonin-1 dsRNA were used to quantify the decrease in axonin-1 expression (A). Tubulin was used as a control for loading equal amounts of proteins. A specific decrease of axonin-1 (B) but not of the related molecules NrCAM (C) and NgCAM (D) has been demonstrated by immunohistochemistry. Arrowheads in (B) indicate axonin-1 staining in the left (control) half of the spinal cord. Levels of axonin-1 are lower on the right (electroporated) side (open arrowheads). Equal staining of the dorsal root entry zone (asterisk) serves as an internal staining control. The staining in the dorsal root entry zone is due to sensory axons from the dorsal root ganglia. These neurons were not targeted under the experimental conditions used here. Therefore, equal levels of axonin-1 staining are expected. To show that in ovo RNAi is specifically silencing the targeted gene without affecting structurally related genes, we stained the same sections as shown in (B) with antibodies against NrCAM (C). No difference in NrCAM expression between the two halves of the spinal cord was detectable (compare staining intensity in the dorsal spinal cord indicated by arrowheads). No difference in staining intensity also was seen in the adjacent section stained for NgCAM (D). The staining of commissural axons was equal for both sides of the spinal cord (arrowheads).

appropriate spatial and temporal expression pattern, we could use the same vector that we had used for the generation of the *in situ* probes for the production of dsRNA. Both strands were transcribed and annealed *in vitro*. The dsRNA was injected into chicken embryos *in ovo* using the same technique that was used for our loss-of-function approaches at the protein level using function-blocking antibodies (Perrin and Stoeckli, 2000). The application of short square-wave pulses resulted in efficient targeting of the injected dsRNA into the selected area of the neural tube. Using a plasmid encoding YFP (yellow fluorescent protein) under the control of the beta-acting promoter, we determined that on average 60% of the cells in the electroporated area of the spinal cord took up the injected nucleic acid (Figure 16.3B). The position of the electrode and the variation of embryonic stages can be used to target distinct areas or cell population of the neural tube (Figure 16.3).

In order to demonstrate the specificity of the method, we used dsRNA derived from the full-length cDNAs encoding axonin-1, NrCAM, and NgCAM for *in ovo*

RNAi (Pekarik et al., 2003). The function of these three cell adhesion molecules in commissural axon guidance had been established previously by *in vivo* loss-of-function studies at the protein level (see above; Stoeckli and Landmesser, 1995; Stoeckli et al., 1997; reviewed in Stoeckli and Landmesser, 1998). The injection of dsRNA followed by *in ovo* electroporation reproduced the distinct phenotypes seen after the injection of function-blocking antibodies (Figure 16.2). Using RT-PCR, Western blotting, and immunohistochemistry, the decrease of the targeted gene product without an effect on non-targeted but related proteins was demonstrated (Figure 16.4). Furthermore, we showed that the localization of the cDNA fragment that was used to generate the dsRNA with respect to the full-length sequence did not matter. We did not get any differences in the phenotype or its penetrance when we cut the full-length cDNAs of axonin-1, NrCAM, or NgCAM into three different fragments (Pekarik et al., 2003). *In ovo* RNAi appears to put no constraints on the length of the injected dsRNA fragments. We successfully used fragments of different sizes that were between 200 and 2000 bp long.

General applicability of *in ovo* RNAi

Taken together the characteristics of *in ovo* RNAi make this method a powerful tool for functional genomics. Based on inherent methodological traits many screens identify candidate genes in the form of cDNA fragments. Cloned into vectors with flanking promoters on either side of the insert the synthesis of *in situ* probes and dsRNA is straightforward. Because genes can be silenced in a temporally and spatially controlled manner *in ovo* RNAi allows for the analysis of loss-of-function phenotypes during the development of the nervous system also for genes that would be lethal at early embryonic stages due to additional function, for instance in heart development, a problem frequently encountered in mice generated with conventional gene knock-out strategies (Porter, 1998; Müller, 1999). We took advantage of the fact that *in ovo* RNAi is temporally controllable when we silenced shh (sonic hedgehog) during later stages of spinal cord development, allowing us to study shh's effect on axon guidance without the problem of aberrant spinal cord patterning induced by the absence of shh during early stages of neural tube development (Pekarik et al., submitted).

Because *in ovo* RNAi does not limit the choice of assays used to identify the resulting phenotype a variety of methods can be selected (Bronner-Fraser, 1996). In some cases *in ovo* RNAi even offers the additional advantage that the phenotype is induced only in one half of the nervous system allowing for a direct comparison between the two halves of the body representing the experimental and the control side in the same animal (Figure 16.3). The expression of YFP or another marker encoded by a plasmid co-injected with the dsRNA allows for the identification of the transfected cells. In other cases, as shown in Figure 16.3F, both sides may be affected by gene silencing when dsRNA is injected during early stages of neural tube development to study neural crest cells. We found that neural crest cells are most effectively transfected before they have migrated away from the neural tube closure site. At these early stages neural crest cells still have the choice to grow either way resulting in a bilateral distribution of cells with the target gene silenced.

The choice of parameters used for *in ovo* RNAi determines the cell populations in which the target gene is silenced. One important factor is the time of dsRNA injection and electroporation. As an example, lateral motorneurons were spared when *in ovo* electroporation was used on older embryos although voltage, pulse number, and position of the electrodes were not changed (compare Figure 16.3B and D). Similarly, a change in the position of the electrodes can be used to specifically select for dorsal cell populations (Figure 16.3C).

Although we have used *in ovo* RNAi only for the analysis of axon guidance so far, this method can easily be adapted to other biological processes, such as the analysis of cell migration, cell survival, or cell proliferation. A particularly efficient method is the analysis of whole-mounts after staining of all neurites or the use of YFP expression to identify the trajectories of the axons lacking a particular gene in comparison to control-electroporated, YFP-expressing axons (Figure 16.3G).

Finally, it should be pointed out that *in ovo* RNAi may be a good choice also for scientists who are not focusing on the chicken embryo as a model system. With the completion of the chicken genome (Spring 2004; Burt and Pourquie, 2003) it has become very easy to identify orthologs of mouse genes. The availability of ESTs of chicken genes will allow for an initial assessment of gene function during development in the chick (Boardman et al., 2002; Brown et al., 2003). If a gene is found to play a role early in development, one may not succeed in producing a living knock-out mouse with a conventional strategy. Knowledge of the developmental expression of the target gene and its *in vivo* function discovered in the chicken embryo by *in ovo* RNAi could be used to select an appropriate strategy of gene ablation in a mouse before a lot of time is wasted with a non-successful conventional approach, because one may opt for a tissue-specific or inducible knock-out strategy. Thus, the chicken embryo may be the model organism of choice for functional genomics addressing developmental processes.

Acknowledgements

Work in the lab of E.T.S. is supported by the Swiss National Science Foundation, the Human Frontier Science Program Organization, and the Olga Mayenfisch Stiftung.

REFERENCES

Adams, M. D. and Sekelsky, J. J. (2002). From sequence to phenotype: Reverse genetics in *Drosophila melanogaster*. *Nature Reviews Genetics*, **3**, 189–198.

Bell, E. J. and Brickell, P. M. (1997). Replication-competent retroviral vectors for expressing genes in avian cells *in vitro* and *in vivo*. *Molecular Biotechnology*, **7**, 289–298.

Boardman, P. E., Sanz-Ezquerro, J., Overton, I. M., Burt, D. W., Bosch, E., Fong, W. T., Tickle, C., Brown, W. R. A., Wilson, S. A. and Hubbard, S. J. (2002). A comprehensive collection of chicken cDNAs. *Current Biology*, **12**, 1965–1969.

Bourikas, D. and Stoeckli, E. T. (2003). New tools for gene manipulation in chicken embryos. *Oligonucleotides*, **13**, 411–419.

Bronner-Fraser, M. (1996). Methods in avian embryology. In *Methods in cell biology* (ed. M. Bronner-Fraser), Academic Press, San Diego, CA.

Brown, W. R. A., Hubbard, S. J., Tickle, C. and Wilson, S. A. (2003). The chicken as a model for large-scale analysis of vertebrates gene function. *Nature Reviews Genetics*, **4**, 87–98.

Burstyn-Cohen T., Tzarfaty, V., Frumkin, A., Feinstein, Y., Stoeckli, E. T. and Klar, A. (1999). F-spondin is required for accurate pathfinding of commissural axons at the floor plate. *Neuron*, **23**, 233–246.

Burt, D. and Pourquie, O. (2003). Genetics. Chicken genome – science nuggets to come soon. *Science*, **300**, 1669.

Cepko, C. (1988). Retrovirus vectors and their applications in neurobiology. *Neuron*, **1**, 345–353.

Chang, C. and Hemmati-Brivanlou, A. (1998). Cell fate determination in embryonic ectoderm. *Journal of Neurobiology*, **36**, 128–151.

Driever, W. (1995). Axis formation in zebrafish. *Current Opinions in Genetics and Development*, **5**, 610–618.

Dynes, J. L. and Ngai, J. (1998). Pathfinding of olfactory neuron axons to stereotyped glomerular targets revealed by dynamic imaging in living zebrafish embryos. *Neuron*, **20**, 1081–1091.

Fekete, D. M. and Cepko, C. L. (1993). Replication-competent retroviral vectors encoding alkaline phosphatase reveal spatial restriction of viral gene expression/transduction in the chick embryo. *Molecular Cell Biology*, **13**, 2604–2613.

Finley, K. R., Davidson, A. E. and Ekker, S. C. (2001). Three-color imaging using fluorescent proteins in living zebrafish embryos. *BioTechniques*, **31**, 66–70.

Fire, A., Xu, S., Montgomery, M. K., Kostas, S. A., Driver, S. E. and Mello, C. C. (1998). Potent and specific genetic interference by double-stranded RNA in *Caenorhabditis elegans*. *Nature*, **391**, 806–811.

Fitzli, D., Stoeckli, E. T., Kunz, S., Siribour, K., Rader, C., Kunz, B., Kozlov, S. V., Buchstaller, A., Lane, R. P., Suter, D. M., Dreyer, W. J. and Sonderegger, P. (2000). A direct interaction of axonin-1 with NgCAM-related cell adhesion molecule (NrCAM) results in guidance, but not growth of commissural axons. *Journal of Cell Biology*, **149**, 951–968.

Heasman, J. (2002). Morpholino oligos: Making sense of antisense? *Developmental Biology*, **243**, 209–214.

Hudson, D. F., Morrison, C., Ruchaud, S. and Earnshaw, W. C. (2002). Reverse genetics of essential genes in tissue-culture cells: 'Dead cells talking.' *Trends in Cell Biology*, **12**, 281–287.

Itasaki, N., Bel-Vialar, S. and Krumlauf, R. (1999). 'Shocking' developments in chick embryology: Electroporation and in ovo gene expression. *Nature Cell Biology*, **1**, E203–207.

Jackson, I. J. (2001). Mouse mutagenesis on target. *Nature Genetics*, **28**, 198–200.

Jontes, J. D., Buchanan, J. and Smith, S. J (2000). Growth cone and dendrite dynamics in zebrafish embryos: Early events in synaptogenesis imaged *in vivo*. *Nature Neurosciences*, **3**, 231–237.

Karlstrom, R. O., Trowe, T. and Bonhoeffer, F. (1997). Genetic analysis of axon guidance and mapping in the zebrafish. *Trends in Neurosciences*, **20**, 3–8.

Kennerdell, J. R. and Carthew, R. W. (1998). Use of dsRNA-mediated genetic interference to demonstrate that frizzled and frizzled 2 act in the wingless pathway. *Cell*, **95**, 1017–1026.

Lebedeva, I. and Stein, C. A. (2001). Antisense oligonucleotides: Promise and reality. *Annual Reviews of Pharmacology and Toxicology*, **41**, 403–419.

Le Douarin, N. M. (1973). A Feulgen-positive nucleolus. *Experimental Cell Research*, **77**, 459–468.

Le Douarin, N. M. and Teillet, M. A. (1973). The migration of neural crest cells to the wall of the digestive tract in avian embryo. *Journal of Embryology & Experimental Morphology*, **30**, 31–48.

Lee, S. S., Lee, R. Y., Fraser, A. G., Kamath, R. S., Ahringer, J. and Ruvkun, G. (2003). A systematic RNAi screen identifies a critical role for mitochondria in *C. elegans* longevity. *Nature Genetics*, **33**, 40–48.

Le Lievre, C. S. and Le Dourain, N. M. (1975). Mesenchymal derivatives of the neural crest: Analysis of chimaeric quail and chick embryos. *Journal of Embryology & Experimental Morphology*, **341**, 25–54.

Li, Y.-X., Farrell, M. J., Liu, R., Mohanty, N. and Kirby, M. L. (2000). Double-stranded RNA injection produces null phenotypes in zebrafish. *Developmental Biology*, **217**, 394–405.

Logan, M. and Tabin, C. (1998). Targeted gene misexpression in chick limb buds using avian replication-competent retroviruses. *Methods*, **14**, 407–420.

Morales, A. V. and de Pablo, F. (1998). Inhibition of gene expression by antisense oligonucleotides in chick embryos *in vitro* and *in vivo*. *Current Topics in Developmental Biology*, **36**, 37–49.

Morgan, B. A. and Fekete, D. M. (1996). Manipulating gene expression with replication-competent retroviruses. *Methods in Cell Biology*, **51**, 185–218.

Müller, U. (1999). Ten years of gene targeting: Targeted mouse mutants, from vector design to phenotype analysis. *Mechanisms in Development*, **82**: 3–21.

Muramatsu, T., Mizutani, Y., Ohmori, Y. and Okumura, J. (1997). Comparison of three nonviral transfection methods for foreign gene expression in early chicken embryos in ovo. *Biochemical Biophysical Research Communications*, **230**, 376–380.

Nasevicius, A. and Ekker, S. C. (2001). The zebrafish as a novel system for functional genomics and therapeutic development applications. *Current Opinions in Molecular Therapy*, **3**, 224–228.

Neuhauss, S. (2003). Behavioral genetic approaches to visual system development and function in zebrafish. *Journal of Neurobiology*, **54**, 148–160.

Oates, A. C., Bruce, A. E. E. and Ho, R. K. (2000). Too much interference: Injection of double-stranded RNA has nonspecific effects in the zebrafish embryo. *Developmental Biology*, **224**, 20–28.

O'Malley, D. M., Zhou, Q. and Gahtan, E. (2003). Probing neural circuits in the zebrafish: A suite of optical techniques. *Methods*, **30**, 49–63

Pekarik, V., Bourikas, D., Miglino, N., Joset, P., Preiswerk, S. and Stoeckli, E. T. (2003). Screening for gene function in chicken embryo using RNAi and electroporation. *Nature Biotechnology*, **21**, 93–96.

Perrin, F. E. and Stoeckli, E. T. (2000). Use of lipophilic dyes in studies of axonal pathfinding *in vivo*. *Microscopy Research Techniques*, **48**, 25–31.

Perrin, F. E., Rathjen, F. G. and Stoeckli, E. T. (2001). Distinct subpopulations of sensory afferents require F11 or axonin-1 for growth to their target layers within the spinal cord of the chick. *Neuron*, **30**, 707–723.

Petropoulos, C. J. and Hughes, S. H. (1991). Replication-competent retrovirus vectors for the transfer and expression of gene cassettes in avian cells. *Journal of Virology*, **65**, 3728–3737.

Porter, A. (1998). Controlling your losses: Conditional gene silencing in mammals. *Trends in Genetics*, **14**, 73–79.

Selleck, M. A. J. (1996). Culture and microsurgical manipulation of the early avian embryo. In *Methods in avian embryology* (ed. M. Bronner-Fraser), Academic Press, San Diego, CA.

Simmer, F., Moorman, C., Van Der Linden, A. M., Kuijk, E., Van Den Berghe, P. V., Kamath, R., Fraser, A. G., Ahringer, J. and Plasterk, R. H. (2003). Genome-wide RNAi of C. elegans using the hypersensitive rrf-3 strain reveals novel gene function. *PLoS Biology*, **1**, E12.

St. Johnston, D. (2002). The art and design of genetic screens: *Drosophila melanogaster*. *Nature Reviews Genetics*, **3**, 176–188.

Stoeckli, E. T. (2003). RNAi in avian embryos. In *RNAi: A Guide to Gene Silencing*. (ed. G.J. Hannon) Cold Spring Harbor Laboratory Press, Cold Spring Harbor, NY.

Stoeckli, E. T. and Landmesser, L. T. (1995). Axonin-1, Nr-CAM, and Ng-CAM play different roles in the *in vivo* guidance of chick commissural neurons. *Neuron*, **14**, 1165–1179.

Stoeckli, E. T. and Landmesser, L. T. (1998). Guidance at choice points. *Current Opinions in Neurobiology*, **8**, 73–79.

Stoeckli, E. T., Sonderegger, P., Pollerberg, G. E. and Landmesser, L. T. (1997). Interference with axonin-1 and NrCAM interactions unmasks a floor-plate activity inhibitory for commissural axons. *Neuron*, **18**, 209–221.

Summerton, J. and Weller, D. (1997). Morpholino antisense oligomers: Design, preparation, and properties. *Antisense Nucleic Acids & Drug Development*, **7**, 187–195.

Svoboda, P., Stein, P., Hayashi, H., and Schultz, R. M. (2000). Selective reduction of dormant maternal mRNAs in mouse oocytes by RNA interference. *Development*, **127**, 4147–4156.

Swartz, M., Eberhart, J., Mastick, G. S. and Krull, C. E. (2001). Sparking new frontiers: Using *in vivo* electroporation for genetic manipulations. *Developmental Biology*, **233**, 13–21.

Wargelius, A., Ellingsen, S. and Fjose, A. (1999). Double-stranded RNA induces specific developmental effects in zebrafish embryos. *Biochemical Biophysical Research Communications*, **263**, 156–161.

Wianny, F. and Zernicka-Goetz, M. (2000). Specific interference with gene function by double-stranded RNA in early mouse development. *Nature Cell Biology*, **2**, 70–75.

Zhao, Z., Cao, Y., Li, M. and Meng, A. (2001). Double-stranded RNA injection produces nonspecific defects in zebrafish. *Developmental Biology*, **229**, 215–223.

Gene silencing in
model organisms

17 Practical applications of RNAi in *C. elegans*

Karen E. Stephens, Olivier Zugasti, Nigel J. O'Neil, and Patricia E. Kuwabara

I. The discovery of RNAi in *C. elegans*

This chapter presents an overview of the methods and practical applications of RNA interference (RNAi) in *C. elegans*. Major exploitation of RNAi in *C. elegans* dates from the observation by Fire et al. (1998) that the injection of sequence-specific double-stranded (ds)RNA into the germ line of *C. elegans* was capable of silencing the activity of its cognate gene. The resulting mutant phenocopy closely resembled the phenotype produced by a genetically null or reduced-function mutant. Amazingly, the effect was found to be cell non-autonomous – RNAi is capable of spreading across cell boundaries and affecting tissues beyond the germ line in *C. elegans* (Fire et al., 1998). This phenomenon is related to post-transcriptional gene silencing (PTGS) in plants and quelling in fungi, all of which are believed to be natural defence mechanisms against viral invasion or transposon jumping (Plasterk, 2002 and references therein). RNAi is now employed routinely across phyla as a tool for systematically analysing gene function in most organisms with complete genome sequences (Hammond et al., 2001). Gene silencing by RNAi holds great promise as a therapeutic for treating human disease and genetic disorders. Considerable progress has been made in understanding the molecular mechanisms underlying the process of RNAi and the reader is referred to other chapters in this book for details.

C. elegans has established itself as an excellent model organism for biomedical research, in part because of its complete cell lineage, transparency, short generation time and ease of genetic manipulation (Riddle et al., 1997). The genome sequence of *C. elegans* was essentially completed in 1998 and carries an estimated 19,000 protein coding genes (*C. elegans* Sequencing Consortium, 1998). The curated genome sequence is available on the *C. elegans* database, WormBase (http://www.wormbase.org; Stein et al., 2001). Over half of predicted *C. elegans* genes have homologues in other organisms and about 75% of known human disease genes have homologues in the worm (*C. elegans* Sequencing Consortium, 1998; Kuwabara and O'Neil, 2001). The power of RNAi combined with the genome sequence and biology of *C. elegans* provides a potent mix for exploring gene function and gaining insights into basic biology and human disease.

II. Methods for performing RNAi in *C. elegans*

RNAi is now used routinely in *C. elegans* as a tool for gaining rapid insights into gene function; the jump from gene sequence to function can often be accomplished within a week. Many researchers have turned to the worm to gain an understanding of gene homologues that are also found in other organisms, such as the human presenilin gene, which is mutated in some patients with Alzheimer disease (Levitan et al., 1996).

What factors does one need to consider when planning an RNAi experiment in *C. elegans*? Although the technique is relatively simple, there are a number of parameters that need to be considered. The basics for performing RNAi are outlined below.

Template design

First, assuming that you have already identified a gene of interest, for example, *yfg-1* (your favourite gene), it is necessary to design a template for synthesising dsRNA. Unlike mammalian cells, the use of dsRNA molecules >30 nt in *C. elegans* does not induce a Type I interferon antiviral response (Zamore, 2002). Hence, DNA templates of 200 nucleotides are used routinely; 81 nucleotides is the minimum size dsRNA segment capable of inducing a potent RNAi response (Parrish et al., 2000). It is possible to use a genomic fragment containing both exon and intron sequences as a template; however, cDNA or exon-only templates have been shown to be more effective (Kraemer et al., 1999; Subramaniam and Seydoux, 1999). Once a suitable sequence complementary to the target gene has been selected, BLASTn should be performed to ensure that RNAi response will be gene-specific. Genes sharing greater than 70 to 80% nucleotide identity are likely to be simultaneously inhibited (Zhang et al., 1997). This property of RNAi can be beneficially exploited to eliminate the activities of redundant genes that are closely related by sequence; it is more difficult and time-consuming to generate null mutations in two closely related genes. Finally, the strength of RNAi induction can be region-dependent, so if the region of choice does not induce an RNAi response, it might be worthwhile trying other gene regions (Parrish et al., 2000; Tabara et al., 1998).

If a transgenic gene fusion such as *yfg-1::gfp* is used, it is necessary to be aware of the transitive RNAi effect when selecting a complementary target sequence. Transitive RNAi occurs when secondary siRNAs are generated to sequences that are located 5′ to the initial target sequence. For example, in the case of *yfg-1::gfp*, dsRNA complementary to the *gfp* region alone is sufficient to trigger the formation of siRNAs to *yfg-1* and inhibit any gene bearing sequence identity to *yfg-1* (Sijen et al., 2001).

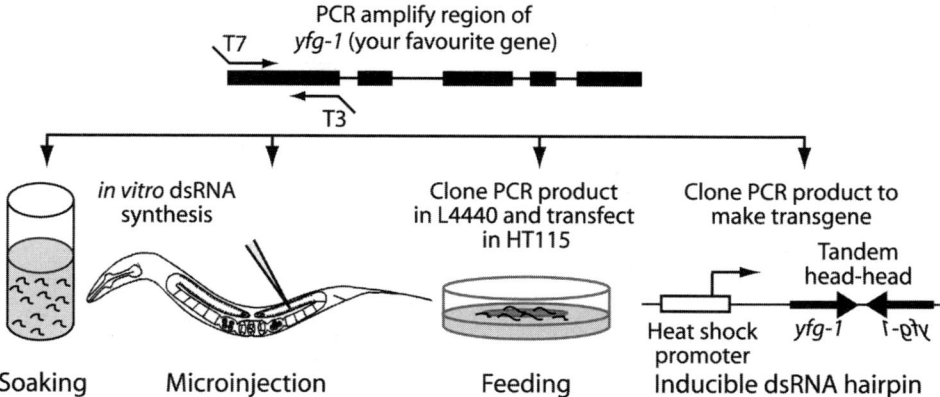

Figure 17.1. Flow diagram depicting methods for performing RNAi in _C. elegans_. After identifying _yfg-1_ (your favourite gene), gene-specific primers with added promoter sequences for T7 or T3 RNA polymerase are selected and used to generate a PCR product. In the example provided, the first exon is PCR amplified from genomic DNA. Subsequently, the PCR product can be used as a template to generate _in vitro_ synthesised dsRNA for RNAi by soaking or microinjection, cloned into the RNAi feeding vector L4440, or cloned into a vector that can generate an inducible dsRNA hairpin in transgenic animals (Fire et al., 1998; Tavernarakis et al., 2000).

Methods of delivery

There are four standard methods for delivering dsRNA in _C. elegans_: germline injection, soaking, feeding and _in vivo_ expression of heritable inverted repeat genes (Figure 17.1). Each method will be discussed in turn.

Historically, germline microinjection was the first method used to administer dsRNA because this is how transgenes are introduced into _C. elegans_ (Fire et al., 1998). Microinjection is technically the most demanding of the methods, but many investigators believe it is also the most effective. However, the precision of injection might not be as critical as initially believed because the RNAi response spreads from the site of injection, whether it is the germ line or intestine (Fire et al., 1998). A 1 to 2 mg/ml solution of dsRNA is injected into each gonad arm and at least three well-injected animals are analysed.

Feeding worms with bacteria expressing dsRNAs from inducible promoters has become a popular method for inducing RNAi because of the ease of its application (Timmons and Fire, 1998). A suitable template DNA (see above) is introduced into a multiple cloning site flanked by T7 promoters in the L4440 vector and transfected into an RNase III deficient _E. coli_ feeding host HT115 (DE3), which ameliorates the stability of RNA (Box 17.2) (Timmons et al., 2001; Timmons and Fire, 1998). The bacterial dsRNAs are ingested through the intestine and the RNAi effect is propagated through the worm (Fire et al., 1998). These bacterial strains provide an easily renewable dsRNA-producing resource, amenable for high-throughput RNAi.

Soaking worms in a solution of dsRNA is another simple, yet less economical method for inducing RNAi. Generally, larval worms are cleared of their normal bacterial food source OP50 prior to soaking in 4 μl of a 4 mg/ml solution

1. After identifying your gene of interest, design sequence-specific primers (Tm ~60°C) capable of PCR amplifying a suitable plasmid-based or genomic DNA template 0.2 kb for dsRNA synthesis (see text for further details). Promoter sequences for either T3 or T7 RNA polymerase (T3, 5′ aattaaccctcactaaagg; T7, 5′ tatacgactcactatgg) should be added to the 5′ ends of the primers.

2. Set up a PCR to contain 50 ng N2 genomic DNA; 5 μl 10× Taq buffer (100 mM Tris-HCl (pH 8.3), 500 mM potassium chloride, 0.01% (w/v) gelatin and 15 mM magnesium chloride); 5 μl of 5 μM Forward primer; 5 μl of 5 μM Reverse primer; 5 μl of 2.5 mM dNTP mix; 0.25 μl of 5U/μl Taq polymerase and water to a total volume of 50 μl.

3. Perform the PCR as follows: 1 cycle at 92°C for 3 minutes; 35 cycles at 92°C for 20 seconds, 58°C for 40 seconds, 72°C for 1 minute per kb; 1 cycle at 72°C for 6 minutes, end. Precipitate at −70°C for 2 hours with 100μl of 95% ethanol and 5μl of 1 M NaCl. Pellet DNA by centrifugation at maximum speed for 7 minutes, wash the pellet with 200μl of ice-cold 70% ethanol and resuspend in 7μl of DEPC-treated TE buffer.

4. If using the Ambion Megascript™ RNA synthesis kit, set up separate sense and antisense transcription reactions, as instructed by the manufacturer: 1 μl of the above PCR product, 0.5 μl of 10× reaction buffer, 2 μl of rNTPs, 1 μl of RNase-free water and 0.5 μl of T3 or T7 enzyme mix and incubate at 37°C for 4 hours.

5. Clean RNA using Qiagen RNeasy™ columns: add 47 μl of RNase-free water to each reaction, remove 2 μl from each tube for gel analysis and combine the remaining T3 and T7 reactions for each gene. Add 350 μl of RLT buffer, mix, add 250 μl of 95% ethanol, mix. Apply the sample to an RNeasy column, centrifuge and discard flow-through. Add 700 μl of RW1 buffer, centrifuge and discard flow-through. Wash RNA in 500 μl of RPE buffer, centrifuge, discard flow-through and repeat. Elute RNA into a total volume of 130 μl RNase-free water, in 2 even steps, collecting in a fresh 1.5 ml tube by centrifugation.

6. Anneal 50 μl sense and antisense ssRNAs by heating to 68°C for 10 minutes then incubating at 37°C for 30 minutes with 10 μl of a 6× injection buffer (40 mM potassium phosphate pH 7.5, 6 mM potassium citrate pH 7.5 and 4% polyethylene glycol 6000) or to a final concentration of 10.9 mM di-sodium hydrogen phosphate, 5.5 mM potassium di-hydrogen phosphate, 2.1 mM sodium chloride, 4.7 mM ammonium chloride, 3 mM spermidine and 0.05% gelatin for soaking (Maeda et al., 2001). A final dsRNA concentration of 1 mg/ml for injection or 4 mg/ml for soaking is required. Assay both ssRNA and dsRNA on a 1% agarose gel.

Box 17.1. Generation of *in vitro* transcribed dsRNA.

1. The most commonly used feeding vector is L4440. A sequence $\geq$0.2 kb that is complementary to the target gene of interest is inserted into the multiple cloning site of L4440, which is flanked by convergent T7 RNA polymerase promoters (Timmons and Fire, 1998). This plasmid is usually transfected into the RNase III deficient *E. coli* strain HT115 to stabilise dsRNAs (Timmons et al., 2001). It should be noted that plasmid recovery from the *E. coli* strain HT115 is not recommended.

2. Inoculate a culture of 2x TY medium containing 50 μg/ml ampicillin and 15 μg/ml tetracycline with a single colony of HT115 carrying the L4440 plasmid with your gene of interest and grow overnight. Seed 5 cm NGM agar plates containing antibiotics and 0.4 mM isopropyl-β-D-thiogalactopyranoside (IPTG) with about 50 ml of the overnight culture. Allow bacteria to grow overnight at 20°C before adding worms.

3. Introduce five L4 hermaphrodites onto a feeding plate and grow at 20°C. After 12 to 18 hours, transfer each worm onto a separate plate. Continue clonal transfer of single worms to fresh plates at 12 to 18 hour intervals, or more frequently, until the worm ceases to produce fertilised eggs.

4. After the parental worm has been transferred, examine the plate 24 hours later for the presence of dead eggs.

5. Continue to score the RNAi phenotype of the P_0s over the next 1 to 5 days. F1 progeny can be transferred onto additional RNAi plates to maintain depletion of a gene of interest. At no time should the worms be allowed to starve.

Box 17.2. RNAi feeding protocol (adapted from Timmons et al., 2001).

of dsRNA for 24 hours (Tabara et al., 1998). Addition of spermidine to the soaking solution increases the efficiency and reproducibility of the RNAi effect (Maeda et al., 2001). After soaking, worms are transferred to a fresh feeding plate for phenotypic scoring.

The three methods described above all produce transient induction of RNAi. A method for stable inheritance of the RNAi response is available that involves the integration of transgenes expressing heritable inverted repeats under the control of an inducible promoter, such as the heat-shock promoter [(Figure 17.1) (Tavernarakis et al., 2000)]. Cloning of hairpin structures and sequence validation is more problematic than conventional cloning. This technique has been beneficially employed to deplete the activities of neuronal genes that are refractory to other methods of RNAi (Tavernarakis et al., 2000).

Phenotypic analysis

Many researchers have noted that the response of individual worms to RNAi is variable, so it is good practice to examine the progeny produced from more than three RNAi-treated worms per experiment. Worms should be maintained under normal laboratory conditions at 20°C and not allowed to starve.

The types of phenotypes that can be scored are very much dependent on the training of the investigator and the nature of the investigation. A basic guide is presented below, but this should only be considered a first approach. The effect of RNAi is detected in the F1 progeny of hermaphrodites that have been subjected to RNAi. This is usually achieved by transferring the RNAi treated P_0 mother to a fresh plate every 6 to 12 hours; the first 15 to 20 F1 progeny to develop will be unaffected by RNAi. A basic screen for embryonic lethality, which is an indicator that a gene has an essential function, can be undertaken by counting unhatched (dead) embryos 24 hours after the P_0 hermaphrodite has been transferred to a fresh plate. Microscopy can help to determine the cause of death and might provide insights into the cell biological processes affected. For genes that prove to be essential for embryonic viability, it is possible to bypass the embryo stage and to address whether these genes are also required at later stages of development. This can be achieved by performing RNAi using L1 stage larvae, which can be obtained as synchronous cultures (Wood, 1988). In this situation, RNAi feeding and soaking are the methods of choice. Throughout post-embryonic development worms should be examined for defects in morphogenesis, motility, size, molting and developmental timing, to name but a few possible mutant phenotypes (Riddle, 1997). Adult animals can be scored for sterility and other germline defects that can affect their progeny.

To validate the success of gene inhibition by RNAi, especially in the absence of a detectable phenotype, it is desirable, whenever possible, to examine mRNA levels by Northern analysis and protein by Western blotting or immunocytochemistry.

Whole genome RNAi screens

A major goal of any functional genomics project is to annotate the function of all genes. From this perspective, RNAi performed by injection, feeding or soaking was demonstrated by a number of laboratories to be amenable to high-throughput analysis, thereby providing a rapid method for determining whether the inhibition of a gene produces a mutant phenocopy, which in turn can then be used to infer the normal wild-type function of a gene (Fraser et al., 2000; Gonczy et al., 2000; Maeda et al., 2001; Piano et al., 2000). It was found that about 10.3% of all genes displayed an RNAi mutant phenocopy by feeding, compared with 12.9% of genes on chromosome III by injection and 13.9% of genes on chromosome I by feeding (Fraser et al., 2000; Gonczy et al., 2000; Kamath et al., 2003).

In addition, directed screens have been employed to identify numerous genes that might participate in a similar biological process or pathway. The availability of an RNAi feeding library has made it possible to identify genes affecting chromosome morphogenesis, lifespan, nucleotide excision repair, fat storage and transposon-gene silencing (Ashrafi et al., 2003; Colaiacovo et al., 2002; Lee et al., 2003; Pothof et al., 2003; Vastenhouw et al., 2003). The feeding library is a renewable resource that covers 86% of all predicted *C. elegans* genes (Kamath et al., 2003).

III. Extended practical applications of RNAi in *C. elegans*

Genetic enhancers and suppressors of RNAi

Genetic screens have been performed to identify genes that can enhance or are deficient in RNAi. Here we discuss how two such mutants, *rrf-1* and *rrf-3*, which encode members of the RNA-directed RNA polymerase family, can be used to advantage in RNAi based screens. *rrf-3* is an enhancer of RNAi; when RNAi is performed with an *rrf-3* mutant, mutant phenocopies have been detected for genes that previously failed to show an obvious RNAi mutant phenocopy, including neuronal genes believed to be refractory to RNAi (Simmer et al., 2002). In a pilot study testing 80 *C. elegans* dsRNA expressing bacteria, 26 bacteria elicited mutant phenocopies in N2 animals and an additional 23 mutant phenocopies were detected using the *rrf-3* mutant (Simmer et al., 2002). However, it should be noted that *rrf-3* worms produce fewer progeny because of defects in spermatogenesis and that they also produce more males than wild-type animals (Sijen et al., 2001). We have also noted that *rrf-3* appears to synergise with, instead of enhance, some genes that are active in the germline (N. J. O'Neil and P. E. Kuwabara, unpublished data).

Our laboratory has performed an RNAi screen of 24 *C. elegans patched-related (ptr)* genes, which encode membrane proteins containing sterol-sensing domains (SSDs) (Kuwabara et al., 2000); *patched* is a multipass membrane protein that is a receptor for Hedgehog (Ingham and McMahon, 2001). The SSD family of proteins are involved in processes connected with lipids, sterols or sterol-modified proteins and cholesterol homeostasis, and are implicated in steroid hormone, lipoprotein and bile acid metabolism in mammals (Kuwabara and Labouesse, 2002). After performing feeding RNAi with N2 wild-type animals, we identified 7 *ptr* genes that were involved in molting; however, when feeding RNAi was repeated with an *rrf-3* mutant, we identified an additional 10 *ptr* genes involved in molting (O. Zugasti and P. E. Kuwabara, unpublished data). The *rrf-3* genetic background also increased the expressivity of the mutant phenocopy, making it possible to detect more severe molting defects at earlier stages of development.

rrf-1 mutants are partially deficient in RNAi. Animals carrying a mutation in *rrf-1* remain sensitive to the effects of germline RNAi, but are deficient in somatic RNAi (Sijen et al., 2001). These properties of *rrf-1* allowed Miller et al. (2003) to perform genetic mosaic analysis on *vab-1*, a gene that encodes the *C. elegans* ephrin receptor. By performing *vab-1(RNAi)* using the *rrf-1* mutant, it was found that *vab-1* is required in different tissues to mediate different aspects of its activity (Miller et al., 2003).

Multigenerational RNAi

Basic whole-genome RNAi screens would not be expected to detect the entire range of potential mutant phenotypes, especially those corresponding to genes that have no immediately obvious visible phenotype. For example, in the absence of genotoxic stress, sterility arising from a mutation in *mrt-2*, a gene that promotes genomic stability, is not evident until after the tenth generation (Ahmed and

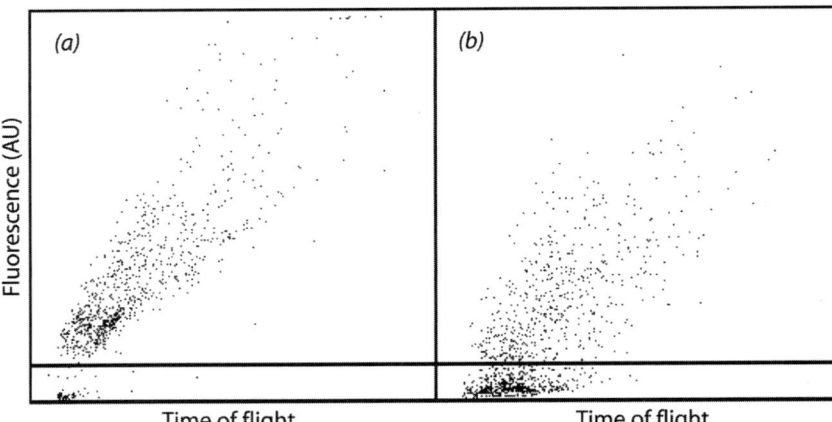

Figure 17.2. COPAS Biosort detects a reduction in the expression of a *myo-2::gfp* reporter by *gfp(RNAi)* in liquid culture. Matched populations of young larvae and embryos expressing a *myo-2::gfp* reporter were fed concentrated (*a*) NA22 bacteria or (*b*) bacteria expressing *gfp* dsRNA. The COPAS Biosort tracks worms based on their length (X-axis, time-of-flight) and fluorescence (Y-axis, arbitrary units). Each point represents a single worm. The horizontal line depicts a fluorescence threshold: worm above the horizontal line are fluorescent whereas those below are not. RNAi induced by *gfp* dsRNA reduces the number of fluorescent worms from 89% (840/944) to 34% (541/1592).

Hodgkin, 2000). For such genes, we have shown that multigenerational RNAi, which involves the clonal transfer of progeny from one generation to the next onto freshly prepared RNAi feeding plates, can help to uncover long-term defects. The most important consideration when using multigenerational RNAi is that worms should never be allowed to starve between transfers, because this negates the previous inhibition (N. J. O'Neil and P. E. Kuwabara, unpublished data). Using this technique we have shown that *mrt-2(RNAi)* animals begin to show genomic instability after the sixth generation, despite the lack of an immediate defect.

Semi-automated phenotypic screens

The COPAS Biosort (Union Biometrica) is a flow sorter that can sort and dispense worms based on size, density and fluorescent properties. We have taken advantage of the Biosort and shown that it is possible to quantify a reduction in the expression of a GFP fluorescent reporter caused by *gfp(RNAi)*. An example of such an experiment is described here. Similarly staged worms expressing a *myo-2::gfp* reporter were grown in liquid containing an empirically determined concentration of either wild-type NA22 bacteria or HT115 bacteria expressing a *gfp* RNAi feeding construct, and then passed through the COPAS Biosort for analysis. As shown in Figure 17.2, the COPAS Biosort can detect a reduction in the number of worms expressing the *myo-2::gfp* reporter after *gfp(RNAi)*. In panel (*a*), 89% of worms fed NA22 (n = 944) express GFP, whereas this number is reduced after *gfp(RNAi)* to 34% (n = 1592) (Figure 17.2(*b*)). Therefore the Biosort can be used for semi-automated high-throughput RNAi screening to detect genes capable of directly affecting the transcription of a *myo-2::gfp* transcriptional reporter,

Table 17.1. Synthetic multivulval phenotypes (Muv) are detected in *rrf-3* mutants by feeding worms bacteria expressing hybrid dsRNA

RNAi feeding construct	Worm strain	Number of worms scored	Number of Muv worms
lin-15	N2	921	0
lin-15	rrf-3	479	1
lin-35	N2	963	0
lin-35	rrf-3	530	0
lin-15 + lin-35	N2	936	0
lin-15 + lin-35	Rrf-3	484	0
lin-15:lin-35	N2	1004	0
lin-15:lin-35	rrf-3	552	452

or other appropriate reporter, which would produce either reduced or ectopic GFP expression.

Synthetic screens by dual feeding

Synthetic lethal screens in yeast have proved to be powerful tools for the identification of functional gene interactions (e.g. Mullen et al., 2001). In *C. elegans*, synthetic phenotypes resulting from the simultaneous reduction of two independent gene activities have also been described (Thomas et al., 2003). It has been noted that the introduction of dsRNA corresponding to more than one gene sometimes fails to elicit the same RNAi response as introduction of dsRNA corresponding to a single gene (Fraser et al., 2000; Gonczy et al., 2000; Kamath et al., 2001). To facilitate the detection of synthetic mutant phenotypes using RNAi, we have shown that the use of hybrid dsRNA molecules is a successful strategy. We have constructed hybrid templates by inserting sequences of about 200 to 500 bp from each of two individual genes into a single copy of the L4440 vector. This dual-feeding method was tested by simultaneously disrupting the activities of *lin-15* and *lin-35* by RNAi, using the enhancer *rrf-3* because the vulva is a tissue that can be refractory to the effects of RNAi. The *lin-15;lin-35* double-mutant produced a synthetic multivulval mutant phenotype (Muv), which was not observed in either a *lin-15* or *lin-35* single mutant (Lu and Horvitz, 1998). Similarly, we found by RNAi feeding that 90% of *lin-15:lin-35(RNAi)* animals were Muv (Table 17.1). In contrast, *lin-15(RNAi)* or *lin-35(RNAi)* animals each failed to develop ectopic vulvae. Moreover, mixing the individually grown *lin-15* and *lin-35* feeding bacteria on plates failed to elicit the formation of ectopic vulvae. Preliminary studies indicate that mixing concentrated bacteria expressing *lin-15* or *lin-35* dsRNA and performing RNAi in liquid can also generate Muv animals.

Phylogenetic studies of gene function using RNAi

The whole genome of *C. briggsae*, a nematode estimated to have diverged from *C. elegans* about 50 million years ago, has recently been sequenced (http://www.WormBase.org). This makes it possible to explore the evolutionary conservation of genetic control regions and gene function between *C. briggsae* and *C. elegans*. Unfortunately, very few genetic mutants are available for *C. briggsae*; however, it has been demonstrated that RNAi is an effective tool for eliminating gene activity in *C. briggsae* (Kuwabara, 1996). Hence, it is anticipated that RNAi will prove to be an invaluable tool for gaining insights into the conservation of gene function between these closely related nematodes.

IV. Future validation

It should be stressed that RNAi alone does not replace the need for obtaining a genetic mutant when exploring the role of a gene. The mutant phenocopy produced by RNAi does not always mimic a null mutant, although it usually resembles a reduced-function mutation, nor does it always show the same penetrance. For example the lifespan extension seen in an *age-1* mutant is 100% and in *age-1(RNAi)* animals it is only 30% (Lee et al., 2003; Morris et al., 1996). Identification of genetic suppressors and enhancers can be dependent on having a reproducible non-null level of gene inhibition, which is not always the case with RNAi. Nonetheless, RNAi is unsurpassed as a method for obtaining rapid insights into gene function and has opened doors that will lead to the development of new therapeutics.

REFERENCES

Ahmed, S. and Hodgkin, J. (2000). MRT-2 checkpoint protein is required for germline immortality and telomere replication in C. elegans. *Nature*, **403**, 159–164.

Ashrafi, K., Chang, F. Y., Watts, J. L., Fraser, A. G., Kamath, R. S., Ahringer, J. and Ruvkun, G. (2003). Genome-wide RNAi analysis of *C. elegans* fat regulatory genes. *Nature*, **421**, 268–272.

C. elegans Sequencing Consortium (1998). Genome sequence of the nematode *C. elegans*: A platform for investigating biology. *Science*, **282**, 2012–2018.

Colaiacovo, M. P., Stanfield, G. M., Reddy, K. C., Reinke, V., Kim, S. K. and Villeneuve, A. M. (2002). A targeted RNAi screen for genes involved in chromosome morphogenesis and nuclear organization in the *C. elegans* germline. *Genetics*, **162**, 113–128.

Fire, A., Xu, S., Montgomery, M. K., Kostas, S. A., Driver, S. E. and Mello, C. C. (1998). Potent and specific genetic interference by double-stranded RNA in *C. elegans*. *Nature*, **391**, 806–811.

Fraser, A. G., Kamath, R. S., Zipperlen, P., Martinez-Campos, M., Sohrmann, M. and Ahringer, J. (2000). Functional genomic analysis of *C. elegans* chromosome I by systematic RNA interference. *Nature*, **408**, 325–330.

Gonczy, P., Echeverri, C., Oegema, K., Coulson, A., Jones, S. J., Copley, R. R., Duperon, J., Oegema, J., Brehm, M., Cassin, E., Hannak, E., Kirkham, M., Pichler, S., Flohrs, K., Goessen, A., Leidel, S., Alleaume, A. M., Martin, C., Ozlu, N., Bork, P. and Hyman, A. A. (2000). Functional genomic analysis of cell division in *C. elegans* using RNAi of genes on chromosome III. *Nature*, **408**, 331–336.

Hammond, S. M., Caudy, A. A. and Hannon, G. J. (2001). Post-transcriptional gene silencing by double-stranded RNA. *Nature Reviews in Genetics*, **2**, 110–119.

Ingham, P. W. and McMahon, A. P. (2001). Hedgehog signaling in animal development: paradigms and principles. *Genes & Development*, **15**, 3059–3087.

Kamath, R. S., Fraser, A. G., Dong, Y., Poulin, G., Durbin, R., Gotta, M., Kanapin, A., Le Bot, N., Moreno, S., Sohrmann, M., Welchman, D. P., Zipperlen, P. and Ahringer, J. (2003). Systematic functional analysis of the _Caenorhabditis elegans_ genome using RNAi. _Nature_, **421**, 231–237.

Kamath, R. S., Martinez-Campos, M., Zipperlen, P., Fraser, A. G. and Ahringer, J. (2001). Effectiveness of specific RNA-mediated interference through ingested double-stranded RNA in _Caenorhabditis elegans_. _Genome Biology_, **2**, RESEARCH0002.

Kraemer, B., Crittenden, S., Gallegos, M., Moulder, G., Barstead, R., Kimble, J. and Wickens, M. (1999). NANOS-3 and FBF proteins physically interact to control the sperm-oocyte switch in _Caenorhabditis elegans_. _Current Biology_, **9**, 1009–1018.

Kuwabara, P. E. (1996). Interspecies comparison reveals evolution of control regions in the nematode sex-determining gene _tra-2_. _Genetics_, **144**, 597–607.

Kuwabara, P. E. and Labouesse, M. (2002). The sterol-sensing domain: Multiple families, a unique role? _Trends in Genetics_, **18**, 193–201.

Kuwabara, P. E., Lee, M. H., Schedl, T. and Jefferis, G. S. (2000). A _C. elegans_ patched gene, _ptc-1_, functions in germ-line cytokinesis. _Genes & Development_, **14**, 1933–1944.

Kuwabara, P. E. and O'Neil, N. (2001). The use of functional genomics in _C. elegans_ for studying human development and disease. _Journal of Inheritable and Metabolic Diseases_, **24**, 127–138.

Lee, S. S., Lee, R. Y., Fraser, A. G., Kamath, R. S., Ahringer, J. and Ruvkun, G. (2003). A systematic RNAi screen identifies a critical role for mitochondria in _C. elegans_ longevity. _Nature Genetics_, **33**, 40–48.

Levitan, D., Doyle, T. G., Brousseau, D., Lee, M. K., Thinakaran, G., Slunt, H. H., Sisodia, S. S. and Greenwald, I. (1996). Assessment of normal and mutant human presenilin function in _Caenorhabditis elegans_. _Proceedings of the National Academy of Sciences USA_, **93**, 14940–14944.

Lu, X. and Horvitz, H. R. (1998). _lin-35_ and _lin-53_, two genes that antagonize a _C. elegans_ Ras pathway, encode proteins similar to Rb and its binding protein RbAp48. _Cell_, **95**, 981–991.

Maeda, I., Kohara, Y., Yamamoto, M. and Sugimoto, A. (2001). Large-scale analysis of gene function in _Caenorhabditis elegans_ by high-throughput RNAi. _Current Biology_, **11**, 171–176.

Miller, M. A., Ruest, P. J., Kosinski, M., Hanks, S. K. and Greenstein, D. (2003). An Eph receptor sperm-sensing control mechanism for oocyte meiotic maturation in _Caenorhabditis elegans_. _Genes & Development_, **17**, 187–200.

Morris, J. Z., Tissenbaum, H. A. and Ruvkun, G. (1996). A phosphatidylinositol-3-OH kinase family member regulating longevity and diapause in _Caenorhabditis elegans_. _Nature_, **382**, 536–539.

Mullen, J. R., Kaliraman, V., Ibrahim, S. S. and Brill, S. J. (2001). Requirement for three novel protein complexes in the absence of the Sgs1 DNA helicase in _Saccharomyces cerevisiae_. _Genetics_, **157**, 103–118.

Parrish, S., Fleenor, J., Xu, S., Mello, C. and Fire, A. (2000). Functional anatomy of a dsRNA trigger: Differential requirement for the two trigger strands in RNA interference. _Molecular Cell_, **6**, 1077–1087.

Piano, F., Schetter, A. J., Mangone, M., Stein, L. and Kemphues, K. J. (2000). RNAi analysis of genes expressed in the ovary of _Caenorhabditis elegans_. _Current Biology_, **10**, 1619–1622.

Plasterk, R. H. (2002). RNA silencing: The genome's immune system. _Science_, **296**, 1263–1265.

Pothof, J., van Haaften, G., Thijssen, K., Kamath, R. S., Fraser, A. G., Ahringer, J., Plasterk, R. H. and Tijsterman, M. (2003). Identification of genes that protect the _C. elegans_ genome against mutations by genome-wide RNAi. _Genes & Development_, **17**, 443–448.

Riddle, D. L., Blumenthal, T., Meyer, B. J. and Priess, J. R. (1997). _C. elegans II_. Cold Spring Harbor Laboratory Press, Cold Spring Harbor, NY.

Sijen, T., Fleenor, J., Simmer, F., Thijssen, K. L., Parrish, S., Timmons, L., Plasterk, R. H. and Fire, A. (2001). On the role of RNA amplification in dsRNA-triggered gene silencing. *Cell*, **107**, 465–476.

Simmer, F., Tijsterman, M., Parrish, S., Koushika, S. P., Nonet, M. L., Fire, A., Ahringer, J. and Plasterk, R. H. (2002). Loss of the putative RNA-directed RNA polymerase RRF-3 makes *C. elegans* hypersensitive to RNAi. *Current Biology*, **12**, 1317–1319.

Stein, L., Sternberg, P., Durbin, R., Thierry-Mieg, J. and Spieth, J. (2001). WormBase: network access to the genome and biology of *Caenorhabditis elegans*. *Nucleic Acids Research*, **29**, 82–86.

Subramaniam, K. and Seydoux, G. (1999). *nos-1* and *nos-2*, two genes related to *Drosophila nanos*, regulate primordial germ cell development and survival in *Caenorhabditis elegans*. *Development*, **126**, 4861–4871.

Tabara, H., Grishok, A. and Mello, C. C. (1998). RNAi in *C. elegans*: Soaking in the genome sequence. *Science*, **282**, 430–431.

Tavernarakis, N., Wang, S. L., Dorovkov, M., Ryazanov, A. and Driscoll, M. (2000). Heritable and inducible genetic interference by double-stranded RNA encoded by transgenes. *Nature Genetics*, **24**, 180–183.

Thomas, J. H., Ceol, C. J., Schwartz, H. T. and Horvitz, H. R. (2003). New genes that interact with *lin-35 Rb* to negatively regulate the *let-60 ras* pathway in *Caenorhabditis elegans*. *Genetics*, **164**, 135–151.

Timmons, L., Court, D. L. and Fire, A. (2001). Ingestion of bacterially expressed dsRNAs can produce specific and potent genetic interference in *Caenorhabditis elegans*. *Gene*, **263**, 103–112.

Timmons, L. and Fire, A. (1998). Specific interference by ingested dsRNA. *Nature*, **395**, 854.

Vastenhouw, N. L., Fischer, S. E., Robert, V. J., Thijssen, K. L., Fraser, A. G., Kamath, R. S., Ahringer, J. and Plasterk, R. H. (2003). A genome-wide screen identifies 27 genes involved in transposon silencing in *C. elegans*. *Current Biology*, **13**, 1311–1316.

Wood, B. (1988). *The nematode C. elegans*. Cold Spring Harbor Laboratory Press, Cold Spring Harbor, NY.

Zamore, P. D. (2002). Ancient pathways programmed by small RNAs. *Science*, **296**, 1265–1269.

Zhang, B., Gallegos, M., Puoti, A., Durkin, E., Fields, S., Kimble, J. and Wickens, M. P. (1997). A conserved RNA-binding protein that regulates sexual fates in the *C. elegans* hermaphrodite germ line. *Nature*, **390**, 477–484.

18 Inducible RNAi as a forward genetic tool in *Trypanosoma brucei*

Mark E. Drew, Shawn A. Motyka, James C. Morris, Zefeng Wang, and Paul T. Englund

Introduction to RNAi in Trypanosomes

In December 1998 the Ullu lab at Yale published the first report of dsRNA-mediated mRNA degradation in *Trypanosoma brucei* (Ngo et al., 1998). These experiments used either electroporation of *in vitro* synthesized dsRNA or transient *in vivo* expression of single-strand RNA that forms a stem–loop structure capable of inducing RNAi. At that time, the arsenal of genetic techniques available to the trypanosome researcher was limited (Clayton, 1999). For example, gene knockout was possible through homologous recombination, although the diploid genome of *T. brucei* complicated this approach. Furthermore, essential genes were difficult to examine by knockout, necessitating complex strategies in which inducible ectopic expression needed to be maintained while the genomic knockouts were generated (Wirtz et al., 1999). RNAi raised hopes for a powerful and convenient genetic approach for these eukaryotes.

Our lab has made extensive use of RNAi in studying gene function in *T. brucei*, the eukaryotic parasite that causes African sleeping sickness. The purpose of this chapter is to review the steps our lab has taken in developing an inducible RNAi system that has allowed us to achieve the goal of *bona fide* RNAi-based forward genetics in *T. brucei*. In addition, this chapter will review our development of an easy-to-use, inducible RNAi system, presenting a few examples of how this approach has allowed us to gain new insights into gene function, especially in the case of essential genes. We will also describe how we have exploited this inducible system to do genome-wide forward genetics in this parasite. For current reviews of the status of RNAi in other pathogenic organisms see (Beverley, 2003; Cottrell and Doering, 2003).

Utility of inducible RNAi

Electroporation of dsRNA is an effective means of carrying out RNAi in *T. brucei*. However, for many genes, long-term exposure to dsRNA is needed for a phenotype to become evident because not only must there be a reduction of mRNA but protein levels must also fall below a functional "threshold." Thus, a stable RNAi expression system is necessary. The Ullu group furthered RNAi technology in *T. brucei* through development of a stable and inducible RNAi approach (Shi et al., 2000). This method utilized an expression vector in which two identical gene fragments, arranged in opposite orientation and separated by "stuffer" DNA, were inserted downstream of a tetracycline (tet)-inducible promoter. This construct was then linearized and transfected into a transgenic line of *T. brucei* that stably expressed the tet repressor, resulting in tet-mediated expression of a stem–loop RNA transcript. Our lab made a similar vector expressing a stem–loop RNA (Wang et al., 2000).

A major goal of our lab has been to study how a trypanosome replicates its unusual mitochondrial DNA, known as kinetoplast DNA (kDNA), and RNAi has been invaluable in these studies. kDNA is a DNA network containing several thousand catenated DNA minicircles (for comprehensive reviews on kDNA see (Klingbeil et al., 2001; Morris et al., 2001). The topology of kDNA resembles that of medieval chain mail. Given its network structure, topoisomerase II (topo II) must certainly play key roles in the replication and maintenance of kDNA. One such enzyme from *T. brucei* had been cloned (Strauss and Wang, 1990; Wang and Englund, 2001) and shown (in a related species) to localize to two antipodal sites flanking the condensed kDNA (Melendy and Ray, 1989). To test its function by RNAi, we cloned a fragment of the enzyme's coding region into the stem–loop expression vector (Figure 18.1A) and integrated this construct into the genome (Wang and Englund 2001). Within three days after induction of RNAi, the topo II protein had nearly disappeared and by day 6 the cells had stopped growing and begun to die (Figure 18.1B). This experiment provided strong evidence that this topoisomerase is in fact essential for the parasite's survival. In this case, the utility of inducible, heritable RNAi allowed us to examine daily the biochemical status of kDNA and its replication intermediates with the goal of understanding the precise function of the topo II. From these studies we found, remarkably, that after RNAi induction the kDNA network gradually shrank in size until it finally disappeared. As shown in the DAPI-stained trypanosomes in Figure 18.1C, each cell in an uninduced culture (– tet) has a nucleus (N) and a prominent kinetoplast (K). However, after RNAi silencing of the topo II for 9 days (+ tet), the kDNA network had either shrunk in size or, in most cases, disappeared. To understand how topo II knockdown could cause the loss of the kDNA network, we examined minicircle replication intermediates. It was known that during replication, covalently closed minicircles are released from the network (by a topo II), and after replication as free minicircles, their progeny in gapped form are attached to the network [(also by a topo II) (Englund, 1979)]. We found that after 2 days of induction of topo II silencing, there was a pronounced increase in the relative abundance of gapped-free minicircle replication intermediates. Because these intermediates accumulated, attachment onto the network must have been suppressed by RNAi. Therefore, a function of the topo II must be to attach these gapped progeny minicircles back onto the network. Without sustained minicircle attachment, the network shrinks and disappears, as shown in Figure 18.1C.

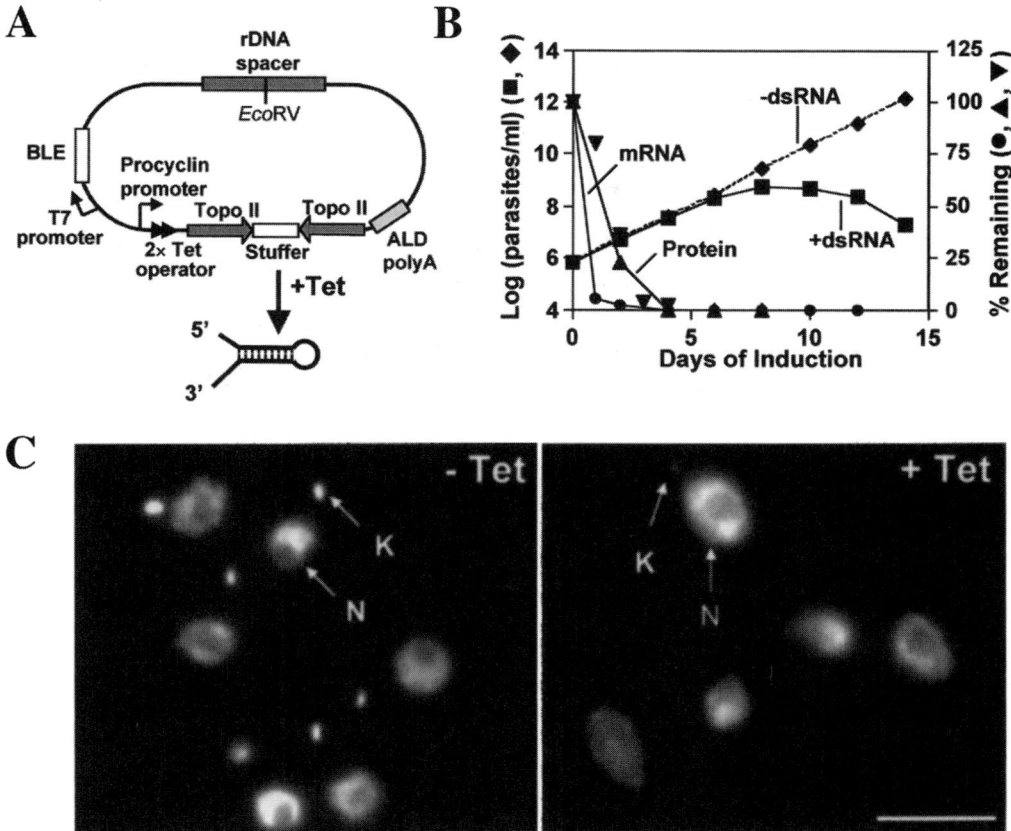

Figure 18.1. Expression of stem-loop dsRNA causes loss of topo II mRNA and protein and loss of kDNA. A) Stem–loop construct for topoisomerase RNAi: topoisomerase gene fragment (Topo II); phleomycin resistance gene (*BLE*); aldolase poly(A) addition sequence (*ALD polyA*). The construct is not drawn to scale. B) Effect of topoisomerase RNAi on cell growth. RNAi was uninduced (filled diamonds) or induced with 1 μg/ml tet (filled squares). The graph also shows topoisomerase mRNA levels (filled circles, evaluated by a northern blot, not shown) and topoisomerase protein levels (filled upright triangles and filled inverted triangles, evaluated by western blot, not shown). C) Cells expressing topoisomerase dsRNA lose kDNA. Cells were fixed and treated with DAPI to stain the nucleus (N) and kinetoplast (K). Left panel shows uninduced cells, and right panel shows cells after 9 days of RNAi. Scale bar, 5 μm. (Reprinted from Wang, Z. and Englund, P. T. (2001). RNA interference of a trypanosome topoisomerase II causes progressive loss of mitochondrial DNA. *EMBO Journal*, 20, 4674–4683 with permission of the publisher).

From these and other studies, RNAi clarified the function of this mitochondrial topo II.

Development of an inducible, dual-promoter RNAi vector: pZJM

The topoisomerase experiment demonstrated the power of heritable, inducible RNAi as a reverse genetic tool. It allowed us to deplete the cell of an essential protein, which in turn allowed analysis of the progression of the lethal phenotype. One hindrance to this approach was that construction of each stem–loop construct required considerable effort. With this in mind, we modified the expression vector by inserting opposed T7 promoters, each under the control of

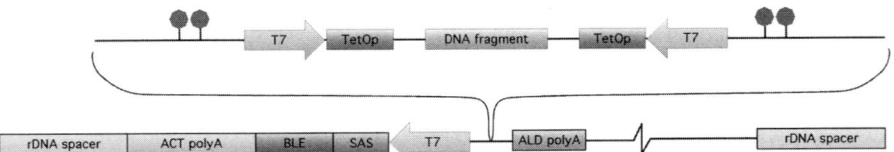

Figure 18.2. The pZJM RNAi vector. The tet operator (*Tet Op*), dual T7 terminators (red octagons), tet-inducible T7 promoters (*T7 arrows*), ribosomal DNA spacer (*rDNA*), actin poly(A) addition sequence (*ACT polyA*), phleomycin resistance gene (*BLE*), splice acceptor site (*SAS*), aldolase poly(A) addition sequence (*ALD polyA*). The plasmid is shown in linearized form, after cleavage in the rDNA spacer, and is not drawn to scale. (See color section.)

tet operator sequences (Figure 18.2). This method of transcribing a dsRNA through the use of opposed T7 promoters had been already developed by the Fire group at the Carnegie Institution (Timmons and Fire, 1998). Additionally we added T7 terminator sequences upstream of each of the T7 promoters. This vector, designated pZJM, allowed us to insert a fragment of DNA between the opposing promoter sequences and then transfect the construct into trypanosomes engineered to express T7 RNA polymerase and the tet repressor [(developed by Elisabeth Wirtz and George Cross, Rockefeller University) (Wirtz et al., 1999)]. We could then induce production of dsRNA by addition of tet to the culture medium (Wang et al., 2000). This RNAi vector strategy has also been adopted by the Donelson lab at University of Iowa resulting in construction of a vector, p2T7^{Ti}, that is very similar to pZJM (LaCount et al., 2002).

Reverse genetics using pZJM

pZJM is effective for most genes tested. An example is shown in Figure 18.3. This experiment involves the gene for ornithine decarboxylase (ODC), an enzyme that converts ornithine to putrescine (a step in polyamine biosynthesis). In this experiment, we inserted a 500-bp fragment of the ODC gene into pZJM and stably transfected the construct into the transgenic *T. brucei* (Wang et al., 2000). When cultured in the absence of tet, these cells grew normally. However, in as few as 40 hours of exposure to tet ODC mRNA is silenced, and within 3 days the cells began to show a severe growth phenotype. Furthermore (as seen when ODC was knocked out using a conventional gene replacement approach (Li et al., 1996)), addition of putrescine to the culture medium bypassed the need for this enzyme and restored normal growth, thus confirming the specificity of ODC silencing. The induction of RNAi was also reversible, as demonstrated by the restoration of normal growth after removal of tet from the medium at day 13.

Multiple genes can be targeted by a single pZJM construct

The ease of cloning a single gene fragment into pZJM, coupled with the ability to induce and maintain prolonged expression of dsRNA, made this vector a valuable tool for reverse genetic experiments in *T. brucei* (for two of many examples, see

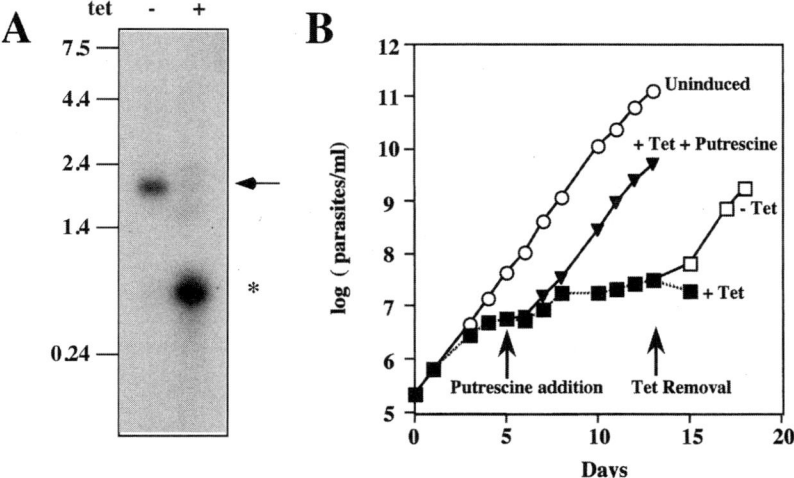

Figure 18.3. Cells induced to express ornithine decarboxylase (ODC) dsRNA exhibit loss of mRNA and fail to grow, but can be rescued by addition of the enzyme's product, putrescine. A. Northern analysis of total RNA from uninduced (−) or 40 h tet-induced (+) cells transfected with pZJM(ODC). ODC mRNA (arrow) and dsRNA (asterisk) are indicated. B. Growth of pZJM(ODC) cells that were either uninduced (open circles) or induced with tet for 5 days (filled squares). The induced culture was then split with one-half grown in the presence of both tet and 1mM putrescine (triangles). At day 13, the tet was removed from a portion of the induced culture (open squares), demonstrating the reversibility of RNAi induction. (Reprinted from Wang, Z., Morris, J. C., Drew, M. E. and Englund, P. T. (2000). Inhibition of *Trypanosoma brucei* gene expression by RNA interference using an integratable vector with opposing T7 promoters. *Journal of Biological Chemistry*, 275, 40174–40179 with permission of the publisher).

Grams et al., 2002; Li and Wang, 2002). We could also silence multiple genes with a single pZJM construct. For example, we created a chimeric pZJM insert composed of ~600 bp each of genes for a mitochondrial topoisomerase II and a mitochondrial DNA polymerase β (Wang et al., 2000). After transfection and RNAi induction, the cells not only exhibited a growth phenotype [(as was predicted from experiments that individually targeted the genes) (Figure 18.4A)] but also each gene's mRNA was significantly reduced (Figure 18.4B).

Forward genetics using an RNAi library

We have been especially interested in developing pZJM for forward genetics, to discover new genes or unknown functions for known genes. Our strategy has been to construct an RNAi library by random fragmentation of purified genomic DNA (average size ~660 bp; nearly all trypanosome genes do not contain introns), followed by ligation into the pZJM vector (Morris et al., 2002). After amplification of the library in *E. coli*, we then transfected a sufficient amount of library DNA into *T. brucei* to achieve a ~5-fold coverage of the genome. We maintained the library parasites in the absence of tet.

We proved the effectiveness of this library in two different experiments. In the first study, we induced the transfected cells with tet for seven days, and then screened for cells with reduced binding to the lectin concanavalin A [(ConA)

A

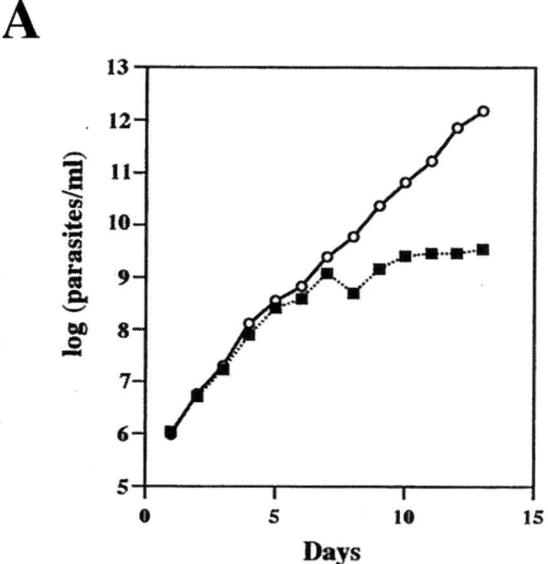

B

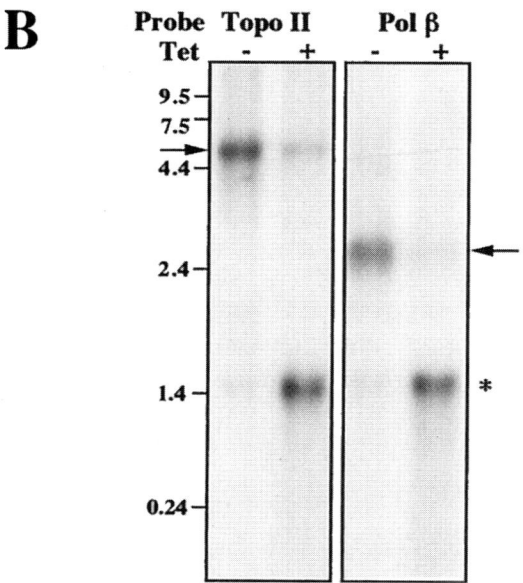

Figure 18.4. Multiple genes can be targeted by a single pZJM insert. A. RNAi of cells transfected with pZJM(topoII/polβ), containing a chimeric insert with sequences of both genes, stops cell growth. Cultures were grown in the absence (open circles) or presence of 1 μg/ml tet (closed squares). B. Northern analysis of total RNA from uninduced (–) and induced (+) cells. The blot was probed for topoisomerase mRNA (*rightward arrow*), then stripped and re-probed for polβ mRNA (*leftward arrow*). The *asterisk* indicates dsRNA. (Reprinted from Wang, Z., Morris, J. C., Drew, M. E. and Englund, P. T. (2000). Inhibition of *Trypanosoma brucei* gene expression by RNA interference using an integratable vector with opposing T7 promoters. *Journal of Biological Chemistry*, 275, 40174–40179 with permission of the publisher.)

A

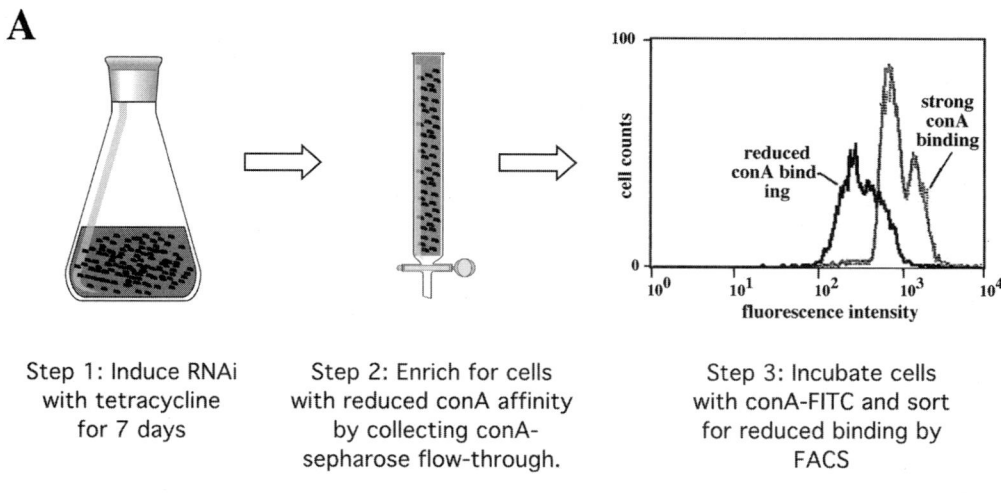

Step 1: Induce RNAi
with tetracycline
for 7 days

Step 2: Enrich for cells
with reduced conA affinity
by collecting conA-
sepharose flow-through.

Step 3: Incubate cells
with conA-FITC and sort
for reduced binding by
FACS

B

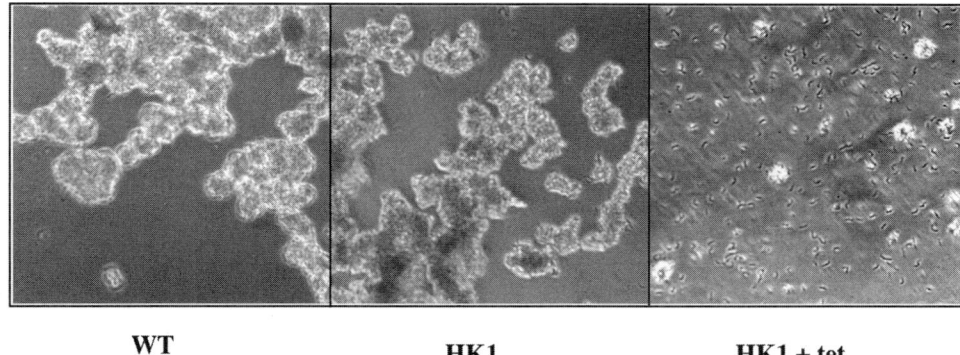

WT   **HK1**   **HK1 + tet**

Figure 18.5. Forward genetics using the *T. brucei* RNAi library. A. Scheme for screening for trypanosomes that do not bind ConA. Library cells, induced for RNAi, were screened for reduced ConA
binding by successive rounds of ConA-Sepharose chromatography and ConA-FITC cell sorting. B.
RNAi of the hexokinase gene *(HK1)* prevents agglutination by ConA. Wildtype (WT), uninduced
pZJM(HK1), and tet-induced (14 days) pZJM(HK1) cells were incubated with 5 μg/ml ConA for 24
h and then visualized by phase microscopy. RNAi of HK1 inhibits ConA-induced cell clumping.
(Reprinted from Morris, J. C., Wang, Z., Drew, M. E. and Englund, P. T. (2002). Glycolysis modulates
trypanosome glycoprotein expression as revealed by an RNAi library. *EMBO Journal*, 21, 4429–4438
with permission of the publisher).

(Morris et al., 2002)]. ConA binds to and induces killing of trypanosomes expressing a specific form of the surface glycoprotein EP-procyclin (Pearson et al., 2000).
Through multiple rounds of ConA chromatography and fluorescence-activated
cell sorting, we isolated a clone that did not bind ConA (Figure 18.5A). We identified the gene fragment responsible for its RNAi-dependant phenotype by amplifying the pZJM insert using primers flanking the T7 promoter sequences. Surprisingly, no open reading frame was apparent in the fragment. However, upon
analyzing its genomic locus, it appeared that the fragment was within a potential
3′ untranslated region of a gene encoding the hexokinase gene, *HK1*. Further analysis by Northern blotting revealed that this fragment did in fact silence the HK1

gene, causing loss of ability of ConA to agglutinate (Figure 18.5B). Subsequent experiments confirmed the role of HK1 in affecting expression levels of the procyclins. See (Morris et al., 2002) for discussion of the significance of this finding; the Morris lab is currently investigating the molecular mechanisms involved.

The second test of pZJM-based forward genetics involved a selection of library-transfected trypanosomes for cells resistant to killing by tubercidin, a toxic adenosine analog (Drew et al., 2003). In this experiment, we induced RNAi for seven days and then added tubercidin to the cell culture. We were successful in isolating a clone that was resistant to tubercidin only upon induction of RNAi. The pZJM insert responsible for this phenotype was a fragment of a hexose transporter gene, *THT-1*, that is a member of a tandemly arrayed cluster of nearly identical hexose transporter genes. Northern analysis revealed a loss of mRNA of this gene, as well as the other members of the cluster. Furthermore, RNAi nearly abolished glucose transport activity as measured by whole-cell glucose uptake. We had hypothesized *a priori* that silencing of genes such as those encoding a membrane-localized nucleoside transporter could give rise to a resistance phenotype, and therefore we were surprised to find that a hexose transporter was responsible. Further analysis revealed that RNAi silencing of some glycolytic enzymes also confered tubercidin resistance. These findings suggest that a target of tubercidin (or a phosphorylated metabolite) is glucose metabolism, and in fact we found that tubercidin triphosphate inhibited the glycolytic enzyme phosphoglycerate kinase. This finding presented a dilemma. Why should tubercidin inhibition of glycolysis kill trypanosomes but RNAi silencing of glucose transport or glycolysis does not? The answer is based on the finding that trypanosomes can switch their metabolism from a glucose-requiring state (in which glycolysis is essential) to a state dependent on amino acids (ter Kuile, 1997). They adapt to the latter state when glucose is removed from the medium. Tubercidin must kill the parasites quickly, giving them no time to adapt to amino-acid based metabolism. In contrast, RNAi silencing of the hexose transporter or a glycolytic enzyme occurs over several days, providing time for adaptation. Consistent with this explanation, we found that cells growing in glucose-depleted medium were resistant to tubercidin, even cells never transfected with pZJM. Thus, the RNAi library provided information not only into a mechanism of tubercidin toxicity, but it also gave insight into metabolic regulation in this parasite.

Concluding remarks

The field of biomedical research abounds with examples of how seminal biological discoveries are quickly adapted into powerful tools for the researcher. Restriction endonucleases, *Taq* polymerase, and the green fluorescent protein are but a few of the more prominent examples. RNAi should be added to this list. As a tool for the trypanosome researcher, the relative ease of reverse genetics using inducible-RNAi has significantly advanced the status of research throughout the field. Equally exciting is the application of true forward genetics using an RNAi library, a strategy that we continue to develop.

Acknowledgements

The authors would like to thank the members of the Englund group for their continued support and helpful discussions throughout this project. We also thank Viiu Klein and Theresa Westhead for invaluable technical assistance.

REFERENCES

Beverley, S. M. (2003). Protozomics: Trypanosomatid parasite genetics comes of age. *Nature Reviews of Genetics*, **4**, 11–19.

Clayton, C. E. (1999). Genetic manipulation of kinetoplastida. *Parasitology Today*, **15**, 372–378.

Cottrell, T. R. and Doering, T. L. (2003). Silence of the strands: RNA interference in eukaryotic pathogens. *Trends in Microbiology*, **11**, 37–43.

Drew, M. E., Morris, J. C., Wang, Z., Wells, L., Sanchez, M., Landfear, S. M. and Englund, P. T. The adenosine analog tubercidin inhibits glycolysis in *Trypanosoma brucei* as revealed by an RNA interference library. *Journal of Biological Chemistry*, **278**, 46596–46600.

Englund, P. T. (1979). Free minicircles of kinetoplast DNA in *Crithidia fasciculata*. *Journal of Biological Chemistry*, **254**, 4895–4900.

Grams, J., Morris, J. C., Drew, M. E., Wang, Z., Englund, P. T. and Hajduk, S. L. (2002). A trypanosome mitochondrial RNA polymerase is required for transcription and replication. *Journal of Biological Chemistry*, **277**, 16952–16959.

Klingbeil, M. M., Drew, M. E., Liu, Y., Morris, J. C., Motyka, S. A., Saxowsky, T. T., Wang, Z. and Englund, P. T. (2001). Unlocking the secrets of trypanosome kinetoplast DNA network replication. *Protist*, **152**, 255–262.

LaCount, D. J., Barrett, B. and Donelson, J. E. (2002). *Trypanosoma brucei* FLA1 is required for flagellum attachment and cytokinesis. *Journal of Biological Chemistry*, **277**, 17580–17588.

Li, F., Hua, S. B., Wang, C. C. and Gottesdiener, K. M. (1996). Procyclic *Trypanosoma brucei* cell lines deficient in ornithine decarboxylase activity. *Molecular and Biochemical Parasitology*, **78**, 227–236.

Li, Z. and Wang, C. C. (2002). Functional characterization of the 11 non-ATPase subunit proteins in the trypanosome 19 S proteasomal regulatory complex. *Journal of Biological Chemistry*, **277**, 42686–42693.

Melendy, T. and Ray, D. S. (1989). Novobiocin affinity purification of a mitochondrial type II topoisomerase from the trypanosomatid *Crithidia fasciculata*. *Journal of Biological Chemistry*, **264**, 1870–1876.

Morris, J. C., Drew, M. E., Klingbeil, M. M., Motyka, S. A., Saxowsky, T. T., Wang, Z. and Englund, P. T. (2001). Replication of kinetoplast DNA: An update for the new millennium. *International Journal for Parasitology*, **31**, 453–458.

Morris, J. C., Wang, Z., Drew, M. E. and Englund, P. T. (2002). Glycolysis modulates trypanosome glycoprotein expression as revealed by an RNAi library. *European Molecular Biology Organization Journal*, **21**, 4429–4438.

Ngo, H., Tschudi, C., Gull, K. and Ullu, E. (1998). Double-stranded RNA induces mRNA degradation in *Trypanosoma brucei*. *Proceedings of the National Academy of Sciences USA*, **95**, 14687–14692.

Pearson, T. W., Beecroft, R. P., Welburn, S. C., Ruepp, S., Roditi, I., Hwa, K. Y., Englund, P. T., Wells, C. W. and Murphy, N. B. (2000). The major cell surface glycoprotein procyclin is a receptor for induction of a novel form of cell death in African trypanosomes *in vitro*. *Molecular and Biochemical Parasitology*, **111**, 333–349.

Shi, H., Djikeng, A., Mark, T., Wirtz, E., Tschudi, C. and Ullu, E. (2000). Genetic interference in *Trypanosoma brucei* by heritable and inducible double-stranded RNA. *RNA*, **6**, 1069–1076.

Strauss, P. R. and Wang, J. C. (1990). The TOP2 gene of *Trypanosoma brucei*: A single-copy gene that shares extensive homology with other TOP2 genes encoding eukaryotic DNA topoisomerase II. *Molecular and Biochemical Parasitology*, **38**, 141–150.

ter Kuile, B. H. (1997). Adaptation of metabolic enzyme activities of *Trypanosoma brucei* promastigotes to growth rate and carbon regimen. *Journal of Bacteriology*, **179**, 4699–4705.

Timmons, L. and Fire, A. (1998). Specific interference by ingested dsRNA. *Nature*, **395**, 854.

Wang, Z. and Englund, P. T. (2001). RNA interference of a trypanosome topoisomerase II causes progressive loss of mitochondrial DNA. *European Molecular Biology Organization Journal*, **20**, 4674–4683.

Wang, Z., Morris, J. C., Drew, M. E. and Englund, P. T. (2000). Inhibition of *Trypanosoma brucei* gene expression by RNA interference using an integratable vector with opposing T7 promoters. *Journal of Biological Chemistry*, **275**, 40174–40179.

Wirtz, E., Leal, S., Ochatt, C. and Cross, G. A. (1999). A tightly regulated inducible expression system for conditional gene knock-outs and dominant-negative genetics in Trypanosoma brucei. *Molecular and Biochemical Parasitology*, **99**, 89–101.

19 RNA-mediated gene silencing in fission yeast

Greg M. Arndt

Introduction

Different forms of RNA-mediated gene silencing, namely antisense RNA, ribozymes and double-stranded RNA (dsRNA), act in naturally occurring mechanisms of gene regulation and provide tools for artificially silencing specific genes (Brantl, 2002). Early work in bacteria indicated that RNA could act as a key regulator of complex biological systems (Itoh and Tomizawa, 1980). These observations led to the use of antisense RNA as a regulator of gene expression in both prokaryotes and eukaryotes (Izant and Weintraub, 1984; Pestka et al., 1984). Equally, characterization of plant viruses and viroids uncovered the presence of RNA sequences with the ability to catalyze degradation of homologous RNA (Forster and Symons, 1987). Adaptation of these catalytic RNAs, or ribozymes, to cleave any target sequence led to applications in both plant biotechnology and medicine (Peurta-Fernandez et al., 2003). The recent discovery of dsRNA as a key effector of RNA-directed gene silencing, in a wide range of different organisms, has revolutionized studies of gene function (Fire et al., 1998). In addition, it has provided evidence for the existence of a multi-component protein complex capable of using dsRNA signals to mediate post-transcriptional and transcriptional gene silencing (Hannon, 2002).

Model organisms have been instrumental in advancing our understanding of the mechanisms underlying different forms of gene regulation. In the area of gene silencing directed by RNA, this is best exemplified by the genetic and biochemical studies in nematodes, plants and flies, examining the mechanism of dsRNA-mediated gene regulation or RNA interference (RNAi). Genetic analysis has identified host genes essential for transcriptional and post-transcriptional gene silencing, while biochemical studies have begun to elucidate the molecular machinery involved (Hannon, 2002). These results highlight the merits of a well-developed model system and its utility in better understanding the underlying mechanism(s) of RNA-mediated gene silencing.

In this review, we describe the development and use of a single-cell eukaryotic model system, in the fission yeast *Schizosaccharomyces pombe*, for examining different types of RNA-mediated gene silencing. Specifically, this model has been used to assess the parameters affecting antisense RNA and dsRNA regulation, the molecular mechanism(s) involved and the transferability of these findings to other eukaryotic cell types. Furthermore, results obtained using this yeast model support the existence of a common pathway for gene suppression by antisense and dsRNA. In addition, we also summarize pioneering studies using fission yeast to demonstrate the role of RNA in controlling the formation of *silent* heterochromatin.

Yeast model for RNA-mediated gene silencing

In an effort to develop a simple model system for examining RNA-mediated gene silencing, we used the fission yeast *Schizosaccharomyces pombe* as the host. Prior to development of this yeast model, numerous laboratories attempted to develop similar systems in the budding yeast *Saccharomyces cerevisiae* with little success (Atkins et al., 1994). At the time, it was predicted that budding yeast either contains cellular factors that interfere with artificial RNA-mediated gene silencing or lacks the essential factors for effective gene suppression. Completion of the genome sequence of both budding and fission yeast has permitted a comparison of the genes in common and different between these two yeast species. Interestingly, budding yeast lacks the genes encoding the components of the RNA induced silencing complex (RISC), the multi-protein complex responsible for mediating RNAi (Aravind et al., 2000). In contrast, fission yeast has retained these proteins and recent studies have indicated that this machinery is required for heterochromatin formation at centromeric repeats (Hall et al., 2002; Reinhart and Bartel, 2002; Volpe et al., 2002). It has been hypothesized that budding yeast may have lost these components, as it became less dependent on DNA methylation to silence endogenous gene expression.

In the model generated, yeast strains stably expressing the *E. coli lacZ* reporter gene under constitutive promoter control were constructed using targeted gene integration [(Arndt et al., 1995) (Figure 19.1)]. All *lacZ* antisense genes were maintained on episomal plasmids under control of the thiamine-repressible *nmt1* promoter. In this way, expression of the *lacZ*-specific complementary RNAs could be conditionally controlled by the presence or absence of thiamine. The *lacZ* target gene offered specific advantages including the availability of a rapid and sensitive reporter protein assay and production of a blue colony color on medium containing a specific chromogenic substrate. Each of the test constructs could be screened for impact on β-galactosidase activity and the *lacZ*-encoded cellular phenotype. Using this model, antisense RNA regulation was shown to be conditional and reversible, with the level of *lacZ* gene silencing being dependent on the level of antisense RNA. In addition, the extent of suppression mediated by the antisense RNA correlated with the degree of colony color, providing a phenotypic indicator of the efficacy of gene silencing.

A primary goal in developing the fission yeast model was to examine factors impacting on the efficiency of RNA silencing. At the time of establishing this

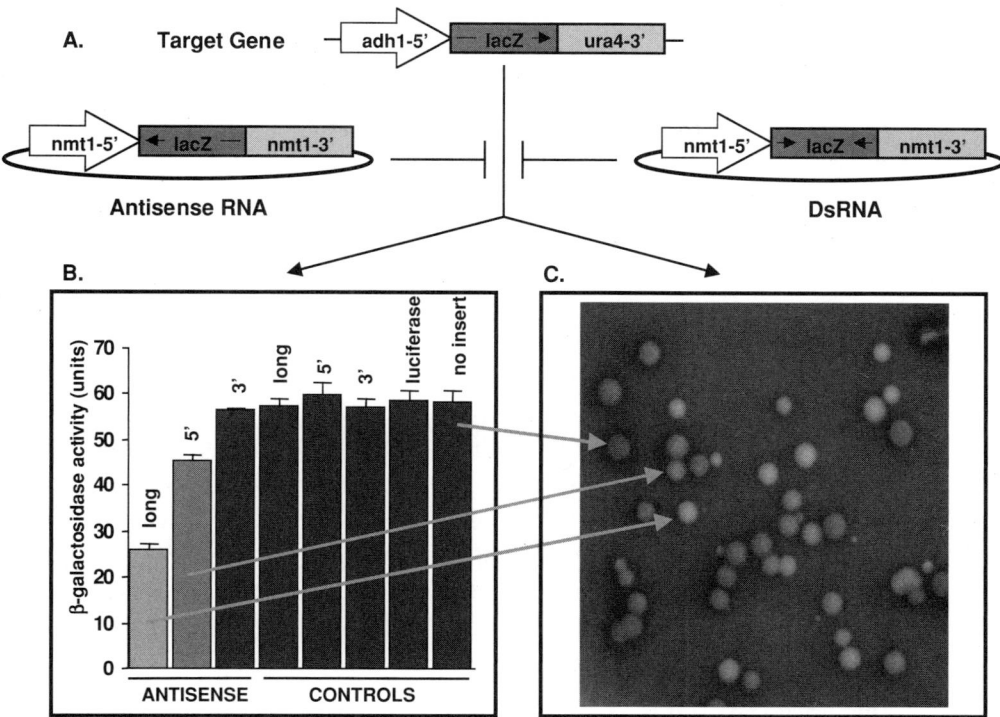

Figure 19.1. Yeast model of RNA-mediated gene silencing. (A) The *lacZ* target gene is integrated into the fission yeast genome under control of the constitutive *adh1* promoter and *ura4* 3' terminator sequences. The genes expressing antisense RNA (left) or dsRNA (right) are driven by a thiamine-repressible *nmt1* promoter and contained on episomal plasmids. (B) β-galactosidase activity in yeast transformants containing different plasmid constructs. The abbreviations used are as follows: long − long *lacZ* antisense or sense RNA; 5' − *lacZ* antisense or sense RNA from 5' end of target RNA; 3' − *lacZ* antisense or sense RNA from 3' end of target RNA; luciferase − luciferase sense RNA; no insert − empty vector control. (C) Blue colony color phenotypes of yeast colonies with different levels of β-galactosidase activity. The orange arrows indicate the resident constructs contained in specific colonies.

model, it was evident that antisense RNA design was largely empirical and that the critical factors for efficacy were undefined. From a number of studies, it was shown that the length, target site, dose and localization of the antisense RNA contribute to its efficacy (Brantl, 2002). Using the yeast model, we examined both the length and dose of antisense RNA. Complementary RNAs of differing sizes were tested and, in general, the longer antisense RNA sequences conferred the greatest activity. Antisense RNA dose studies showed that antisense RNA regulation displayed a dose dependency with suppression greater with higher steady-state levels of antisense RNA. In addition to studies on the length and dose of the antisense RNA, we have also extensively examined target site selection for short and long complementary RNAs *in vivo*. Finally, this model has also been used to determine the influence of antisense gene location, relative to the target gene, on the efficacy of gene silencing (Raponi et al., 2000).

The concept of antisense RNA-mediated gene suppression involves the formation of a double-stranded RNA between the antisense RNA and its complementary target RNA. As such, it is possible that antisense RNA operates through a RNAi-like pathway resulting in target RNA degradation (Di Serio et al., 2001; Martinez et al., 2002). In an attempt to understand the mode of action of antisense RNA, we examined the fate of the *lacZ* target RNA in yeast transformants expressing *lacZ* antisense RNA and showing reduced β-galactosidase activity (Raponi and Arndt, 2003). Northern analyses indicated that *lacZ* RNA was reduced in the presence of effective antisense RNA and that the level of β-galactosidase activity correlated with target RNA. This indicated that antisense RNA-mediated gene silencing in fission yeast occurred through target RNA destruction, a mechanism consistent with dsRNA-mediated gene silencing. It has recently been shown that gene regulation by antisense RNA in plants occurs through the production of small RNAs (DiSerio et al., 2001). To further decipher the mechanism of antisense RNA-mediated silencing, we have shown that *lacZ*-specific small RNAs are produced (Raponi and Arndt, 2003). In addition, as discussed in a future section of this review, we have demonstrated that a specific host factor can enhance both forms of RNA-mediated gene regulation. All of these observations support the hypothesis that antisense RNA and dsRNA either operate through the same mechanism or share cellular components.

Screening for RNAs capable of effective gene silencing

In all forms of RNA-mediated gene silencing, the accessibility to the target mRNA is considered a key hurdle to effective gene suppression. This accessibility is dependent on the cellular localization of the two complementary RNAs and their ability to hybridize. *In vitro* strategies for the identification of accessible antisense sequences have been developed (Kretschmer-Kazemi Far and Sczakiel, 2003). However, the *in vivo* efficacy of these sequences remains difficult to predict. To overcome some of the limitations associated with *in vitro* assays and *in silico* prediction programs, we adapted the fission yeast model to screen for effective antisense sequences to any target RNA within the cellular environment (Arndt et al., 2000). In this generic system, target sequences were fused to the 5′ end of the *lacZ* reporter gene and the fusion gene used to construct a yeast strain with all the characteristics associated with the *lacZ* protein (Figure 19.2). The target sequence was then randomly fragmented to construct a conditional gene-specific expression library of 1×10^5 members. These libraries were screened in the target-*lacZ* gene fusion strain for constructs that modified the blue colony color phenotype. Previous optimization indicated that as little as a 20% reduction in β-galactosidase activity resulted in a visually detectable decrease in blue color.

This model has been used to identify the most effective RNA-mediated silencing constructs against a variety of genes including *lacZ*, *ura4*, c-myc and Chk1. In most instances these constructs were shown to be equally effective at reducing target gene expression in human cells, suggesting conservation of the cellular machinery responsible for gene silencing between fission yeast and human cells. Alignment of the sequences identified, to the target sequences, indicated the presence of gene-specific patterns for the antisense RNAs, thus delineating sub-regions

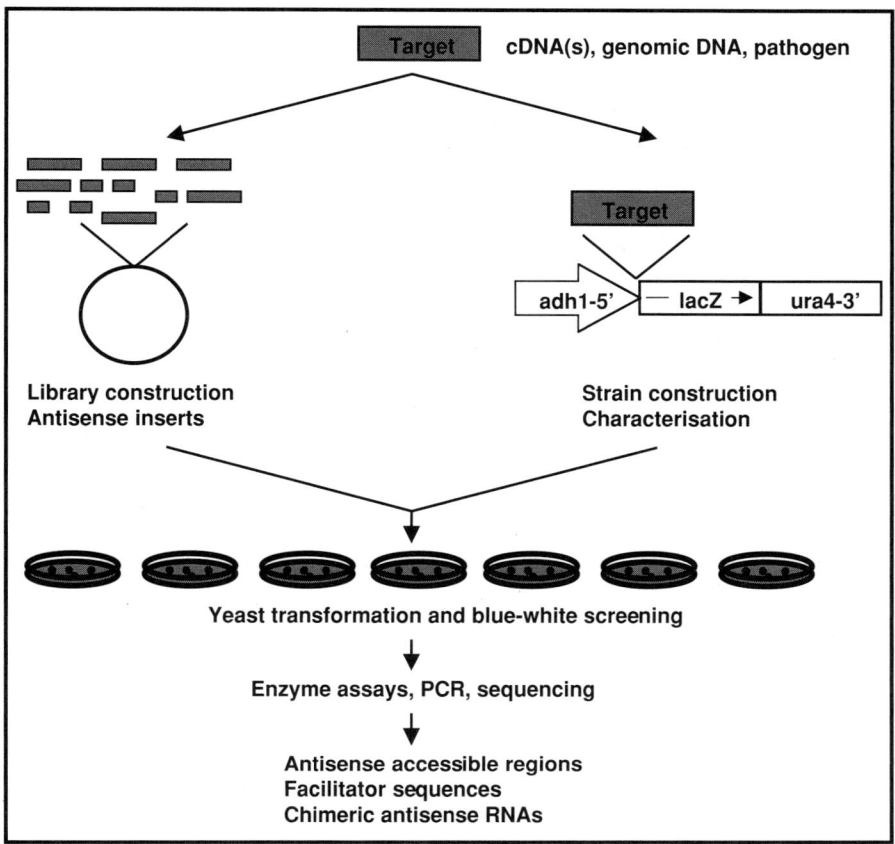

Figure 19.2. *In vivo* genetic screen for identification of effective RNA inhibitors. A target sequence, derived from cDNA or genomic DNA, is fused to the *lacZ* reporter gene to construct a target-*lacZ* fusion gene (right). This sequence is also fragmented to generate an expression library (left) that is delivered to a yeast strain containing the target-*lacZ* fusion gene. Following blue/white colony color screening, plasmids are recovered and analysed for insert sequences. This *in vivo* screen permits the identification of target RNA accessible regions, antisense facilitator sequences and novel RNA inhibitors.

of the target RNA accessible to RNA-mediated gene silencing. Interestingly, use of the fission yeast screen identified a class of novel *trans*-acting RNA modulators referred to as chimeric antisense RNAs (Arndt et al., 2000). These RNAs have the potential to hybridize with different regions of the target RNA or to form alternative intramolecular structures, including dsRNA, which may enter the RNAi pathway. Alternatively, sequences within these hybrid RNAs may act as facilitators to increase the accessibility of other complementary sequences to the target RNA. The identification of these unique RNA inhibitor sequences emphasizes the distinct advantages of using an *in vivo* genetic screen.

The possible applications of this genetic screen in the post-genomic era are numerous. Firstly, based on the simplicity of this system it can be used to identify a number of constructs for a large number of target genes. The transferability of these constructs to human cells allows for rapid testing in gene function studies. Secondly, this screen can be used in functional genomics to assist in assigning function to orphan genes. Finally, the constructs identified may encode RNAs that operate through novel post-transcriptional or transcriptional regulatory pathways. These may be used to better understand the underlying cellular mechanisms of how RNAs direct gene repression.

DsRNA-mediated gene regulation in fission yeast

The discovery that dsRNA can be used to mediate gene silencing in a wide variety of organisms has provided a tool for genetic control of gene expression and the basis for understanding the mechanism(s) of RNA-mediated gene silencing (Hannon, 2002). Genetic and biochemical studies using model organisms such as *C. elegans* and *Drosophila* have identified some of the host-encoded genes involved and the cellular machinery required for gene-specific silencing by dsRNA (Denli and Hannon, 2003). It is now believed that longer dsRNA is processed by the RNase III-like endonuclease Dicer into small 21–25 nucleotide dsRNA species and that these smaller RNAs are the effectors of gene silencing. The identification of these mediators has provided an essential tool for studying gene function in mammalian cells (McManus and Sharp, 2002). Studies in model organisms have identified the key components of RISC, which uses the small RNAs to hydrolyze complementary target RNA (Caudy et al., 2003; Hammond et al., 2003; Liu et al., 2003).

The presence of the RNAi cellular machinery in fission yeast makes it an ideal model for examining this form of gene silencing. In this regard, we have used our model to show that dsRNA can regulate the expression of target sequences [(Raponi and Arndt, 2003) (Figure 19.3)]. Preliminary observations in our laboratory suggested that antisense RNA-mediated gene suppression is enhanced by the expression of additional non-coding sense RNA. Using this information, we proceeded to demonstrate that co-expression of sense and antisense RNAs, in the presence of homologous target mRNA, enhanced gene silencing. Furthermore, expression of a *lacZ* panhandle RNA induced silencing, indicating that the generation of dsRNA through either intramolecular or intermolecular hybridisation was central to RNA-mediated gene regulation (Raponi and Arndt, 2003).

In many ways the results obtained with co-expression of sense and antisense RNA were unexpected in that the delivery of excess sense RNA would be predicted to titrate antisense RNA away from the target mRNA leading to reduced suppression. Consistent with gene silencing in other systems, the enrichment was associated with a concomitant reduction in *lacZ* target mRNA (Raponi and Arndt, 2003). The specificity of this form of dsRNA-mediated gene inhibition was demonstrated by showing that a c-myc-*lacZ* fusion gene could be regulated by co-expression of sense and antisense c-myc RNAs. The co-expression of antisense and non-coding sense RNAs as a strategy for inducing dsRNA-mediated gene silencing has been used in other organisms (Waterhouse et al., 1998; Aoki et al., 2003; Wang et al., 2003). In addition, the co-existence of naturally occurring complementary RNAs in different organisms suggests that eukaryotic cells may use this as a mechanism of gene regulation (Yelin et al., 2003).

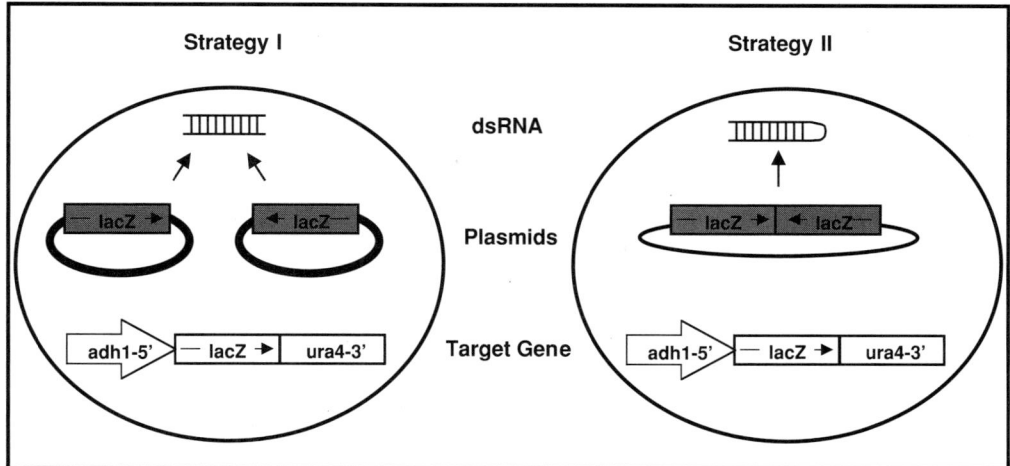

Figure 19.3. Strategies for dsRNA-mediated gene silencing. Strategy I involves the co-transformation of the *lacZ* gene-expressing yeast strain with plasmids encoding antisense RNA and non-coding sense RNA. Intermolecular hybridization of these sequences produces a dsRNA species. Strategy II uses a single plasmid encoding an inverted repeat derived from the *lacZ* target gene. Expression from this plasmid produces RNA capable of forming dsRNA by intramolecular hybridization.

An alternative strategy for the intracellular expression of dsRNA involves using inverted repeat sequences derived from the target gene (Waterhouse et al., 1998; Paul et al., 2002). Transcribed RNA has the capacity to self-anneal and form a panhandle or hairpin structure, which presumably can act as a substrate for Dicer-mediated cleavage into small effector RNAs. Using our model, we tested this approach by producing a vector encoding a 6.2–kb transcript, with a 1–kb loop and 2.5-kb of self-complementarity (Raponi and Arndt, 2003). Gene silencing mediated by this RNA was shown to be dose-dependent and associated with the production of *lacZ*-specific small RNAs, further establishing the system as an ideal model for studying dsRNA-mediated gene silencing.

Use of the fission yeast model to examine both antisense and dsRNA-mediated gene silencing has indicated that these two forms of RNA-directed gene regulation may operate through similar mechanisms, with the production of dsRNA being a key factor. At least two observations support this conclusion. Firstly, both forms of regulation are associated with the production of small gene-specific RNAs, suggesting that the dsRNA formed by intermolecular hybridization between antisense RNA and target mRNA or by intramolecular hybridization are converted into effector RNAs similar in size to those implicated in RNAi. Secondly, over-expression of co-factors of antisense RNA regulation also enhanced gene silencing by dsRNA. In particular, co-expression of a previously identified antisense enhancing sequence (aes), *aes2* (EF-Tu), with the *lacZ* panhandle RNA stimulated dsRNA-mediated gene silencing (Raponi and Arndt, 2003). Furthermore, over-expression of the ATP-dependent RNA helicase, *ded1*, with characteristics of helicases proposed to play a role in RNAi also enhanced antisense RNA efficacy. The *ded1* helicase acts during translation initiation to unwind RNA hybrids (Grallert et al., 2000). *aes2* encodes for a version of the mitochondrial elongation factor EF-Tu that lacks the localization domain essential for uptake into the mitochondrion (Glick and Schatz, 1991). EF-Tu transports tRNA to the translating ribosome by recognizing the stem structure of the tRNA (Condeelis, 1995). It is proposed that these factors may be rate limiting in the mechanism of RNA-mediated gene silencing in fission yeast. In addition, the sub-cellular localisation of these factors to the cytoplasm, and more precisely the translational machinery, is consistent with the mechanism of RNAi (Djikeng et al., 2003) and the association of components of the RISC complex with ribosomes (Bernstein et al., 2001). The utilization of common mechanisms for gene silencing by antisense RNA and dsRNA is further supported by studies in plants and flies (Di Serio et al., 2001; Martinez et al., 2001).

Antisense and dsRNA enhancing factors

Genetic studies involving mutagenesis have been used to understand the mechanism of RNA-induced gene silencing, particularly RNAi (Bosher and Labouesse, 2000). This has led to the identification of factors that, upon loss of function, reduce or eliminate gene silencing. These genes include those encoding RNA helicases, RNA-dependent RNA polymerase, RecQ DNA helicase, Dicer, Argonaute-like proteins, and other accessory proteins (Hannon, 2002). A common feature of these approaches is the identification of genes that are non-essential for cell growth and viability. This is not surprising given that the genetic screens select against genes essential for cell growth and development.

Dominant genetics is an alternative approach to deciphering gene function and involves increasing the cellular dose of specific proteins and determining the impact of gene over-expression on cellular phenotype(s) (Ramer et al, 1992). Using the fission yeast model, we developed a novel over-expression strategy to identify unique host-encoded factors that enhance antisense RNA-mediated gene silencing (Raponi and Arndt, 2002). This method allows for the identification of both essential and non-essential host genes. Furthermore, the factors identified are, in theory, rate limiting and provide genes that upon over-expression may be used to enhance RNA-mediated silencing in recalcitrant cell systems. Using the colony color screen described earlier, a fission yeast cDNA expression library was delivered to *lacZ* gene-containing yeast, expressing *lacZ* antisense RNA, and colonies screened for reduced blue color. Four novel *aes* factors were identified that enriched antisense RNA-directed gene inhibition by up to 50%. In addition, the ATP-dependent RNA helicase *ded1* was shown to boost the efficacy of partially effective *lacZ* antisense RNA. Of the five factors identified as enhancing RNA-mediated gene silencing, three [*aes2* (EF-Tu), *aes4* (L7a) and *ded1*] can associate with the translational machinery. Moreover, at least one of these factors, *aes2* (EF-Tu), was shown to promote gene silencing by dsRNA. Possible roles for the *aes* factors in the RNAi pathway are summarised in Figure 19.4.

The observation that the majority of factors identified in the fission yeast screen have an association with components of the translational machinery suggests that RISC is likely associated with ribosomes. Previous studies on the purification of the protein complex have indicated its physical association with

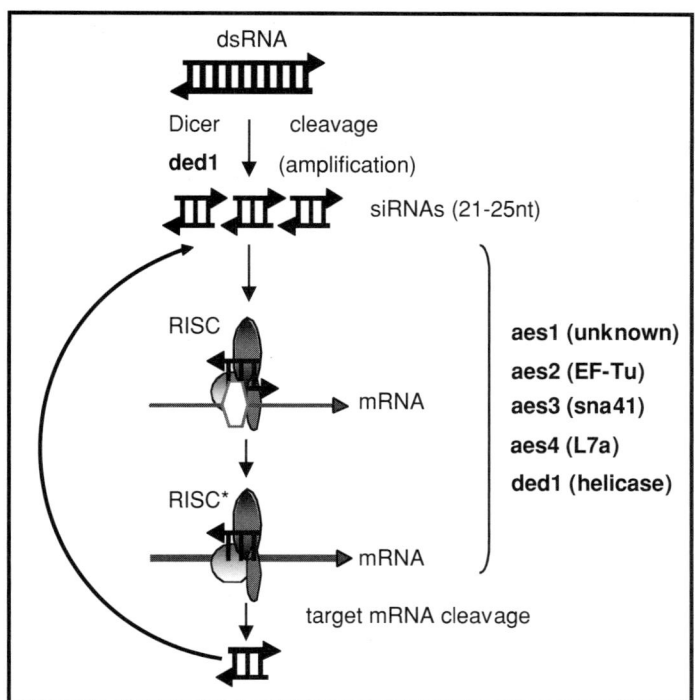

Figure 19.4. *Aes* factors involved in RNA-mediated gene silencing. Long dsRNA generated by intermolecular or intramolecular hybridization undergoes cleavage by Dicer to produce small 21-25 nucleotide (nt) RNAs. These RNAs are associated with inactive RISC (RISC) to produce active RISC (RISC*). This complex containing a guide RNA strand binds homologous target mRNA and directs its turnover. The stages in this pathway at which the aes factors (*aes1-4*) and *ded1* may play a functional role are indicated.

ribosomes (Bernstein et al., 2001) and, more recently, siRNA ribonucleoprotein complexes have been shown to co-sediment with polyribosomes (Djikeng et al., 2003). Characterization of proteins associated with RISC have identified the homolog of the Fragile X mental retardation protein (FMR1), an RNA binding protein that binds translating ribosomes and acts as a negative regulator of translation (Ishizuka et al., 2002). In addition to this close association between RISC and the translational machinery, there is also evidence that nonsense-mediated decay components are associated with ribosomes (Atkin et al., 1997). Furthermore, genetic studies have identified a link between RNAi-mediated gene silencing and nonsense-mediated decay (Domeier et al., 2000).

This interplay between RISC, polyribosomes and the nonsense RNA decay machinery suggests that siRNAs may be part of a complex surveillance system to trigger degradation of mRNA. Given that translation can occur both in the nucleus and the cytoplasm (Iborra et al., 2001), this surveillance mechanism would provide a rapid response system for destroying aberrant RNAs or for controlling mRNA stability and turnover within both sub-cellular compartments. In the context of this model, the *aes* factors identified may be additional accessory components of RISC, ribosomes or mRNA decay complexes that are rate limiting or rate determining in the surveillance mechanism. Further work will be required to understand the precise mechanism of action of the host-encoded factors identified using the fission yeast system.

The over-expression strategy for dissecting host-encoded genes involved in RNA silencing complements mutagenesis and overcomes some of the limitations associated with these loss-of-function studies. Furthermore, it is possible to co-express these factors with different forms of gene silencing, such as sense RNA, antisense RNA or dsRNA to enrich gene inhibition. Finally, these host-encoded factors may be involved in the mechanism of silencing, either directly or indirectly, and therefore assist in understanding how antisense RNA or dsRNA direct specific gene suppression.

RNA-mediated heterochromatin formation in fission yeast

The formation of heterochromatin is essential for a wide range of gene silencing phenomena including imprinting, X-inactivation, genomic integrity, and centromere and telomere function (Grewal and Moazed, 2003). Factors contributing to the creation of heterochromatin domains include the presence of repeat sequences and modifications to associated histones. A series of recent studies in fission yeast have begun to understand the mechanism underlying the formation of this *silent* chromatin with the evidence implicating dsRNA as the effector (Hall et al., 2002; Volpe et al., 2002; Schramke and Allshire, 2003).

A link between dsRNA and chromatin silencing was suggested by studies in plants in which dsRNA homologous to a target transgene directed methylation of the homologous DNA (Mette et al., 2000). However, it was not clear from these studies how dsRNA directed chromatin re-modelling at a specific genomic region. Fission yeast has no DNA methylation and heterochromatin is marked by modifications of histone H3, specifically methylation of lysine 9. More recently, it has been shown that deletion of genes involved in RNAi, specifically those encoding Dicer, Argonaute and RNA-dependent RNA polymerase, leads to loss of gene silencing at centromeres and the mating-type locus in fission yeast (Volpe et al., 2002; Hall et al., 2002). Furthermore, it was demonstrated that non-coding transcripts from centromeric repeat sequences, with the capacity to form dsRNA, undergo processing to siRNAs that direct histone modification and sequester chromatin silencing factors at the centromere (Reinhart and Bartel, 2002; Volpe et al., 2002). Most recently, Schramke and Allshire (2003) showed that RNA-mediated gene silencing occurs at non-centromeric regions by expressing a small hairpin RNA (shRNA) specific for the *ura4* locus. Silencing was dependent on the RNAi machinery and the silenced locus exhibited features associated with centromeric heterochromatin. These authors went on to show that silencing of meiosis-specific genes in vegetative cells is controlled by flanking retrotransposon long terminal repeats (LTR) that produce non-coding transcripts. It is likely that this mechanism of gene repression is dependent on convergent transcription across the repeat sequences and processing of the LTR transcripts by the RNAi pathway. The above studies suggest that RNAi-dependent formation of silent chromatin can occur at centromeres, meiosis-induced genes, and from shRNA expressed *in trans* in fission yeast.

Noncoding RNAs control imprinting and X-inactivation in mammalian cells, the latter of which occurs through histone H3 lysine 9 methylation following the expression of a long antisense RNA *in cis* (Wutz, 2003). This structure is similar to the proposed model for heterochromatin formation at repeat sequences in fission yeast. Several outstanding questions remain to be answered. Firstly, is RNAi-mediated heterochromatin formation universal to other organisms? Secondly, what are the conditions by which siRNAs direct gene silencing at the DNA level and what are the rules governing this phenomenon? Thirdly, can this naturally occurring mechanism of gene silencing be exploited for applications? Single-cell fission yeast will remain instrumental in understanding this gene-silencing phenomenon and providing answers to these key questions.

Summary

RNAs have a role in naturally occurring mechanisms of gene silencing in eukaryotic cells and have been exploited to control specific gene expression for both gene functional studies and applications in medicine and biotechnology. Variability in the efficacy of these approaches, particularly antisense RNA, led us to develop a single-cell yeast model for examining factors impacting on RNA-mediated gene silencing. In this review, we have highlighted the utility of this yeast for studying key parameters, defining host genes and identifying the most effective RNA species for silencing. It is anticipated that fission yeast will continue to act as an ideal model for artificial RNA-mediated gene silencing and understanding the mechanisms by which endogenous RNAs control gene expression.

REFERENCES

Aoki, Y., Cioca, D. P., Oidaira, H., Kamiya, J. and Kiyosawa, K. (2003). RNA interference may be more potent than antisense RNA in human cancer cell lines. *Clinical and Experimental Pharmacology and Physiology*, **30**, 96–102.

Aravind, L., Watanabe, H., Lipman, D. J. and Koonin, E. V. (2000). Lineage-specific loss and divergence of functionally linked genes in eukaryotes. *Proceedings of the National Academy of Sciences USA*, **97**, 11319–11324.

Arndt, G. M., Atkins, D., Patrikakis, M. and Izant J. G. (1995). Gene regulation by antisense RNA in the fission yeast *Schizosaccharomyces pombe*. *Molecular & General Genetics*, **248**, 293–300.

Arndt, G. M., Patrikakis, M. and Atkins, D. (2000). A rapid genetic screening system for identifying gene-specific suppression constructs for use in human cells. *Nucleic Acids Research*, **28**, e15.

Atkin, A. L., Schenkman, L. R., Eastham, M., Dahlseid, J. N., Lelivelt, M. J., and Culbertson, M. R. (1997). Relationship between yeast polyribosomes and Upf proteins required for nonsense mRNA decay. *Journal of Biological Chemistry*, **272**, 22163–22172.

Atkins, D., Arndt, G. M. and Izant J. G. (1994). Antisense gene expression in yeast. *Biol. Chem. Hoppe-Seyler*, **375**, 721–729.

Bernstein, E., Caudy, A. A., Hammond, S. M., and Hannon, G. J. (2001). Role for a bidentate ribonuclease in the initiation step of RNA interference. *Nature*, **409**, 363–366.

Bosher, J. M. and Labouesse, M. (2000). RNA interference: genetic wand and genetic watchdog. *Nature Cell Biology*, **2**, E31–E36.

Brantl, S. (2002). Antisense-RNA regulation and RNA interference. *Biochemica et Biophysica Acta*, **1575**, 15–25.

Caudy, A. A., Ketting, R. F., Hammond, S. M., Denli, A. M., Bathoorn, A. M. P., Tops, B. B. J., Silva, J. M., Myers, M. M., Hannon, G. J. and Plasterk, R. H. A. (2003). A micrococcal nuclease homologue in RNAi effector complexes. *Nature*, **425**, 411–414.

Condeelis, J. (1995). Elongation factor 1α, translation and the cytoskeleton. *Trends in Biochemical Sciences*, **20**, 169–170.

Denli, A. M. and Hannon, G. J. (2003). RNAi: An ever-growing puzzle. *Trends in Biochemical Sciences*, **28**, 196–201.

DiSerio, F., Schob, H., Iglesias, A., Tarina, C., Bouldoires, E., and Meins, F. Jr (2001). Sense- and antisense-mediated gene silencing in tobacco is inhibited by the same viral suppressors and is associated with accumulation of small RNAs. *Proceedings of the National Academy of Sciences USA*, **98**, 6506–6510.

Djikeng, A., Shi, H., Tschudl, C., Shen, S. and Ullu, E. (2003). An siRNA ribonucleoprotein is found associated with polyribosomes in *Trypanosoma brucei*. *RNA* **9**, 802–808.

Domeier, M. E., Morse, D. P., Knight, S. W., Portereiko, M., Bass, B. L. and Mango, S. E. (2000). A link between RNA interference and nonsense-mediated decay in *C. elegans*. *Science*, **289**, 1928–1931.

Fire, A., Xu, S., Montegomery, M., Kostas, S., Driver, S. and Mello, C. (1998). Potent and specific genetic interference by double-stranded RNA in *C. elegans*. *Nature*, **391**, 806–811.

Forster, A. and Symons, R. (1987). Self-cleavage of plus and minus RNAs of a virusoid and a structural model for the active sites. *Cell*, **49**, 211–220.

Glick, B. and Schatz, G. (1991). Import of proteins into mitochondria. *Annual Review of Genetics*, **25**, 21–44.

Grallert, B., Kearsey, S., Lenhard, M., Carlson, C., Nurse, P., Boye, E. and Labib, K. (2000). A fission yeast general translation factor reveals links between protein synthesis and cell cycle control. *Journal of Cell Science*, **113**, 1447–1458.

Grewal, S. I. S. and Moazed, D. (2003). Heterochromatin and epigenetic control of gene expression. *Science*, **301**, 798–802.

Hall, I. M., Shankaranarayana, G. D., Noma, K., Ayoub, N., Cohen, A., and Grewal, S. I. S. (2002). Establishment and maintenance of a heterochromatin domain. *Science*, **297**, 2232–2237.

Hammond, S. M., Boettcher, S., Caudy, A. A., Kobayashi, R. and Hannon, G. J. (2003). Argonaute2, a link between genetic and biochemical analyses of RNAi. *Science*, **293**, 1146–1150.

Hannon, G. J. (2002). RNA interference. *Nature*, **418**, 244–251.

Iborra, F. J., Jackson, D. A. and Cook, P. R. (2001). Coupled transcription and translation within nuclei of mammalian cells. *Science*, **293**, 1139–1142.

Ishizuka, A., Siomi, M. C. and Siomi, H. (2002). A Drosophila Fragile X protein interacts with components of RNAi and ribosomal proteins. *Genes & Development*, **16**, 2497–2508.

Itoh, T. and Tomizawa, J. (1980). Formation of an RNA primer for initiation of replication of ColE1 DNA by ribonuclease H. *Proceedings of the National Academy of Sciences USA*, **77**, 2450–2454.

Izant, J. and Weintraub, H. (1984). Inhibition of thymidine kinase gene expression by antisense RNA: A molecular approach to genetic analysis. *Cell*, **36**, 1007–1015.

Kretschmer-Kazemi Far, R. and Sczakiel, G. (2003). The activity of siRNA in mammalian cells is related to structural target accessibility: a comparison with antisense oligonucleotides. *Nucleic Acids Research*, **31**, 4417–4424.

Liu, Q., Rand, T. A., Kalidas, S., Du, F., Kim, H., Smith D. P. and Wang, X. (2003). R2D2, a bridge between the initiation and effector steps of the Drosophila RNAi pathway. *Science*, **301**, 1921–1925.

Martinez, J., Patkaniowska, A., Urlaub, H., Luhrmann, R., and Tuschl, T. (2002). Single-stranded antisense RNAs guide target RNA cleavage in RNAi. *Cell*, **110**, 563–574.

McManus, M. T. and Sharp, P. A. (2002). Gene silencing in mammals by small interfering RNAs. *Nature Reviews Genetics*, **3**, 737–747.

Mette, M. F., Aufsatz, W., van der Winden, J., Matzke, M. A. and Matzke, A. J. (2000). Transcriptional and promoter methylation triggered by double-stranded RNA. *European Molecular Biology Organization Journal*, **19**, 5194–5201.

Paul, C. P., Good, P. D., Winder, I. and Engelke, D. R. (2002). Effective expression of small interfering RNA in human cells. *Nature Biotechnology*, **20**, 505–508.

Pestka, S., Daugherty, B., Jung, V., Hotta, K. and Pestka, R. (1984). Anti-mRNA: Specific inhibition of translation of single mRNA molecules. *Proceedings of the National Academy of Sciences USA*, **81**, 7525–7528.

Peurta-Fernandez, E., Romero-Lopez, C., Barroso-de Jesus, A. and Berzal-Herranz, A. (2003). Ribozymes: Recent advances in the development of RNA tools. *Federation of European Microbiology Society Microbiology Reviews*, **27**, 75–97.

Ramer, S., Elledge, S. and Davis, R. (1992). Dominant genetics using a yeast genomic library under the control of a strong inducible promoter. *Proceedings of the National Academy of Sciences USA*, **89**, 11589–11593.

Raponi, M., Atkins, D., Dawes, I. W. and Arndt, G. M. (2000). The influence of antisense gene location on target gene suppression in the fission yeast *Schizosaccharomyces pombe*. *Antisense & Nucleic Acid Drug Development*, **10**, 29–34.

Raponi, M. and Arndt, G. M. (2002). Dominant genetic screen for cofactors that enhance antisense RNA-mediated gene silencing in fission yeast. *Nucleic Acids Research*, **30**, 2546–2554.

Raponi, M. and Arndt, G. M. (2003). Double-stranded RNA-mediated gene silencing in fission yeast. *Nucleic Acids Research*, **31**, 4481–4489.

Reinhart, B. J. and Bartel, D. P. (2002). Small RNAs correspond to centromere heterochromatic repeats. *Science*, **297**, 1831.

Schramke, V. and Allshire, R. (2003). Hairpin RNAs and retrotransposon LTR effect RNAi and chromatin-based gene silencing. *Science*, **301**, 1069–1074.

Volpe, T. A., Kidner, C., Hall, I. M., Teng, G., Grewal, S. I. S, and Martienssen, R. A. (2002). Regulation of heterochromatic silencing and histone H3 lysine-9 methylation by RNAi. *Science*, **297**, 1833–1837.

Wang, J., Tekle, E., Oubrahim, H., Mieyal, J.J., Stadtman, E. R. and Chock P. B. (2003). Stable and controllable RNA interference: Investigating the physiological function of glutathionylated actin. *Proceedings of the National Academy of Sciences USA*, **100**, 5103–5106.

Waterhouse, P., Graham, M. and Wang, M. (1998). Virus resistance and gene silencing in plants can be induced by simultaneous expression of sense and antisense RNA. *Proceedings of the National Academy of Sciences USA*, **95**, 13959–13964.

Wutz, A. (2003). *Xist* RNA associates with chromatin and causes gene silencing. In *Noncoding RNAs: Molecular Biology and Molecular Medicine*, ed. J. Barciszewski and V.A. Erdmann, pp. 49–65, New York, NY: Kluwer Academic/Plenum Publishers.

Yelin, R., Dahary, D., Sorek, R., Levanon, E. Y., Goldstein, O., Shoshan, A., Diber, A., Biton, S., Tamir, Y., Khosravi, R., Nemzer, S., Pinner, E., Walach, S., Bernstein, J., Savitsky, K. and Rotman, G. (2003). Widespread occurrence of antisense transcription in the human genome. *Nature Biotechnology*, **21**, 379–386.

20 RNA silencing in filamentous fungi: *Mucor circinelloides* as a model organism

Rosa M. Ruiz-Vázquez

Introduction

RNA silencing, a nucleotide sequence-specific RNA degradation mechanism that results in the suppression of gene expression, has emerged over the past decade as a topic of interest for the genetic manipulation of eukaryotes. This process, which is manifested in organisms ranging from protozoa to vertebrates, is triggered by double-stranded RNA (dsRNA) molecules, which are processed into 21–25 nucleotide (nt)-long RNA duplexes by an RNaseIII enzyme named Dicer. These small interfering RNAs (siRNAs) are incorporated into a multiprotein complex, the RNA-induced silencing complex (RISC), which specifically degrades all mRNA sharing sequence identity with the siRNAs. (For recent reviews see Cerutti, 2003; Denli and Hannon, 2003.)

The dsRNA molecules that trigger the silencing response are found naturally in an eukaryotic cell as replicative intermediates upon virus infection, or during the replication process of a transposable element. Experimentally, RNA silencing can be triggered by the deliberate introduction of dsRNA molecules or inverted repeat transgenes, the latter inducing gene silencing through transcription into hairpin dsRNAs in the nucleus. However, transgenes transcribing only sense RNA are also able to activate the silencing mechanism (Meins, 2000). Exactly how these sense transgenes are able to produce dsRNA molecules is still an open question, although it has been proposed that abnormally processed RNA ("aberrant" RNA) generated from these transgenes is the signal that triggers the silencing mechanism. These hypothetical aberrant RNAs would be converted to dsRNA through the action of a RNA-dependent RNA polymerase (RdRP), which has been identified in several organisms as an essential element for gene silencing. The silencing state can be transmitted between distant cells and through the syncytium of filamentous fungi, in a process known as systemic RNA silencing (Cogoni et al., 1996; Voinnet and Baulcombe, 1997).

RNA silencing has been described as a conserved mechanism that defends the genome against exogenous pathogenic and endogenous parasitic nucleic acids, such as viruses and transposons. However, recent evidence supporting the

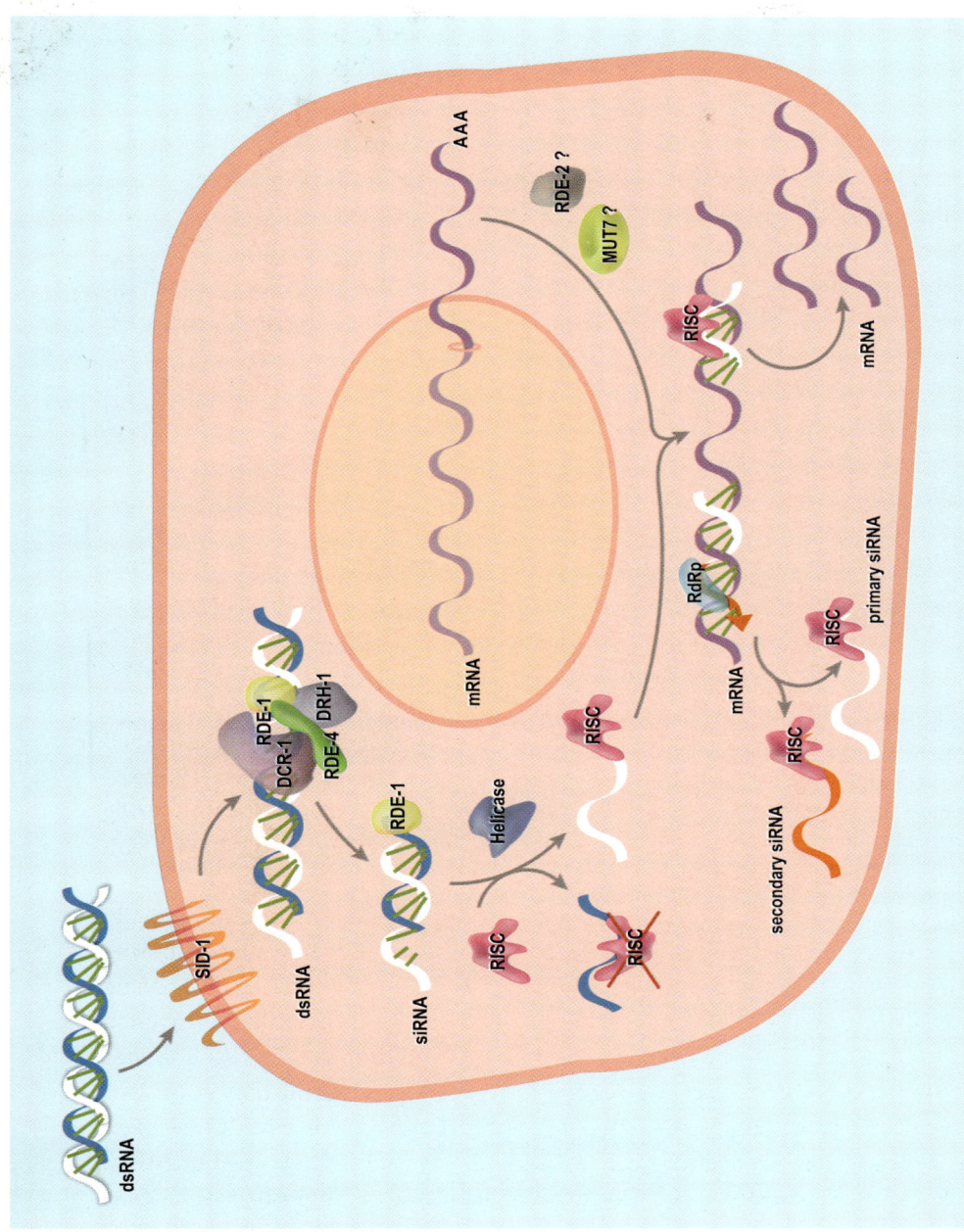

Figure 1.1. Model of RNAi pathway in *C. elegans*. Transmembrane protein SID-1 allows dsRNA to enter the cell. In the cytoplasm, dsRNA gets processed by DCR-1, existing in a complex with RDE-4, RDE-1 and DRH-1. The resulting siRNA is shown bound to RDE-1. Double stranded siRNA is unwound by a helicase, allowing RISC complex to bind single strands of siRNA. RDE-2 and MUT-7 might help in bringing together target mRNA and antisense siRNA (white) bound to RISC. RISC, guided by siRNA (white), degrades mRNA. Antisense siRNA also serves as a primer for RdRp, which uses mRNA as a template and produces more antisense siRNAs (orange). Sense siRNAs (blue) do not accumulate in *C. elegans*.

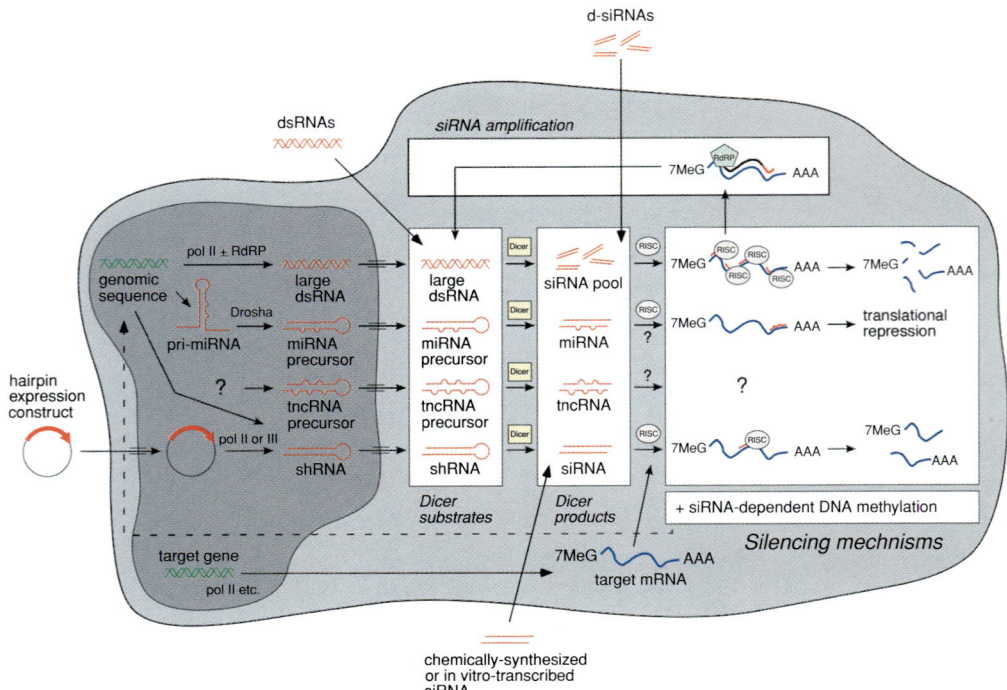

Figure 2.1. Sequence dependent regulation of gene expression. Expression of particular genomic sequences produces various dsRNAs. Dicer is responsible for processing the dsRNAs into small RNAs. The small RNA is then incorporated into RISC, guiding the protein complex to a specific mRNA. RISC enforces specific gene suppression either by cleavage of the mRNA or inhibition of translation. The degree of complementarity between the small RNA and the mRNA determines the fate of the mRNA; perfect or near perfect complementarity generally results in cleavage of the mRNA, whereas, a few mismatches results in suppression of translation. In a sequence specific manner, small RNAs also regulate chromatin structure and hence gene expression. Some of the processes depicted here may not be present in all organisms or cell types. From a practical standpoint small RNAs can be used to investigate gene function; several types of small dsRNAs can either be introduced into the cell or produced inside the cell. In all cases the small RNA is funneled into the RNAi pathway and triggers specific gene suppression.

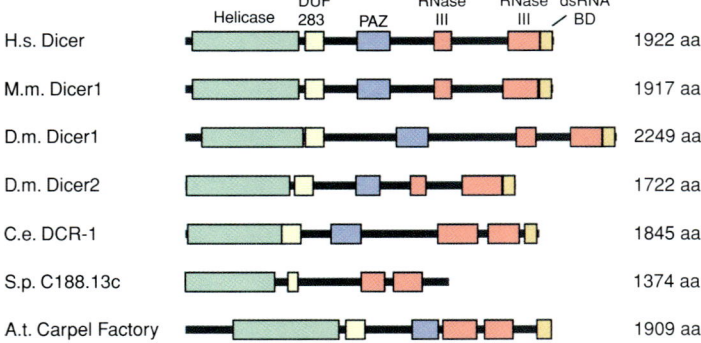

Figure 2.2. Alignment of Dicer homologs. Most Dicer proteins contain six conserved domains: a DExH helicase domain; a domain of unknown function (DUF283); a PAZ domain; two RNase III catalytic domains; and finally a double stranded RNA binding domain. Spacing between the domains is different for the homologs, which may explain the different size classes of small RNAs found in some species.

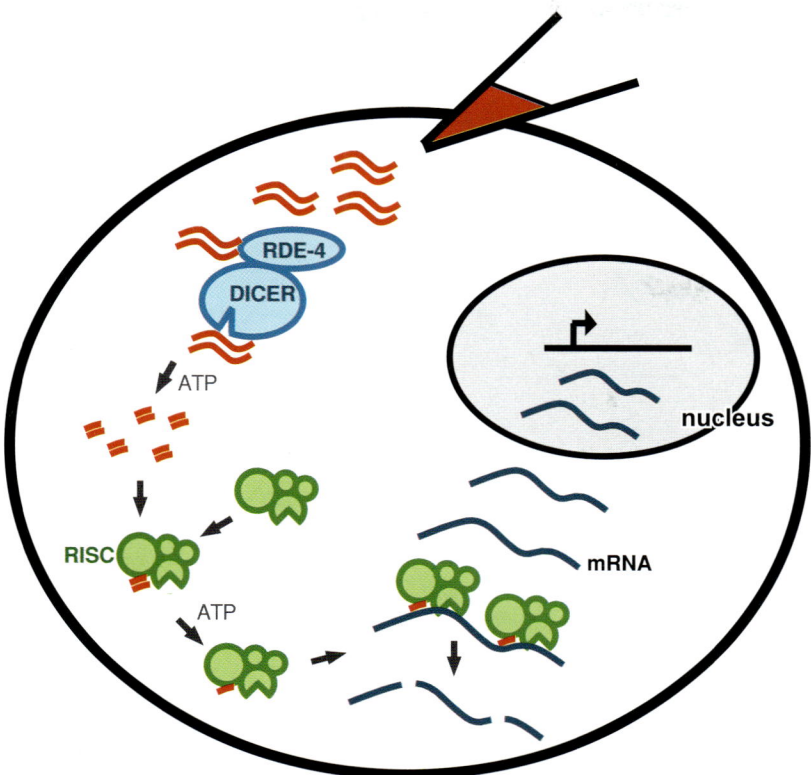

Figure 3.1. Model for mRNA degradation in the cytoplasm by RNAi. Introduced dsRNAs (red) are recognized by RDE-4/R2D2, a dsRNA binding protein. These dsRNAs are then processed by Dicer into 21-23nt duplexes that can associate with an enzyme complex called RISC. After unwinding of the siRNAs, RISC becomes competent to target homologous mRNA transcripts for degradation.

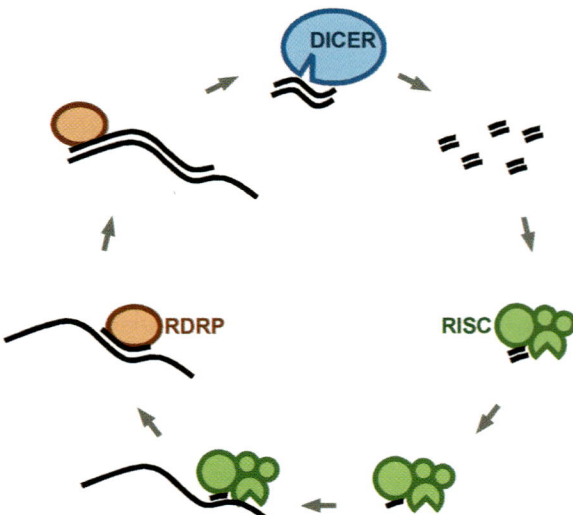

Figure 3.2. Amplification of dsRNA by an RNA-dependent RNA polymerase. In certain organisms, new dsRNAs can be generated by RDRPs, primed by siRNAs on mRNA targets. The new dsRNAs can be used subsequently by Dicer to create more siRNAs, which can lead to additional rounds of amplification.

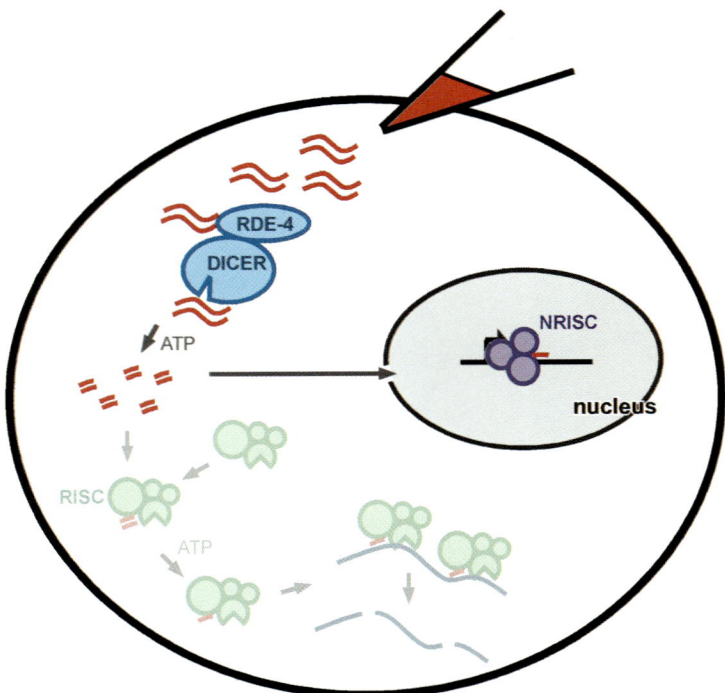

Figure 3.3. Model for gene silencing in the nucleus by RNAi. RNAi can also silence the transcription of targeted genes in certain organisms. In this model a signal can direct a putative nuclear RNAi silencing complex (NRISC), composed of chromatin modifying proteins, to the targeted locus, silencing gene expression at the level of transcription.

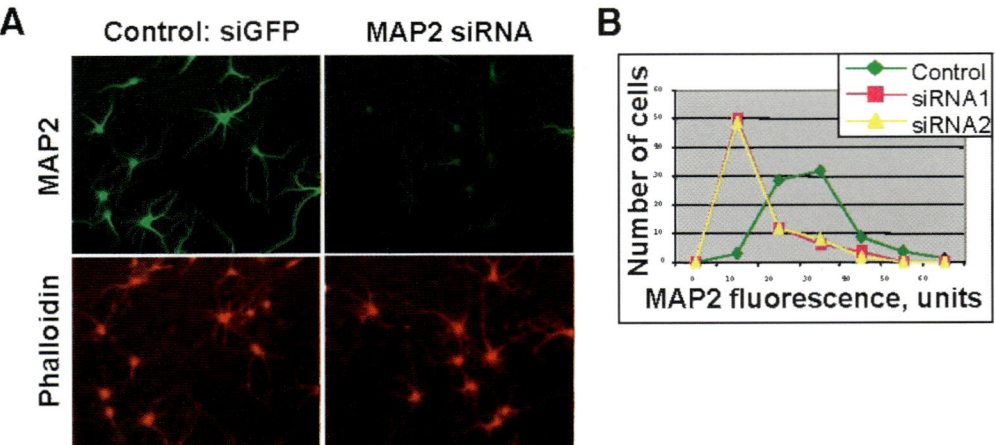

Figure 5.4. Microtubule Associated Protein 2 (MAP2) suppression in primary cortical neurons by cognate 21nt-siRNAs. A. Double fluorescence staining of neurons transfected with non-specific siRNA or with MAP2-siRNA. Upper panels-staining with MAP2 monoclonal antibody (green); lower panels-staining with actin-bound toxin phalloidin (red). B. Distribution of MAP2 expression levels in control and targeted cells, two different siRNA (siRNA1 and siRNA2) show a very similar effect. In each experiment, at least 70 random neurons per experimental condition were analyzed and gene expression was quantified in both control and targeted cells. The figure is reprinted from: Krichevsky, A. M. and Kosik, K. S. "RNAi functions in cultured mammalian neurons." *Proc Natl Acad Sci U S A.*, **99**(18):11926–9 (2002).

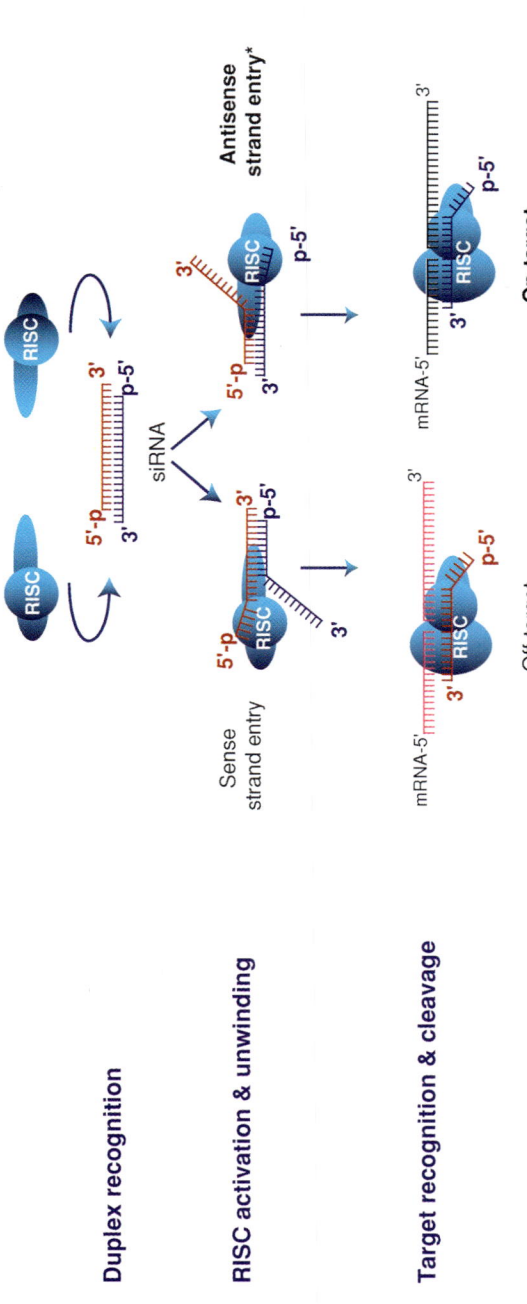

Figure 6.2. A model for the siRNA-mediated RNAi mechanism depicting major steps in (1) initial duplex recognition by the pre-RISC complex; (2) ATP-dependent RISC activation and siRNA unwinding, (3) target recognition, (4) target cleavage and product release. The desired outcome is directed by antisense strand entry* into the RISC to produce "on-target" silencing effects. Undesirable off-target effects are thought to be directed by sense strand identity to unrelated sequences but can be minimized using rational design coupled with comprehensive sequence analyses.

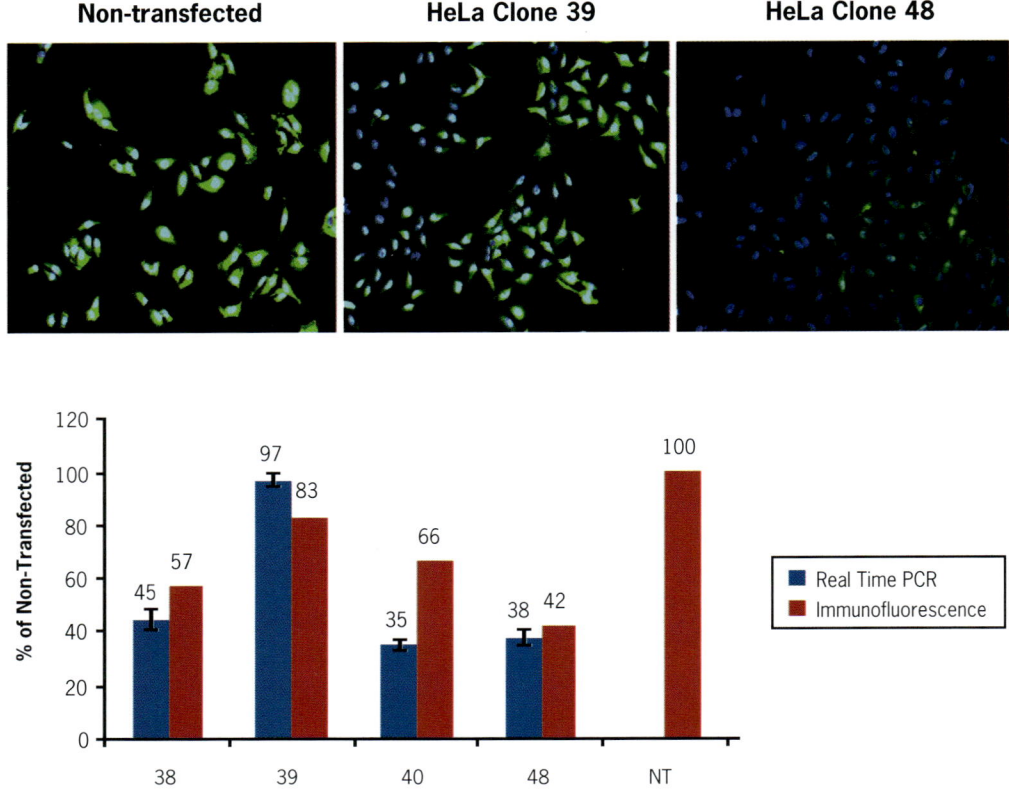

Figure 10.3. Long Term Silencing of GAPDH with CMV Puro Plasmid. HeLa cells were transfected with a CMV puro plasmid expressing GAPDH-specific siRNAs. The cells were cloned, and clonal populations were selected in 2.5 μg/ml puromycin. Three weeks after selection, GAPDH expression was analyzed by (A) RT-PCR or (B) immunofluorescence. Expression levels of several cell clones are shown. Green: GAPDH. Blue: DAPI stained nuclei.

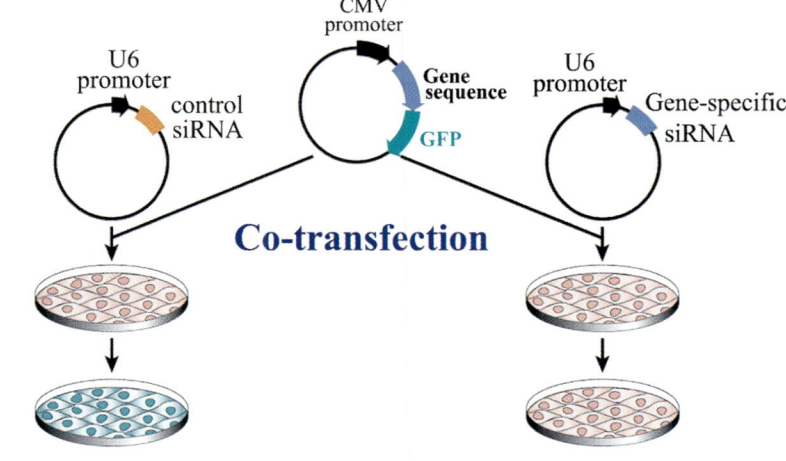

Cells fluorescence green   **Fluorescence is lost**

Figure 11.4. Method for identifying effective shRNA sequences. To screen for siRNAs that are effective against a gene of interest, the gene to be targeted is cloned into an expression vector as a translational fusion to a fluorescent protein. This construct is then co-transfected with test and control siRNA sequences against the gene. If the siRNA sequence is effective, then expression of the fusion protein will be reduced, resulting in a loss of fluorescence.

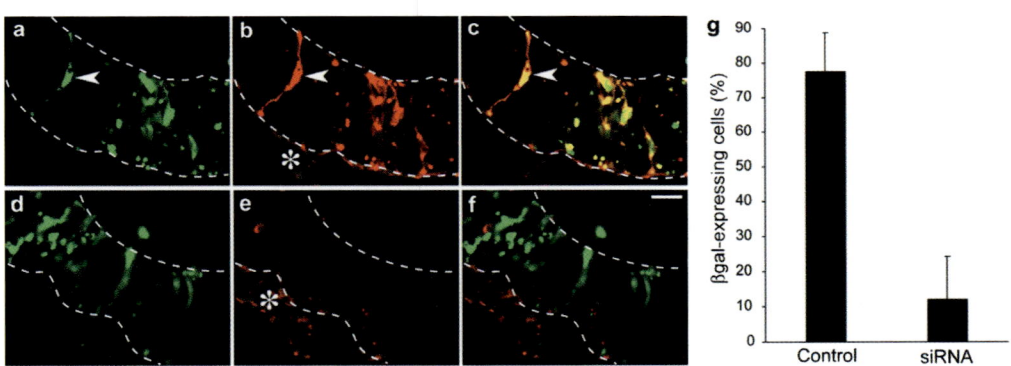

Figure 15.4. RNAi in the neuroepithelium of E10 mouse embryos. E10 mouse embryos were injected, into the lumen of the telencephalic neural tube, with the two reporter plasmids pEGFP-N2 (for GFP) and pSVpaXD (for βgal), either without (**a–c** and **g**, **Control**) or with (**d-f** and **g**, **siRNA**) βgal-directed esiRNAs, followed by directional electroporation and whole embryo culture for 24 hours. (**a-f**) Horizontal cryosections through the targeted region of the telencenphalon were analysed by double fluorescence for expression of GFP (green; **a** and **d**) and βgal immunoreactivity (red; **b** and **e**). Co-expression of GFP and βgal in neuroepithelial cells appears yellow in the merge (**c** and **f**, **arrowheads**). Note the lack of βgal expression in neuroepithelial cells in the presence of βgal-directed esiRNAs. Upper and lower dashed lines indicate the lumenal (apical) surface and basal border of the neuroepithelium, respectively. Asterisks in (**b** and **e**) indicate signal due to the cross-reaction of the secondary antibody used to detect βgal with the basal lamina and underlying mesenchymal cells. Scale bar in (**f**), 20 μm. (**g**) Quantitation of the percentage of GFP-expressing neuroepithelial cells that also express βgal without (**Control**) or with (**siRNA**) application of βgal-directed esiRNAs. Data are the mean of three embryos analyzed as in (**a-f**); bars indicate S.D. (Reprinted figure with permission from PNAS).

Figure 16.1. Chicken embryos are a good model system for developmental studies due to their accessibility. Chicken embryos can be accessed in ovo (A) through a window in the eggshell that can be resealed after manipulations with a coverslip and melted paraffin. As an alternative approach, chicken embryos can be used as ex ovo cultures (B). With both methods embryos can be kept alive throughout embryonic development.

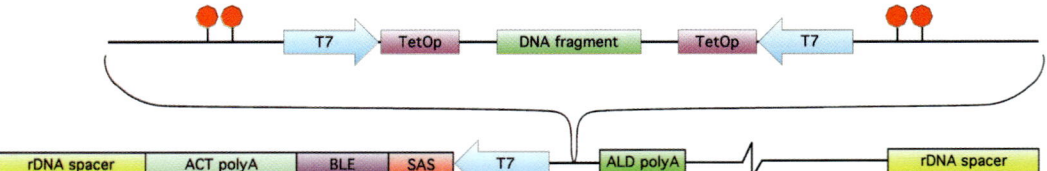

Figure 18.2. The pZJM RNAi vector. The tet operator (*Tet Op*), dual T7 terminators (red octagons), tet-inducible T7 promoters (*T7 arrows*), ribosomal DNA spacer (*rDNA*), actin poly(A) addition sequence (*ACT polyA*), phleomycin resistance gene (*BLE*), splice acceptor site (*SAS*), aldolase poly(A) addition sequence (*ALD polyA*). The plasmid is shown in linearized form, after cleavage in the rDNA spacer, and is not drawn to scale.

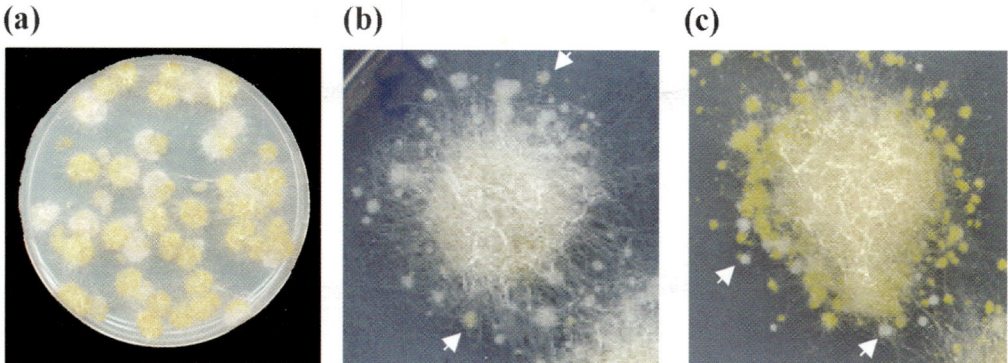

Figure 20.1. (a) Albino and wild type (yellow) colonies obtained by transformation of the wild type strain with *carB* sequences. Segregation of albino (b) and wild-type (c) transformants after a cycle of vegetative growth. Colonies showing different phenotypes (arrows) are obtained from spores of the original transformants. Photographs were taken after illumination with blue light for 24 hours.

Figure 21.1. ACMV-[CM]-infected *N. benthamiana* showing recovery phenotype. *N. benthamiana* plants imaged at 2-weeks post inoculation [(WPI) (control-A-left; Infected-A-right)] and at 5-WPI (control-B-left; Infected-B-right).

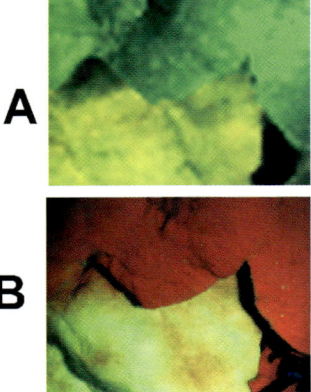

Figure 21.2. ACMV-[CM]-infected GFP silenced GFP-transgenic *N. benthamiana* (line 16C). Plant photographed using dissecting microscope (A) Normal light and (B) UV filter. Symptom-less recovered leaves appeared red under UV light.

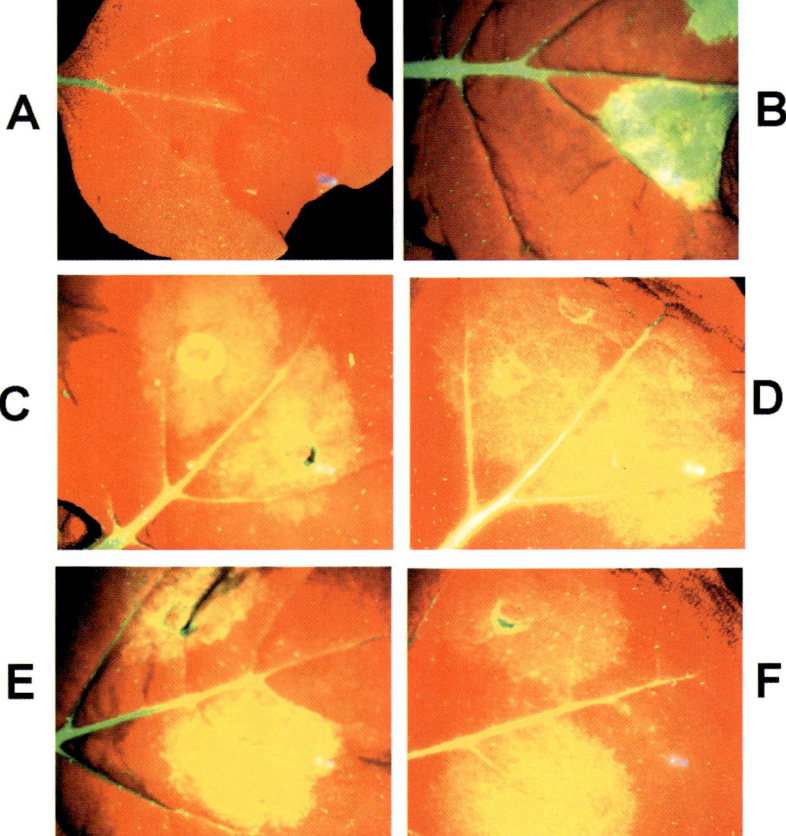

Figure 21.3. Effect of anti-PTGS activity of AC2 gene of EACMCV and ICMV; and AC4 gene of ACMV-[CM] and SLCMV. Leaf of GFP-transgenic *N. benthamiana* (line 16C) plant agroinfiltrated with pBin-GFP alone (A), or bacterial mixture harboring pBin-GFP along with the following viral gene constructs, P1/HC-Pro of TEV (B); AC4 of ACMV-[CM] (C), AC2 of EACMCV (D), AC4 of SLCMV (E) and AC2 of ICMV (F). Leaves were photographed 7 days after infiltration using a dissecting microscope.

Three Stage Activity Nanoparticles

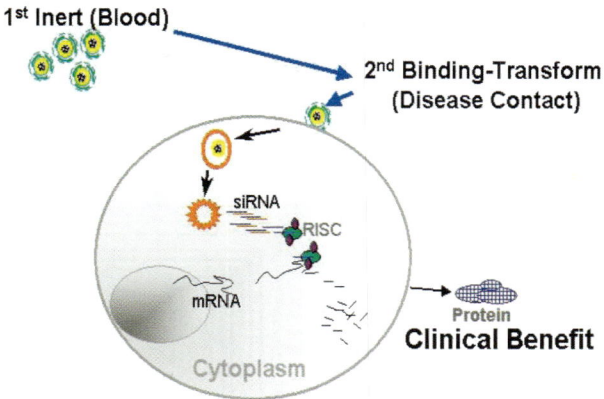

Figure 22.2. Ideal System for systemic delivery of siRNA.

HCV dependent translation initiation

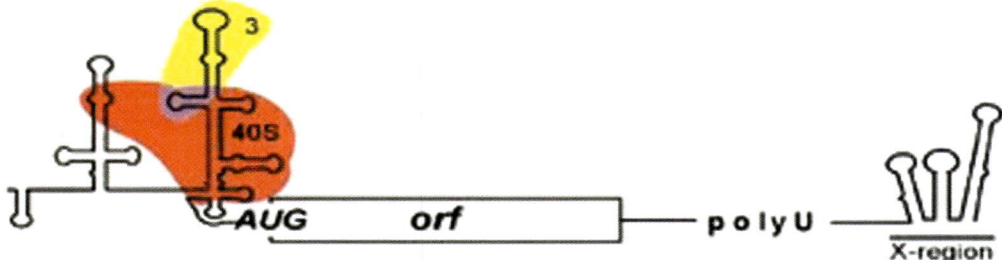

cap mediated translation initiation

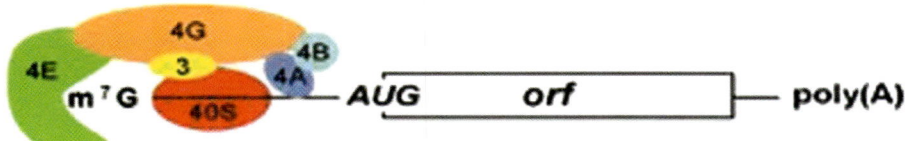

Figure 23.1. Comparison of HCV-IRES- and 5′ cap- dependent translation A) The HCV-IRES driven internal translation initiation, starts with the direct recruitment of the 40S ribosomal subunit and the eukaryotic initiation factor eIF 3. In cap-dependent translation initiation, the recruitment of the 40S ribosomal subunit requires the recognition of the m⁷GpppG cap at the mRNA 5′end, by the initiation factor eIF4E at which the eIF 4F complex, consisting of the initiation factors eIF 4E, eIF 4G and eIF 4A, is assembled. The recruitment of the 40S ribosomal subunit takes place via eIF 3 which binds to eIF4G.

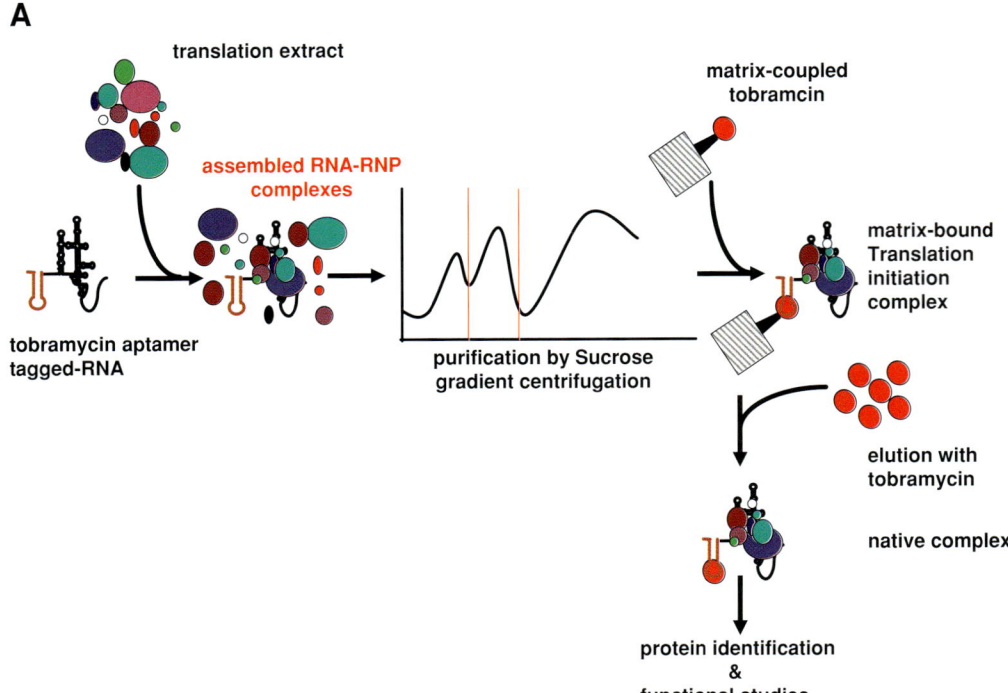

A

translation extract

matrix-coupled
tobramcin

assembled RNA-RNP
complexes

matrix-bound
Translation
initiation
complex

tobramycin aptamer
tagged-RNA

purification by Sucrose
gradient centrifugation

elution with
tobramycin

native complex

protein identification
&
functional studies

Figure 23.2. Tobramycin-tag affinity chromatography of translation initiation complex A) The HCV RNA 5′UTR bearing the HCV-IRES fused to the tobramycin aptamer was incubated with cytoplasmic HeLa cell extract under physiological conditions. Protein complexes assembled on the 5′UTR sequence were loaded on a 5–25% sucrose gradient and distributed according to their molecular weight by ultracentrifugation. The resulting 48S ribosomal peak fractions were subsequently incubated with the sepharose-coupled aminoglycoside tobramycin. Following extensive washing of the affinity matrix, proteins were eluted from the complex assembled at the HCV-IRES. Proteins eluted from the HCV-IRES were isolated from a silver stained polyacrylamide gel and identified by mass spectrometry (LC-MS) according to representative peptides.

A

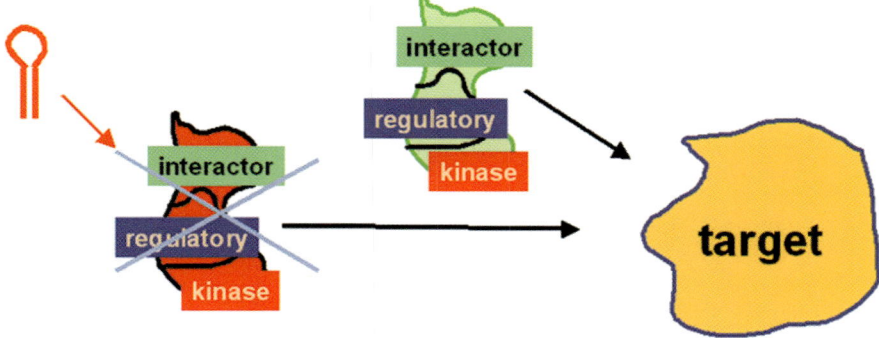

B

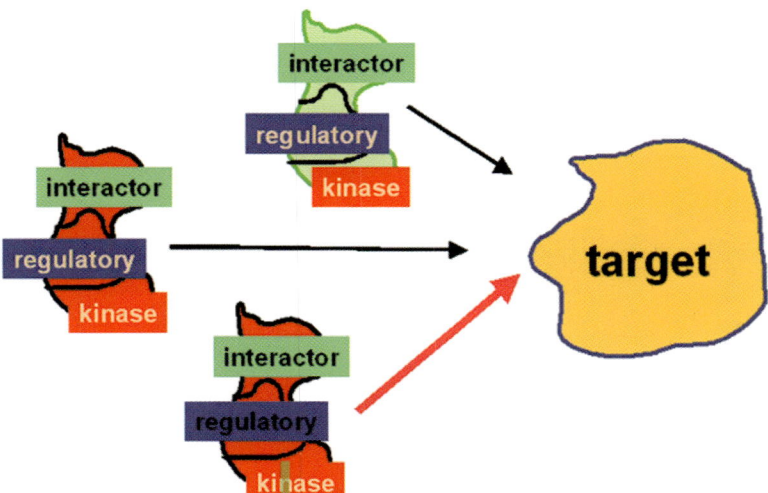

Figure 24.4. *Schematic models of interference of protein kinases by RNAi knockdown and dominant negative mutant.* Two kinases are depicted as interacting with the same target. Inhibition by RNAi or dominant negative is depicted in red in each part of the figure. A. Knockdown of a kinase by shRNA removes the targeted kinase from the cell. B. Inhibition of kinases by expression of a dominant negative mutant.

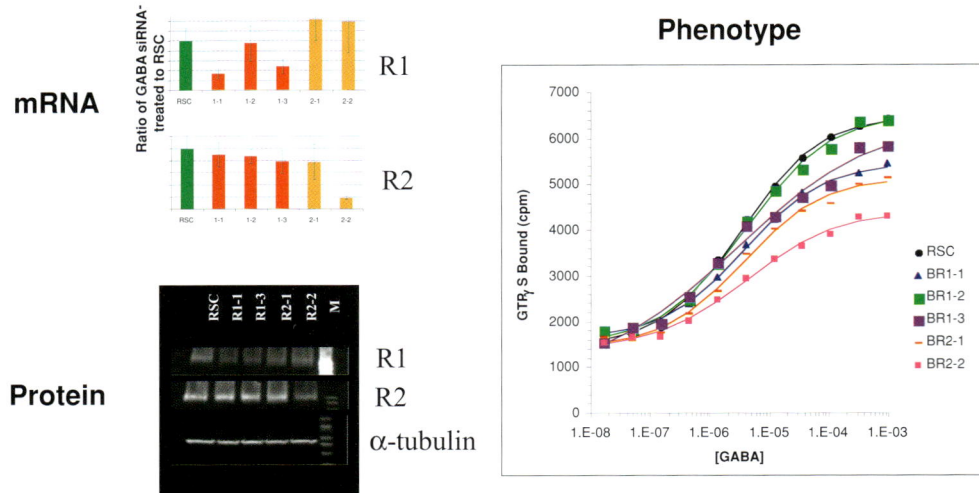

Figure 25.3. Consequences of RNAi-mediated knockdown onGABA$_B$R1 and GABA$_B$R2 receptors. mRNA levels of GABA$_B$R1 and GABA$_B$R2 were determined 24 hours post transfection with specific siRNAs. The green bar indicates the random sequence control, the red bars (1−1, 1−2 and 1−3) are siRNAs specific for GABA$_B$R1, the orange bars (2−1 and 2−2) show siRNAs specific for GABA$_B$R2. **Protein** knockdown of GABA$_B$R1 and GABA$_B$R2 was analysed after treatment of cells for 72 hours with siRNAs, and quantified by immunoblot relative to the internal control, α-tubulin. The **phenotype** associated with the knockdown of GABA$_B$R1 and GABA$_B$R2 was characterised using a ^{35}S GTPγassay.

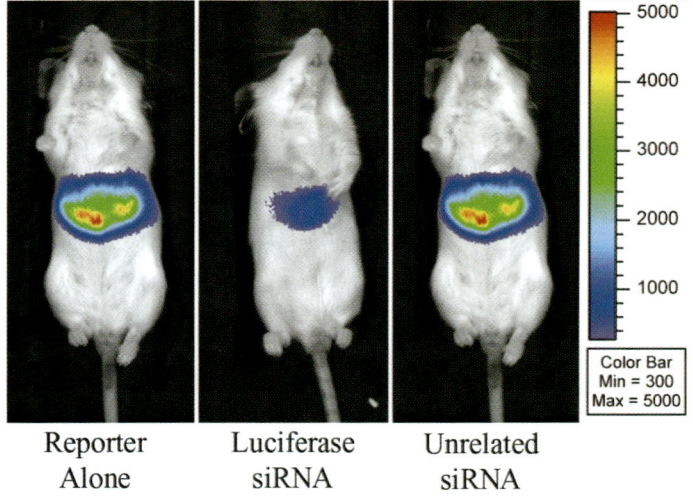

Figure 27.2. Light emitted from living mice as the result of luciferase expression is significantly reduced in the presence of luciferase siRNAs. Representative images of mice co-transfected with the luciferase plasmid pGL3-Control and either no siRNA (left), luciferase siRNA (middle) or unrelated siRNA (right). A pseudocolor image representing intensity of emitted light (red most and blue least intense) superimposed on a grayscale reference image (for orientation) shows that.RNAi functions in adult mammals. Forty μg of annealed 21-mer siRNAs (Dharmacon) were hydrodynamically transfected into livers of mice with the 2 μg of pGL3-Control DNA. Seventy two hours after transfection, mice were anesthetized and given 3 mg of luciferin intraperitoneally 15 min prior to imaging with a cooled CCD camera. IVIS imaging system (Xenogen, Alameda, CA) courtesy of Dr. Christopher Contag, Stanford University. Image reprinted with permission from McCaffrey et al., 2004.

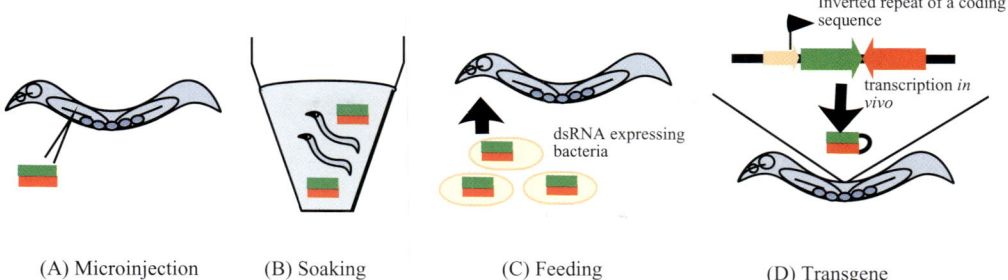

(A) Microinjection   (B) Soaking   (C) Feeding   (D) Transgene

Figure 30.1. dsRNA delivery methods. (A) Microinjection of the *in vitro* synthesized dsRNA. (B) Soaking in the dsRNA solution. (C) Feeding bacteria that express dsRNA. (D) *In vivo* transcription of hairpin RNAs from the transgene. By choosing promoters that control the expression of hairpin RNAs, inducible- or tissue-specific RNAi can be elicited.

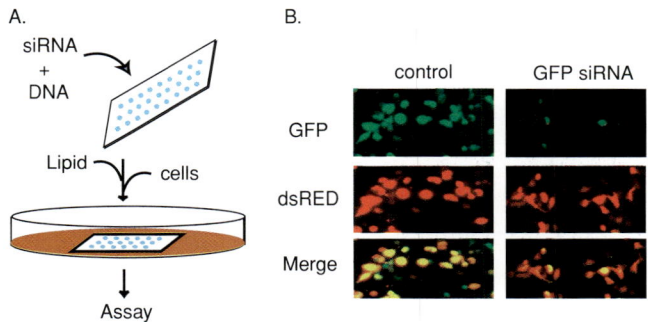

Figure 33.1. siRNA microarray for gene silencing. (A) Experimental strategy for siRNA microarrays. The desired cDNA and siRNAs are printed as individual spots on glass slides and exposed briefly to lipid before placing HEK293 cells on the printed slides in culture dish. Transfected cells are visualized using fluorescent microscopy and evaluated for the effect of RNAi. (B) Parallel RNAi on microarrays. Fluorescence photomicrograph of cells after reverse transfection of the indicated siRNA and cDNAs is shown.

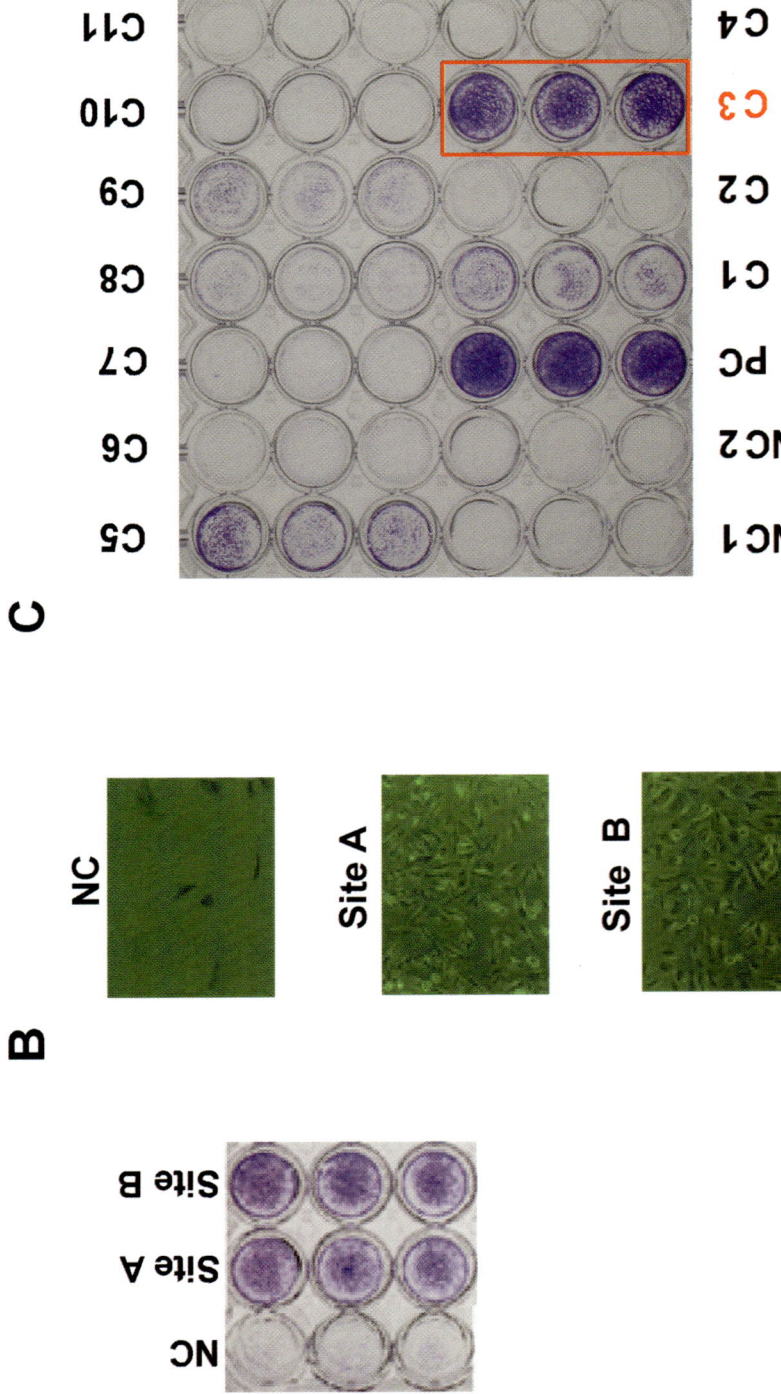

Figure 34.6. Identification of a previously unidentified gene that is involved in apoptosis in human cells. (A) shRNAi vectors directed against two sites (A and B) in the transcript of the gene for protein kinase R (PKR) prevented apoptosis in response to dsRNA. After the induction of apoptosis and fixation, HeLa S3 cells were stained with crystal violet, which only stains living cells. NC, Negative control. (B) Images of cells shown in (A) prior to staining. (C) The shRNAs designated C1 through C11 were tested for their ability to prevent apoptosis. These shRNAs were designed to target the transcripts of genes for kinases or transcription factors or genes that might be related to apoptosis. Each shRNA was tested in triplicate and cells that had been challenged with dsRNA were fixed and subjected to staining with crystal violet. PC, Positive control (shRNA directed against the gene for PKR; NC1 and NC2, negative controls; C1 through C11, shRNAs directed against specific genes. The shRNA designated C3 prevented apoptosis and the cells were stained purple with crystal violet.

involvement of RNA silencing and related pathways in functions other than defence, such as endogenous gene regulation and heterochromatin formation (Denli and Hannon, 2003), have injected new interest into this field. In addition, the possibility of obtaining "loss of function" phenotypes by means of the introduction of cognate-specific sequences has recently opened the way to the functional genomic analysis of several organisms (Fraser et al., 2000).

RNA silencing in filamentous fungi

In filamentous fungi, RNA silencing was first described and molecularly characterised in the Ascomycete *Neurospora crassa*, where it was referred to as "quelling" (reviewed in Pickford et al., 2002). One of the most relevant contributions derived from analysis of the gene-silencing mechanism in this fungus has been the identification of the first genetic elements known to be involved in this phenomenon, the *qde* (quelling defective) genes (Cogoni and Macino, 1997). Isolation of these genes helped identify the first components of a large list of genes and proteins involved in the silencing mechanism in different organisms: *qde-1* gene, which encodes an RNA-dependent RNA polymerase (Cogoni and Macino, 1999a), *qde-2*, which codes for a piwi-PAZ domain protein (Catalanoto et al., 2000) belonging to the Argonauta family of proteins that has been shown to be part of the RISC complex, and *qde-3*, which encodes a DNA helicase (Cogoni and Macino, 1999b). Of great significance was the discovery of the first RdRP enzyme as a determinant of quelling. Indeed, similar proteins, it has been subsequently demonstrated, are required for transgene-induced gene silencing in *Caenorhabditis elegans* (Smardon et al., 2000) and *Arabidopsis thaliana* (Dalmay et al., 2000). RdRP enzymes were also shown to be required in the gene silencing triggered directly by dsRNA, suggesting an additional role for these proteins in the silencing mechanism. In fact, the Qde-1-homologous enzymes seem to be involved in an amplification step of the silencing signal that multiplies the effect of gene silencing (see below). Additional screening for quelling-defective mutants in *N. crassa* did not reveal any other genes involved in the silencing mechanism (Pickford et al., 2002). It is possible that some genes exist that could be redundant in the *N. crassa* genome or could be involved in essential functions other than quelling, which would render non-viable mutations.

The 25-nt siRNAs that are specifically associated to the action of the Dicer nuclease are also accumulated in silenced transformants of *N. crassa* (Catalanoto et al., 2002), suggesting that a Dicer-like enzyme must participate in the silencing machinery of this fungus. Two Dicer-homologous proteins can be deduced from the genomic DNA sequence of *N. crassa*, although no individual Dicer gene has yet been identified as an essential element for gene silencing in this organism. The DNA-helicase encoded by the *qde-3* gene seems to act upstream of the production of the 25-nt siRNAs, as any accumulation of the small RNAs is prevented in the *qde-3* mutants. Similarly, such accumulation is also hindered in the *qde-1* mutants, as is to be expected from the supposed role of the corresponding gene product. However, the production of the 25-nt RNAs does not depend on a functional *qde-2* gene, indicating that the piwi-PAZ domain protein is involved in a downstream step of the silencing mechanism. Indeed, 25-nt RNAs bound to a QDE-2 protein complex have been isolated, which agrees with the idea that this protein could be a part of the *N. crassa* RISC complex (Catalanoto et al., 2002).

Besides *N. crassa*, transgene-induced post-transcriptional gene silencing has been described in the filamentous fungi *Cladosporium fulvum* (Hamada and Spanu, 1998) and, recently, *Magnaporthe oryzae* (Kadotani et al., 2003). However, the new findings concerning RNA silencing in the Zygomycete *Mucor circinelloides* (Nicolás et al., 2003) make this fungus a suitable organism for investigating some unresolved questions about the silencing mechanism.

Gene silencing in *Mucor circinelloides*

A gene-silencing mechanism in *M. circinelloides* was postulated to explain the unexpected phenotype of transformants carrying additional copies of the regulatory gene *crgA*, a repressor of light-induced carotenogenesis (Navarro et al., 2000; Navarro et al., 2001). *M. circinelloides* responds to blue light by activating carotenoid biosynthesis (Navarro et al., 1995). The white-yellowish colonies of the wild-type strain turn bright yellow upon illumination due to the accumulation of β-carotene. Lack of the *crgA* function results in over-accumulation of carotenoids both in the dark and the light, which unequivocally indicates that this gene acts as a negative regulator of photocarotenogenesis. Unexpectedly, transformants harbouring several copies per nucleus of complete or 3′-truncated *crgA* sequences also accumulated large amounts of β-carotene in dark and light conditions, which led us to suggest that a gene-silencing mechanism was operative in the *crgA* multicopy transformants, resulting in the suppression of CrgA protein synthesis.

The existence of a silenced mechanism was confirmed using the *carB* gene as a visual reporter of silencing. Gene *carB* codes for the phytoene dehydrogenase enzyme (Velayos et al., 2000), which is involved in the biosynthesis of the first coloured carotenoid. Thus, inactivation of the *carB* function results in albino colonies, which accumulate the colourless precursor phytoene. When sense transgenes containing complete or 3′-truncated *carB* sequences are introduced into the wild-type *M. circinelloides* strain, 3–10% of the transformants show an albino phenotype instead of a bright yellow wild-type colour (Figure 20.1a). The silenced phenotype is the result of the degradation of the mature *carB* mRNA and is not associated with DNA rearrangement or methylation of the coding region of transgenic or endogenous *carB* sequences (Nicolás et al., 2003).

The frequency of silencing observed in these experiments is much lower than that reported for different transgene-induced silencing systems, possibly due to the fact that, unlike in other organisms, transgenes do not integrate into the *M. circinelloides* genome, but exist as autonomous molecules as a result of the self-replicative condition of the vectors used for transformation. Thus, transgene expression in *M. circinelloides* is not affected by position effects or host regulatory

(a)   (b)   (c)

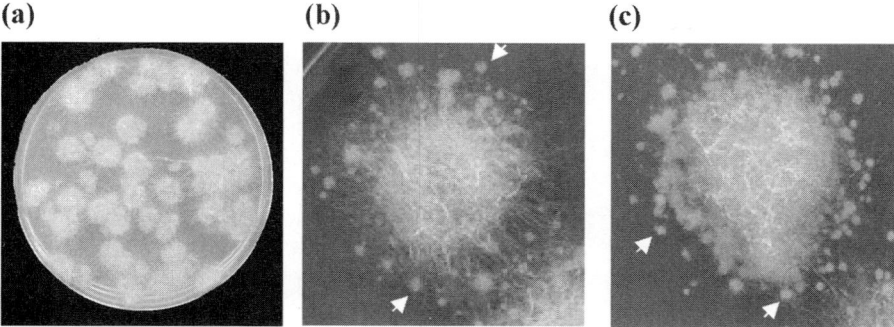

Figure 20.1. (a) Albino and wild type (yellow) colonies obtained by transformation of the wild type strain with *carB* sequences. Segregation of albino (b) and wild-type (c) transformants after a cycle of vegetative growth. Colonies showing different phenotypes (arrows) are obtained from spores of the original transformants. Photographs were taken after illumination with blue light for 24 hours. (See color section.)

sequences at insertion sites, both of which are thought to be involved in the production of truncated RNAs, antisense RNAs and longer than full-length RNAs. These molecules may be detected as "aberrant" RNA by the RdRP enzyme, leading to dsRNA formation. However, the exact "aberrant" RNA quality that confers a signal which can be recognised by the RdRP enzyme in the gene silencing induced by sense transgenes is still unknown. The self-replicative condition of transforming DNA in *M. circinelloides*, which allows easy recovery of transgenes from the silenced strains, make this organism a suitable model for analysing this question.

Most *M. circinelloides* transformants contain a low number of transgenic sequence copies, which may also be responsible for the observed low silencing efficiency, since a relationship between silenced phenotype and the number of copies of transgenes has been demonstrated (Nicolás et al., 2003). This could mean that a high local concentration of "aberrant" RNA is required to activate the putative RdRP-mediated copying, as has been suggested in other systems (Tang et al., 2003). Alternatively, the low silencing frequency may indicate the low efficiency of the enzymes involved in the production of small RNAs.

Both albino and wild-type transformants have an unstable phenotype, since they segregate into a range of different phenotypes after a cycle of vegetative growth (Figure 20.1b and 1c). Cytoplasmic and nuclear segregation into the multinucleated spores of *M. circinelloides* may explain this instability. The cytoplasmatically diffusible molecule thought to be involved in propagation of the silenced state must reach an adequate concentration to achieve efficient target mRNA degradation (Cogoni et al., 1996). Thus, different local concentrations of the triggering molecule, together with different proportions of transformed nuclei into the spores, would probably result in different intensities of the silenced phenotype in the progeny of albino transformants. In support of this hypothesis, the frequency of reversion decreases when spores are recovered from colonies grown under conditions that minimise cytoplasmic segregation (our unpublished results).

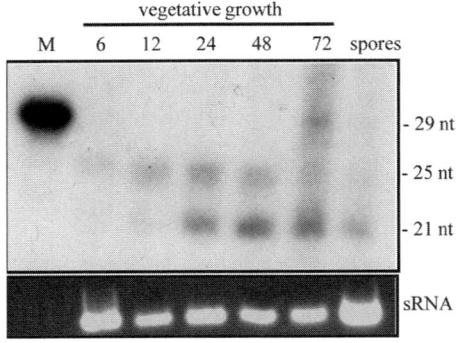

Figure 20.2. Differential accumulation of 21-nt and 25-nt antisense RNAs during vegetative growth of silenced strains. RNA samples isolated from a silenced transformant at different times of vegetative growth (hours) were hybridised with a *carB* antisense-specific riboprobe. M: size marker. The predominant RNA species in the RNA samples (sRNA) are shown as control for loading. (Modified from Nicolás et al., 2003).

Two classes of small antisense RNAs in RNA silencing

Gene silencing in *M. circinelloides* is associated with the presence of small sense and antisense RNAs, as occurs in all RNA silencing systems. Unusually, two size classes of small antisense RNAs, which are differentially accumulated during the vegetative growth, have been identified in this fungus (Nicolás et al., 2003). Long (25-nt) and short (21-nt) antisense RNAs can be detected in silenced transformants, the 25-nt class being more abundant at the beginning of vegetative growth, whereas the 21-nt RNAs are predominantly produced late in the vegetative life cycle (Figure 20.2). The short antisense RNA is transmitted through the spores of the silenced transformants, and it may be responsible for initiating silencing after germination.

Different size classes of siRNAs have also been described in plants (Hamilton et al., 2002) and, recently, in the fungus *M. oryzae* (Kadotani at al., 2003). In light of the effect that viral suppressors of silencing have on the production of the two classes of antisense RNA, it has been suggested that the 21-nt class in plants could be involved in the sequence-specific mRNA degradation process that takes place in the cytoplasm of the silenced strains. The same experiments indicate that the long 25-nt RNA class could participate in the systemic spreading of RNA silencing through the plant and in the sequence-specific methylation of homologous DNA sequences, a process often associated with systemic RNA silencing. These two classes of small RNAs seem to be produced by two different Dicer-like enzymes in plants, as can be deduced from the *in vitro* silencing assays based on wheat germ extracts (Tang et al., 2003). Four Dicer-like proteins are encoded by the *Arabidopsis* genome, two of them containing putative nuclear localisation signals, which could indicate a nuclear pathway for processing dsRNA in plants. Indeed, recent results show that at least one of the Dicer-like enzymes of *Arabidopsis* is a nuclear protein required for processing dsRNA precursors to the 21-nt micro RNAs involved in developmental regulation (Papp et al., 2003). These experiments also indicate that at least some 21-nt siRNAs are processed in the nucleus by an unknown nuclear Dicer-like enzyme.

The nucleotide sequence of the *M. circinelloides* genome has not been determined. No Dicer gene(s) have been yet identified to be central components of the silencing mechanism in this organism. However, two Dicer-homologous genes can be deduced from the genomic sequence of the fungus *N. crassa*. Thus, we can assume that the two different size classes of siRNAs identified in *M. circinelloides* are produced by two different Dicer enzymes. If this is the case, the differential accumulation of the 21-nt and 25-nt RNAs could be due to different expression patterns of the corresponding Dicer genes and/or to the different sub-cellular location of the corresponding protein products. The early phase of vegetative growth in *M. circinelloides* is characterised by rapid elongation of the hyphae and an active metabolism, the maximum biomass being reached around the time of glucose depletion (in our conditions, 24–30 hours). At the end of the filamentous growth, there is a change in the internal structure of the hyphae and nuclei become the prevalent component that can be observed (McIntyre et al., 2002). This is even more evident in the spores, which contain, on average, 6–7 nuclei. Thus, it is tempting to speculate that the 21-nt RNA molecules accumulated late during the vegetative growth are mainly processed by a nuclear Dicer enzyme, whereas the 25-nt RNAs are probably derived from cytoplasmic Dicer activity. Identification of the *M. circinelloides* Dicer-like genes and generation of null mutants by gene replacement will allow us to prove this hypothesis.

Amplification of silencing signal and transitive RNA silencing

Genetic analysis suggests that an amplification step may be required for efficient gene silencing in plants and animals (Cerutti, 2003). This step would produce secondary siRNAs using the target mRNA transcripts as a template in a reaction catalysed by a putative RdRP enzyme. These secondary siRNAs can also be detected from sequences of the target gene upstream and downstream of the initial triggering molecule, a process that has been named transitive silencing. In *C. elegans*, transitive silencing shows upstream directionality, while in plants, there is a bi-directional spreading from the initiator region into the adjacent 5′ and 3′ sequences of the target gene. Primer-dependent and primer-independent synthesis by the RdRP enzyme has been proposed to explain the 3′ → 5′ and 5′ → 3′ directionality of transitive silencing. We have also observed bi-directional spreading of the silencing signal in *M. circinelloides*. When a restricted portion of the *carB* gene is used to induce silencing, small antisense RNAs corresponding to upstream and downstream sequences of the initial triggering molecules (secondary siRNAs) are detected in silenced transformants [(Nicolás et al., 2003; our unpublished results) (Figure 20.3)].

Transitive silencing in *M. circinelloides* is associated with both size classes of antisense RNAs. The 21-nt and 25-nt secondary RNAs are both preferentially generated from the 3′ region of the endogenous *carB* gene (Figure 20.3), which may be explained by the preferential transcription from the 3′-end of the target mRNA by the putative RdRP combined with the low processivity of the enzyme activity. Our results also indicate that the 21/25 nt molecules detected in silenced transformants mainly correspond to the amplification of the silencing signal on the target mRNA. Interestingly, both 21 nt- and 25 nt-long RNAs are associated with the amplification and spreading of silencing signals in *M. circinelloides*,

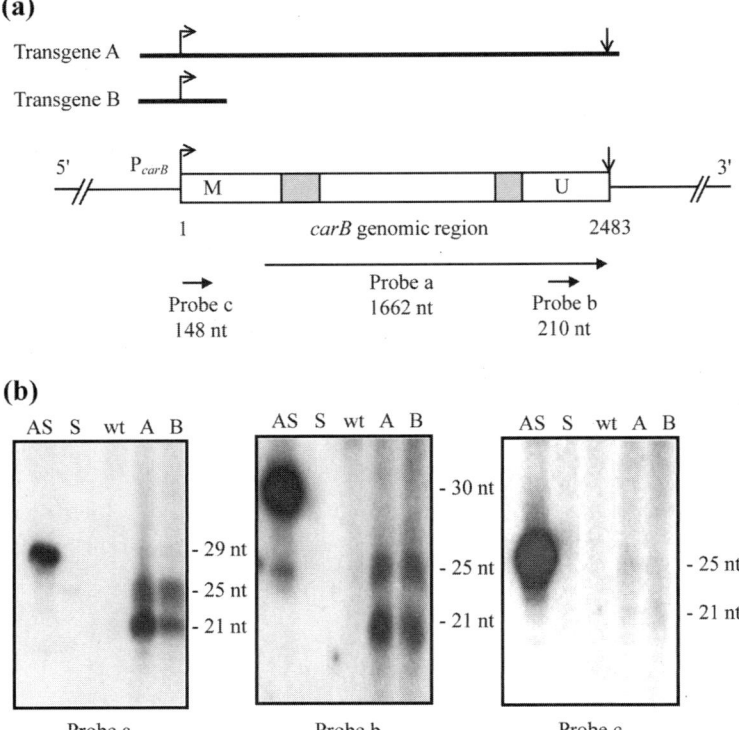

Figure 20.3. Secondary sense and antisense RNAs in transgene-induced gene silencing. (a) Schematic representation of the *carB* genomic region. The transcription start and polyadenylation sites (arrows), the translation start (M) and termination (U) codons and the exons and introns (open and shaded boxes) are indicated. Above the scheme, *carB* constructs used for transformation. Below, *carB* riboprobes used to detect small antisense RNAs. (b) Northern blot analyses showing that silenced transformants (lane B) obtained with a 5′ portion of the *carB* gene (transgene B) accumulate secondary 21/25 nt antisense RNAs corresponding to downstream sequences of the initial triggering molecules (probe a). Hybridisation with probes b and c indicates that both size classes of small antisense RNAs preferentially represent the 3′ portion of the target gene. The wild-type non-silenced strain (wt) and a silenced transformant obtained with the full-length transgene A (lane A) has been included as controls. Antisense (AS) and sense (S) oligonucleotides are used as size markers and specificity controls. (Modified from Nicolás et al., 2003.)

contrary to what happen in plants, where the *in vitro* production of dsRNA from a single-strand RNA molecule as a result of the RdRP activity seems to be coupled to the 25-nt siRNAs (Tang et al., 2003). Considering that two different Dicer-like enzymes may process the two different size classes of antisense RNAs in *M. circinelloides*, our results indicate that the dsRNA produced by the RdRP activity is delivered to both processing Dicer-like enzymes. Figure 20.4 shows a hypothetical model for the processing of the two size classes of siRNAs involved in gene silencing, and the amplification and spreading of the silencing signal in *M. circinelloides*.

Future prospects

M. circinelloides is clearly a suitable organism for investigating some unresolved questions concerning RNA silencing. The self-replicative nature of transgenes and

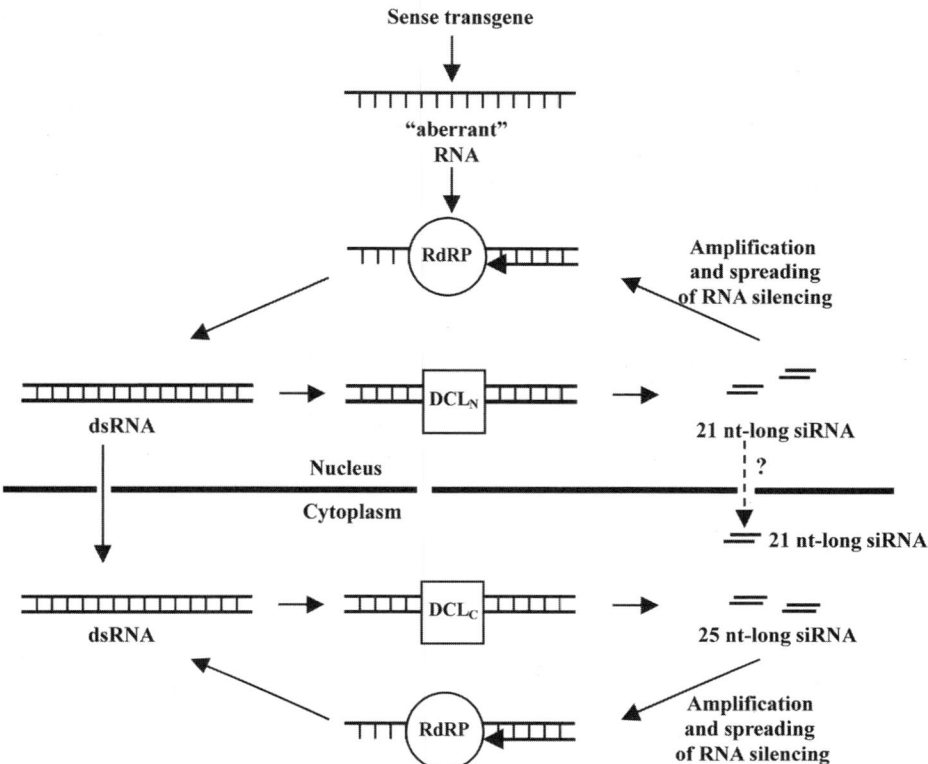

Figure 20.4. Hypothetical model of different pathways involved in the processing of 21 nt-long and 25 nt-long siRNAs in *M. circinelloides* RNA silencing. Self-replicative transgenes transcribe "aberrant" RNAs in the nucleus, which are recognised by a putative RNA-dependent RNA polymerase (RdRP) enzyme and converted to dsRNA. This dsRNA is processed by a putative nuclear Dicer-like enzyme (DCL$_N$) into 21 nt-long siRNAs duplexes, which can be transported to the cytoplasm. Precursors of dsRNA may also be located to the cytoplasm, where they are processed into 25 nt-long siRNAs by a putative cytoplasmic Dicer-like activity (DCL$_C$). Both size classes of small RNAs are associated with the amplification and spreading of the silencing signal, suggesting that the RdRP enzyme may deliver the dsRNA precursor to both processing Dicer enzymes.

the possibility that they can be recovered from silenced strains will provide a useful tool for identifying the molecular events that differentiate the triggering molecules in silenced and non-silenced transformants. Furthermore, *M. circinelloides* is the only organism, to date, in which the differential accumulation of two size classes of antisense RNAs has been described. The ease with which this fungus can be manipulated and the availability of a variety of molecular tools will eventually permit identification of the Dicer-homologous genes, analysis of their expression patterns and the study of the sub-cellular distribution of their corresponding protein products. In addition, bi-directional transitive silencing has been demonstrated to occur in this fungus, and it is possible that this mechanism could operate in others. This feature must be investigated before using RNA silencing as a genetic tool for inactivating any given single gene, since secondary siRNAs may target other mRNA species that contain sequences homologous to these secondary RNAs.

Large-scale analysis of gene function by homologous-dependent gene silencing has not been extensively explored in fungi. However, besides *N. crassa* and

M. circinelloides, other filamentous fungi have been demonstrated to be suitable for this analysis, since efficient gene silencing triggered by antisense RNA or dsRNA molecules has recently been reported in several pathogenic and non-pathogenic fungi (Cottrell and Doering, 2003). The application of RNA silencing technology to pathogenic fungi is of special interest as a strategy for identifying genes that are essential for growth – a crucial step in the development of antifungal drugs. To date, RNA silencing has been successfully reported in the human pathogen *Cryptococcus neoformans* (Liu et al., 2002) and in the rice pathogenic fungus *Magnaporthe oryzae* (Kadotani et al., 2003), and efforts are under way to establish RNA silencing in other pathogenic fungi, such as *Aspergillus fumigatus* (Cottrell and Doering, 2003). The results of such research will open up the way for this new tool to be used for genome-wide analysis in these, and possibly other, fungal organisms.

Note added in proofs: During the publishing process of this manuscript it was reported that two Dicer genes are involved in transgene-induced gene silencing in *N. crassa* (Catalanotto et al., 2004).

REFERENCES

Catalanotto, C., Azzalin, G., Macino, G. and Cogoni, C. (2000). Gene silencing in worms and fungi. *Nature*, **404**, 245.

Catalanotto, C., Azzalin, G., Macino, G. and Cogoni, C. (2002). Involvement of small RNAs and role of the *qde* genes on the gene silencing pathway in *Neurospora crassa*. *Genes & Development*, **16**, 790–795.

Catalanotto, C., Pallotta, M., ReFalo, P., Sachs, M. S., Vayssie, L., Macino, G. and Cogoni, C. (2004). Redundancy of the two dicer genes in transgene-induced posttranscriptional gene silencing in Neurospora crassa. *Molucular and Cellular Biology*, **24**: 2536–2545.

Cerutti, H. (2003). RNA interference: Traveling in the cell and gaining functions? *Trends in Genetics*, **19**, 39–46.

Cogoni, C. and Macino, G. (1997). Isolation of quelling-defective (*qde*) mutants impaired in post-transcriptional transgene-induced gene silencing in *Neurospora crassa*. *Proceedings of the National Academy of Sciences USA*, **94**, 10233–10238.

Cogoni, C. and Macino, G. (1999a). Gene silencing in *Neurospora crassa* requires a protein homologous to RNA-dependent RNA polymerase. *Nature*, **399**, 166–169.

Cogoni, C. and Macino, G. (1999b). Posttranscriptional gene silencing in *Neurospora by* a RecQ DNA helicase. *Science*, **286**, 2342–2344.

Cogoni, C., Ireland, J. T., Schumacher, M., Schmidhauser, T., Selker, E. U. and Macino, G. (1996). Transgene silencing of the *al-1* gene in vegetative cells of *Neurospora* is mediated by a cytoplasmic effector and does not depend on DNA-DNA interactions or DNA methylation. *European Molecular Biology Organization Journal*, **15**, 3153–3163.

Cottrell, T. R. and Doering, T.L. (2003). Silence of the strands: RNA interference in eukaryotic pathogens. *Trends in Microbiology*, **11**, 37–43.

Dalmay, T., Hamilton, A., Rudd, S., Angell, S. and Baulcombe, D. C. (2000). An RNA-dependent RNA polymerase gene in *Arabidopsis* is required for post-transcriptional gene silencing mediated by a transgene but not by a virus. *Cell*, **101**, 543–553.

Denli, A. M. and Hannon, G. J. (2003). RNAi: An ever-growing puzzle. *Trends in Biochemical Sciences*, **28**, 196–201.

Fraser, A. G., Kamath, R. S., Zipperlen, P., Martinez-Campos, M., Sohrmann, M. and Ahringer, J. (2000). Functional genomic analysis of *C. elegans* chromosome I by systemic RNA interference. *Nature*, **408**, 325–330.

Hamada, W. and Spanu, P. D. (1998). Co-suppression of the hydrophobin gene *HCf-1* is correlated with antisense RNA biosynthesis in *Cladosporium fulvum*. *Molecular and General Genetics*, **259**, 630–638.

Hamilton, A., Voinnet, O., Chappell, L. and Baulcombe, D. (2002). Two classes of short interfering RNA in RNA silencing. *European Molecular Biology Organization Journal*, **21**, 4671–4679.

Kadotani, N., Nakayashiki, H., Tosa, Y. and Mayama, S. (2003). RNA silencing in the phytophatogenic fungus *Magnaporthe oryzae*. *Molecular Plant-Microbe Interactions*, **16**, 769–776.

Liu, H., Cottrell, T. R., Pierini, L. M., Goldman, W. E. and Doering, T. L. (2002). RNA interference in the pathogenic fungus *Cryptococcus neoformans*. *Genetics*, **160**, 463–470.

McIntyre, M., Breum, J., Arnau, J. and Nielsen, J. (2002). Growth physiology and dimorphism of *Mucor circinelloides* (*syn. racemosus*) during submerged batch cultivation. *Applied Microbial Biotechnology*, **58**, 495–502.

Meins, F., Jr. (2000). RNA degradation and models for post-transcriptional gene-silencing. *Plant Molecular Biology*, **43**, 261–273.

Navarro, E., Sandmann, G. and Torres-Martínez, S. (1995). Mutants of the carotenoid biosynthetic pathway of *Mucor circinelloides*. *Experimental Mycology*, **19**, 186–190.

Navarro, E., Ruiz-Pérez, V. L. and Torres-Martínez, S. (2000). Overexpression of the *crgA* gene abolishes light requirement for carotenoid biosynthesis in *Mucor circinelloides*. *European Journal of Biochemistry*, **267**, 800–807.

Navarro, E., Lorca-Pascual, J. M., Quiles-Rosillo, M. D., Nicolás, F. E., Garre, V., Torres-Martínez, S. and Ruiz-Vázquez, R. M. (2001). A negative regulator of light-inducible carotenogenesis in *Mucor circinelloides*. *Molecular and General Genomics*, **266**, 463–470.

Nicolás, F. E., Torres-Martínez, S. and Ruiz-Vázquez, R. M. (2003). Two classes of small antisense RNAs in fungal RNA silencing triggered by non-integrative transgenes. *The European Molecular Biology Organization Journal*, **22**, 3983–3991.

Papp, I., Mette, M. F., Aufsatz, W., Daxinger, L., Schauer, E., Ray, A., van der Winden, J., Matzke, M. and Matzke, A. J. M. (2003). Evidence for nuclear processing of plant micro RNA and short interfering RNA precursors. *Plant Physiology*, **132**, 1382–1390.

Pickford, A. S., Catalanotto, C., Cogoni, C. and Macino, G. (2002). Quelling in *Neurospora crassa*. *Advances in Genetics*, **46**, 277–303.

Smardon, A., Spoerke, J. M., Stacey, S. C., Klein, M. E., Mackin, N. and Maine, E. M. (2000). EGO-1 is related to RNA-directed RNA polymerase and functions in germ-line development and RNA interference in *C. elegans*. *Current Biology*, **10**, 169–178.

Tang, G., Reinhart, B. J., Bartel, D. P. and Zamore, P. D. (2003). A biochemical framework for RNA silencing in plants. *Genes & Development*, **17**, 49–63.

Velayos, A., Blasco, J. L., Alvarez, M. I., Iturriaga, E. A. and Eslava, A. P. (2000). Blue-light regulation of phytoene dehydrogenase (*carB*) gene expression in *Mucor circinelloides*. *Planta*, **210**, 938–946.

Voinnet, O. and Baulcombe, D. C. (1997). Systemic signaling in gene silencing. *Nature*, **389**, 553.

21 RNAi and gene silencing phenomena mediated by viral suppressors in plants

Ramachandran Vanitharani, Padmanabhan Chellappan,
and Claude M. Fauquet

Introduction

Posttranscriptional gene silencing (PTGS) is a natural, universal mechanism that degrades both cellular and viral mRNA in a homology-dependant manner in diverse eukaryotes, and has now become a major area of research and development in both plants and animals. It was first discovered in plants (Napoli et al., 1990), while a mechanistically similar phenomenon is known to occur in a wide range of organisms, including *Caenorhabditis elegans*, *Drosophila melanogaster* and mammals termed RNA-interference (RNAi) (Fire et al., 1998, Hammond et al., 2000) and in *Neurospora crassa* termed quelling (Cogoni and Macino, 1997). Transgenes and viruses can induce PTGS in plants, and it is now recognized as a natural defense mechanism against virus infection (Hamilton and Baulcombe, 1999). Recent studies at the molecular level revealed that all of these phenomena are considered as manifestations of a general RNA-targeting pathway (Vance and Vaucheret, 2001). The mechanism by which a virus infection triggers PTGS in plants is not fully understood, but it is evident that dsRNA is a strong inducer of PTGS (Waterhouse et al., 2001). Such dsRNA molecules are produced during RNA virus replication using their own RNA-dependent RNA polymerase (RdRP), or alternatively host RdRPs convert any "aberrant" ssRNA in the cell, from viral origin or cell origin, into dsRNA (Dalmay et al., 2000; Ahlquist, 2002). The biology and biochemistry of RNAi was discussed in detail in Section I of this book.

In plants, gene silencing generates a nucleic acid–based mobile signal that moves from cell to cell through plasmodesmata and systemically via vascular tissues and can trigger PTGS in distant tissues and across a graft union (Palauqui et al., 1997). The signal would move together with, or in advance of, the virus, and mediates silencing of the viral RNA in the newly infected cells. Consequently, the infection would progress slowly or would be arrested (Voinnet et al., 1999). Another remarkable feature of RNA silencing in plants is that it is associated with intracellular signaling between the cytoplasm and the nucleus. Evidence for intracellular signaling is that RNA degradation is typically correlated with methylation of cognate DNA in the nucleus, even when silencing is induced by viruses that replicate exclusively in the cytoplasm (Jones et al., 1998). Another class of small RNAs called as micro RNAs (miRNAs) encoded by the host genome exist in higher life forms, from worms to humans including plants, and play a major role in regulating development (Mallory et al., 2002; Palatnik et al., 2003), suggesting that they arose early in eukaryotic evolution as did the RNA silencing mechanism. Unlike siRNAs, which interfere with gene expression at the level of transcription, the siRNAs interfere with gene expression at the translational level.

Geminiviruses and gene silencing

This chapter focuses on ssDNA virus-induced PTGS and suppression of RNA silencing. Geminiviruses are characterized by small geminate particles (18 × 20 nm) containing single-stranded circular DNA molecules (Stanley, 1983) that cause economically significant diseases in a wide range of cereal, vegetables, and fiber crops worldwide. These viruses amplify their genomes in the nuclei of host cells by rolling circle replication mechanism that uses double-stranded DNA intermediates as the replication and transcription templates (Hanley-Bowdoin et al., 1999). Cassava is an important food staple for 600 million people in Africa, Asia and South America and is highly prone to mosaic disease. Cassava mosaic disease (CMD) is a complex disease associated with eight different species of geminiviruses, belonging to the genus *Begomovirus* of the family *Geminiviridae* (Fauquet and Stanley, 2003). Cassava-infecting geminiviruses are transmitted by whiteflies (*Bemicia tabaci*) and also spread through infected cuttings, which are the usual mode of cassava propagation. The genome of these viruses is divided into two components defined as DNA-A and DNA-B and both the components are required for infectivity in plants (Stanley, 1983). Previously, it was reported that severe mosaic disease in cassava in the field is associated with the presence of double-infection with African cassava mosaic virus (ACMV) and East African cassava mosaic virus [(EACMV) (Harrison et al., 1997)] and found that these two viruses interact synergistically causing severe symptoms in cassava (Fondong et al., 2000; Pita et al., 2001). Molecular understanding of the cassava-infecting geminiviruses, in terms of their ability to induce and suppress gene silencing, would be very useful to develop gene silencing based strategies to control these devastating viruses.

I. RNA silencing as a defense against viruses

More than 90% of plant RNA viruses have ssRNA genomes that are replicated using virus-encoded RNA-dependent RNA polymerase (RdRP) through the formation of dsRNA intermediates, the potential inducer of PTGS. Plants protect themselves using their intrinsic defense system to tackle these viruses. The remaining 10% of the plant virus genomes are encoded by ssDNA or dsDNA while others have dsRNA genome. Plant viruses are both inducers and targets of PTGS;

therefore, defending plants against viruses is recognized to be a role of PTGS. Consequently, many plant viruses have evolved to encode a suppressor of PTGS to combat this resistance mechanism. Upon infection, viruses induce the plant's intrinsic defense system, which in turn eventually targets the viruses resulting with the production of virus-derived siRNAs as shown for RNA viruses such as *Potato virus X* (PVX) and *Cymbidium ringspot virus* [(CymRSV) (Hamilton and Baulcombe, 1999; Szittya et al., 2003)] infected plants. Geminiviruses possess ssDNA genome and it does not comprise a dsRNA phase in its replication cycle. However, recently these nuclear replicating ssDNA containing monopartite geminivirus such as *Tomato yellow leaf curl virus* [(TYLCV) (Lucioli et al., 2003)] and five distinct species of cassava-infecting bipartite geminiviruses, namely *African cassava mosaic virus* from Cameroon (ACMV-[CM]), *East African cassava mosaic Cameroon virus* (EACMCV), EACMV from Uganda (EACMV-[UG]), *Sri-Lankan cassava mosaic virus* (SLCMV) and *Indian cassava mosaic virus* (ICMV) were shown to trigger the PTGS system in infected plants with the production of virus-specific siRNAs (Chellappan et al., 2004a). RNA virus-induced PTGS is more quickly revealed with the presence of virus-derived siRNAs even at 4 days after inoculation as shown in PVX-infected plants (Hamilton and Baulcombe, 1999). But it seems that ssDNA viruses are much slower in triggering the host's PTGS system as detectable amounts of virus-derived siRNAs are produced only at 7 days after inoculation as noticed in the case of cassava-infecting geminiviruses (Chellappan et al., 2004a).

(i) Virus-induced gene silencing and endogenous gene inactivation

The intimate interaction between viruses and RNA silencing in plants is being exploited to create a novel high-throughput technology for functional genomics in plants. Virus-induced gene silencing (VIGS) is a type of RNA silencing that is initiated by virus vectors carrying portions of host genes that can suppress the gene expression in plants by degrading the homologous transcripts in a sequence-specific manner (Baulcombe, 1999). A number of nuclear endogenous genes and transgenes have been targeted successfully by VIGS (Jones et al., 1998), although it is not clear whether endogenous mRNA is involved in the triggering and/or maintenance of RNA silencing. VIGS has been demonstrated for a number of RNA and DNA plant viruses. Viral vectors of *Tobacco mosaic virus* [(TMV) (Kumagai et al., 1995), PVX (Ruiz et al., 1998)] and *Tobacco rattle virus* (TRV) (Ratcliff et al., 2001) that carried host-related inserts such as phytoene desaturase (PDS), an enzyme required for carotenoid pigment production or a gene encoding a component of the magnesium chelatase complex required for chlorophyll production, caused homology-based silencing of gene expression in plants. In geminiviruses, *Tomato golden mosaic virus* (Kjemtrup et al., 1998; Peele et al., 2001) and *Cabbage leaf curl virus* (Turnage et al., 2002) have been shown as viral vectors to silence endogenous genes (PDS and magnesium chelatase) in plants. Use of viral-vectors to silence host genes is advantageous over making transgenic plants which consumes an enormous amount of time and energy for studying especially functional genomics.

(ii) Virus-induced PTGS and recovery phenotype

Classical literature shows that some virus–host interactions naturally lead to host recovery, which is characterized by the initial development of systemic symptoms followed by symptom recovery. The natural recovery responses induced by a nepovirus, with an ssRNA genome (Ratcliff et al., 1997), and a caulimovirus, with a dsDNA genome (Covey et al., 1997), were shown to be similar to RNA-mediated virus resistance (RMVR) found in transgenic plants. A similarity between viral defense and gene silencing has been predicted for a nepovirus, as recovered plants showed cross-protection to the same virus or to closely related viruses, but not to unrelated viruses (Ratcliff et al., 1997). Additionally, the recovery of *Nicotiana benthamiana* from infection by a tobravirus expressing GFP was also shown to be associated with RMVR activation in the recovered leaves (Ratcliff et al., 1999). Furthermore, it was demonstrated that induced resistance mechanism is functionally equivalent to PTGS because it could also target homologous mRNA transiently transcribed from the T-DNA of an *Agrobacterium* strain delivered by leaf infiltration (Ratcliff et al., 1999). It appears, therefore that higher plants contain a novel cellular mechanism to recognize and degrade, with a high specificity, the RNAs of invading viruses.

This symptom recovery phenomenon is quite unusual for geminiviruses (ss-DNA viruses). However, we observed that ACMV-[CM]-infected *N. benthamiana* and cassava plants showed severe symptoms of yellow mosaic and leaf distortion in the systemically infected leaves until 2 weeks after inoculation; later, the plants started to show recovery (Figure 21.1). This symptom-recovery phenomenon prompted us to look into the possible involvement of virus-induced PTGS in ACMV-[CM] infected plants. We found that ACMV-[CM]-induced PTGS is associated with the production of virus-derived siRNAs starting from week one post inoculation, gradually increased over time and was abundant in the newly developed symptom-recovered leaves, both in *N. benthamiana* and cassava. This increase in virus-derived siRNA accumulation was accompanied by reduction in the accumulation levels of both viral DNA and mRNA, indicating the presence of a low level of virus load in the recovered leaves. Moreover, ACMV-[CM]-induced PTGS targeted the AC1 gene. The explanation would be that in ACMV-[CM]-infected plants, the presence of virus-specific siRNAs promoted the degradation of the corresponding mRNAs in a sequence-specific manner, which in turn affected both the viral replication and transcription. As a result, there was a reduction in virus titer and fewer symptoms in newly developed leaves. The role of viral suppressor is important in determining recovery phenotype (Chellappan et al., 2004a).

To investigate the capacity of ACMV-[CM] in relation to its ability to suppress gene silencing, the infectious clones of these viruses were inoculated on to *N. benthamiana* plants in which GFP transgene was silenced. GFP transgenic *N. benthamiana* line 16C plants (kindly provided Dr. Baulcombe) accumulate high levels of GFP mRNA and GFP protein, so that the leaves and stem appear green under UV illumination (Ruiz et al., 1998). Infiltrating the lower leaf of the seedling with a solution of *A. tumefaciens* harboring a binary plasmid containing a GFP transcription

Figure 21.1. ACMV-[CM]-infected *N. benthamiana* showing recovery phenotype. *N. benthamiana* plants imaged at 2-weeks post inoculation [(WPI) (control-A-left; Infected-A-right)] and at 5-WPI (control-B-left; Infected-B-right). (See color section.)

cassette induced silencing of GFP transgene. PTGS induced in the infiltrated patch is followed by systemic spread throughout the plant. Tissue that had undergone silencing appears red under UV light due to the fluorescence of the chlorophyll (Voinnet and Baulcombe, 1997). When inoculated to a host plant containing a post-transcriptionally silenced transgene, the virus that possess anti-PTGS activity can reverse silencing that is already established and/or prevent its onset in the new growth (Beclin et al., 1998; Brigneti et al., 1998). Infection of GFP-silenced 16C plants with ACMV-[CM] showed a complete reversal of GFP-silencing and became highly GFP fluorescent in infected tissue concomitant with the occurrence of virus symptoms from 3 dpi to 21 dpi (Figure 21.2). But eventually the GFP-silencing was not blocked in the subsequent newly emerging leaves, and as a consequence these leaves appeared red under UV-illumination with higher amount of GFP-specific siRNAs and very low level of GFP mRNA accumulation. In addition, viral suppression of PTGS correlated well with the physical presence of

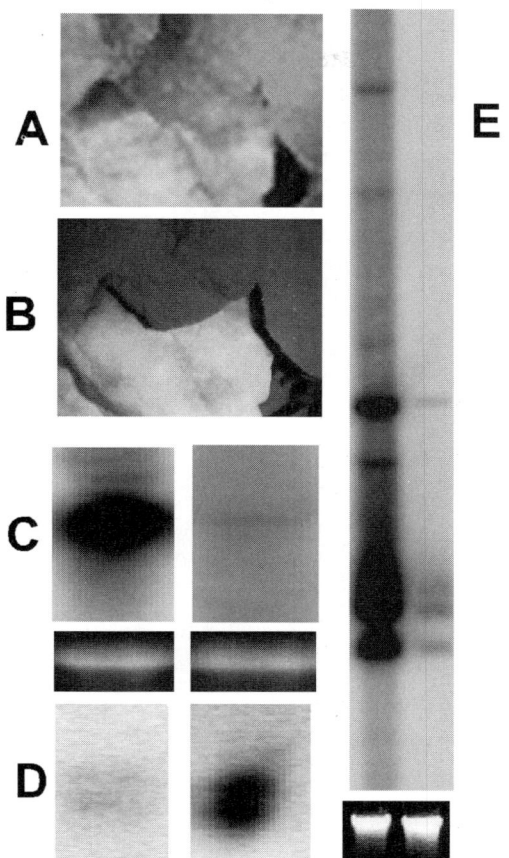

Figure 21.2. ACMV-[CM]-infected GFP silenced GFP-transgenic *N. benthamiana* (line 16C). Plant photographed using dissecting microscope (A) Normal light and (B) UV filter. Symptom-less recovered leaves appeared red under UV light. (C) Northern blot to detect GFP mRNA accumulation in lower symptomatic (left) and (right) upper silenced leaves. (D) Detection level of GFP-specific siRNA accumulation in lower symptomatic (left) and (right) upper silenced leaves. GFP-specific probe was used in panel (C) and (D). (E) Detection of viral DNA accumulation in the lower symptomatic (left) and (right) upper recovered leaves. ACMV-[CM] DNA-A specific probe was used. (See color section.)

the virus in the symptomatic leaves compared to low level of viral DNA accumulation in the symptom recovered upper leaves in which GFP-silencing persisted (Figure 21.2). Our results clearly show that ACMV-[CM] encodes a protein that suppresses the PTGS and is active locally in the virus-infected leaves (Vanitharani et al., 2004). In the ACMV-Kenyan strain (ACMV-[KE]), a related strain to ACMV-[CM], the AC2 gene has been shown to be a mild suppressor of PTGS (Voinnet et al., 1999).

(iii) Virus-induced PTGS and non-recovery phenotype

Recovery is not a usual plant response to virus infection. Virus-derived siRNAs implicating virus-induced PTGS has been shown for RNA and DNA viruses, such as PVX, CymRSV and TYLCV, despite the fact that the infected plants do not recover. These findings indicate that the ongoing silencing of viral RNAs may occur in many other persistent and successful virus infections and that active viral RNA silencing may not be sufficient to lead to a recovery. Non-recovery type cassava-infecting geminiviruses (EACMCV, EACMV-[UG] and ICMV) were also able to trigger the host's PTGS system with the production of virus-derived siRNAs, however it was not sufficient to induce recovery phenotype in infected plants. The possible explanation is that either these viruses might encode a strong suppressor of PTGS or these viruses may accumulate to such a high level that the plants still show symptoms despite the action of PTGS.

(iv) Transgene-induced gene silencing

In plants, PTGS not only protects the host against virus infection but also down-regulates transgene expression, and may also be an important component in the control of development. The first evidence for RNA silencing was from petunia plant transformed with an additional copy of a gene (*chalcone synthase*) required for floral pigment production. The transgene activated RNA silencing and as a consequence both transgene and the endogenous gene were suppressed resulting in 'white flowers' instead dark purple flowers. The phenomenon was then termed as co-suppression, literally linking that sense transgene-mediated suppression of the gene expression of both transgene and homologous endogenous gene (Napoli et al., 1990). Gene silencing in transgenic plants is manifested as decreased accumulation of specific mRNAs and occurs most often when there are multiple copies of a particular sequence present in the genome (Jorgensen, 1992; Depicker and Montagu, 1997). The multiple copies may consist of coding sequences, promoter regions, or both. Depending on the type of repeat sequence whether promoter or coding regions, transgene-induced gene silencing could operate at the level of transcription (TGS) or/and post-transcription (PTGS). Both TGS and PTGS are nucleotide sequence homology dependent; however, genes silenced transcriptionally are homologous in their promoter regions, whereas genes targeted for PTGS share homology in transcribed regions. TGS results from decreased transcription of the affected transgene(s) and is associated with increased promoter methylation that occurs in the nucleus (Mette et al., 1999). In contrast, PTGS has no apparent effect on transcription of the target gene but promotes a rapid and specific turnover of RNA transcripts in the cytoplasm. Both types are associated with the presence of repeated DNA sequences and result in low levels accumulation of the transgene mRNA.

Several lines of research suggest a link between PTGS in transgenic plants and virus resistance, because transgene-induced PTGS causes resistance against viruses that have nucleotide sequences similar to that of the transgene (English et al., 1996; Goodwin et al., 1996). When plants transformed with virus-derived transgenes, designed to provide protection through a protein-mediated mechanism (Abel et al., 1986), gave protection against viruses even when little or no transgene protein (transprotein) was produced indicating the involvement of an RNA-mediated virus resistance [(RMVR) (Lindbo and Dougherty, 1992; Pinto et al., 1999)]. RMVR was further evidenced using untranslatable transgenes in plants that provided resistance only through the RNA level (Waterhouse et al., 1998).

Moreover, simultaneous expression of sense- and anti-sense RNA of a transgene can activate PTGS in plants, and this phenomenon is recognized to be responsible for immunity to viral infection in many transformed plants that carry homologous viral transgene sequences (Waterhouse et al., 2001). However, the level of pathogen-derived resistance in transgenic plants was specific to the same virus or to several strains of the same species as demonstrated in the case of *Cucumber mosaic virus* (CMV), *Tobacco etch virus* (TEV) and *Rice yellow mottle virus* [(RYMV) (Zaitlin et al., 1994; Goodwin et al., 1996; Waterhouse et al., 1998; Pinto et al., 1999).

We developed transgenic cassava plants harboring multiple copies of the full-length *AC1* gene (codes for replication-associated protein) of ACMV-[KE], out of which two lines that produced transgene-derived siRNAs conferred resistance not only to homologous virus, but also to heterologous cassava-infecting virus species of geminiviruses (Chellappan et al., 2004b). The cross protection to other species of cassava-infecting geminiviruses such as EACMCV and SLCMV could be because the *AC1* gene of these viruses shares a considerable amount of sequence homology. EACMCV shares 66% and SLCMV shares 77% nucleotide sequence homology with the *AC1* gene of ACMV. Thus, transgenes that are derived from viral DNA are able to induce gene silencing, thereby suppress the accumulation of viruses that are similar in nucleotide sequence and impart resistance.

II. Suppression of virus-induced gene silencing and viral disease

(i) Suppression of gene silencing and synergism

In virus–host interactions, plants have evolved to posses PTGS a natural antiviral defense to resist virus infection. As a counter-defense, viruses have been co-evolved to encode proteins, which can suppress the PTGS, thus enabling the pathogen to become established and complete its life cycle in the host (Vance and Vaucheret, 2001). Mixed infection leading to synergistic viral diseases in plants has been known for a long time, and recently it has been implicated that it is due to suppression of host-surveillance mechanism by the interacting viruses; suggesting that the two phenomena may be linked. The systematic analysis of the classical synergism between PVX and potyviruses led to the identification of the potyviral helper component proteinase (HC-Pro) as the synergy determinant and a broad-range virulence enhancer (Pruss et al., 1997). Subsequent studies established that HC-Pro is able to suppress PTGS, providing a direct link between the ability to suppress RNA silencing and thereby permitting viruses to accumulate beyond the normal host-mediated limits leading to the severity of viral diseases (Anandalakshmi et al., 1998; Kasschau and Carrington, 1998). The next identified viral suppressor of PTGS was the 2b protein of *Cucumber mosaic virus* (CMV) (genus *Cucumovirus*) in *N. benthamiana* (Brigneti et al., 1998). The 2b protein of *Tomato aspermy virus* (TAV-2b), also a cucumovirus, induced a synergistic disease in seven host species when it was expressed from in the CMV genome in the place of CMV-2b. This is probably because TAV-2b is a more efficient and a different type of suppressor of PTGS than CMV-2b, as demonstrated in *N. benthamiana*, for example, by the fact that suppression of a transgene RNA silencing by TAV-2b happened at least three days earlier than that achieved by CMV-2b (Li et al., 1999). Furthermore, TAV-2b, but not CMV-2b, triggered a typical hypersensitive virus resistance in related host species (*N. tabacum*), characterized by the formation of necrotic lesions and the transcriptional induction of pathogenesis-related protein genes indicating that TAV-2b is a different type of suppressor (Li et al., 1999).

(ii) Mechanism of viral suppressor action on RNA silencing

Over the last three years, 22 RNA silencing-suppressor proteins have been identified in several plant viruses and 1 from an Flock house insect virus [(FHV) (Li et al., 2002)], and 1 from a host plant that interacts with the HC-Pro of TEV [(Anandalakshmi et al., 2000) (Table 21.1)]. However, the mode of action of these suppressor proteins in the RNA silencing pathway is not well understood. Moreover, these identified anti-PTGS suppressor proteins do not share any common sequence and presumably act at different levels of the PTGS system. The size of these viral suppressor proteins ranges from 95 to 469 amino acids, their pHi spans from 5.08 to 10.41 and they do not share any obvious conserved domain. Their viral function is mostly unknown, but even in known cases; it varies from one virus to another. With regards to plant RNA viruses, HC-Pro of the TEV has been shown to interfere with the maintenance of silencing at a step coincident with, or upstream of siRNA production, because it did not prevent the RNA silencing signal from becoming systemic (Llave et al., 2000). On the other hand, the 2b protein of CMV prevents initiation of PTGS in newly emerging tissues (Brigneti et al., 1998; Guo and Ding, 2002). A third step in the silencing pathway, the systemic signaling, is targeted by the suppressor protein p25 of PVX that belongs to the genus *Potexvirus* (Voinnet et al., 2000). The viral suppressor protein p19 of *Cymbidium ringspot virus* has been shown to bind directly to siRNAs, thus acting to deplete the specificity determinants and preventing further amplification of the PTGS (Silhavy et al., 2002). Recently, *Turnip crinkle virus* coat protein has been shown to suppress PTGS at an early initiation step, most likely by interfering with the function of the Dicer-like RNase in plants (Qu et al., 2003). The mode of action of these proteins has been poorly understood. In the case of the *Peanut clump virus* p15 protein, it has been shown that it works as a dimer, but the dimerization domain itself is not responsible for the PTGS suppression (Dunoyer et al., 2002). More research on the HC-Pro revealed that it regulates the accumulation of miRNAs that are involved in plant development (Kasschau et al., 2003; Mallory et al., 2002) and in addition, a host-protein that interacts with the HC-Pro called rgs-CAM (regulator of gene silencing Calmodulin-like protein) has been cloned from tobacco. This protein resembles Calmodulin and shares a significant level of sequence homology with it. It has also been shown that the rgs-CAM protein has a suppressor function similar to HC-Pro when it is over-expressed in transgenic tobacco plants (Anandalakshmi et al., 2000). It is puzzling that why plants themselves have PTGS suppressor proteins and that does not show phenotype otherwise over-expressed. Suppression of PTGS is not restricted to plant viruses;

as evidenced that Flock house virus infection requires suppression of RNA silencing by an FHV-encoded protein called B2 in *Drosophila* cells, indicating that RNA silencing and suppression are universal phenomena (Li et al., 2002).

(iii) Geminiviral suppressors of gene silencing

In plant ssDNA viruses, the *AC2* gene that encodes a transcriptional activator protein (TrAP) of ACMV-[KE] and *C2* (a positional homologue of AC2 in monopartite geminiviruses) of *Tomato yellow leaf curl China virus* (TYLCCNV) have been identified as suppressors of PTGS (Van Wezel et al., 2002; Voinnet et al., 1999). Recently, the requirement of NLS for anti-PTGS activity has been demonstrated for C2, in TYLCCNV (Dong et al., 2003). In our laboratory, we have identified another 4 viral suppressors belonging to geminiviruses [(Table 21.1) (Vanitharani et al., 2004)]. Two of them are of the same type of AC2 protein and were cloned from the *East African cassava mosaic Cameroon virus* (EACMCV) and from the *Indian cassava mosaic virus* (ICMV) respectively. We also identified a second type of suppressor protein, called AC4 from ACMV-[CM] and from *Sri Lankan cassava mosaic virus* [(SLCMV) (Figure 21.3)] The mode of action of these silencing suppressor proteins is under investigation.

In the case of cassava mosaic disease scenario, previous studies have shown that natural dual infection of ACMV and EACMV together in cassava plants in the field results in unusually severe synergistic disease that correlated well with higher amounts of viral DNA accumulation of both viruses, and that were reproducible in *N. benthamiana* using infectious clones (Fondong et al., 2000; Harrison et al., 1997; Pita et al., 2001). Further experiments based on the level of viral DNA accumulation in BY2 protoplasts as a tool indicated that synergism between these two viruses is a two-way process wherein the DNA-A component of both viruses are involved. More detailed analysis revealed that AC4 of ACMV-[CM] and AC2 of EACMCV are responsible for enhanced pathogenicity by increasing the level of both viral DNA accumulation (Vanitharani et al., 2004). Later, *Agrobacterium tumefaciens* infiltration assay revealed that these two genes have anti-PTGS activity due to their ability to suppress locally induced PTGS (Figure 21.3). Therefore, we concluded that synergism between ACMV-[CM] and EACMCV is mediated by the differential and complementary PTGS suppression ability of the AC4 and AC2 genes of ACMV-[CM] and EACMCV, respectively, in a temporal and spatial manner (Vanitharani et al., 2004). It is worth noting that PTGS suppression by viral-encoded proteins is non-specific, therefore it is an advantage for the viruses in mixed infection to establish severe disease by shutting down the host's intrinsic defense system.

III. siRNA-mediated down-regulation of virus infection

Gene silencing mediated by dsRNA is a sequence-specific RNA degradation mechanism highly conserved in eukaryotes that serves as an antiviral defense pathway in both plants and *Drosophila*. In plants, viruses are known to trigger PTGS with the subsequent production of virus-derived siRNAs, the molecular markers

Table 21.1. List of PTGS suppressors known to date

Virus	Protein	AA#	Mr (Da)	PHi	Viral Function	Suppressor type	Reference
ssDNA viruses *Geminivirus*							
ACMV-[CM]*	AC4	100	11,086	10.41	?	Strong in *N. ben*	(Vanitharani et al., 2004)
SLCMV	AC4	100	10,756	9.08	?	Strong in *N. ben*	(Vanitharani et al., 2004)
EACMCV	AC2	135	15,484	9.61	Trans. activator	Mild in *N. ben*	(Vanitharani et al., 2004)
ICMV	AC2	135	15,262	8.51	Trans. activator	Mild in *N. ben*	(Vanitharani et al., 2004)
ACMV-[KE]	AC2	135	15,160	8.82	Trans. activator	Mild in *N. ben*	(Voinnet et al., 1999)
TYLCCNV	C2	134	15,179	8.83	?	Mild in *N. ben*	(Van Wezel et al., 2002)
ssRNA viruses							
CMV *Cucumovirus*	2b	110	12,759	6.27	Movement	Mild in *N. ben*, amplification level	(Brigneti et al., 1998)
PCV *Pecluvirus*	P15	124	14,574	8.13	?	Med In *N. ben*	(Dunoyer et al., 2002)
BNYVV *Benyvirus*	P14	129	14,704	8.6	?	Med In *N. ben*	(Dunoyer et al., 2002)
RYMV *Sobemovirus*	P1	157	17,821	5.08	Movement	Med in *N. ben*	(Voinnet et al., 1999)
TBSV *Tombusvirus*	P19	172	19,396	5.28	Symptom determinant	Very strong in *N. ben*,	(Voinnet et al., 1999)
BYV *Closterovirus*	P21	177	20,559	8.0	Replication	Med *N. ben*	(Reed et al., 2003)
PVX *Potexvirus*	P25	226	24,267	6.7	Movement	Mild in *N. ben*, amplification level	(Voinnet et al., 2000)
BWYV *Polerovirus*	P0	238	27,299	9.25	?	Strong in *N. ben*	(Pfeffer et al., 2002)
CABYV *Polerovirus*	P0	238	27,263	10.1	?	Med in *N. ben*	(Pfeffer et al., 2002)
PLRV *Polerovirus*	P0	247	28,002	7.37	?	Suppressor of PTGS	(Pfeffer et al., 2002)
TCV *Carmovirus*	CP	351	38,068	9.38	Coat protein	Strong in *N. ben*	(Qu et al., 2003)
TEV *Potyvirus*	HC-Pro	459	51,857	7.74	protease	Very strong in *N. ben* maintenance level	(Anandalakshmi et al., 1998)
BSMV *Hordeivirus*	δ[b] protein				?	Suppressor of PTGS	(Yelina et al., 2002)
Negative ssRNA viruses							
RHBV *Tenuivirus*	NS3	203	23,206	8.6	?	Mild in *N. ben*	(Bucher et al., 2003)
TSWV *Tospovirus*	NSs	464	52,454	8.92	?	Mild in *N. ben*	(Bucher et al., 2003)
Insect virus							
FHV *Alphanodavirus*	B2	106	11,647	7.24	?	Mild *N. ben*	(Li et al., 2002)
Plant							
Tobacco	Rgs-CAM	192	21,496	4.65	Calmodulin – like protein	Strong in tobacco	(Anandalakshmi et al., 2000)

* Abbreviations: *N. ben* for *Nicotiana benthamiana*, *Med* for medium, *Trans. activator* for Transcription activator, ? – viral function not known.

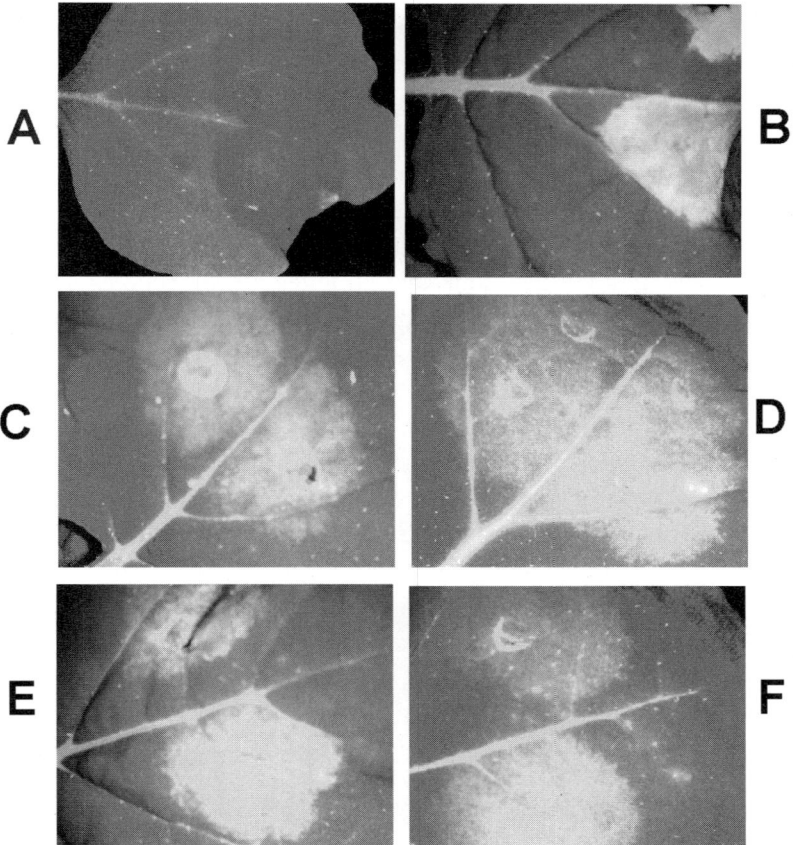

Figure 21.3. Effect of anti-PTGS activity of AC2 gene of EACMCV and ICMV; and AC4 gene of ACMV-[CM] and SLCMV. Leaf of GFP-transgenic *N. benthamiana* (line 16C) plant agroinfiltrated with pBin-GFP alone (A), or bacterial mixture harboring pBin-GFP along with the following viral gene constructs, P1/HC-Pro of TEV (B); AC4 of ACMV-[CM] (C), AC2 of EACMCV (D), AC4 of SLCMV (E) and AC2 of ICMV (F). Leaves were photographed 7 days after infiltration using a dissecting microscope. (See color section.)

of gene silencing, as shown in PVX (Hamilton and Baulcombe, 1999), CymRSV (Szittya et al., 2003), and geminiviruses such as TYLCV (Lucioli et al., 2003) and in cassava-infecting geminiviruses (Chellappan et al., 2004a). This induced defense mechanism is characterized by sequence-specific resistance against virus infection. Earlier studies demonstrated that introduction of additional gene copies or long dsRNA constructs through *Agrobacterium* or particle bombardment can trigger gene silencing in plants. The use of siRNAs, an intermediate in the gene silencing pathway, has become a powerful tool to specifically down-regulate gene expression, has been successfully demonstrated in a wide variety of mammalian cells (Elbashir et al., 2001; Fire et al., 1998; Gitlin et al., 2002), but not in plants, with the exception of a transient phenotype on GFP plants bombarded with siGFP [(siRNA designed to target GFP transgene) (Klahre et al., 2002)]. We observed that geminivirus-induced PTGS is associated with the production of virus-derived siRNAs leading to recovery phenotype in *N. benthamiana* and in cassava

(Chellappan et al., 2004a). This led us to ask the question whether gene silencing can be used as a tool to control geminivirus infection.

(i) Delivery of siRNA into protoplasts

Protoplasts provide a method for rapid and quantitative analysis of transgene expression before committing to the process of transgenic plants. We developed and utilized a protoplast system to facilitate rapid and quantitative analysis of synthetic double-stranded siRNA-mediated interference on gene expression (Vanitharani et al., 2003a). By electroporation of *N. tabacum* BY-2 protoplasts, we demonstrated that siRNAs could be delivered into protoplasts, making it possible to evaluate the effect of siRNAs targeted against reporter genes such as GFP and DsRed. Basically, protoplasts were transfected with both the reporter plasmids (GFP and DsRed) along with the cognate siRNA targeting GFP or DsRed separately. In both cases, the presence of the siRNA in the plant cells resulted in a sequence specific, down-regulation of the target gene. Transient expression of GFP and DsRed in transformed protoplasts was measured by using a fluorometer 24 hours post-transfection. The use of siRNAs in protoplasts to specifically down-regulate gene expression provides a rapid way to study gene regulation and has certain advantages over plant viral vectors, some of which act as inducers and suppressors of PTGS.

(ii) Specificity of siRNAs interference on gene expression

The GFP and DsRed coding sequences are 36% homologous in nucleic acid sequence. The siRNA designed to target GFP was 50% identical to the analogous sequence in DsRed, whereas the siRNA for DsRed was only 33% identical to the analogous sequence in GFP. Each strand of the siRNAs was synthesized separately, and pairs were annealed to create the duplex dsRNA with the characteristics of siRNA. Protoplasts coelectroporated with plasmids encoding GFP and DsRed plus siRNA cognate to GFP showed a significant reduction of GFP fluorescence to 58% of the level in the absence of siRNA. In contrast, siGFP did not interfere with DsRed fluorescence, indicating that interference is sequence-specific. Similarly, protoplasts co-delivered with both reporter plasmids along with siRNA targeted to DsRed showed a 47% reduction in DsRed expression compared with control, but had no effect on GFP expression. This study showed that co-delivery of both reporter genes (GFP and DsRed) in combination with either one of the siRNA specifically inhibited the respective targeted gene expression without affecting the expression of the other gene. In addition, it is clear from our results that introduced siRNA can induce gene silencing in a "transitive manner" in cultured plant cells, i.e., siRNA targeted to one sequence in a gene results in degradation of the entire or most of the mRNA to short polynucleotides (secondary siRNAs) outside the introduced siRNA targeted region (Vanitharani et al., 2003). This phenomenon has been demonstrated with the production of siRNAs on 5'-side of the targeted gene in *C. elegans* (Sijen et al., 2001) and on both sides of the targeted gene in transgenic GFP plants (Klahre et al., 2002).

(iii) siRNA interference with viral DNA accumulation

There has been no direct proof that siRNAs could specifically down-regulate gene expression and/or viral replication, particularly in plant cell cultures and specifically for ssDNA viruses. The knowledge that siRNA can down-regulate GFP or DsRed expression in protoplasts raised the possibility that such molecules could be used to suppress geminivirus accumulation in plant cells. In geminiviruses, the replication-associated protein (AC1) is a multi-functional protein and is indispensable for viral DNA replication. Protoplasts were coinoculated with the combination of (i) infectious clones of ACMV-[CM] DNA-A and DNA-B (Figure 21.4A) with or without siAC1 (Figure 21.4B) or (ii) infectious clones of DNA-A and DNA-B of EACMCV with or without siAC1. Total DNA isolated from protoplasts 36 and 48 hours post-transfection as well as viral DNA accumulation was quantified using Southern blot hybridization using virus-specific probes. We observed a 65–68% reduction in viral DNA accumulation in reactions with siAC1 compared with the control at 36 and 48 hours post-transfection, respectively, indicating that siRNAs inhibited ACMV-DNA accumulation (Figure 21.4D). However, siAC1 did not interfere with accumulation of EACMCV DNA, another species of cassava-infecting geminiviruses, indicating that this siAC1 was highly specific in action. The siAC1 sequence of ACMV-[CM] shares 67% homology to its counterpart in EACMCV. Next, we examined the effect of siRNA on accumulation of viral mRNA. Total RNA isolated from protoplasts inoculated with infectious clones of DNA-A and DNA-B of ACMV-[CM] with or without siAC1 was analyzed on Northern blot using AC1-specific probe. The effect of siAC1 on accumulation of AC1 mRNA was very dramatic, and showed a 90–92% reduction in mRNA levels compared with the control at 36 and 48 hours posttransfection, respectively (Figure 21.4C). This result demonstrates that siAC1 specifically degraded the polycistronic transcript, which contains the AC1 mRNA. Based on the results of siGFP-induced RNA silencing, we predicted that introduction of siAC1 would result in degradation of AC1 mRNA, and/or might serve as a primer to synthesize dsRNA using the AC1 viral mRNA as the template to amplify the PTGS signal (Tuschl et al., 1999; Vance and Vaucheret, 2001). Once synthesized, AC1 protein is stable and it is therefore not surprising that the effect of PTGS is less on DNA accumulation than on accumulation of mRNA level. Moreover, once viral DNA is produced, other viral single-stranded DNA binding proteins including the coat protein or nuclear shuttle protein BV1 protects it. Therefore, even though 90% of mRNA was degraded the remaining 10% is probably sufficient to enable continued accumulation of viral DNA. As in the case of the GFP and DsRed reporter genes, siRNA-mediated gene silencing was specific against the parent sequence only, with the siAC1 proving ineffective for suppressing accumulation of EACMCV. The nucleotide sequence of the siAC1 targeted in the *AC1* gene of ACMV-[CM] differs by 33% compared with its counterpart in EACMCV. ACMV siAC1 differs from EACMCV-AC1 counterpart by seven nts, revealing the specificity of siRNAs. We conclude that the inhibitory effect requires a very high degree of identity between the siRNAs and the target mRNA sequences (Vanitharani et al., 2003).

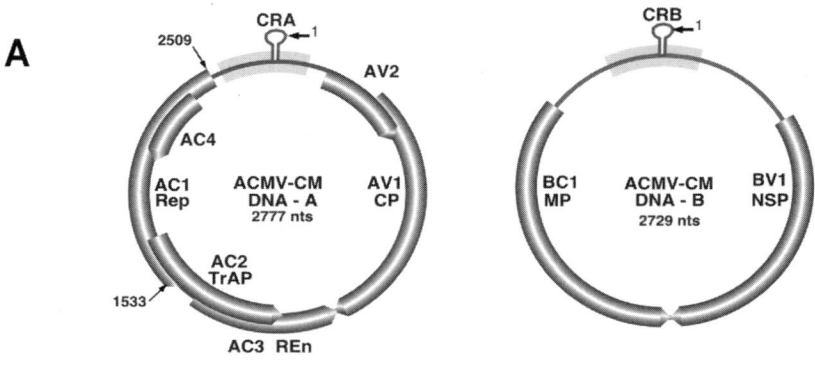

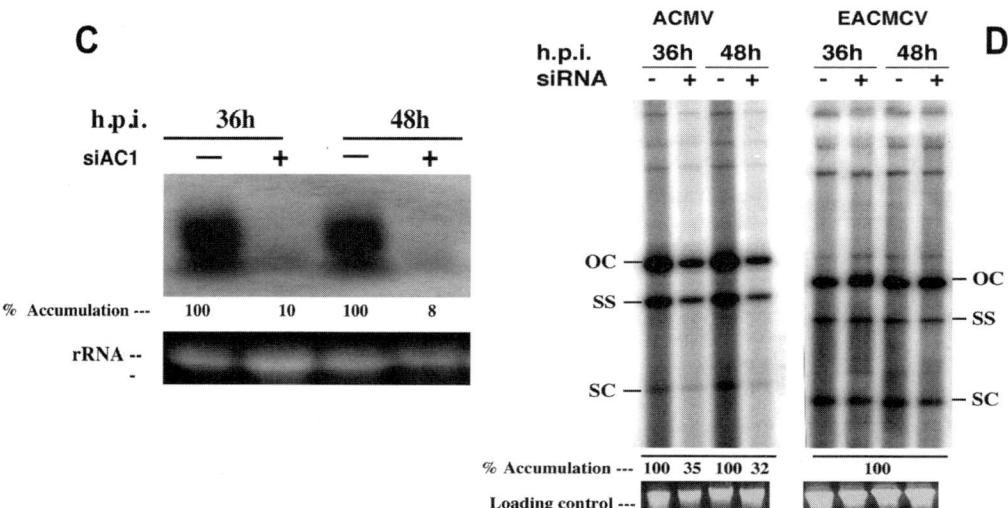

Figure 21.4. (A) Genome organization of DNA-A and DNA-B of ACMV-[CM]. DNA-A contains six ORFs AC1, AC2, AC3, AC4, AV1 and AV2. DNA-B contains BV1 and BC1 ORFs. V-represents virion-sense, and C-represents complementary-sense strands. AC1 encodes replication-associated protein (Rep). (B) Schematic representation of *AC1* coding sequence of ACMV-[CM] DNA-A. siAC1 sequence (co-ordinates 161-181 nts) designed to target AC1 gene of ACMV-[CM]. (C) Northern blot analysis of AC1-specific mRNA level in transfected BY2 tobacco protoplasts. Total RNA isolated from protoplasts inoculated with the infectious clones of (DNA-A and DNA-B) ACMV-[CM] along with (+) or without (−) siRNA specific for AC1 (siAC1) of ACMV-[CM] was loaded in each lane. AC1-specific [α-^{32}P]dCTP-labelled DNA was used as the probe. Hours post inoculation (h.p.i.). Ethidium bromide staining of ribosomal RNA (rRNA) shows equal loading of the samples. (D) Southern blots showing relative levels of viral DNA accumulation at 36–48 hrs post-transfection in BY2 tobacco protoplasts. For co-transfection, either infectious clones (DNA-A and DNA-B) of ACMV-[CM] or EACMCV along with (+) or without (−) siAC1 were used. The blots were hybridized with [α-^{32}P]dCTP-labelled ACMV-[CM]-specific (left panel) or EACMCV-specific probes (right panel). Ethidium bromide stained gel to show loading control. Hours post inoculation (h.p.i.).

(iv) Advantage of siRNA/protoplast system

Use of a cell-based system such as described here for rapid evaluation and study of siRNA and the ability to down-regulate gene expression will be a valuable tool for investigating gene regulation in plants. In addition, we provide evidence that siRNA can interfere with and suppress accumulation of the economically important single-stranded DNA geminiviruses (Vanitharani et al., 2003). Another advantage of using siRNA-induced gene silencing over long dsRNA is that it dissects the gene-silencing pathway at least into two steps: (i) the initiation and (ii) the maintenance steps. As a counter-defense mechanism, though a number of suppressor proteins have been identified from plant RNA and DNA viruses, insect virus, and one from host plant, their exact mode of action on the RNA-silencing pathway is yet to be determined. These identified silencing suppressor proteins may act at different steps in the PTGS pathway. Our results suggest that use of siRNA in protoplasts system may be a valuable tool to understand the mechanism of gene silencing in plants as well as a means to determine the point in the silencing pathway at which plant viral suppressors act (Vanitharani et al., 2003).

Summary

RNA silencing is an ancient cellular defense mechanism conserved among different kingdoms of organisms (Plasterk, 2002). It is therefore understandable that viruses utilizing RNA during replication and gene expression would need to develop strategies to evade this defense system. It is now becoming evident that many plant viruses have evolved an active mechanism to counteract silencing by encoding suppressor proteins that interfere with the process (Li et al., 2002, Vance and Vaucheret, 2001). All plant RNA viruses replicate through dsRNA intermediate, a potential inducer of PTGS. Geminiviruses do not encounter dsRNA in their life cycle. Nevertheless, these viruses are able to trigger the host's PTGS with the production of virus-derived siRNAs as reported recently for monopartite, TYLCV (Lucioli et al., 2003) and also in bipartite cassava-infecting geminiviruses (Chellappan et al., 2004a). Furthermore, recovery from symptoms in ACMV-[CM]-infected plants is associated with higher accumulation of virus-derived siRNAs, especially from the C-terminus of AC1. Moreover, it appears that suppressor protein in ACMV-[CM] acts in a temporal/spatial manner; as a consequence it is active only at certain time and may be weak compared to suppressor proteins of non-recovery type geminiviruses (Vanitharani et al., 2004). Therefore, ACMV-[CM]-infected plants showed recovery from symptoms at the later stage of infection cycle. The specificity of siRNA-mediated interference of gene expression in protoplasts allows quick screening of genes transiently before committing to the process of transgenic plants. This system would be very useful tool to identify the action of different suppressor proteins in the RNA silencing pathway, whether it targets the upstream steps before the siRNA synthesis or downstream steps after siRNA synthesis. The fact that one could use siRNAs to down-regulate viral DNA accumulation provides a way to identify the best candidate viral sequence which would target more than one virus will be useful to control the devastating viruses such as geminiviruses (Vanitharani et al., 2003).

Acknowledgements

We thank Dr. Vicki Vance (University of South Carolina, USA) for P1/HC-Pro construct, and Dr. David Baulcombe (John Innes Center, UK) for GFP-transgenic *N. benthamiana* line 16C plants. This work was funded by Donald Danforth Plant Science Center.

REFERENCES

Abel, P. P., Nelson, R. S., De, B., Hoffmann, N., Rogers, S. G., Fraley, R. T. and Beachy, R. N. (1986). Delay of disease development in transgenic plants that express the tobacco mosaic virus coat protein gene. *Science*, **232**, 738–743.

Ahlquist, P. (2002). RNA-dependent RNA polymerases, viruses, and RNA silencing. *Science*, **296**, 1270–1273.

Anandalakshmi, R., Marathe, R., Ge, X., Herr, J. M., Jr., Mau, C., Mallory, A., Pruss, G., Bowman, L. and Vance, V. B. (2000). A calmodulin-related protein that suppresses post-transcriptional gene silencing in plants. *Science*, **290**, 142–144.

Anandalakshmi, R., Pruss, G. J., Ge, X., Marathe, R., Mallory, A. C., Smith, T. H. and Vance, V. B. (1998). A viral suppressor of gene silencing in plants. *Proceedings of the National Academy of Sciences USA*, **95**, 13079–13084.

Baulcombe, D. C. (1999). Fast forward genetics based on virus-induced gene silencing. *Current Opinion in Plant Biology*, **2**, 109–113.

Beclin, C., Berthome, R., Palauqui, J. C., Tepfer, M. and Vaucheret, H. (1998). Infection of tobacco or Arabidopsis plants by CMV counteracts systemic post-transcriptional silencing of nonviral (trans) genes. *Virology*, **252**, 313–317.

Brigneti, G., Voinnet, O., Li, W., Ji, L., Ding, S. and Baulcombe, D. C. (1998). Viral pathogenicity determinants are suppressors of transgene silencing in Nicotiana benthamiana. *European Molecular Biology Organization Journal*, **17**, 6739–6746.

Bucher, E., Sijen, T., Haan, P. de, Goldbach, R. and Prins, M. (2003). Negative-strand tospoviruses and tenuiviruses carry a gene for a suppressor of gene silencing at analogous genomic positions. *Journal of Virology*, **77**, 1329–1336.

Chellappan, P., Vanitharani, R. and Fauquet, C. M. (2004a). Short-interfering RNA accumulation correlates with host recovery in DNA virus infected hosts and gene silencing targets specific viral sequences. *Journal of Virology*, **78**, 7465–7477.

Chellappan, P., Masona, M. V., Vanitharani, R., Taylor, N. J. and Fauquet, C. M. (2004b). Broad spectrum resistance to ssDNA viruses associated with transgene-induced gene silencing in cassava. *Plant Molecular Biology*, In press.

Cogoni, C. and Macino, G. (1997). Isolation of quelling-defective (qde) mutants impaired in posttranscriptional transgene-induced gene silencing in Neurospora crassa. *Proceedings of the National Academy of Sciences USA*, **94**, 10233–10238.

Covey, S. N., Al-Kaff, N. S., Langara, A. and Turner, D. S. (1997). Plants combat infection by gene silencing. *Nature*, **385**, 781–782.

Dalmay, T., Hamilton, A., Rudd, S., Angell, S. and Baulcombe, D. C. (2000). An RNA-dependent RNA polymerase gene in Arabidopsis is required for posttranscriptional gene silencing mediated by a transgene but not by a virus. *Cell*, **101**, 543–553.

Depicker, A. and Montagu, M. V. (1997). Post-transcriptional gene silencing in plants. *Current Opinion in Cell Biology*, **9**, 373–382.

Dong, X., van Wezel, R., Stanley, J. and Hong, Y. (2003). Functional characterization of the nuclear localization signal for a suppressor of posttranscriptional gene silencing. *Journal of Virology*, **77**, 7026–7033.

Dunoyer, P., Pfeffer, S., Fritsch, C., Hemmer, O., Voinnet, O. and Richards, K., E. (2002). Identification, subcellular localization and some properties of a cysteine-rich suppressor of gene silencing encoded by peanut clump virus. *Plant Journal*, **29**, 555–567.

Elbashir, S. M., Harborth, J., Lendeckel, W., Yalcin, A., Weber, K. and Tuschl, T. (2001). Duplexes of 21-nucleotide RNAs mediate RNA interference in cultured mammalian cells. *Nature*, **411**, 494–498.

English, J. J., Mueller, E. and Baulcombe, D. C. (1996). Suppression of virus accumulation in transgenic plants exhibiting silencing of nuclear genes. *Plant Cell*, **8**, 179–188.

Fauquet, C. M. and Stanley, J. (2003). Geminivirus classification and nomenclature: Progress and problems. *Annals of Applied Biology*, **142**, 165–189.

Fire, A., Xu, S., Montgomery, M. K., Kostas, S. A., Driver, S. E. and Mello, C. C. (1998). Potent and specific genetic interference by double-stranded RNA in *Caenorhabditis elegans*. *Nature*, **391**, 806–811.

Fondong, V. N., Pita, J. S., Rey, M. E., de Kochko, A., Beachy, R. N. and Fauquet, C. M. (2000). Evidence of synergism between African cassava mosaic virus and a new double-recombinant geminivirus infecting cassava in Cameroon. *Journal of General Virology*, **81**, 287–297.

Gitlin, L., Karelsky, S. and Andino, R. (2002). Short interfering RNA confers intracellular antiviral immunity in human cells. *Nature*, **418**, 430–434.

Goodwin, J., Chapman, K., Swaney, S., Parks, T. D., Wernsman, E. A. and Dougherty, W. G. (1996). Genetic and biochemical dissection of transgenic RNA-mediated virus resistance. *Plant Cell*, **8**, 95–105.

Guo, H. and Ding, S. (2002). A viral protein inhibits the long range signaling activity of the gene silencing signal. *European Molecular Biology Organization Journal*, **21**, 398–407.

Hamilton, A. J. and Baulcombe, D. C. (1999). A species of small antisense RNA in posttranscriptional gene silencing in plants. *Science*, **286**, 950–952.

Hammond, S. M., Bernstein, E., Beach, D. and Hannon, G. J. (2000). An RNA-directed nuclease mediates post-transcriptional gene silencing in *Drosophila* cells. *Nature*, **404**, 293–6.

Hanley-Bowdoin, L., Settlage, S. B., Orozco, B. M., Nagar, S. and Robertson, D. (1999). Geminiviruses: Models for plant DNA replication, transcription, and cell cycle regulation. *Critical Reviews in Plant Sciences*, **18**, 71–106.

Harrison, B. D., Zhou, X., Otim-Nape, G. W., Liu, Y. and Robinson, D. J. (1997). Role of a novel type of double infection in the geminivirus-induced epidemic of severe cassava mosaic in Uganda. *Annals of Applied Biology*, **131**, 437–448.

Jones, A. L., Thomas, C. L. and Maule, A. J. (1998). De novo methylation and co-suppression induced by a cytoplasmically replicating plant RNA virus. *European Molecular Biology Organization Journal*, **17**, 6385–6393.

Jorgensen, R. (1992). Silencing of plant genes by homologous transgenes. *Agbiotech News and Information*, **4**, 265N-273N.

Kasschau, K., D., Xie, Z., Allen, E., Llave, C., Chapman, E. J., Krizan, K. A. and Carrington, J. C. (2003). P1/HC-Pro, a viral suppressor of RNA silencing, interferes with Arabidopsis development and miRNA function. *Developmental Cell*, **4**, 205–217.

Kasschau, K. D. and Carrington, J. C. (1998). A counterdefensive strategy of plant viruses: suppression of posttranscriptional gene silencing. *Cell*, **95**, 461–470.

Kjemtrup, S., Sampson, K. S., Peele, C. G., Nguyen, L. V., Conkling, M. A., Thompson, W. F. and Robertson, D. (1998). Gene silencing from plant DNA carried by a geminivirus. *The Plant Journal: for Cell and Molecular Biology*, **14**, 91–100.

Klahre, U., Crete, P., Leuenberger, S. A., Iglesias, V. A. and Meins, F., Jr. (2002). High molecular weight RNAs and small interfering RNAs induce systemic posttranscriptional gene silencing in plants. *Proceedings of the National Academy of Sciences USA*, **99**, 11981–11986.

Kumagai, M. H., Donson, J., Della-Cioppa, G., Harvey, D., Hanley, K. and Grill, L. K. (1995). Cytoplasmic inhibition of carotenoid biosynthesis with virus-derived RNA. *Proceedings of the National Academy of Sciences USA*, **92**, 1679–1683.

Li, H., Li, W. X. and Ding, S. W. (2002). Induction and suppression of RNA silencing by an animal virus. *Science*, **296**, 1319–1321.

Li, H., Lucy, A. P., Guo, H., Li, W., Ji, L., Wong, S. and Ding, S. (1999). Strong host resistance targeted against a viral suppressor of the plant gene silencing defence mechanism. *European Molecular Biology Organization Journal*, **18**, 2683–2691.

Lindbo, J. A. and Dougherty, W. G. (1992). Pathogen-derived resistance to a potyvirus: Immune and resistant phenotypes in transgenic tobacco expressing altered forms of a potyvirus coat protein nucleotide sequence. *Molecular Plant Microbe Interactions*, **5**, 144–153.

Llave, C., Kasschau, K. D. and Carrington, J. C. (2000). Virus-encoded suppressor of posttranscriptional gene silencing targets a maintenance step in the silencing pathway. *Proceedings of the National Academy of Sciences USA*, **97**, 13401–13406.

Lucioli, A., Noris, E., Brunetti, A., Tavazza, R., Ruzza, V., Castillo, A. G., Bejarano, E. R., Accotto, G. P. and Tavazza, M. (2003). Tomato yellow leaf curl Sardinia virus rep-derived resistance to homologous and heterologous geminiviruses occurs by different mechanisms and is overcome if virus-mediated transgene silencing is activated. *Journal of Virology*, **77**, 6785–6798.

Mallory, A. C., Reinhart, B. J., Bartel, D., Vance, V. B. and Bowman, L. H. (2002). A viral suppressor of RNA silencing differentially regulates the accumulation of short interfering RNAs and micro-RNAs in tobacco. *Proceedings of the National Academy of Sciences USA*, **99**, 15228–15233.

Mette, M. F., Winden, J. v. d., Matzke, M. A. and Matzke, A. J. M. (1999). Production of aberrant promoter transcripts contributes to methylation and silencing of unlinked homologous promoters in trans. *European Molecular Biology Organization Journal*, **18**, 241–248.

Napoli, C., Lemieux, C. and Jorgensen, R. (1990). Introduction of a chimeric chalcone synthase gene into petunia results in reversible co-suppression of homologous genes in trans. *Plant Cell*, **2**, 279–289.

Palatnik, J. F., Allen, E., Wu, X., Schommer, C., Schwab, R., Carrington, J. C. and Weigel, D. (2003). Control of leaf morphogenesis by microRNAs. *Nature*, **425**, 257–263.

Palauqui, J. C., Elmayan, T., Pollien, J. M. and Vaucheret, H. (1997). Systemic acquired silencing: transgene-specific post-transcriptional silencing is transmitted by grafting from silenced stocks to non-silenced scions. *European Molecular Biology Organization Journal*, **16**, 4738–4745.

Peele, C., Jordan, C. V., Muangsan, N., Turnage, M., Egelkrout, E., Eagle, P., Hanley-Bowdoin, L. and Robertson, D. (2001). Silencing of a meristematic gene using geminivirus-derived vectors. *The Plant Journal: for Cell and Molecular Biology*, **27**, 357–366.

Pfeffer, S., Dunoyer, P., Heim, F., Richards, K. E., Jonard, G. and Ziegler-Graff, V. (2002). P0 of beet Western yellows virus is a suppressor of posttranscriptional gene silencing. *Journal of Virology*, **76**, 6815–6824.

Pinto, Y. M., Kok, R. A. and Baulcombe, D. C. (1999). Resistance to rice yellow mottle virus (RYMV) in cultivated African rice varieties containing RYMV transgenes. *Nature Biotechnology*, **17**, 702–707.

Pita, J. S., Fondong, V. N., Sangare, A., Otim-Nape, G. W., Ogwal, S. and Fauquet, C. M. (2001). Recombination, pseudorecombination and synergism of geminiviruses are determinant keys to the epidemic of severe cassava mosaic disease in Uganda. *Journal of General Virology*, **82**, 655–665.

Plasterk, R. H. (2002). RNA silencing: The genome's immune system. *Science*, **296**, 1263–1265.

Pruss, G., Ge, X., Shi, X. M., Carrington, J. C. and Vance, V. B. (1997). Plant viral synergism: The potyviral genome encodes a broad-range pathogenicity enhancer that transactivates replication of heterologous viruses. *Plant Cell*, **9**, 859–868.

Qu, F., Ren, T. and Morris, T. J. (2003). The coat protein of Turnip crinkle virus suppresses posttranscriptional gene silencing at an early initiation step. *Journal of Virology*, **77**, 511–522.

Ratcliff, F., Harrison, B. D. and Baulcombe, D. C. (1997). A similarity between viral defense and gene silencing in plants. *Science*, **276**, 1558–1560.

Ratcliff, F., Martin-Hernandez, A. M. and Baulcombe, D. C. (2001). Tobacco rattle virus as a vector for analysis of gene function by silencing. *Plant Journal*, **25**, 237–245.

Ratcliff, F. G., MacFarlane, S. A. and Baulcombe, D. C. (1999). Gene silencing without DNA:RNA-mediated cross-protection between viruses. *Plant Cell*, **11**, 1207–1215.

Reed, J. C., Kasschau, K. D., Prokhnevsky, A. I., Gopinath, K., Pogue, G. P., Carrington, J. C. and Dolja, V. V. (2003). Suppressor of RNA silencing encoded by Beet yellows virus. *Virology*, **306**, 203–209.

Ruiz, M. T., Voinnet, O. and Baulcombe, D. C. (1998). Initiation and maintenance of virus-induced gene silencing. *Plant Cell*, **10**, 937–946.

Sijen, T., Fleenor, J., Simmer, F., Thijssen, K. L., Parrish, S., Timmons, L., Plasterk, R. H. A. and Fire, A. (2001). On the role of RNA amplification in dsRNA-triggered gene silencing. *Cell*, **107**, 465–476.

Silhavy, D., Molnar, A., Lucioli, A., Szittya, G., Hornyik, C., Tavazza, M. and Burgyan, J. (2002). A viral protein suppresses RNA silencing and binds silencing-generated, 21- to 25-nucleotide double-stranded RNAs. *European Molecular Biology Organization Journal*, **21**, 3070–3080.

Stanley, J. (1983). Infectivity of the cloned geminivirus genome requires sequences from both DNAs. *Nature*, **305**, 643–645.

Szittya, G., Silhavy, D., Molnar, A., Havelda, Z., Lovas, A., Lakatos, L., Banfalvi, Z. and Burgyan, J. (2003). Low temperature inhibits RNA silencing-mediated defence by the control of siRNA generation. *European Molecular Biology Organization Journal*, **22**, 633–640.

Turnage, M. A., Muangsan, N., Peele, C. G. and Robertson, D. (2002). Geminivirus-based vectors for gene silencing in Arabidopsis. *The Plant Journal: for Cell and Molecular Biology*, **30**, 107–114.

Tuschl, T., Zamore, P. D., Lehmann, R., Bartel, D. P. and Sharp, P. A. (1999). Targeted mRNA degradation by double-stranded RNA *in vitro*. *Genes & Development*, **13**, 3191–3197.

Van Wezel, R., Dong, X., Liu, H., Tien, P., Stanley, J. and Hong, Y. (2002). Mutation of three cysteine residues in Tomato yellow leaf curl virus-China C2 protein causes dysfunction in pathogenesis and posttranscriptional gene-silencing suppression. *Molecular Plant-Microbe Interactions: Mpmi* **15**, 203–208.

Vance, V. and Vaucheret, H. (2001). RNA silencing in plants – defense and counterdefense. *Science*, **292**, 2277–2280.

Vanitharani, R., Chellappan, P. and Fauquet, C. M. (2003). Short interfering RNA-mediated interference of gene expression and viral DNA accumulation in cultured plant cells. *Proceedings of the National Academy of Sciences USA*, **100**, 9632–9636.

Vanitharani, R., Chellappan, P., Pita, J. S. and Fauquet, C. M. (2004). Differential roles of AC2 and AC4 of cassava geminiviruses in mediating synergism and posttranscriptional gene silencing suppression. *Journal of Virology*, In press.

Voinnet, O. and Baulcombe, D. C. (1997). Systemic signalling in gene silencing. *Nature*, **389**, 553.

Voinnet, O., Lederer, C. and Baulcombe, D. C. (2000). A viral movement protein prevents spread of the gene silencing signal in Nicotiana benthamiana. *Cell*, **103**, 157–167.

Voinnet, O., Pinto, Y. M. and Baulcombe, D. C. (1999). Suppression of gene silencing: a general strategy used by diverse DNA and RNA viruses of plants. *Proceedings of the National Academy of Sciences USA*, **96**, 14147–14152.

Waterhouse, P. M., Graham, M. W. and Wang, M. (1998). Virus resistance and gene silencing in plants can be induced by simultaneous expression of sense and antisense RNA. *Proceedings of the National Academy of Sciences USA*, **95**, 13959–13964.

Waterhouse, P. M., Wang, M. and Lough, T. (2001). Gene silencing as an adaptive defence against viruses. *Nature*, **411**, 834–842.

Yelina, N. E., Savenkov, E. I., Solovyev, A. G., Morozov, S. Y. and Valkonen, J. P. (2002). Long-distance movement, virulence, and RNA silencing suppression controlled by a single protein in hordei- and potyviruses: complementary functions between virus families. *Journal of Virology*, **76**, 12981–12991.

Zaitlin, M., Anderson, J. M., Perry, K. L., Zhang, L. and Palukaitis, P. (1994). Specificity of replicase-mediated resistance to cucumber mosaic virus. *Virology*, **201**, 200–205.

Drug target validation

22 Delivering siRNA *in vivo* for functional genomics and novel therapeutics

Patrick Y. Lu and Martin C. Woodle

Introduction

Use of siRNA *in vivo* to down regulate expression of a specific gene requires knowledge of target sequence accessibility, target tissue deliverability, and, for most applications, siRNA stability in both extracellular and intracellular environments (McManus and Sharp, 2002). Unlike *in vitro* transfection of siRNA into cells, *in vivo* delivery of siRNA into targeted tissue of animal models is much more complicated, involving physical, chemical and biological approaches, and in some cases their combination (Lu et al., 2003). A consequence of the fast-growing literature on using siRNA as a research tool for functional genomics is an emerging interest in siRNA as a therapeutic. Therapeutic applications, however, clearly depend upon optimized local and systemic delivery of siRNA *in vivo*. Therefore, delivering siRNA into targeted tissues and maintaining its activity within targeted cells and on the targeted gene sequence are the key aspects considered here, in order to fulfill the goals of both functional genomic research and therapeutic development.

Currently, using siRNA to characterize gene function and to explore therapeutic potential is spreading over almost every field in biomedical research. This phenomenon results from two basic realities: 1) siRNA is proving to be a very potent, robust, and easy to use inhibitor activated by a natural process and 2) down regulation of individual genes is a powerful tool for understanding their biological functions and may generate therapeutic benefits by reversing the pathological effects caused by over-expression of those genes. Not surprisingly, cancer research has become the most dynamic and exciting area for application of siRNA inhibitor. Other therapeutic areas are also attractive for siRNA application either for use as a research tool for validation of gene functions or to explore novel therapeutics.

Although increasing numbers of studies on target identification and validation using siRNA *in vitro* have been reported, limited reports of *in vivo* studies have indicated a lack of effective delivery methods for siRNA agents. The key to *in vivo* application is a delivery system that transports siRNA duplex into the target tissue and cells and keeps the siRNA functioning within the cell. Advancement of the

303

siRNA delivery systems will determine the successful range of applications of this unique inhibitor *in vivo*. Thanks to more than a decade of effort on development of nucleic acid delivery for gene therapy, currently, there are many of such delivery tools able to be adapted for siRNA delivery, including biological-based vectors, synthetic chemical–based and physical-based systems.

Delivering siRNA *in vivo*

The effectiveness of using siRNA as a functional genomic tool has been demonstrated by targeting different genes in various cell types. Transfection of siRNA into the cultured cells is relatively easy (McManus and Sharp, 2002) compared with other forms of nucleic acid, e.g. plasmid. The concept for using siRNA delivery to modulate gene expression as a means to investigate gene function in pathological tissue is illustrated in Figure 22.1. Delivery of siRNA duplex specifically targeting certain genes in the cell or tissue typically has been performed with siRNA targeting an individual gene, but multiples or groups of genes are possible. The beneficial or detrimental effects of induced phenotypic changes in pathological status can be analyzed with various means including biochemical, pharmacological and histological assays. Overall, this approach clearly depends upon the effectiveness of the siRNA delivery to modulate expression of specific genes and as a result induce physiological or pharmacological effects. A wide variety of nucleic acid delivery systems have been developed, including viral vectors and "non-viral" approaches, achieving efficient and significant modulation of gene expression for many types of cells. Regardless of whether the systems are biologics such as viral vectors or synthetic agents such as polymer formulations of chemically synthesized oligonucleotides, siRNA delivery can be used to decrease gene expression, and – if the gene is important – result in phenotypic changes. Nonetheless, further improvement remains a critical need for application of siRNA delivery to drug target discovery and validation, and potential clinical application.

Viral vectors

Viruses are natural biological agents that have been attractive as a starting point for gene delivery by removal of replication capability and adaptation for delivery of nucleic acids. Their development has been very successful, leading to extensive studies both as research tools and as therapeutic agents in clinical trials. One of the earliest systems studied extensively for therapeutic applications, retroviral vectors, have been adapted for expression of libraries for genomic studies (Devroe and Silver, 2002), although many viral vectors have been studied for this purpose (Xia et al., 2002; Tiscornia et al., 2003). Nonetheless, most viral vector studies of delivery of siRNA utilize only a few types: retroviral vectors, adenoviral vectors and most recently lentiviral vectors. These viral vectors offer two distinct classes of vectors with properties that in most respects are polar opposites.

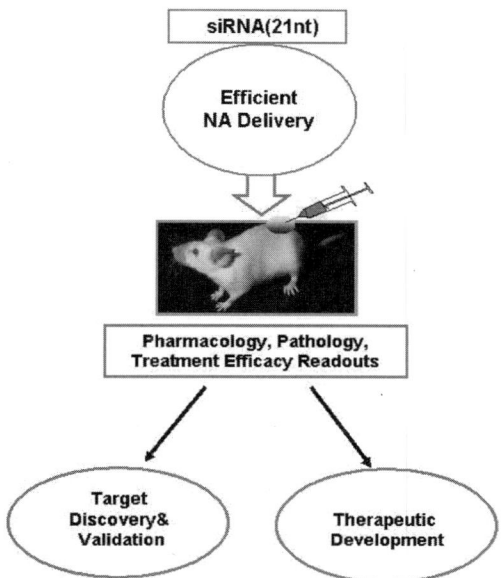

Figure 22.1. Applications of *in vivo* delivery of siRNA

Their differences include form of nucleic acid, chromosomal integration, gene size capacity, capsid form, requirement for cell division, and tissue tropism. Consequently, the biological application often mandates choice of one or the other. Nonetheless, adjustments to the vectors can be made that override an inherent preference. For example, retroviral vectors that deliver RNA should be preferred for delivery of RNA-based ribozymes, but in fact engineering of the expression cassette in adenoviral vectors to express mRNA forming these active agents is becoming common. In fact, the first studies reporting expression of siRNA by viral vectors utilized adenoviral vectors.

Xia and colleagues (Xia et al., 2002) have described an adenoviral-mediated delivery mechanism that results in specific silencing of targeted genes through expression of small interfering RNA (siRNA). They established proof of principle by markedly diminishing expression of exogenous and endogenous genes *in vitro* and *in vivo* in brain and liver, and further applied this strategy to a model system of a major class of neurodegenerative disorders, the polyglutamine diseases, to show reduced polyglutamine aggregation in cells. Devroe and Silver have developed a retroviral siRNA delivery system based on commercially available vectors (Devroe and Silver, 2002). They found that cells infected with the anti-NDR siRNA virus dramatically down regulate NDR expression, whereas control viruses have no effect on total NDR levels. The experiments also demonstrated that retrovirus-delivered siRNA provides significant advancement over previous methods by providing efficient, uniform delivery and immediate selection of stable "knockdown" cells. They believed that this development would provide a method to rapidly assess gene function in established cell lines, primary cells, or animals.

To test different viral vector systems, Tiscornia and colleagues (Tiscornia et al., 2003) used lentiviral vectors to express siRNAs and to knockdown the expression of specific genes *in vitro* and *in vivo*. This lentiviral vector expressing GFP-siRNA was capable of inhibiting GFP expression in 293T-GFP cell lines and eggs from GFP-positive transgenic mice. More interestingly, pups from F(1) progeny, which expressed siGFP, showed considerably diminished fluorescence and decreased GFP. They proposed that an approach of combining transgenesis by lentiviral vectors expressing siRNAs could be used successfully to generate a large number of mice in which the expression of a specific gene(s) was down regulated substantially. They also suggested that this approach of generating "knockdown" mice would aid functional genomics. Regardless of the differences between adenoviral, retroviral and lentiviral vector systems, their efficiency and prolonged expression has resulted in widespread acceptance, especially as a research tool, but therapeutic use remains unproven.

The viral vectors are an effective means to deliver nucleic acids, but have several limitations. One of the greatest challenges of using viral vectors is that the preparations contain essential viral proteins that frequently introduce biological effects of their own. One such effect is generating immune responses that limit applications in animals. Added steps are needed to create each viral vector construction once the siRNA sequences have been identified and obtained. On the other hand, viral vector–based deliveries of siRNA agents are also quite effective, if toxicity, immune response, and long-term safety are not major concerns. Caution must be exercised since under certain circumstances, viral infection may induce unexpected complication that may mask the RNAi true effect.

"Non-viral" vectors

A wide range of nucleic acid delivery systems have been devised without adaptation of virus and thus fall into the broad category called "non-viral" vectors. Lumped together in this poorly defined class are many different systems covering vastly different forms of nucleic acid. Likewise, this class covers a wide variety of methods to formulate and deliver the nucleic acid to enable intracellular delivery and exertion of the gene activity including cationic complexes with lipids or polymers or both, PLGA microsphere depot formulations, hydrophilic protective polymers, physical force based delivery such as bombardment with nucleic acid coated gold particles or electroporation, and many other delivery methods. Note that the common feature conveyed by the non-viral category allows for many different forms of the nucleic acid, spanning forms generated biologically, e.g. plasmids, to forms generated synthetically, e.g. phosphorothioate antisense and siRNA oligonucleotides. In addition, the success of each delivery *in vivo* largely depends on the route and targeted tissues of the delivery, e.g. the hydrodynamic approach is effective for liver delivery, electroporation is suitable to muscle delivery and targeted systemic delivery is able to reach neovasculature.

One of the important non-viral delivery methods is based on the use of cationic transfection agents to form complexes with oligonucleotides. The most commonly used transfection reagents have been cationic lipids, producing lipoplex complexes. Sørensen et al. (2003) showed that systemic delivery of siRNAs using cationic liposome-based intravenous injection in mice of plasmid encoding the green fluorescent protein (GFP) with its cognate siRNA inhibited GFP gene expression in various organs. Furthermore, intraperitoneal injection of anti-TNF-alpha siRNA inhibited lipopolysaccharide-induced TNF-alpha gene expression, whereas secretion of IL1-alpha was not inhibited. Similarly, a number of cationic polymers also have been developed for delivery of plasmids, forming the analogous polyplex class, with many commercial reagents also available. Delivery of siRNA into mouse liver has also been achieved using large volumes of aqueous solution injected rapidly creating a high pressure in the vascular circulation (Lewis et al., 2002; McCaffrey et al., 2002; Song et al., 2003; Zender et al., 2003). This approach, often called hydrodynamic delivery, has indicated that siRNA is effective at inhibiting the expression of a transgene expressed from the genome and imply that delivery of siRNA to the liver results in uptake of siRNA by at least a majority of hepatocytes. This procedure will allow the use of siRNA in mice for gene function and drug target validation studies (Sen et al., 2003; Zender et al., 2003).

As found with viral vectors, development of non-viral gene delivery methods that achieve tumor targeted nucleic acid delivery and activity directly within the tumor from an intravenous administration has proven to be difficult. Recent results obtained using ligand targeting to tumor neovasculature suggests that systemic tumor therapeutics using gene delivery is a possibility (Woodle et al., 2001; Ogris et al., 2003) but limited to gene expression in endothelial cells of tumor blood vessels, not the tumor cells. For systemic targeted delivery of siRNA we have developed a layered nanoplex system that combines tissue-targeted nanoparticles giving cytoplasmic delivery with highly potent siRNA active in the cytoplasm. We designed a modular chemical conjugation that allowed separation of functional aspects of the nanoparticle: self-assembly, formation of a steric polymer surface layer, and exposed ligands. For targeting tumors, we coupled polyethylene glycol (PEG) with an Arg-Gly-Asp (RGD)-motif peptide ligand on one end to binding αv-integrins on activated endothelial cells to a polycation agent, polyethyleneimine (PEI) with the other end (Woodle et al., 2001). The reagent binds siRNA and forms layered nanoplexes. The particles self-assemble with the PEG located on the surface and expose peptides that target neovasculature. These advances in nanoparticle systems appear promising for therapeutic application of siRNA for metastatic cancer and many other angiogenesis related diseases.

Non-viral vectors have many advantages, especially for therapeutics, over the viral vectors. For example, immunogenicity of viral vectors has precluded multiple administrations and resulted in severe toxicity limitations. Chromosomal integration, and the resulting safety concerns, is also avoided by non-viral vectors. While potential safety issues are not a limitation for functional genomics, immunogenicity and other side effects can obscure the results and need to be avoided. Clearly, effective siRNA delivery is crucial for *in vivo* functional genomics, i.e. inhibition of gene expression in a significant number of cells so that the biological effect that a drug targeting the same protein can be emulated. This means that approximately the same percentage of cells and the cells in the same location within the target tissue reached by the drug should be affected by the siRNA delivery. As mentioned for viral vectors, an equally important requirement for non-viral vectors is delivery without significant background activity from the delivery method itself. Non-viral delivery methods tend to have much lower levels of biological effects, other than that from the gene, relative to viral vectors with their many protein components.

Delivering siRNA through local and systemic administrations

Delivering siRNA molecules *in vivo* basically falls into two approaches: local and systemic administrations. These two approaches can be uniquely suitable for a particular tissue type, or can be used together to reach the same targeted organs. For example, skin and muscle can be better accessed using local delivery, while lung and tumor can be reached efficiently by both local and systemic deliveries. The choice between local and systemic deliveries largely depends on the targeting tissues and cell types, and the expected outcome for siRNA-mediated gene knockdown. In addition, this choice usually is only a part of the entire consideration of study design involving vector carriers, administration routes and approaches for siRNA delivery *in vivo*. There are increasing data from our group and others supporting the conclusion that siRNA duplex is a very potent sequence specific inhibitor in many tissue types.

Delivery of siRNA into tumor

Although primary tumors grow locally, malignant tumors grow fast, penetrating and destroying local tissues, and spread throughout the body via blood or the lymphatic system. Their unpredictable and uncontrolled growth makes malignant cancer dangerous, and fatal in many cases. The tumors established at distant locations are not morphologically typical of the original tissue and are not encapsulated. Therefore, therapeutics using local delivery of siRNA may be limited to certain primary tumor types, such as head-and-neck cancer and skin cancer. Systemic delivery of siRNA, on the other hand, through intravenous administration, is clinically feasible for many types of the malignancies. Systemic delivery requires siRNA oligo stability in the blood stream and in the local environment before reaching the target cell. In addition, it requires stability and means for the siRNA oligo to pass through multiple tissue barriers until reaching the target cell, without loss of activity once within the target cells. To meet these requirements, ligand-directed nanoparticle formulations are an ideal system for systemic anti-cancer siRNA delivery, especially for therapeutic application. An emergence of such nanoparticle systems initially for plasmid gene delivery suggests that similar systems can be adapted for siRNA oligonucleotides. Viral vectors also are being evaluated for tumor-targeted gene delivery and ultimately may prove effective for expression siRNA or other forms such as short hairpin RNA (shRNA), but lack of repeated administration remains an impasse. For functional validation of the tumorigenic genes, though, intratumoral delivery of siRNA is still a very effective approach, since preclinical human xenograft tumor models are assessable and the determining step for clinical studies. Local delivery can be carried out in several ways, including use of cationic lipid and polymer complexes, direct injection of formulated siRNA solution, viral vector expression vectors, and potentially using physical enhancement methods.

Delivery of siRNA into liver

Both viral and non-viral gene delivery methods have been developed for effective liver expression and applied to siRNA. Perhaps as a result of an initial success and ready availability, many groups have adapted hydrodynamic delivery to deliver siRNA into the hepatocytes with relative high efficiency first for reporter studies (Lewis et al., 2002; McCaffrey et al., 2002) and subsequently for therapeutic studies (Song et al., 2003; Zender et al., 2003). Although this particular method is not clinically feasible for the human patient, it is an effective approach for gene functional validation. Song et al. (2003) injected *Fas* siRNA and achieved down regulation of *Fas* mRNA levels and protein in mouse hepatocytes, and the effects persisted without diminution for 10 days. Zender and colleagues (Zender et al., 2003) delivered 21-nt siRNAs against caspase 8 systemically resulting in inhibition of caspase 8 gene expression in the liver, thereby preventing *Fas* (CD95)-mediated apoptosis. As mentioned before, adenovirus has also been applied for liver delivery of siRNA. Although both hydrodynamic delivery and adenoviral vector are not clinical acceptable methods, these two systems have a relatively high hepatocyte efficiency that is a very powerful tool for hepatocyte functional genomic studies. Targeted systems for siRNA delivery into the liver have also been a very attractive approach and are under development (Ren et al., 2001). Clearly, such liver targeted delivery systems are more clinically feasible for development of siRNA-based therapeutics for treatment of various liver-related metabolic and hepatitis viral diseases.

Delivery of siRNA into ocular tissue

There are an increasing number of clinical protocols for treating eye diseases with nucleic acid drugs such as short antisense oligo or long single-stranded RNA aptamer. These clinically feasible delivery approaches are well suited for siRNA administration but to date depend upon local administration. Using a model of retinal neovascularization induced by laser damage, Reich and colleagues (Reich et al., 2003) delivered non-formulated siRNA specific to murine VEGF to the subretinal space and observed significant reduction of the eye angiogenesis. Using a different murine model with herpes simplex virus DNA induced angiogenesis on corneal surface, in collaboration with Prof. Barry Rouse at the University of Tennessee, we have investigated both local and systemic delivery of siRNA for VEGF pathway genes formulated with polymer nanoparticle carriers. Using an intra-subjunctival injection delivery, the siRNA was able to knockdown expression of the pro-angiogenesis genes and significantly inhibit the ocular neo-vascularization. Using an intravenous administration of the siRNA packaged in a ligand-directed nanoparticle with the same murine model, we observed a similar anti-angiogenesis effect (manuscript submitted). Clearly, delivery of siRNA through different routes with different formulations can be used to knockdown targeted genes sufficiently for gene function studies and potentially for therapeutic applications. For ocular neovascularization diseases, both local and systemic delivery approaches are very likely to be clinically feasible.

Pulmonary delivery of siRNA

Delivery of nucleic acid into the airway has been a very active area, in part due to intense efforts for development of therapy to treat Cystic Fibrosis patients using various viral and non-viral vector systems. Because of an overwhelming immune response to adenoviral vectors, efforts for pulmonary nucleic acid delivery have turned to non-viral carriers to draw upon their generally reduced toxicity and immune response. In comparison with cationic lipoplexes, natural lung surfactant products, e.g. Survanta and Infasurf, have been studied and found to provide improved delivery efficiency without sign of toxicity. There is a line of evidence that siRNA can effectively knockdown viral proteins of a group of RNA viruses, e.g. influenza (Ge et al., 2003) and SARS coronavirus, resulting in significant effects of anti-viral infection in various mammalian cell systems. To evaluate these *in vitro* proven siRNAs *in vivo*, we have developed a pulmonary delivery system for siRNA administration using a surfactant formulation, resulting in efficient siRNA delivery as measured by knockdown of a co-delivered reporter gene expression plasmid (data not shown). In murine models, we have relied upon an oral-tracheal delivery method for siRNA duplex and other nucleic acid for effective gene expression manipulations in the lung. The formulations are applicable using this method of delivery include surfactant, lipid and polymer nanoparticles, and viral vectors. Nasal delivery and other means of airway instillations can also be applied to achieve effective siRNA delivery to the specific tissue of interest. These pulmonary delivery approaches are very suitable for using siRNA, including inhibition of viral infection and replication in the airway, such as treatment of SARS.

Delivery of siRNA into skin, joint and muscle

Skin, joints and muscle are relatively accessible for local administration methods. Effective topical administration of siRNA to the skin depends upon use of established nucleic acid delivery methods to transport the siRNA across the stratus corneium to the active dermal tissues or into hair follicles. Direct injection of siRNA molecules formulated with cationic lipids or polymers can be used to achieve local delivery efficiency. The key to achieving therapeutic efficacy using designed siRNA agent-sequences to any of gene targets implicated in rheumatic diseases, musculoskeletal diseases or skin diseases is the intracellular delivery of siRNA agents at the diseased site, and more specifically the targeted cellular components of the musculoskeletal system and skin. We have effectively delivered siRNA duplexes into mouse skin and muscle with local delivery of siRNA duplex followed by electroporation enhancement (data not shown).

Delivering siRNA as a functional genomic tool

The traditional approach to identifying genes involved in specific biological processes begins with determination of which genes are involved or correlate with the process, i.e. which genes are either up or down regulated in the cells and tissues of the biological system of interest. When applied to pathological processes, this approach generates pools of candidate gene targets for therapeutic intervention. However, it lacks the power to validate these candidates as drug targets since correlation is insufficient for causation (whether the target is a cause of the diseases or a result), or more importantly for therapeutic intervention, the key is control associated with efficacy. Consequently, a tool that can selectively down regulate individual genes within the cells and tissues of a pathological process is widely recognized as a valuable means for understanding gene function, especially from the perspective of facilitating discovery of the protein targets for drugs to inhibit. To this end many years have gone into use of antisense and ribozyme methods for obtaining such selective gene inhibition. Since siRNA is proving to be a more potent and robust means for this important objective it is being rapidly adopted as a gene function tool.

Target identification and validation

One of the key hurdles for drug target discovery or validation directly in animal disease models has been a lack of effective *in vivo* nucleic acid delivery methods. While antisense activity in animals has been limited in general, intravenous administration of aqueous formulations has been used to inhibit liver gene function. Using siRNA in animals (Lewis et al., 2002; McCaffrey et al., 2002; Sen et al., 2003, and our unpublished observations) may provide substantial further improvements for gene inhibition. A key element of the challenge is often a large amplitude of inhibition (or over-expression) of the gene for a significant effect on the phenotype to be observed. Requirements for large phenotypic effects appear to be a bigger challenge for therapeutics than for gene function studies, particularly for several classes of genes and proteins.

In cancer, the tumorigenesis process is thought to be the result of abnormal over-expression of oncogenes, growth factors, and mutant tumor suppressors even though under-expression of other proteins also plays critical roles. Efforts to identify and validate tumorigenic targets have been focused mainly on those targets over-expressed in the tumor tissues and promoting tumorigenesis as a means to enable development of small molecule and antibody anti-cancer drugs acting through an inhibitor mechanism. Studies designed to reveal whether a gene target plays a tumorigenic role use siRNA duplexes specifically targeting its mRNA sequence to knockdown its expression and observe whether the effect on pathology is a direct and specific effect. We have developed a unique target identification approach (Lu et al., 2003), named Efficacy-First[TM] discovery. This method utilizes efficient delivery of nucleic acid into the xenograft tumor model to induce phenotypic effects on tumor growth rate and then identification of genes significantly up or down regulated, those correlating with induced tumor growth behaviors. Using this method, we have selected a pool of gene targets remarkably over-expressed in tumors having accelerated growth after treatment with bFGF expression plasmid. To differentiate which of these gene targets are playing disease-controlling roles for tumor growth, we designed and administered siRNAs specifically targeting their mRNA sequences by intratumoral delivery of the siRNA duplexes. By measuring tumor growth rates following repeated administrations of siRNA duplexes, inhibition of a few of the candidate targets induces efficacy, e.g. those genes have been validated as inhibitor drug targets. This validation process based on tumor growth inhibition efficacy by siRNA *in vivo* delivery with clinically relevant xenograft tumor models has been named Disease-Control[TM] Validation (Lu et al., 2003).

The power of the method is derived from data sets derived from induced disease dynamics, the key for therapeutic effects. Content identified for gene and proteins associated with efficacy represents the best information needed to identify candidate drug targets. In other words, the method generates gene and protein data sets from animal models enriched for answers to the question, "what genes and proteins are in pathways that control the pathology?" In order for therapeutic intervention to be successful, it must control the pathological process or at least the symptoms that result. To fulfill that search criteria, the data set needs to contain, if not be enriched in, data that relates to controlling pathological processes including inhibition of undesirable processes and enhancement of desirable processes. To be effective, this method requires high efficiency of *in vivo* delivery applicable for the disease tissue of interest. Importantly, this approach offers an ability to gain insight into the genes and proteins associated with the later stages of pathology since it can be applied to established disease tissues.

A new way to characterize tumorigenic targets

The emergence of siRNA-selective inhibition of gene expression has changed the landscape of cancer research on tumorigenic genes. This powerful inhibitory agent is able to down regulate target gene expression in a sequence-specific manner in various biological systems, from cell to tissue and even to whole organs. There are dramatically increasing reports demonstrated that siRNA and small hairpin RNA (shRNA expressed by viral or non-viral vectors) are able to knockdown tumorigenic genes both *in vitro* and *in vivo* (Lu et al., 2003). The cancer-causing targets include genes involving in cell cycle regulation, growth factors activation, protein kinases signaling and DNA repair, etc.

Due to strong interest in the p53 tumor suppressor protein, numerous studies have been conducted using siRNA-mediated inhibitions. One interesting study has shown (Martinez et al., 2002) that utilizing synthetic siRNAs to suppress gene expression in mammalian cells was able to discriminate a single base difference in siRNAs between mutant and WT p53 in cells expressing both forms, resulting in the restoration of wild-type p53 protein function. This result indicates that siRNAs may be used to suppress expression of point-mutated genes and suggest the potential for selective and personalized antitumor therapy. Another study (Jiang and Milner, 2002, 2003) using siRNA to silence pathogenic viral genes, E6 and E7 of human papillomavirus type 16, resulted in selective degradation of E6 and E7 mRNA. This silencing was sustained for at least 4 days following a single dose of siRNA. E6 silencing induced accumulation of cellular p53 protein, versus E7 silencing induced apoptotic cell death. This was the first demonstration that siRNA can induce selective silencing of exogenous viral genes in mammalian cells and, importantly, that the process of siRNA interference does not interfere with the recovery of cellular regulatory systems previously inhibited by viral gene expression. Taken together, using siRNA to reveal tumorigenic properties of newly found targets and their relationship with p53 relevant pathways represents one of the most active research areas for siRNA as a tool to study tumor gene function. Many of these studies provided not only better understanding of the mechanism of action of the validated gene targets, but also potential means for siRNA cancer therapeutics. When bcr-abl oncogene causing chronic myeloid leukemia (CML) and bcr-abl-positive acute lymphoblastic leukemia (ALL) was targeted by chemically synthesized siRNA (Scherr et al., 2003), the mRNA was reduced up to 87% in bcr-abl-positive cell lines and in primary cells from CML patients. This mRNA reduction was specific for bcr-abl because c-abl and c-bcr mRNA levels remained unaffected. In addition, protein expression of BCR-ABL and of laminA/C was reduced by the specific siRNAs up to 80% in bcr-abl-positive and normal CD34(+) cells, respectively. These data indicated that siRNA could specifically and efficiently interfere with the expression of an oncogenic fusion gene in hematopoietic cells.

Growth factors are gene products that play important roles in the regulation of cell division and tissue proliferation. Activation and over-expression of various growth factors, e.g. EGF, FGF, PDGF, TGF, IGF and VEGF, etc., in tumor cells indicate very active tumor growth. Each growth factor has a specific cell-surface receptor. Binding of specific growth factors at the cell surface activates that growth factor receptor. The activated receptor, in turn, activates an intracellular protein, or substrate, typically part of an intracellular signaling pathway. Among the well-known cancer cell growth factors, VEGF and VEGF R2 represent two of the most widely recognized and highly validated targets. Rather than using siRNA in cell culture, we used siRNA-mediated down regulation of endogenous genes in clinically relevant animal models, human breast carcinoma cell produced xenograft tumor models. Marked tumor growth inhibition was observed following repeated delivery of the siRNAs specific to hVEGF and mVEGFR2, accompanied by knockdown of expression of the growth factor at both mRNA and protein levels. These results indicate the feasibility of using siRNA delivery in animal tumor models for drug target validation according to its ability to achieve efficacy. This is the first evidence of siRNA-mediated anti-tumorigenesis in the xenograft model (Lu et al., 2002, and unpublished observations). Using siRNA to down regulate epidermal growth factor receptor (EGFR or erbB1), a tyrosine kinase specific for the epidermal growth factor (EGF), one study (Nagy et al., 2003) showed that endogenous erbB1 can be specifically and extensively suppressed in A431 human epidermoid carcinoma cells. As a consequence, EGF-induced tyrosine phosphorylation was inhibited and cell proliferation was reduced due to induction of apoptosis. Selective inhibition of the expression of the fusion protein was achieved with a siRNA specific for the EGFP mRNA, whereas the erbB1-specific siRNA inhibited the expression of both molecules. This result indicated that siRNA-mediated inhibition of erbB1 and other erbB tyrosine kinases might constitute a useful therapeutic approach in the treatment of human cancer.

These examples have demonstrated the power of siRNA as a tool for functional genomics. The inhibitory effect of siRNA is not only a sequence specific effect, but also sustainable and obtained with relatively few side effects (low noise level). The output of these studies provides biological function of the target gene and suggests a therapeutic potential of the specific siRNA agent used in the study.

Delivering siRNA as therapeutics

The specificity and potency of siRNA duplexes suggest that they may be useful therapeutic agents. However, while development of siRNA as a therapeutic agent faces a number of challenges, the two most critical hurdles are maintaining its stability *in vivo* and delivery to disease tissue and cells. To increase biological stability, medicinal chemistry originally developed for oligonucleotides and small-molecule conjugates are being studied to improve delivery of siRNA drugs. However, use of chemical modification alone to reduce nuclease digestion doesn't address requirements for better pharmacokinetics and tissue distribution, in large part a result of fast urinary excretion. With conjugation of lipophilic residues to increase serum protein binding, improved pharmacokinetics of the oligonucleotide, alteration of its biodistribution and reduced urinary excretion were observed. Ultimately, though, such chemical modification must also address requirements for entering targeted cells, overcoming endosomal and other intracellular barriers, and retain activity with the cellular RNAi machinery.

It can be predicated that chemically manipulating the siRNA backbone or modifying siRNA residues will increase the stability and biodistribution while maintaining at least some activity. However, this approach may also bring in some unpredictable changes and increase side effects, e.g., the specificity to the target sequence, the pharmacokinetics, the biodistribution and the cellular locations, etc. We are taking a different approach by using delivery formulations of siRNA with minimum chemical modification. The fundamental thinking behind this approach is that RNA interference is a natural process involving a complicated cellular mechanism where siRNA is the predominant intermediate playing a sequence-specific silencing function. Therefore, preserving the biochemical authenticity of siRNA and improving its *in vivo* delivery efficiency will have the best chance for success in therapeutic application. To this end, we also believe

Three Stage Activity Nanoparticles

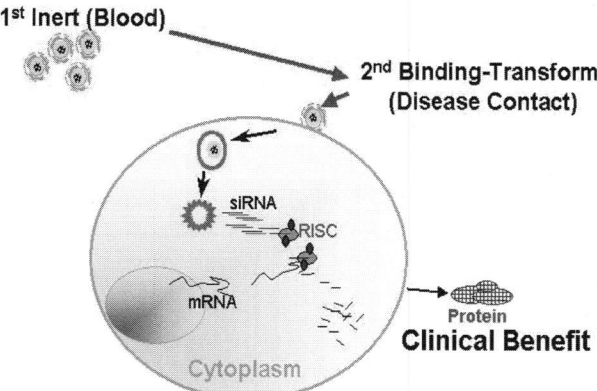

Figure 22.2. Ideal system for systemic delivery of siRNA. (See color section.)

that many routes and carries are suitable for siRNA delivery either locally or systemically. Although local and topical deliveries of siRNA may fit well for certain disease applications, the systemic delivery of siRNA will have much broader therapeutic applications. The ideal system for siRNA systemic delivery should be a nanoparticle with three stages of action (Figure 22.2). This three-stage system should be able to first protect the siRNA duplex from degradation in body fluid such as blood and at the same time avoid aggregation and non-specific binding of the nanoparticles. The system should also have targeting capability to reach disease tissue specifically. When the particle binds and enters the targeted cells, the siRNA content should be released for action. Since siRNA specifically inhibits expression of the targeted disease-causing gene and protein, the clinical benefit will be achieved. As described earlier, we have developed a layered nanoplex system that combines tissue-targeted nanoparticles giving cytoplasmic delivery with highly potent siRNA active in the cytoplasm. This nanoparticle system appears promising for therapeutic application of siRNA for metastatic cancer and many other angiogenesis related diseases.

Although the hydrodynamic delivery of siRNA duplexes into mouse liver was quite efficient and has facilitated many proof-of-concept studies for the potential therapeutic uses of siRNA, it is not clinically feasible in human studies. HIV, HBV and HCV are the infectious viruses in particular being first targeted using specific siRNAs in either chemically synthesized or vector expressed formats with the hydrodynamic deliveries (Lewis et al., 2002; McCaffrey et al., 2002). Using the same approach, one study has shown that siRNA-mediated knockdown of cyclin E resulted in significant inhibition of hepatocellular carcinoma tumor on nude mice. As we expected, using RNAi for therapy is not limited to directly silencing pathogenic genes or disease-causing mutant genes. As disease mechanisms become increasingly clear, its application can be expanded to silence genes involved in known pathogenic pathways. For example, an obvious target for treatment of Alzheimer's disease (AD) is the β-site APP-cleaving enzyme BACE, which is required for the production of Aβ peptide and is present at elevated levels in the cortex of people with AD.

Considerable evidence shows that many types of human diseases result from over-expression of multiple disease causing genes. Thus the potent and specific siRNA inhibitory agent holds tremendous potential for treatment of those diseases. Additionally, siRNA provides a unique opportunity as a therapeutic agent: using a combination of siRNA duplexes (siRNA oligo cocktail) targeting multiple genes involved in the disease pathology to achieve much better therapeutic efficacy. This approach is based on two important facts: although inhibitory siRNA duplexes are sequence specific, all use an identical chemistry (dsRNA oligonucleotides); on the other hand, many human diseases are the result of over-expression of multiple endogenous and exogenous disease causing genes. Using an siRNA oligo cocktail targeting multiple disease causing genes represents an advantageous therapeutic approach with a synergistic effect. Nevertheless, even as a single agent siRNA has tremendous therapeutic potential that will be realized when clinical feasible deliveries are developed.

REFERENCES

Devroe, E. and Silver, P. A. (2002). Retrovirus-delivered siRNA. *BMC Biotechnology*, **28**, 15.

Ge, Q., McManus, M. T., Nguyen, T., Shen, C. H., Sharp, P. A., Eisen, H. N. and Chen, J. (2003). RNA interference of influenza virus production by directly targeting mRNA for degradation and indirectly inhibiting all viral RNA transcription. *Proceedings of the National Academy of Sciences USA*, **100**, 2718–23.

Hood, J. D., Bednarski, M., Frausto, R., Guccione, S., Reisfeld, R. A., Xiang, R. and Cheresh, D. A. (2002). Tumor regression by targeted gene delivery to the neovasculature. *Science*, **296**, 2404–2407.

Jiang, M. and Milner, J. (2002). Selective silencing of viral gene expression in HPV-positive human cervical carcinoma cells treated with siRNA, a primer of RNA interference. *Oncogene*, **21**, 6041–6048.

Jiang., M. and Milner, J. (2003). Bcl-2 constitutively suppresses P53-dependent apoptosis in colorectal cancer cells. *Genes & Development*, **17**, 832–837.

Lewis, D. L., Hagstrom, J. E., Loomis, A. G., Wolff, J. A. and Herweijer, H. (2002). Efficient delivery of siRNA for inhibition of gene expression in postnatal mice. *Nature Genetics*, **32**, 107–108.

Lu, Y. P., Xie, Y. F. and Woodle, M. C. (2003). siRNA-mediated antitumorgenesis for drug target validation and therapeutics. *Current Opinion in Molecular Therapeutics*, **5**, 225–234.

Lu, P., Xie, F., Scaria, P. and Woodle, M. (2003). From correlation to causation to control: Utilizing preclinical disease models to improve cancer target discovery. *Preclinica*, **1**, 31–42.

Lu, P., Frank, X., Quinn, T., Jun, X., Scaria, P., Zhou, Q. and Woodle, M. (2002). Tumor inhibition by RNAi-mediated VEGF and VEGFR2 down regulation in xenograft models. *Cancer Gene Therapy*, **10**, Supplement 1, 011.

Martinez, L. A., Naguibneva, I., Lehrmann, H., Vervisch, A., Tchenio, T., Lozano, G. and Harel-Bellan, A. (2002). Synthetic small inhibiting RNAs: Efficient tools to inactivate oncogenic mutations and restore P53 pathways. *Proceedings of the National Academy of Sciences USA*, **99**, 14849–14854.

McCaffrey, A. P., Meuse, L., Pham, T. T., Conklin, D. S., Hannon, G. J. and Kay, M. A. (2002). RNA interference in adult mice. *Nature*, **418**, 6893, 38–39.

McManus, M. T. and Sharp, P. A. (2002). Gene silencing in mammals by small interfering RNAs. *Nature Medicine*, **3**, 737–747.

Nagy, P., Arndt-Jovin, D. J. and Jovin, T. M. (2003). Small interfering RNAs suppress the expression of endogenous and GFP-fused epidermal growth factor receptor (erbB1) and Induce apoptosis in erbB1-overexpressing cells. *Experimental Cell Research*, **285**, 39–49.

Ogris, M., Walker, G., Blessing, T., Kircheis, R., Wolschek, M. and Wagner, E. (2003). Tumor-targeted gene therapy: Strategies for preparation of ligand-polyethylene glycol-polyethylenimine/DNA complexes. *Journal of Controlled Release*, **91**, 173–181.

Reich, S. J., Fosnot, J., Kuroki, A., Tang, W., Yang, X., Maguire A. M., Bennett, J. and Tolentino, M. J. (2003). Small interfering RNA (siRNA) targeting VEGF effectively inhibits ocular neovascularization in a mouse model. *Molecular Vision*, **9**, 210–216.

Ren, T., Zhang, G. and Liu, D. (2001). Synthesis of galactosyl compounds for targeted gene delivery. *Bioorganic & Medicinal Chemistry*, **9**, 2969–2978.

Scherr, M., Battmer, K., Winkler, T., Heidenreich, O., Ganser, A. and Eder, M. (2003). Specific inhibition of bcr-abl gene expression by small interfering RNA. *Blood*, **101**, 1566–1569.

Sen, A., Steele, R., Ghosh, A. K., Basu, A., Ray, R. and Ray, R. B. (2003). Inhibition of hepatitis C virus protein expression by RNA interference. *Virus Research*, **96**(1–2), 27–35.

Scherr, M., Battmer, K., Winkler, T., Heidenreich, O., Ganser, A. and Eder, M. (2003). Specific inhibition of bcr-abl gene expression by small interfering RNA. *Blood*, **101**, 1566–1569.

Song, E., Lee., S. K., Wang, J., Ince, N., Ouyang, N., Min, J., Chen, J., Shankar, P. and Lieberman, J. (2003).RNA interference targeting Fas protects mice from fulminant hepatitism. *Nature Medicine*, **9**, 347–351.

Sørensen, D. R., Leirdal, M. and Sioud, M. (2003). Gene silencing by systemic delivery of synthetic siRNAs in adult mice. *Journal of Molecular Biology*, **4**, 761–766.

Tiscornia, G., Singer, O., Ikawa, M. and Verma, I. M. (2003). A general method for gene knockdown in mice by using lentiviral vectors expressing small interfering RNA. *Proceedings of the National Academy of Sciences USA*, **100**, 1844–1848.

Woodle, M. C., Scaria, P., Ganesh, S., Subramanian, K., Titmas, R., Cheng, C., Pan, Y., Weng, K., Gu, C. and Torkelson, S. (2001). Sterically stabilized polyplex: ligand-mediated activity. *Journal of Controlled Release*, **74**, 309–311.

Xia, H., Mao, Q., Paulson., H. L. and Davidson, B. L. (2002). Sirna-mediated gene silencing *in vitro* and *in vivo*. *Nature Biotechnology*, **20**, 1006–1010.

Zender, L., Hütker, S., Liedtke, C., Tillmann, H. L., Zender, S., Mundt, B., Waltemathe, M., Gösling, T., Flemming, P., Malek, N. P., Trautwein, C., Manns, M. P., Kühnel, F. and Kubicka, S. (2003).Caspase 8 small interfering RNA prevents acute liver failure in mice. *Proceedings of the National Academy of Sciences USA*, **100**, 7797–7802.

23 The role of RNA interference in drug target validation: Application to Hepatitis C

Antje Ostareck-Lederer, Sandra Clauder-Münster, Rolf Thermann,
Maria Polycarpou-Schwarz, Marc Gentzel, Matthias Wilm, and Joe D. Lewis

Introduction

The Hepatitis C Virus (HCV) is the main causative agent of non-A, non-B hepatitis in humans and a major cause of mortality and morbidity in the world. At this time there is no effective vaccination or cure for Hepatitis C, an infection affecting at least 170 million people worldwide. This slow-processing disease is transmitted through contaminated blood transfusions and needle sharing, and frequently leads to liver cirrhosis and cancer (Cohen, 1999).

The genetic pattern associated with HCV consists of a positive-sense-strand RNA genome of ~9600 nucleotides (nts) that contains a single large open-reading frame. The structure and organization of the HCV genome is similar to those of members of the pestivirus and flavivirus genera of the family *Flaviviridae* (Takamizawa et al., 1991). HCV is now classified as a distinct genus of this family, with at least six major genotypes that differ from each other in their nucleotide sequence by up to 35%. The 341-nts long 5′untranslated region (5′UTR) and the adjacent core protein coding sequence are highly conserved (Simmonds, 1995; Smith et al., 1995). The HCV RNA 5′UTR contains a highly structured internal ribosome entry site (IRES) that mediates initiation of translation of the viral polyprotein by a 5′ cap-independent mechanism that is unprecedented in eukaryotes [(Jackson and Kaminski, 1995; Reynolds et al., 1995) (Figure 23.1)]. The first step in translation initiation is the assembly of a 43S preinitiation complex consisting of the eukaryotic initiation factors (eIF) 3, eIF 2, GTP, the initiator tRNA and a 40S ribosomal subunit. The HCV-IRES recruits this 43S complex and directs its precise attachment at the initiation codon to form a 48S stable complex with the RNA by internal initiation independent of the eIFs 4A, 4B or 4F, which are required for the initiation of eukaryotic mRNA translation (Figure 23.1A). The 48S complex is competent to complete the remaining steps to initiate protein synthesis. The assembly of 48S complexes is enhanced by, but does not require, eIF 3. In the final stage of initiation, eIF 3 is required for the subsequent joining of the 60S subunit to the 48S complex to form an 80S ribosome capable

A

HCV dependent translation initiation

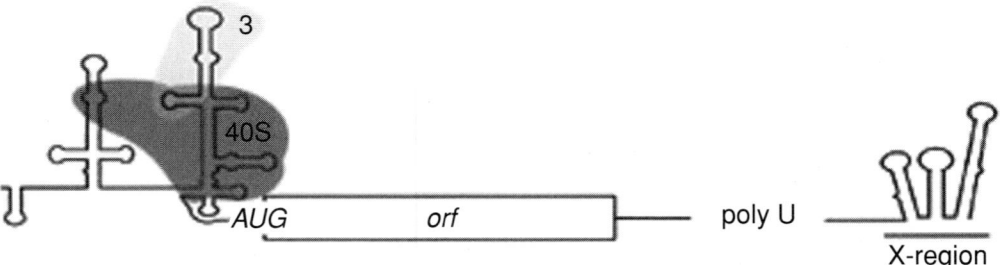

cap mediated translation initiation

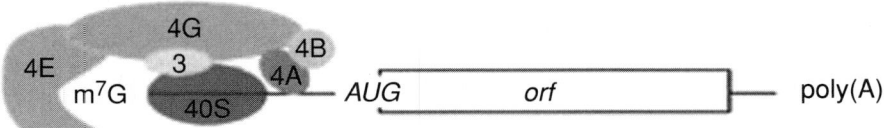

Figure 23.1. Comparison of HCV-IRES- and 5′ cap- dependent translation A) The HCV-IRES driven internal translation initiation, starts with the direct recruitment of the 40S ribosomal subunit and the eukaryotic initiation factor eIF 3. In cap-dependent translation initiation, the recruitment of the 40S ribosomal subunit requires the recognition of the m^7GpppG cap at the mRNA 5′end, by the initiation factor eIF4E at which the eIF 4F complex, consisting of the initiation factors eIF 4E, eIF 4G and eIF 4A, is assembled. The recruitment of the 40S ribosomal subunit takes place via eIF 3 which binds to eIF4G. (See color section.)

of polyprotein synthesis (Pestova et al., 1998). The factor requirement for the initiation process on the HCV-IRES is simpler than the 5′ cap dependent translation of cellular mRNAs and allows the viral start-codon to be recognized in a 5′ cap- and 3′ poly(A) independent way. Therefore HCV polyprotein synthesis is initiated by a mechanism that is fundamentally distinct from most host-cell mRNAs. Other cellular proteins, like the autoantigen La and the pyrimidine tract binding protein (PTB), which are not required for the cap-dependent initiation of translation, have been found to interact with the highly structured HCV-IRES (Bartenschlager and Lohmann, 2000).

The unique mechanism of HCV RNA translation initiation appears to be possible only because the IRES itself plays an important role, first, in recruiting translation components to form a 43S pre-initiation complex and second, in orienting this complex so that nucleotides flanking the initiation codon can enter the mRNA-binding cleft of the 40S subunit and the initiation codon can establish a productive interaction with the anticodon of initiator tRNA in the ribosomal "P" site (Hellen and Pestova, 1999). Mutational studies combined with biochemical probing analysis revealed that the HCV-IRES consists of three major structural

B

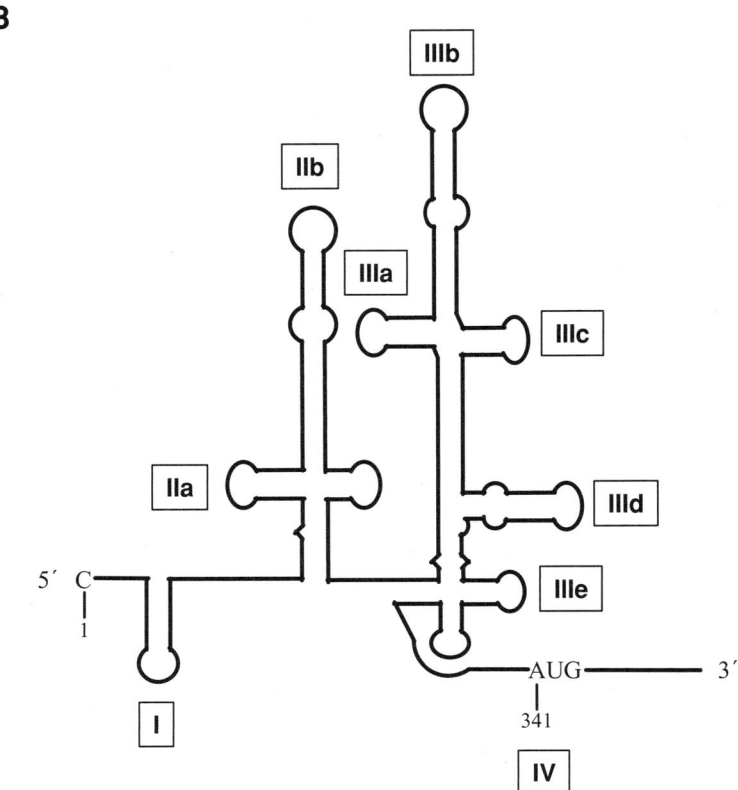

Figure 23.1. (*cont.*) B) The depicted model of the HCV-IRES secondary structure shows the position and numbering of the different loops identified by biochemical probing analysis as well as the translation initiation codon (Honda et al., 1996; Honda et al., 1999).

domains (II, III, and IV) radiating from a complex pseudoknot [(Brown et al., 1992; Honda et al., 1996; Honda et al., 1999) (Figure 23.1B)]. Reconstitution of the 40S/HCV-IRES at ~20-Å resolution produced by cryo-electron microscopy reveals a single elongated shape. Domains IIIa-IIIc and II extend in opposite directions from a small central domain that includes stem loops and junctions IIIe-IIIf (Spahn et al., 2001). Enzymatic and chemical footprinting and domain deletion experiments have identified the ribosome- and eIF 3-binding sites within the HCV-IRES. The ribosomal 40S subunit binds to subdomains IIb, IIIa, IIIc, IIId, IIIe, IIIf and IV and junction IIIabc (Pestova et al., 1998; Kolupaeva et al., 2000; Lukavsky et al., 2000; Kieft et al., 2001), while eIF 3 binding is localized to subdomains IIIa and IIIb and junction IIIabc [(Buratti et al., 1998; Pestova et al., 1998; Sizova et al., 1998; Odreman-Macchioli et al., 2000; Kieft et al., 2001) (see Figure 23.1A)]. The structure of the HCV-IRES is different from the structure of other viral and cellular IRESs (Hellen and Sarnow, 2001), making this RNA motif and its complexes with 40S and eIF 3 attractive targets for the discovery of new antiviral agents because of the potential for selectivity.

To identify specific HCV-IRES interacting proteins required for the initiation of HCV RNA translation as candidates for drug target development, we made use

A

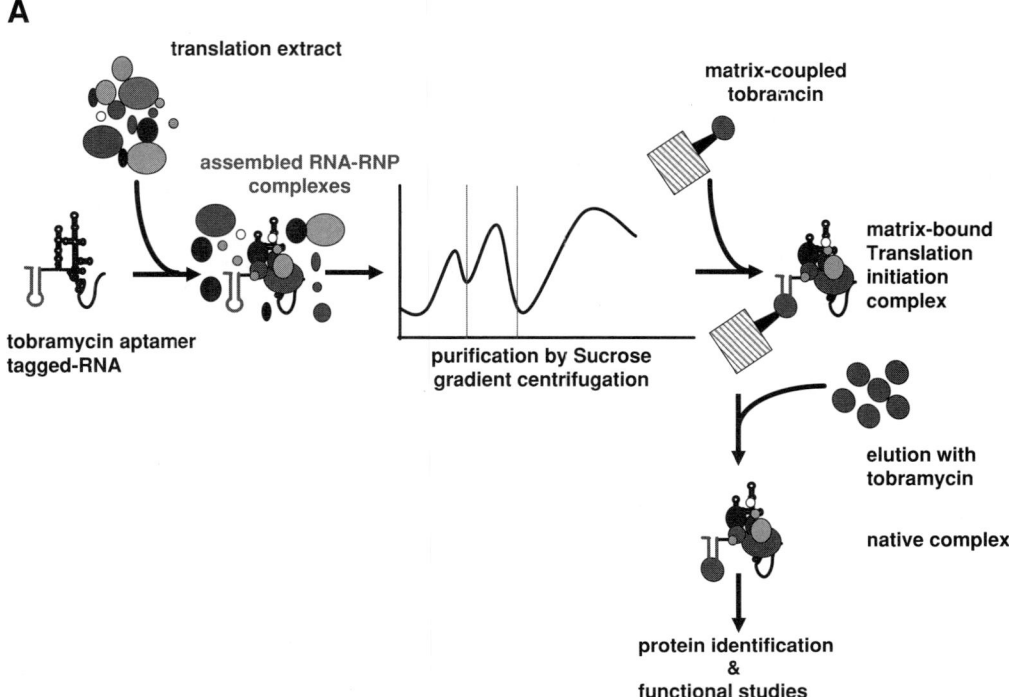

Figure 23.2. Tobramycin-tag affinity chromatography of translation initiation complex A) The HCV RNA 5′UTR bearing the HCV-IRES fused to the tobramycin aptamer was incubated with cytoplasmic HeLa cell extract under physiological conditions. Protein complexes assembled on the 5′UTR sequence were loaded on a 5–25% sucrose gradient and distributed according to their molecular weight by ultracentrifugation. The resulting 48S ribosomal peak fractions were subsequently incubated with the sepharose-coupled aminoglycoside tobramycin. Following extensive washing of the affinity matrix, proteins were eluted from the complex assembled at the HCV-IRES. Proteins eluted from the HCV-IRES were isolated from a silver stained polyacrylamide gel and identified by mass spectrometry (LC-MS) according to representative peptides. (See color section.)

of our Riboproteomics™ approach. This technique consists of three consecutive procedures to identify and validate interacting factors:

(1) The highly specific purification of large native ribonuclear protein complexes (RNPs) assembled at the HCV-IRES under gentle conditions employing the tobramycin-tag affinity purification assay (Hartmuth et al., 2002).

(2) The identification of the protein components by polyacrylamide electrophoresis and mass spectrometry followed by the analysis of identified peptides using RIBObase™

(3) The characterization and validation of the identified protein:HCV-IRES interactions *in vivo* by RNA interference [(RNAi) (Elbashir et al., 2000] in combination with functional cell based expression and translation assays.

Identification of HCV-IRES interacting proteins

To purify protein complexes under native conditions at a large scale, we made use of the tobramycin-tag affinity technique (Figure 23.2A), which has proved

B

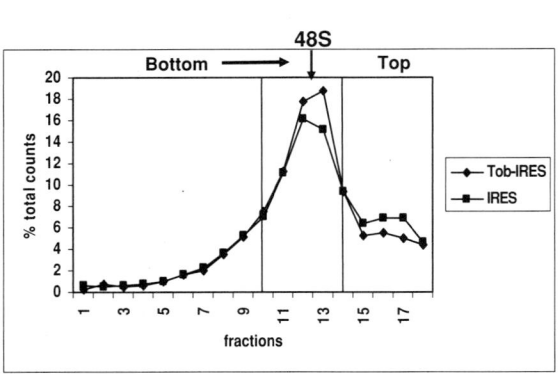

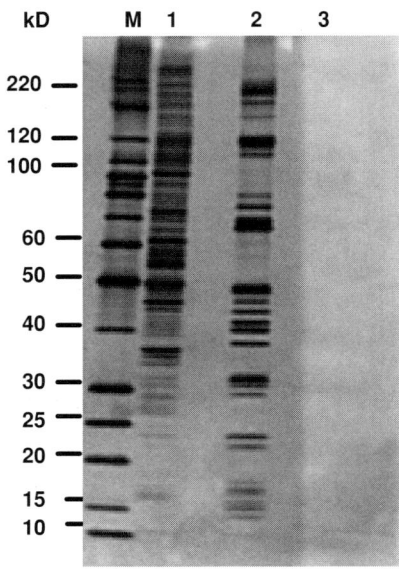

Figure 23.2. (*cont.*) B) The purification of the 48S translation pre-initiation complex formed on the HCV-IRES in a 5–25% sucrose gradient is shown at the left panel. The tobramycin-tag does not affect the distribution of protein complexes in the sucrose gradient. The typical protein pattern from the sucrose-gradient 48S peak fractions (10–14) is presented at the right panel: lane M: marker proteins, lane 1: HeLa cell lysate used as input material for the translation initiation reaction, lane 2: protein pattern obtained from the 48S peak fractions following tobramycin-tag affinity purification of the protein complexe assembled in the HeLa cell lysate at the HCV-IRES under translation initiation conditions, lane 3: specificity control using the non-tagged HCV-IRES in the HeLa cell lysate translation initiation reaction. Protein bands isolated from a silver stained gel were analysed by mass spectrometry.

to be highly efficient and specific in the analysis of the protein composition of pre-spliceosomal complexes (Hartmuth et al., 2002; Jiang and Patel, 1998). For this purpose, a RNA aptamer (Tob), which specifically interacts with the aminoglycoside tobramycin, was cloned in *cis* to the HCV-IRES RNA sequence (Tob-HCV-IRES) (Thermann et al., in preparation). The Tob-HCV-IRES was incubated with cytoplasmic HeLa cell extract (Bergamini et al., 2000) under translation initiation conditions to form 48S IRES ribosomal complexes. The presence of the non-hydrolysable GTP analog GMP-PNP inhibits the further processing of the translation initiation reaction and the joining of the 60S ribosomal subunit and results in an enrichment of 48S complexes. The resulting 48S complexes were purified by sucrose gradient sedimentation and the appropriate fractions were pooled (Figure 23.2B, left panel). In the next step, the complexes were affinity purified by binding to tobramycin-sepharose from the pooled sucrose gradient fractions. Protein fractions specifically eluted from the tobramycin affinity matrix were analyzed subsequently by SDS-PAGE and silver staining (Figure 23.2B, right panel). As shown on the gel, proteins ranging in molecular weight from <10 KDa to >150 kDa can be observed; significantly, the background level of binding is very low (Figure 23.2B, lanes 2 and 3). Mass spectrometry analysis (LC-Tandem MS) was performed to identify specifically eluted proteins from the silver stained gel (compare Figure 23.2B, right panel, lanes 1–3). The LC-Tandem MS revealed 44 proteins that could be identified in databases and these could be categorized into at least 3 groups: 40S ribosomal subunit, eIF3 subunit and other proteins. In summary, 23 of the 32 ribosomal 40S subunit proteins were found and no ribosomal 60S subunit proteins. The absence of the ribosomal 60S subunit proteins reinforces the conclusion that the purity of the 48S translation initiation complex is high. In addition to the ribosomal proteins all subunits of eIF3 and one protein associated with the eIF3 subunit were detected. Most importantly, 11 other known and unknown interacting proteins were found in the isolated 48S complexes, e.g. La and nucleolin (Thermann et al., in preparation). These proteins were prioritized for testing in cell based assays in combination with RNAi using Anadys' proprietary database, RIBObase.

Development of an approach combining RNAi and a HeLa cell-based assay to study specific protein requirements for HCV-IRES dependent translation

In order to analyze the requirement of the identified proteins for HCV-IRES dependent translation initiation, we performed RNAi experiments in combination with a HeLa cell-based translation assays. To establish the method, we have chosen the eukaryotic translation initiation factor eIF 4E, which is essential for the 5′ cap dependent initiation of translation of cellular mRNAs, but which would not be expected to be required for translation of proteins initiated at the HCV-IRES (Figure 23.1A). A second protein, PTB, which had been identified as an HCV-IRES interacting cellular protein (Bartenschlager and Lohmann, 2000), was used to validate cell viability when expression of a cellular protein that does assemble with the HCV-IRES is reduced by RNAi.

First we screened for small interfering RNA duplexes (siRNAs) effective in decreasing eIF 4E and PTB protein expression. Three (PTB) or two (eIF 4E) different siRNAs were screened by HeLa cell transfection. The siRNA that most efficiently decreased PTB and eIF 4E expression, as analyzed by Northern and Western blotting, was chosen for further functional analysis. Surprisingly, we found that the strong reduction of PTB and eIF 4E expression did not significantly affect cell viability and growth rate (Clauder-Münster et al., in preparation).

At the time point at which the expression of PTB and eIF4E was reduced most efficiently, we transfected *in vitro* transcribed reporter mRNAs coding for Luciferase bearing either the HCV-IRES in 5′ position and the authentic HCV RNA 3′UTR or a 5′ cap structure and a 3′ poly (A_{98}) tail respectively (Figure 23.3A). As shown in Figure 23.3B, duplicates in a Western blot, PTB (lanes 1–8) and eIF 4E (lanes 9–16) protein levels were significantly decreased by the transfected siRNAs (lanes 5–8 and 13–16) compared to control cells (lanes 1–4 and 9–12). At the 96-hour time point, the reporter RNAs were transfected and incubated for three hours, then LUC expression levels were determined (Figure 23.3C). Translation initiated at the 5′ cap (Figure 23.3C, left panel) was decreased only to a minor extent

A

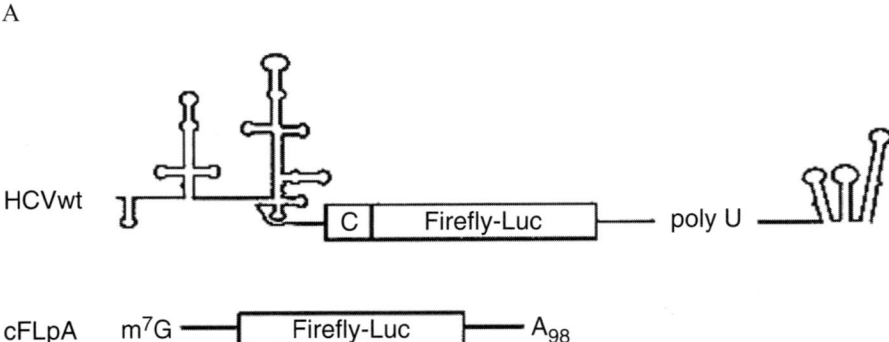

Figure 23.3. RNAi against PTB and eIF 4E in HeLa cells. A) The reporter RNA constructs used in the HeLa cell-based translation assay are shown: HCV$_{wt}$ represents the firefly Luciferase (LUC) open reading frame (orf) bearing at the 5′end the HCV-IRES and the authentic 3′UTR of the HCV RNA. As a control for translation initiation complex formation on a 5′UTR characteristic for a cellular mRNA, the cFLpA transcript, consisting of the LUC-orf, a 5′UTR starting with an m^{7}G cap structure and a 3′UTR is followed by a poly tail (A$_{98}$) was used in the HeLa cell translation assay.

when PTB expression was significantly reduced (Figure 23.3B, lanes 5 and 6). In contrast, the strong reduction of eIF 4E (Figure 23.3B, lanes 13 and 14) caused a significantly lower LUC expression from the 5′ cap LUC mRNA compared to control cells (Figure 23.3C, left panel). HCV-IRES mediated LUC translation (Figure 23.3C, right panel) was significantly enhanced when eIF4E expression was reduced (Figure 23.3B, lanes 15 and 16), whereas the reduction of PTB protein (Figure 23.3B, lanes 7 and 8) had no effect on HCV-IRES mediated translation.

The results obtained for PTB and eIF 4E show that the expression of both proteins can be reduced significantly in HeLa cells using RNAi. Using a cell-based translation assay we could show that 5′ cap dependent translation is reduced to 30% of control levels when eIF 4E expression is decreased, whereas under those conditions HCV-IRES mediated translation is enhanced three-fold. This effect is likely due to an increased availability of ribosomal subunits and other components required for translation initiation at the HCV-IRES. The reduction of PTB expression had no effect on translation initiated at a 5′ cap structure or the HCV-IRES, indicating that there is no direct requirement for both types of translation initiation. This is consistent with data obtained in rabbit reticulocyte lysate, where the activity of the HCV-IRES is unaffected by the depletion of PTB (Kaminski et al., 1995). However there are low remaining levels of PTB and it remains a possibility that PTB cannot be made limiting for translation of HCV-IRES using these siRNAs.

Figure 23.3. (*cont.*) B) HeLa cells were transfected with siRNAs against PTB (left panel, lanes 5–8) or eIF 4E (right panel, lanes 13–16). The RNAi effect on protein expression compared to control cells not transfected with siRNAs (lanes 1–4 and 9–12) was analysed by Western blotting using antibodies against PTB and eIF4E, as well as GAPDH to monitor the specificity of RNA interference. C) To quantify the influence of reduced PTB and eIF4E expression on the 5′ cap-dependent (left panel) and HCV-IRES-mediated translation (right panel), LUC expression from the constructs presented in Fig. 3A was analysed in HeLa cells.

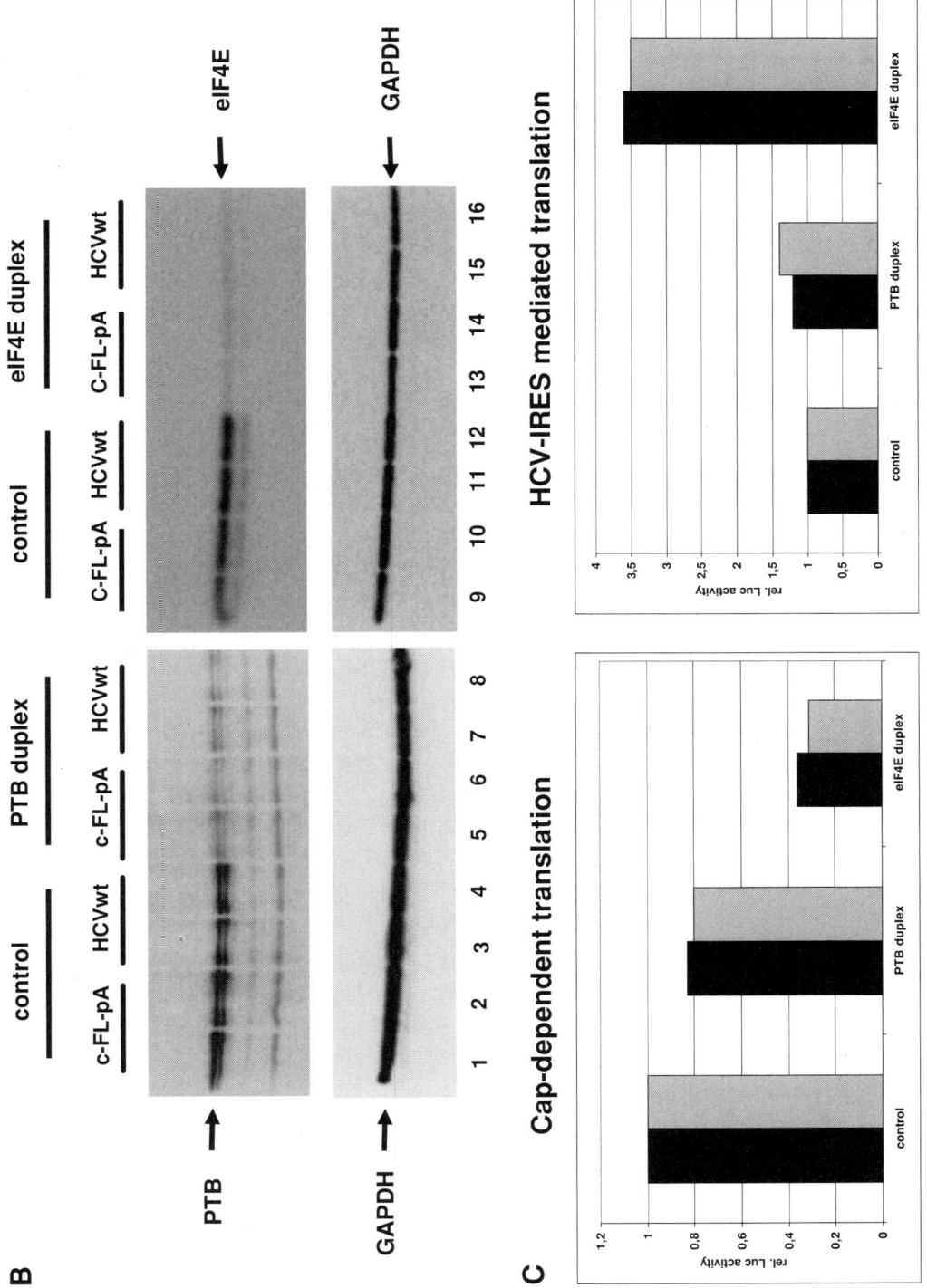

325

A

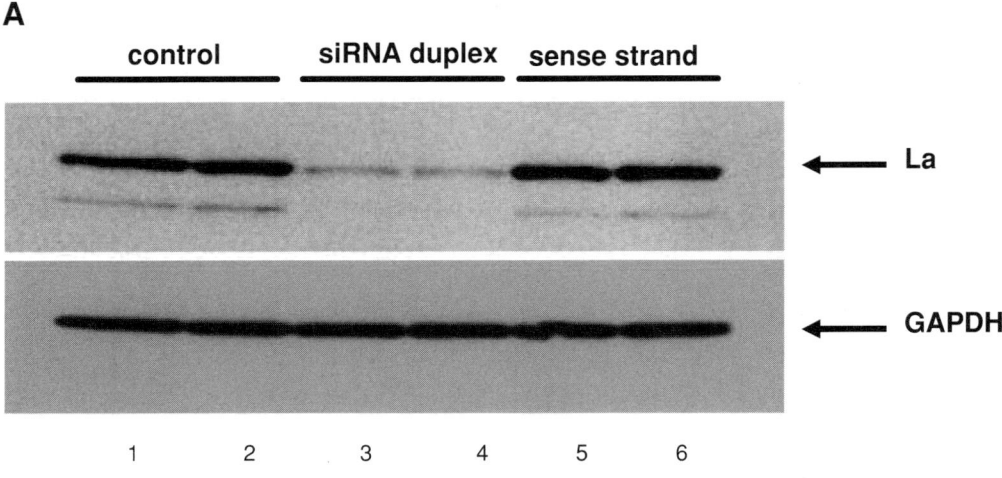

B

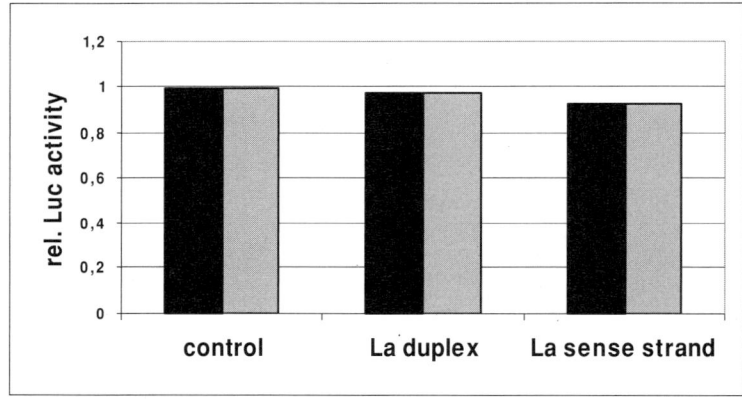

Figure 23.4. RNAi against La A) HeLa cells were transfected with siRNAs against La (lanes 3 and 4) or sense strand as control (lanes 5 and 6). The RNAi effect on protein expression compared to control cells not transfected with siRNAs (lanes 1 and 2) was analysed by Western blotting using an antibody against La, as well as GAPDH to monitor the specificity of RNA interference. B) To quantify the influence of reduced La expression on HCV-IRES-mediated translation LUC expression from the construct presented in Fig. 3A was analysed in HeLa cells.

RNAi-based functional characterization of HCV-IRES interacting proteins identified in the 48S IRES complexes: La and p40

One of the proteins identified in purified 48S IRES complexes was the autoantigen protein La. La protein has been described previously as an HCV-IRES interacting

Figure 23.5. RNAi against p40 A) HeLa cells were transfected with siRNAs against p40 (lane 2) or the sense strand as a control (lane 3). The RNAi effect on RNA compared to control cells not transfected with siRNAs (lane 1) was analysed by Northern blotting using DNA fragments against p40, as well as GAPDH to monitor the specificity of RNA interference. B) To quantify the influence of reduced p40 expression on 5′ cap-dependent or HCV-IRES-mediated translation, LUC expression from the constructs presented in Fig. 3A was analyzed in HeLa cells.

A Northern Blot analysis

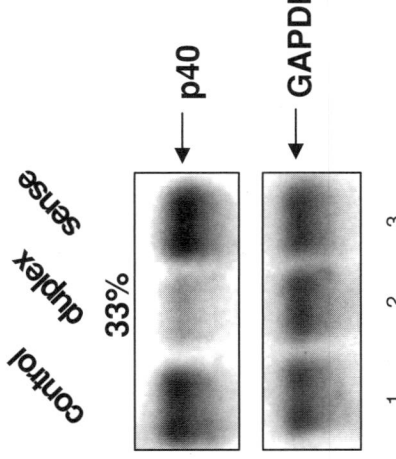

B

HCV-IRES mediated translation

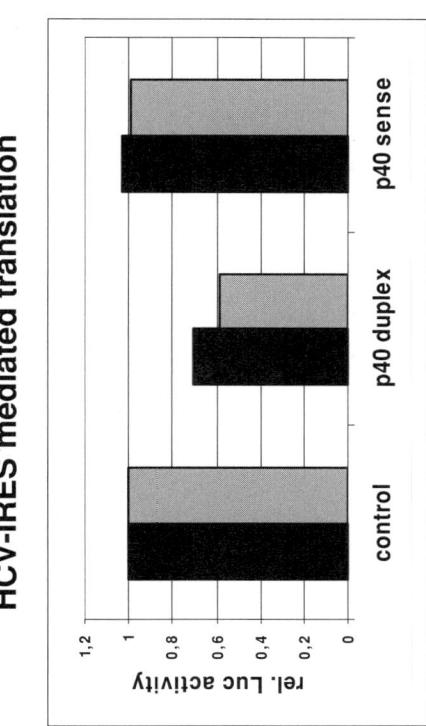

Cap-dependent translation

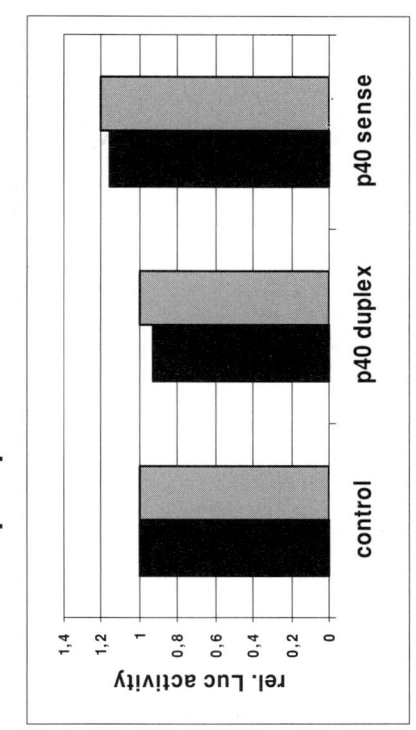

327

protein (Ali and Siddiqui, 1997). In addition to La, another purified protein (p40) was chosen for further functional characterization. In Figure 23.4A the effect of RNAi using a functional La mRNA-specific siRNA on La protein expression in HeLa cells is detected immunologically. The La protein level is significantly reduced when HeLa cells were transfected with the functional La siRNA (Figure 23.4A, lanes 3 and 4), compared to nontransfected cells (Figure 23.4A, lanes 1 and 2) or Hela cells transfected with the sense-strand siRNA as a control (Figure 23.4A, lanes 5 and 6). The reduced level of La protein does not cause a decrease in HCV-IRES–dependent Luciferase expression in the cell-based translation assay (Figure 23.4B), indicating that the binding of La protein to the HCV-IRES has no consequence for its function in translation initiation (Hellen and Sarnow, 2001).

Transfection of HeLa cells using a functional siRNA against p40 mRNA reduced the p40 mRNA level significantly (Figure 23.5A). Luciferase expression levels obtained at day two post-transfection show that LUC expression is reduced to 60% from the HCV-IRES LUC mRNA, compared to a slight reduction of LUC expression from the LUC mRNA bearing a 5′ cap (Figure 23.5B). These results suggest that HCV-IRES dependent translation is more dependent on the function of the cellular protein p40 than the 5′cap-dependent translation initiation. Further validation of this finding, i.e. mapping of the HCV-IRES interaction sites and complementation experiments need to be performed to qualify p40 as a potential target for HCV-directed drug development.

Conclusions

In order to identify *bona fide* host cell factors potentially involved in the formation of HCV-IRES-mediated translation initiation complexes, we developed a novel RNA-affinity purification assay based on the tobramycin aptamer tag affinity technique. Using a combination of sucrose gradient sedimentation and tobramycin affinity purification we were able to isolate highly enriched 48S complexes. We have analyzed the composition of the complex using mass spectrometry. As expected we isolated components of the 40S ribosomal subunit as well as eIF 3 and several associated proteins. Importantly, we have identified several proteins that have not been previously implicated in HCV-IRES mediated translation initiation and may provide entry points to developing small-molecule therapeutics. We have analyzed the role of three host cell proteins in the HCV-IRES-mediated translation *in vivo* that were identified using our Riboproteomics™ approach. By RNA interference we could significantly reduce the levels of La or PTB in HeLa cells without affecting the overall grow rate of the cells. At the levels we could observe the depletion of La and PTB had no influence on the HCV-IRES-mediated translation *in vivo*. This indicates that although PTB has been shown to bind efficiently to HCV genomic sequences it does not contribute to an increased translation efficiency under the conditions applied. In contrast to La and PTB, the *in vivo* depletion of protein p40 results in a reproducible significant reduction in translation mediated *via* the HCV-IRES. 5′ cap-dependent translation that represents the major pathway of protein expression from cellular mRNAs was affected only to a minor extent under the conditions used. Protein p40 is therefore a potential target for the identification of small molecule therapeutics that specifically affect HCV-IRES-mediated translation initiation. The other proteins identified are currently being analyzed to determine whether their *in vivo* depletion has a specific consequence for the HCV-IRES-mediated translation.

The next step: Currently no functional assays are available for protein p40 which has been validated as required for HCV-IRES mediated translation. However, even in the absence of a functional assay, we can take advantage of Anadys' proprietary affinity screening technology, ATLAS™, to identify small molecules that bind specifically to p40. Specificity of the small molecule/protein interaction can then be addressed *in vivo* using the HCV-IRES- and 5′ cap-dependent cell-based translation assay systems that we have developed and implemented in these studies.

RNAi combined with cell-based assays should also be a powerful tool to investigate *in vivo* drug targets where the EC50 of a compound can be determined in both the wild-type and RNAi-treated cells. If the targeted protein is required for HCV-IRES-mediated translation, one would expect the EC50 to be more potent when protein levels are reduced.

Acknowledgment

We thank Iain Mattaj, Reinhard Lührmann and Matthias Hentze for valuable suggestions during the experimental work. We are grateful to Dirk H. Ostareck for critical reading of the manuscript. The antibody against La was kindly provided by Ger J.M. Pruijn. The work was supported by a grant from the Bundesministerium für Bildung und Technologie (031U215C).

REFERENCES

Ali, N. and Siddiqui, A. (1997). The La antigen binds 5′ non-coding region of the hepatitis C virus RNA in the context of the initiator AUG codon and stimulates internal ribosome entry site-mediated translation. *Proceedings of the National Academy Sciences USA*, **94**, 2249–2254.

Bartenschlager, R. and Lohmann, V. (2000). Replication of hepatitis C virus. *Journal of General Virology*, **7**, 1631–1648.

Bergamini, G., Preiss, T. and Hentze, M. W. (2000). Picornavirus IRESes and the poly(A) tail jointly promote cap-independent translation in a mammalian cell-free system. *RNA*, **6**, 1781–1790.

Brown, E. A., Zhang, H., Ping, L.-H. and Lemon, S. M. (1992). Secondary structure of the 5′ nontranslated regions of hepatitis C virus and pestivirus genomic RNAs. *Nucleic Acids Research*, **20**, 5041–5045.

Buratti, E., Tisminetzky, S., Zotti, M. and Baralle, F. M. (1998). Functional analysis of the interaction between HCV 5′UTR and putative subunits of eukaryotic translation initiation factor eIF3. *Nucleic Acids Research*, **26**, 3179–3187.

Cohen, J. (1999). The scientific challenge of hepatitis C. *Science*, **285**, 26–30.

Elbashir, S. M., Harboth, J., Lendeckel, W., Yalcin, A., Weber, K. and Tuschl, T. 2001. Duplexes of 21-nucleotide RNAs mediate RNA interference in mammalian cell culture, *Nature*, **411**, 494–98.

Hartmuth, K., Urlaub, H., Vornlocher, H.-P., Will, C. L., Gentzel, M., Wilm, M. and Lührmann, R. (2002). Protein composition of human prespliceosomes isolated by a to-bramycin affinity-selection method. *Proceedings of the National Academy Sciences USA*, 99,16719–16724.

Hellen, C. U. T. and Pestova, T. V. (1999). Translation of hepatitis C virus RNA. *Journal of Viral Hepatitis*, 6, 79–87.

Hellen, C. U. T. and Sarnow, P. (2001). Internal ribosome entry sites in eukaryotic mRNA molecules. *Genes & Development*, 15, 1593–1612.

Honda, M., Beard, M. R., Ping, L.-H. and Lemon, S. M. (1999). A phylogenetically conserved stem-loop structure at the 5′ border of the internal ribosome entry site of hepatitis C virus is required for cap-independent viral translation. *Journal of Virology*, 73, 1165–1174.

Honda, M., Ping, L.-H., Rijnbrand, R. C., Amphlett, E., Clarke, B., Rowlands, D. and Lemon, S. M. (1996). Structural requirements for initiation of translation by internal ribosome entry within genome-length hepatitis C virus RNA. *Virology*, 222, 31–42.

Jackson, R. J. and Kaminski, A. (1995). Internal initiation of translation in eukaryotes: The picornavirus paradigm and beyond. *RNA*, 1, 985–100.

Jiang, L. and Patel, D. J. (1998). Solution structure of the tobramycin-RNA aptamer complex. *Nature Structural Biology*, 5, 769–774.

Kaminski, A., Hunt, S. L., Patton, J. G. and Jackson, R. J. (1995). Direct evidence that the polypyrimidine tract-binding protein (PTB) is essential for internal initiation of transla-tion of encephalomyocarditis virus RNA. *RNA*, 1, 924–938.

Kieft, J., Zhou, K., Rubin, R. and Doudna, J. A. (2001). Mechanism of ribosome recruitment by hepatitis C IRES RNA. *RNA*, 7, 194–206.

Kolupaeva, V. G., Pestova, T. V. and Hellen, C. U. T. (2000). An enzymatic footprinting analysis of the interaction of 40S ribosomal subunits with the internal ribosomal entry site of hepatitis C virus. *Journal of Virology*, 74, 6242–6250.

Lukavsky, P. J., Otto, G. A., Lancaster, A. M., Sarnow, P. and Puglisi, J. D. (2000). Structures of two RNA domains essential for hepatitis C virus internal ribosome entry site function. *Nature Structural Biology*, 7, 1105–1110.

Odreman-Macchioli, F. E., Tisminetzky, S. G., Zotti, M., Baralle, F. E. and Buratti, E. (2000). Influence of correct secondary and tertiary RNA folding on the binding of cellular factors to the HCV-IRES. *Nucleic Acids Research*, 28, 875–885.

Pestova, T. V., Shatsky, I. N., Fletcher, S. P, Jackson, R. J. and Hellen, C. U. T. (1998). A prokaryotic-like mode of cytoplasmic eukaryotic ribosome binding to the initiation codon during internal translation initiation of hepatitis C and classical swine fever virus RNAs. *Genes & Development*, 12, 67–83.

Reynolds, J. E., Kaminski, A., Kettinen, H. J., Grace, K., Clarke, B. E., Carroll, A. R., Rowlands, D. J. and Jackson, R. J. (1995). Unique features of internal initiation of hepatitis C virus RNA translation. *European Molecular Biology Organization Journal*, 14, 6010–6020.

Simmonds, P. (1995). Variability of hepatitis C virus. *Hepatology*, 21, 570–583.

Sizova, D. V., Kolupaeva, V. G., Pestova, T. V., Shatsky, I. N. and Hellen, C. U. T. (1998). Specific interaction of eukaryotic translation initiation factor 3 with the 5′ nontranslated regions of hepatitis C virus and classical swine fever virus RNAs. *Journal of Virology*, 72, 4775–4782.

Smith D. B., Mellor, J. Jarvis, L. M., Davidson, F., Kolberg, J., Urdea, M., Yap, P. L. and Simmonds, P. (1995). Variations of the hepatitis C virus 5′ non-coding region: Implications for secondary structure, virus detection and typing. *Journal of General Virology*, 78, 1749–1761.

Spahn, C. M., Kieft, J. S., Grassucci, R. A., Penczek, P. A., Zhou, K., Doudna, J. A. and Frank, J. (2001). Hepatitis C virus IRES RNA-induced changes in the conformation of the 40s ribosomal subunit. *Science*, 291, 1959–1962.

Takamizawa, A., Mori, C., Fuke, I., Manabe, S., Murakami, S., Fujita, J., Onishi, E., Andoh, T., Yoshida, I. and Okayama, H. (1991). Structure and organization of the hepatitis C virus genome isolated from human carriers. *Journal of Virology*, 65, 1105–1113.

24 **RNAi and the drug discovery process**

Steven A. Haney, Peter Lapan, Jeff Aalfs, Chris Childs, Paul Yaworsky, and Chris Miller

Introduction

Drug discovery is a complex process that seeks to identify therapeutics for treating human disease. It has very high failure rate, and by one estimate, the total cost for a therapeutic successfully brought to market is 803 million dollars (DiMasi et al., 2003). Failure occurs at all points in the process, with failures at the pre-development stages being the most common, and failures at the clinical stages being the most costly. Scientists working in drug discovery are continually challenged to identify ways to improve the process. Current efforts are largely "target-based" approaches. Once chosen, the target may be studied *in vitro* for more than a year, and in model systems of the disease for up to four years. Errors in determining whether a given target is truly effective in treating a disease may not be detected until Phase II or Phase III clinical studies, which follow many years of study and a financial investment of tens or hundreds of millions of dollars. As such, target validation is a critical aspect of the drug discovery process.

The pharmaceutical industry has invested heavily in genomics because of its promise to provide a continuing supply of drug targets (Wiley, 1998; Ohlstein et al., 2000; Baba, 2001). Implicit in this investment has been the expectation that the targets provided by genomics are highly validated (Debouck and Metcalf, 2000). Thus, genomics has grown broadly across the drug discovery process, from target identification to phamacogenomics. Genomics efforts in the drug discovery process are currently focused on genes that are mostly likely to be involved in disease. This strategy is to look only at the "druggable targets," those gene products for which the intervention by small-molecule therapeutics has already been demonstrated (Hopkins and Groom, 2002). These druggable targets are a group of functionally defined families, such as enzymes (including protein kinases) and receptors (such as GPCRs), that naturally function by binding small molecules. This is because the task of finding a new small molecule that interferes with the function of the target protein which already binds a small molecule is simpler (or at least better understood) than that of developing a small molecule to protein that does not, such as protein-protein interactions of scaffolding proteins. However, both types of proteins play critical roles in disease (Balakin et al., 2002; Golemis et al., 2002; Klabunde and Hessler, 2002; Sawyers, 2002; Williams and Mitchell, 2002)]. As therapeutics against new classes of proteins are developed, new families will be added to the group of druggable targets.

Using a pharmaceutically defined subset of the genome greatly reduces the efforts needed to develop comprehensive target validation technologies. Furthermore, working within a single gene family allows for validation assays to be set up that leverage specific characteristics of the family, while limiting the screening to a manageable number of genes, ranging from thirty to several hundred. In these family-based assays, a common strategy may be employed over the entire family. Examples include the systematic mutational analysis of protein kinases and GPCRs in cell-based assays (Bishop et al., 2000; Chalmers and Behan, 2002). A number of gene-based target validation technologies exist as well. These include DNA antisense (Dean, 2001; Taylor, 2001; Sternberger et al., 2002), morpholino oligos (Iversen, 2001), ribozymes (Rhoades and Wong-Staal, 2003), peptide and nucleic acid aptamers (Colas et al., 1996; Drolet et al., 1999; Geyer et al., 1999; Fisch, 2001; James, 2001) and arrayed cDNA expression libraries (Michiels et al., 2002). All have been shown to work, but most are difficult to implement, for reasons of cost, intellectual property restrictions, or because they are technically challenging.

It is for the reasons stated above that the rapid progress made in developing RNAi for the study of biological processes has been greeted so enthusiastically by the pharmaceutical industry (O'Neil et al., 2001; Appasani, 2003). RNAi has been used by many investigators and has been shown to be relatively simple to use, highly effective, economical and relatively free of restrictive intellectual property constraints (Shi, 2003; Voorhoeve and Agami, 2003). The technology has advanced very quickly, and these advances have greatly increased the direct and potential applications in basic biology and drug discovery (Elbashir et al., 2001; Elbashir et al., 2002; Krichevsky and Kosik, 2002; Lee et al., 2002; McManus et al., 2002; Miyagishi and Taira, 2002; Paddison et al., 2002; Paul et al., 2002; Shi et al., 2002). Indeed, while a significant impact of RNAi on target validation is immediately evident, additional steps in the drug development process can also benefit from RNAi technologies.

However, while RNAi has opened many potential avenues for improving the drug discovery process, these remain only potential opportunities until the RNAi technologies themselves are validated, shown to be robust, and sources of experimental artifacts characterized. Our work described here was based on an interest in determining how RNAi compared with other target validation technologies, and its potential for scalability of target validation studies.

Materials and methods

Cell culture

All cell lines were obtained from the American Type Culture Collection (ATCC, Mannassas, VA), and cultured under standard conditions. SW480 cells were cultured in RPMI (Invitrogen, Carlsbad, CA) supplemented with 10% Fetal Bovine

Serum, 1X MEM nonessential amino acids and 20 μg/mL gentamycin (all from Invitrogen) and seeded at 20000 cells per well of a 96-well plate. HeLa cells are cultured in Dulbecco's Modified Medium supplemented with 10% Fetal Bovine Serum and 20 μg/mL gentamycin, and seeded at 4000 per well of a 96-well plate. U2OS (osteosarcoma) cells and HEK-293 (Transformed kidney epithelium) cells were maintained using standard procedures. U2OS cells were grown in McCoy's 5A medium with 10% FBS. 293 cells were grown in Dulbecco's Modified Eagle Medium (DMEM, Invitrogen), supplemented with 10% FBS and Penicillin/Streptomycin.

Wnt3A was obtained from conditioned medium from normal L cells, or from L cells stably transfected with the Wnt3A gene (Shibamoto et al. 1998). Briefly, cells were split and allowed to grow for 4 days in the growth medium. The medium was collected, replaced, and the cells were allowed to grow another 4 days. At the end of the 4 days, the medium was collected and the two batches combined and stored at 4°C. For experiments, the conditioned medium was diluted 1:5 or 1:10 in the appropriate growth medium before use.

Design and synthesis or construction of siRNAs and shRNAs

Duplex siRNAs were designed using Tuschl's rules (Shibamoto et al., 1998; Elbashir et al., 2001; Harborth et al., 2001; Elbashir et al., 2002). Oligos were synthesized and annealed by Dharmacon (Lafayette, CO). Smartpools™ against specific genes were designed and synthesized by Dharmacon from the published sequence using 2' ACE chemistry by Dharmacon, Inc. Pools are a collection of four 21mer siRNA duplexes. Due to space considerations, oligonucleotide sequences are not listed here but are available as a supplemental table from the corresponding author.

Transfections

Transfections using Lipofectamine2000 and Opti-MEM (Invitrogen) were performed using manufacturer's instructions at siRNA concentrations of 20–200 nM and 3.2 ng/μl (reporter plasmids, at a TOPFLASH: pRL-SV40 ratio of 40:1, for U2OS cells, 10:1 for SW480 cells). Reporter plasmids were pTOPFLASH (TCF-Firefly Luciferase, Upstate, Waltham, MA) and pRL-SV40 (SV40-Renilla Luciferase, Promega, Madison, WI). Cells were allowed to recover 24 hr before further treatments. 293 cells were treated in the same way as described for the U2OS cells, with the following modifications: the wells were coated with poly-D-lysine in water prior to plating, and the tranfection complexes were diluted by half.

For the experiments in Figure 24.3: After 18–24 hours exposure to Wnt3A, GSKi, or control medium, the cells were rinsed with 100μl/well PBS. 20 μl Passive Lysis Buffer (Promega) was added to each well, and the plates were rocked gently for 20 min. Assays were performed on a Fluroskan AscentFL luminometer (Thermo Labsystem, Vantaa, Finland), using the Dual-Luciferase assay reagent kit from Promega.

RNA expression analysis by TaqMan

Total RNA was harvested 24 hours post-transfection. FAM-MGB target primers and probes were purchased from the Assay-on-demand collection (Applied

Table 24.1. mRNA level reductions by siRNAs

Target	RNA level
β-catenin	0.31
LRP6	0.25
Dkk1	0.31

As measured by TaqMan.

Biosystems, Inc., Foster City, CA), and reference genes (GAPDH or β-2-micro-globulin) were VIC-MGB labeled ABI Predeveloped Assay Reagents (PDAR, Applied Biosystems). Sequences of TaqMan amplicons designed for this study are listed in Table 24.1, except for those designed by ABI through the Assay-on-Demand service (AoD). AoD reagents used in these studies are for β-catenin (cat# 4318286T) and CDK2 (Hs00176475_m1).

Twenty-five microliter amplifications were performed using the ABI one-step RT-PCR master mix. Samples were analyzed using the comparative method (ABI User Bulletin #2) with 6 treatment replicates. For CDK2 analysis of probe location effects, RNA samples for treatment were pooled prior to TaqMan analysis. Cycling conditions were 48°C 30 minutes, 95°C 10 minutes, 40 cycles of 95°C 15 seconds, 60°C 1 minute.

For experiments in U2OS cells: RNA was harvested 24–48 hours after transfection. 5 μl of RNA from each well was then added to a 25-μl TaqMan RT-PCR reaction and monitored through 40 cycles in the ABI7000 gene analyzer (Applied Biosystems). RNA level were assessed by comparing amplification profiles of target genes to those of reference genes in transfected and mock-transfected cells.

Western blotting

Cells growing in 96-well plates were solubilized in 30 μL of 1× Protein Sample Buffer (2% SDS, 10% Glycerol, 75 μM Tris pH6.8, 0.04% Bromphenol Blue, 150μM μM DTT); triplicate wells were pooled and homogenized with an 18G needle. Samples were held at 95°C for 5 minutes, spun, and 10 μL was loaded on a 4–20% Novex gradient gel. Electrophoresis was performed at 70 V constant for 3 hours. The gel is transferred to Hybond ECL nitrocellulose (Amersham, Piscataway, NJ) in 1× Tris Glycine SDS/20% methanol buffer (Biorad, Hercules, CA) at 30 V constant for 2 hours. Blots were prepared by standard methods using HRP-conjugated secondary antibodies and ECL reagents (Amersham).

Results and discussion

RNAi against β-catenin in SW480 cells

Initially, we were interested in determining how robust and facile siRNA was for knocking down a target. As a pilot study, we found β-catenin to be a good choice for a proof-of-principle study since it is part of a well-characterized signal transduction pathway, and subject to rapid turnover at the protein level (Moon et al., 2002). We began our studies by synthesizing 9 RNA duplexes that correspond to some of the sites chosen by Roh and coworkers for knockdown by DNA anti-sense (Roh et al., 2001), as at this time we were considering both technologies. 7 duplexes target the β-catenin gene. One duplex synthesized was CA5sc, which was a scrambled sequence of CA5. A second scrambled sequence was chosen *de novo*. Both scramble sequences were searched by BLAST (Altschul et al., 1990) against GenBank, and no significant matches were identified. The relative positions of the sequences are shown in Figure 24.1A. These duplexes were used to treat SW480 colon carcinoma cells at 200 nM. No significant toxicity was associated with transfection of the duplexes at this concentration, but in subsequent experiments, we have observed that they are still effective at much lower concentrations (25 nM, see next section). Most of these duplexes were characterized in three assays. In the first, the extent of message knockdown was determined by TaqMan (Figure 24.1B and 1D), Western blot (Figure 24.1D) and functional inhibition of β-catenin signaling using a TCF4-dependant luciferase construct (Figure 24.1C and 1D). In the last assay, reporter plasmids were used to measure activitation of TCF by β-catenin. In each of these assays, several siRNA duplexes were shown to be effective, including CA5, the one sequence that was also shown to be effective as a DNA antisense oligo. Figure 24.1D compares how the duplexes performed in each of the three assays (primary Western blot data were omitted from the figure for clarity). In general, there was good agreement between each of the measures of β-catenin inhibition, with CA5 being the most effective in the TaqMan assay and CA2 showing the strongest correlation between assays overall. The most notable exception was the CAS duplex which did not have a significant effect on message level (Figure 24.1B) but was effective in Western blots and in the activity assay (Figure 24.1C). Such discordance between measures of message level, and functional consequences at the protein and activity levels, may limit the utility of TaqMan in measuring the effectiveness of an RNAi duplex.

Characterization of target site effectiveness on mRNA level as determined by TaqMan

Measurements of both mRNA and protein levels following RNAi treatment are important for demonstrating that the treatment with RNAi does in fact cause knockdown of the intended target. Taqman offers the advantage of requiring the gene sequence *in silico* only, as does RNAi itself, allowing previously uncharacterized genes, and even putative genes, to be screened. We were interested in determining if the location of the TaqMan primer and probe (covering a section of the gene which we refer to as the amplicon) influenced the assessment of the target knockdown. To test this, we used an siRNA duplex that we previously had shown to be effective in knocking down CDK2 (NM_001798) at position 399 in several assays (data not shown). Using this reagent we measured CDK2 mRNA levels using 7 independent TaqMan primer/probe sets at distinct sites along the CDK2 mRNA. A schematic showing the relationship between the CDK2–399 duplex and the TaqMan amplicons is shown in Figure 24.2A. One amplicon was designed by Applied Biosystems, as part of their Assay-on-Demand service, and its sequence is not known to us. We compared all of these amplicons against four

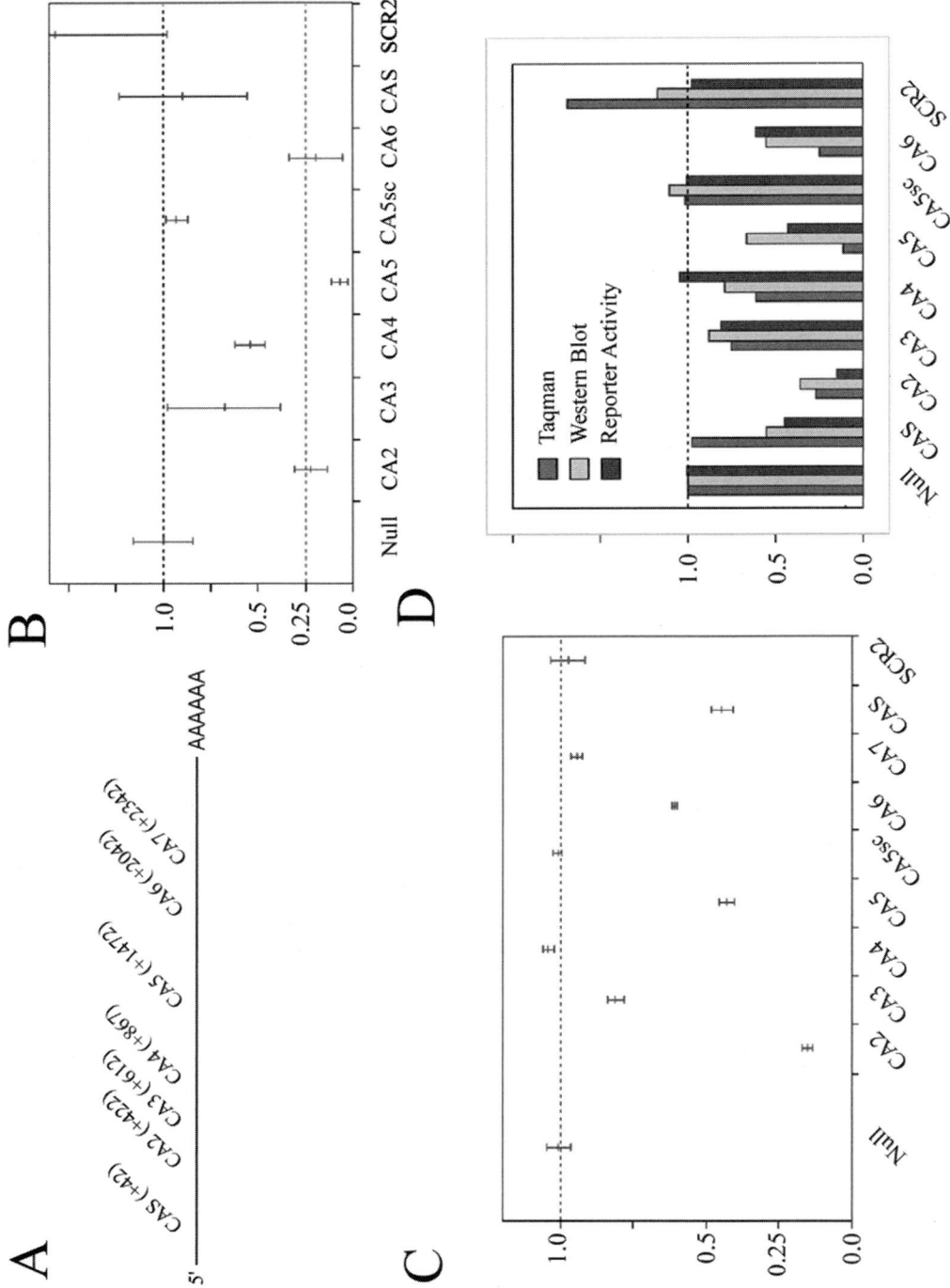

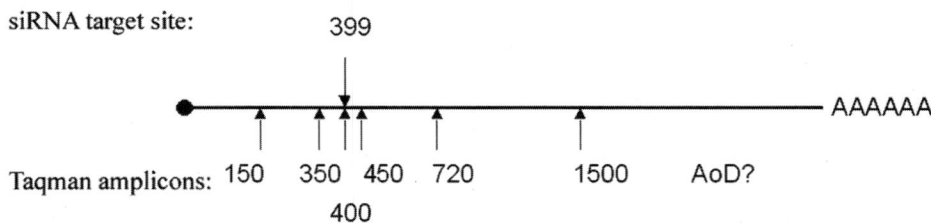

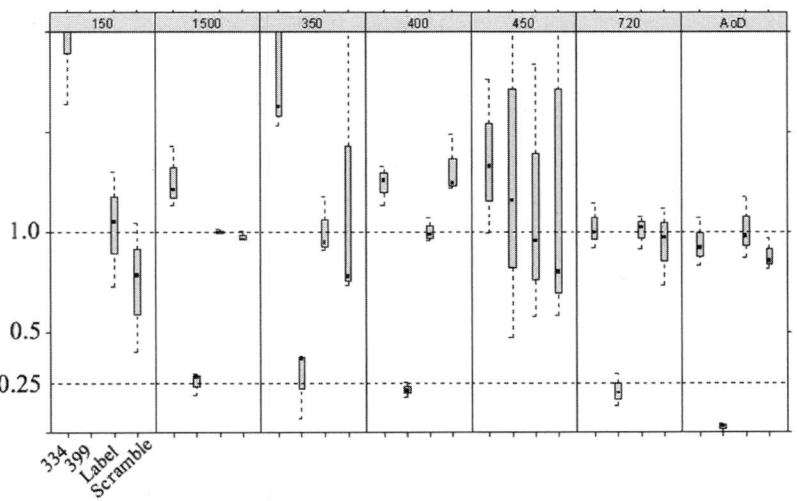

Figure 24.2. *Influence of TaqMan amplicon location on the determination of mRNA levels following siRNA treatment.* siRNA duplexes targeting CDK2 at either position 334 or 399 were used to knockdown CDK2 expression. A labeled siRNA duplex was used for measuring relative mRNA levels, and a second scrambled sequence control also used. Following treatment with each of these regents, mRNA levels were measured using 7 different primer/probe amplicons. Sequences of the amplicons are indicated by the first nucleotide of the amplicon, except for the Assay-on-Demand amplicon designed by Applied Biosystems, for which the location in not known. A. Schematic representation of the siRNA duplex showing the most consistent results (399, top of part A), and of the locations of the amplicons used to measure mRNA levels (bottom of part A). B. Results of different amplicons used to measure CDK2 mRNA levels, presented as box plots. Amplicons are indicated at the top of each of the panels. Duplex siRNAs are indicated at the bottom of the first panel, the pattern is consistent for the remaining panels.

Figure 24.1. *Characterization of siRNAs against β-Catenin in SW480 colon carcinoma cells.* Data is presented as the mean +/− the standard error, and is measured relative to null samples, as described in the text. A. Schematic representation of the β-catenin message showing relative positions of target sequences used in this study. B. Extent of mRNA level reduction following treatment with indicated siRNA duplexes. C. Inhibition of TCF4-mediated luciferase expression following treatment with siRNA duplexes. D. Concordance analysis of siRNAs against β-Catenin from experiments measuring messages levels (in A), protein levels as measured by Western Blots (not shown) and reporter assay levels (in C).

treatments: CDK2–334, a duplex that shows weak and non-reproducible results; CDK2–399, described above; and two duplexes that do not target CDK2: a labeled duplex, and the scramble duplex used in the experiments shown in Figure 24.1. In Figure 24.2B the results are shown. Overall, there was good consistency for all of the amplicons, confirming CDK2–399 to be very effective for knockdown of CDK2 message. The major exception was the amplicon at base 450 of the CDK2 message, which failed to show any knockdown of CDK2 message. The other amplicons verify that the duplex used is effective, so this failure of the 450 amplicon is clearly an artifact. Since it is the amplicon closest to the target site of the siRNA duplex, on the 3′ side of the target site, it is possible that this portion of the message could be masked by the cellular RNAi machinery. CDK2–400 is unique among the amplicons tested in this study, because it actually spans the siRNA duplex sequence. Further testing with additional examples of effective siRNAs against different sites in the CDK2 message, and with studies of additional genes, will clarify whether there is a functional relationship between the target site and the amplicon that should be taken into consideration when designing experiments.

Knockdown of β-catenin in U2OS cells and HEK293 cells by siRNA and shRNA

Having demonstrated that chemically synthesized siRNAs can be effective reagents, we wanted to extend these findings in two directions. First, we wanted to compare shRNAs to siRNAs. shRNAs (Brummelkamp et al., 2002; Lee et al., 2002; Paddison et al., 2002; Paul et al., 2002; Sui et al., 2002; Yu et al., 2002) allow different experiments to be performed, including *in vivo* (xenograft) studies of modified cell lines (Hemann et al., 2003), and regulated expression of gene knockdown (Wang et al., 2003). If siRNA function can be predictive of shRNA function, for the same sequence, it allows us to extend experiments performed with siRNAs into shRNA-based systems directly, allowing for persistent or regulated knockdown of a target gene. Second, we wanted to determine if the reduced message levels we had observed actually carried functional consequences in the transfected cells; one could imagine that even partial knockdown of a gene might significantly affect its activity in the context of the cell.

We used the Wnt pathway to address both of these questions (Figure 24.3A). The Wnt pathway responds to extracellular secreted peptides by blocking the activity of the Axin/APC/GSK-3B complex, or β-catenin destruction complex (Cadigan and Nusse, 1997; Polakis, 2000). The destruction complex phosphorylates β-catenin, triggering its degradation by the ubiquitinase pathway. In the presence of certain Wnt proteins, the complex is inactivated through a pathway involving the LRP5/6 co-receptors and the Frizzled (Fz) receptor, and the cytoplasmic protein Disheveled (DVL). As a result, β-Catenin accumulates and translocates to the nucleus, where it dimerizes with the TCF transcription factor to activate a family of Wnt-induced genes. β-catenin plays a crucial role in this pathway, and its elimination would be expected to inhibit signaling.

To assay signaling directly, we used the TCF-luciferase assay, in which luciferase is placed under the control of TCF binding sites; this construct has been shown

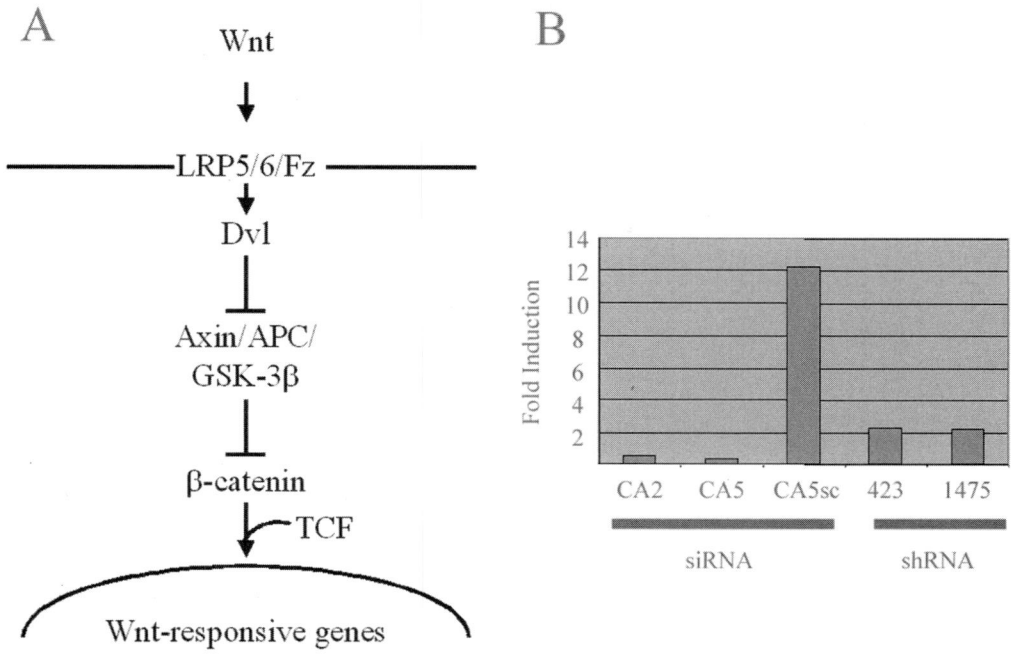

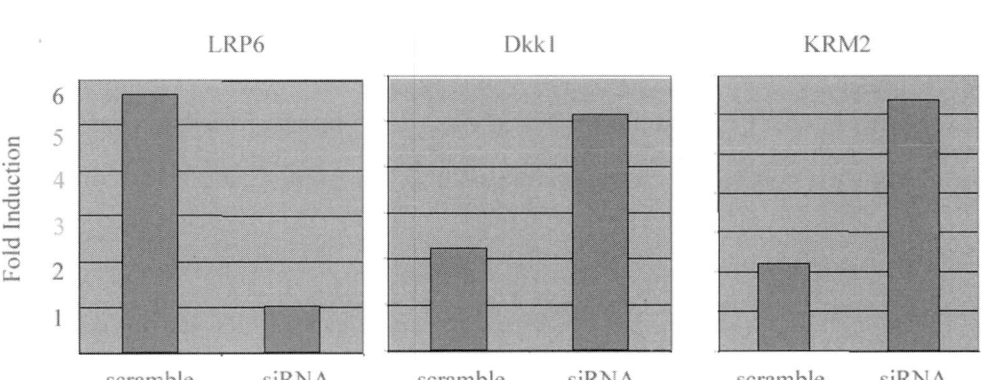

Figure 24.3. *Inhibition of Wnt-pathway signaling by siRNAs and shRNAs.* A. Schematic representation of the Wnt/β-catenin pathway. Wnts are soluble peptide ligands for the receptor complex of Frizzled (Fz) and LRP5 or LRP6. Activation of this complex activates Disheveled, which in turn ultimately inhibits glycogen synthase kinase 3β (GSK-3β). GSK-3β phosphorylates β-catenin, promoting degradation through the ubiquitin pathway. Inhibition of GSK-3β increases the active β-catenin, resulting in accumulation in the nucleus, and activation of TCF-dependent target genes. B. Inhibition of a TCF-dependent reporter by siRNAs and shRNAs against β-catenin in U2OS cells, following activation of the pathway using a small molecule inhibitor of GSK-3β. CA2, CA5 and CA5sc siRNAs used in Figure 1, shRNAs designated 423 and 1475 are 29 bp shRNA hairpins that correspond to CA2 and CA5, respectively. C. Modulation of the Wnt/β-catenin pathway by siRNAs, following stimulation with Wnt-3A. Wnt-3A added to U2OS cells treated with siRNAs targeting different components of the pathway. LRP6 is described above, Dikkopf protein 1 (Dkk1) and Kremen2 (KRM2) are extracellular and membrane proteins, respectively, that function as negative regulators of the pathway.

to respond to Wnt signals (Cadigan and Nusse, 1997; Moon et al., 2002). In this experiment, cells were stimulated with an inhibitor of GSK-3B (Coghlan et al., 2000); the same results can be seen when cells are stimulated with Wnt3A-conditioned medium (not shown). As shown in Figure 24.3B, the siRNAs targeting β-catenin effectively abolish Wnt pathway signaling, as compared with a control in which cells are transfected with a scrambled-sequence siRNA. Importantly, two shRNAs, 1475–29 (designated 1475 in the figure) and 423–29 (designated 423) also strongly affect activity. These vector-based shRNA constructs correspond to the CA2 and CA5 sequences, respectively. Several other shRNAs, designed against other sites within the β-catenin gene, were less effective at reducing mRNA levels or signaling (not shown); this result suggests that siRNAs can be instructive in the design of shRNAs against the same gene.

We next targeted other members of the Wnt pathway, using the TCF-Luciferase reporter as a functional assay (Figure 24.3C). The co-receptor LRP6 is a critical effector of the Wnt pathway; as expected, its removal reduces TCF-luciferase activity to background levels, under conditions where control cells are strongly stimulated (Figure 24.3C, left panel). The human Dickkopf homolog 1 (Dkk1) and the Kremen 2 (KRM2) genes are antagonists of the Wnt pathway; Dkk1 factor is a secreted peptide, while KRM2 is a cell-surface receptor (Mao et al., 2002). siRNAs targeting either of these mRNAs enhance the activity of the Wnt pathway (Figure 24.3C, center and right panels), demonstrating that siRNAs can have significant effects on the activity of a signaling pathway, even when they do not completely eliminate the target mRNA (Table 24.1).

RNAi for target validation and drug lead characterization

RNAi is an effective method for (high-throughput) target validation

Our analysis of RNAi as a general method for target validation has shown it to be effective and relatively facile for such purposes. This makes RNAi a viable technology for target validation, including at the high-throughput level. We have found siRNAs to function by several methods, and that the sequence of an effective siRNA duplex frequently functions effectively as an shRNA. Both siRNAs and shRNAs have been used to knock down the expression of target genes, as demonstrated by mRNA, protein and activity assays. Furthermore, the use of RNAi to investigate signaling in the Wnt/β-catenin pathway has been demonstrated by the use of these reagents to block signaling by a protein kinase inhibitor and a secreted signaling protein. The characterization of other members of the Wnt/β-catenin pathway by RNAi demonstrates that RNAi can be used for general target validation studies.

We have found that TaqMan is typically a very efficient means for determining how effectively an RNAi reagent is functioning. However, sites exist in genes that are slow to disappear following knockdown, as was observed in Figure 24.2 for CDK2. This could be because the sequence targeted by this amplicon in CDK2 is unusually stable, or its proximity to the target sequence of the siRNA indicates that a 3′ proximal region of 50–100 bases is protected by the RISC machinery.

Whatever the reason, it is clear that occasional problems may occur in relying exclusively on TaqMan for validating reagents. However, it usually performs well, and in cases where an antibody or cDNA is not available (including cases of novel or predicted genes), it can be used.

RNAi is compatible with other methods of target validation

While capable of screening many genes rapidly, RNAi does not explicitly model the actions of a small molecule inhibitor. Suppressing the expression of a candidate target is similar to inhibiting its function, such as with a small molecule, but the distinctions between the two can be important. Many drug targets maintain complex interactions with other proteins, and it is foreseeable that interactions with these other proteins may be a significant component of the therapeutic function of a small molecule inhibitor. The best characterized examples are the protein kinases, which can be converted to dominant negative isoforms by mutations that render the protein unable to bind ATP. These mutations more explicitly model the effect of a small molecule inhibitor. Blocking the interaction of a kinase with its associated proteins by mutationally inactivating it may leave the signal transduction pathway nonfunctional, whereas suppressing the expression of the kinase may allow a related kinase to substitute. Schematic examples are depicted in Figure 24.4. Additionally, multiple kinases and associated proteins exist in balanced relationships that are affected differently when one is targeted by either a dominant negative mutant or an inhibitory RNA. Unique consequences to different mechanisms of preventing kinase function have been observed in model systems such as yeast (Madhani et al., 1997), where different phenotypes are observed when a protein kinase is deleted, versus when it is mutationally inactivated. This has also been observed for CDK2 in colon cancer cells (Tetsu and McCormick, 2003).

How does a situation such as this affect a genomic-level target validation project? Pharmaceutically speaking, it may not be all that important that CDK2 activity is dispensable in colon cancer, as long as a small molecule inhibitor of CDK2 mimics the dominant negative form of the protein, and the chemical agent ultimately treats colon cancer. Does a distinction between the observations between these two methods carry any limitations for either one in target validation? While it is true that an inhibitor of CDK2 may effectively treat colon cancer despite CDK2 dispensability, a failure to identify CDK2 by RNAi is not a false negative, since it is not the explicit target. Instead, another CDK may be the true target, and a genomics-level project that examines CDK2 would presumably also look at other CDKs. Thus, one would expect that depletion of CDKs would identify one or more targets for treating colon cancer. Instead, the above example highlights the idea that target selection is a complex decision (Knowles and Gromo, 2003). Target validation can leverage the results of high-throughput screens for gene function using RNAi, and then apply those findings to a more focused and rigorous study. A thorough understanding of a candidate target in several assays will provide a clear set of expectations about how an effective drug candidate should affect each model system. Additional techniques will also be critical in this

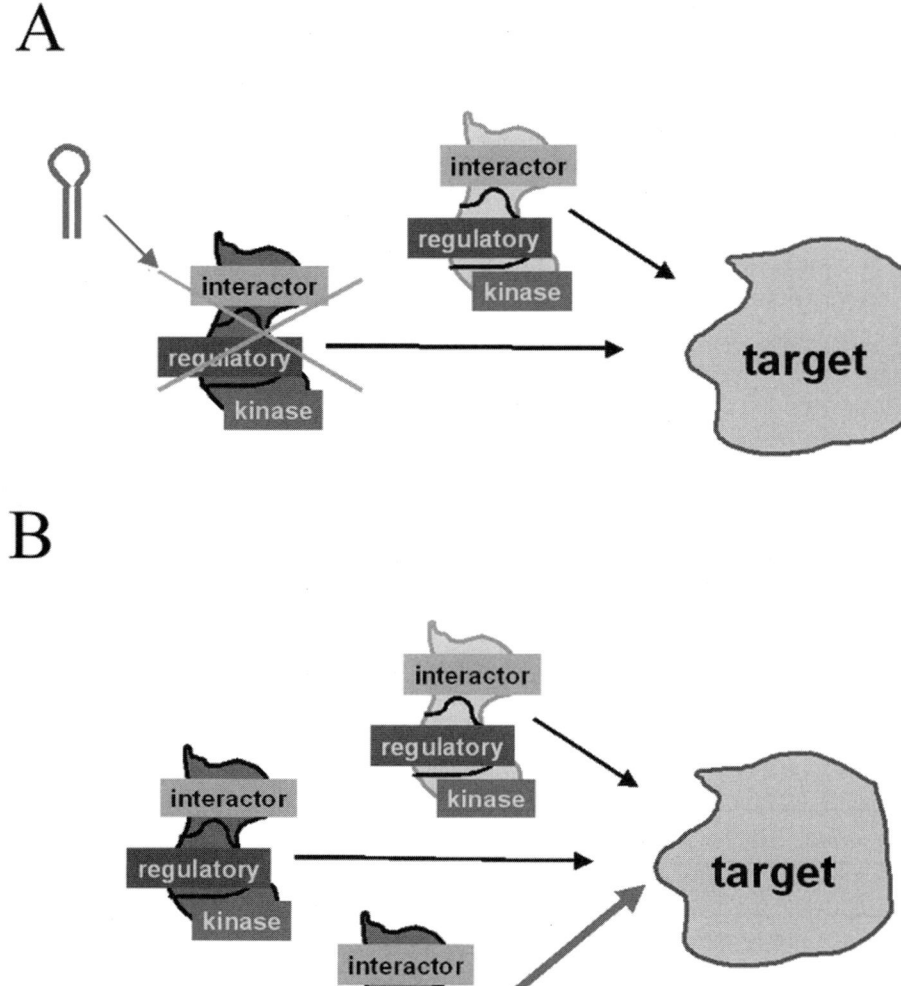

Figure 24.4. *Schematic models of interference of protein kinases by RNAi knockdown and dominant negative mutant.* Two kinases are depicted as interacting with the same target. Inhibition by RNAi or dominant negative is depicted in red in each part of the figure. A. Knockdown of a kinase by shRNA removes the targeted kinase from the cell. B. Inhibition of kinases by expression of a dominant negative mutant. (See color section.)

process, including dominant negative mutants (discussed above) and cell permeable peptides (Fawell et al., 1994; Fahraeus et al., 1996; Nagahara et al., 1998).

RNAi can be important for validating small molecule inhibitors and model systems

Beyond the use of RNAi in Target Validation, we see uses for it in additional steps of the drug discovery process. Two that we can readily identify are compound validation and model system validation. Compound validation encompasses the series of secondary assays used to determine whether a compound functions specifically on the target in several assays. RNAi provides an opportunity to continually check whether the compound is acting as expected in each of the subsequent tests. For example, a small molecule inhibitor of a protein kinase or GPCR may be tested in several cell lines in order to determine how broadly effective it is. However, not all of these lines may have been tested thoroughly prior to their use as secondary assays, and unexpected behavior may be indicative of either unanticipated behavior of the compound or uncharacterized issues with the model system. The potential for this occurring increases as the targets become increasingly novel. RNAi of the target (as well as dominant negatives and cell permeable peptides) allows a rigorous "reality check" of the model systems used.

RNAi provides an opportunity to develop specific cell models for further characterizing a small molecule inhibitor. In this case, RNAi would not be used to mimic the inhibition of the target, but to increase the number of model systems available to characterize the compound. RNAi would be used to establish clearly defined genetic backgrounds for determining the mechanism with which a compound functions *in vivo*. Currently, creating isogenic cell lines is very labor intensive, so only limited sets of strains can be established (Torrance et al., 2001). Such lines are critical in defining the genetic context in which a therapeutic can be expected to function. For example, the rapamycin-like inhibitor of mTor function, CCI-779, is significantly more effective in PTEN(–/–) cell lines and tumor xenografts than into PTEN wild-type lines and tumor xenografts (Neshat et al., 2001). PTEN mutations are observed in 50% of classified cancers, including many of the major cancers, such as of the prostate (Li et al., 1997). An important part of establishing this relationship is showing it can be observed in many cell lines. Yu and coworkers were able to do so using several pairs of well-characterized breast cancer cell lines (Yu et al., 2001). Now that this relationship between PTEN status and CCI-779 sensitivity has been established, it has been observed in multiple myeloma (Shi et al., 2002) and in prostate cancer cells (Grunwald et al., 2002). Once again, moving forward into the study of new and even novel drug targets, the challenge will be to rigorously characterize the target and its inhibitor throughout the preclinical phase of drug discovery. RNAi can play an important role in the creation of lines that have precisely defined genetic backgrounds, which will result in a better understanding of how the compounds function going into clinical studies.

REFERENCES

Altschul, S. F., Gish, W., Miller, W., Myers, E. W. and Lipman, D. J. (1990). Basic local alignment search tool. *Journal of Molecular Biology*, **215**, 403–410.

Appasani, K. (2003). RNA interference: A technology platform for target validation, drug discovery and therapeutic development. *Drug Discovery World*, Summer Issue, 61–68.

Baba, Y. (2001). Development of novel biomedicine based on genome science. *European Journal of Pharmaceutical Sciences*, **13**, 3–4.

Balakin, K. V., Tkachenko, S. E., Lang, S. A., Okun, I., Ivashchenko, A. A. and Savchuk, N. P. (2002). Property-based design of GPCR-targeted library. *Journal of Chemical Informatics and Computer Sciences*, **42**, 1332–1342.

Bishop, A. C., Ubersax, J. A., Petsch, D. T., Matheos, D. P., Gray, N. S., Blethrow, J., Shimizu, E., Tsien, J. Z., Schultz, P. Z., Rose, M. D., Wood, J. L., Morgan, D. O. and Shokat, K. M. (2000). A chemical switch for inhibitor-sensitive alleles of any protein kinase. *Nature*, **407**, 395–401.

Brummelkamp, T. R., Bernards, R. and Agami, R. (2002). A system for stable expression of short interfering RNAs in mammalian cells. *Science*, **296**, 550–553.

Cadigan, K. M. and Nusse, R. (1997). Wnt signaling: A common theme in animal development. *Genes & Development*, **11**, 3286–3305.

Chalmers, D. T. and Behan, D. P. (2002). The use of constitutively active GPCRs in drug discovery and functional genomics. *Nature Reviews Drug Discovery*, **1**, 599–608.

Coghlan, M. P., Culbert, A. A., Cross, D. A., Corcoran, S. L., Yates, J. W., Pearce, N. J., Rausch, O. L., Murphy, G. J., Carter, P. S., Cox, L. R., Mills, D., Brown, M. J., Haigh, D., Ward, R. W., Smith, D. G., Murray, K. J., Reith, A. D. and Holder, J. C. (2000). Selective small molecule inhibitors of glycogen synthase kinase-3 modulate glycogen metabolism and gene transcription. *Chemical Biology*, **7**, 793–803.

Colas, P., Cohen, B., Jessen, T., Grishina, I., McCoy, J. and Brent, R. (1996). Genetic selection of peptide aptamers that recognize and inhibit cyclin-dependent kinase 2. *Nature*, **380**, 548–550.

Dean, N. M. (2001). Functional genomics and target validation approaches using antisense oligonucleotide technology. *Current Opinion in Biotechnology*, **12**, 622–625.

Debouck, C. and Metcalf, B. (2000). The impact of genomics on drug discovery. *Annual Reviews of Pharmacology and Toxicology*, **40**, 193–207.

DiMasi, J. A., Hansen, R. W. and Grabowski, H. G. (2003). The price of innovation: new estimates of drug development costs. *Journal of Health Economics*, **22**, 151–185.

Drolet, D. W., Jenison, R. D., Smith, D. E., Pratt, D. and Hicke, B. J. (1999). A high throughput platform for systematic evolution of ligands by exponential enrichment (SELEX). *Combinatorial Chemistry and High Throughput Screening*, **2**, 271–278.

Elbashir, S., Harborth, M., J., Lendeckel, W., Yalcin, A., Weber, K. and Tuschl, T. (2001). Duplexes of 21-nucleotide RNAs mediate RNA interference in cultured mammalian cells. *Nature*, **411**, 494–498.

Elbashir, S., Harborth, M. J., Weber, K. and Tuschl, T. (2002). Analysis of gene function in somatic mammalian cells using small interfering RNAs. *Methods*, **26**, 199–213.

Fahraeus, R., Paramio, J. M., Ball, K. L., Lain, S. and Lane, D. P. (1996). Inhibition of pRb phosphorylation and cell-cycle progression by a 20-residue peptide derived from p16CDKN2/INK4A. *Current Biology*, **6**, 84–91.

Fawell, S., Seery, J., Daikh, Y., Moore, C., Chen, L. L., Pepinsky, B. and Barsoum, J. (1994). Tat-mediated delivery of heterologous proteins into cells. *Proceedings of the National Academy of Sciences USA*, **91**, 664–668.

Fisch, I. (2001). Peptide display in functional genomics. *Combinatorial Chemistry and High Throughput Screening*, **4**, 157–169.

Geyer, C. R., Colman-Lerner, A. and Brent, R. (1999). "Mutagenesis" by peptide aptamers identifies genetic network members and pathway connections. Proceedings of the *National Academy of Sciences USA*, **96**, 8567–8572.

Golemis, E. A., Tew, K. D. and Dadke, D. (2002). Protein interaction-targeted drug discovery: Evaluating critical issues. *BioTechniques*, **32**, 636–638.

Grunwald, V., DeGraffenried, L., Russel, D., Friedrichs, W. E., Ray, R. B. and Hidalgo, M. (2002). Inhibitors of mTOR reverse doxorubicin resistance conferred by PTEN status in prostate cancer cells. *Cancer Research*, **62**, 6141–6145.

Harborth, J., Elbashir, S. M., Bechert, K., Tuschl, T. and Weber, K. (2001). Identification of essential genes in cultured mammalian cells using small interfering RNAs. *Journal of Cell Science*, **114**, 4557–4565.

Hemann, M. T., Fridman, J. S., Zilfou, J. T., Hernando, E., Paddison, P. J., Cordon-Cardo, C., Hannon, G. J. and Lowe, S. W. (2003). An epi-allelic series of p53 hypomorphs created by stable RNAi produces distinct tumor phenotypes *in vivo*. *Nature Genetics*, **33**, 396–400.

Hopkins, A. L. and Groom, C. R. (2002). The druggable genome. *Nature Reviews Drug Discovery*, **1**, 727–730.

Iversen, P. L. (2001). Phosphorodiamidate morpholino oligomers: Favorable properties for sequence-specific gene inactivation. *Current Opinion in Molecular Therapy*, **3**, 235–238.

James, W. (2001). Nucleic acid and polypeptide aptamers: A powerful approach to ligand discovery. *Current Opinion in Pharmacology*, **1**, 540–546.

Klabunde, T. and Hessler, G. (2002). Drug design strategies for targeting G-protein-coupled receptors. *Chemicalbiochemistry*, **3**, 928–944.

Knowles, J. and Gromo, G. (2003). A guide to drug discovery: Target selection in drug discovery. *Nature Reviews Drug Discovery*, **2**, 63–69.

Krichevsky, A. M. and Kosik, K. S. (2002). RNAi functions in cultured mammalian neurons. *Proceedings of the National Academy of Sciences USA*, **99**, 11926–11929.

Lee, N. S., Dohjima, T., Bauer, G., Li, H., Li, M. J., Ehsani, A., Salvaterra, P. and Rossi, J. (2002). Expression of small interfering RNAs targeted against HIV-1 rev transcripts in human cells. *Nature Biotechnology*, **20**, 500–505.

Li, J., Yen, C., Liaw, D., Podsypanina, K., Bose, S., Wang, S. I., Puc, J., Miliaresis, C., Rodgers, L., McCombie, R., Bigner, S. H., Giovanella, B. C., Ittmann, M., Tycko, B., Hibshoosh, H., Wigler, M. H. and Parsons, R. (1997). PTEN, a putative protein tyrosine phosphatase gene mutated in human brain, breast, and prostate cancer. *Science*, **275**, 1943–1947.

Madhani, H. D., Styles, C. A. and Fink, G. R. (1997). MAP kinases with distinct inhibitory functions impart signaling specificity during yeast differentiation. *Cell*, **91**, 673–684.

Mao, B., Wu, W., Davidson, G., Marhold, J., Li, M., Mechler, B. M., Delius, H., Hoppe, D., Stannek, P., Walter, C., Glinka, A. and Niehrs, C. (2002). Kremen proteins are Dickkopf receptors that regulate Wnt/beta-catenin signalling. *Nature*, **417**, 664–667.

McManus, M. T., Petersen, C. P., Haines, B. B., Chen, J. and Sharp, P. A. (2002). Gene silencing using micro-RNA designed hairpins. *RNA*, **8**, 842–850.

Michiels, F., van Es, H., van Rompaey, L., Merchiers, P., Francken, B., Pittois, K., van der Schueren, J., Brys, R., Vandersmissen, J., Beirinckx, F., Herman, S., Dokic, K., Klaassen, H., Narinx, E., Hagers, A., Laenen, W., Piest, I., Pavliska, H., Rombout, Y., Langemeijer, E., Ma, L., Schipper, C., Raeymaeker, M. D., Schweicher, S., Jans, M., van Beeck, K., Tsang, I. R., van de Stolpe, O. and Tomme, P. (2002). Arrayed adenoviral expression libraries for functional screening. *Nature Biotechnology*, **20**, 1154–1157.

Miyagishi, M. and Taira, K. (2002). U6 promoter-driven siRNAs with four uridine 3′ overhangs efficiently suppress targeted gene expression in mammalian cells. *Nature Biotechnology*, **20**, 497–500.

Moon, R. T., Bowerman, B., Boutros, M. and Perrimon, N. (2002). The promise and perils of Wnt signaling through beta-catenin. *Science*, **296** 1644–1646.

Nagahara, H., Vocero-Akbani, A. M., Snyder, E. L., Ho, A., Latham, D. G., Lissy, N. A., Becker-Hapak, M., Ezhevsky, S. A. and Dowdy, S. F. (1998). Transduction of full-length TAT fusion proteins into mammalian cells: TAT-p27Kip1 induces cell migration. *Nature Medicine*, **4**, 1449–1452.

Neshat, M. S., Mellinghoff, I. K., Tran, C., Stiles, B., Thomas, G., Petersen, R., Frost, P., Gibbons, J. J., Wu, H. and Sawyers, C. L. (2001). Enhanced sensitivity of PTEN-deficient tumors to inhibition of FRAP/mTOR. *Proceedings of the National Academy of Sciences USA*, **98**, 10314–10319.

Ohlstein, E. H., Ruffolo, Jr. R. R. and Elliott, J. D. (2000). Drug discovery in the next millennium. *Annual Reviews of Pharmacology and Toxicology*, **40**, 177–191.

O'Neil, N. J., Martin, R. L., Tomlinson, M. L., Jones, M. R., Coulson, A. and Kuwabara, P. E. (2001). RNA-mediated interference as a tool for identifying drug targets. *American Journal of Pharmacogenomics*, **1**, 45–53.

Paddison, P. J., Caudy, A. A., Bernstein, E., Hannon, G. J. and Conklin, D. S. (2002). Short hairpin RNAs (shRNAs) induce sequence-specific silencing in mammalian cells. *Genes & Development*, **16**, 948–958.

Paul, C. P., Good, P. D., Winer, I. and Engelke, D. R. (2002). Effective expression of small interfering RNA in human cells. *Nature Biotechnology*, **20**, 505–508.

Polakis, P. (2000). Wnt signaling and cancer. *Genes & Development*, **14**, 1837–1851.

Rhoades, K. and Wong-Staal, F. (2003). Inverse Genomics as a powerful tool to identify novel targets for the treatment of neurodegenerative diseases. *Mechanisms of Ageing and Development*, **124**, 125–132.

Roh, H., Green, D. W., Boswell, C. B., Pippin, J. A. and Drebin, J. A. (2001). Suppression of beta-catenin inhibits the neoplastic growth of APC-mutant colon cancer cells. *Cancer Research*, **61**, 6563–6568.

Sawyers, C. L. (2002). Rational therapeutic intervention in cancer: Kinases as drug targets. *Current Opinion Genetics and Development*, **12**, 111–115.

Shi, Y. (2003). Mammalian RNAi for the masses. *Trends in Genetics*, **19**, 9–12.

Shi, Y., Gera, J., Hu, L., Hsu, J. H., Bookstein, R., Li, W. and Lichtenstein, A. (2002). Enhanced sensitivity of multiple myeloma cells containing PTEN mutations to CCI-779. *Cancer Research*, **62**, 5027–5034.

Shibamoto, S., Higano, K., Takada, R., Ito, F., Takeichi, M. and Takada, S. (1998). Cytoskeletal reorganization by soluble Wnt-3a protein signalling. *Genes & Cells*, **3**, 659–670.

Sternberger, M., Schmiedeknecht, A., Kretschmer, A., Gebhardt, F., Leenders, F., Czauderna, F., Von Carlowitz, I., Engle, M., Giese, K., Beigelman, L. and Klippel, A. (2002). GeneBlocs are powerful tools to study and delineate signal transduction processes that regulate cell growth and transformation. *Antisense Nucleic Acid Drug Development*, **12**, 131–143.

Sui, G., Soohoo, C., Affar el, B., Gay, F., Shi, Y. and Forrester, W. C. (2002). A DNA vector-based RNAi technology to suppress gene expression in mammalian cells. *Proceedings of the National Academy of Sciences USA*, **99**, 5515–55120.

Taylor, M. F. (2001). Target validation and functional analyses using antisense oligonucleotides. *Expert Opinion in Therapeutic Targets*, **5**, 297–301.

Tetsu, O. and McCormick, F. (2003). Proliferation of cancer cells despite CDK2 inhibition. *Cancer Cell*, **3**, 233–245.

Torrance, C. J., Agrawal, V., Vogelstein, B. and Kinzler, K. W. (2001). Use of isogenic human cancer cells for high-throughput screening and drug discovery. *Nature Biotechnology*, **19**, 940–945.

Voorhoeve, P. M. and Agami, R. (2003). Knockdown stands up. *Trends in Biotechnology*, **21**, 2–4.

Wang, J., Tekle, E., Oubrahim, H., Mieyal, J. J., Stadtman, E. R. and Chock, P. B. (2003). Stable and controllable RNA interference: Investigating the physiological function of glutathionylated actin. *Proceedings of the National Academy of Sciences USA*, **100**, 5103–5106.

Wiley, S. R. (1998). Genomics in the real world. *Current Pharmaceutical Design*, **4**, 417–422.

Williams, D. H. and Mitchell, T. (2002). Latest developments in crystallography and structure-based design of protein kinase inhibitors as drug candidates. *Current Opinion Pharmacology*, **2**, 567–573.

Yu, J. Y., DeRuiter, S. L. and Turner, D. L. (2002). RNA interference by expression of short-interfering RNAs and hairpin RNAs in mammalian cells. *Proceedings of the National Academy of Sciences USA*, **99**, 6047–6052.

Yu, K., Toral-Barza, L., Discafani, C., Zhang, W. G., Skotnicki, J., Frost, P. and Gibbons, J. J. (2001). mTOR, a novel target in breast cancer: the effect of CCI-779, an mTOR inhibitor, in preclinical models of breast cancer. *Endocrinology Related Cancer*, **8**, 249–258.

25 RNA interference technology in the discovery and validation of druggable targets

Neil J. Clarke, John E. Bisi, Caretha L. Creasy, Michael K. Dush, Kris J. Fisher, John M. Johnson III, Christopher J. A. Ring, and Mark R. Edbrooke

The role of gene suppression technologies in pharmaceutical drug development

The completion of the human genome sequence (International Human Genome Sequencing Consortium, 2001; Venter et al., 2001) has ushered in a new era for the pharmaceutical industry. With access to a comprehensive list of candidate drug targets, together with a growing understanding of their association to signalling and metabolic pathways and human disease, there would seem to be endless opportunity for novel drug development. From a clinical perspective, this most basic of genetic insights should enhance the efficacy of existing drugs, while at the same time bring new drugs to market for many new therapeutic indications. From a corporate vantage, utilising the information embedded in the human genome for improved target validation should effectively reduce the attrition of candidate drug targets and therefore translate to a "competitive advantage."

With thousands of candidate drug targets to select from, it is necessary to invest in strategic methods in order to navigate the complex maze of genetic information. Indeed, the identification and validation of targets that map to chemically tractable gene families (the 'druggable' genome, or 'pharmome') is now recognised as a fundamental challenge to the entire pharmaceutical industry (Figure 25.1). In an effort to achieve this goal, molecular technologies that suppress the expression of a candidate drug target (either in *cis*- or *trans*-) have become fixtures in most pharmaceutical R&D programmes, and there are many examples where this has been successful. To date, perhaps the most visible successes along these lines come from mouse knockouts (Zambrowicz and Sands, 2003). However, in a R&D environment where time is of the essence, mouse knockouts are not necessarily the fastest route to target validation. For this reason, technologies that model the simplicity of chemical tool compounds in application have emerged as an attractive alternative. DNA-based antisense, that uses modified oligonucleotides to target cognate mRNAs for destruction by an RNaseH mechanism (Zamenick and Stephenson, 1978), has been available for a number of years. More recently, RNAi, and in particular siRNA, has taken centre stage as an

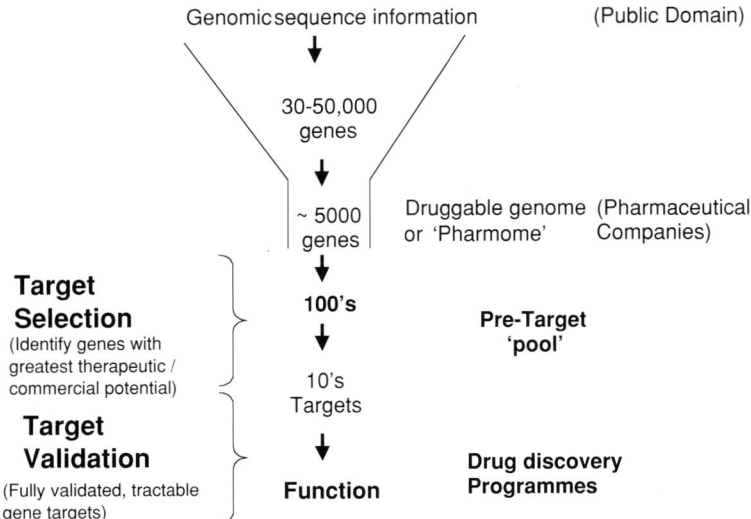

Figure 25.1. Turning sequence into function: the target validation 'funnel'. There is intense commercial pressure from drug discovery programmes to identify and select the most therapeutically relevant target genes from the chemically tractable gene families (known collectively as the 'pharmome').

effective, biologically relevant mechanism for post-transcriptional gene suppression (Elbashir et al., 2001a, 2001b).

Although antisense and siRNA operate by very different biological mechanisms, they are very similar from a research operations standpoint. Both rely on transient transfection procedures for cell delivery, which unfortunately currently precludes their use in some cell types. Because the potency of any given reagent is a function of intracellular delivery, optimising the transfection efficiency to the target cell population is a critical parameter for both reagents. Another common property of antisense and siRNA is the relatively rapid rate with which they are cleared from transfected cells, resulting in a transient (24–96 h) gene suppression phenotype. This property becomes noticeably problematic for cell-based assays that require sustained knockdown e.g. where differentiation of cells is involved. However, some of these issues may be overcome by the ability to develop RNAi in the context of an expression system that encodes a short-hairpin derivative (shRNA) that is processed to the active siRNA conformation by cellular factors. The significance of this approach is that a shRNA expression cassette can be packaged into virtually any conventional gene transfer system, including high-efficiency viral vectors e.g. adenovirus. This offers the prospect for long-term gene suppression (>72 h), delivery to transfection-resistant cells via viral vector infection, and application to animal models i.e. germ line transmission or systemic delivery by viral vectors.

For RNAi to be an effective resource for the pharmaceutical industry, it should be expandable to a working scale that spans the pharmome. Examples of existing genome-wide technology platforms include transcriptomic and proteomic analysis, expression profiling (quantitative and qualitative), and the use of organismal and animal model systems (e.g. *C. elegans*, *Drosophila*, Mouse) to promote the assignment of phenotypes to experimentally induced genotypes (e.g. knockouts). To what extent RNAi can follow this lead and contribute phenotypic information on candidate drug targets, at scale, from arrayed cell-based assays has yet to be fully investigated. Regardless, the compelling efficacy that appears to define RNAi-mediated gene suppression, combined with the prospect for novel delivery options via shRNA, would appear to be consistent with a system that is compatible with pharmaceutical high-throughput operations platforms.

Use of systematic RNAi knockdown in model organisms for target identification

The discovery of the phenomenon of RNAi in the nematode worm *Caenorhabditis elegans* as a response to double-stranded RNA (dsRNA), and the later discovery that sequence-specific gene silencing occurs in worms fed *E. coli* expressing dsRNA (Fire et al., 1998), have given rise to large-scale RNAi approaches. By constructing a library consisting of bacterial clones, each capable of expressing dsRNA designed to correspond to a single gene, loss-of-function phenotypes can be revealed by systematic RNAi analysis to 86% of the worm genome (Kamath et al., 2003; Ashrafi et al., 2003; Timmons and Fire, 1998; Timmons et al., 2001; Kamath et al., 2001). The stable and heritable expression of the dsRNA can be used reproducibly to study numerous phenotypes such as sterility, embryonic or larval lethality, slow post-embryonic growth or post-embryonic defect, (Kamath et al., 2003), to name just a few. In certain cases these phenotypes can reflect cellular functions which are highly conserved, or at least provide clues to possible functions in other species such as human.

An application of this approach for drug discovery can be to analyse the genes identified in the *C. elegans* fat regulatory study to identify potential obesity targets (Ashrafi et al., 2003, Figure 25.2). In this study, 305 genes were identified that resulted in decreased fat storage upon inactivation and 112 gene inactivations increased fat storage. Many of the identified genes are known to be involved in mammalian fat, lipid or sterol metabolism. For instance, increased fat storage was observed upon inactivation of C43H6.8, a nuclear hormone receptor whose potential orthologue is Nhlh-2/Nscl-2, a gene that when knocked out in the mouse causes obesity. By simply identifying the human orthologues to each protein encoded by these genes, and choosing those which encode potential drug targets (e.g. GPCRs, nuclear receptors, enzymes, transporters and ion channels) potential targets can be identified. Literature searches with the human orthologue and other model organism orthologues (e.g. flies, zebrafish, and yeast) can be used to prioritise the list followed by target validation studies on those proteins of most interest. These studies could include mRNA and protein expression studies, mammalian cellular assays using RNAi and other functional assays, mouse KO studies, and pathway expansion studies via yeast two-hybrid, proteomics and model organisms.

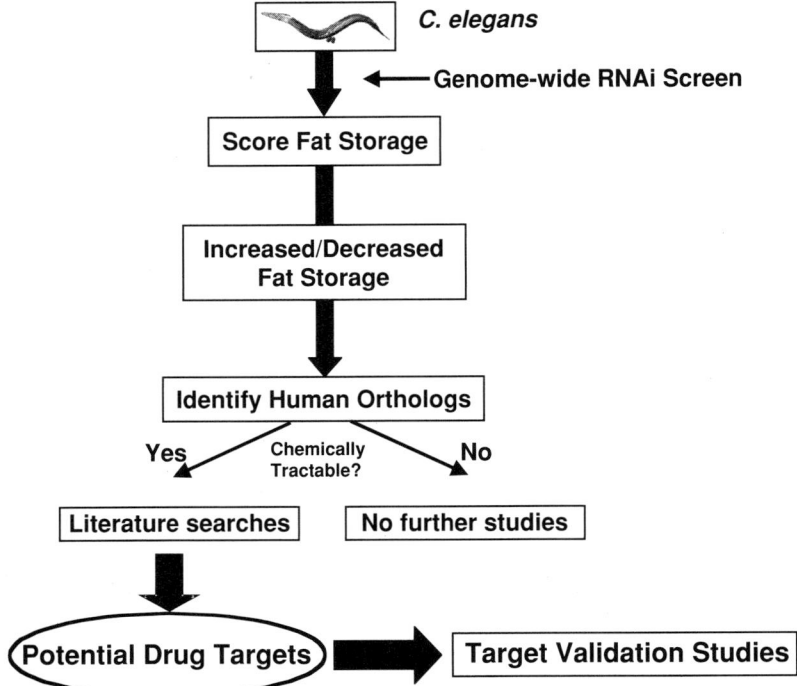

Figure 25.2. Systematic RNAi knockdown in *C. elegans*. Scheme used for systematic RNAi studies of target orthologues of human genes in *C. elegans* (as illustrated for genes involved in fat metabolism).

While *C. elegans* is the only multicellular organism for which genome-wide RNAi is possible at the moment, systematic studies have been conducted in cultured cells from *Drosophila melanogaster* (Somma et al., 2002; Lum et al., 2003). One approach utilised a reporter gene to identify components of the *Hedgehog* pathway via systematic RNAi to ~43% of the predicted *Drosophila* genes including every kinase and phosphatase (Lum et al., 2003). With this approach, four new putative components of the *Hedgehog* pathway were identified. Using a similar approach, with either all of the potential targets or with a few key protein families, potential targets can be examined in disease relevant functional assays to prioritise which targets to move forward to high throughput drug screening.

Use of RNAi for target validation in mammalian cells *in vitro*: Knockdown of GABA$_B$R1 and GABA$_B$R2 receptor expression

The recent, highly significant discovery that the use of 21–22 nucleotide short interfering RNAs (siRNAs), homologous to a target gene sequence, could lead to specific gene silencing in mammalian cells (Elbashir et al., 2001a; 2001b), has ensured their widespread use for *in vitro* modulation of gene expression studies (Harborth et al., 2001; Elbashir et al, 2002; McManus et al., 2002; Novina et al., 2002; Spankuch-Schmitt et al., 2002). Indeed we, and others in the pharmaceutical industry, have evaluated this approach to modulate the expression of a range of genes from the chemically tractable gene families. Historically, the G-protein coupled receptors (GPCRs) represent the most tractable target class of proteins for drug intervention. Although gene knockdown approaches, such as antisense (Zamenick and Stephenson, 1978; Izant and Weintraub, 1985), have been applied in the study of GPCR function (Rohrer and Kobilka, 1998; Van Oekelen et al., 2003), RNAi has not so far been extensively used for this purpose in mammalian cells. Thus, we set out to establish if an RNAi knockdown approach could be successfully applied to this target family. To do this, a number of different targets within the GPCR target class were considered based on the following criteria:

- The receptor needs to be liganded (i.e. not an orphan), unless it was constitutively active.
- The receptor should be expressed in a cell type or cell line to which reagent delivery (i.e. siRNA) is possible e.g. successful transfection to >70% cells in the population.
- A cell line in which the receptor is endogenously expressed is advantageous (as this may more accurately reflect the normal intracellular signalling processes for the receptor).
- TaqMan™ probes and primers, which are specific for the mRNA of the gene being targeted, should be available.
- Robust antibodies should be available to the protein product of the target gene.
- A robust phenotypic assay, by which the consequences of the gene knockdown can be measured, must be available.

The GABA$_B$ receptor (Blein et al., 2000; Bowery and Enna, 2000; Jones et al., 2000; Couve et al., 2000) was chosen as a GPCR of interest, and one that fulfilled the requirements listed above. It is a G-protein coupled receptor that mediates inhibition throughout the central and peripheral nervous systems, and could be a therapeutic target in conditions as diverse as epilepsy and hypertension. The functional receptor is a heterodimer composed of homologous 7-transmembrane domain subunits (called GABA$_B$R1 and GABA$_B$R2), both of which are required for the formation of a fully functional GABA$_B$ receptor (White et al., 1998).

Signalling through the active, fully functional heterodimeric receptor can be measured in a membrane-based, high-throughput GTPγS binding assay (Wieland and Jakobs, 1994) in a recombinant CHO cell line, where a number of different GPCRs are known to be expressed endogenously. We tested the hypothesis that siRNA mediated knockdown of the mRNA for either subunit in the GABA$_B$ heterodimer should lead to a specific decrease in signalling through this receptor in the presence of the ligand (GABA) in this assay. It was already known that delivery of siRNAs to CHO cells is possible. Hence, the combination of this particular target gene in such a well-defined system was thought to be an excellent starting point in which to assess successful in-house design, specificity, and functional outcome of *in vitro* RNAi knockdown as a target validation tool against genes in this particular target class.

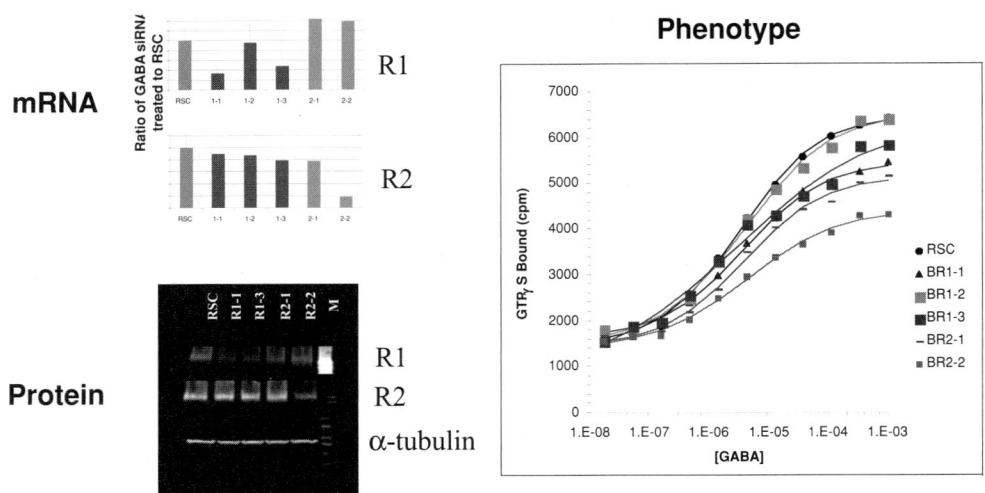

Figure 25.3. Consequences of RNAi-mediated knockdown on GABA$_B$R1 and GABA$_B$R2 receptors. mRNA levels of GABA$_B$R1 and GABA$_B$R2 were determined 24 hours post transfection with specific siRNAs. The green bar indicates the random sequence control, the red bars (1−1, 1−2 and 1−3) are siRNAs specific for GABA$_B$R1, the orange bars (2−1 and 2−2) show siRNAs specific for GABA$_B$R2. **Protein** knockdown of GABA$_B$R1 and GABA$_B$R2 was analysed after treatment of cells for 72 hours with siRNAs, and quantified by immunoblot relative to the internal control, α-tubulin. The **phenotype** associated with the knockdown of GABA$_B$R1 and GABA$_B$R2 was characterised using a ^{35}S GTPγ assay. (See color section.)

There was a good correlation between the level of knockdown at the mRNA, protein and phenotypic response for siRNAs targeting GABA$_B$R1 (see Figure 25.3). Both active siRNA motifs, 1.1 and 1.3, knocked down mRNA levels (by 67% and 52% respectively) and protein levels (by approximately 45%) after 48 hours of treatment of the cells. Both siRNAs yielded a similar response in the GTPγ[^{35}S] ATP assay for the functional heterodimer. This is in contrast with the inactive siRNA motif, 1.2 (which did not give significant knockdown of the mRNA), and closely mirrored the response of the random sequence control (RSC) siRNA included in the phenotypic assay. Significantly, none of the siRNAs specific to GABA$_B$R1 produced any knockdown of GABA$_B$R2 mRNA or protein, indicating good specificity of the knockdown tools.

With respect to GABA$_B$R2, the levels of mRNA and protein knockdown given by siRNA motif 2.2 were 82% and >50% respectively. Motif 2.1 did not appear to have any significant effect. However, both siRNAs produced a significant knockdown in phenotypic response, and this was greater than that seen with either of the active siRNAs used to target GABA$_B$R1. We assume that this is due to the fact that GABA$_B$R2 is expressed at lower levels than GABA$_B$R1 in these cells.

From these observations, we were able to conclude that an RNAi-mediated knockdown approach was able to suppress GABA$_B$ receptor mRNA and protein levels, concomitant with a decrease in receptor function. The knockdown was highly specific (R1 tools did not knock down R2 levels and vice versa), and there was a good correlation between the level of knockdown of mRNA and the suppression of function. In addition, we found that a relatively modest suppression

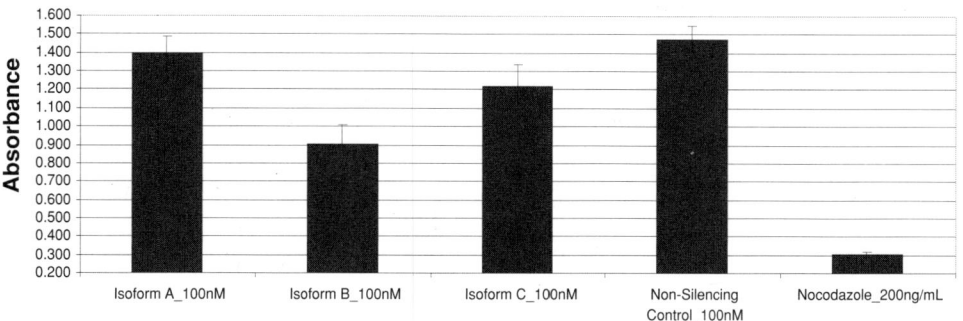

Figure 25.4. Inhibition of specific isoforms of kinases using siRNAs. HT-29 cells were transfected with 100nM of isoform specific siRNAs. An MTS based cell proliferation assay was performed at 96 hours post-transfection to measure the phenotypic response elicited by selectively inhibiting each isoform. Changes in cell proliferation were compared to the non-silencing siRNA negative control and the positive control compound, Nocodazole.

in protein level e.g. 45% for R1 and 50% for R2, caused a significant suppression of function, but the magnitude and order of the changes seen were entirely consistent with what is known about the biology of the system being tested. Moreover, we found that the phenotypic data obtained in the GTPγS assay was very similar to that generated using specific antagonists ('tool' compounds; data not shown). Hence, we conclude that RNAi can be used to modulate expression and function of members of the GPCR target class for target validation purposes.

Current alternative uses of RNAi in mammalian cells in the drug discovery process

Compound validation

As demonstrated by the data above, one of the virtues of RNAi is its specificity. Studies have indicated that single base mismatches between a siRNA and its corresponding mRNA can deleteriously affect the ability of the siRNA to recognise and promote the degradation of that mRNA. Indeed, Brummelkamp et al., (2002) have shown that siRNA designed to an activated allele of Ras (Ras V12) does not affect the levels of wild-type Ras mRNA. Thus, by designing siRNAs or shRNAs to target a specific mRNA, it is possible to investigate the function of a single member of a family of closely related members in a particular biological process without perturbing the expression of other members. This is especially important in drug discovery, as it is important to know whether a drug can (or should) recognise a particular family member, or whether multiple family members are required for a particular cellular process, thus requiring that that drug recognise and inhibit multiple family members.

We have evaluated this specificity of siRNA by elucidating the roles of specific isoforms of kinases in cell proliferation. siRNAs, specific to each of three members of a family of kinases implicated in tumourigenesis, were transfected into cells and then screened for their effects on proliferation. As shown in Figure 25.4, isoforms A and C have no effect on proliferation even though protein levels are reduced approximately 90 percent (not shown), whereas isoform B has a pronounced effect on proliferation. These results have important applications in our efforts to design therapeutic molecules against this family of kinases.

Assay development

RNAi can be used in the engineering of new cell lines, either to improve their properties by removing the expression of proteins that are deleterious to the growth and division of the cells, or by removing the expression of a protein that is a target for drug screening. In this latter approach, a 'negative control' cell line can be generated for a high-throughput screen, obviating the requirement for the generation and isolation of 'knockout' cells from another source e.g. mutant human cells or from KO mice.

Future perspectives

Platform RNAi-mediated knockdown approaches for target identification *in vitro*

The ability to perform large-scale, genome-wide loss-of-function screens using dsRNA in *C. elegans* was described in detail above. The relative ease and low-level costs associated with this approach, together with its demonstrated utility in permitting identification of genes of potential therapeutic interest, will ensure an ongoing role for this in the drug discovery process. However, since the ultimate aim of drug discovery is to prosecute targets of therapeutic interest in man, the ability to carry out similar, large-scale, loss-of-function screens against human target genes in disease relevant pathway or mechanistic assays is of paramount importance. The development of siRNAs and, more recently, the functionally equivalent short-hairpin RNAs (shRNAs: Paddison et al., 2002; McManus et al., 2002), has provided the tools with which this goal can be achieved at large scale and at high throughput in mammalian primary cells and cell lines (see Figure 25.5). In a similar manner to that described for the *C. elegans* work, libraries consisting of large numbers of siRNA or shRNA motifs are targeted against members of various gene families (or even the complete genome). These are being generated by various academic laboratories and biotech and pharmaceutical companies. Indeed, a number of commercial organisations are offering such collections of siRNAs (including Dharmacon, Qiagen, and Ambion), or shRNAs expressed in arrayed adenoviral libraries (for example, Galapagos Genomics).

We have conducted a pilot study on the feasibility of using an arrayed shRNA library to identify motifs that will knock down the expression of members of a family of therapeutically relevant targets. The objective of this study was to develop the methods for prosecuting shRNA motifs in a 96-well format, to identify bottlenecks, and to determine the costs, both in terms of reagents and manpower, of a library approach. Three shRNA motifs to each of our chosen targets were cloned into a vector containing a human U6 small nuclear RNA promoter, and transfected into cells, along with a vector that expresses the mRNA recognised by the shRNA. Our results indicate that it is relatively straightforward to

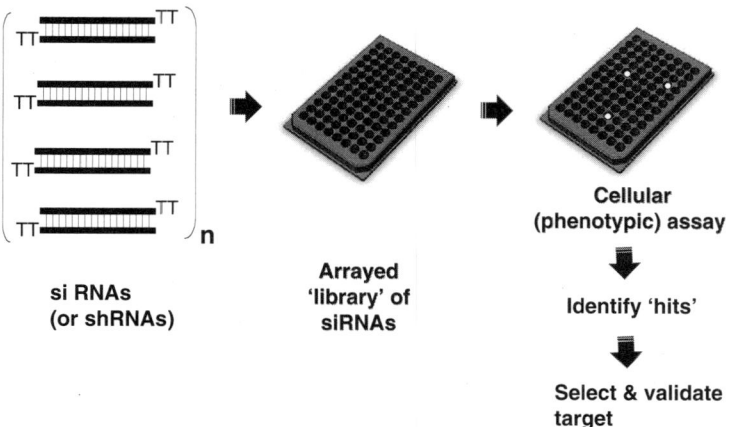

Figure 25.5. Pharmome-scale target identification using RNAi. Libraries of arrayed siRNAs or shRNAs could be used to investigate the role of defined gene targets in mammalian cells using disease pathway or disease mechanism-relevant phenotypic assays.

scale-up the cloning and screening of shRNA motifs. As shown in Figure 25.6, 60 percent of the targets had at least one shRNA motif that knocked down mRNA expression.

Whichever format is used for this approach however, the requirement for i) high quality reagents, ii) cells to which the reagents can be delivered, and iii) the availability of robust, disease relevant, assays, is of paramount importance. It is only the successful combination of these factors which will allow correlation of a particular molecular target with the underlying disease relevant phenotype, and therefore allow the selection of biological targets for development of potential new medicines.

TaqMan Analysis of selected Targets

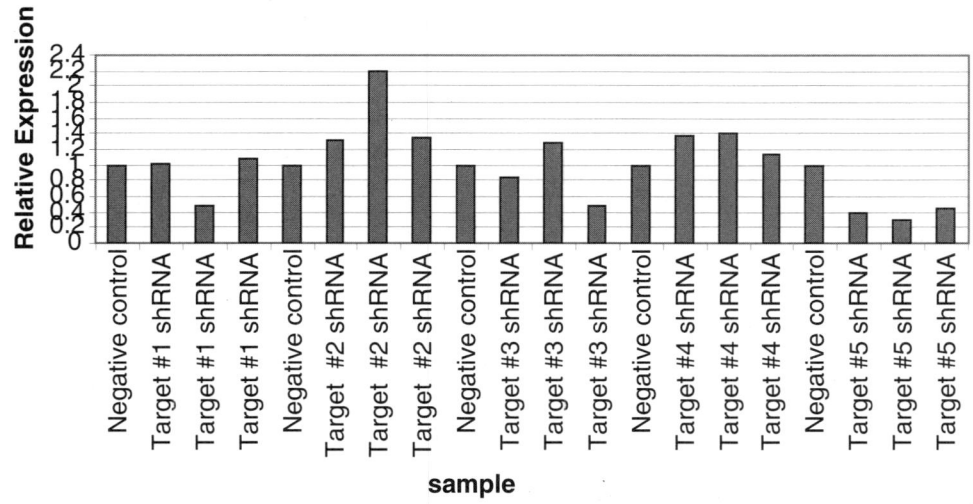

Figure 25.6. Generation and functional validation of a shRNA library.

Viral vectors for expression of shRNA

Sustained or regulated gene suppression requires the expression of shRNA within cells and the subsequent processing of the shRNA to siRNA. shRNA expression cassettes can be delivered to cells in plasmid form or alternatively as genetically engineered viruses. The choice of virus depends upon the application, for example the cell type in which expression of the shRNA is desired and the required duration of expression. Adenoviruses are able to infect a wide variety of cell types. However expression of transgenes is transient since viral genome integration is not part of their replication cycle. In contrast, retroviruses and a group of parvoviruses, termed adeno-associated viruses, are able to integrate their genomes into that of their hosts facilitating long-term expression of the shRNA (Walther and Stein, 2000). A number of recent reports have described the use of lentiviruses, not only for expression of shRNA in somatic cells, but also for the generation of transgenic rodents (Lois et al., 2002; Pfeifer et al., 2002; Rubinson et al., 2003). Lentiviruses possess a number of key features that lend their use to these applications including the ability to infect a very wide variety of both dividing and non-dividing cell types including embryonic stem cells and to efficiently integrate their genomes into that of their host cells (Walther and Stein, 2000).

RNAi *in vivo*

Animal model systems, especially those based on rodents such as mouse and rat, have played a role in the drug discovery process for many years. These have consisted of models of disease or disease mechanisms, toxicological and drug metabolic testing, and target validation studies. Transgenic models, especially gene 'knockout' (KO) mice have played a key role in the target validation strategies for many drug targets as they permit the detailed *in vivo* investigation of the consequences of loss of expression of a particular gene. Although useful, these types of model suffer from the significant length of time, resource and cost associated with their generation. As a consequence of an unpredicted role in development, many gene knockouts are also lethal at the embryonic stage, such that the exact function in the adult animal remains unclear.

Since the initial accounts of RNA interference in mouse (Wianny and Zernicka-Goetz, 2000; Svoboda et al., 2000), there have been a number of reports demonstrating that both transient (siRNA/shRNA mediated; McCaffrey et al., 2002; Lewis et al., 2002; Sørensen et al., 2003; Xia et al., 2002) and stable (germline transmission of a dsRNA or shRNA-encoding transgene; Stein et al., 2003; Hasuwa et al., 2002; Carmell et al., 2003; Hemann et al., 2003; Kunath et al., 2003; Shinagawa and Ishii, 2003) knockdown of gene expression is possible in both mouse and rat. Though in their infancy, these approaches (in particular that of germline stable knockdown) have a number of particular attractions compared to the conventional knockout mouse. These are based mainly on the reduced timescales associated with making a 'knockdown' transgenic compared to a conventional knockout mouse. However, there are other potential benefits. Firstly, as KO mice are diploid it is usually necessary to target both alleles of a particular target gene. However, a dominant loss-of-function approach, such as RNAi, will inhibit both alleles simultaneously. Secondly, it should be possible to generate an 'allelic series' of RNAi mutants, with varying degrees of ablation of the expression of the target gene. Also, it should be possible to generate constructs where the expression of the RNAi transgene can be regulated by exogenous factors e.g. drugs including tetracycline, leading to a similar variation in gene knockdown but also avoiding the complication of gene knockdown during embryonic development. Additionally, there is the possibility that a gene knockdown approach might be more predictive than a complete KO for target validation, efficacy, and drug side effect determinations as the levels of 'gene interference' are not absolute. Finally, the mouse is often not the optimum choice for animal studies and frequently rat or other rodent models are preferred. Until now, the ability to carry out systematic gene knockout or knockdown studies in these other species has not been possible. It should therefore be noted that lentiviruses have been shown to deliver shRNA expression cassettes to mouse, cynomolgus monkey and human embryonic stem cells (Asano et al., 2002; Gropp et al., 2003; Ma et al., 2003; Rubinson et al., 2003), and lentiviruses have been used to generate transgenic rats as well as mice (Lois et al., 2002).

RNAi therapeutics

Hand-in-hand with the development of gene knockdown tools for target identification and validation is the prospect of using the same reagents as therapeutic agents in their own right. With this in mind, a number of companies have been established recently with the primary aim of developing RNAi as a way of treating disease (companies including Alnylam and Sirna). These companies will be able to build on the lessons learnt from the ongoing development of antisense reagents as therapeutics, including the use of novel chemistries to aid efficacy and drug half-life, and the extensive knowledge of pharmacodynamics, drug toxicology, and delivery. Only time will tell whether RNAi drugs will become mainstream therapeutics in their own right.

Conclusions

There is intense competitive pressure to identify and validate novel therapeutic targets from the chemically tractable subset of genes in the genome, or 'pharmome.' Pharmaceutical companies have embraced the full range of currently available functional genomic technologies, both in small scale and as 'platforms,' in order to help expedite this process. The recent discovery of RNA interference and subsequent examples of its use as a successful tool for knocking down gene expression in model organisms, mammalian cells and *in vivo* have ensured its place in the target identification and validation toolbox. The recent developments in RNAi directed to mammalian cells allow its use in bespoke analyses for target validation directed against single targets of interest, as exemplified by the work directed towards the GABA$_B$ receptor, a member of the important, chemically tractable gene family of G-protein coupled receptors. Large-scale knockdown of gene expression in the nematode worm *C. elegans* demonstrates the power that high-throughput efforts can have on prioritising targets. Recent developments offer promise that similar large-scale loss-of-function screens implemented in mammalian cells will allow the direct identification of human target genes of possible therapeutic interest. Finally, researchers are now beginning to utilise RNAi for *in vivo* target identification and validation approaches. If this application of the technology becomes fully established, it may offer significant time savings for drug discovery.

The rapidity with which RNAi technology has been 'embraced' by the scientific community is perhaps the greatest testament to the current need for a simple, cost-effective, yet robust, gene knockdown technology. The simplicity of the approach and relatively low cost are evidenced by the fact that, aside from the knowledge of the sequences of the genes to be knocked down, little in the way of expensive additional laboratory equipment is required and most of the major costs are associated with reagents. Whether the technology and range of applications for RNAi meet our full expectations in the future, and prove to be robust enough approaches to help deliver and validate high quality therapeutic targets to the drug discovery business, remains to be seen.

REFERENCES

Asano, T., Hanazono, Y., Ueda, Y., Muramatsu, S.-I., Kume, A., Suemori, H., Suzuki, Y., Harii, K., Hasegawa, M., Nakatsuji, N. and Ozawa, K. (2002). Highly efficient gene transfer into primate embryonic stem cells with a simian lentivirus vector. *Molecular Therapy*, **6**, 162–168.

Ashrafi, K., Chang, F. Y.,Watts, J. L., Fraser, A. G., Kamath, R. S., Ahringer, J. and Ruvkun, G. (2003). Genome-wide RNAi analysis of *Caenorhabditis elegans* fat regulatory genes. *Nature*, **421**, 268–272.

Blein, S., Hawrot, E. and Barlow, P. (2000). The metabotropic GABA receptor: molecular insights and their functional consequences. *Cellular and Molecular Life Sciences*, **57**, 635–650.

Bowery, N. G. and Enna, S. J. (2000). γ-Aminobutyric acid$_B$ receptors: First of the functional metabotropic heterodimers. *Journal of Pharmacology and Experimental Therapeutics*, **292**, 2–7.

Brummelkamp, T. R., Bernards, R. and Agami, R. (2002). A system for stable expression of short interfering RNAs in mammalian cells. *Science*, **296**, 550–553.

Carmell, M. A., Zhang, L., Conklin, D. S., Hannon, G. J. and Rosenquist, T. A. (2003). Germline transmission of RNAi in mice. *Nature Structural Biology*, **2**, 91–92.

Couve, A., Moss, S. J. and Pangalos, M. N. (2000). GABA$_B$ receptors: A new paradigm in G-protein signalling. *Molecular and Cellular Neuroscience*, **16**, 296–312.

Elbashir, S. M., Harborth, J., Lendeckel. W., Yalcin, A., Weber, K. and Tuschl, T. (2001a). Duplexes of 21 nucleotide RNAs mediate RNA interference in cultured mammalian cells. *Nature*, **411**, 494–498.

Elbashir, S. M, Lendeckel, W. and Tuschl, T. (2001b). RNA interference is mediated by 21- and 22- nucleotide RNAs. *Genes & Development*, **15**, 188–200.

Elbashir, S. M., Harborth, J., Weber, K. and Tuschl, T. (2002). Analysis of gene function in somatic mammalian cells using small interfering RNAs. *Methods*, **26**, 199–213.

Fire, A., Xu, S., Montgomery, M. K., Kostas, S. A, Driver, S. E. and Mello, C. C. (1998). Potent and specific gene interference by double-stranded RNA in *Caenorhabditis elegans*. *Nature*, **391**, 806–811.

Gropp, M., Itsykson, P., Singer, O., Ben-Hur, T., Reinhartz, E., Galun, E. and Reubinoff, B. E. (2003). Stable genetic modification of human embryonic stem cells by lentiviral vectors. *Molecular Therapy*, **7**, 281–287.

Harborth, J., Elbashir, S. M., Bechert, K., Tuschl, T. and Weber, K. (2001). Identification of essential genes in cultured mammalian cells using small interfering RNAs. *Journal of Cell Science*, **114**, 4557–4565.

Hasuwa, H., Kaseda, K., Einarsdottir, T. and Okabe, M. (2002). Small interfering RNA and gene silencing in transgenic mice and rats. *Federation of European Biochemical Society Letters*, **532**, 227–230.

Hemann, M., T., Fridman, J. S., Zilfou, J. T., Hernando, E., Paddison, P. J., Cordon-Cardo, C., Hannon, G. J. and Low, S. W. (2003). An epi-allelic series of p53 hypomorphs created by stable RNAi produces distinct tumor phenotypes *in vivo*. *Nature Genetics*, **33**, 396–400.

International Human Genome Sequencing Consortium. (2001). Initial sequencing and analysis of the human genome. *Nature*, **409**, 860–921.

Izant, J. G. and Weintraub, H. (1985). Constitutive and conditional suppression of exogenous and endogenous genes by anti-sense RNA. *Science*, **229**, 345–352.

Jones, K. A., Tamm, J. A., Craig, D. A., Yao, W.-J. and Panico, R. (2000). Signal transduction by GABA$_B$ receptor heterodimers. *Neuropsychopharmacology*, **23** (S4), S42–49.

Kamath, R. K., Martinez-Campos, M., Zipperlen, P., Fraser, A. G. and Ahringer, J. (2001). Effectiveness of specific RNA-mediated interference through ingested double-stranded RNA in *C. elegans*. *Genome Biology*, **2**, 1–10.

Kamath, R. S., Fraser, A. G., Dong, Y., Durbin, R., Gotta, M., Kanapin, A., Le Bot, N., Moreno, S., Sohrmann, M., Welchman, D. P., Zipperlen, P. and Ahringer, J. (2003). Systematic functional analysis of the *Caenorhabditis elegans* genome using RNAi. *Nature*, **421**, 231–237.

Kunath, T., Gish, G., Lickert, H., Jones, N., Pawson, T. and Rossant, J. (2003). Transgenic RNA interference in ES cell-derived embryos recapitulates a genetic null phenotype. *Nature Biotechnology*, **21**, 559–561.

Lewis, D. L., Hagstrom, J. E., Loomis, A. G., Wolff, J. A. and Herweijer, H. (2002). Efficient delivery of siRNA for inhibition of gene expression in postnatal mice. *Nature Genetics*, **32**, 107–108.

Lois, C., Hong, E.J., Pease, S., Brown, E. J. and Baltimore, D. (2002). Germline transmission and tissue-specific expression of transgenes delivered by lentiviral vectors. *Science*, **295**, 868–872.

Lum., L., Yao, S., Mozer, B., Rovescalli, A., Von Kessler, D., Nirenberg, M. and Beachy, P. A. (2003). Identification of Hedgehog pathway components by RNAi in *Drosophila* cultured cells. *Science*, **299**, 2039–2045.

Ma, Y., Ramezani, A., Lewis, R., Hawley, R. G. and Thomson, J. A. (2003). High level sustained transgene expression in human embryonic stem cells using lentiviral vectors. *Stem Cells*, **21**, 111–117.

McCaffrey, A. P., Meuse, L., Pham, T. T., Conklin, D. S., Hannon, G. J. and Kay, M. A. (2002). RNA interference in adult mice. *Nature*, **418**, 38–39.

McManus, M. T., Haines, B. B., Chen, J. and Sharp, P. A. (2002a). siRNA-mediated silencing in T-cells. *Journal of Immunology*, **169**, 5754–5760.

McManus, M. T., Petersen, C. P., Haines, B. B., Chen, J. and Sharp, P. (2002b). Gene silencing using micro-RNA designed hairpins. *RNA*, **8**, 842–850.

Novina, C. D., Murray, M. F., Dykxhoorn, D. M., Beresford, J., Reiss, S. K., Lee, S. K., Collman, R. G., Lieberman, J., Shankar, P. and Sharp, P. A. (2002). siRNA-directed inhibition of HIV infection. *Nature Medicine*, **8**, 681–686.

Paddison, P. J., Caudy, A. A., Bernstein, E., Hannon, G. J. and Conklin, D. S. (2002). Short hairpin RNAs (shRNAs) induce sequence-specific silencing in mammalian cells. *Genes & Development*, **16**, 948–958.

Pfeifer, A., Ikawa, M., Dayn, Y. and Verma, I. (2002). Transgenesis by lentiviral vectors: Lack of gene silencing in mammalian embryonic stem cells and preimplantation embryos. *Proceedings of the National Academy of Sciences USA*, **99**, 2140–2145.

Rohrer, D. K. and Kobilka, B. K, (1998). G protein-coupled receptors: Functional and mechanistic insights through altered gene expression. *Physiological Reviews*, **78**, 35–52.

Rubinson, D. A., Dillon, C. P., Kwiatkowski, A. V., Sievers, C., Yang, L., Kopinja, J., Zhang, M., McManus, M. T., Gertler, F. B., Scott, M. L. and Van Parijs, L. (2003). A lentivirus-based system to functionally silence genes in primary mammalian cells, stem cells and transgenic mice by RNA interference. *Nature Genetics*, **33**, 401–406.

Shinagawa, T. and Ishii, S. (2003). Generation of ski-knockdown mice by expressing a long double strand RNA from an RNA polymerase II promoter. *Genes & Development*, **17**, 1340–1345.

Somma, M. P., Fasulo, B., Cenci, G., Cundari, E. and Gatti, M. (2002). Molecular dissection of cytokinesis by RNA interference in *Drosophila* cultured cells. *Molecular Biology of the Cell*, **13**, 2448–2460.

Sørensen, D. R., Leirdal, M. and Sioud, M. (2003). Gene silencing by systemic delivery of synthetic siRNAs in adult mice. *Journal of Molecular Biology*, **327**, 761–766.

Spankuch-Schmitt, B., Bereiter-Hahn, J., Kaufmann., M. and Strebhardt, K. (2002). Effect of RNA silencing of polo-like kinase-1 (PLK-1) on apoptosis and spindle formation in human cancer cells. *Journal of the National Cancer Institute*, **94**, 1863–1876.

Stein, P., Svoboda, P. and Schultz, R. M. (2003). Transgenic RNAi in mouse oocytes: A simple and fast approach to study gene function. *Developmental Biology*, **256**, 187–193.

Svoboda, P., Stein, P., Hayashi, H. and Schultz, R. M. (2000). Selective reduction of dominant maternal mRNAs in mouse oocytes by RNA interference. *Development*, **127**, 4147–4156.

Timmons, L. and Fire, A. (1998). Specific interference by ingested dsRNA. *Nature*, **395**, 854.

Timmons, L., Court, D. L. and Fire, A. (2001). Ingestion of bacterially expressed dsRNAs can produce specific and potent genetic interference in *Caenorhabditis elegans*. *Gene*, **263**, 103–112.

Van Oekelen, D., Luyten, W. H. M. L. and Leysen, J. (2003). Ten years of antisense inhibition of brain G-protein coupled receptor function. *Brain Research Reviews*, **42**, 123–142.

Venter, J. C. et al. (2001). The sequence of the human genome. *Science*, **291**, 1304–1351.

Walther, W. and Stein, U. (2000). Viral vectors for gene transfer: A review of their use in the treatment of human diseases. *Drugs*, **60**, 249–271.

White, J., Wise, A., Main, M. J., Green, A., Fraser, N. J., Disney, G. H., Barnes, A., Emson, P., Foord, S. M. and Marshall, F. (1998). Heterodimerization is required for the formation of a functional GABA$_B$ receptor. *Nature*, **396**, 679–682.

Wianny, F. and Zernicka-Goetz, M. (2000). Specific interference with gene function by double-stranded RNA in early mouse development. *Nature Cell Biology*, **2**, 70–75.

Wieland, T. and Jakobs, K. H. (1994). Measurement of receptor stimulated-guanosine 5'-O-(γ-thio) triphosphate binding by G proteins. *Methods in Enzymology*, **237**, 3–13.

Xia, H., Mao, Q., Paulson, H. L. and Davidson, B. L. (2002). siRNA-mediated gene silencing *in vitro* and *in vivo*. *Nature Biotechnology*, **20**, 1006–1010.

Zambrowicz, B. P. and Sands, A. T. (2003). Knockouts model the 100 best-selling drugs – will they model the next 100? *Nature Reviews Drug Discovery*, **2**, 38–51.

Zamenick, P. C. and Stephenson, M. L. (1978). Inhibition of Rous sarcoma virus replication and cell transformation by a specific oligodeoxynucleotide. *Proceedings of the National Academy of Sciences USA*, **75**, 280–284.

SECTION SIX

Therapeutic and drug development

26 RNAi-mediated silencing of viral gene expression and replication

Derek M. Dykxhoorn

Introduction

RNA interference (RNAi) is a gene-silencing technique first described by Fire and colleagues in *Caenorhabditis elegans* (Fire et al., 1998). They found that the introduction of long dsRNA into the nematode *C. elegans* led to the targeted disruption of the homologous mRNA. RNAi is related to a variety of other gene-silencing phenomena including posttranscriptional gene silencing (PTGS) in plants and quelling in the fungus *Neurospora crassa* (Jorgensen, 1990; Romano and Macino, 1992). The link between these seemingly divergent processes is the presence of dsRNA homologous to the silenced gene (Bernstein et al., 2001b; Waterhouse et al., 2001).

The addition of long dsRNAs into *Drosophila melanogaster* embryo extracts was shown to silence gene expression in a sequence-specific manner, recapturing the RNAi reaction *in vitro* (Tuschl et al., 1999). Although initiated by long dsRNA, the effector molecules that guide the mRNA degradation were found to be small (21- to 23-nt) dsRNA species, termed small interfering (si)RNAs, produced by the cleavage of the long dsRNAs (Zamore et al., 2000). The introduction of chemically synthesized siRNAs into the extracts was found to be sufficient for targeting mRNA degradation (Elbashir et al., 2001a). Similarly, small RNA species were found *in vivo* in a wide variety of organisms and cells undergoing dsRNA-mediated gene silencing including plants, *Drosophila* embryos, *Drosophila* Schneider 2 (S2) cells and *C. elegans* (Hamilton and Baulcombe, 1999; Hammond et al., 2000; Parrish et al., 2000, Yang et al., 2001).

When analyzed, these short RNAs had the characteristic cleavage pattern of an RNase III type enzyme with a 21–23-nt dsRNA duplex, symmetric 2- to 3-nt $3'$ overhangs and $5'$-phosphate and $3'$-hydroxyl groups (Elbashir et al., 2001a). This led to the identification of the highly conserved Dicer family of RNase III enzymes as the mediator of the long dsRNA cleavage into siRNAs (Bernstein et al., 2001a). In addition to cleaving long dsRNAs, Dicer can cleave hairpin RNA structures, including short hairpin (sh) RNAs and endogenously expressed micro (mi) RNA precursors (Hütvagner and Zamore, 2002; Paddison et al., 2002). To direct the

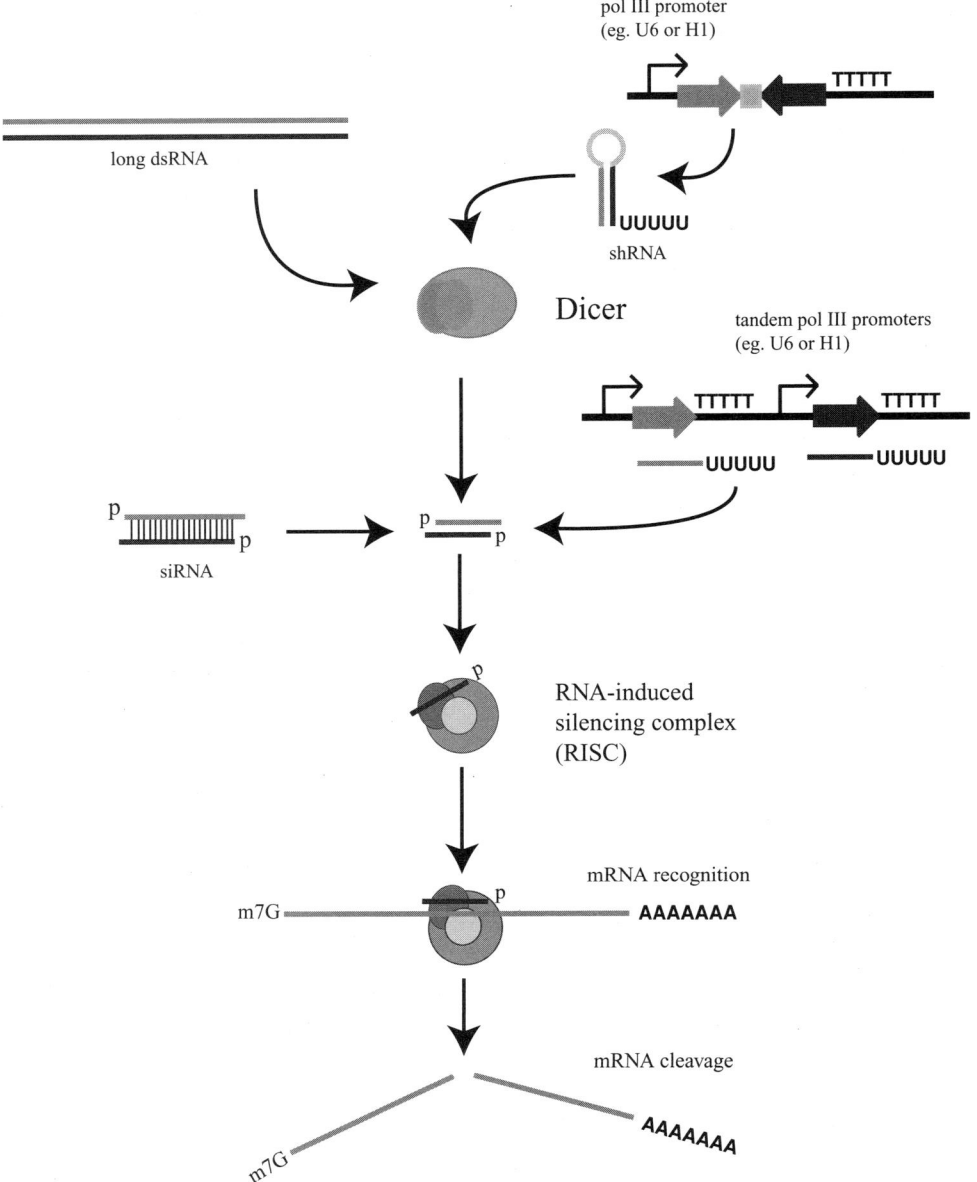

Figure 26.1. The siRNA pathway. In the siRNA pathway, long dsRNAs are cleaved by the RNase III family member, Dicer, into short interfering RNAs. The siRNAs are incorporated into the RNA-induced silencing complex (RISC), unwound in a reaction which reqires ATP hydrolysis, and the antisense strand of the siRNA guides RISC to the homologous site on the target mRNA (Hütvagner and Zamore, 2002). The endonuclease activity of RISC cleaves the mRNA in the center of the duplexed region, 10-nt from the 5′ end of the siRNA (Elbashir et al., 2001b). Short hairpin RNA can be expressed from a DNA vector that contains a single RNA polymerase III promoter. shRNAs can serve as substrates for cleavage by Dicer to produce an active siRNA. In a Dicer independent reaction, chemically synthesized siRNAs can become incorporated into RISC and guide mRNA degradation. Similarly, siRNAs that silence gene expression can also be produced by the expression of the individual sense and antisense strands from tandem RNA polymerase III promoters on DNA vectors. The sense and the antisense strand can associate in *trans*, directly forming a functional siRNA without the need for Dicer processing.

cleavage of the homologous mRNA, the siRNAs are taken up in a multisubunit complex called RISC (RNA-induced silencing complex). The antisense strand of the siRNA guides RISC to the homologous site on the mRNA leading to the endonucleoytic cleavage of the mRNA (Figure 26.1) (recently reviewed in Sharp, 2001; Hannon, 2002; McManus and Sharp, 2002; Zamore, 2002; Dykxhoorn et al., 2003).

The characterization of small RNA species from a variety of organisms and cell types has shed light on the endogenous role(s) of the RNAi pathway and the role small RNA molecules play in the regulation of gene expression. These roles include the regulation of development and the preservation of genome integrity by the suppression of mobile genetic elements including transposons, retrotransposons and viruses. One class of small RNAs that has been cloned or identified computationally are the short, single stranded micro(mi)RNAs (Pasquinelli, 2002; Pasquinelli and Ruvkun, 2002; Lim et al., 2003a,b). The function of the majority of miRNAs remains unknown. The archetype miRNAs, the short temporal (st)RNAs let-7 and lin-4, have been shown to be involved in the regulation of *C. elegans* development. These miRNAs appear to function by binding to sites on the mRNA with which they have only partial sequence complementarity leading to the repression of translation. The cloning of miRNAs from different developmental stages in *D. melanogaster* and different stages of differentiation of mouse embryonic stem cells has shown that miRNAs are expressed in a specific spatial and temporal pattern further supporting their role in regulating development (Aravin et al., 2003; Houbaviy et al., 2003).

In addition to being involved in the regulation of development, it appears that small RNAs play a role in the preservation of the integrity of the organism's genome by the suppression of invading genetic material. The cloning of small RNAs from *D. melanogaster* and *Trypanosoma brucei* led to the discovery of siR-NAs corresponding to regions of repetitive DNA, termed repeat-associated siRNAs (rasiRNAs), including transposons, retrotransposons, satellite and microsatellite DNA (Elbashir et al., 2001b; Djikeng et al., 2001; Aravin et al., 2003). Genetic studies from *C. elegans* and *Schizosacchromyces pombe* have demonstrated a link between the RNAi pathway and suppression of mobile genetic elements. Mutations in certain components of the RNAi pathway in *C. elegans* have been shown to increase the frequency of transposable element mobilization (Tabara et al., 1999). Similarly, the maintenance of retrotransposon silencing in *S. pombe* was shown to be dependent on an intact RNAi pathway (Schramke and Allshire, 2003).

The role of RNAi in the suppression of viral infection

The most extensive studies of endogenous silencing of viral genes by short dsRNA species have been in plant systems. It has long been known that a plant will be protected from the adverse effects of a more pathogenic virus if previously infected by a mild strain of a closely related virus (Waterhouse et al., 2001). Although protection is seen in mammalian systems (this forms the basis of many vaccination strategies), this protection is based on the priming of the host immune system against an invading viral pathogen. Plants do not have such an adaptive immune system. Instead, the protection that is seen is based on the similarity of the nucleic acid sequences. Plants defend themselves by exploiting the requirement of most plant viruses to go through a dsRNA intermediate during their replication cycle. In fact, the presence of dsRNA in plants has long been used as a way of diagnosing a viral infection since healthy plants do not contain dsRNA or extensively self-complementary ssRNA (Morris and Dodds, 1979). These dsRNAs can serve as substrates for siRNA production that are available to silence viral gene expression leading to a 'recovery' phenotype in the plants. In addition, the plants appear to use the existing siRNAs to protect themselves from infection by more pathogenic, but related, plant viruses (Waterhouse et al., 2001).

RNAi as a gene-silencing tool

RNA interference mediated by the introduction of long dsRNA has been used as a gene-silencing tool in a wide variety of organisms including *C. elegans*, plants, *Drosophila*, mosquito, and mouse oocytes (Fire et al., 1998; Kennerdell and Carthew, 1998; Baulcombe, 1999; Svoboda et al., 2000; Wianny and Zerricka-Goetz, 2000; Caplen et al., 2002;). However, the application of RNAi-based technologies was limited in vertebrate cells due to the nonsequence-specific suppression of gene expression by the interferon response. In the presence of long dsRNA, $2'–5'$ oligoadenylate synthase is activated producing oligoadenylates ($pppA(2'p5'A)_n$) which in turn activate RNase L, triggering the nonspecific degradation of mRNA. In addition, long dsRNA activates the protein kinase PKR which phosphorylates the translation initiation factor eIF2α, leading to a global suppression of mRNA translation. The activation of these pathways often causes the cells to undergo programmed cell death [(apoptosis) (Stark et al., 1998)].

In order to silence gene expression in mammalian cells, Tuschl and colleagues introduced siRNAs into mammalian cells by liposome-mediated transfection. They found that the introduction of dsRNAs shorter than 30 nt appears to bypass the activation of the interferon pathway while still permitting effective, sequence-specific gene silencing [(Figure 26.1) (Elbashir et al, 2001b)]. This has led to the widespread use of siRNA technologies as a tool for reverse genetic experiments in mammalian cells and has hinted at the therapeutic potential of this technology. One exciting application has been the use of siRNA-based gene silencing technologies in the inhibition of viral replication and infection. To date, there have been no reports of endogenous siRNAs being produced in virally infected mammalian cells. However, mammalian cells have been shown to possess the machinery for the cleavage of long dsRNA into siRNAs and the degradation of target mRNA. The introduction of chemically synthesized siRNAs into mammalian cells has been shown to be an effective mechanism for the suppression of viral replication and confers a degree of protection to the host cells. This chapter will look at several clinically relevant examples of the application of RNAi-based technologies for the silencing of viral gene expression and the inhibition of viral replication, as well as discussing some of the challenges of using RNAi-based technology against viral targets.

The use of RNAi to target viruses

Viruses represent a diverse group of pathogens that have widely divergent modes of replication, life cycles and genomic organization, from single- and double-stranded RNA viruses to DNA viruses, from viruses with all their genetic information on a single genomic unit to viruses with segmented genomes, from cytoplasmically replicating to nuclear replicating viruses, from viruses whose life cycle requires passage through a secondary host to viruses which are passed directly from one infected host to a new host. Although the various viral families differ extensively from one another, protein expression by all viruses is dependent on the translation of virally encoded mRNAs. Therefore, all viruses will be susceptible, to one degree or another, to the targeted disruption of their mRNA by siRNAs. The most widely used approach to silence viruses has been the direct targeting of viral protein expression and replication in permissive host cells. RNAi-based approaches have also been used to target the cellular receptor(s) (eg. CCR5 and HIV-1) used by viruses to gain entry into host cells and has been used to silence viral replication in a secondary host (eg. mosquito cells).

The inhibition of viral replication by siRNAs

Models of viral hepatitis

The Hepatitis C virus (HCV) represents a major global health concern. Approximately 3% of the world population (~270 million people) is chronically infected with HCV. Between 40 and 60% of infected individuals will progress to chronic liver disease that can lead to cirrhosis of the liver and hepatocellular carcinoma, with many infected patients requiring liver transplants (Detre et al., 1996; Alter et al., 1992; Bradley, 2000). Currently, patients with HCV are being treated with a combination of interferon (IFN) and ribavirin (Christie et al., 1999; Lanford and Bigger, 2002; McCaffrey et al., 2002). However, interferon therapy in particular has a poor response rate with greater than 60% of patients failing to clear the virus (McHutchinson et al., 1998). This lack of responsiveness to IFN may be a result of the interaction of the HCV E2 and NS5A proteins with PKR, interfering with its kinase activity and functionally eliminating that arm of the IFN pathway (Taylor et al., 1999; Abid et al., 2000).

HCV is a member of the *flaviviridae* family of viruses with a single stranded, positive polarity, cytoplasmically replicating RNA genome [(Figure 26.2A) (Lindenbach and Rice, 2001)]. The viral proteins are translated as a single polyprotein that is cleaved posttranslationally into 10 viral proteins. One of the difficulties in studying HCV has been the lack of a tissue culture model system that can sustain a productive viral infection. Therefore, an HCV replicon – a cDNA copy of the HCV genome – system has been developed which allows for the HCV replication and protein expression when stably transfected into human hepatoma cell line

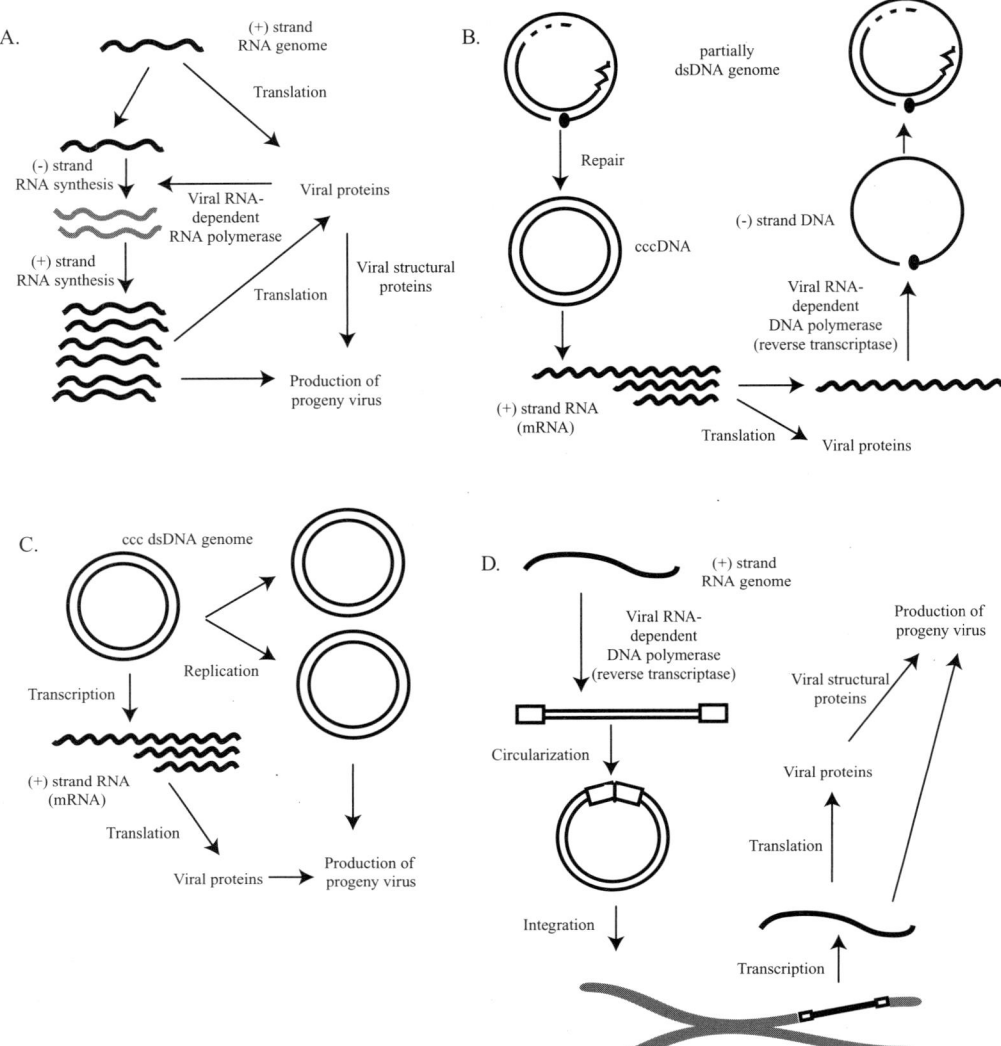

Figure 26.2. Viral replication cycles. A) General schematic of the replication of single stranded positive polarity RNA viruses, for example, poliovirus and the flaviviruses, Dengue virus and Hepatitis C virus. The single stranded RNA genome can serve as the template for viral protein expression (translation), as well as the template for the synthesis of the replication intermediate, (−) strand RNA, produced by the virally encoded RNA-dependent RNA polymerase. The (−) stranded RNA is then used as a template for the synthesis of new (+) stranded RNA genomes. Although the (+) polarity single stranded viruses replicate their genomes in the same general manner, there are differences between the viral families, for example replication of the poliovirus genome occurs in the nucleus of the host cell, whereas the replication of the flaviviral genome is completely cytoplasmic. B) Replication of the Hepatitis B viral (HBV) genome. The hepatitis B virus has a partially double stranded DNA genome. When HBV infects a host cell, the partially dsDNA genome is brought to the nucleus where the breaks in the dsDNA strands are repaired, forming a covalently closed circular DNA. Unlike most DNA viruses that undergo conservative replication, HBV replicates from an RNA intermediate through the action of a virally encoded RNA-dependent DNA polymerase (reverse transcriptase). Briefly, the repaired DNA genome is transcribed into mRNAs that fall into two size classes, the 3.5-kb class and the 2.4-/2.1-kb class. Each of the mRNAs is translated to produce different viral proteins. In addition, the 3.5-kb mRNA, representing the full genome length, serves as a template for the

[(Huh-7) (Lohmann et al., 1999)]. The targeting of HCV by siRNAs against two regions of the HCV genome, the NS3 and NS5A regions, resulted in a dramatic inhibition of HCV replication ($\sim$21- to 23-fold decrease in transcript levels) and decreased HCV protein expression. Comparison of the levels of three IFN-induced genes, MxA, PKR and OAS, showed that siRNA transfection did not activate known IFN-inducible pathways. In addition, no alteration of Huh-7 cell cycle progression was seen (Kapadia et al., 2003).

Another extensively studied agent of acute and chronic hepatitis is the Hepatitis B virus (HBV). It is estimated that 350 million individuals are chronically infected with HBV worldwide (Kao and Chen, 2002). Although similar clinical symptoms (increased risk of chronic liver disease and hepatocellular carcinoma) are associated with HBV and HCV infections, the viruses themselves are completely unrelated (Ganem and Varmus, 1987). HBV is the prototypical member of the *Hepadnaviridae* family of viruses (Ganem and Schneider, 2001). The viral genome is composed of a 3.2-kb relaxed, circular, partially double-stranded DNA genome containing four open reading frames: P (viral polymerase-reverse transcriptase), C (core structural protein of the viral nucleocapsid), S (viral surface glycoprotein) and X. Although HBV is a DNA virus, it replicates using an RNA intermediate as the template for the virally encoded reverse transcriptase activity (Figure 26.2B). There are two size classes of transcripts expressed from the HBV genome, the 3.5-kb genomic transcript and the 2.1/2.4-kb subgenomic transcripts. The 3.5-kb transcript serves as both the template for the core protein/HBeAg and polymerase-reverse transcriptase translation and as the template for replication of the HBV genome. The 2.4/2.1-kb transcript serves as the template for the viral surface glycoprotein (HBsAg) translation. Although a good tissue culture model system is lacking, the transfection of the human hepatoma cell lines, Huh-7 and HepG2 cells, with an HBV replicon can recapitulate many of the steps of HBV replication (Sells et al., 1987; Yang et al., 2002). The targeting of the core region of the HBV genome by siRNAs led to a substantial decrease in the amount of the 3.5-kb mRNA, resulting in an $\sim$5-fold decrease in the expression of HBeAg, without altering the level of HBsAg (Hamasaki et al., 2003). In addition, there was a dramatic inhibition of viral replication in siRNA treated cells as measured by a decrease in the amount of open circular and single-stranded HBV DNA replication intermediates.

Figure 26.2. *(cont.)* reverse transcription reaction producing a single stranded (−) polarity DNA. From the (−) stranded DNA the (+) DNA strand is synthesized. C) Replication of the human papillomaviral (HPV) genome. HPV has a covalently closed circular double stranded (ccc dsDNA) genome that is episomally maintained in the nucleus of the infected host cell. The degree of DNA replication depends on the state of the infection. D) Replication of retroviral genomes. Retroviruses, such as HIV-1, have a single stranded (+) polarity RNA genome. The (+) stranded RNA genome is used as the template for the synthesis of a linear dsDNA version of the genome. The linear dsDNA genome becomes circularized and transported to the nucleus of the host cell. A virally encoded integrase facilitates the incorporation of the viral genome into the host chromosome, referred to as the proviral form. Transcription from the HIV-1 LTR promoter produces a 9.2-kb mRNA that can be used for the expression of viral proteins (either from the full length mRNA or spliced forms of the mRNA), as well as, serving as the viral genome for the formation of progeny viruses.

Both HCV and HBV can cause chronic hepatitis by inducing damage to the liver. Although these viruses vary drastically in their modes of replication and genomic organization, viral protein expression and replication was effectively silenced by treatment with siRNAs. These results also demonstrate that the use of siRNAs against HCV did not induce changes in the cell cycle progression of the siRNA transfected cells nor did it induce a non-specific interferon response.

Human Papilloma virus

Human Papilloma virus (HPV) is a member of a group of small DNA viruses that preferentially infect squamous epithelial cells (Figure 26.2C). The most common outcomes of HPV infection are benign lesions of the skin (papillomas or warts) and mucous membranes [(condylomas) (Howley and Lowy, 2001)]. However, in some cases, HPV infection can lead to cellular transformation and malignancy. In fact, over 90% of human cervical cancers are positive for HPV (Jiang and Milner, 2002). There are a variety of HPV strains that have been categorized by their ability to cause cellular transformation from low-risk types (eg. HPV-6 and HPV-11) to high-risk types (eg. HPV-16 and HPV-18). High-risk types of HPV are causally linked with the initiation and malignant progression of human cervical carcinomas (Lowy and Howley, 2001). The expression of several viral oncoproteins, including E6 and E7, appears to promote this malignant transformation. E6 and E7 function by indirectly or directly binding to cellular proteins involved in cell growth regulation. Normally these genes are involved in the induction of proliferation of the epithelial cells associated with the benign lesions seen in HPV infections. However, upon immortalization it appears that regions of the HPV genome, including those that span the E6 and E7 genes, become integrated into the hosts genome leading to a sustained, misregulated expression of E6 and E7.

The targeting of E6 in two immortalized human cervical carcinoma cell lines positive for a high risk type HPV, HPV-16, resulted in the accumulation of the tumor suppressor p53 which led to arrest in the G1 phase of the cell cycle and a reduction in the rate of cell growth. Immortalized cells in which E7 was targeted appeared to undergo programmed cell death (Jiang and Milner, 2002). Similar results were found in HeLa cells that are stably infected with the high risk HPV strain, HPV-18 (Hall and Alexander. 2003). These results demonstrate that the targeting of HPV by siRNA effectively inhibited the oncogenic effects of E6 and E7 leading to a restriction in the growth of the immortalized human cervical carcinoma cells.

Poliovirus

Poliovirus is a rapidly replicating, cytoplasmic, single stranded, positive polarity RNA virus from the enterovirus family of *picornaviridae* [(Figure 26.2A) (Pallansch and Roos, 2001)]. It is highly cytolytic, inducing apoptosis of the host cells and the release of viral progeny within 6–8 h after the initial infection. HeLa cells were transfected with siRNAs against either the capsid protein (siC) or the polymerase (siP) regions of the genome, followed by an infection with live poliovirus

(Gitlin et al., 2002). The siRNAs against the poliovirus reduced the viral titer to ~1–3% of the control siRNA transfected or untransfected but infected cells and promoted the clearance of the virus from most of the cells. Although the control cells were completely lysed by 8 h post-infection, the siRNA-treated cells (siC or siP) were protected from lysis for up to 26 h post-infection. The transfection of cells with both the siC and siP siRNAs extended the protection of the cells even further to 54 h post-infection. To ensure that the viral silencing and protection were not a result of the induction of the nonspecific antiviral interferon response, mouse fibroblasts deficient in PKR and RNase L were transfected with siC or siP and subsequently infected with poliovirus. Similar to the HeLa cell results, the siRNA-transfected mouse fibroblasts showed a robust inhibition of poliovirus replication. These results show that siRNAs were capable of efficiently and specifically silencing a rapidly replicating, highly cytolytic virus without the need for a functional interferon pathway.

Vector-mediated RNAi

Chemically synthesized siRNAs have been used to effectively inhibit gene expression through the targeted degradation of the homologous mRNA. However, this approach is limited by the transient nature of the siRNA-induced gene silencing (Dykxhoorn et al., 2003). Unlike *C. elegans* and plants which have a means of amplifying the RNAi response through the activity of an endogenous RNA-dependent RNA polymerase, mammalian cells can not amplify the RNAi response (Dalmay et al., 2000; Sijen et al., 2001; Chiu and Rana, 2002; Stein et al., 2003). Therefore, as cells divide the effective concentration of the siRNAs in each cell decreases until the level of siRNAs falls below a threshold concentration under which effective silencing can not be maintained. To get around this limitation, DNA vector-based approaches have been developed. These vectors fall into two categories, those which express the individual strands of an siRNA in *trans* from tandem promoters or those which express an siRNA precursor, such as short hairpin (sh) RNAs, which can be cleaved by Dicer to produce active siRNAs (Figure 26.1). These expression units have been successfully incorporated into a variety of vector including plasmid, retroviral and adenoviral systems (see Dykxhoorn et al., 2003 for a recent review of DNA vector-mediated RNAi).

Vector-mediated RNAi has been used as a method to extend the temporal length of silencing of viral gene expression and inhibition of viral replication. Targeting of the Hepatitis C virus (HCV) by DNA vectors, either by the expression of sense and antisense strands of the siRNA from tandem promoters (Wilson et al., 2003) or the expression of short hairpin (sh) RNA (Yokota et al., 2003) against HCV led to the inhibition of HCV replication and extended the protective effects of the siRNAs from a few days for siRNA treated cells to longer than three weeks (Wilson et al., 2003). Similarly, DNA vectors were used to express shRNAs against Hepatitis B virus (HBV) leading to a decrease in viral protein expression and a reduction in the formation of replication intermediates in a tissue culture model system (Shlomai and Shaul, 2003; McCaffrey et al., 2003).

In vivo delivery of siRNAs

Recently, hydrodynamic injection of siRNAs has been used as a method for the *in vivo* delivery of siRNAs to a variety of mouse tissues and organs. This technique involves the rapid injection of large volumes of a physiological solution containing siRNAs into the tail vein of mice (Liu et al., 1999; Zhang et al., 2001; Lewis et al., 2002; McCaffrey et al., 2002; Song et al., 2003a&b). Several groups have shown that the coinjection of siRNAs or vectors that express shRNAs with a luciferase expression plasmid inhibited luciferase expression in mouse liver (Lewis et al., 2002; McCaffrey et al., 2002). In addition, Song et al (2003a) injected siRNAs against endogenous Fas receptor leading to decreased expression in the liver and protection of the mice from fulminant hepatitis.

Although not targeting HCV directly, McCaffrey et al (2002) demonstrated that the hydrodynamic injection of siRNAs against a region of the HCV genome could be a potential method for the inhibition of the virus. They showed a substantial reduction in the level of luciferase activity from a vector expressing an HCV NS5B-luciferase fusion protein when cotransfected with an anti-HCV NS5B shRNA expression vector in mouse livers.

In a more extensive analysis, shRNA expression vectors were introduced into immunocompetent (C57BL/6J) and immunocompromised (NOD/SCID; NOD/LtSz-Prhdc^scid/J) mice with the HBV genomic DNA by hydrodynamic injection (McCaffrey et al., 2003). Northern blot analysis probing for HBV mRNA showed that injection of the shRNA construct against HBV led to a reduction in the amount of HBV mRNA (77% and 92% less HBV mRNA in the immunocompetent and immunocompromised host, respectively). Although the shRNA will directly target only the 3.5-kb message, all the viral transcripts were suppressed presumably due to the silencing of important factors involved in viral replication and the expression of viral proteins. Similar effects were observed by Novina et al (2002) in RNAi-mediated inhibition of HIV-1. In addition, mice treated with shRNA showed no HBV replicative intermediates, implying an inhibition of viral replication. The levels of HBsAg in the serum were also substantially decreased (reduced by 88% relative to the control). *In situ* staining of cells in the liver showed a 99% decrease in the number of HBV capsid protein positive cells relative to the control transfections.

Inhibition of viral infection by blocking viral entry

The majority of studies using RNAi-mediated gene silencing to inhibit viral replication have directly targeted the viral mRNA. Although effective at inhibiting viral replication and protein expression, the virus is still permitted to enter the host cell. Recently, several groups have attempted to inhibit viral infection by preventing entry of the virus into the host cell by silencing the cellular receptor(s) required for viral uptake. For this to be feasible, the cellular receptor to which the virus binds must be known and the targeted disruption of the receptor must have no deleterious effects on the host.

SiRNA-mediated gene silencing has been used extensively to inhibit HIV-1 replication by targeting the virus directly [(Figure 26.2D) (Capodici et al., 2002, Coburn and Cullen, 2002; Jacque et al., 2002; Lee et al., 2002; Novina et al., 2002)] and by inhibiting the cellular receptors that facilitate viral entry into the cells (Martinez et al., 2002; Qin et al., 2003; Song et al., 2003b). Due to the critical role the primary HIV-1 receptor, the CD4 molecule, plays in maintaining immune cell function this molecule is not a good target for silencing. However, infection by HIV-1 requires additional coreceptors, CCR5 or CXCR4, which facilitate infection by macrophage- and lympho-tropic strains of HIV-1 respectively. The loss of the CCR5 receptor appears to have no deleterious effects on immune cell function or host viability (Nansen et al., 2002). Targeting the CCR5 receptor in terminally differentiated primary macrophages by siRNAs led to the silencing of CCR5 and the effective protection from infection of a CCR5-tropic strain of HIV-1 for 21 days post-infection (Song et al., 2003b). Infection of peripheral blood T lymphocytes with a lentiviral vector expressing a shRNA against CCR5 lead to a 10-fold decrease in CCR5 expression on the cells and a 3–7-fold decrease in HIV-1 infected cells when challenged with a CCR5-tropic virus (Qin et al., 2003). Similar effects were seen with the silencing of the CXCR4 receptor in the context of lymphotropic strains of HIV-1 (Martinez et al., 2002).

Blocking viral infection by blocking viral replication in the secondary host

One aspect in the control of virus spreading and the containment of viral infection has been in the control of the vectors that transmit the virus from one host to another. Arthropods, such as mosquitoes, have been shown to transmit a wide variety of human pathogens including the viral pathogens, Dengue, West Nile and Yellow Fever viruses [(Figure 26.3A and B)(Nathanson, 2001)]. In an effort to determine whether the use of dsRNAs against viral targets was feasible in mosquito cells, Caplen et al. (2002) transfected the *Aedes albopictus* (mosquito) C6/36 cell line with siRNAs against two different mosquito-borne viruses, Semliki Forest virus (SFV) and serotype 1 Dengue virus (DEN1). SFV is a member of the alphavirus family (Griffin, 2001). Its genome is a single-stranded, positive polarity RNA that is both the genome of the virus and is translated to give a polyprotein which is cleaved into four nonstructural proteins. SFV replicates in the cytoplasm of the host cell producing a full-length minus strand from which both genomic and subgenomic RNAs can be produced. The subgenomic RNAs expressed from an internal promoter produce the SFV structural proteins. SFV can infect a variety of mammals and under some conditions can cause lethal encephalitis in humans. Since mosquito cells lack an interferon response, longer dsRNAs (78 nt) were used to transfect the cells and were found to potently reduce the expression of a GFP transgene expressed from a SFV replicon (the cloned cDNA version of SFV). Although these results suggest that siRNAs can be used to effectively silence the expression of viral genes, the replicon system does not recapitulate all the steps in a viral infection.

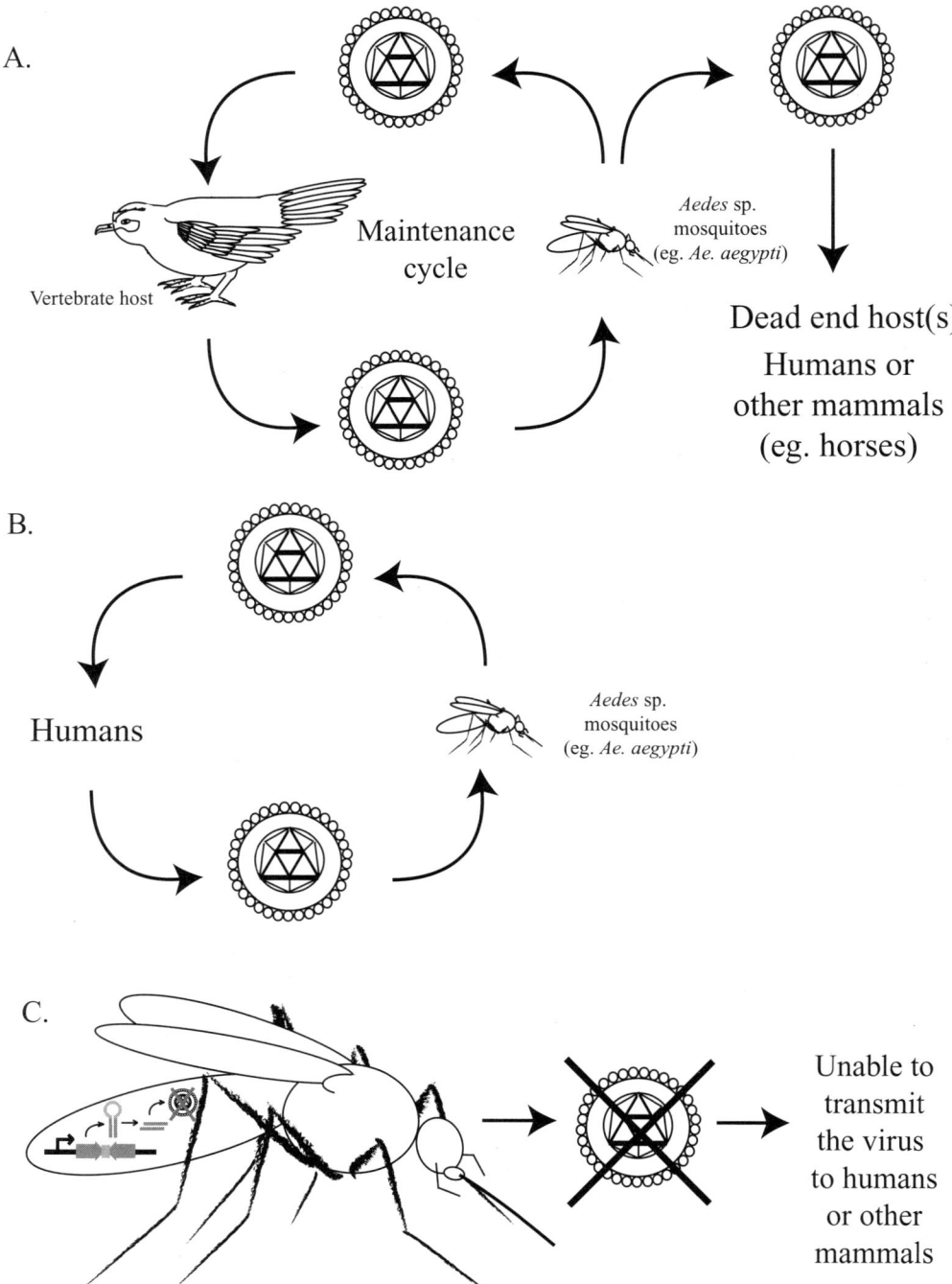

Figure 26.3. Transmission of arthropod-borne viruses. A) The transmission of many arthropod viruses, for example West Nile virus, requires the amplification and maintenance of the virus in a vertebrate host, such as birds and in some cases mammals (eg. pigs). The infection of human beings leads to a dead end infection, the virus is capable of infecting and exerting its pathogenic effects but there is insufficient viral replication to produce high enough viral titers for transmission of the virus from humans to mosquitoes (Nathanson, 2001). B) Transmission of Dengue virus. Humans infected with Dengue virus can produce sufficient viral titers to allow for the uptake and transmission

To test whether cells treated with dsRNA are capable of inhibiting viral protein expression and replication in infected cells, C6/36 mosquito cells were transfected with dsRNA and infected with Dengue virus serotype 1 (DEN1). Dengue virus is the causative agent of dengue fever, a self-limiting febrile illness that is often accompanied by minor hemorrhagic manifestations or hemorrhagic fever–dengue shock syndrome. Dengue virus infections represent the most prevalent arthropod-borne viral disease affecting ~100 million people every year (Burke and Monath, 2001). Like HCV, Dengue virus is a member of the *flaviviridae* family of viruses, having a single-stranded RNA genome. Although highly cytopathic in vertebrate cells, Dengue virus can actively replicate in the mosquito host cells and maintain a persistant, noncytopathic infection. Transfection with dsRNA against DEN1 and subsequent infection of C6/36 cells with live DEN1 virus led to a 1–2 log decrease in the amount of viral RNA and a log decrease in the amount of virus released to the culture supernatant compared to control infections.

Although this technique effectively inhibited viral expression in mosquito cells, the effects were only transient. To overcome this limitation, Adelman and colleagues (2002) developed transgenic C6/36 cell lines that expressed a long hairpin RNA (290 bp dsRNA stem) homologous to the Matrix protein (prM) region of the Dengue type 2 RNA genome. Clonal cell lines were obtained which effectively suppressed viral replication, with nearly undetectable levels of the DEN2 RNA genome, and viral protein expression, as monitored by the staining of the cells for the DEN2 E antigen. These cell lines were resistant to infection even after 40 to 50 passages of the cells. This silencing was specific because the expression of the hairpin RNA against the DEN2 prM in the clonal cell lines did not inhibit infection by a related flavivirus (West Nile virus, NY99 strain).

The development of tools that allow for the creation of transgenic mosquitoes, such as transposition-based gene transformation methods, combined with RNAi-based silencing technologies, may serve to be an effective method for the development of pathogen resistant mosquitoes (Figure 26.3C). This would decrease the number of mosquitoes that can serve as vectors for the infection of humans with various pathogenic agents and could potentially reduce the incidence of human disease and the impact of these diseases on human populations.

Viral variation and escape

There are a number of important issues to consider when dealing with the treatment of viral infections, including the high degree of sequence variability within

Figure 26.3. *(cont.)* of the virus by mosquitoes directly to another human without passing through a maintenance cycle in another host (Nathanson, 2001). C) RNAi-mediated inhibition of viral replication in transgenic mosquitoes. Transgenic techniques can potential be used to create mosquitoes that stably express siRNA from hairpin RNA precursors and actively silence viral gene expression and replication (Adelman et al., 2002). The transgenic mosquitoes would be able to take up the virus, but would be unable to replicate and transmit the virus, thereby decreasing the likelihood that the virus could be transmitted to uninfected humans or other animals.

certain types of viruses and the rapidly evolving nature of viruses. There are predominately two main causes that lead to the high degree of viral diversity, mutation of the viral genome and population size (DeFilippis and Villarreal, 2001). As both of these factors increase so does the amount of viral diversity. Viral mutations may be the result of insertions, deletions or mismatches which subtly alter the viral genome or it may involve larger scale changes such as the acquisition of new sequences by recombination of the viral genetic material with additional genetic material (eg. from the host genome) or reassortment of viral genetic material from several viruses coinfecting the same host cell. This incorporation of mutations is particularly acute in RNA viruses whose replication requires an RNA-dependent RNA polymerase and retroviruses that require an RNA-dependent DNA polymerase (reverse transcriptase) for replication, many of which lack proofreading activities. In addition, most viruses have a rapid rate of replication that allows for the production of large populations of viral progeny. It is this diversity in the viral progeny population that enables viruses to survive and replicate under selective pressure. This selective pressure most often involves the avoidance of the antiviral defense mechanisms of the host. Different viruses have adopted different strategies of evasion, from the infection of privileged sites that are protected from the host's immune system (e.g. Herpes simplex viruses) to the continual changing of the viral antigens presented to the host (eg. HIV-1).

Influenza A is one of the viruses that use the mutation of viral antigens to escape the host's protective immunity. Influenza A is the most prevalent respiratory infection in humans, infecting ~10–20% of the population and causing up to 40,000 deaths in the United States each year (Thompson et al., 2003). Despite the availability of vaccines against influenza, it remains a public health problem. This is largely due to the continual appearance of viruses with mutations in the viral antigens, in particular the hemagglutinin (HA) and neuraminidase (NA) glycoproteins (the viral antigens that elicit protective antibodies), and the emergence of new viral strains. Influenza A virus (*orthomyxoviridae* family) contains a single-stranded, negative polarity segmented RNA genome. Similar to other RNA viruses, the replication of influenza A requires an RNA-dependent RNA polymerase that has a low degree of fidelity and therefore is continually introducing mutations into the viral genome, referred to as antigenic drift (Lamb and Krug, 2001). However, unlike other RNA viruses, such as HCV (*flaviviridae*) and poliovirus (*picornaviridae*), Influenza A has a segmented genome (8 RNA segments). This allows for new viral strains to emerge due to the reassortment or mixing of RNA segments, particularly the HA and NA segments, from viruses of different species, referred to as antigenic shift. These processes produce viruses which are recognized as foreign even to a host which has been previously infected with Influenza A or has been vaccinated against a different strain of Influenza A. How then can researchers design effective therapeutic approaches against a continually changing virus? In order to get around these changes researchers have used bioinformatic approaches to design siRNAs that will target regions of the virus that are the most conserved between strains of the virus, with the assumption that the most conserved regions will tolerate less diversity.

Ge et al. (2003) used this strategy to design siRNAs against Influenza A and to test their effectiveness in suppressing viral replication and viral gene expression. Choosing siRNA sequences against the most conserved regions of the influenza A genome led to the identification of several siRNAs that strongly inhibited the replication of the virus in a tissue culture model system and a developing chicken embryo model of Influenza A infection. Similar levels of inhibition were seen against a related strain of Influenza A.

Although the selective pressure that viruses face usually involves avoidance of the antiviral defense mechanisms of the host, the introduction of antiviral agents also applies selective pressure to the virus. The same population dynamics, the frequent rate of mutation and the rapid rate of replication, which generate the diversity that allows viruses to evade the adaptive immune system of the host, also allow for the development of resistance to antiviral agents. A variety of antiviral therapies have been developed which inhibit viral replication. Although these agents have shown many clinical benefits, their application in some cases has been hampered by the emergence of drug-resistant viruses as seen in the antiviral therapy to HIV-1. For example, resistance of HIV viral reverse transcriptase to nucleoside analog reverse transcriptase inhibitors (NRTI) may be the result of a single nucleotide substitution or the accumulation of serial mutations (Larder and Kemp, 1989). These same principles should also hold true for the development of resistance to siRNAs.

It is known that siRNAs will tolerate only limited mismatches in their recognition sites especially around the site of RISC-mediated cleavage (middle of the recognition site). Just as viruses can undergo mutations which allow them to become resistant to antiviral drugs, it would be assumed that viruses could also mutate the recognition site for the siRNA and produce viruses which are resistant to particular siRNAs. In fact, this was seen by Gitlin and colleagues (2003) in their studies on poliovirus. They found that prolonged exposure of HeLa cells transfected with siRNA targeting the poliovirus capsid (siC) to high titers of poliovirus led to cytopathic effects. This cell lysis was not due to a decrease in the activity of the siRNAs within the time frame of the experiment. Instead they found that virus isolated from the siC-treated cells were no longer susceptible to siC siRNAs (viral escape siRNA mutants). Sequencing of these viruses revealed that they had acquired a point mutation in the middle of the siC siRNA recognition site. Interestingly, this was a silent point mutation affecting the mRNA sequences, but not the amino acid sequence, thereby making the virus resistant to the siRNA but preserving protein function and the viability of the virus.

Whether the escape siRNA mutant virus was produced as a result of the poliovirus replication in the siRNA treated cells or was present at a low level in the original poliovirus inoculums, the siRNAs present in the cells effectively selected for the escape mutants. Therefore, it is fairly clear that viruses will be able to mutate in order to escape targeting by siRNAs. However, due to the ease with which siRNA targets can be selected and the number of potentially effective siRNAs for any target, viruses can be easily targeted with a number of siRNAs to decrease the chances that viral escape mutants will emerge.

Limitations and challenges

RNAi-based technologies have been effective at silencing viral gene expression and replication in tissue culture model systems. However, their application in *in vivo* situations has been limited by the need for effective delivery methods. Viral vector–based approaches, particularly the use of HIV-1-based lentiviral vectors, appear to be the most attractive vehicles for the delivery of siRNAs. The HIV-1 based lentiviral vectors allow for the infection of a variety of cell types, including nondividing cells and germ cells, leading to the establishment of a stable phenotype due to their insertion into the host genome as proviral DNA. However, the benefits of stable silencing which are offered by lentiviral delivery systems are balanced by the concern that the integration of the virus in particular genomic locations could potentially lead to the activation of gene expression and the possibility of oncogenic transformation. In fact, this was found to occur in patients where lentiviral vectors were used for the delivery of proteins as a therapy for X-linked severe combined immunodeficiency [(X-SCID) (Hacien-Bay-Abina et al., 2003; Marshall, 2003)].

The use of hydrodynamic injection has been shown to deliver siRNAs to various tissues in mice when injected into the tail vein. Although this technique has been used to successfully silence various transgenes (e.g. luciferase) and endogenous genes (e.g. the Fas receptor), there are several challenges that need to be overcome before it can be applied therapeutically. Due to the inefficiencies in the uptake of siRNAs by the various tissues, large amounts of costly siRNAs are used in these experiments with the majority of the siRNAs being excreted by the animal. The hydrodynamic injection of siRNAs leads to the non-tissue specific silencing of gene expression in several tissues including the liver, kidney, spleen, lung, and pancreas (Lewis et al., 2002). The conjugation of agents that would allow for the more efficient uptake of the siRNAs would greatly improve the efficiency of RNAi *in vivo* and reduce the amounts of material needed for efficient silencing. In addition, agents that could selectively target siRNAs to specific tissues and cell types would help to reduce the non-specific targeting seen with hydrodynamic injection. Protocols that could efficiently deliver siRNAs without the need for high-pressure, high-volume injections would also help to facilitate the use of this technique in humans.

The rapid pace of technical developments in RNAi-based technologies and the increased understanding of the biology of RNA interference will help to overcome the barriers which prevent the use of RNAi as effective therapeutic modalities in humans.

Acknowledgements

Due to the large number of papers that have used RNAi-based technologies for the silencing of viral targets, only selected papers have been discussed in this chapter, I apologize for any oversights. I would like to thank L. Maliszewski, S. Lee and E. Song for the critical reading of this manuscript.

REFERENCES

Abid, K., Quadri, R. and Negro, F. (2000). Hepatitis C virus, the E2 envelope protein, and alpha-interferon resistance. *Science*, **287**, 1555.

Adelman, Z. N., Sanchez-Vargas, I., Travanty, E. A., Carlson, J. O., Beaty, B. J., Blair, C. D. and Olson, K. E. (2002). RNA silencing of dengue virus type 2 replication in transformed C6/36 mosquito cells transcribing an inverted-repeat RNA derived from the virus genome. *Journal of Virology*, **76**, 12925–12933.

Alter, M. J., Margolis, H. S., Krawczynski, K., Judson, F. N., Mares, A., Alexander, W. J., Hu, P. Y., Miller, J. K., Gerber, M. A., Sampliner, R. E., et al. (1992). The natural history of community-acquired hepatitis C in the United States. The Sentinel Counties Chronic non-A, non-B Hepatitis Study Team. *New England Journal of Medicine*, **327**, 1899–1905.

Aravin, A. A., Lagos-Quintana, M., Yalcin, A., Zavolan, M., Marks, D., Snyder, B., Gaasterland, T., Meyer, J. and Tuschl, T. (2003). The small RNA profile during *Drosophila melanogaster* development. *Developmental Cell*, **5**, 337–350.

Baulcombe, D. (1999). Viruses and gene silencing in plants. *Archives in Virology Supplement*, **15**, 189–201.

Bernstein, E., Caudy, A. A., Hammond, S. M. and Hannon, G. J. (2001a). Role for a bidentate ribonuclease in the initiation step of RNA interference. *Nature*, **409**, 363–366.

Bernstein, E., Denli, A. M. and Hannon, G. J. (2001b). The rest is silence. *RNA*, **7**, 1509–1521.

Bradley, D. W. (2000). Studies of non-A, non-B hepatitis and characterization of the hepatitis C virus in chimpanzees. *Current Topics in Microbiology and Immunology*, **242**, 1–23.

Burke, D. S. and Monath, T. P. (2001). Flaviviruses. In *Fields' Virology*, 4th edn. pp. 1043–1125. Philadelphia, PA; Lippincott, Williams and Wilkins.

Caplen, N. J., Zheng, Z., Falgout, B. and Morgan, R. A. (2002). Inhibition of viral gene expression and replication in mosquito cells by dsRNA-triggered RNA interference. *Molecular Therapy*, **6**, 243–251.

Capodici, J., Kariko, K. and Weissman, D. (2002). Inhibition of HIV-1 infection by small interfering RNA-mediated RNA interference. *Journal of Immunology*, **169**, 5196–5201.

Chiu, Y. L. and Rana, T. M. (2002). RNAi in human cells: basic structural and functional features of small interfering RNA. *Molecular Cell*, **10**, 549–561.

Christie, J. M., Chapel, H., Chapman, R. W. and Rosenberg, W. M. (1999). Immune selection and genetic sequence variation in core and envelope regions of hepatitis C virus. Hepatology, **30**, 1037–1044.

Coburn, G. A. and Cullen, B. R. (2002). Potent and specific inhibition of human immunodeficiency virus type 1 replication by RNA interference. *Journal of Virology*, **76**, 9225–9231.

Dalmay, T., Hamilton, A., Rudd, S., Angell, S. and Baulcombe, D. C. (2000). An RNA-dependent RNA polymerase gene in Arabidopsis is required for posttranscriptional gene silencing mediated by a transgene but not by a virus. *Cell*, **101**, 543–553.

Defilippis, V. R. and Villarreal, C. P. (2001). Virus Evolution. In *Fields' Virology*, 4th edn. pp. 1043–1125. Philadelphia, PA; Lippincott, Williams and Wilkins.

Detre, K. M., Belle, S. H. and Lombardero, M. (1996). Liver transplantation for chronic viral hepatitis. *Viral Hepatitis Review*, **2**, 219–228.

Djikeng, A., Shi, H., Tschudi, C. and Ullu, E. (2001). RNA interference in Trypanosoma brucei: cloning of small interfering RNAs provides evidence for retroposon-derived 24–26-nucleotide RNAs. *RNA*, **7**, 1522–1530.

Dykxhoorn, D. M., Novina, C. D. and Sharp, P. A. (2003). Killing the messenger: Short RNAs that silence gene expression. *Nature Reviews Molecular Cell Biology*, **4**, 457–467.

Elbashir, S. M., Harborth, J., Lendeckel, W., Yalcin, A., Weber, K. and Tuschl, T. (2001a). Duplexes of 21-nucleotide RNAs mediate RNA interference in cultured mammalian cells. *Nature*, **411**, 494–498.

Elbashir, S. M., Lendeckel, W. and Tuschl, T. (2001b). RNA interference is mediated by 21- and 22-nucleotide RNAs. *Genes & Development*, **15**, 188–200.

Fire, A., Xu, S., Montgomery, M. K., Kostas, S. A., Driver, S. E. and Mello, C. C. (1998). Potent and specific genetic interference by double-stranded RNA in *Caenorhabditis elegans*. *Nature*, **391**, 806–811.

Ganem, D. and Schneider, R. J. (2001). Hepadnaviridae: The viruses and their replication. In *Fields' Virology*, 4th edn, pp.2923–2969. Philadelphia, PA; Lippincott, Williams and Wilkins.

Ganem, D. and Varmus, H. E. (1987). The molecular biology of the hepatitis B viruses. *Annual Reviews of Biochemistry*, **56**, 651–693.

Ge, Q., McManus, M. T., Nguyen, T., Shen, C. H., Sharp, P. A., Eisen, H. N. and Chen J. (2003). RNA interference of influenza virus production by directly targeting mRNA for degradation and indirectly inhibiting all viral RNA transcription. *Proceedings of the National Academy of Sciences USA*, **100**, 2718–2723.

Gitlin, L., Karelsky, S. and Andino, R. (2002). Short interfering RNA confers intracellular antiviral immunity in human cells. *Nature*, **418**:430–434.

Griffin, D. (2001). Alphaviruses. In *Fields' Virology*, 4th edn. pp. 917–962. Philadelphia, PA; Lippincott, Williams and Wilkins.

Hacien-Bay-Abina, S., von Kalle, C., Schmidt, M. and LeDeist, F. (2003). A serious adverse event after successful gene therapy for X-linked severe combined immunodeficiency. *New England Journal of Medicine*, **348**, 255–256.

Hall, A. H. S. and Alexander, K. A. (2003). RNA interference of human papillomavirus type 18 E6 and E7 induces senescence in HeLa cells. *Journal of Virology*, **77**, 6066–6069.

Hamasaki, K., Nakao, K., Matsumoto, K., Ichikawa, T., Ishikawa, H. and Eguchi, K. (2003). Short interfering RNA-directed inhibition of hepatitis B virus replication. *Federation of European Biochemical Society Letters*, **543**, 51–54.

Hamilton, A. J. and Baulcombe, D. C. (1999). A species of small antisense RNA in posttranscriptional gene silencing in plants. *Science*, **286**, 950–952.

Hammond, S. M., Bernstein, E., Beach, D. and Hannon, G. J. (2000). An RNA-directed nuclease mediates post-transcriptional gene silencing in *Drosophila cells*. *Nature*, **404**, 293–296.

Hannon, G. J. (2002). RNA interference. *Nature*, **418**, 244–251.

Houbaviy, H. B., Murray, M. F. and Sharp, P. A. (2003). Embryonic stem cell-specific MicroRNAs. *Developmental Cell*, **5**, 351–358.

Howley, P. M. and Lowy, D. R. (2001). Papillomaviruses and their replication. In *Fields' Virology*, 4th edn. pp. 2197–2229. Philadelphia, PA; Lippincott, Williams and Wilkins.

Hütvagner, G. and Zamore, P. D. (2002). A microRNA in a multiple turnover RNAi enzyme complex. *Science*, **297**, 2056–2060.

Jacque, J. M., Triques, K. and Stevenson, M. (2002). Modulation of HIV-1 replication by RNA interference. *Nature*, **418**, 435–438.

Jiang, M. and Milner, J. (2002). Selective silencing of viral gene expression in HPV-positive human cervical carcinoma cells treated with siRNA, a primer of RNA interference. *Oncogene*, **21**, 6041–6048.

Jorgensen, R. (1990). Altered gene expression in plants due to trans interactions between homologous genes. *Trends in Biotechnology*, **8**, 340–344.

Kao, J. H. and Chen, D. S. (2002). Global control of hepatitis B virus infection. *Lancet Infectious Diseases*, **2**, 395–403.

Kapadia, S. B., Brideau-Andersen, A. and Chisari, F. V. (2003) Interference of hepatitis C virus RNA replication by short interfering RNAs. *Proceedings of the National Academy of Sciences USA*, **100**, 2014–2018.

Kennerdell, J. R. and Carthew, R. W. (1998). Use of dsRNA-mediated genetic interference to demonstrate that frizzled and frizzled 2 act in the wingless pathway. *Cell*, **95**, 1017–1026.

Lamb, R. A. and Krug, R. M. (2001). Orthomyxoviridae: The viruses and their replication. In *Fields' Virology*, 4th edn. pp. 1487–1531. Philadelphia, PA; Lippincott, Williams and Wilkins.

Lanford, R. E. and Bigger, C. (2002). Advances in model systems for hepatitis C virus research. *Virology*, **293**, 1–9.

Larder, B. and Kemp. S. (1989). Multiple mutations in HIV-1 reverse transcriptase confer high-level resistance to zidovudine (AZT). *Science*, **246**, 1155–1158.

Lee, N. S., Dohjima, T., Bauer, G., Li, H., Li, M. J., Ehsani, A., Salvaterra, P., and Rossi, J. (2002). Expression of small interfering RNAs targeted against HIV-1 rev transcripts in human cells. *Nature Biotechnology*, **20**, 500–505.

Lewis, D. L., Hagstrom, J. E., Loomis, A. G., Wolff, J. A. and Herweijer, H. (2002). Efficient delivery of siRNA for inhibition of gene expression in postnatal mice. *Nature Genetics*, **32**, 107–108.

Lim, L. P., Glasner, M. E., Yekta, S., Burge, C. B. and Bartel, D. P. (2003a). Vertebrate microRNA genes. *Science*, **299**, 1540.

Lim, L. P., Lau, N. C., Weinstein, E. G., Abdelhakim, A., Yekta, S., Rhoades, M.W., Burge, C. B. and Bartel, D. P. (2003b). The microRNAs of *Caenorhabditis elegans*. *Genes & Development*, **17**, 991–1008.

Lindenbach, B. D. and Rice, C. M. (2001). Flaviviridae: the viruses and their replication. In *Fields' Virology*, 4th edn. pp. 999–1041. Philadelphia, PA; Lippincott, Williams and Wilkins.

Liu, F., Song, Y. and Liu, D. (1999). Hydrodynamic-based transfection in animals by systemic administration of plasmid DNA. *Gene Therapy*, **6**, 1258–1266.

Lohmann V., Korner, F., Koch, J., Herian. U., Theilmann, L. and Bartenschlager R. (1999). Replication of subgenomic hepatitis C virus RNAs in a hepatoma cell line. *Science*, **285**, 110–113.

Lowy, D. R. and Howley, P. M. (2001). Papillomaviruses. In *Fields' Virology*, 4th edn. pp. 2231–2264. Philadelphia, PA; Lippincott, Williams and Wilkins.

Marshall, E. (2003). Gene therapy: Second child in French trial is found to have leukemia. *Science*, **299**, 320.

Martinez M. A., Gutierrez, A., Armand-Ugon, M., Blanco, J., Parera M., Gomez, J., Clotet, B. and Este J. A. (2002). Suppression of chemokine receptor expression by RNA interference allows for inhibition of HIV-1 replication. *AIDS*, **16**, 2385–2390.

McCaffrey, A. P., Nakai, H., Pandey, K., Huang, Z., Salazar, F. H., Xu, H., Wieland, S. F., Marion, P. L. and Kay, M. A. (2003). Inhibition of hepatitis B virus in mice by RNA interference. *Nature Biotechnology*, **21**, 639–644.

McCaffrey, A. P., Ohashi, K., Meuse, L., Shen, S., Lancaster, A. M., Lukavsky, P. J., Sarnow, P. and Kay, M. A. (2002). Determinants of hepatitis C translational initiation *in vitro*, in cultured cells and mice. *Molecular Therapy*, **5**, 676–684.

McHutchison, J. G., Gordon, S. C., Schiff, E. R., Shiffman, M. L., Lee, W. M., Rustgi, V. K., Goodman, Z. D., Ling, M. H., Cort, S. and Albrecht, J. K. (1998). Interferon alfa-2b alone or in combination with ribavirin as initial treatment for chronic hepatitis C. Hepatitis Interventional Therapy Group. *New England Journal of Medicine*, **339**, 1485–1492.

McManus, M. T. and Sharp, P. A. (2002). Gene silencing in mammals by small interfering RNAs. *Nature Reviews Genetics*, **3**, 737–747.

Morris, T. J. and Dodds, J. A. (1979). Isolation and analysis of double stranded RNA from virus infected plant and fungal tissue. *Phytopathology*, **69**, 854–858.

Nansen, A., Christensen, J. P., Andreasen, S. O., Bartholdy, C., Christensen, J. E. and Thomsen, A. R. (2002). The role of CC chemokine receptor 5 in antiviral immunity. Blood, **99**,1237–1245.

Nathanson, N. (2001). Epidemiology. In *Fields' Virology*, 4th edn. pp.371–392. Philadelphia, PA; Lippincott, Williams and Wilkins.

Novina, C. D., Murray, M. F., Dykxhoorn, D. M., Beresford, P. J., Riess, J., Lee, S. K., Collman, R. G., Lieberman, J., Shankar, P. and Sharp, P. A. (2002) siRNA-directed inhibition of HIV-1 infection. *Nature Medicine*, **8**, 681–686.

Paddison, P. J., Caudy, A. A., Bernstein, E., Hannon, G. J. and Conklin, D. S. (2002). Short hairpin RNAs (shRNAs) induce sequence-specific silencing in mammalian cells. *Genes & Development*, **16**, 948–958.

Pallansch, M. A. and Roos, R. P. (2001). Enteroviruses: Polioviruses, Coxsackie viruses and newer enteroviruses. In *Fields' Virology*, 4th edn. pp.723–775. Philadelphia, PA; *Lippincott, Williams and Wilkins*.

Parrish, S., Fleenor, J., Xu, S., Mello, C. and Fire, A. (2000). Functional anatomy of a dsRNA trigger: Differential requirement for the two trigger strands in RNA interference. *Molecular Cell*, **6**, 1077–1087.

Pasquinelli, A. E. (2002). MicroRNAs: Deviants no longer. *Trends in Genetics*, **18**, 171–173.

Pasquinelli, A. E. and Ruvkun, G. (2002). Control of developmental timing by microRNAs and their targets. *Annual Reviews of Cell and Developmental Biology*, **18**, 495–513.

Qin, X. F., An, D. S., Chen, I. S. and Baltimore, D. (2003). Inhibiting HIV-1 infection in human T cells by lentiviral-mediated delivery of small interfering RNA against CCR5. *Proceeding of the National Academy of Sciences USA*, **100**, 183–188.

Romano, N. and Macino, G. (1992). Quelling: Transient inactivation of gene expression in *Neurospora crassa* by transformation with homologous sequences. *Molecular Microbiology*, **6**, 3343–3353.

Schramke, V. and Allshire, R. (2003). Hairpin RNAs and retrotransposon LTRs effect RNAi and chromatin-based gene silencing. *Science*, **301**, 1069–1074.

Sells, M. A., Chen, M. L. and Acs, G. (1987). Production of hepatitis B virus particles in Hep G2 cells transfected with cloned hepatitis B virus DNA. *Proceedings of the National Academy of Sciences USA*, **84**, 1005–1009.

Sharp, P. A. (2001). RNA interference – 2001. *Genes & Development*, **15**, 485–490.

Shlomai, A. and Shaul, Y. 2003. Inhibition of Hepatitis B virus expression and replication by RNA interference. *Hepatology*, **37**, 764–770.

Sijen, T., Fleenor, J., Simmer, F., Thijssen, K. L., Parrish, S., Timmons, L., Plasterk, R. H. and Fire, A. (2001). On the role of RNA amplification in dsRNA-triggered gene silencing. *Cell*, **107**, 465–476.

Song, E., Lee, S. K., Dykxhoorn, D. M., Novina, C., Zhang, D., Crawford, K., Cerny, J., Sharp, P. A., Lieberman, J., Manjunath, N. and Shankar, P. (2003a). Sustained small interfering RNA-mediated human immunodeficiency virus type 1 inhibition in primary macrophages. *Journal of Virology*, **77**, 7174–7181.

Song, E., Lee, S. K., Wang, J., Ince, N., Ouyang, N., Min, J., Chen, J., Shankar, P. and Lieberman, J. (2003b). RNA interference targeting Fas protects mice from fulminant hepatitis. *Nature Medicine*, **9**, 347–351.

Stark, G. R, Kerr, I. M., Williams, B. R., Silverman, R. H. and Schreiber, R. D. (1998). How cells respond to interferons. *Annual Reviews of Biochemistry*, **67**, 227–264.

Stein, P., Svoboda, P., Anger, M. and Schultz, R. M. (2003). RNAi: Mammalian oocytes do it without RNA-dependent RNA polymerase. *RNA*, **9**, 187–192.

Svoboda, P., Stein, P., Hayashi, H. and Schultz, R. M. (2000). Selective reduction of dormant maternal mRNAs in mouse oocytes by RNA interference. *Development*, **127**, 4147–4156.

Tabara, H., Sarkissian, M., Kelly, W. G., Fleenor, J., Grishok, A., Timmons, L., Fire, A. and Mello, C. C. (1999). The rde-1 gene, RNA interference, and transposon silencing in *C. elegans*. Cell, **99**, 123–132.

Taylor, D. R., Shi, S. T., Romano, P. R., Barber, G. N. and Lai, M. M. (1999). Inhibition of the interferon-inducible protein kinase PKR by HCV E2 protein. *Science*, **285**, 107–110.

Thompson, W. W., Shay, D. K., Weintraub, E., Brammer, L., Cox, N., Anderson, L. J. and Fukuda, K. (2003). Mortality associated with influenza and respiratory syncytial virus in the United States. *Journal of the American Medical Association*, **289**, 179–186.

Tuschl, T., Zamore, P. D., Lehmann, R., Bartel, D. P. and Sharp, P. A. (1999). Targeted mRNA degradation by double-stranded RNA *in vitro*. *Genes & Development*, **13**, 3191–3197.

Waterhouse, P. M., Wang, M. B. and Lough T. (2001). Gene silencing as an adaptive defence against viruses. *Nature*, **411**, 834–842.

Wianny, F. and Zernicka-Goetz, M. (2000). Specific interference with gene function by double-stranded RNA in early mouse development. *Nature Cell Biology*, **2**, 70–75.

Wilson, J. A., Jayasena, S., Khvorova, A., Sabatinos, S., Rodrigue-Gervais, I. G., Arya, S., Sarangi, F., Harris-Brandts, M., Beaulieu, S. and Richardson C. D. (2003). RNA interference blocks gene expression and RNA synthesis from hepatitis C replicons propagated in human liver cells. *Proceedings of the National Academy of Sciences USA*, **100**, 2783–2788.

Yang, D., Lu, H. and Erickson, J. W. (2001). Evidence that processed small dsRNAs may mediate sequence-specific mRNA degradation during RNAi in *Drosophila* embryos. *Current Biology*, **10**, 1191–1200.

Yang, P. L., Althage, A., Chung, J., Chisari, F. V. (2002). Hydrodynamic injection of viral DNA: A mouse model of acute hepatitis B virus infection. *Proceedings of the National Academy of Sciences USA*, **99**, 13825–13830.

Yokota, T., Sakamoto, N., Enomoto, N., Tanabe, Y., Miyagishi, M., Maekawa, S., Yi, L., Kurosaki, M., Taira, K., Watanabe, M. and Mizusawa, H. (2003). Inhibition of intracellular hepatitis C virus replication by synthetic and vector-derived small interfering RNAs. *European Molecular Biology Organization Reports*, **4**, 602–608.

Zamore, P. D., Tuschl, T., Sharp, P. A. and Bartel, D. P. (2000). RNAi: Double-stranded RNA directs the ATP-dependent cleavage of mRNA at 21 to 23 nucleotide intervals. *Cell*, **101**, 25–33.

Zamore, P. D. (2002). Ancient pathways programmed by small RNAs. *Science*, **296**, 1265–1269.

Zhang, G., Budker, V., Williams, P., Subbotin, V. and Wolff, J. A. (2001). Efficient expression of naked DNA delivered intraarterially to limb muscles of nonhuman primates. *Human Gene Therapy*, **12**, 427–438.

27 RNAi in drug development: Practical considerations

Dmitry Samarsky, Margaret Taylor, Mark A. Kay, and Anton P. McCaffrey

Introduction

The development of pharmaceuticals is a complex process that requires a tremendous investment of human and financial resources. In the post-genomic era, drug development has traditionally followed an established protocol that begins with the identification of a new target, proceeds with the identification, optimization and clinical evaluation of small molecule inhibitors of that target and culminates with the successful launch of a new pharmaceutical entity (Figure 27.1). The estimated cost of bringing a new drug to the market place is approximately $900 million over an 8+ year timeframe. RNAi technology can be applied at multiple steps in the drug development process and has the potential to change the pattern of drug development altogether. Indeed, RNAi has been widely employed as a powerful target discovery and validation tool *in vitro* and is currently being applied *in vivo* for target validation using whole organisms. In addition several companies are pursuing the clinical development RNAi compounds for the treatment of various diseases.

Mediators of RNAi fall into two main categories: synthetic RNAi compounds, *e.g.* short interfering RNA (or siRNA) and Invitrogen's *STEALTH RNAi*™ compounds, that can be delivered exogenously; and RNAi molecules, *e.g.* short hairpin RNA, or shRNA, that are expressed endogenously from plasmid or viral vectors. Both of these formats have merits as well as limitations, and have application in cultured cells (*in vitro*) and in whole organisms (*in vivo*). Synthetic RNAi compounds are of a defined composition and are easily synthesized under GMP conditions, making them more appealing to the pharmaceutical industry and regulators than endogenously expressed RNAi triggers. Since the effects of synthetic RNAi are transient, however, they may require repeated administrations for many *in vitro* and *in vivo* applications. Gene silencing with expressed RNAi-inducing molecules is longer lasting and, in the case of integrated expression cassettes, can be transmitted through the germ line to progeny. An advantage of plasmid expression vectors is that they can be produced easily and cheaply since

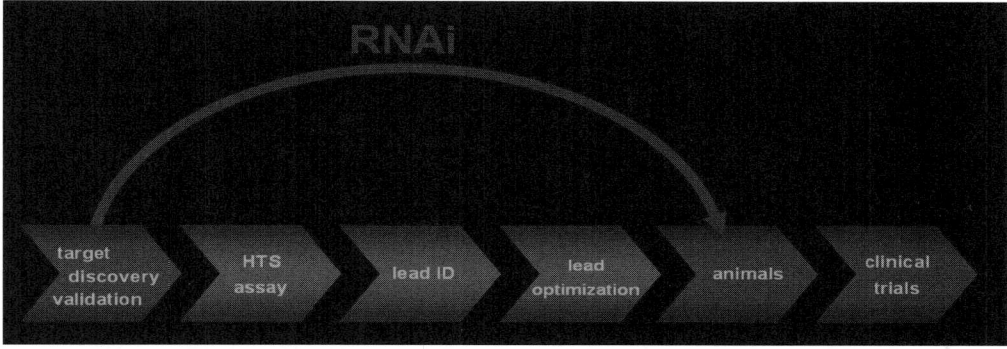

Figure 27.1. At the early stages (often referred as *"in vitro* drug target discovery and validation") potential drug targets are selected using a range of criteria and then validated using *in vitro* test systems. Then high-throughput assays are developed and these assays are used to screen small molecule libraries. At later stages, the most promising compounds are optimized to provide better performance, tested in animal systems for their activity and ADMET (absorbance, distribution, metabolism, excretion and toxicity) properties, and evaluated in clinical trials.

the infrastructure for their large scale production already exists. Production of viral vectors is more costly and labor intensive. The primary advantages associated with the use of custom made viruses are the ability to transduce many cell types (*e.g.* primary cells) that are difficult to transfect using non-viral methods, and also to efficiently transduce many target tissues *in vivo.*

The emergence of RNAi technology has significantly affected the way in which scientists approach both the early and late stages of the drug development process. In early stage discovery and pre-clinical experiments, RNAi is a powerful target discovery and validation tool that may be applied to *in vitro* and *in vivo* models of disease. After validation in pre-clinical models, RNAi compounds can be directly evaluated in the clinic as potential therapeutic modalities, bypassing the need for the development of traditional small molecule or biologic drug candidates. The successful application of RNAi in the drug development process requires that certain criteria be met according to the specifics of the disease system. In this chapter we will discuss various factors that are important determinants for RNAi success in cell culture and the whole organism.

RNAi *in vitro*

Several factors play key roles in the successful application of RNAi to *in vitro* systems, including the hit rate, specificity, delivery and longevity of silencing activity. The relative importance of each factor depends upon the specific system in which the RNAi is being applied. For example, if a large number of genes has to be analyzed at the early target discovery stage, high hit rate (*i.e.* number of compounds that must be tested to identify one yielding sufficient knock-down) is a critical factor. In another example, when the effect of the gene knock-down must be monitored many days after the knock-down, or if the targeted protein has a long half-life, the longevity of the silencing is important.

Generating validated reagents

Many scientists are employing RNAi technology to develop a toolbox of reagents that can be used to manipulate the expression of target genes implicated in the regulation of key signaling and biochemical pathways. In order for that toolbox to be useful, the reagents in it must effectively and specifically target the intended genes. To generate such an arsenal of reagents, a panel of RNAi duplexes must be screened to determine those that most effectively inhibit the target. Effective RNAi compounds can then be used to evaluate key pathways across numerous disease models. Depending upon the number of target genes, time and cost can become prohibitive factors in generating validated RNAi duplexes; therefore, the hit rate of the duplexes becomes a critical factor as it impacts both the cost and length of the screening process. To increase the throughput of the screening process, it is desirable to employ a cell line that is easy to deal with and for which reliable, reproducible and robust transfection protocols are available. Of course, the gene(s) of interest must be expressed in the cell line chosen for screening. Identification of effective duplexes is usually based on inhibition of the target mRNA at 24–48 hours post-transfection. Evaluation of inhibition at the level of protein translation is another option, but is more complicated than the mRNA knockdown, due to the lack of good antibodies for some targets. In addition, the half-life of some proteins is long, precluding the evaluation of knock-down within 48 hours of the transfection. Rather than conducting these screening steps in their own labs, scientist are turning to suppliers of so-called validated or pre-screened RNAi compounds (*e.g.* Invitrogen, Qiagen, Ambion, Dharmacon, Eurogentec). Although there are no uniform criteria for acceptable efficacy, in most cases RNAi compounds yielding more than 70% knock-down of the target transcript are considered "functional."

High-throughput applications

In a different setting, researchers may be interested in knocking down a large number of genes (hundreds or thousands) within a disease model (*e.g.* apoptosis, cell proliferation, inflammation, etc.). RNAi is well suited for such applications because the compounds can be readily designed based solely on target sequence information. In high-throughput settings, one ideally relies on a phenotypic readout to identify the target genes whose down-regulation by RNAi impacts the pathway under investigation; therefore, high hit rate of the duplexes and a well-planned phenotypic assay are critical for success. The assay should be reliable and robust with an easily quantifiable endpoint. Robust and straightforward phenotypic assays may include the evaluation of the expression of downstream marker genes by Q-RT-PCR or ELISA (Invitrogen, unpublished). The need for duplexes with a high hit rate (>90%) is critical to avoid false negatives. Several companies are in the process of developing genome-wide libraries of validated compounds (*e.g.* Invitrogen, Dharmacon, Qiagen, Ambion); however, the use of such libraries is not yet widely accepted (most likely due to partiality and high cost). In addition, substantial efforts are underway to improve the hit rate of duplexes by developing RNAi sequence design algorithms that accurately predict the sequence of active compounds (*e.g.* Ambion, Dharmacon, Invitrogen, Compugen). The recent discovery that siRNA compounds are asymmetrical in terms of recognition by RISC complex and the identification of the factors (oligonucleotide composition at the ends of the RNAi complexes) which seem to improve the incorporation of the RNAi compound antisense strand into RISC complex (Khvorova et al., 2003; Schwarz et al., 2003) have a material impact on our ability to predict the sites of effective duplexes.

There are some alternative approaches that increase the probability of successfully inhibiting the target gene of interest with just one treatment per gene. One such approach, the generation of mixtures of targeted compounds, was introduced by Sequitur (now Invitrogen) using antisense technology (unpublished), and has been actively used by Dharmacon with RNAi in the form of so-called SMARTpools. Mixing multiple targeting compounds in a single treatment essentially guarantees hit rates higher than 90%. Another approach is to use long double-stranded RNA (dsRNA) that has been processed into small pieces by enzyme Dicer, which essentially mimics the natural processing of long dsRNAs *in vivo*. The concern with both approaches is that the introduction of multiple compounds in a single treatment may increase the potential for generating nonspecific effects. In the case of mixtures of known compounds, these concerns may be resolved by additional testing of individual siRNAs after initial screens.

Difficult cell lines

As mentioned in the previous sections, the generation of validated reagents and high throughput screens are certainly facilitated by employing a robust cell line that is readily transfectable with RNAi. Immortalized cell lines are often used for this reason. It is important to note, however, that the genetic programs of immortalized cells are significantly altered compared to those of primary cells. Thus, the appropriate evaluation of a specific pathway or model system frequently requires the use of cell lines that are more difficult to transfect such as monocytes, lymphocytes, neurons and other primary cells. In these types of systems, delivery becomes the critical criterion that must be addressed in order for successful application of RNAi. Non-adherent (suspension) cells, which are used extensively in the study of inflammation and immune regulation, frequently require the use of specialized delivery protocols, such as electroporation. While many adherent cell types can be transfected using cationic lipids or non-cationic delivery vehicles, some, including neurons and other primary cells, may require the development of novel delivery protocols for RNAi. It is critical that the delivery protocols that are developed for these cell types are compatible with the phenotypic readout. In order to clearly evaluate the effect of target gene knock-down, the signal-to-noise ratio must be such that differences in the phenotypic response (which may be small differences) are perceptible. For example, the release of cytokines by human monocytes (THP-1) in response to stimulation is blunted upon repeated delivery of RNAi by electroporation (Dr. Joanne Kamens, Abbott Laboratories, personal communication).

In cases where sufficient delivery is not achievable within the context of the phenotypic assay, several alternative approaches can be suggested: 1) spiking RNAi reagent with fluorescent compound, followed by sorting (problems: stressful for cells, simple labeling of standard siRNA compounds interferes with gene silencing); 2) co-injection of RNAi compounds with expression vectors providing the target (problem: not always physiologically relevant); 3) creation of stable cell lines (problem: not good for primary cells); 4) infection with viral expression vectors (problems: time consuming, cell type-dependent and non-specific effects due to the viral vector).

Assays with extended timeframe

The use of RNAi to evaluate the roles of target genes within certain biological systems is complicated by the duration of the assay. For example, in some systems it is essential that the target gene be knocked down for seven days or longer. In such cases, specificity of the RNAi compounds as well as the duration of the silencing effect of RNAi is critical to the success of RNAi in the system. The duration of the effect becomes particularly important if the cells are actively dividing over the course of the assay (thereby diluting the effective concentration of RNAi per cell). In our experience, one way to enhance the probability of success is to simplify the readout protocol as much as possible (reducing the timing from transfection to readout, simple vs. complex assays) in order to reduce the noise. For example, if the endpoint of an assay is a biochemical or protein readout at day 7 of differentiation, it is often possible to assess the expression levels of the mRNA encoding the protein of interest 2–3 days earlier (Invitrogen, unpublished observations). Such a reduction in timeframe may obviate the need for repeated transfection of the cells, thereby limiting the potential for transfection related side effects.

The development of RNAi compounds with improved stability and enhanced duration of activity may also serve to eliminate the need for repeated transfection protocols. Work is in progress by several groups to generate chemically modified RNAi compounds that exhibit improved stability inside the cell while retaining high levels of activity. The specificity of the compounds in long-term assays is also critical since side effects can be cumulative over time and create the "noise" that obscures the actual signal. An attractive alternative to the use of endogenously delivered RNAi compounds is generation of stable cell lines, expressing RNAi-triggering molecules from integrated cassettes. Such approach, however, is time consuming and requires the use of integrating viral vectors in non-immortalized cells.

RNAi *in vivo*

There are two main *in vivo* applications for RNAi in the drug development process: target validation within the context of the whole organism and as a therapeutic modality. Delivery of RNAi to target cells and tissues in mammalian organism is considerably more difficult than in cultured cells. This step is likely to be a critical bottleneck in the *in vivo* application of RNAi. Thus, initial therapeutic targets and target validation models will probably be chosen based on the accessibility of various organs to RNAi triggers. It is also important to keep in mind that RNAi is particularly suitable to disease systems where the knock-down of a single gene (monogenic diseases) or small group of genes results in a therapeutic phenotype.

After a brief review of current delivery paradigms and progress in the field, we will give examples of tissue types that can be transfected/transduced relatively easily *in vivo* and are the site of important dominant, monogenic diseases. Examples of attractive disease targets in these accessible organs will be provided.

Delivery sets the rule

Strategies for *in vivo* delivery of nucleic acids are divided into two broad categories, non-viral and viral. Prior to the discovery of RNAi, a considerable amount of work was devoted to plasmid and non-viral antisense delivery as well as gene therapy using viral vectors and plasmid based approaches. This work is likely to be instructive for RNAi practitioners, although it should be noted that there could be important differences in delivery of RNAi compounds *vs.* traditional antisense.

In classical non-viral delivery systems, nucleic acids are complexed with cationic lipids or cationic polymers (reviewed in Davis, 2002). While these methods can be quite effective in cultured cells, their use *in vivo* results in delivery to only a subset of tissues and is often accompanied by toxic side effects. Intravenous administration of cationic lipid/oligonucleotide complexes results in gene expression predominantly in the lung with lower levels of gene expression in the liver and the heart (Iyer et al., 2002).

More recently, delivery of "naked" nucleic acids has received considerable attention (Herweijer and Wolff, 2003). In mammals, naked nucleic acids enter cells very inefficiently. In limited settings, physical methods such as hydrodynamic transfection and *in vivo* electroporation can significantly enhance uptake. *In vivo* electroporation has been successfully used to facilitate the uptake of siRNAs. Transduction of cells *in vivo* by viral vectors is much more effective than by non-viral methods. The target tissues that can be efficiently transduced, depending upon the type of viral vector used (reviewed in Lundstrom, 2003), include liver, muscle, retina, CNS, lymphocytes and tumors. Unfortunately, viral vectors can also have unwanted side effects such as toxicity and in some cases insertional mutagenesis. Currently no non-viral or viral vector system possesses ideal properties. Nevertheless, the examples of *in vivo* efficacy with RNAi cited below may provide motivation for renewed studies on the improvement of *in vivo* delivery methodologies.

Early experiments using RNAi *in vivo*

Initial proof-of-principle experiments in mouse oocytes microinjected with dsRNA first demonstrated that RNAi was functional in mammals (Svoboda et al., 2000; Wianny and Zernicka-Goetz, 2000) but the effect was transient, precluding studies in adult mice. More recently, hydrodynamic transfection has been used

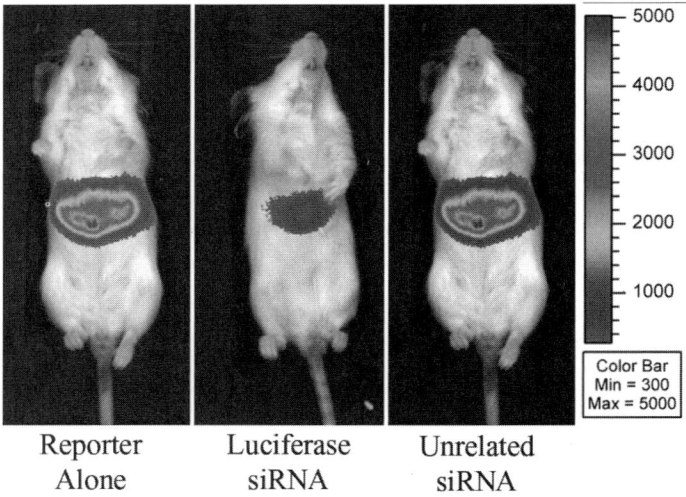

Reporter   Luciferase   Unrelated
Alone   siRNA   siRNA

Figure 27.2. Light emitted from living mice as the result of luciferase expression is significantly reduced in the presence of luciferase siRNAs. Representative images of mice co-transfected with the luciferase plasmid pGL3-Control and either no siRNA (left), luciferase siRNA (middle) or unrelated siRNA (right). A pseudocolor image representing intensity of emitted light (red most and blue least intense) superimposed on a grayscale reference image (for orientation) shows that.RNAi functions in adult mammals. Forty μg of annealed 21-mer siRNAs (Dharmacon) were hydrodynamically transfected into livers of mice with the 2 μg of pGL3-Control DNA. Seventy two hours after transfection, mice were anesthetized and given 3 mg of luciferin intraperitoneally 15 min prior to imaging with a cooled CCD camera. IVIS imaging system (Xenogen, Alameda, CA) courtesy of Dr. Christopher Contag, Stanford University. Image reprinted with permission from McCaffrey et al., 2004. (See color section.)

to introduce siRNAs and shRNAs into the livers of mice [(Figure 27.2) (McCaffrey et al., 2002; Lewis et al., 2002)] and less efficiently into other organs. This method is highly efficient, delivering hundreds or thousand of molecules per hepatocyte, and the level of inhibition observed can exceed 95% (McCaffrey et al., 2002). While this approach is likely to be applicable in very limited clinical settings, efforts are underway to extend this approach to primates (reviewed in Herweijer and Wolff, 2003). Nevertheless, hydrodynamic transfection may be a valuable tool for *in vivo* target validation in the liver. In another important development, Hasuwa et al. demonstrated that it is possible to create transgenic knock-down mice by microinjecting shRNA expression cassettes (Hasuwa et al., 2002).

Electroporation has also been used to introduce siRNAs into chick and mouse embryos. Hu et al. first used electroporation to show that Rous sarcoma virus (RSV) pro-viral DNA could be co-delivered with RSV siRNAs into chick embryos. RSV siRNAs, but not control siRNAs were able to inhibit RSV replication and enhance survival (Hu et al., 2002). Because of its easy accessibility, Stoeckli's colleagues have suggested that *in ovo* electroporation of chick embryos could be used for functional genomics (Pekarik et al., 2003; Bourikas et al. Chapter 16 of this book). They showed that RNA interference against axonal guidance cues produced the expected developmental phenotype. They then used this method to screen candidate genes to identify new developmental regulators. Extending this electroporation approach to mice, Calegari et al. introduced endoribonuclease-prepared short interfering RNA into neuroepithelial cells of post-implantation mouse embryos (Calegari et al., 2002).

Recently, viral vectors have been used to express shRNAs from a variety of viral vectors *in vitro* (reviewed in Dykxhoorn et al., 2003; Dykxhoorn, Chapter 26). In the first *in vivo* proof of principle experiment with shRNAs expressed from a virus, Xia et al. showed that a first generation adenoviral vectors that expressed a GFP shRNA reduced the levels of this protein after direct injection of the virus into the brains of mice (Xia et al., 2002). They also showed that tail vein injection of an adenovirus expressing an shRNA could reduce the levels of β-glucuronidase in the livers of mice. In contrast to most reports on expressed shRNAs, these researchers used a Pol II expression cassette rather than the more common Pol III cassette, and significant optimization was required to produce functional shRNAs. Rubinson et al. and Hemann et al. used lentiviral and retroviral vectors, respectively, to transduce hematopoietic stem cells (HSCs) *ex vivo* before reintroducing them into lethally irradiated mice (Rubinson et al., 2003; Hemann et al., 2003). Rubinson infected murine HSCs with a CD8$^+$ targeting lentivirus. After sorting for GFP positive (transduced) cells, they injected these cells into lethally irradiated congenic mice. After 8 weeks, the infected HSCs had reconstituted all blood cell lineages and the frequency of splenocytes expressing CD8 was reduced by 10 fold. Hemann and co-workers used retroviral vectors to express three different shRNAs that suppressed the *Trp53* tumor suppressor gene to varying degrees. Interestingly, transplantation of infected HSCs into mice resulted in a range of tumorogenic phenotypes that correlated with the degree of p53 suppression. In contrast to gene knockout mice, the authors suggested that it is possible to construct an "epi-allelic series of hypomorphic mutations" using a series of shRNAs with various potencies. Clearly *ex vivo* infection of lymphocytes or HSCs with shRNA expression vectors, followed by reimplantation into patients, has promise for the treatment of diseases such as HIV/AIDS. Studies in our laboratory (A. P. M. and M. A. K.) and those of others are currently underway to validate the use of helper-dependent (HD) adenoviral vectors and adeno-associated viruses (AAV) for inducing RNAi.

Clearly, if RNAi is to be effective, siRNAs or shRNA expression vectors must be able to reach the appropriate sub-cellular locale within the organ of interest. Given the many hurdles for the application of RNAi it seems prudent to select disease targets based on the transfectablility/transducability of the organ in which the disease manifests itself.

Liver

The liver is an attractive target organ because it is one of the most easily transfectable tissues using non-viral delivery methods and it is easily transduced using common viral vectors. Potential target diseases in the liver include hepatitis viruses as well as multiple metabolic genetic diseases.

One in every sixteen people world-wide is infected with hepatitis B virus (HBV) and one in every forty is infected with hepatitis C virus (HCV). Although a vaccine exists for HBV, currently there are no effective treatments for millions of people chronically infected with HBV or HCV. A number of groups have shown that RNAi can inhibit the replication of HCV replicons in cultured cells (Kapadia et al., 2003; Randall et al., 2003; Seo et al., 2003).

Recently, we (A. P. M and M. A. K.) initiated an HBV viral replication cycle in mice by transfection with plasmids containing the HBV genome. Co-transfection with HBV shRNA expression plasmids resulted in a marked reduction in the level of HBV RNAs, replicated genomes and viral proteins in this mouse model (McCaffrey et al., 2003). Thus, RNAi may represent a novel strategy for the treatment of these viral diseases *in vivo*. Viral hepatitis is a target being currently pursued by a number of RNAi companies (*e.g.* Avocel, Sirna and Alnylam). RNAi has also been used to treat non-viral hepatitis *in vivo*. Hydrodynamic transfection of Fas siRNAs reduced liver damage in a mouse model of fulminant hepatitis (Song et al., 2003).

The liver is also affected by a number of inherited genetic disorders in which accumulation of an aberrant protein causes liver damage. These patients would not only benefit from replacement of the missing protein, but could avoid liver damage if production of the flawed protein could be diminished. One such group of diseases is broadly known as lysosomal storage diseases (reviewed in Cheng and Smith, 2003). Another such disease is alpha-1 antitrypsin (AAT) deficiency, which is one of the most common lethal hereditary disorders in Caucasians of European descent. Previously, Zern et al. used ribozymes to reduce levels of the defective AAT mRNA by as much as 85% while at the same time expressing a copy of the gene that was immune to cleavage (Ozaki et al., 1999). A similar approach using RNAi could reduce the amount of defective mRNA even further.

The eye

Target tissues that are easily accessed, such as the eye, may also be attractive targets for RNAi therapies since they can be treated by local rather than systemical injection. The eye has the additional advantage of being an organ of immune privilege, which may facilitate the use of viral vectors. Age-related macular degeneration (AMD) causes blindness in a large portion of the elderly. AMD is caused by neovascularization in the eye and one possible treatment for this disease is inhibition of vascular endothelial growth factor (VEGF). Recently siRNAs targeting VEGF were shown to effectively inhibit ocular neovascularization in a mouse model (Reich et al., 2003). These authors compared the degree of silencing observed when expressing shRNAs using the U6 and H1 promoters. They found no significant differences between these two popular systems. RNAi for the treatment of AMD is currently being pursued by at least one RNAi company (Sirna Therapeutics).

Vasculature – angiogenesis

Inhibition of angiogenesis has also been suggested as a method for inhibiting the growth of tumors. VEGF has been implicated as a critical signaling molecule in the promotion of tumor development (Carmeliet and Jain, 2000) and there is considerable interest in blocking VEGF in order to cut off the blood supply to tumors. Zhang et al. have used RNAi to knock down specific isoforms of VEGF in cultured cells (Zhang et al., 2003). In the future, localized delivery of VEGF RNAi to tumors may be another strategy to suppress tumor growth. Other potential disease targets associated with the vasculature include tissue death after heart failure and restenosis.

The brain

There are many dominant neurodegenerative diseases affecting the brain which could, in theory, be treated using RNAi. However, the brain is protected by the formidable blood–brain barrier. In general, non-viral methods are not very effective at transversing this barrier or transfecting neurons in culture. On the other hand, many viruses can effectively transduce cells in the brain if the blood–brain barrier can be circumvented by retrograde transport or direct injection (reviewed in Davidson and Breakefield, 2003). As mentioned above, RNAi has been induced in the brain by direct injection of an adenoviral vector expressing a GFP shRNA. In culture, RNAi has been used to target a gene involved in a neurodegenerative disorder caused by a polyglutamine expansion (Xia et al., 2002). Furthermore, in cultured neurons, RNAi was used to target a missense mutation in the Tau protein that is involved in frontotemporal dementia (Miller et al., 2003). It seems likely that viral delivery of shRNAs for the treatment of neurological disorders will soon be tested in animals. Other possible targets for RNAi in the brain include Parkinson's disease, stroke, spinal cord injury, lysosomal storage diseases and brain tumors.

Ex vivo strategies

Ex vivo treatment of cells or organs may be another promising area for therapeutic development. *Ex vivo* treatments may be advantageous because they avoid systemic delivery problems. For some diseases such as HIV/AIDS and leukemia, it may be possible to harvest cells, treat them with siRNAs or viral vectors expressing shRNAs and then reintroduce them into patients. Numerous studies in cultured cells have shown that RNAi targeting either HIV or HIV co-receptors can inhibit viral replication or infection of cells by HIV. Immunization of a subset of cells with RNAi may provide a selective advantage, allowing these cells to repopulate the immune system of infected individuals. As described above, Rubinson et al. and Hemann et al. used lentiviral and retroviral vectors, respectively, to transduce hematopoietic stem cells before re-introducing them into mice (Hemann et al., 2003; Rubinson et al., 2003) paving the way for HIV/AIDS gene therapy with RNAi.

Conclusions

RNAi applications *in vitro* have clearly gained a prominent position in the molecular biological toolbox for gene function analysis. Yet a number of factors, discussed in this chapter, still limit the use of RNAi for some applications. In the next several years, scientists and product developers from both academia and industry will have to concentrate on improvement and perfection on such properties of RNAi compounds as deliverability, longevity of effect, stability, specificity and hit rate.

In vivo application of RNAi has a dual role, both as an *in vivo* target validation tool, and as a therapeutic modality. RNA interference provides an exciting new therapeutic strategy for the treatment of a broad range of diseases. While significant challenges remain before this technology is applied in the clinic, there has been rapid progress in the field. One major remaining obstacle is the efficient delivery of RNAi triggers to target tissues *in vivo*. Thus it seems prudent to select disease targets that are manifest in organs that are accessible to transfection with siRNAs or shRNA expression plasmids or are easily transduced with viruses that express shRNAs. These target organs include but are not limited to liver, eye, vasculature, brain and lymphocytes. In the coming years, improvements in formulation chemistry and delivery methods are expected to expand this list of potential target organs and to pave the way for more widespread use of RNAi *in vivo*.

REFERENCES

Calegari, F., Haubensak, W., Yang, D., Huttner, W. B. and Buchholz, F. (2002). Tissue-specific RNA interference in postimplantation mouse embryos with endoribonuclease-prepared short interfering RNA. *Proceedings of the National Academy of Sciences USA*, **99**, 14236–14240.

Carmeliet, P. and Jain, R. K. (2000). Angiogenesis in cancer and other diseases. *Nature*, **407**, 249–257.

Cheng, S. H. and Smith, A. E. (2003). Gene therapy progress and prospects: Gene therapy of lysosomal storage disorders. *Gene Therapy*, **10**, 1275–1281.

Davidson, B. L. and Breakefield, X. O. (2003). Viral vectors for gene delivery to the nervous system. *Nature Reviews in Neuroscience*, **4**, 353–364.

Davis, M. E. (2002). Non-viral gene delivery systems. *Current Opinions in Biotechnology*, **13**, 128–131.

Dykxhoorn, D. M., Novina, C. D. and Sharp, P. A. (2003). Killing the messenger: Short RNAs that silence gene expression. *Nature Reviews in Molecular Cell Biology*, **4**, 457–467.

Hasuwa, H., Kaseda, K., Einarsdottir, T. and Okabe, M. (2002). Small interfering RNA and gene silencing in transgenic mice and rats. *FEBS Letters*, **532**, 227–230.

Hemann, M. T., Fridman, J. S., Zilfou, J. T., Hernando, E., Paddison, P. J., Cordon-Cardo, C., Hannon, G. J. and Lowe, S. W. (2003). An epi-allelic series of p53 hypomorphs created by stable RNAi produces distinct tumor phenotypes *in vivo*. *Nature Genetics*, **33**, 396–400.

Herweijer, H. and Wolff, J. A. (2003). Progress and prospects: Naked DNA gene transfer and therapy. *Gene Therapy*, **10**, 453–458.

Hu, W. Y., Myers, C. P., Kilzer, J. M., Pfaff, S. L. and Bushman, F. D. (2002). Inhibition of retroviral pathogenesis by RNA interference. *Current Biology*, **12**, 1301–1311.

Iyer, M., Berenji, M., Templeton, N. S. and Gambhir, S. S. (2002). Noninvasive imaging of cationic lipid-mediated delivery of optical and PET reporter genes in living mice. *Molecular Therapy*, **6**, 555–562.

Kapadia, S. B., Brideau-Andersen, A. and Chisari, F. V. (2003). Interference of hepatitis C virus RNA replication by short interfering RNAs. *Proceedings of the National Academy of Sciences USA*, **100**, 2014–2018.

Khvorova, A., Reynolds, A. and Jayasena, S. D. (2003). Functional siRNAs and miRNAs exhibit strand bias. *Cell*, **99**, 133–141.

Lewis, D. L., Hagstrom, J. E., Loomis, A. G., Wolff, J. A. and Herweijer, H. (2002). Efficient delivery of siRNA for inhibition of gene expression in postnatal mice. *Nature Genetics*, **32**, 107–108.

Lundstrom, K. (2003). Latest development in viral vectors for gene therapy. *Trends in Biotechnology*, **21**, 117–122.

McCaffrey, A. P., Meuse, L., Pham, T. T., Conklin, D. S., Hannon, G. J. and Kay, M. A. (2002). RNA interference in adult mice. *Nature*, **418**, 38–39.

McCaffrey, A. P., Nakai, H., Pandey, K., Huang, Z., Salazar, F. H., Xu, H., Wieland, S. F., Marion, P. L. and Kay, M. A. (2003). Inhibition of hepatitis B virus in mice by RNA interference. *Nature Biotechnology*, **21**, 639–644.

Miller, V. M., Xia, H., Marrs, G. L., Gouvion, C. M., Lee, G., Davidson, B. L. and Paulson, H. L. (2003). Allele-specific silencing of dominant disease genes. *Proceedings of the National Academy of Sciences USA*, **100**, 7195–7200.

Ozaki, I., Zern, M. A., Liu, S., Wei, D. L., Pomerantz, R. J. and Duan, L. (1999). Ribozyme-mediated specific gene replacement of the alpha1-antitrypsin gene in human hepatoma cells. *Journal of Hepatology*, **31**, 53–60.

Pekarik, V., Bourikas, D., Miglino, N., Joset, P., Preiswerk, S. and Stoeckli, E. T. (2003). Screening for gene function in chicken embryo using RNAi and electroporation. *Nature Biotechnology*, **21**, 93–96.

Randall, G., Grakoui, A. and Rice, C. M. (2003). Clearance of replicating hepatitis C virus replicon RNAs in cell culture by small interfering RNAs. *Proceedings of the National Academy of Sciences USA*, **100**, 235–240.

Reich, S. J., Fosnot, J., Kuroki, A., Tang, W., Yang, X., Maguire, A. M., Bennett, J. and Tolentino, M. J. (2003). Small interfering RNA (siRNA) targeting VEGF effectively inhibits ocular neovascularization in a mouse model. *Molecular Vision*, **9**, 210–216.

Rubinson, D. A., Dillon, C. P., Kwiatkowski, A. V., Sievers, C., Yang, L., Kopinja, J., Rooney, D. L., Ihrig, M. M., McManus, M. T., Gertler, F. B., Scott, M. L. and Van Parijs, L. (2003). A lentivirus-based system to functionally silence genes in primary mammalian cells, stem cells and transgenic mice by RNA interference. *Nature Genetics*, **33**, 401–406.

Schwartz, D. S., Hütvagner, G., Du, T., Xu, Z., Aronin, N. and Zamore, P. D. (2003). Asymmetry in the assembly of the RNAi enzyme complex. *Cell*, **115**, 199–208.

Seo, M. Y., Abrignani, S., Houghton, M. and Han, J. H. (2003). Small interfering RNA-mediated inhibition of hepatitis C virus replication in the human hepatoma cell line Huh-7. *Journal of Virology*, **77**, 810–812.

Song, E., Lee, S. K., Wang, J., Ince, N., Ouyang, N., Min, J., Chen, J., Shankar, P. and Lieberman, J. (2003). RNA interference targeting Fas protects mice from fulminant hepatitis. *Nature Medicine*, **9**, 347–351.

Svoboda, P., Stein, P., Hayashi, H. and Schultz, R. M. (2000). Selective reduction of dormant maternal mRNAs in mouse oocytes by RNA interference. *Development*, **127**, 4147–4156.

Wianny, F. and Zernicka-Goetz, M. (2000). Specific interference with gene function by double-stranded RNA in early mouse development. *Nature Cell Biology*, **2**, 70–75.

Xia, H., Mao, Q., Paulson, H. L. and Davidson, B. L. (2002). siRNA-mediated gene silencing *in vitro* and *in vivo*. *Nature Biotechnology*, **20**, 1006–1010.

Zhang, L., Yang, N., Mohamed-Hadley, A., Rubin, S. C. and Coukos, G. (2003). Vector-based RNAi, a novel tool for isoform-specific knock-down of VEGF and anti-angiogenesis gene therapy of cancer. *Biochemical and Biophysical Research Communications*, **303**, 1169–1178.

28 RNA interference studies in liver failure

Lars Zender, Michael P. Manns, and Stefan Kubicka

Clinical features and molecular mechanisms of acute liver failure

Both acute and chronic hepatic failure present diseases associated with high mortality. Acute liver failure (ALF) is a dramatic clinical syndrome in which a previously normal liver fails within days or weeks, resulting in haemorrhage, electrolyte and metabolic disturbance, cardiovascular instability, renal failure, cerebral oedema and encephalopathy. Worldwide the most frequent cause of acute liver failure is viral hepatitis. Advances in intensive care monitoring and the establishment of liver transplantation programmes have made a significant impact on survival of patients with ALF in the last 30 years. Depending on the time course three subgroups of ALF have been proposed: hyperacute, acute and subacute liver failure (O'Grady et al., 1993). Patients with hyperacute disease have the most favorable prognosis, while survival of patients with acute and subacute liver failure is still less than 15% (Plevris et al., 1998). Currently for the majority of the patients with acute and subacute liver failure orthotopic liver transplantation is the only curative therapy with long-term survival rates of approximately 90%. Organ availability, however, remains the most important limiting factor for liver transplantation. Although some patients may recover without liver transplantation, most of the patients die while waiting for a donor liver. Attempts have been made to remove toxins and to provide important liver products by the use of bioartificial supports systems. The objectives of these bioarteficial support system is to bridge patients to transplantation or recovery. Recently a metaanalysis of 12 randomized clinical studies demonstrates that artificial support systems may reduce mortality in acute-on-chronic liver failure compared with standard medical therapy. However, these support systems did not appear to affect mortality in acute liver failure (Kjaergard et al., 2003).

Is appears to be naive to expect artificial support systems to completely replace the synthetic, metabolic and excretory functions that are maintained by a healthy liver in an organism. In addition it has been shown that the dying hepatocytes itself may contribute to the clinical syndrome and the mortality of ALF. Both animal and human studies have shown that total hepatectomy significantly improves (for a short time) the clinical stability of the subjects in ALF (Ejlersen et al., 1994). Consequently the protection of hepatocytes should be an urgent aim in the treatment of ALF.

In acute and chronic liver diseases death of hepatocytes is predominantly mediated by apoptosis. The molecular mechanisms involved in apoptosis of hepatocytes are well known, providing interesting molecular targets for protection of hepatocytes during ALF. After injury of hepatocytes by viruses or toxins, transcription factors in the nucleus or signals released from the cell membrane trigger suicide pathways, leading to the activation of caspase cascades, which subsequently execute apoptotic death. In animal models of acute viral hepatitis, apoptosis of hepatocytes is mediated by transcription factors, such as p53 (Kubicka et al., 1999) and NFκB (Kuhnel et al., 2000) and death receptors, such as Fas (CD95), TNFR-1 and TRAIL-DR5 (Kondo et al., 1997; Kuhnel et al., 2000; Mundt et al., 2003). Studies in humans showed that apoptosis of hepatocytes in Wilson's disease, viral hepatitis and toxic liver damage is mainly triggered through Fas (CD95) (Galle et al., 1995; Strand et al., 1998). However, the contribution of TNF- and TRAIL-receptors in suicide of hepatocytes has also been demonstrated in human viral hepatitis (Mundt et al., 2003; Streetz et al., 2000). Consequently a reasonable strategy to treat patients with ALF would be to inhibit death-receptor mediated apoptosis to maintain liver function and to save the organ.

Inhibition of apoptosis in animal models of acute liver failure by caspase-inhibitors and antisense-oligodesoxyribonucleotides

Inhibition of apoptosis has not been investigated so far in clinical trials with patients with liver diseases. But it has been shown in different mouse models that systemic administration of caspase inhibitors are capable to prevent apoptosis of hepatocytes *in vivo*, resulting in improvement of survival in different models of liver injury (Guillot et al., 2001; Kim et al., 2000; Kubo et al., 1998; Rodriguez et al., 1996a). In a clinical situation patients present with ongoing or even advanced diseases. Therefore it is important that treatment with caspase-inhibitors was effective even when delayed until liver damage was already detectable, promising also successful therapeutic effects in patients with ongoing acute liver failure. However, general inhibition of all caspases in all tissues is a very unspecific approach that may have severe side effects in clinical trials. Prevention of apoptosis in lymphocytes may trigger autoimmunity and inhibition of caspase activity in distinct tissues may result in carcinogenesis. Surprisingly, it has been revealed that systemic application of the general caspase inhibitor zVAD-fmk did not protect mice from TNF-induced shock syndrome, but rather sensitized them to TNF-R1-mediated toxicity by enhancing oxidative stress and mitochondrial damage, which led to hyperacute cardiovascular collaps, renal injury and mortality (Cauwels et al., 2003). For successful molecular therapy of patients with acute liver failure approaches have to be established that target specifically essential mediators of apoptosis in the liver, but not in extrahepatic tissues.

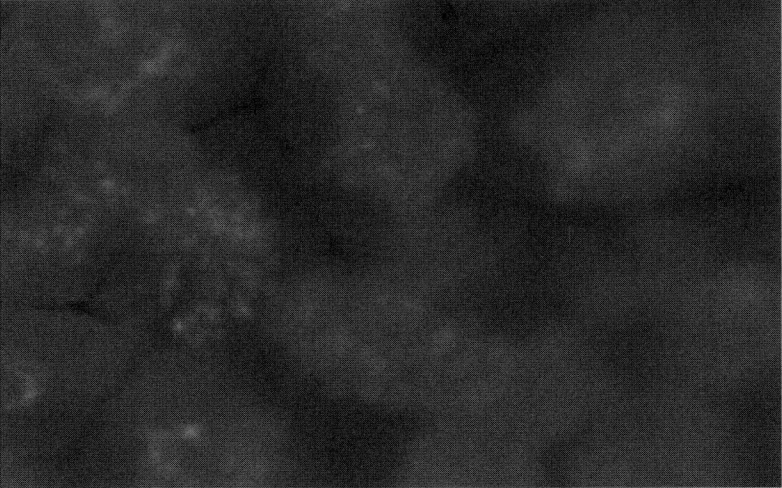

Figure 28.1. Hydrodynamic injection leads to efficient uptake of naked DNA into approximately 10% of the hepatocytes. Visualisation of hepatocytes transfected with EGFP-expressing plasmids by fluorescence microscopy of the liver surface.

Inhibition of gene expression by antisense oligodesoxynucleotides or siRNAs may be a more specific approach compared to pharmacological inhibition of caspase (protease) activity. As with all gene therapy methods, effective and specific uptake of antisense oligodesoxynucleotides or siRNAs into hepatocytes *in vivo* is critical. It appears that a high fraction of peripherally administered plasmid-DNA, oligonuclotides or siRNA is absorbed by the liver after high-volume "hydrodynamic" injection (Herweijer and Wolff, 2003). Delivery of a large volume presumably results in a short term right heat failure and in a return flow of a large volume into the liver, whereby the therapeutic liquid may be squeezed through the endothelial layer of the sinusoids. The optimal volume for hydrodynamic transfection is approximately 10% of the animal's body weight injected within 5–7 seconds. By this method 10–15% of hepatocytes can be transfected with plasmid-DNA (Figure 28.1). Although this method is not feasible for clinical application, it allows the investigation of potential therapeutic oligonuclotides or siRNA sequences in animal models of liver diseases.

In human liver diseases the Fas/FasL-system plays a major role in cell death of hepatocytes. Therefore the inhibition of the FasL/Fas-system is a reasonable strategy to prevent apoptosis in liver injury and has been investigated in many mouse models of acute liver failure. Administration of an anti-Fas oligonucleotide in mice reduced Fas expression in the liver by 90%, resulting in complete protection from fulminant hepatitis, induced by the agonistic Fas antibody Jo-2 (Zhang et al., 2000). Interestingly, oligonucleotide-mediated supression of Fas reduced also the severity of acetaminophen-mediated fulminant hepatitis, but was without any effect on concanavalin A (Con-A)-mediated liver injury. Acetaminophen-mediated liver injury is (together with viral hepatitis and mushroom poisoning) one of the frequent causes of acute liver failure in humans. However, the physiological

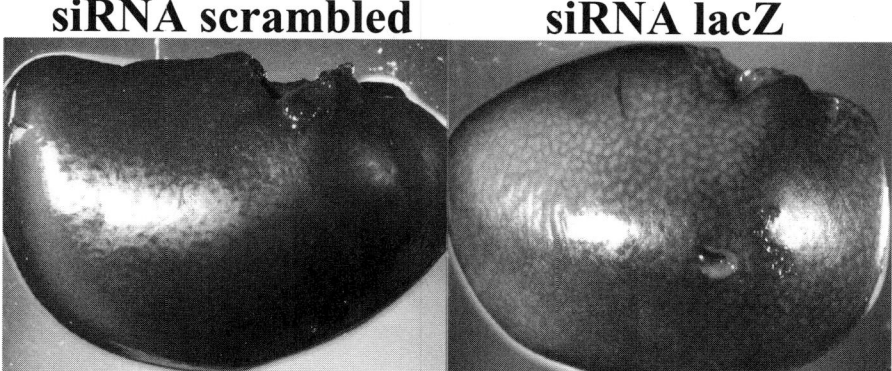

siRNA scrambled **siRNA lacZ**

Figure 28.2. LacZ-staining of the whole liver shows downregulation of x-gal gene expression after hydrodynamic injection of siRNAlacZ in lacZ-transgenic mice (C57BL/6J-TgN(MtnlacZ)).

relevance of Con-A-mediated liver damage is completely unclear. The Con-A model of hepatitis is a very complex and poorly understood model of liver injury, in which activated immune cells (such as T lymphocytes and natural killer cells) mediate apoptosis and necrosis of "healthy" hepatocytes. There are contradictory results about the real contribution of Fas in Con-A mediated liver injury (Seino et al., 1997; Watanabe et al., 1996), which may explain the therapeutic failure of the anti-Fas oligonucleotide in the Con-A model.

Therapeutic utility of siRNA in mouse models of liver diseases

It appears likely that in gene therapy of liver diseases several hurdles of gene delivery can be overcome by the use of siRNA. Compared to plasmid DNA many more hepatocytes can be transfected with siRNA by hydrodynamic injection (Song et al., 2003; Zender et al., 2003), resulting in significant RNA interference mediated silencing of gene expression in the whole liver (Figure 28.2). The number of injections appears to influence the uptake of siRNA into hepatocytes and the effectiveness of gene silencing. Approximately 70% of LacZ-transgenic hepatocytes showed a significant downregulation of LacZ gene expression after single injection of LacZ siRNA (Zender et al., 2003), while 86% (+/– 6%) of hepatocytes took up siRNA labeled with the flurophore Cy5 following three injections (Song et al., 2003). The very effective uptake of siRNA by hepatocytes may be explained by the better pharmacological properties of siRNA compared to plasmid-DNA. siRNAs are much smaller molecules than plasmid-DNA, and therefore extravasation through the endothelial layers and uptake into the hepatocytes may be easier.

Compared to antisense oligodesoxynucleotides, gene silencing by siRNA appears to be more specific and much more effective (Bertrand et al., 2002). Recently the therapeutic utility of siRNA in acute liver failure was demonstrated by Song et al. and our group (Song et al., 2003; Zender et al., 2003). Song et al. used 50 μg Fas-siRNA dissolved in 1 ml PBS per injection to silence Fas-expression in the liver. Hydrodynamic tail vein injections with 50 μg Fas-siRNA were repeated

8 and 24 hours later. Three injections of Fas-siRNA suppressed the expression of Fas in the liver to a similar degree as did the intraperitoneal injection of antisense oligodeoxynucleotides (Zhang et al., 2000). However, Zhang et al. treated the mice with a 14-fold higher dosage of nucleic acids (6 mg/kg of anti-Fas oligodesoxynucleotide for 12 consecutive days), indicating the less effective capacity of oligodesoxynucleotide to silence gene expression *in vivo*. In agreement with the results from Zhang et al., inhibition of Fas expression by RNA interference protected the liver against Jo-2 and resulted in significantly improved survival in the Jo-2 model of acute liver failure (Song et al., 2003). In contrast to Fas-antisense oligodesoxynucleotide, application of Fas-siRNA was also capable to protect the animals against ConA-mediated acute hepatitis. To determine whether siRNA administration is successful after the noxious insult, Song et al. used a chronic model of liver injury by weekly application of lower Con-A doses. Treatment with Fas-siRNA was delayed until 24 hours after the second of six weekly Con-A injections. All control animals developed bridging fibrosis in the liver, while no hepatic fibrosis was seen in mice treated with Fas-siRNA, indicating that Fas-siRNA provides protection even after the initiation of chronic liver injury.

Since it has been shown that in viral hepatitis, in addition to FasL, TNFα and TRAIL are also important mediators of apoptosis (Mundt et al., 2003; Streetz et al., 2000), an essential early downstream mediator of all death receptors appears to be a suitable target to achieve therapeutic success in clinical relevant viral hepatitis animal models. Thus we directed siRNA against caspase 8, which is a key enzyme in death receptor–mediated apoptosis. Single application of 0.45–0.6 nmol/g caspase 8-siRNA resulted in very effective inhibition of caspase 8 expression in the liver, thereby attenuating acute liver damage induced by Jo2 or adenovirus expressing Fas ligand (Ad-FasL). To evaluate the therapeutic efficacy of caspase 8-siRNA in an ongoing viral hepatitis, we applied the siRNA at a time point during Ad-FasL and adenovirus wild-type–mediated hepatitis with already elevated liver transaminsases (i.e. 8 h after viral infection). Improvement of survival due to RNA interference was significant even when caspase 8 siRNA was applied during an ongoing viral-mediated acute liver failure. It is of particular interest that the therapy was not only successful in animal models with Fas-specific liver injury but also in acute liver failure mediated by adenovirus wild type, which is an animal model reflecting multiple molecular mechanisms involved in human acute viral hepatitis.

In acute viral hepatitis it is reasonable not only to prevent apoptosis of hepatocytes but also to inhibit viral replication. Several studies demonstrates that siRNA targeted to viral RNA sequences can effectively protect human cells against RNA-viruses such as HIV (Jacque et al., 2002; Lee et al., 2002; Novina et al., 2002), poliovirus (Gitlin et al., 2002) or hepatitis C virus (Randall et al. 2003) In addition, two recently published studies show effective inhibition of replication initiation of the hepatitis B virus in mouse models *in vivo* (Klein et al., 2003; McCaffrey et al., 2003). These findings indicate that RNA interference provides an attractive therapeutic strategy for the inhibition of many molecular targets that are involved in acute or chronic viral hepatitis.

Delivery and therapeutic effects of siRNA in extrahepatic organs

Recently Lewis et al. (2002). demonstrated that systemic coinjection of luciferase reporter plasmids with siRNA-luciferase results in strong inhibition of luciferase activity not only in the liver, but also in kidney, spleen, lung and pancreas of postnatal mice, suggesting that siRNA is effectively delivered into many organs after systemic application in mammals. Systemic Fas (CD95) engagement induced by the agonistic antibody Jo2 has been reported to cause apoptosis in lymphocytes (Nishimura et al., 1997), endothelial cells (Janin et al., 2002; Wanner et al., 1999), alveolar epithelial cells of the lung (Matute-Bello et al., 2001) as well as glomerular and mesangial cells in the kidney (Gonzalez-Cuadrado et al., 1997). The investigation of extrahepatic organs of Jo-2 treated animals was of particular importance, because there is evidence that mice with Fas-resistant hepatocytes die after Jo2-treatment due to stimulation of Fas receptors present on other organs (Rodriguez et al., 1996b). To evaluate the therapeutic utility of siRNA in extrahepatic organs, we investigated inhibition of caspase 8 activity following Jo-2 application in lung and kidney. Caspase 8 activity in kidney and lung of mice was significantly inhibited by caspase 8-siRNA, confirming the uptake of siRNA into these organs after systemic application (Zender et al., 2003). But in contrast to the liver we did not observe a reduced rate of Fas (CD95)-mediated apoptosis in these organs following caspase 8-siRNA treatment, indicating that delivery of siRNA into lung and kidney is less effective compared to the liver. The less therapeutic effect of siRNA in extrahepatic organs may explain, at least in part, some of the deaths in the group of mice treated with caspase 8-siRNA and Jo-2 (Zender et al., 2003). Since application of Jo-2 did not change the histomorphology of kidney and lung in our model of acute liver failure (presumptively because the animals died too rapidly before tissue damage in lung and kidney was detectable), our experiments were not suitable for answering the question whether systemic treatment with naked siRNA is capable of providing real therapeutic effects in extrahepatic organs.

Future perspectives of the treatment of liver diseases by siRNA

The liver was the appropriate organ to demonstrate the therapeutic utility of siRNA, because >70% of hepatocytes take up siRNA after hydrodynamic injection. However, high-volume hydrodynamic injection is not feasible in clinical trials. Consequently, the next rational step in siRNA therapy of liver diseases is the optimization of siRNA delivery into hepatocytes *in vivo*. Effective and non-toxic delivery methods of siRNA have to be established in animal models before clinical trials in human can be initiated. One useful approach may be the regional application of siRNA instead of systemic injection. It has been shown that efficient transgene expression in hepatocytes throughout the liver can be obtained following delivery of naked plasmid DNA via the portal vein, the hepatic vein, or the bile duct in mice and rats (Herweijer and Wolff, 2003). In our study we investigated whether administration of smaller volumes of caspase 8-siRNA with 10%

lipiodol into the portal vein (without hydrodynamic injection) would also result in protection of the liver against Jo-2-mediated liver injury. Indeed, compared to systemic hydrodynamic injection, inhibition of liver cell apoptosis seemed more effective after regional treatment in mice (Zender et al., 2003), indicating that portal vein injection may be a suitable method for siRNA-therapy of liver diseases in clinical trials. The normal liver gets approximately 70% of the blood flow through the portal vein and 30% through the hepatic artery. However, in acute liver failure and chronic liver diseases, the portal vein resistance is elevated and the blood flow is shifted to the hepatic artery so that regional delivery of siRNA through the hepatic artery instead of the portal vein may be more effective in patients with advanced liver diseases.

In addition to the route of delivery specific binding of siRNA to hepatocytes have to be improved for clinical trials. It has been shown that conjugation of oligodeoxynucleotides with cholesterol moieties results in significantly enhanced and almost selective uptake by the liver, but endothelial cells were the major site of uptake, whereas hepatocytes accounted only for approximately 25% of the liver uptake (Bijsterbosch et al., 2002). By linking a specific peptide to an antisense oligonucleotide, specific delivery in ErbB2-positive cancer cells was achieved (Shadidi and Sioud, 2003). One promising approach for effective and non-toxic delivery of siRNA into hepatocytes may be the linkage of the therapeutic siRNAs with small peptides that bind specifically, or preferentially, to hepatocytes.

It has been demonstrated in mice that the RNA interference effect lasts for at least 10 days in quiescent hepatocytes following injection of siRNA (Klein et al., 2003; Song et al., 2003). The effective prevention of apoptosis for an interval of 10–14 days may be an appropriate method for preserving the function of the liver in patients with acute liver diseases, allowing time for recovery and regeneration of the organ. But long-term side effects of siRNA-mediated gene silencing have to be investigated in animal models before the onset of clinical trials. Currently gene therapeutic approaches for non-malignant diseases are under critical consideration because they may result in genetic instability and enhanced cancer development (Williams and Baum, 2003). Although siRNA-mediated RNA interference is a timely limited process, one cannot rule out the possibility that even short-term inhibition of critical genes, such as apoptosis-inducing genes, may result in enhanced cancer development, according to the hit-and-run hypothesis of carcinogenesis (Galloway and McDougall, 1983).

Taken together the current results raise new hopes for the development of future effective siRNA-therapies for many liver diseases. But delivery of siRNA into hepatocytes have to be optimized and more information about long-term side effects is needed before siRNA-therapy can be safely used in clinical trials in humans.

REFERENCES

Bertrand, J. R., Pottier, M., Vekris, A., Opolon, P., Maksimenko, A. and Malvy, C. (2002). Comparison of antisense oligonucleotides and siRNAs in cell culture and *in vivo. Biochemical and BiophysicalResearch Communications*, **296**, 1000–1004.

Bijsterbosch, M. K., Manoharan, M., Dorland, R., Van Veghel, R., Biessen, E. A., and Van Berkel, T. J. (2002). bis-Cholesteryl-conjugated phosphorothioate oligodeoxynucleotides are highly selectively taken up by the liver. *Journal of Pharmacology and Experimental Therapeutics*, **302**, 619–626.

Cauwels, A. Janssen, A. B., Waeytens, A., Cuvelier, C. and Brouckaert, P. (2003). Caspase inhibition causes hyperacute tumor necrosis factor-induced shock via oxidative stress and phospholipase A2. *Nature Immunology*, **4**, 387–393.

Ejlersen, E., Larsen, F. S., Pott, F., Gyrtrup, H. J., Kirkegaard, P. and Secher, N. H. (1994). Hepatectomy corrects cerebral hyperperfusion in fulminant hepatic failure. *Transplantion Proceedings*, **26**, 1794–1795.

Galle, P. R., Hofmann, W. J., Walczak, H., Schaller, H., Otto, G., Stremmel, W., Krammer, P. H. and Runkel, L. (1995). Involvement of the CD95 (APO-1/Fas) receptor and ligand in liver damage. *Journal of Experimental Medicine*, **182**, 1223–1230.

Galloway, D. A. and McDougall, J. K. (1983). The oncogenic potential of herpes simplex viruses: Evidence for a 'hit-and-run' mechanism. *Nature*, **302**, 21–24.

Gitlin, L., Karelsky, S. and Andino, R. (2002). Short interfering RNA confers intracellular antiviral immunity in human cells. *Nature*, **418**, 430–434.

Gonzalez-Cuadrado, S., Lorz, C., Garcia, d. M., O'Valle, F., Alonso, C., Ramiro, F. Ortiz-Gonzalez, A., Egido, J. and Ortiz, A. (1997). Agonistic anti-Fas antibodies induce glomerular cell apoptosis in mice *in vivo*. *Kidney International*, **51**, 1739–1746.

Guillot, C., Coathalem, H., Chetritt, J., David, A., Lowenstein, P., Gilbert, E., Tesson, L., van Rooijen, N., Cuturi, M. C., Soulillou, J. P. and Anegon, I. (2001). Lethal hepatitis after gene transfer of IL-4 in the liver is independent of immune responses and dependent on apoptosis of hepatocytes: a rodent model of IL-4-induced hepatitis. *Journal of Immunology*, **166**, 5225–5235.

Herweijer, H. and Wolff, J. A. (2003). Progress and prospects: Naked DNA gene transfer and therapy. *Gene Therapy*, **10**, 453–458.

Jacque, J. M., Triques, K. and Stevenson, M. (2002). Modulation of HIV-1 replication by RNA interference. *Nature*, **418**, 435–438.

Janin, A., Deschaumes, C., Daneshpouy, M., Estaquier, J., Micic-Polianski, J., Rajagopalan-Levasseur, P., Akarid, K., Mounier, N., Gluckman, E., Socie, G. and Ameisen, J. C. (2002). CD95 engagement induces disseminated endothelial cell apoptosis *in vivo*: Immunopathologic implications. *Blood*, **99**, 2940–2947.

Kim, K. M., Kim, Y. M., Park, M., Park, K., Chang, H. K., Park, T. K., Chung, H. H. and Kang, C. Y. (2000). A broad-spectrum caspase inhibitor blocks concanavalin A-induced hepatitis in mice. *Clinical Immunology*, **97**, 221–233.

Kjaergard, L. L., Liu, J., Als-Nielsen, B. and Gluud, C. (2003). Artificial and bioartificial support systems for acute and acute-on-chronic liver failure: A systematic review. *Journal of American Medical Association*, **289**, 217–222.

Klein, C., Bock, C. C., Wedemeyer, H., Wustefeld, T., Locarnini, S., Dienes, H. H., Kubicka, S., Manns, M. M. and Trautwein, C. (2003). Inhibition of hepatitis B virus replication *in vivo* by nucleoside analogues and siRNA. *Gastroenterology*, **125**, 9–18.

Kondo, T., Suda, T., Fukuyama, H., Adachi, M. and Nagata, S. (1997). Essential roles of the Fas ligand in the development of hepatitis. *Nature Medicine*, **3**, 409–413.

Kubicka, S., Kuhnel, F., Zender, L., Rudolph, K. L., Plumpe, J., Manns, M. and Trautwein, C. (1999). p53 represses CAAT enhancer-binding protein (C/EBP)-dependent transcription of the albumin gene. A molecular mechanism involved in viral liver infection with implications for hepatocarcinogenesis. *Journal of Biological Chemistry*, **274**, 32137–32144.

Kubo, S., Sun, M., Miyahara, M., Umeyama, K., Urakami, K., Yamamoto, T., Jakobs, C., Matsuda, I. and Endo, F. (1998). Hepatocyte injury in tyrosinemia type 1 is induced by fumarylacetoacetate and is inhibited by caspase inhibitors. *Proceedings of the National Academy of Sciences USA*, **95**, 9552–9557.

Kuhnel, F., Zender, L., Paul, Y., Tietze, M. K., Trautwein, C., Manns, M. and Kubicka, S. (2000). NFkappaB mediates apoptosis through transcriptional activation of Fas (CD95) in adenoviral hepatitis. *Journal Biological Chemistry*, **275**, 6421–6427.

Lee, N. S., Dohjima, T., Bauer, G., Li, H., Li, M. J., Ehsani, A., Salvaterra, P. and Rossi, J. (2002). Expression of small interfering RNAs targeted against HIV-1 rev transcripts in human cells. *Nature Biotechnology*, **20**, 500–505.

Matute-Bello, G., Winn, R. K., Jonas, M., Chi, E. Y., Martin, T. R. and Liles, W. C. (2001). Fas (CD95) induces alveolar epithelial cell apoptosis *in vivo*: Implications for acute pulmonary inflammation. *American Journal of Pathology*, **158**, 153–161.

McCaffrey, A. P., Nakai, H., Pandey, K., Huang, Z., Salazar, F. H., Xu, H., Wieland, S. F., Marion, P. L. and Kay, M. A. (2003). Inhibition of hepatitis B virus in mice by RNA interference. *Nature Biotechnology*, **21**, 639–644.

Mundt, B., Kuhnel, F., Zender, L., Paul, Y., Tillmann, H., Trautwein, C., Manns, M. P. and Kubicka, S. (2003). Involvement of TRAIL and its receptors in viral hepatitis. *FASEB Journal*, **17**, 94–96.

Nishimura, Y., Hirabayashi, Y., Matsuzaki, Y., Musette, P., Ishii, A., Nakauchi, H., Inoue, T. and Yonehara, S. (1997). *In vivo* analysis of Fas antigen-mediated apoptosis: Effects of agonistic anti-mouse Fas mAb on thymus, spleen and liver. *International Immunology*, **9**, 307–316.

Novina, C. D., Murray, M. F., Dykxhoorn, D. M., Beresford, P. J., Riess, J., Lee, S. K., Collman, R. G., Lieberman, J., Shankar, P. and Sharp, P. A. (2002). siRNA-directed inhibition of HIV-1 infection. *Nature Medicine*, **8**, 681–686.

O'Grady, J. G., Schalm, S. W. and Williams, R. (1993). Acute liver failure: Redefining the syndromes. *Lancet*, **342**, 273–275.

Plevris, J. N., Schina, M. and Hayes, P. C. (1998). Review article: The management of acute liver failure. *Alimental Pharmacology Therapy*, **12**, 405–418.

Randall, G., Grakoui, A. and Rice, C. M. (2003). Clearance of replicating hepatitis C virus replicon RNAs in cell culture by small interfering RNAs. *Proceedings of the National Academy of Science USA*, **100**, 235–240.

Rodriguez, I., Matsuura, K., Ody, C., Nagata, S. and Vassalli, P. (1996a). Systemic injection of a tripeptide inhibits the intracellular activation of CPP32-like proteases *in vivo* and fully protects mice against Fas-mediated fulminant liver destruction and death. *Journal of Experimental Medicine*, **184**, 2067–2072.

Rodriguez, I., Matsuura, K., Khatib, K., Reed, J. C., Nagata, S. and Vassalli, P. (1996b). A bcl-2 transgene expressed in hepatocytes protects mice from fulminant liver destruction but not from rapid death induced by anti-Fas antibody injection. *Journal of Experimental Medicine*, **183**, 1031–1036.

Seino, K., Kayagaki, N., Takeda, K., Fukao, K., Okumura, K. and Yagita, H. (1997). Contribution of Fas ligand to T cell-mediated hepatic injury in mice. *Gastroenterology*, **113**, 1315–1322.

Shadidi, M. and Sioud, M. (2003). Identification of novel carrier peptides for the specific delivery of therapeutics into cancer cells. *Federation of the American Society for Experimental Biologists Journal*, **17**, 256–258.

Song, E., Lee, S. K., Wang, J., Ince, N., Ouyang, N., Min, J., Chen, J., Shankar, P. and Lieberman, J. (2003). RNA interference targeting Fas protects mice from fulminant hepatitis. *Nature Medicine*, **9**, 347–351.

Strand, S., Hofmann, W. J., Grambihler, A., Hug, H., Volkmann, M., Otto, G., Wesch, H., Mariani, S. M., Hack, V., Stremmel, W., Krammer, P. H. and Galle, P. R. (1998). Hepatic failure and liver cell damage in acute Wilson's disease involve CD95 (APO-1/Fas) mediated apoptosis. *Nature Medicine*, **4**, 588–593.

Streetz, K., Leifeld, L., Grundmann, D., Ramakers, J., Eckert, K., Spengler, U., Brenner, D., Manns, M. and Trautwein, C. (2000). Tumor necrosis factor alpha in the pathogenesis of human and murine fulminant hepatic failure. *Gastroenterology*, **119**, 446–460.

Wanner, G. A., Mica, L., Wanner-Schmid, E., Kolb, S. A., Hentze, H., Trentz, O. and Ertel, W. (1999). Inhibition of caspase activity prevents CD95-mediated hepatic microvascular perfusion failure and restores Kupffer cell clearance capacity. *Federation of the American Society for Experimental Biologists Journal*, **13**, 1239–1248.

Watanabe, Y., Morita, M. and Akaike, T. (1996). Concanavalin A induces perforin-mediated but not Fas-mediated hepatic injury. *Hepatology*, **24**, 702–710.

Williams, D. A. and Baum, C. (2003). Medicine. Gene therapy – new challenges ahead. *Science*, **302**, 400–401.

Zender, L., Hutker, S., Liedtke, C., Tillmann, H. L., Zender, S., Mundt, B., Waltemathe, M., Gosling, T., Flemming, P., Malek, N. P., Trautwein, C., Manns, M. P., Kuhnel, F. and Kubicka, S. (2003). Caspase 8 small interfering RNA prevents acute liver failure in mice. *Proceedings of the National Academy of Sciences USA*, **100**, 7797–7802.

Zhang, H., Cook, J., Nickel, J., Yu, R., Stecker, K., Myers, K. and Dean, N. M. (2000). Reduction of liver Fas expression by an antisense oligonucleotide protects mice from fulminant hepatitis. *Nature Biotechnology*, **18**, 862–867.

29 RNAi applications in living animal systems

Lisa Scherer and John J. Rossi

Introduction

RNA interference, or RNAi, refers to a conserved post-transcriptional gene silencing mechanism in which a small antisense RNA serves as the sequence specific effector molecule. In plants, *Caenorhabditis elegans* and *Drosophila*, where the mechanism was originally elucidated (Bernstein et al., 2001), this effector is derived from a long, double-stranded RNA (dsRNA) trigger, which is processed by the cellular enzyme Dicer into 21–23 nucleotide dsRNAs referred to as short interfering RNAs (siRNAs). The anti-sense strand of the siRNA becomes incorporated into the multi-protein RNA-induced silencing complex (RISC); there, it serves as a guide for cleavage of the mRNA within the target site, when the anti-sense strand is completely complementary to the target.

The RNAi pathway was also shown to occur in mammalian cells subsequent to the discovery of RNAi in lower eukaryotes (Caplen et al., 2001; Elbashir et al., 2001); however, since dsRNA longer than 30 nucleotides triggers the interferon pathway in most mammalian cells, long dsRNA cannot be used to induce RNAi. This difficulty can be overcome by using exogenously introduced or transcribed siRNAs to bypass the Dicer step and directly activate homologous mRNA degradation without initiating the interferon response in most cases. The siRNAs may be in the form of two separate strands, mimicking the natural Dicer product; or may consist of a single hairpin molecule (shRNA) where the sense and anti-sense strands are linked by a short loop.

Since these initial demonstrations, the potential for high specificity and efficiency of targeted gene knockdown have led to a rapidly expanding body of literature reporting success with RNAi as a tool for targeted gene inhibition in mammalian cancer and viral therapeutics, and functional genomics. Initial applications in mammalian systems were primarily in tissue culture and *ex vivo* systems, but uses of RNAi in living organisms are emerging, where RNAi is also being used as a genetic tool to investigate questions in developmental biology. This review is designed to introduce the reader to some of the emerging spectrum of applications of RNAi in whole animal systems.

Generating RNAi knockdown transgenic rodents by RNAi

Undifferentiated ES cells were among the first mammalian cells where RNAi was applied (Yang et al., 2001), since ES cells, like mouse oocytes and pre-implantation embryos, lack the interferon response, allowing the use of long dsRNA as an RNAi trigger. Because of the rapid onset of the interferon response upon differentiation, si/shRNAs are typically used to mediate stable RNAi for long-term gene knockdown in ES cells. One of the general applications of RNAi in this context is the determination of the effects of down-regulation of specific genes in early murine development, particularly embryonic lethals. For example, (Kunath et al., 2003), investigators used this approach to investigate the developmental onset of aberrations due to RasGAP downregulation. While *Rasa 1* homozygous null ES cells have no phenotype, the same mutation in mice is an embryonic lethal and affects the ability of endothelial cells to organize into a highly vascularized network resulting in extensive neuronal cell death. Stable ES cell lines were generated by electroporation of a linearized plasmid carrying a RasGAP shRNA and *neo* transgenes into R1 ES cells. The developmental capacity of four clones with a spectrum of protein suppression levels was assessed in completely ES cell-derived chimeric embryos generated by the tetraploid aggregation method. Day 9.5 embryos derived from RasGAP-silenced ES cells had defective heart, tail and head structures. The severity of phenotypes was variable but correlated with degree of RasGAP suppression. These results illustrate the application of RNAi to investigate early mammalian developmental defects.

These experiments lay the groundwork for using RNAi to produce specific gene hypomorphic animals. Initial reports used expressed siRNAs against GFP to successfully down-regulate GFP in transgenic mouse and rat model systems (Hasuwa et al., 2002), and provided proof of principle that RNAi can be used to generate 'knockdown' transgenic rodents. Several groups were subsequently successful in extending these applications.

In one approach, si/shRNA expression cassettes were introduced by electroporation into ES cells, which were then used to generate knockdown transgenic animals (Carmell et al., 2003). The target gene was *NeiI*, a DNA *N*-glycosylase proposed to have a role in DNA repair. The majority of ES cell lines showed an ~80% reduction in NeiI mRNA and protein, as well as a two-fold increase in sensitivity to ionizing radiation. Two independently isolated ES cell lines were injected into BL/6 blastocysts and high percentage F_0 chimeras then outcrossed to check for germline transmission. The shRNA expression construct was successfully transmitted to 13/27 and 12/26 F_1 progeny derived from the two ES cell lines. Heritable expression of *NeiI* shRNA was verified and also correlated with reduced *NeiI* expression in liver, heart, and spleen as assayed by RT-PCR. These animals can now be assessed for defects in response to specific agents that damage DNA, to define the role of *NeiI* in DNA repair *in vivo*.

Interestingly, these same authors initially tried to produce transgenic founder animals by direct injection of shRNA expression plasmids into pronuclei. This method was not successful in their hands, although not for lack of attempts. Seven different, well-studied endogenous genes with obvious hypomorphic phenotypes were targeted with combinations of three different shRNA expression plasmids each. The expected phenotypes were not observed in any case, even though the presence of the transgene was verified in some animals. The reason for this failure is unclear given the success of this method in other hands (see below); these authors did not mention whether the effectiveness of the shRNAs was tested in preliminary tissue culture experiments. Happily, two groups recently reported success with two methods that do not require ES cell lines for generating transgenic knockdown animals using RNAi.

In the first case (Hasuwa et al., 2002), investigators injected linearized plasmids bearing transgenes directly into the pronuclei of heterozygous EGFP-expressing fertilized eggs, derived from fertilizing wild-type eggs *in vitro* with sperm from homozygous EGFP mice. The dual cassette plasmids carried an HI promoter driving an shRNA against the target EGFP, as well as an HcRed1 reporter gene. Expression of the red fluorescent protein (RFP) monitors genomic integration of the plasmid; the degree of concomitant reduction of green fluorescence, the effectiveness of the shRNA. Consequently, cells or embryos that take up the shRNA plasmid are expected to express both transgenes, and show a 'switch' from red to green fluorescence. Controls carrying the HcRed parental plasmid (no shRNA cassette) fluoresce with both colors. Embryos with the expected phenotypes were observed by day 3 of *in vitro* culture.

Having verified the experimental approach in pre-implantation embryos, mouse eggs were transplanted immediately after injection into oviducts of pseudopregnant females. Analysis of several transgenic day 10.5 founder embryos showed decreases in EGFP fluorescence and protein levels in all organs of day 10.5, long after silencing produced by injected synthetic siRNAs wanes. Newborn founder mice were identified by their red fluorescence and confirmed by genomic PCR analysis. Mice derived from the HcRed parental plasmid alone showed no decrease in EGFP fluorescence relative to the original EGFP transgenic mice. However, those carrying the EGFP shRNA showed a decrease, though not complete elimination, of green fluorescence relative to the parental murine line in all tissues examined, which persisted into adulthood. Most importantly, fluorescent and immunoblot analysis of day 10.5 and 12.5 F_1 embryos derived from backcrossing a doubly transgenic male to wild-type female showed that the RNAi silencing effect was heritable. Finally, the same methodology was used successfully to generate transgenic rats, an important contribution since no ES rat cell lines exist.

Alternatively, lentiviral vectors may be used to introduce shRNA expression cassettes in mice (Tiscornia et al., 2003). Lentiviral vectors can transduce non-dividing cells, and have been used to generate transgenic rodents by *in vitro* transduction of fertilized eggs at preimplantation stages (Pfeifer et al., 2002; Yu et al., 2002). Preliminary experiments used packaged provirus derived from lentiviral vectors carrying HI promoter driven GFP shRNA cassettes in the 3' LTR to transduce a 293 EGFP stable cell line at different multiplicities of infection.

Fluorescence and Western analyses demonstrated dose-dependent decreases in GFP expression. Parallel transductions with lentivirus carrying the p53 shRNA gene did not decrease GFP levels.

To create transgenic mice, fertilized eggs were collected from females previously mated to transgenic EGFP males carrying multiple copies of GFP on both alleles. Two-cell stage fertilized eggs were transduced for 48 hours after removal of the zona pellucida. GFP reduction could be seen as early as 48 hours post-transduction. Blastocysts were transferred into the uteri of pseudopregnant females. The hemizygous pups were prescreened for reduced levels of whole-body GFP fluorescence; presence of integrated lentivirus and retention of the GFP gene were confirmed by PCR. Two F_0 EGFP-shRNA positive females were back-crossed to wild-type males and both day-13 embryos and live pups from the second mating analyzed.

Of the nine F_1 embryos examined from the first cross, 2 carried integrated lentivirus, but only 1 also carried the founder line EGFP transgene; protein extracts derived from this embryo showed decreased EGFP (by Western analysis) compared to littermates. One F_1 pup out of 6 born from the second mating carried both transgenes, and had lower total body fluorescence compared to littermates. Interestingly, in the both the dual transgenic animals examined, the average number of lentiviral shRNA integrations in the F_1 progeny (2 and 10) was lower than the average copy numbers in their F_0 mothers (13 and 21, respectively). Thus, this method was also successful RNAi-mediated gene knockdown in mice.

Both direct DNA injection and lentiviral transduction of early embryos show promise as methods for producing hypomorphic transgenic animals, with the potential to target genes expressed anywhere in the animal. The range of possible applications overlaps considerably with transgenic animals produced by standard ES methods; however, these newer protocols allow generation of genetically modified animals in those species where ES cell lines are not established and homologous recombination has not been demonstrated. In addition, since shRNA can reduce gene products in *trans*, the effects of gene knockdown can be examined in hemizygous individuals. One disadvantage of this system is the current lack of tissue- and developmentally specific promoters for expressing shRNAs, though these are under development. More problematic is the potential for position effects. Since retroviral integration is known to preferentially occur at sites of open chromatin, there is a possibility of adversely affecting normal gene expression near the insertion site. Multiple integration sites within a cell are likely to exacerbate the problem, as well as introduce dosage effects. Reciprocally, adjacent sequences at an insertion site may also affect transgene expression, which is well documented for P-element transformation in *Drosophila*. Though the strong pol III promotors typically used for shRNA expression may be less susceptible to position effects, this may not be true for inducible, stage- and tissue-specific promoters that may be used in the future. Analysis of multiple lines may be required to ensure that the observed effects are due to transgene knockdown. Nonetheless, where ES cell lines are available, it may still be advantageous to perform preliminary analysis in RNAi knockdown lines, which can be derived relatively rapidly, and follow up in transgenic animals generated by standard methods.

In another development, there is one report of a method using long dsRNAs to create knockdown mice without activating the interferon response (Shinagawa et al., 2003). The conceptual basis for the approach was that if long dsRNA activates the interferon response only when present in the cytoplasm, long dsRNA that is retained in the nucleus of mammalian cells might be able to mediate RNAi without inducing an interferon response. To this end, a plasmid expression cassette, pDECAP (Deletion of Cap structure and polyA), was designed to produce intronless transcripts lacking the 5' cap and polyA tail required for efficient nuclear export. A CMV promoter drives transcription of a leader sequence with a *cis*-acting hammerhead ribozyme directed against site upstream between the 5' end of the leader and the ribozyme to remove the m^7G cap structure that directs mRNA export post-transcriptionally. Sequences coding for approximately 500 nucleotides of the target sequence as a sense–12 nucleotide loop–antisense inverted repeat can be inserted downstream of the leader. The transcriptional unit ends with a pol II transcriptional pause site, to allow the nascent transcript more time to form a hairpin and the ribozyme to remove the cap; the polyA site is omitted. The authors hypothesized that nuclear DICER-like activity, perhaps used in endogenous microRNA pathways, could process the dsRNA to siRNAs, which could then be exported to the cytoplasm and enter RISC.

Preliminary experiments targeting firefly luciferase in this manner verified the technique in 293T cells; there was no interferon response, as assayed by $eIF2\alpha$ phosphorylation. An endogenous transcriptional corepressor, Ski, was chosen as a target to generate knockdown mice using a derivative of pDECAP (pDECAP-ski); one advantage of targeting Ski was that a traditionally derived transgenic knockout mouse, $Ski^{-/-}$, already existed for comparison. Linearized pDECAP-ski or pDECAP-β-gal (as a control) DNAs were injected into fertilized mouse oocytes and embryos analyzed for successful integration by Southern blot. Northern analysis showed the presence of small 21–22 nt Ski RNAs were generated only in transgenic, not wild-type embryos. Neural tube and heart defects were present in 5/13 stage E9-E12 pDECAP-ski embryos: eye formation abnormalities were found in abnormalities in 2/15 E14-E16 animals. In contrast, no (0/9) transgenic pDECAP-β-gal -derived embryos showed defects. *In situ* hybridization of tissues from pDECAP-ski transgenic embryos, $Ski^{-/-}$ transgenic mice, wild-type mice (parental line) and pDECAP-β-gal-derived embryos were compared. Tissues from pDECAP embryos and $Ski^{-/-}$ transgenic mice showed similar reductions in Ski mRNA. Reduction of Ski mRNA in $Ski^{-/-}$ mice results in ectopic expression of ornithine decarboxylase (*Odc*) in nueral epithelial cells leading to increased apoptosis, and is responsible for defects in neural tube closure. Ectopic ODC expression was also observed in pDECAP-Ski transgenic mice. Taken together, these results indicate that pDECAP-Ski expression induces a very similar pattern of abnormalities as seen in $Ski^{-/-}$ transgenic mice.

The use of a long dsRNA to target a gene circumvents the difficulty of finding an effective single mRNA target sequence; however, potential caveats to this method deserve consideration. First, there was only about 30% penetrance of Ski-defects in pDECAP-ski embryos; this may possibly have been due to interactions between integration site and the CMV promoter, but further experiments will be required to firmly establish the cause. Secondly most, but not all, of the defects typical of *Ski*⁻/⁻ transgenic mice were observed in pDECAP-ski embryos; this could be due to promoter effects as well, but suggests that a certain amount of caution must be used in interpreting data from animals generated by this method. Third, the use of long dsRNA increases the potential of off-target effects. In this particular case, expression from pDECAP-Ski did not down-regulate the related Sno mRNA, which is encouraging in this instance, but off-target effects would be expected to vary depending on the target sequence. Also, the effects on other transcripts were not determined. Next, no adult mice were generated using pDECAP-Ski or pDECAP-β-gal, so the long-term effects of long dsRNA expression and heritability are unknown. Finally, it is not clear whether mRNA suppression of Ski is due to a true siRNA effect or a gene silencing mechanism; indeed, that is unclear in many RNAi experiments, but is more of a question here given the methodology.

RNAi delivery and gene silencing in post-natal animals

The potential of RNAi to down-regulate gene expression in adult animals is also enormous, and has implications for clinical applications. High-pressure, tail-vein injection of a large volume of physiological solution into the tail vein (Liu et al., 1999; Zhang et al., 1999) was the first method used to achieve efficient delivery of synthetic siRNA in adult mice (Lewis et al., 2002). Preliminary experiments co-injecting a synthetic siRNA with a plasmid expressing target gene demonstrated feasibility; a 80–90% specific reduction of co-injected P-luciferase reporter mRNA was observed in all tissues examined. The same method was used successfully to reduce expression of EGFP in mice expressed from an endogenous transgene. In particular, a substantial reduction of EGFP was observed in most hepatocytes, though the authors mention that *in vivo* uptake of fluorescently labeled siRNA is not equal in all liver cells. Nonetheless, the successful targeting of hepatocytes by this method has made it attractive for the study of applications of RNAi in murine liver model systems of injury and disease.

The first endogenous gene targeted by this method was Fas, which is over-expressed during liver inflammation in damaged hepatocytes and can lead to apoptosis (Song et al., 2003). Eighty percent of mice infused with Fas siRNA prior to induction of autoimmune hepatitis were protected from liver failure and death compared to untreated controls. More importantly, mice already suffering from auto-immune hepatitis improved after Fas siRNA treatment. Fas has also been a siRNA target for reducing hepatocyte apoptosis upon allogenic hepatocyte trans-plantation (Wang et al., 2003). Reduction of Fas in virally infected hepatocytes could potentially be used to prevent or reduce disease-induced liver damage.

RNAi is also being considered to target liver virus replication directly. RNAi has been used successfully against HCV replicon intermediates, but only in tissue cul-ture (Kapadia et al., 2003; Randall et al., 2003; Sen et al., 2003; Seo et al., 2003;

Wilson et al., 2003; Yokota et al., 2003) as an animal HCV viral replication model does not yet exist. As an alternative, one group (McCaffrey et al., 2002) used a HCV-luciferase fusion reporter construct to assay the effects of RNAi-mediated down-regulation of HCV. The reporter construct and synthetic siRNAs directed against the HCV portion of the fusion, luciferase, or an irrelevant sequence were introduced by hydrodynamic transfection into mice, and levels of luciferase monitored by quantitative whole body imaging. The HCV siRNAs were more effective even than the luciferase siRNAs at specifically reducing luciferase levels in murine liver.

RNAi has also been used to reduce HBV replication in a murine model system (McCaffrey et al., 2003). Many steps of hepatitis B virus (HBV) replication occur in mice after hydrodynamic transfection with plasmids containing the HBV genome (Yang et al., 2001). A combination of U6-HBV shRNA expression constructs, prescreened for efficacy in tissue culture, HBV replicon and secreted human 1-antitrypsin (hAAT) as a transfection control was introduced into immunocompetent and immunodeficient mice by hydrodynamic transfection; analyses were done only in mice having comparable serum hAAT levels. HBV mRNA levels were reduced up to 90% by HBV shRNAs in both normal and NOD/SCID mice. Viral single-stranded DNA and double-stranded DNA replicative intermediates as well as serum antigen levels were also reduced in livers of NOD/SCID mice, indicating that the effects are not due to immunological responses. Immunohistochemical staining of liver sections showed a substantial decrease in the number of cells HBV antigen positive cells as well. These studies show the potential for using hydrodynamic transfection to study disease and pathological processes in the mammalian liver, as well as the feasibility of using RNAi. Care must be taken to monitor side effects associated with systemic delivery, particularly since down-regulation of the target gene in organs besides the liver could complicate interpretation of results.

Synthetic siRNAs also have the potential to alleviate short-term, acute conditions due to endogenous gene expression. For instance, overproduction of TNF-α by activated macrophages can cause many pathogenic responses including septic shock, cachexia and rheumatoid arthritis; the ability to control TNF-α levels would have potential therapeutic applications. A chemically synthesized siRNA targeting endogenous TNF-α specifically reduced TNF-α production and increased survival when introduced by intraperitoneal injection in a murine model system of LPS-induced septic shock, secondary to overproduction of TNF-α by peritoneal macrophages (Sørensen et al., 2003). siRNAs may also be introduced by subretinal delivery to investigate the role of candidate genes involved in retinal disease and development. For instance, retinal pigment epithelial (RPE) cells, through abnormal VEGF production, are thought to initiate the process that leads to pathologic neovascularization and blindness in animal models of age-related macular degeneration (AMD) that induce choroidal neovascularization (CNV). In this system, subretinal injection of a synthetic human VEGF siRNA specifically reduced the expression of VEGF protein from a co-injected VEGF expression construct and reduced the size of VEGF-induced neovascularization by 75% relative to controls 36 hours post-injection (Reich et al., 2003).

A modification of this technique employs *in vivo* electroporation after subretinal DNA injection to transduce retinal cells of neonatal mouse and rat pups (Matsuda and Cepko, 2004). *In vivo* electroporation generates transgenic retinal tissue faster, more conveniently, and with better efficacy compared to conventional retinal transduction methods using viral vectors such as MMLV, tested in parallel. Introduction of plasmid vectors bearing siRNA expression constructs targeting transcription factors Crx or Nrl into newborn mice induced retinal phenotypes at day 20 similar to those observed in the respective conventional knockout murine lines. While the expression from plasmid constructs are relatively short term compared to those mediated by integrating viral vectors, the effects are long enough for loss-of-function analysis in rodent retinal physiological processes and development.

In vivo electroporation is also currently being used to deliver synthetic or expressed siRNAs in other mammalian tissues. RNAi-mediated knockdown in postnatal rat cerebellum was part of a group of experiments demonstrating the role of Cdh1 in controlling axonal growth and patterning (Konishi et al., 2004) and the role of the tyrosine kinase MuSK in muscle synapse assembly (Kong et al., 2004).

Conclusions

In a relatively short time, RNAi has moved from the realm of an esoteric oddity, peculiar to a few lower eukaryotes, to a broadly conserved mechanism involved in development and basic cellular processes as well as a powerful addition to the tools for anti-sense mediated gene-specific targeting. The research summarized in this review indicates that mammalian RNAi applications are moving rapidly from the *in vitro* and *ex vivo* and into the *in vivo* realms, with the potential to facilitate our understanding of mammalian gene function and disease. We expect that as methods of siRNA expression and delivery expand, so will the spectrum of applications.

REFERENCES

Bernstein, E., Denli, A. M. and Hannon, G. J. (2001). The rest is silence. *RNA*, 7, 1509–1521.

Caplen, N. J., Parrish, S., Imani, F., Fire, A. and Morgan, R. A. (2001). Specific inhibition of gene expression by small double-stranded RNAs in invertebrate and vertebrate systems. *Proceedings of the National Academy of Sciences USA*, **98**, 9742–9747.

Carmell, M. A., Zhang, L., Conklin, D. S., Hannon, G. J. and Rosenquist, T. A. (2003). Germline transmission of RNAi in mice. *Nature Structural Biology*, **10**, 91–92.

Elbashir, S. M., Harborth, J., Lendeckel, W., Yalcin, A., Weber, K. and Tuschl, T. (2001). Duplexes of 21-nucleotide RNAs mediate RNA interference in cultured mammalian cells. *Nature*, **411**, 494–498.

Hasuwa, H., Kaseda, K., Einarsdottir, T. and Okabe, M. (2002). Small interfering RNA and gene silencing in transgenic mice and rats. *Federation of European Biochemical Society Letters*, **532**, 227–230.

Kapadia, S. B., Brideau-Andersen, A. and Chisari, F. V. (2003). Interference of hepatitis C virus RNA replication by short interfering RNAs. *Proceedings of the National Academy of Sciences USA*, **100**, 2014–2018.

Kong, X. C., Barzaghi, P. and Ruegg, M. A. (2004). Inhibition of synapse assembly in mammalian muscle *in vivo* by RNA interference. *European Molecular Biology Organization Reports*, **5**, 183–188.

Konishi, Y., Stegmuller, J., Matsuda, T., Bonni, S. and Bonni, A. (2004). Cdh1-APC controls axonal growth and patterning in the mammalian brain. *Science*, **303**, 1026–1030.

Kunath, T., Gish, G., Lickert, H., Jones, N., Pawson, T. and Rossant, J. (2003). Transgenic RNA interference in ES cell-derived embryos recapitulates a genetic null phenotype. *Nature Biotechnology*, **21**, 559–561.

Lewis, D. L., Hagstrom, J. E., Loomis, A. G., Wolff, J. A. and Herweijer, H. (2002). Efficient delivery of siRNA for inhibition of gene expression in postnatal mice. *Nature Genetics*, **32**, 107–108.

Liu, F., Song, Y. and Liu, D. (1999). Hydrodynamics-based transfection in animals by systemic administration of plasmid DNA. *Gene Therapy*, **6**, 1258–1266.

Matsuda, T. and Cepko, C. L. (2004). Electroporation and RNA interference in the rodent retina *in vivo* and *in vitro*. *Proceedings of the National Academy of Sciences USA*, **101**, 16–22.

McCaffrey, A. P., Meuse, L., Pham, T. T., Conklin, D. S., Hannon, G. J. and Kay, M. A. (2002). RNA interference in adult mice. *Nature*, **418**, 38–39.

McCaffrey, A. P., Nakai, H., Pandey, K., Huang, Z., Salazar, F. H., Xu, H., Wieland, S. F., Marion, P. L. and Kay, M. A. (2003). Inhibition of hepatitis B virus in mice by RNA interference. *Nature Biotechnology*, **21**, 639–644.

Pfeifer, A., Ikawa, M., Dayn, Y. and Verma, I. M. (2002). Transgenesis by lentiviral vectors: Lack of gene silencing in mammalian embryonic stem cells and preimplantation embryos. *Proceedings of the National Academy of Sciences USA*, **99**, 2140–2145.

Randall, G., Grakoui, A. and Rice, C. M. (2003). Clearance of replicating hepatitis C virus replicon RNAs in cell culture by small interfering RNAs. *Proceedings of the National Academy of Sciences USA*, **100**, 235–240.

Reich, S. J., Fosnot, J., Kuroki, A., Tang, W., Yang, X., Maguire, A. M., Bennett, J. and Tolentino, M. J. (2003). Small interfering RNA (siRNA) targeting VEGF effectively inhibits ocular neovascularization in a mouse model. *Molecular Vision*, **9**, 210–216.

Sen, A., Steele, R., Ghosh, A. K., Basu, A., Ray, R. and Ray, R. B. (2003). Inhibition of hepatitis C virus protein expression by RNA interference. *Virus Research*, **96**, 27–35.

Seo, M. Y., Abrignani, S., Houghton, M. and Han, J. H. (2003). Small interfering RNA-mediated inhibition of hepatitis C virus replication in the human hepatoma cell line Huh-7. *Journal of Virology*, **77**, 810–812.

Shinagawa, T. and Ishii, S. (2003). Generation of Ski-knockdown mice by expressing a long double-strand RNA from an RNA polymerase II promoter. *Genes & Development*, **17**, 1340–1345.

Song, E., Lee, S. K., Wang, J., Ince, N., Ouyang, N., Min, J., Chen, J., Shankar, P. and Lieberman, J. (2003). RNA interference targeting Fas protects mice from fulminant hepatitis. *Nature Medicine*, **9**, 347–351.

Sørensen, D. R., Leirdal, M. and Sioud, M. (2003). Gene silencing by systemic delivery of synthetic siRNAs in adult mice. *Journal of Molecular Biology*, **327**, 761–766.

Tiscornia, G., Singer, O., Ikawa, M. and Verma, I. M. (2003). A general method for gene knockdown in mice by using lentiviral vectors expressing small interfering RNA. *Proceedings of the National Academy of Sciences USA*, **100**, 1844–1848.

Wang, J., Li, W., Min, J., Ou, Q. and Chen, J. (2003). Fas siRNA reduces apoptotic cell death of allogeneic-transplanted hepatocytes in mouse spleen. *Transplantation Proceedings*, **35**, 1594–1595.

Wilson, J. A., Jayasena, S., Khvorova, A., Sabatinos, S., Rodrigue-Gervais, I. G., Arya, S., Sarangi, F., Harris-Brandts, M., Beaulieu, S. and Richardson, C. D. (2003). RNA interference blocks gene expression and RNA synthesis from hepatitis C replicons propagated in human liver cells. *Proceedings of the National Academy of Sciences USA*, 25275–25279.

Yang, P. L., Althage, A., Chung, J. and Chisari, F. V. (2002). Hydrodynamic injection of viral DNA: A mouse model of acute hepatitis B virus infection. *Proceedings of the National Academy of Sciences USA*, **99**, 13825–13830.

Yang, S., Tutton, S., Pierce, E. and Yoon, K. (2001). Specific double-stranded RNA interference in undifferentiated mouse embryonic stem cells. *Molecular Cell Biology*, **21**, 7807–7816.

Yokota, T., Sakamoto, N., Enomoto, N., Tanabe, Y., Miyagishi, M., Maekawa, S., Yi, L., Kurosaki, M., Taira, K., Watanabe, M. and Mizusawa, H. (2003). Inhibition of intracellular hepatitis C virus replication by synthetic and vector-derived small interfering RNAs. *European Molecular Biology Organization Reports*, **4**, 602–608.

Yu, J. Y., DeRuiter, S. L. and Turner, D. L. (2002). RNA interference by expression of short-interfering RNAs and hairpin RNAs in mammalian cells. *Proceedings of the National Academy of Sciences USA*, **99**, 6047–6052.

Zhang, G., Budker, V. and Wolff, J. A. (1999). High levels of foreign gene expression in hepatocytes after tail vein injections of naked plasmid DNA. *Human Gene Therapy*, **10**, 1735–1737.

High-throughput genome-wide RNAi analysis

30 High-throughput RNAi by soaking in *Caenorhabditis elegans*

Asako Sugimoto

1. Introduction

The phenomenon of RNA-mediated interference (RNAi) was first discovered in the nematode *Caenorhabditis elegans* (Fire et al., 1998), in which introduction of double-stranded RNA (dsRNA) causes specific inactivation of genes with corresponding sequences. RNAi was soon recognized as an experimentally simple and technically undemanding method for gene knockdown in this organism. The emergence of this new technology, which coincided with the completion of the sequencing of the *C. elegans* genome (The *C. elegans* Sequencing Consortium, 1998), has brought about a dramatic shift in the experimental strategies adopted in the study of this organism, and greatly expanded our understanding of gene function at the whole-genome level. Our group have established an optimized "soaking method" for dsRNA delivery into worms (Maeda et al., 2001), and applied this technique for large-scale analysis of gene function. In this paper, I describe the characteristics of the "soaking method," and its application in the study of developmental processes in *C. elegans*.

2. RNAi in *C. elegans*

After the finding of RNAi in *C. elegans*, a series of subsequent analyses demonstrated that wide range of animals, plants, and fungi has RNAi-related gene regulatory mechanisms (Cogoni and Macino, 2000). Biochemical and genetic approaches, conducted mainly in *C. elegans* and *Drosophila*, have revealed the evolutionarily conserved nature of the molecular mechanism of RNAi (see other chapters in this book for detail). Although the "core RNAi mechanism" appears to be conserved among diverse organisms, some aspects of RNAi observed in *C. elegans* are not universally found in other species. First, while long (>100 base pairs) dsRNA molecules are potent effectors of gene-specific silencing in *C. elegans* and other invertebrate species (Fire et al., 1998), in mammals, long dsRNA induces the activation of sequence-nonspecific dsRNA responses, such as the RNA-dependent protein kinase (PKR) pathway that repress general translation. Because of this, siRNAs or short-hairpin RNAs (shRNAs) must be used instead of long dsRNAs to knock down genes in mammalian cells (Elbashir et al., 2001; Paddison et al., 2002). Secondly, in *C. elegans* and some other invertebrates and plants, gene silencing is observed even in body regions remote from the site of the initial dsRNA delivery. This phenomenon is called "systemic RNAi" (Winston et al., 2002). Another intriguing aspect of RNAi in *C. elegans* is the amplification of RNAi signals. Findings from genetic and biochemical analyses suggest that siRNAs prime dsRNA synthesis though the action of RNA-dependent RNA polymerase (RdRP), using the target mRNAs as templates (Sijen et al., 2001). Generated new dsRNAs are processed into "secondary siRNAs" that can silence the expression of the target genes. The amplification of siRNAs and their efficient intercellular transport contribute to the robustness and persistence of the RNAi effect in *C. elegans*.

Although RNAi in *C. elegans* is a powerful method for the inactivation of gene function, it does have several limitations. There are some tissue-specific and gene-specific differences in sensitivity to RNAi, the underlying mechanisms of which are not fully understood. In particular, genes that function in the nervous system have proven refractory to RNAi (Tavernarakis et al., 2000). In addition, there is significant interexperimental variability in RNAi results, which may be due to subtle differences in experimental conditions, such as the developmental stage at which the worms are treated or the concentration of dsRNA (Simmer et al., 2003). Researchers should therefore interpret RNAi results carefully, taking into account the penetrance and variability of the RNAi effects.

3. RNAi by soaking

Optimization of the soaking method

In the RNAi experiments of *C. elegans*, dsRNA can be delivered into the body of worms by four different methods (Figure 30.1). Originally dsRNA was delivered by microinjection (Fire et al., 1998). Timmons and Fire showed that feeding the worms *E. coli* expressing dsRNA is sufficient to induce the RNAi effect (Timmons and Fire, 1998). Tabara et al. reported that simply soaking the worms in dsRNA could also elicit interference (Tabara et al., 1998). Finally, hairpin RNA can be expressed from the transgene, which enable tissue-specific, or inducible, RNAi by using specific promoters (Tavernarakis et al., 2000). Early reports of the latter three delivery methods note that the effect of RNAi is less potent compared to those obtained by direct microinjection.

Since the soaking method is much less labor intensive compared to microinjection and does not require special equipment, we optimized the conditions aiming to induce a more stable RNAi effect. To determine the optimal soaking conditions, a number of parameters are examined, including concentration of dsRNA and buffer ingredients and the length of soaking time.

The RNAi effect was generally concentration dependent, but minimum concentration to elicit phenotypes varies dependent on the genes. Some genes (such as *cyk-4*) were effectively knocked down with as low as 1–5 ng/μl of dsRNA, but other genes require higher concentrations to obtain reproducible phenotypes. Currently we generally use 0.5–1 μg/μl dsRNA for our experiments.

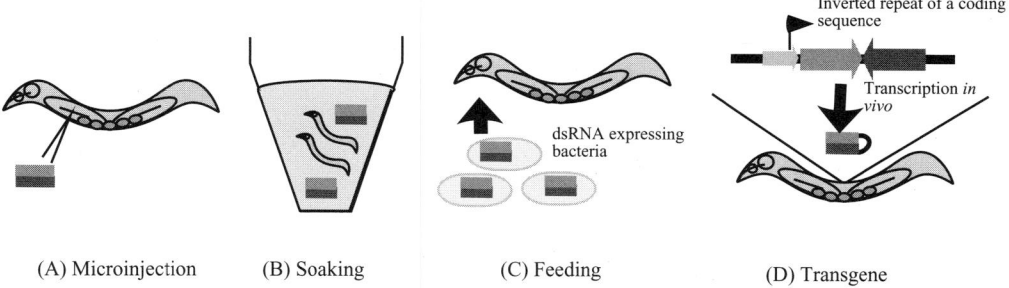

(A) Microinjection   (B) Soaking   (C) Feeding   (D) Transgene

Figure 30.1. dsRNA delivery methods. (A) Microinjection of the *in vitro* synthesized dsRNA. (B) Soaking in the dsRNA solution. (C) Feeding bacteria that express dsRNA. (D) *In vivo* transcription of hairpin RNAs from the transgene. By choosing promoters that control the expression of hairpin RNAs, inducible- or tissue-specific RNAi can be elicited. (See color section.)

The most significant modification we made for the soaking method is the addition of spermidine in the soaking solution (Maeda et al., 2001). We found that the addition of spermidine (final concentration 3 mM) in the dsRNA solution significantly improved the efficiency and reproducibility of RNAi experiments by the soaking method. For example, in the case of the *unc-22* gene that functions in muscle, RNAi by soaking without spermidine caused the expected "twitcher" phenotypes in 32% of the progeny of the soaked worms, whereas with spermidine the phenotypes were observed in 100% of the progeny (Maeda et al., 2001). Although the reason for the positive effect of spermidine for the soaking method is unclear, one possibility is that the positive charge of spermidine neutralizes the negative charge of dsRNA, which might facilitate the uptake of dsRNA through the intestinal membrane.

In the series of control experiments, expected phenotypes were observed at the nearly complete penetrance for genes involved in early embryogenesis (*glp-1* and *mei-1*), and germline development (*mes-3*, *fem-1*, *gld-1*, and *daz-1*) (Maeda et al., 2001). For some genes that function in adults (*unc-22*, *gld-1*, and *daz-1*), mutant phenotypes were also observed in the P0 animals that were soaked in the RNA solution (Maeda et al., 2001). Thus, the modified soaking method can be applied for the genes that function at various stages in the development.

To inhibit gene functions from early embryogenesis, L4 larva are soaked for 24 hours (Figure 30.2A). During the soaking process, the worms develop to adulthood, but their oogenesis arrests in the starved condition in the soaking buffer, although the RNAi process appears to progress during this period. Therefore, a strong RNAi effect can be observed generally from the first embryo laid after the recovery of the worms to the agar plates with bacteria as food, and at most ~300 RNAi-affected embryos per worm can be obtained. It should be noted that some genes take a longer period of time to exhibit RNAi phenotypes than others (in such a case, embryos laid 24–48 hours after the recovery show phenotypes at higher penetrance than 0–24 hours). In the case of injection into adult worms or feeding L4 larva, genes were not effectively silenced in the embryos laid in the first several to 24 hours either by injection or feeding method (Fire et al., 1998; Kamath et al., 2001). It has been observed that the silencing will become more effective over time, thus producing many unaffected progeny. Therefore, for genes

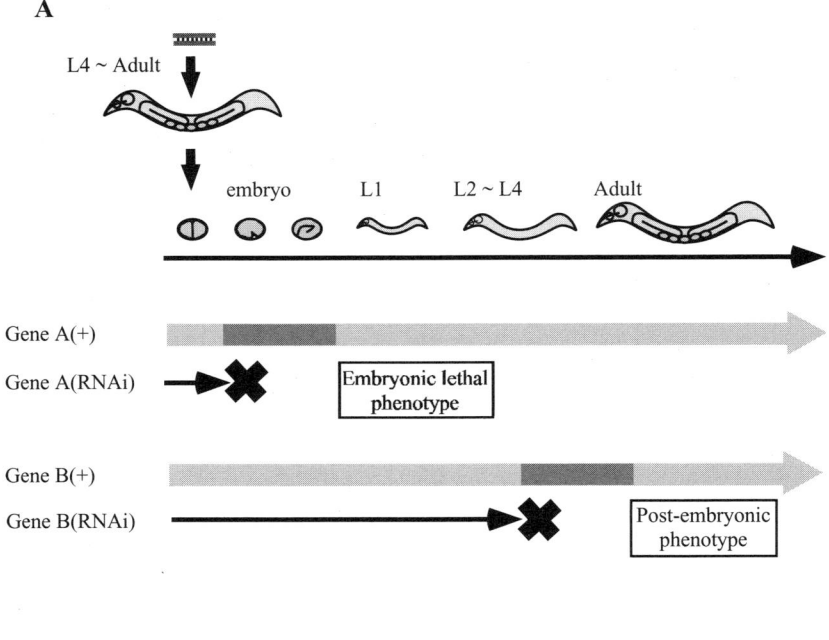

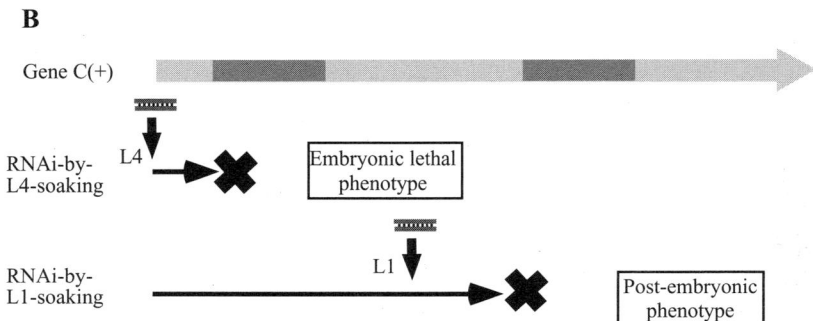

Figure 30.2. RNAi experiments in *C. elegans*. (A) In general, dsRNA corresponding to the target gene is introduced into the L4 or adult worm. The dsRNA can be delivered by microinjection, soaking or feeding method. The phenotypes by reduction of gene function are observed mainly in the next generation. (B) Developmental stage specific RNAi. Some genes function at multiple developmental stages. In such a case, lethal phenotypes at the embryonic stage would prevent the observation of the phenotypes at later stages. To inhibit the post-embryonic gene functions, dsRNA can be delivered into the worms at the first larval stage using the soaking methods so that post-embryonic gene functions are specifically knocked down.

that take longer time to be silenced, the soaking method may elicit more severe or highly penetrant phenotypes than other methods, because animals are exposed to dsRNA for a longer time before laying eggs.

Post-embryonic RNAi by L1-soaking method

By allowing investigators to select the developmental stage for dsRNA delivery, the soaking method can be used to conduct stage-specific RNAi experiments (Figure 30.2B). To perform post-embryonic specific knockdowns, L1 larvae (instead of L4 or adults) are soaked in dsRNA solution (Kuroyanagi et al., 2000). While soaking (i.e., without food), the development of L1 larvae is arrested but the RNAi response continues, making it possible to suppress gene function from the onset of post-embryonic development. Feeding method can also be used for post-embryonic RNAi, but the RNAi effect is elicited gradually during post-embryonic development even if L1 larvae are fed. Therefore, the feeding method is better suited to the analysis of late larval development and of phenotypes that require a persistent RNAi effect, such as analyses of behavior and longevity (Dillin et al., 2002; Lee et al., 2003).

The L1-soaking method is particularly useful to analyze multiple roles of the essential genes throughout development. In general, early phenotypes (e.g., embryonic lethality) prevent the analysis of the potential phenotypes at later phenotypes, if any. By inhibiting the gene function after embryogenesis by RNAi-by-L1 soaking, the post-embryonic role of the embryonic lethal gene can be analyzed. For example, the *glp-1* gene, a homolog of Notch (Austin and Kimble, 1989), is required for cell fate determination in early embryogenesis (Priess et al., 1987), as well as for the regulation of germ cell proliferation (Austin and Kimble, 1987): RNAi-by-L4 soaking of the *glp-1* gene results in embryonic lethality due to abnormal differentiation, whereas L1 soaking causes sterility in adults due to the dramatically reduced number of germ cells.

In the L4-soaking RNAi experiments, phenotypes are generally exhibited in the progeny of soaked worms, but a significant fraction of the genes causes sterility of the soaked worms. This is because during the L4 stage through adulthood, at which worms are soaked in dsRNA, germ cells are the only cell type that actively proliferate and differentiate, and RNAi inhibition of the genes involved in these processes causes germline defects. Such "P0 sterile" genes may be germline-specific, or may function both in germline and soma. These two possibilities can also be distinguished by post-embryonic stage specific RNAi: If L1-soaking of "P0 sterile" genes causes sterility, but not other phenotypes, indicating that the gene is involved solely in germline development. On the other hand, if that results of L1-soaking in larval lethality of other post-embryonic defect, in addition to sterility, it may indicate that the gene is essential both in germline and soma. Thus, RNAi by soaking is a powerful method to characterize genes that play multiple roles in different developmental stages, by simply selecting the stage of the worms to be soaked.

4. High-throughput analysis by RNAi by soaking

Since the RNAi effect of the modified soaking method is nearly equivalent to that by microinjection, we applied this method for high-throughput RNAi.

To perform genome-wide or large-scale RNAi, a gene library is required to provide the templates for dsRNA synthesis. Our group chose a non-redundant cDNA set established by Y. Kohara. Thus far, Kohara's *C. elegans* EST project has isolated cDNAs corresponding to ~10,000 genes, representing more than half of the predicted number of genes (Reboul et al., 2001). They have established a non-redundant cDNA set by choosing one clone per gene. M. Vidal's group recently published the *C. elegans* "ORFeome" consisting of ~12,000 cloned ORFs (Reboul et al., 2003). This "ORFeome" set has not been used for genome-wide RNAi analysis, but does provide an alternative source for the *in vitro* synthesis of dsRNA for use for injection or soaking.

Our rationale for using the non-redundant cDNA set is as follows: (i) Since mature mRNAs are the target molecules in RNAi (Montgomery et al., 1998), cDNA would be the better template to synthesize dsRNA, compared to the PCR-amplified genomic fragment that contain intron sequences. Since many of the gene structure predictions are computer-based, mis-prediction of gene structure would be detrimental for the efficacy of the RNAi response. Indeed, a comparison of cDNA sequence and gene prediction using the GeneFinder software revealed that more than 50% of predicted genes needed corrections to their exon–intron structures (Reboul et al., 2003). In addition, 9% of known genes were not predicted by the computer analysis of the genome sequence, but were rather identified from isolated cDNAs (Reboul et al., 2001). (ii) cDNA templates tend to be longer than PCR-based genomic DNA templates, which make them more likely to produce stronger RNAi effects. In the PCR-based gene set, PCR primers are designed to amplify the genomic regions that cover the coding region (thus in many cases, not all exons are included), and the average size of the PCR-based gene sets (see below) are ~1 kb (Gonczy et al., 2000; Kamath and Ahringer, 2003). On the other hand, majority of the cDNA clones in the non-redundant cDNA set are nearly full-length, and the average size is ~1.5 kb. (iii) Each cDNA is cloned in a pBluescript vector that contains promoter sequences for T3 and T7 RNA polymerases, which is convenient for RNA synthesis *in vitro*. On the other hand, a disadvantage of the cDNA libraries currently available is that they cover at most ~60% of the genes in the genome (Reboul et al., 2003), and do not contain genes expressed at extremely low levels.

Two groups have used genomic PCR products based on predicted gene structures to perform functional genomic RNAi analyses. While T. Hyman's group used *in vitro* synthesized dsRNA for injection (Gonczy et al., 2000), J. Ahringer's group cloned the PCR products into a plasmid vector expressing dsRNA in *E. coli*. This bacterial strain library, which contains 86% of the genes predicted for *C. elegans*, is now available to the public. The use of these two types of gene libraries (cDNA sets and PCR-based genomic library), considering their relative merits, should provide complementary coverage of the genome-wide RNAi analysis.

5. Functional genomics by RNAi by soaking

The procedure for the large-scale RNAi by soaking is schematically shown in Figures 30.3 and 30.4, which was modified from our original published protocol. All steps are performed with a 96-well format so that an automated liquid handling system (Biomek 2000, Beckman) can be used. Each cDNA clone was amplified by PCR with a primer pair both of which contain a T7 promoter site. The amplified products were used as templates for *in vitro* transcription by the T7 RNA

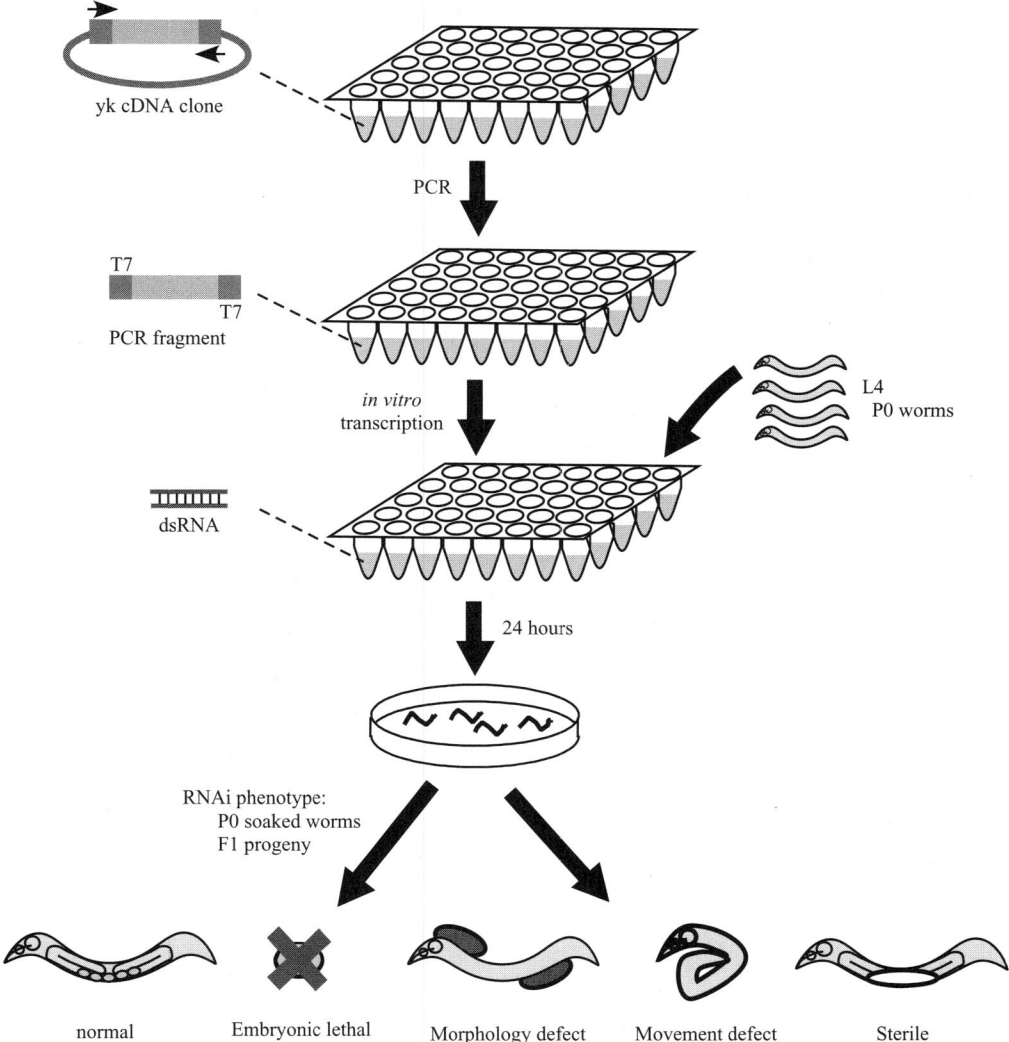

yk cDNA clone

PCR

T7

T7
PCR fragment

in vitro
transcription

L4
P0 worms

dsRNA

24 hours

RNAi phenotype:
P0 soaked worms
F1 progeny

normal   Embryonic lethal   Morphology defect   Movement defect   Sterile

Figure 30.3. Procedures for the high-throughput RNAi-by-soaking. Template PCR, *in vitro* transcription of dsRNA, and soaking are performed with 96-well format plates. Phenotypes of recovered worms (P0 generation) and their progeny (F1 generation) are observed for detectable phenotypes by dissecting microscope.

polymerase. Since both ends of the PCR products have T7 promoters, dsRNA can be produced in a single reaction. After the treatment by DNase I to digest the template DNA, dsRNA are purified using Wizard Plus SV 96 DNA binding plate (Promega, #A2271). The dsRNA solution is concentrated using an evaporator to 0.5–1 μg/μl.

Six L4 larva were put in the soaking solution [containing 0.5–1 μg dsRNA, 0.25 $\times$ M9 (without Mg^{2+}), 3mM spermidine, 0.05% gelatine], and kept at 20°C for 24 hours. The soaked worms (P0 generation) developed to adults while soaking, which were then recovered on a NGM plate (Plate 1) with food (*E. coli* OP50) and incubated at 25°C, and allowed to lay fertilized eggs onto the plate (Figure 30.4).

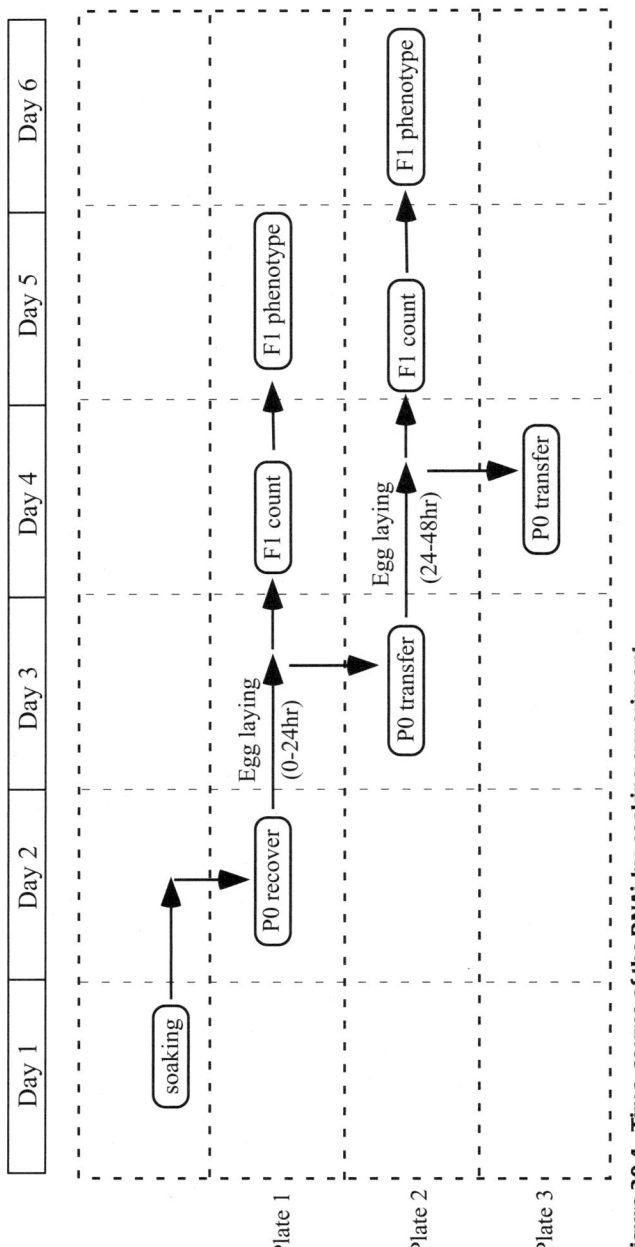

Figure 30.4. Time-course of the RNAi-by-soaking experiment.

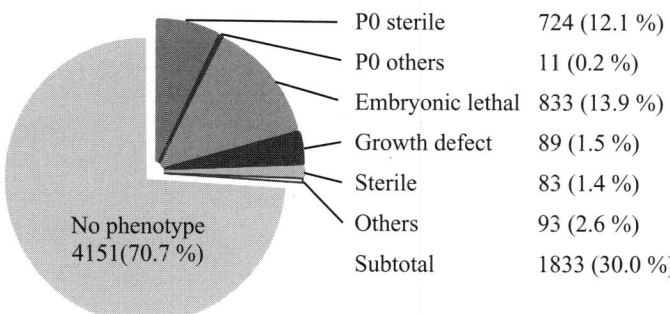

P0 sterile	724 (12.1 %)
P0 others	11 (0.2 %)
Embryonic lethal	833 (13.9 %)
Growth defect	89 (1.5 %)
Sterile	83 (1.4 %)
Others	93 (2.6 %)
Subtotal	1833 (30.0 %)

No phenotype 4151 (70.7 %)

5984 cDNA clones examined

Figure 30.5. Summary of phenotypes by RNAi-by-L4-soaking.

After 24 hours, P0 worms were transferred to a new plate (Plate 2) with food to let them lay eggs. After another 24 hours, P0 worms were removed from Plate 2. Phenotypes of P0 worms (throughout the procedure), F1 progeny on Plate 1 (eggs laid 0–24 hours after recovery) and Plate 2 (eggs laid 24–48 hours after recovery) were observed under a dissecting microscope. Observed phenotypes include lethality (embryonic, larval, adult), sterility (in P0 or F1), and defects in growth, morphology, and locomotion.

To date, ~6000 cDNA were examined for RNAi phenotypes. Of these, ~30% exhibited detectable phenotypes (Figure 30.5). The most frequently observed phenotype was embryonic lethality (13.9%): the genes in this category are regarded to be essential for various aspects of embryogenesis. The second-most frequent phenotype was sterility/small fecundity in the soaked P0 animals (12.1%). The genes caused this phenotype should play some role in germline proliferation/development/differentiation in adults. In addition, some fraction of this group is likely to function in somatic tissues as well as germline. Indeed, many genes that caused P0 sterility exhibited other phenotypes in the escapers, such as F1 embryonic and larval lethality. On the other hand, the genes caused F1 sterility with no apparent somatic phenotypes (1%) are likely to play a germline-specific function.

Several groups have performed functional genomic analyses by high-throughput RNAi. Ahringer's group used the feeding method to perform RNAi analysis for ~86% of predicted genes (Fraser et al., 2000; Kamath et al., 2003). More recently, Plasterk's group performed another round of genome-wide RNAi with the same feeding library using an RNAi-hypersensitive mutant *rrf-3* as the host strain (Simmer et al., 2003). The *rrf-3* gene encodes one of the four RdRP genes, and its mutants exhibit higher RNAi sensitivity (Simmer et al., 2002). Hyman's group has been performing systematic RNAi by injection, using the PCR-amplified genomic fragments as templates, and has published the results of their analysis of chromosome III (Gonczy et al., 2000). Piano and Kemphues's group has also performed a smaller scale analysis using an ovary cDNA library (Piano et al., 2000). For the most part, the results of large-scale RNAi conducted by the various groups are consistent with each other, but there has also been some lack of agreement. Specifically, our group's results show the highest percentage in the genes that showed phenotypes. This is partly because each of the laboratories used different definitions of phenotypes (e.g., our group recorded >10% embryonic lethality, whereas Hyman's group found >90%). In addition, we examined the phenotypes at 25°C, a more stringent condition to worms than the conventional culture condition at 20°C, by which more severe phenotypes are expected to be observed. In addition, genes whose cDNA has been isolated, meaning those which are relatively highly expressed during normal development, were significantly more likely to show RNAi phenotypes (Fraser et al., 2000). Therefore, our analysis using the cDNA set naturally produces more phenotypes than PCR-based genomic libraries.

These high-throughput RNAi analyses use dissecting microscopes for the first round of phenotype analysis to analyze a wide variety of developmental abnormalities, e.g., embryonic and larval lethality, sterility, and morphological or locomotion defects. However, since neuronal cells are less sensitive to RNAi (Tavernarakis et al., 2000), many genes that play essential functions in neurons are likely to have been missed in these analyses. In addition, phenotypes that require specific assays (e.g., aging, stress response) and male-specific phenotypes (male-specific morphogenesis, mating behavior, etc.) were not examined, and will have to be screened separately.

Collectively, these large-scale RNAi results have provided important insights into the genome organization of *C. elegans*. The percentage of genes that caused detectable RNAi phenotypes ranged between 10 and 25% in the five screens; such genes are thought to play essential roles in viability and proper development (Fraser et al., 2000; Gonczy et al., 2000; Kamath et al., 2003; Maeda et al., 2001; Piano et al., 2000; Simmer et al., 2003). Genes that have orthologues in other organisms were found to be more likely to produce phenotypes than nonconserved genes (Fraser et al., 2000; Gonczy et al., 2000; Piano et al., 2002). In addition, genes whose cDNA has been isolated, meaning those which are relatively highly expressed during normal development, were significantly more likely to show RNAi phenotypes (Fraser et al., 2000).

Our large-scale RNAi analysis revealed that genes required for viability are significantly underrepresented on the X chromosome compared to those on the autosomes (Maeda et al., 2001). The incidence of the clones that caused detectable phenotypes was nearly the same among the autosomes; it averaged 29.3% (ranging from 25.6% of chromosome V to 34.0% of chromosome I). In contrast, the incidence on the X chromosome was 16.4%, which was significantly lower than that of autosomes. This finding was later confirmed by the genome-wide RNAi analysis by the feeding method (Kamath et al., 2003). They further revealed the biased distribution of the essential genes within chromosomes. In the autosomes, essential genes are enriched twofold in the central "cluster" region of each chromosome where rates of recombination are low. Furthermore, genes that show viable post-embryonic phenotypes are present in greater frequency in the center of the X chromosome (Kamath et al., 2003). There are two possible explanations for this underrepresentation of essential genes on X chromosome. It may be that mutations in the essential genes on the X chromosome directly lead to lethality or sterility in XO males, with the result that essential genes tend to be removed from the X chromosome. It has also been reported that the X chromosome is transcriptionally silenced in the germline during mitosis and early meiosis (Kelly et al., 2002), which would also contribute to a reduced incidence of genes required in germline development and early embryogenesis on this chromosome.

6. Use of functional genomic RNAi data for in-depth characterization

The accumulated functional genomic RNAi data is a useful source for the pre-selection of genes for further in-depth analyses. From our large-scale RNAi analysis, we selected genes that caused F1 sterility with no apparent somatic phenotypes, which are expected to play a germline-specific function (Maeda et al., 2001). As the secondary analysis, the germline of these RNAi-affected animals was examined with DIC microscopy for cellular morphology, and with DAPI staining for nuclear morphology, which revealed various kinds of germline aberration (a defect in the proliferation of germ cells, germline sex determination, oogenesis, and fertilization), and gonadogenesis as well (Maeda et al., 2001).

We are currently focusing on the genes that resulted in embryonic lethality. We record the terminally arrested embryonic phenotypes observed by DIC microscopy, and also examining the post-embryonic phenotypes of these embryonic lethal genes by the L1-soaking method described above. These detailed RNAi phenotype profiles will significantly expand the understanding of the role(s) of each gene in the development of *C. elegans*.

7. Perspectives

Large-scale RNAi analyses have dramatically increased the amount of functional information for each gene of the *C. elegans* genome. Since the data from large-scale functional genomic RNAi analyses are available and searchable from WormBase (Stein et al., 2001) and other Internet-based resources, researchers can now perform the first round of gene screening for particular RNAi phenotypes "*in silico*," before beginning actual work at the bench.

In order to make it possible to effectively analyze the large-scale RNAi phenotype data, each RNAi phenotype needs to be described in a manner that can be easily converted into a format suitable for computer-assisted analysis. As Piano et al. demonstrated in their study of early embryogenesis (Piano et al., 2002), the development of a clearly defined vocabulary for the systematic description of phenotypes is required. Our group now aims to establish a systematic method for the profiling of embryonic and post-embryonic phenotypes. Piano et al. also showed that such systematically recorded RNAi phenotype profiles are readily used for cluster analysis such that genes are clustered based on the relatedness of the phenotypes, which is useful in identifying sets of genes involved in the same developmental process that may consist of genetic pathway or protein complexes (Piano et al., 2002). Furthermore, in conjunction with other genome-wide data, such as expression profiles (Hill et al., 2000; Kim et al., 2001; Reinke et al., 2000) and protein–protein interaction maps (Li et al., 2004), the RNAi phenotype data set promises to provide a rich information source for use in the deciphering of *C. elegans* biology.

Considering that over 40% of *C. elegans* genes have sequence homologies with genes in other organisms (The *C. elegans* Sequencing Consortium, 1998), these functional genomic resources are useful not only to the *C. elegans* research community, but to those studying other organisms as well. With the recent dramatic improvements in RNAi technology, it seems inevitable that large-scale RNAi will be applied to other model organisms in the very near future.

Acknowledgements

We thank members of the Sugimoto lab for helpful discussion; and Naoko Iida for the figures.

REFERENCES

Austin, J. and Kimble, J. (1987). *glp-1* is required in the germ line for regulation of the decision between mitosis and meiosis in *C. elegans*. *Cell*, **51**, 589–599.

Austin, J. and Kimble, J. (1989). Transcript analysis of *glp-1* and *lin-12*, homologous genes required for cell interactions during development of *C. elegans*. *Cell*, **58**, 565–571.

Cogoni, C. and Macino, G. (2000). Post-transcriptional gene silencing across kingdoms. *Current Opinions in Genetics and Development*, **10**, 638–643.

Dillin, A., Hsu, A. L., Arantes-Oliveira, N., Lehrer-Graiwer, J., Hsin, H., Fraser, A. G., Kamath, R. S., Ahringer, J. and Kenyon, C. (2002). Rates of behavior and aging specified by mitochondrial function during development. *Science*, **298**, 2398–2401.

Elbashir, S. M., Harborth, J., Lendeckel, W., Yalcin, A., Weber, K. and Tuschl, T. (2001). Duplexes of 21-nucleotide RNAs mediate RNA interference in cultured mammalian cells. *Nature*, **411**, 494–498.

Fire, A., Xu, S., Montgomery, M. K., Kostas, S. A., Driver, S. E. and Mello, C. C. (1998). Potent and specific genetic interference by double-stranded RNA in *Caenorhabditis elegans*. *Nature*, **391**, 806–811.

Fraser, A. G., Kamath, R. S., Zipperlen, P., Martinez-Campos, M., Sohrmann, M. and Ahringer, J. (2000). Functional genomic analysis of C. elegans chromosome I by systematic RNA interference. *Nature*, **408**, 325–330.

Gonczy, P., Echeverri, C., Oegema, K., Coulson, A., Jones, S. J., Copley, R. R., Duperon, J., Oegema, J., Brehm, M., Cassin, E., *et al.* (2000). Functional genomic analysis of cell division in *C. elegans* using RNAi of genes on chromosome III. *Nature*, **408**, 331–336.

Hill, A. A., Hunter, C. P., Tsung, B. T., Tucker-Kellogg, G. and Brown, E. L. (2000). Genomic analysis of gene expression in *C. elegans*. *Science*, **290**, 809–812.

Kamath, R. S. and Ahringer, J. (2003). Genome-wide RNAi screening in *Caenorhabditis elegans*. *Methods*, **30**, 313–321.

Kamath, R. S., Fraser, A. G., Dong, Y., Poulin, G., Durbin, R., Gotta, M., Kanapin, A., Le Bot, N., Moreno, S., Sohrmann, M., *et al.* (2003). Systematic functional analysis of the *Caenorhabditis elegans* genome using RNAi. *Nature*, **421**, 231–237.

Kamath, R. S., Martinez-Campos, M., Zipperlen, P., Fraser, A. G., and Ahringer, J. (2001). Effectiveness of specific RNA-mediated interference through ingested double-stranded RNA in *Caenorhabditis elegans*. *Genome Biology*, **2**, RESEARCH0002.

Kelly, W. G., Schaner, C. E., Dernburg, A. F., Lee, M. H., Kim, S. K., Villeneuve, A. M. and Reinke, V. (2002). X-chromosome silencing in the germline of *C. elegans*. *Development*, **129**, 479–492.

Kim, S. K., Lund, J., Kiraly, M., Duke, K., Jiang, M., Stuart, J. M., Eizinger, A., Wylie, B. N. and Davidson, G. S. (2001). A gene expression map for *Caenorhabditis elegans*. *Science*, **293**, 2087–2092.

Kuroyanagi, H., Kimura, T., Wada, K., Hisamoto, N., Matsumoto, K. and Hagiwara, M. (2000). SPK-1, a *C. elegans* SR protein kinase homologue, is essential for embryogenesis and required for germline development. *Mechanisms of Development*, **99**, 51–64.

Lee, S. S., Lee, R. Y., Fraser, A. G., Kamath, R. S., Ahringer, J. and Ruvkun, G. (2003). A systematic RNAi screen identifies a critical role for mitochondria in *C. elegans* longevity. *Nature Genetics*, **33**, 40–48.

Li, S., Armstrong, C. M., Bertin, N., Ge, H., Milstein, S., Boxem, M., Vidalain, P. O., Han, J. D., Chesneau, A., Hao, T., Goldberg, D. S., Li, N., Martinez, M., Rual, J. F., Lamesch, P., Xu, L., Tewari, M., Wong, S. L., Zhang, L. V., Berriz, G. F., Jacotot, L., Vaglio, P., Reboul, J., Hirozane-Kishikawa, T., Li, Q., Gabel, H. W., Elewa, A., Baumgartner, B., Rose, D. J., Yu, H., Bosak, S., Sequerra, R., Fraser, A., Mango, S. E., Saxton, W. M., Strome, S., Van Den Heuvel, S., Piano, F., Vandenhaute, J., Sardet, C., Gerstein, M., Doucette-Stamm, L., Gunsalus, K. C., Harper, J. W., Cusick, M. E., Roth, F. P., Hill, D. E. and Vidal, M. (2004). A Map of the Interactome Network of the Metazoan *C. elegans*. *Science*, **303**, 540–543.

Maeda, I., Kohara, Y., Yamamoto, M. and Sugimoto, A. (2001). Large-scale analysis of gene function in *Caenorhabditis elegans* by high-throughput RNAi. *Current Biology*, **11**, 171–176.

Montgomery, M. K., Xu, S. and Fire, A. (1998). RNA as a target of double-stranded RNA-mediated genetic interference in *Caenorhabditis elegans*. *Proceedings of the National Academy of Sciences USA*, **95**, 15502–15507.

Paddison, P. J., Caudy, A. A., Bernstein, E., Hannon, G. J. and Conklin, D. S. (2002). Short hairpin RNAs (shRNAs) induce sequence-specific silencing in mammalian cells. *Genes & Development*, **16**, 948–958.

Piano, F., Schetter, A. J., Mangone, M., Stein, L. and Kemphues, K. J. (2000). RNAi analysis of genes expressed in the ovary of *Caenorhabditis elegans*. *Current Biology*, **10**, 1619–1622.

Piano, F., Schetter, A. J., Morton, D. G., Gunsalus, K. C., Reinke, V., Kim, S. K. and Kemphues, K. J. (2002). Gene clustering based on RNAi phenotypes of ovary-enriched genes in *C. elegans*. *Current Biology*, **12**, 1959–1964.

Priess, J. R., Schnabel, H. and Schnabel, R. (1987). The *glp-1* locus and cellular interactions in early *C. elegans* embryos. *Cell*, **51**, 601–611.

Reboul, J., Vaglio, P., Rual, J. F., Lamesch, P., Martinez, M., Armstrong, C. M., Li, S., Jacotot, L., Bertin, N., Janky, R., *et al.* (2003). *C. elegans* ORFeome version 1.1: Experimental verification of the genome annotation and resource for proteome-scale protein expression. *Nature Genetics*, **34**, 35–41.

Reboul, J., Vaglio, P., Tzellas, N., Thierry-Mieg, N., Moore, T., Jackson, C., Shin-i, T., Kohara, Y., Thierry-Mieg, D., Thierry-Mieg, J., *et al.* (2001). Open-reading-frame sequence tags (OSTs) support the existence of at least 17300 genes in *C. elegans*. *Nature Genetics*, **27**, 332–336.

Reinke, V., Smith, H. E., Nance, J., Wang, J., Van Doren, C., Begley, R., Jones, S. J., Davis, E. B., Scherer, S., Ward, S. and Kim, S. K. (2000). A global profile of germline gene expression in *C. elegans*. *Molecular Cell*, **6**, 605–616.

Sijen, T., Fleenor, J., Simmer, F., Thijssen, K. L., Parrish, S., Timmons, L., Plasterk, R. H. and Fire, A. (2001). On the role of RNA amplification in dsRNA-triggered gene silencing. *Cell*, **107**, 465–476.

Simmer, F., Moorman, C., Van Der Linden, A. M., Kuijk, E., Van Den Berghe, P. V., Kamath, R., Fraser, A. G., Ahringer, J. and Plasterk, R. H. (2003). Genome-wide RNAi of *C. elegans* using the hypersensitive rrf-3 strain reveals novel gene functions. *Public Library of Science Biology*, **1**, E12.

Simmer, F., Tijsterman, M., Parrish, S., Koushika, S. P., Nonet, M. L., Fire, A., Ahringer, J. and Plasterk, R. H. (2002). Loss of the putative RNA-directed RNA polymerase RRF-3 makes *C. elegans* hypersensitive to RNAi. *Current Biology*, **12**, 1317–1319.

Stein, L., Sternberg, P., Durbin, R., Thierry-Mieg, J. and Spieth, J. (2001). WormBase: network access to the genome and biology of *Caenorhabditis elegans*. *Nucleic Acids Research*, **29**, 82–86.

Tabara, H., Grishok, A. and Mello, C. C. (1998). RNAi in *C. elegans*: Soaking in the genome sequence. *Science*, **282**, 430–431.

Tavernarakis, N., Wang, S. L., Dorovkov, M., Ryazanov, A. and Driscoll, M. (2000). Heritable and inducible genetic interference by double-stranded RNA encoded by transgenes. *Nature Genetics*, **24**, 180–183.

The *C. elegans* Sequencing Consortium. (1998). Genome sequence of the nematode *C. elegans*: A platform for investigating biology. *Science*, **282**, 2012–2018.

Timmons, L. and Fire, A. (1998). Specific interference by ingested dsRNA. *Nature*, **395**, 854.

Winston, W. M., Molodowitch, C. and Hunter, C. P. (2002). Systemic RNAi in *C. elegans* requires the putative transmembrane protein SID-1. *Science*, **295**, 2456–2459.

31 Tools for integrative genomics: Genome-wide RNAi and expression profiling in *Drosophila*

Michael Boutros and Marc Hild

Introduction

Recent years have seen the sequencing of the complete genome of human and major model organisms, opening – and demanding – new discovery paradigms. By using a combination of high-throughput methodologies and sophisticated information analysis, several new approaches promise to add function to the raw "sequence" blueprint of life. Currently, high-throughput technologies have been applied to study the transcriptome, by expression profiling (Brown and Botstein, 1999) and ChIP on chip (Wyrick and Young, 2002), the proteome (2D gel-electrophoresis and mass-spectrometry analysis) and protein–protein interactions [(two-hybrid screens, affinity-tag purification coupled with mass-spectrometry, protein chips) (Uetz et al., 2000; Gavin et al., 2002; Drewes and Bouwmeester, 2003; Giot et al., 2003; Li et al., 2004)]. Genome-wide reverse genetic approaches have recently entered the field. In yeast, deletion strains for every gene transcript have been constructed and allow the functional analysis on a genome-wide scale (Wagner et al., 1997; Giaever et al., 2002). Similarly, genome-wide RNAi experiments promise to deliver functional information both in whole-animal assays and focused cell-based screening applications. Typically, hypothesis-driven experiments have relied on combining the strength of different experimental approaches to infer statements about functional properties. It can be similarly expected that in the medium term, a combination of multiple "data-rich" genomic technologies will be required to formulate and test hypotheses about system properties. Ultimately, this should enable us to accurately change physiological systems to prevent or cure disease conditions.

Genome-wide functional screens have become one of the key next challenges to systematically discover gene function. The availability of whole genome sequence of man and major model organisms has opened many new avenues for approaches that test the expression, interaction and depletion of gene products not only through one-gene-at-a-time paradigms but through high-throughput technologies that collect thousands of experimental observations at once or in very short time-frames. In contrast to approaches in the pre-genome era, well-annotated genome sequences will also allow us to make deductions about the "absence" of positive results – resulting in functional maps that cover not only positive but also negative results. The development of high-throughput technologies made screening procedures, such as microarrays and cell-based assays, feasible in many biological replicates, thereby delivering results that can be solidly quantified and rigorously statistically tested. Statistical analyses to assess false negative and false positive rates are an important integral part in any type of large-scale functional genomic approach. In this chapter, we will summarize recent approaches that we believe will be widely used both in academic and industrial discovery programs. Furthermore, we will present an approach to develop a versatile platform that has been used for both microarray and genome-wide RNAi screens.

RNAi and loss-of-function analysis

Large scale loss-of-function approaches have been rather difficult to conduct, largely due to the difficulty of generating collections of "knock-out" genetic backgrounds in metazoans and analyzing phenotypes in formats that are compatible with high-throughput settings. RNA interference (RNAi) has been rapidly filling this gap by permitting silencing of gene expression in diverse metazoan, from worms to humans.

RNAi was first discovered in *C. elegans* by Fire and Mello in 1998 who elegantly showed that introduction of double-stranded RNA molecules homologous to endogenous sequences rapidly leads to destruction of target mRNAs (Fire et al., 1998). Subsequent studies have shown that double-stranded RNA (dsRNA) molecules are both important endogenous regulators of gene expression and a powerful new tool to create transient loss-of-function phenotypes (Hannon, 2002; Montgomery and Fire, 1998). In *C. elegans*, several genome-wide collections of dsRNAs have been applied to study developmental defects, fat storage and other phenotypes (Fraser et al., 2000; Gonczy et al., 2000; Ashrafi et al., 2003; Kamath et al., 2003). *C. elegans in vivo* studies have been facilitated by the ease of dsRNA introduction: feeding of worms with *E. coli* which express double-stranded RNA molecules are sufficient to produce fully penetrant phenotypes. This topic has been detailed by Asako Sugimoto in the previous chapter.

Similarly, dsRNA introduction into *Drosophila* embryos has been successfully used to study gene function. *Drosophila* is one of the best-studied genetic model organisms and has been instrumental for identifying and dissecting conserved pathways that play important roles during development and homeostasis. Most of the knowledge gained on the *Drosophila* genome has relied on the characterization of phenotypes by forward genetic approaches. These studies have provided functional information on approximately 20% of the genes (Nagy et al., 2003). Since the completion of the *Drosophila* genome sequence, it has become increasingly important to develop systematic methods to generate functional information on every gene encoded in the genome. Unlike *C. elegans*, however, feeding animals with dsRNA is not technically feasible and *in vivo* analysis on a genome-wide scale has not yet been conducted. *Drosophila* as a model organism has however several other advantages: many well-established cell lines of different tissue origins are available and have been extensively used in prior studies of many conserved signaling pathways. Further, RNAi experiments in *Drosophila* cells are technically simple: long dsRNA molecules added to the culture are readily internalized by cells without further treatment and lead to fully penetrant phenotypes (Clemens et al., 2000). This makes *Drosophila* cell-based assays an attractive system for high-throughput loss-of-function studies. Cell-based assays using small-interfering RNA have also been developed in mammalian cells and it can be expected that current applications in model organisms will lead the way to similarly powerful approaches to study gene functions in human cells.

Overall, a combination of statistically relevant "wet" biology approaches using recent high-throughput technologies and integration by bioinformatics will increase our capacity to analyze biological processes. It will also require new discussion about "evidence criteria" necessary to deduce biological functions. In the coming years, analyses of integrated genomic datasets will give insights into underlying networks of cellular processes at a level of detail that has not been seen before.

Generating genome-wide reagents

To achieve a high degree of comparability between different functional genomics, we developed an amplicon-based approach that allows the creation of multiple reagent sets used in whole-genome transcriptome and RNAi analysis, *in situ* analysis and possibly antibody technology (Figure 31.1). The amplicon sets are readily "converted" for different usages, such as the production of double-stranded RNA for RNAi experiments and single-stranded RNA for *in situ* hybridization. Such a set also carries the potential for the systematic production of recombinant antibodies to all potential proteins in the genome as most of the amplicons include only exonic sequences and therefore code for specific protein fingerprints. The availability of such a comprehensive antibody collection would open the way for studies of the proteome in fine detail. Versatile amplicon sets are not only very economical but will also allow to better correlation of result data sets obtained by different genomic technologies.

Genome-wide amplicon approach (GAA)

An amplicon set has to fulfill several criteria to make it useful for a spectrum of functional genomic approaches. For *Drosophila*, a median amplicon size of 500 bp was chosen to ensure good hybridization conditions in microarrays but also allow for efficient RNAi, *in situ* hybridizations and sufficient specificity for antibody production. BLAST evaluation against all other predicted transcripts

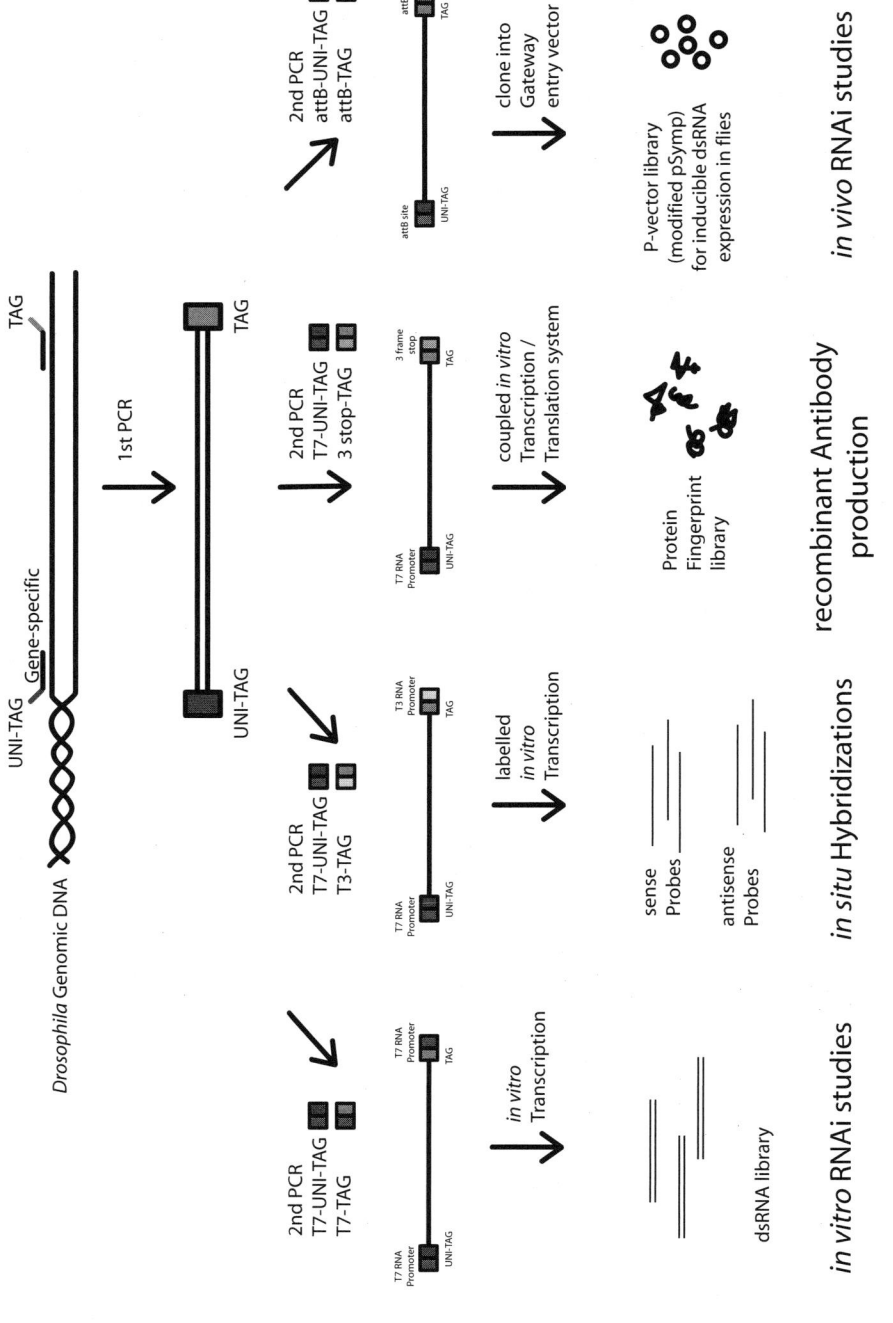

had to be performed to exclude homologies with other genes that would lead to cross-hybridization and off-site effects in RNAi experiments. Such versatility is created by a two-step PCR protocol. First, gene-specific amplicons are generated by PCR from genomic DNA and incorporation of a limited number of TAG sequences at the end of all gene specific primer pairs. Second, the PCR products are re-amplified using specific TAG primers linked to application-specific sequences. Since the main cost drivers for genome-wide reagent sets are gene-specific oligos, this approach is much less expensive than non-TAG approaches. Further, using such a two-step procedure one can easily add required sequences, such as T7 promoters to the TAG specific primers and thereby produce a new set of amplicons which can now serve as templates for large-scale RNA *in vitro* transcription and subsequent dsRNA production. Similar other different sequences can be added to the ends of the amplicon and thereby create the templates for sense and anti-sense probe synthesis used in *in situ* hybridization.

Specifically, based on these requirements the Drosophila GAA design was performed as such:

1. Exclusion of conserved/repeated sequences

All predicted ORF sequences we blasted against each other to detect and exclude regions from the amplicon design which were not unique. Based on the default parameters used in this BLAST search, regions of more than 40 bp length showing more than 70% identity were flagged and consequently excluded from amplicon design. While such a strategy might not exclude regions showing a limited conservation just below the chosen homology criteria, the re-analysis showed that only about 13.5% had a low-stringency second site hit. Based on the expression profiling data we obtained we found that no more than 15% of them showed co-regulation thus demonstrating that less than 2% of the amplicons may be involved in cross-hybridization.

2. Primer selection

For the actual PCR primer design we used the GenomePRIDE software which based on a fuzzy-logic approach automatically selects the optimal position for the gene-specific amplicons as well as the matching PCR primer pairs (Haas et al., 2003). By providing a list of criteria and their corresponding weights for

---

Figure 31.1. Genome-wide Amplicon Approach. Amplicons selected according to the rules outlined in the text are amplified by a 2-step PCR protocol. In a first step gene-specific primers carrying a limited set of TAG sequences at their ends are used to produce 1st PCR products from genomic DNA. Depending on the envisioned downstream application the amplicons are re-amplified using the TAG-specific primers carrying the required, additional sequences at their 5′ ends. For example we added a T7 promoter sequence to the forward UNI-TAG primer as well as to the 9 different reverse TAG primers to produce a template library which can be used in dsRNA synthesis. Alternatively the 2nd PCR product may serve as template for *in situ* probe generation, to produce protein fingerprints as a basis for antibody generation or to generate a library of P-vectors for *in vivo* RNAi studies.

the amplicon selection process the software is easily adjusted to different needs. For our purposes we applied the following rules:

a. Exclude regions of homology
   Based on the all-against-all BLAST search (see above).
b. Exclude most 5′ and 3′ regions from amplicon selection
   Minimizes the effect of false gene models as ab initio gene prediction programs tend to have the most problems in finding the correct start and end of a gene.
c. Homogenous amplicon size
   Ensures comparable hybridization conditions in expression profiling experiments and in in situ *studies. Moreover by choosing a preferred size of 500 bp the amplicon is large enough to ensure efficient RNAi as well as for recombinant antibody production.*
d. Single exon
   All envisioned uses of the amplicon set will benefit if the amplicon consists only of expressed sequences. The presence of introns will lower the efficiency as probes for expression profiling, in situ *hybridization as well as for use in RNAi. Introns will be most harmful for the expression of protein fingerprints and the subsequent use for antibody production.*
e. Frame
   For protein fingerprint expression it will be best to ensure that all amplicons start in the same frame. Alternatively the use of 3 different forward TAG-primers could compensate for lack of this.

3. TAG primer

By using a 2-step PCR protocol we not only improved on the sensitivity and thus success rate of the amplicon generation but also tried to limit the amount of contaminating genomic DNA (used as template for 1st PCR) present in the samples spotted in microarray production. In addition by using a combination of a unique, forward TAG primer in combination with 9 different, reverse TAG primers we could limit cross-well contaminations. Most important the incorporation of these TAG sequences at the end of the gene-specific primer pairs will allow for the very efficient re-amplification of the amplicon set. Based on this limited TAG primers set it is very convenient to add sequences for different down-stream approaches (for example T7 promoter sequences that allow dsRNA production).

4. Validation of the design

Expression profiling using microarrays based on the amplicon set allows for an easy validation of the design. A first test may be the comparison of expression levels/changes in the expression of genes sharing sequence motifs to ensure specificity of the set. Similar genes known to be coordinately regulated or – if available – previously published expression profiling data can be used to show the overall validity of the amplicon selection. In the end after such initial tests, the comparison of the microarray results with *in situ* hybridization data as well as a limited number of RNAi experiments will prove the power of the design.

Microarray analysis using GAA

To include all possible expressed sequences, even sequences that lacked yet sufficient annotation information, we based our genome-wide amplicon design on a combination of the Berkeley *Drosophila* Genome Project (BDGP) *Drosophila* genome annotation with a new *ab initio* gene prediction using the Fgenesh software resulting in a combined total of 21,396 predicted open reading frames (Salamov and Solovyev 2000, Hild et al., 2003; Misra et al., 2002). While the fact that nearly 97% of the BDGP genes were also predicted by Fgenesh validates our overlap criterion, we found another 6,000 predicted genes not represented in the latest BDGP annotation (Misra et al., 2002). Based on the design rules the primer selection software was successful for 21,306 (99.6%) predicted ORFs. Using our 2-step PCR protocol we produced amplicons for 97.7% of all potential genes from genomic DNA.

For microarray analysis, the complete amplicon set was spotted with appropriate controls, resulting in a high-density transcriptome 48,000-spot microarray. To assess the overall quality of our amplicon selection as well as to validate the novel predictions, we performed developmental profiling of the *Drosophila* life-cycle and analyzed the data by correspondence analysis. Our microarray analysis provided evidence for the expression of 9908 (78.3%) of the BDGP release 3.1 genes that are represented on the array and 2636 (42.9%) of the novel genes. A detailed analysis of the data showed that less than 2% of all amplicons may be prone to cross-hybridization (Hild et al., 2003).

Further, previously annotated and newly identified transcripts were analyzed using whole-embryo *in situ* analysis. To this end, the amplicons were modified using a new secondary PCR incorporating either T7 or SP6 promotor sequences to generate anti-sense specific probes using *in vitro* transcription. The various specific patterns observed demonstrated that the subset of novel predicted transcripts not only are expressed, but also validated the GAA design [(Hild et al., 2003) (see also http://hdflyarray.zmbh.uni-heidelberg.de)].

Generating a genome-wide RNAi library

Subsets of all genes in the *Drosophia* genome were successfully used in RNAi screens to identify novel components in various biological processes (Ramet et al., 2002; Goshima and Vale, 2003; Kiger et al., 2003; Lum et al., 2003; Rogers et al., 2003). Following pilot experiments based on a limited number of dsRNAs, we decided to generate a set of genome-wide dsRNAs to conduct high-throughput cell-based screens by silencing every gene in the *Drosophila* genome. Using the amplicon approach and modified secondary primers that incorporated T7-promotors on each end of all PCR products, we synthesized a total of 21,306 dsRNA in a 96-well format covering all 13,600 genes predicted by the *Drosophila* Genome Project plus additional open reading discovered through genome-wide expression profiling.

Each step of the RNA synthesis was adapted and optimized for a high-throughput formats:

1. *Synthesis*: PCR-products were used as templates for *in vitro* transcription reactions. *In vitro* transcription reactions were set up using a robotic liquid handling device and using T7 Megascript reagents (Ambion, TX). To control for consistent RNA quality and add functional controls, each 96-well plate incorporated an additional control PCR fragment that targeted either DIAP-1 (an anti-apoptotic factor), Rho1 (required for cytokinesis), GFP (as a negative control) and an empty well. The well plates were arranged in such a pattern that when reformatted into 384-well plates, each contained a complete set of controls. After a 4-hour incubation time, the reactions were treated with DNase I to digest the template product.

2. *Purification*: Subsequently, RNAs are purified using Millipore filtration plates (MANU3010). Purification by Millipore plates was shown to be superior to other methods, such as glass beads or column purification, which in most cases had a limiting binding capacity and significantly decreased the total yield. RNA products were eluted from the membrane in 10mM Tris pH7 and stored at $-70°$C.

3. *Quality controls*: All RNA products were quality controlled by spectrophotometry to assess the total yield. The size of the products was controlled by gel electrophoresis and documented using an in-house database. On average, a 50-μl *in vitro* transcription reaction yielded approximately 100 μg dsRNA.

4. *Dilution and aliquotting steps*: After purification and quality control steps, the obtained dsRNA were aliquotted into 384-well tissue-culture plates that are ready-to-screen and can be stored for several months at $-70°$C. Using "up-front" batch aliquot procedures allows significant reduction of the time necessary to set up genome-wide RNAi screens and reduces the dependence on high-throughput liquid handling devices.

A one-time production run proved to be sufficient for ~300 genome-wide cell-based screens (depending on the cell and screening format). The total failure rate of PCR and RNA-production was less than 8% during a first run-through synthesis reaction. Figure 31.2 shows the steps required to produce a genome-wide RNAi library (Boutros et al., 2004).

RNAi screening and assays

The principal strategy using this cell-based assay system was to reduce the time required to screen through a genome-wide set of dsRNA. Using defined phenotypic assays, such as cell morphology changes or reporter gene activity, this approach allows us to collect data sets in a very short time frame. High-throughput

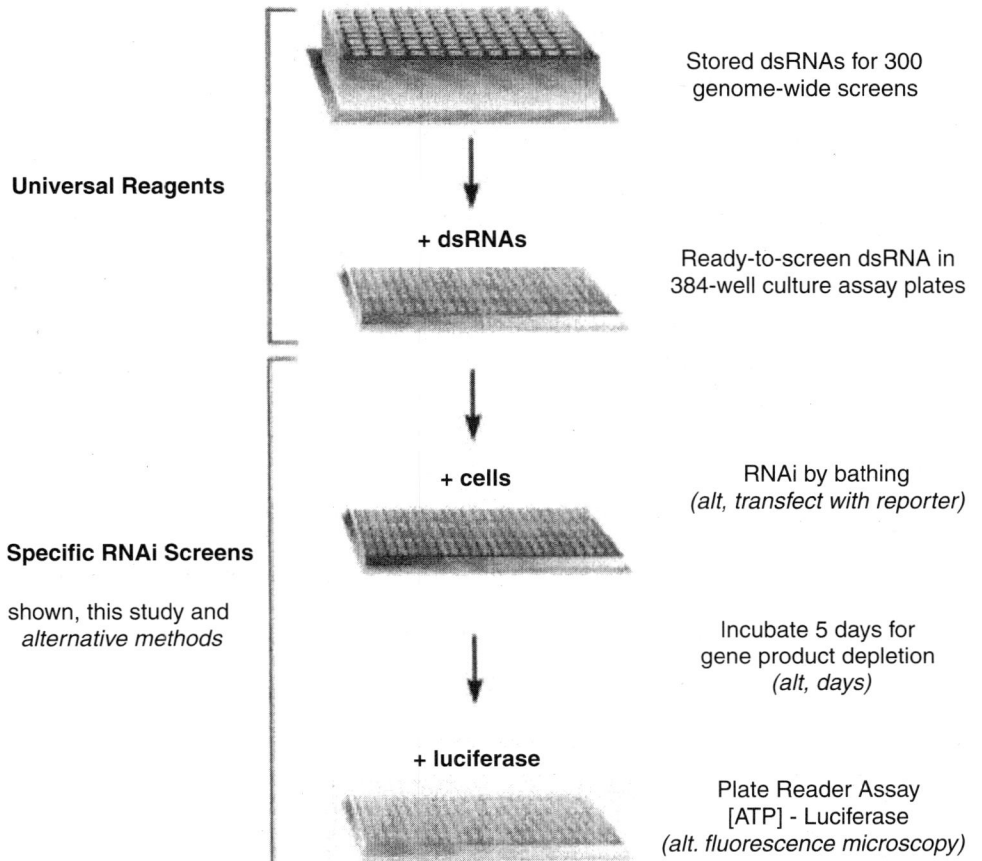

Figure 31.2. Strategy to synthesize and screen a genome-wide RNAi library in *Drosophila*. Synthesized double-stranded (ds) RNA is aliquotted in batches from storage plates into ready-to-use screening plates. These plates are frozen at −80 °C. Once a screen should initiated, plates with dsRNA is thawed and cells are added. After several days growth, specific cellular assays are performed either by monitoring the activity of a reporter gene through luminescent or fluorescent read-out, or by automated microscopy.

cell-based screens using a genome-wide RNAi library are conducted in 384-well plates. Assays were developed that are based on visual cellular phenotypes (to be analyzed by automated microscopes), luminescent or fluorescent readouts (analyzed by plate readers). Computational methods were established to rapidly score phenotypes and potential hits (Boutros et al., 2004). Using high-throughput plate reader systems, for example, luciferase-based assays, for every gene in the genome may be conducted within few days.

   In order to initially characterize RNAi phenotypes in cells, a homogenous luciferase assay that monitors cell death and cell growth phenotypes was used to quantitatively identify factors required for cell proliferation in *Drosophila* macrophage-like cells. First, we quantitatively tested the effect of treating cells either with a dsRNA against GFP or the *Drosophila* homolog of IAP (inhibitor of apoptosis protein). Removal of IAP proteins has been previously shown to induce programmed cell death (Hay et al., 1995). These results indicated that

Kc₁₆₇ Genome-wide RNAi/Screen 1

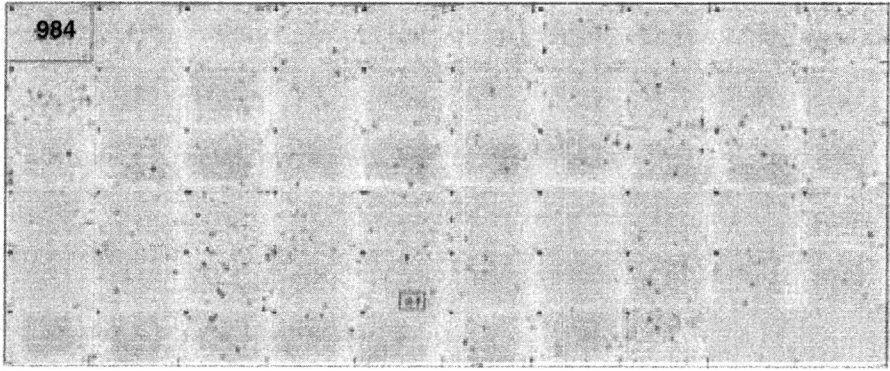

Kc₁₆₇ Genome-wide RNAi/Screen 2

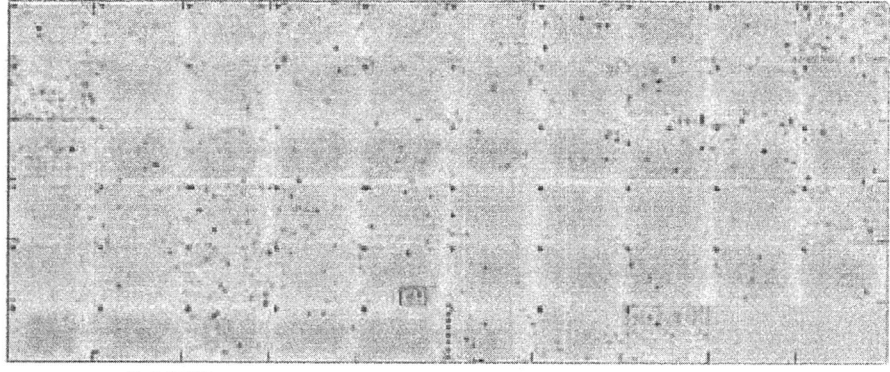

z-score

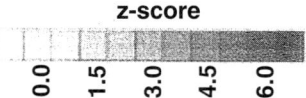

Figure 31.3. Reproducibility of quantitative genome-wide RNAi screens. Shown are duplicate screens for cell proliferation and viability phenotypes. Grey colors indicate quantitative ATP-readout using a luciferin-luciferase assay. Darker values indicate lower ATP-levels, indicative for loss of viability or proliferation phenotypes. Each small box represent a single RNAi experiment. In total, over 20,000 RNAi experiments are performed for each screen. Quantitative analysis showed that duplicate screens had a correlation coefficient of 0.86, indicating that the assays phenotypes are highly reproducible.

we can accurately identify candidate genes which upon removal affect cell death or cell growth phenotypes. Subsequently, all 21 000 dsRNA were assayed in different cell lines. Figure 31.3 shows the results of two genome-wide screens in a Kc167 cell line. In this representation, each dot represents the result of a different dsRNA treatment; on a scale where brightness indicates an increasing number of detected live cells. The analysis of these first data sets showed that the phenotypes are highly reproducible both on qualitative and quantitative levels. Based on added control RNA with known cell death phenotypes, we estimate that the computational analysis has a false negative rate of less than 8%.

When we analyzed different functional groups, we found genes with similar functional properties to give very similar phenotypes. For example, the depletion of ribosomal transcripts leads to a reduced growth level; however they do not induce cell death. In contrast, RNAi against IAP or other survival factors, such as the *Drosophila* homolog of the PDGF/VEGF receptor, showed an increased number of apoptotic cells. The initial screen also allowed us to assess the reproducibility of RNAi-phenotypes, and constitutes a baseline for all pathway specific screens that might be affected by cell growth or cell death phenotypes. These phenotypic hits were further characterized in secondary cell assays by using other apoptotic markers that confirmed the anti-apoptotic function of many novel factors (Boutros et al., 2004).

Quantitative loss-of-function phenotypes on a genome-wide scale showed a high correlation coefficient of 0.86, indicating that this approach is highly reproducible and may allow a quantitative classification over a wide range of readouts. Statistical comparison with expression levels assayed by microarray and RT-PCR showed that dsRNA-induced phenotypes are not correlated with expression levels, indicating that RNAi efficiency is not dependent on the relative expression level of the targeted transcript.

This screening approach allows a rapid assessment for cellular phenotypes by automated microscopy, quantitative fluorescent or luminescent readouts. Using homogenous cell-based assays, such as luminescence reporters, genome-wide screens can be conducted in a few days. Given that the initial screens identified many genes that have been previously shown to act in the same pathways *in vivo*, we expect that this approach can be applied as to many biological pathways that can be represented using a relevant cell-based assays. Many biological assays are already adapted for screening chemical compound libraries and we expect that many cell-based assays are adaptable for RNAi screening applications. The choice of a defined assay that represents a biological pathway coupled with the ability to analyze quantitative readouts might also identify components that are partially redundant or have complex phenotypes *in vivo* and which have been missed in previous genetic screens.

Conclusions

Integrated approaches will become essential in the analysis of genome-wide datasets. Versatile reagents will certainly lead to higher efficiency in screening approaches and provide the flexibility necessary for further technology development. For example, to delineate cellular pathways and deduct systems properties, one would like to comprehensively identify perturbations e.g. by RNAi, then map transcriptional signatures in wild-type and perturbed situations using complementary transcriptome approaches. Subsequent computational analysis of comprehensive data sets will result in quantitative models of complex signal transduction networks and their regulatory circuitry that are at the basis of each living cell. Analysis of the effects of perturbation of these networks predicted using the computer generated models and then compared to the *in vivo* situation will lead to an even more detailed understanding. Multiple reiterations of this experiment–model building circle will finally reveal even the weakest interconnections and allow for an almost complete view on the cellular networks. Additional information will come from studies of the proteome and interaction maps. Large-scale loss-of-function studies will provide important clues about disease-relevant cellular pathways and might more rapidly identify functionally confirmed targets than can be targeted by therapeutics.

REFERENCES

Ashrafi, K., Chang, F. Y., Watts, J. L., Fraser, A. G., Kamath, R. S., Ahringer, J. and Ruvkun, G. (2003). Genome-wide RNAi analysis of *Caenorhabditis elegans* fat regulatory genes. *Nature*, **421**, 268–272.

Boutros, M., Kiger, A. A., Armknecht, S., Kerr, K., Hild, M., Koch, B., Haas, S. A., Paro, R. and Perrimon, N. (2004). Genome-wide RNAi analysis of growth and viability in *Drosophila* cells. *Science*, **303**, 832–835.

Brown, P. O. and Botstein, D. (1999). Exploring the new world of the genome with DNA microarrays. *Nature Genetics*, **21**, 33–37.

Clemens, J. C., Worby, C. A., Simonson-Leff, N., Muda, M., Maehama, T., Hemmings, B. A. and Dixon, J. E. (2000). Use of double-stranded RNA interference in *Drosophila* cell lines to dissect signal transduction pathways. *Proceedings of the National Academy of Sciences USA*, **97**, 6499–6503.

Drewes, G. and Bouwmeester, T. (2003). Global approaches to protein-protein interactions. *Current Opinion in Cell Biology*, **15**, 199–205.

Fire, A., Xu, S., Montgomery, M. K., Kostas, S. A., Driver, S. E. and Mello, C. C. (1998). Potent and specific genetic interference by double-stranded RNA in *Caenorhabditis elegans*. *Nature*, **391**, 806–811.

Fraser, A. G., Kamath, R. S., Zipperlen, P., Martinez-Campos, M., Sohrmann, M. and Ahringer, J. (2000). Functional genomic analysis of *C. elegans* chromosome I by systematic RNA interference. *Nature*, **408**, 325–330.

Gavin, A. C., Bosche, M., Krause, R., Grandi, P., Marzioch, M., Bauer, A., Schultz, J., Rick, J. M., Michon, A. M., Cruciat, C. M. Remor, M., Hofert, C., Schelder, M., Brajenovic, M., Ruffner, H., Merino, A., Klein, K., Hudak, M., Dickson, D., Rudi, T., Gnau, V., Bauch, A., Bastuck, S., Huhse, B., Leutwein, C., Heurtier, M. A., Copley, R. R., Edelmann, A., Querfurth, E., Rybin, V., Drewes, G., Raida, M., Bouwmeester, T., Bork, P., Seraphin, B., Kuster, B., Neubauer, G. and Superti-Furga, G. (2002). Functional organization of the yeast proteome by systematic analysis of protein complexes. *Nature*, **415**, 141–147.

Giaever, G., Chu, A. M., Ni, L., Connelly, C., Riles, L., Veronneau, S., Dow, S., Lucau-Danila, A., Anderson, K., Andre, B. Arkin, A. P., Astromoff, A., El-Bakkoury, M., Bangham, R., Benito, R., Brachat, S., Campanaro, S., Curtiss, M., Davis, K., Deutschbauer, A., Entian, K. D., Flaherty, P., Foury, F., Garfinkel, D. J., Gerstein, M., Gotte, D., Guldener, U., Hegemann, J. H., Hempel, S., Herman, Z., Jaramillo, D. F., Kelly, D. E., Kelly, S. L., Kotter, P., LaBonte, D., Lamb, D. C., Lan, N., Liang, H., Liao, H., Liu, L., Luo, C., Lussier, M., Mao, R., Menard, P., Ooi, S. L., Revuelta, J. L., Roberts, C. J., Rose, M., Ross-Macdonald, P., Scherens, B., Schimmack, G., Shafer, B., Shoemaker, D. D., Sookhai-Mahadeo, S., Storms, R. K., Strathern, J. N., Valle, G., Voet, M., Volckaert, G., Wang, C. Y., Ward, T. R., Wilhelmy, J., Winzeler, E. A., Yang, Y., Yen, G., Youngman, E., Yu, K., Bussey, H., Boeke, J. D., Snyder, M., Philippsen, P., Davis, R. W. and Johnston, M. (2002). Functional profiling of the *Saccharomyces cerevisiae* genome. *Nature*, **418**, 387–391.

Giot, L., Bader, J. S., Brouwer, C., Chaudhuri, A., Kuang, B., Li, Y., Hao, Y. L., Ooi, C. E., Godwin, B., Vitols, E., Vijayadamodar, G., Pochart, P., Machineni, H., Welsh, M., Kong, Y., Zerhusen, B., Malcolm, R., Varrone, Z., Collis, A., Minto, M., Burgess, S., McDaniel, L., Stimpson, E., Spriggs, F., Williams, J., Neurath, K., Ioime, N., Agee, M., Voss, E., Furtak, K., Renzulli, R., Aanensen, N., Carrolla, S., Bickelhaupt, E., Lazovatsky, Y., DaSilva, A., Zhong, J., Stanyon, C. A., Finley, R. L., Jr., White, K. P., Braverman, M., Jarvie, T., Gold, S., Leach, M., Knight, J., Shimkets, R. A., McKenna, M. P., Chant, J. and Rothberg, J. M. (2003). A protein interaction map of *Drosophila melanogaster*. *Science*, **302**, 1727–1736.

Gonczy, P., Echeverri, C., Oegema, K., Coulson, A., Jones, S. J., Copley, R. R., Duperon, J., Oegema, J., Brehm, M., Cassin, E. et al. (2000). Functional genomic analysis of cell division in *C. elegans* using RNAi of genes on chromosome III. *Nature*, **408**, 331–336.

Goshima, G. and Vale, R. D. (2003). The roles of microtubule-based motor proteins in mitosis: Comprehensive RNAi analysis in the *Drosophila* S2 cell line. *Journal of Cell Biology*, **162**, 1003–1016.

Haas, S. A., Hild, M., Wright, A. P., Hain, T., Talibi, D. and Vingron, M. (2003). Genome-scale design of PCR primers and long oligomers for DNA microarrays. *Nucleic Acids Research*, **31**, 5576–5581.

Hannon, G. J. (2002). RNA interference. *Nature*, **418**, 244–251.

Hay, B. A., Wassarman, D. A. and Rubin, G. M. (1995). *Drosophila* homologs of baculovirus inhibitor of apoptosis proteins function to block cell death. *Cell*, **83**, 1253–1262.

Hild, M., Beckmann, B., Haas, S., Koch, B., Solovyev, V., Busold, C., Fellenberg, K., Boutros, M., Vingron, M., Sauer, F., Hoheisel, J. and Paro, R. (2003). An integrated gene annotation and transcriptional profiling approach towards the full gene content of the *Drosophila* genome. *Genome Biology*, **5**, R3.

Kamath, R. S., Fraser, A. G., Dong, Y., Poulin, G., Durbin, R., Gotta, M., Kanapin, A., Le Bot, N., Moreno, S., Sohrmann, M., Welchman, D. P., Zipperlen, P. and Ahringer, J. (2003). Systematic functional analysis of the Caenorhabditis elegans genome using RNAi. *Nature*, **421**, 231–237.

Kiger, A., Baum, B., Jones, S., Jones, M., Coulson, A., Echeverri, C. and Perrimon, N. (2003). A functional genomic analysis of cell morphology using RNA interference. *J. Biology*, **2**, 27.

Li, S., Armstrong, C. M., Bertin, N., Ge, H., Milstein, S., Boxem, M., Vidalain, P. O., Han, J. D., Chesneau, A., Hao, Goldberg, D. S., Li, N., Martinez, M., Rual, J. F., Lamesch, P., Xu, L., Tewari, M., Wong, S. L., Zhang, L. V., Berriz, G. F., Jacotot, L., Vaglio, P., Reboul, J., Hirozane-Kishikawa, T., Li, Q., Gabel, H. W., Elewa, A., Baumgartner, B., Rose, D. J., Yu, H., Bosak, S., Sequerra, R., Fraser, A., Mango, S. E., Saxton, W. M., Strome, S., Van Den Heuvel, S., Piano, F., Vandenhaute, J., Sardet, C., Gerstein, M., Doucette-Stamm, L., Gunsalus, K. C., Harper, J. W., Cusick, M. E., Roth, F. P., Hill, D. E. and Vidal, M. (2004). A map of the interactome network of the metazoan *C. elegans*. *Science*, **303**, 540–543.

Lum, L., Yao, S., Mozer, B., Rovescalli, A., Von Kessler, D., Nirenberg, M. and Beachy, P. A. (2003). Identification of Hedgehog pathway components by RNAi in *Drosophila* cultured cells. *Science*, **299**, 2039–2045.

Misra, S., Crosby, M. A., Mungall, C. J., Matthews, B. B., Campbell, K. S., Hradecky, P., Huang, Y., Kaminker, J. S., Millburn, G. H., Prochnik, S. E., Smith, C. D., Tupy, J. L., Whitfied, E. J., Bayraktaroglu, L., Berman, B. P., Bettencourt, B. R., Celniker, S. E., de Grey, A. D., Drysdale, R. A., Harris, N. L., Richter, J., Russo, S., Schroeder, A. J., Shu, S. Q., Stapleton, M., Yamada, C., Ashburner, M., Gelbart, W. M., Rubin, G. M. and Lewis, S. E. (2002). Annotation of the *Drosophila melanogaster* euchromatic genome: Asystematic review. *Genome Biology*, **3**, RESEARCH0083–3.

Montgomery, M. K. and Fire, A. (1998). Double-stranded RNA as a mediator in sequence-specific genetic silencing and co-suppression. *Trends in Genetics*, 14, 255–258.

Nagy, A., Perrimon, N., Sandmeyer, S. and Plasterk, R. (2003). Tailoring the genome: the power of genetic approaches. *Nature Genetics*, **33** Suppl, 276–284.

Ramet, M., Manfruelli, P., Pearson, A., Mathey-Prevot, B. and Ezekowitz, R. A. (2002). Functional genomic analysis of phagocytosis and identification of a *Drosophila* receptor for E. coli. *Nature*, **416**, 644–648.

Rogers, S. L., Wiedemann, U., Stuurman, N. and Vale, R. D. (2003). Molecular requirements for actin-based lamella formation in *Drosophila* S2 cells. *Journal of Cell Biology*, **162**, 1079–1088.

Salamov, A. A. and Solovyev, V. V. (2000) Ab initio gene finding in *Drosophila* genomic DNA. *Genome Research 2000*, **10**, 516–522

Uetz, P., Giot, L., Cagney, G., Mansfield, T. A., Judson, R. S., Knight, J. R., Lockshon, D., Narayan, V., Srinivasan, M., Pochart, P., Qureshi-Emili, A., Li, Y., Godwin, B., Conover,

D., Kalbfleisch, T., Vijayadamodar, G., Yang, M., Johnston, M., Fields, S. and Rothberg, J. M. (2000). A comprehensive analysis of protein-protein interactions in *Saccharomyces cerevisiae*. *Nature*, **403**, 623–627.

Wagner, U., Brownlees, J., Irving, N. G., Lucas, F. R., Salinas, P. C. and Miller, C. C. (1997). Overexpression of the mouse dishevelled-1 protein inhibits GSK-3beta- mediated phosphorylation of tau in transfected mammalian cells. *Federation of European Biochemical Society Letters*, **411**, 369–372.

Wyrick, J. J. and Young, R. A. (2002). Deciphering gene expression regulatory networks. *Current Opinion in Genetics & Development*, **12**, 130–136.

32 Microarray analysis and RNA silencing to determine genes functionally important in mesothelioma

Maria E. Ramos-Nino and Brooke T. Mossman

Introduction

The most important event in science in this decade has been the publication of the sequence of the human genome by two independent initiatives (Lander et al., 2001; Venter et al., 2001). However, the sequence of the human genome and other species can contribute to our understanding of human biology only if it can be linked with functional information on the roles of encoded proteins. A major goal of science in the "post-genomic era" is to unravel the functions of the many genes discovered by sequencing. Although sequence- or structure-based comparisons are enabling the generation of hypotheses on the biochemical functions of many gene products, determining the role of a large set of genes is still a challenge.

Recently, RNA interference (RNAi) technology has been developed for the down-regulation of selected gene expression in mammalian cells. This technology offers a rapid way to gain insight to loss-of-function phenotypes associated with specific genes. Furthermore, the combination of RNAi with other functional genomic approaches such as gene expression profiling is providing a powerful tool in our efforts to establish the function of genes in the pathogenesis of many diseases. In this chapter, we will summarize the contributions that RNAi and microarrays have had in cancer research, particularly in understanding the mechanisms and pathogenesis of malignant mesothelioma, a devastating tumor of the serosal cells lining the pleural, peritoneal and pericardial cavities (Mossman and Gee, 1989).

Advances in RNAi technology applied to mammalian cells

Different experimental procedures have been used in the field of molecular biology to assign a specific function to a specific gene product. For example, a common method used to investigate the function of a protein in human heritable diseases is a knock-out experiment where a specific protein is mutated or altered to investigate consequent phenotypic or functional changes. This technique has been used in many classical knock-out mouse models which can mimic the phenotypic changes and pathogenesis of several human diseases.

The use of knock-out techniques is a laborious process with sometimes negative results, and establishing reliable methods to reduce or knock out gene expression have been goals of molecular biologists for several years. Many techniques have been tested for this purpose including antisense sequences, aptamers, intramers, ribozymes, chimeric oligonucleotides, etc., with effects difficult to predict, and often weak suppression achieved (Braasch and Corey, 2002). In 1990, two teams lead by Napoli (Napoli et al., 1990) and Stuitje (van der Krol et al., 1990) first reported the co-suppression of an over-expressed *chalcone synthase* in plants. The mechanism of this phenomenon remained a mystery, but it was proposed that the product of degradation of the double-stranded RNA region in the gene might be related to this post-transcriptional gene silencing (Jorgensen et al., 1996; van der Krol et al., 1990). In 1998, Fire's group, at the Carnegie Institute, and Mello at the University of Massachusetts (Fire et al., 1998), demonstrated with the worm *Caenorhabditis elegans* that double-stranded RNA (dsRNA) may specifically and selectively inhibit gene expression, a phenomenon called RNA interference (RNAi). The process of RNA interference is mediated by dsRNA which is cleaved by the enzyme DICER into 21- to 23-nt duplexes containing 2-nt overhang at the 3′ end of each strand. These duplexes are incorporated into a protein complex called the RNA-induced silencing complex (RISC). Directed by the antisense strand of the duplex, RISC recognizes and cleaves the target mRNA [(Bernstein et al., 2001; Elbashir et al., 2001; Sharp, 2001; Zamore, 2001; McManus and Sharp, 2002) (Figure 32.1)]. This process is a highly conserved mechanism believed to serve as an antiviral defense mechanism (Dillin, 2003).

RNAi is extremely active in invertebrate species; however, mammalian cells have developed protective mechanisms against viral infections that may impede the use of this approach in these cells. The presence of dsRNA (>30 bp) of viral origin triggers an interferon response (IFR), and the activation of a dsRNA response protein kinase (PKR). PKR inactivates the translation factor EIF2a which leads to the activation of the 2′, 5′ oligoadenylate synthetase, finally resulting in RNAse L activation. This cascade induces a global non-specific suppression of translation, which in turn triggers apoptosis, reviewed in (Williams, 1997; Gil and Esteban, 2000).

Procedures developed by Ribopharma (now Alnylam Europe AG, for which a patent has been granted) first demonstrated the functionality of RNAi in mammalian cells. This group proved that by using smaller dsRNA (20–24 bp), which they called SIRPLEX™, they could specifically knock out genes without the IFR and subsequently, apoptosis. Similar results confirmed this observation (Caplen et al., 2001; Elbashir et al., 2001).

Further advances in the technique have allowed the induction of stable and heritable gene silencing. Based on the work of a number of investigators (Grishok et al., 2001; Hütvagner et al., 2001; Ketting et al., 2001; Knight and Bass, 2001)

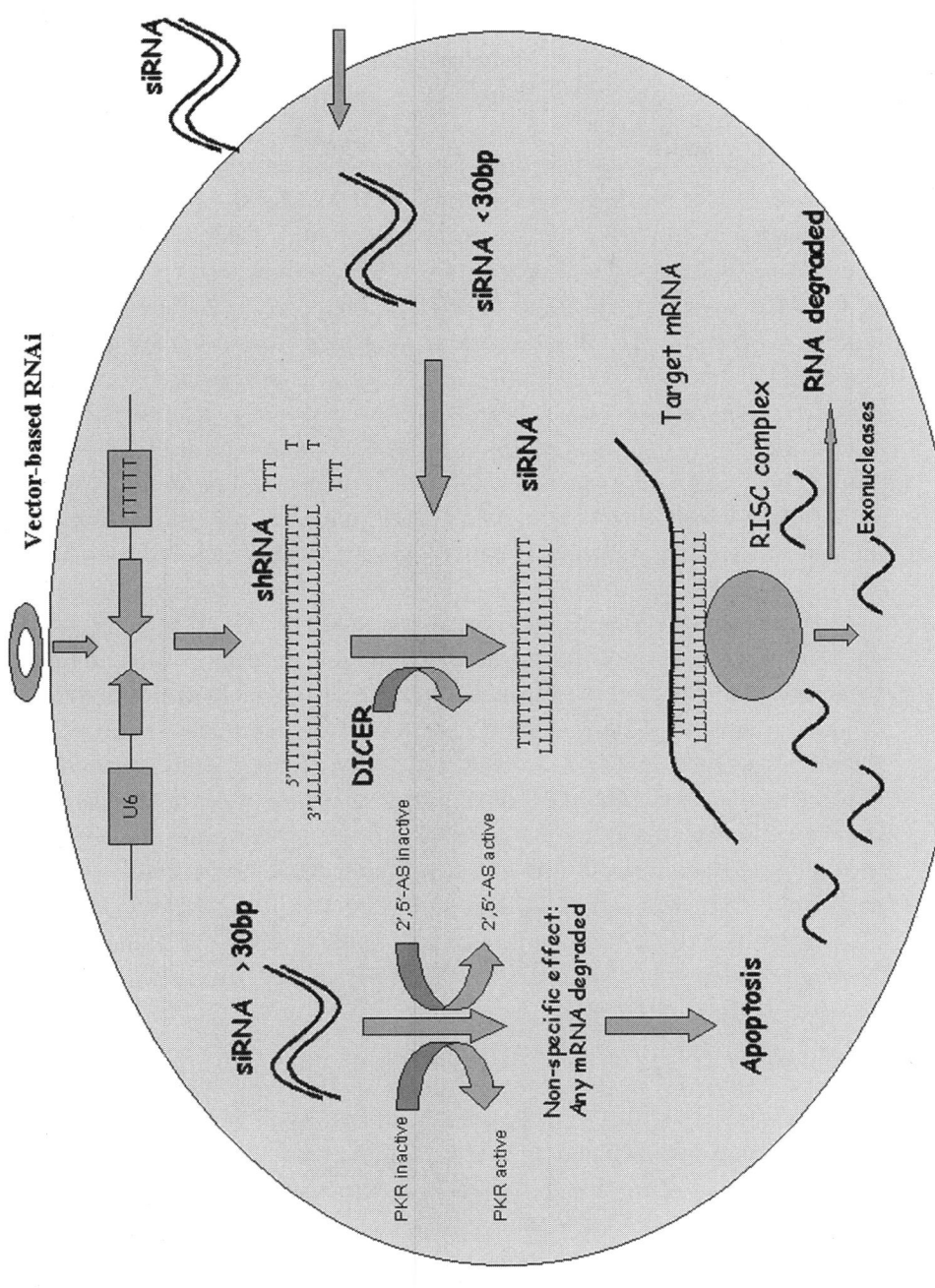

Figure 32.1. Non-specific and specific dsRNA silencing pathways in mammals.

who have shown that endogenously encoded triggers of RNAi-related pathways exist and are transcribed as short hairpin RNAs (generically miRNA). Paddison et al. (2002) studied the use of small RNAs folded in hairpin structures to inhibit the function of genes. A significant disadvantage of siRNA is that effects are transient with phenotypes generated by transfection with such RNAs persisting for ~1 wk. However, the use of short hairpin RNAs (shRNA) synthesized *in vivo* from strong promoters such as polymerase-II promoters like CMV or EF1alpha and RNA polymerase III promoters like U6 or H1 (Brummelkamp et al., 2002; Lee et al., 2002; Miyagishi and Taira, 2002; Paddison et al., 2002; Paul et al., 2002) enabled the creation of continuous cell lines in which suppression of a target gene was stably maintained by RNAi (Yu et al., 2002). RNA polymerase-III expression systems revealed a greater capacity for stable expression of short inhibitory RNAs *in vivo* and *in vitro* (Jennings and Molloy, 1987; Ojwang et al., 1992; Ilves et al., 1996). The use of vector-based shRNA opened the possibility of the use of this technique in *in vivo* work. For example, siRNA-expressing mice and rat transgenic animals which display knock-out phenotypes have been developed (Hasuwa et al., 2002).

The recent demonstrations that synthetic siRNAs can trigger sequence-specific RNAi in mammalian cells have stimulated interest in using this technique to determine gene function in human cells for the development of therapeutic approaches (Carmichael, 2002; Elbashir et al., 2001; Tuschl and Borkhardt, 2002). Although the use of short (20–24 bp) dsRNA does not induce the IFR, the specificity of the gene silencing induced by siRNA in mammalian cells has not been systematically examined. Ideally, the siRNA should not produce any non-specific effects including: 1) cross-hybridization to different mRNAs, 2) binding in a sequence-dependent manner to various cellular proteins as has been shown for antisense oligonucleotides (Branch, 1998; Lebedeva and Stein, 2001), and 3) translational silencing through miRNA effects, etc. (Doench et al., 2003).

Use of microarrays to demonstrate target specificity of siRNAs

In recent studies by Chi et al. (2003), the use of DNA microarrays to profile global gene expression demonstrated the specificity for target mRNA of dsRNA molecules in human kidney cells. This group used a cDNA array containing 36,000 genes to address the question of whether global changes in mRNA levels occurred due to RNAi directed against an exogenous gene, GFP. Results demonstrated no significant changes in global mRNA after 48 and 72 h of siRNA exposure. Other studies have also confirmed the specificity of siRNA. For example, Semizarov and colleagues (Semizarov et al., 2003) tested the hypothesis that if siRNA caused non-specific effects, then comparison of three different siRNA experiments targeting three different genes should share some commonality within their transcriptional profiles. The expression profiles of lung carcinoma cells treated with siRNA directed toward Rb, AKT1 or Plk1 were unrelated, confirming the specificity of the siRNA technique. Furthermore, they observed that the specificity of siRNA was concentration-dependent, showing that nonspecific cross-hybridization occurred at siRNA concentrations ~100 nM.

Microarrays/RNAi and gene function in human cancers

Microarrays hold much promise for the analysis of diseases, and their use has been especially intensive in the field of cancer research. The identification of single gene products that are expressed in tumor cells compared to normal tissue (Alon et al., 1999) is of great pharmacological interest, especially in the establishment of tumor markers for diagnostic purposes, and in the elucidation of target genes for chemo- or immunotherapy. Also, microarrays have been useful in the classification of cancer types in a number of studies (Golub et al., 1999; Alizadeh et al., 2000; Perou et al., 2000; Ramaswamy et al., 2001; van't Veer et al., 2002). Gene expression data can also be combined with analysis of genomic alterations, for example by microarray analysis of comparative genomic hybridization, whereby genomic DNA is hybridized to immobilized bacterial artificial chromosomes (BAC) or cDNA clones to identify gene amplifications or deletions that are involved in tumor development (Pollack et al., 1999). These studies have offered new prospects for refined tumor diagnosis and treatment by identifying different subtypes of tumors that can be targeted with tailored therapies.

The use of RNAi in the field of cancer research is more recent than that of microarrays, but the publications are starting to increase as the technical difficulties are overcome (Spankuch-Schmitt et al., 2002; Butz et al., 2003; Chen et al., 2003; De Schrijver et al., 2003; Harvey et al., 2003; Hemann et al., 2003; Nagy et al., 2003; Sun et al., 2003; Zhang et al., 2003).

The use of microarrays combined with RNAi in cell-based knock-out experiments provides a rapid analysis of phenotypic and functional changes, thus providing insight into the target gene's function in a high-throughput manner. siRNA and microarray analysis also have been proven to be reliable and valuable for the large-scale screening of gene function in drug target identification and validation. For example, in *C. elegans*, RNAi has yielded impressive results in investigating the function of a number of genes implicated in cell division (Chase et al., 2000). Furthermore, the first genome-wide RNAi screens have been performed, covering more than 16,000 genes (Kamath et al., 2003).

Cancer-related studies have reported the success of using RNAi and microarrays to investigate the function of a specific gene in human tumorigenesis. For example, Williams et al. (2003) have identified and validated genes involved in the pathogenesis of colorectal cancer using cDNA microarrays and RNAi. In this study, cDNA microarrays were used to detect differences in gene expression between normal tissue, colon tumors, and polyps isolated from 20 patients. To identify genes that were important in regulating the growth properties of colorectal cancer cells, RNAi was used to disrupt expression of several of the over-expressed genes in a colon tumor cell line, HCT116. The knock out of survivin, a potent inhibitor of apoptosis, severely reduced tumor growth both *in vitro* and in an *in vivo* xenograft model.

Interruption of mitogenic cell signaling pathways as targets for cancer therapy

The evolution of cancer in humans is thought to be a multistep process whereby cancer cells acquire mutant alleles of proto-oncogenes, tumor-suppressor genes, and other genes that control, in one way or another, cell proliferation and progression to malignancy. The general set of genetic and epigenetic alterations that are required to program and maintain the neoplastic phenotype of human cancer cells have been the center of intense research. One of the most interesting pathways thought to be involved in the development and maintenance of neoplastic phenotypes is the mitogen-activated protein kinase pathway (MAPKs).

The MAPKs [Extracellular Signal-Regulated Kinases (ERKs), p38 kinases, and Jun NH$_2$-terminal kinases (JNKs)] are a group of protein kinases that have an important function in mediating the responses of cells to changes in their environment (Figure 32.2), reviewed in (Schaeffer et al., 1999; Rincón et al., 2000). MAPK signaling cascades are generally initiated at the cell surface, but their targets are nuclear transcription factors.

The mammalian ERK1 and ERK2 MAPK are activated by signaling pathways that are initiated often by stimulation of cell surface receptors, a common point of integration being the activation of the small G protein, Ras. Phosphorylated ERKs translocate to the nucleus to phosphorylate transcription factors such as Elk-1 that is essential to transcriptional activation of c-*fos*, a component of the AP-1 transcription factor (Gille et al., 1992; Marais et al., 1993).

Studies designed to determine the functional outcomes of stimulation of individual MAPK pathways have employed pharmacological strategies (PD98059, an inhibitor of MEK1, the upstream activator of ERKs, and pyridinyl imimazole derivatives for inhibition of p38s) and/or transfection approaches. These data show that certain MAPK pathways may be causally linked in specific cell types to cell injury, proliferation, and/or transformation. Functional studies indicate that the activation of ERK1 and ERK2 provides proliferative signals that may contribute to normal growth and the malignant transformation of fibroblasts *in vitro* (Cowley et al., 1994; Mansour et al., 1994). Moreover, recent work shows that expression of dominant negative ERK2 inhibits AP-1 transactivation and neoplastic transformation in epidermal cells (Watts et al., 1998). In other cells, changes in ERK activity are implicated in cellular differentiation and growth arrest (Cowley et al., 1994; Bennett and Tonks, 1997; Whalen et al., 1997). Several laboratories have also shown that ERK1 and 2 activation is linked to the development of apoptotic and necrotic cell injury by stress-inducing agents including asbestos fibers (Zanella et al., 1996; Jimenez et al., 1997; Chin et al., 1998; Bhat and Zhang, 1999; Zanella et al., 1999). A recent report shows that oral administration of a small molecule inhibitor of MEK1 (PD184352) inhibits growth of mouse and human colon tumors after implantation into mice (Sebolt-Leopold et al., 1999). These

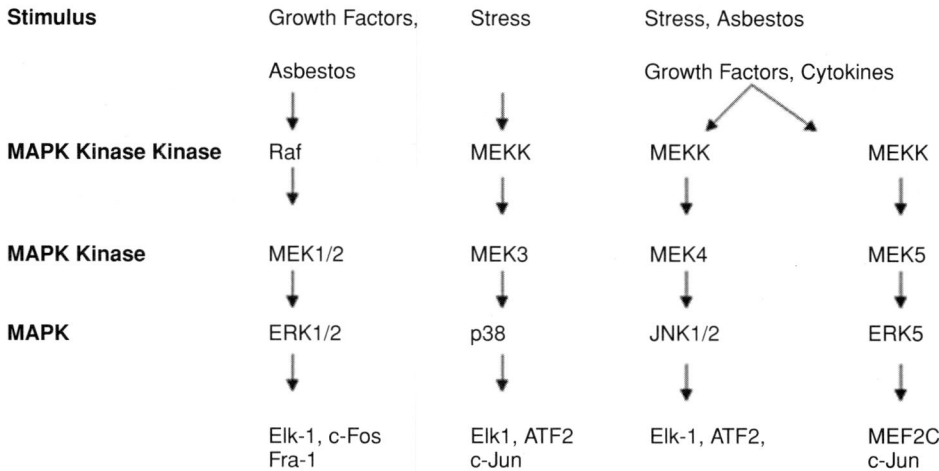

Figure 32.2. A basic diagram of Mitogen-Activated Protein Kinase (MAPK) signal transduction pathways. Modified from Rincón et al., 2000.

results suggest that abrogating the ERK1/2 pathway is a feasible approach to control cell proliferation in colon cancers.

Another MAPK route, the ERK5 pathway, has also been implicated in different cells events including proliferation (Kato et al., 1998; English et al., 1999; Scapoli et al., 2004), transformation, motility/invasion and angiogenesis (Regan et al., 2002; Sohn et al., 2002). This pathway has been shown to be activated by different stimuli including EGF, NGF (Kamakura et al., 1999), hyperosmolarity and oxidative stress (Abe et al., 1996; Yan et al., 1999). Some studies have demonstrated the importance of the ERK5 pathway in tumorigenesis. For example, van Dartel and colleagues (van Dartel et al., 2002) found a very high-level amplification of ERK5 in 17p11.2 of high-grade osteosarcomas, suggesting its contribution to their development. A tyrosine kinase profile of prostate carcinoma demonstrated the expression of MEK5 in derived cell lines and tumor tissue (Robinson et al., 1996). In an investigation of the gene expression profiles of cancer-related genes in hepatoma cell lines, the differential expression of cell cycle/growth regulator in hepatocellular carcinoma showed stronger tendency toward cell proliferation with more than 1.5 fold up-regulation of ERK5 (Liu et al., 2003).

ERK5 participates in neuregulin signal transduction, and is constitutively active in breast cancer cells overexpressing ErbB2 (Esparis-Ogando et al., 2002). ERK5 also activates Cyclin D1, a key step in cell proliferation in breast cancer cells (Mulloy et al., 2003). Overexpression of MEK5, an upstream activator of ERK5, stimulates proliferation, MMP-9 expression and invasion in prostate cells (Mehta et al., 2003). Moreover, ERK5 and ERK2 cooperate to regulate the transcription factor, NF-κB, and cause cell transformation (focus-forming activity) in NIH 3T3 cells (Pearson et al., 2001).

Targets of the MAPK pathways include the early response protooncogenes of the *fos*, (c-*fos*, *fra-1*, *fra-2*, *fosB*) and *jun* (c-*jun*, *junB*, *junD*) families (Ramos-Nino et al., 2002; Reddy and Mossman, 2002). These genes encode protein subunits of the AP-1 family of transcription factors. MAPKs regulate AP-1 transcription factor activity by several mechanisms including transcriptional regulation of AP-1 genes (i.e., c-*fos* and c-*jun*) and phosphorylation of specific Fos and Jun subunit proteins which modulate protein stability and/or transcriptional activity (Karin et al., 1997).

Utilization of different MAPK pathways represents a potential mechanism for the determination of cell-type- and stimuli-specific responses to extracellular agents. More importantly, MAPKs have been identified as a major signaling pathway activated through p21Ras (Denhardt, 1996). Ras proteins are involved in a high percentage of human tumors and are the central point for many signal transduction pathways in the cell. All of the members of the Ras family have GTPase activity and are involved in many different cellular functions including control of transcription, translation, cytoskeleton organization, cell–cell junctions, etc. (Malumbres et al., 1998). Evidence has accumulated demonstrating that Ras signaling has at least 10 effectors including Raf1, MEKK1, phosphatidylinositol 3-kinase (PI3-K), protein kinase zeta, and others (Treinies et al., 1999). In addition, Ras-transformed cells have elevated levels of Fra-1, Fra-2, c-Jun and JunB, which contribute to elevated AP-1 activity (Mechta et al., 1997). Blocking of AP-1 activity in Ras-transformed cells has been shown to suppress transformation (Lloyd et al., 1991).

Fra-1 is also the predominant component of AP-1 complex formation induced by activated Ras which promotes fibroblast transformation (Battista et al., 1998). In support of this hypothesis, a causal role for Fra-1 in cellular transformation has been documented in other systems including esophageal, skin and breast cancers, reviewed by Reddy and Mossman (2002). Overexpression of *fra-1* in fibroblasts results in an anchorage-independent cell growth *in vitro* and tumor formation in nude mice (Wisdom and Verma, 1993; Bergers et al., 1995). Studies in our laboratory have confirmed the requirement of MEK1/2 activation for increased Fra-1 expression in mesotheliomas (Ramos-Nino et al., 2002) and the involvement of PI3-K and c-Src in the regulation of *fra-1* expression (Ramos-Nino and Mossman, manuscript in preparation).

Recently, several lines of evidence generated by various laboratories also indicate a potential role for the Fos family member, Fra-1, in abnormal differentiation and transformation of lung cells. First, a broad overexpression of *fra-1*, but not other AP-1 components, induces some lung tumors in mice (Jochum et al., 2000). Secondly, various toxicants and known carcinogens, such as tobacco smoke (Reddy and Mossman, 2002), silica (Shukla et al., 2001) and asbestos (Heintz et al., 1993; Timblin et al., 1998; Sandhu et al., 2000) persistently activate *fra-1* expression in lung cells both *in vitro* and *in vivo*. Thirdly, Fra-1 positively upregulates gene expression associated with squamous cell metaplasia, a preneoplastic lesion (Patterson et al., 2001; Vuong et al., 2002). Fourthly, the transition from small-cell to non–small-cell lung cancer phenotype induced by H-Ras/c-Myc is associated with the specific induction of *fra-1*, but not other AP-1 family members (Risse-Hackl et al., 1998). Lastly, Fra-1 is a predominant component of the AP-1 complex in asbestos-induced mesotheliomas and proliferating rat mesothelioma cells, and overexpression of the dominant negative Fra-1 mutant inhibits growth of these cells in soft agar (Ramos-Nino et al., 2002). *Fra-1* mRNA is also highly induced in the carcinogen 4-(methyl-*N*-nitroamino)-1-(3-pyridil)-1-butanone (NNK)-induced lung tumors compared to normal lung tissue (Reddy and Mossman, 2002). Taken together, these observations highlight a potential role for Fra-1 in lung and pleural tumorigenesis.

Microarray/siRNA and the study of mesothelioma

The processes involved in the initiation and development of malignant mesothelioma, a tumor derived from pluripotential mesothelial cells which give rise to a variety of tumor phenotypes including epithelial, sarcomatous and mixed varieties (Mossman and Gee, 1989), are under intense investigation (Carbone et al., 2002). This unique tumor has been associated historically with occupational exposures to amphibole types of asbestos (Mossman et al., 1990; Mossman and Gee, 1989) and is increasing in several countries (McDonald and Wagner, 1996). More importantly, the prognosis of patients with mesothelioma is grim, most surviving less than a year after initial diagnosis (Mossman and Gee, 1989; Mossman et al., 1990). Thus, effective therapeutic strategies are desperately needed. In the past few years, simian virus 40 (SV40), a DNA virus, has been linked to the etiology of mesothelioma in multi-institutional studies showing that approximately 50% of human mesotheliomas in the United States contain SV40 large T antigen (Tag) DNA sequences, reviewed by Pass et al. (1998) and Klein et al. (2002).

In an effort to establish possible targets for therapy, our laboratory has used microarray analysis and RNAi technology to elucidate the role of important genes in mesothelioma (Ramos-Nino, 2003). We first characterized, using oligonucleotide microarray analysis, up- or down-regulated gene expression comparatively in 1) rat pleural mesothelial cells (RPM) isolated from the parietal pleura of Fischer 344 rats (Heintz et al., 1993); 2) RPM after acute exposures to crocidolite asbestos fibers $[(Na_2(Fe^{3+})_2(Fe^{2+})_3Si_8O_{22}(OH)_2]$, a high iron-containing amphibole fiber associated with the causation of human mesothelioma (Mossman et al., 1990); and 3) three mesothelioma cell lines: 11, 23 and 52 developed in rats after intraperitoneal injection of crocidolite asbestos (Craighead et al., 1987). The MAPK *erk5* and the transcription factor *fra-1* were upregulated (Table 32.1). After confirming that *fra-1* mRNA was dramatically increased in asbestos-exposed RPM and in human and rat mesotheliomas by Real Time- Quantitative PCR (RT-QPCR), we selected candidate genes following patterns of *fra-1* expression. After confirmation of changes in mRNA levels using real-time Q-PCR, we then used RNAi technology (Figure 32.3) to address the hypothesis that expression of some genes would be *fra-1* dependent. For this purpose, sequence information on *fra*-1 mature mRNA was extracted from the EST database (www.ncbi.nlm.nih.gov). The open frame region from the cDNA sequence around 100 nucleotides downstream of the start codon of the gene was selected to develop the 21-base duplex siRNA molecule (AAGCGCAGA-CACAGACAGUCC). The sequence was BLAST-searched (NCBI database) against EST libraries to ensure the specificity of the siRNA molecule. Transient transfection of cells with the siRNA duplex or a scramble control revealed that expression

Table 32.1. Comparison of Microarray and TaqMan results for different gene expression in asbestos-exposed RPM cells and in three rat mesotheliomas. Fold changes in microarray data are calculated with normalized values against the control (RPM). Fold changes in Taqman data are calculated from normalized ratios (gene expression/HPRT expression) against the control RPM cells. Reprinted with permission from Cancer Research (Ramos-Nino et al., 2003)

GB Acc#:GeneName	Method	RPM	Asb	Meso11	Meso23	Meso52
M19651 *fra-1*	TaqMan	1	9	219	117	311
	Microarray	1	10	289	319	677
M61875 *cd44*	TaqMam	1	17	123	147	279
	Microarray	1	1.5	2	4	5
U65007 *met*	TaqMan	1	3	7	16	9
	Microarray	1	2	30	54	19
X62875 High Mobility Group	TaqMan	1	5	253	432	709
	Microarray	1	3	8	15	23
L32591 *gadd45*	TaqMan	1	5	10	24	32
	Microarray	1	5	4	3	4
X02601 *src*	TaqMan	1	1.3	4	1.5	1.3
	Microarray	1	1.3	2.3	1.8	10.3
AJ005424 *erk5*	TaqMan	1	1.7	17	15	9
	Microarray	1	1.2	1.2	1.2	1.4

RPM: Rat Pleural Mesothelial cells
Asb: Asbestos-treated RPM cells
Meso 11, 23 and 52: Rat mesothelioma cell lines.

of *c-met* and *cd44* genes encoding receptors were linked to *fra-1* expression in asbestos-treated mesothelial cells and mesotheliomas (Figure 32.4).

CD44 is the principal cell surface receptor for the extracellular matrix glycosaminoglycan hyaluronan, which has been found to be increased in mesotheliomas (Li and Heldin, 2001). The binding of CD44 to hyaluronan mediates cell attachment and migration (Lewis et al., 2001). An association between overexpression of CD44s or its alternative spliced variants and aggressiveness or metastasis of a variety of human tumors show the importance of this protein in tumor invasiveness *in vitro* (Faassen et al., 1992; Thomas et al., 1993) and *in vivo* models (Gunthert et al., 1991; Guo et al., 1994). In support of an association between *fra*-1 and *cd44*, p21ras promotes transcription of the *cd44* gene via an AP-1 binding site in the *cd44* (Hofmann et al., 1993). Moreover, Lamb et al. (1997) also found a link between AP-1 and *cd44* induction. Moreover, the ERK pathway regulates alternative pre mRNA splicing variants of CD44 (Weg-Remers et al., 2001).

Our observation that *fra*-1 expression upregulates *c-met* also has mechanistic implications in the development of mesotheliomas by asbestos and SV40T antigens, further supporting the concept that these diverse agents may act through similar cell signaling pathways (Mossman and Gruenert, 2002). The *c-met* protooncogene encodes a transmembrane tyrosine kinase receptor (Met) for hepatocyte growth factor (HGF). HGF or scatter factor is a multifunctional protein involved in tissue repair as well as cancer and metastasis. Others have found a predominant role of HGF in mesothelioma cell invasion and regulation of MMP (matrix metalloproteinases) and TIMP ((MMP-inhibitor) levels (Harvey et al., 2000).

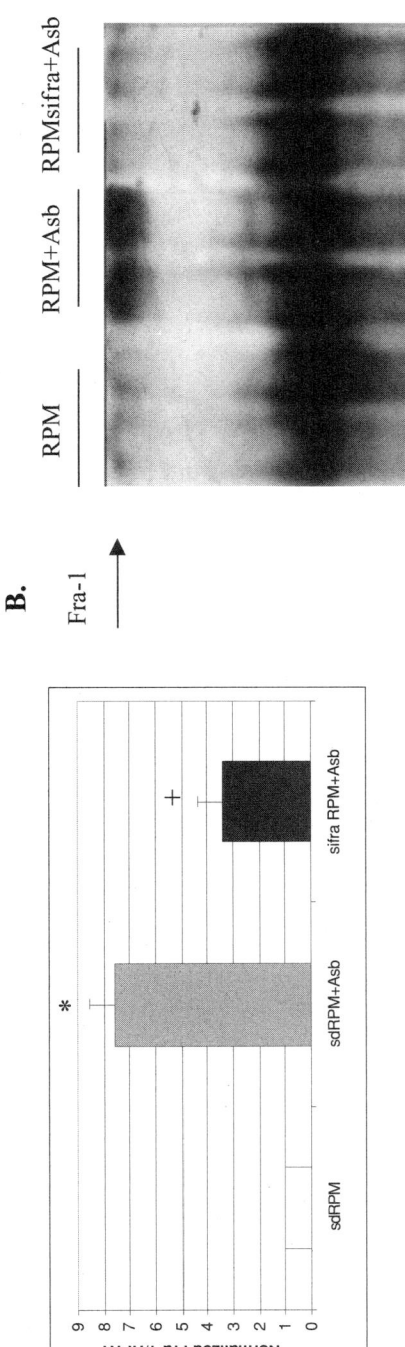

A.

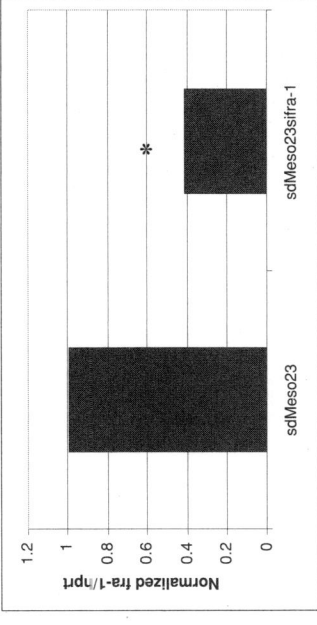

B.

Fra-1

RPM   RPM+Asb   RPMsifra+Asb

C.

Figure 32.3. Silencing of *fra*-1 using a siRNA to target *fra*-1 results in its decreased expression in RPM cells exposed to crocidolite asbestos (5 μg/cm² for 24 h); sd = groups transfected with scrambled duplex (control) and mesothelioma 23. A. RT- QPCR results for RPM and RPM exposed to crocidolite; B. Electrophoretic Mobility Shift Assay (EMSA) super-shift analysis for detection of Fra-1 in AP-1 complexes; C. Real time Q-PCR results for the mesothelioma cell line 23. Reprinted with permission from Cancer Research (Ramos-Nino et al., 2003).

457

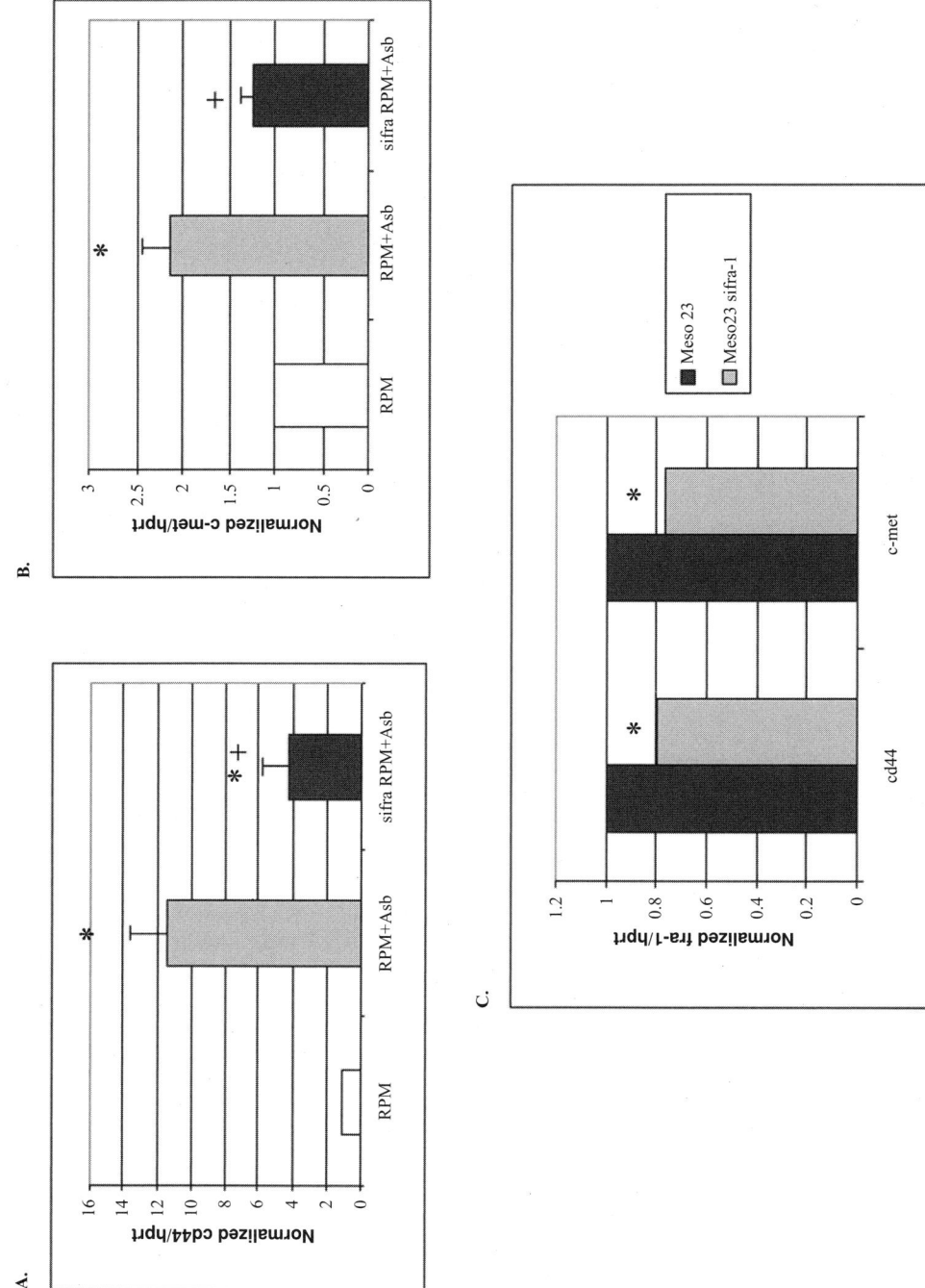

Figure 32.4. RT-QPCR for quantitation of selected genes upregulated in A and B) asbestos-exposed RPM cells and C. mesothelioma 23 after transfection with *sifra*-1 or scrambled duplex (sd). Reprinted with permission from Cancer Research (Ramos-Nino et al., 2003). * = $p < 0.05$; † = $p < 0.05$

A.

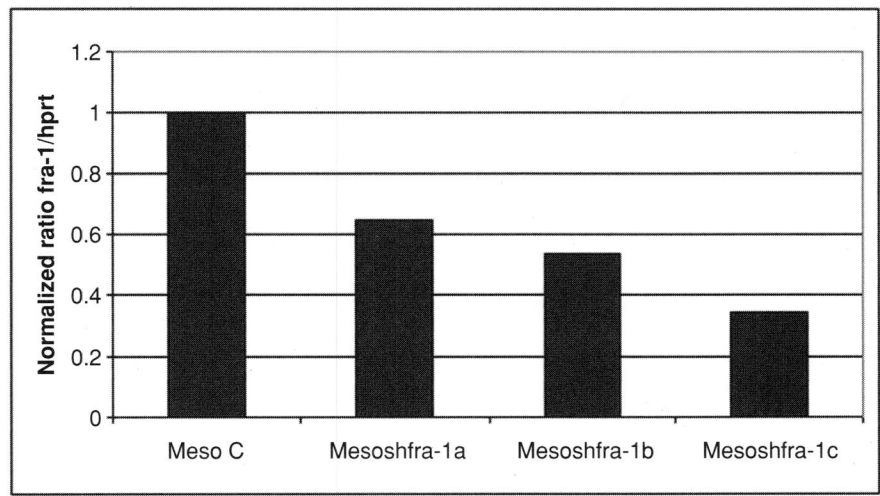

B.

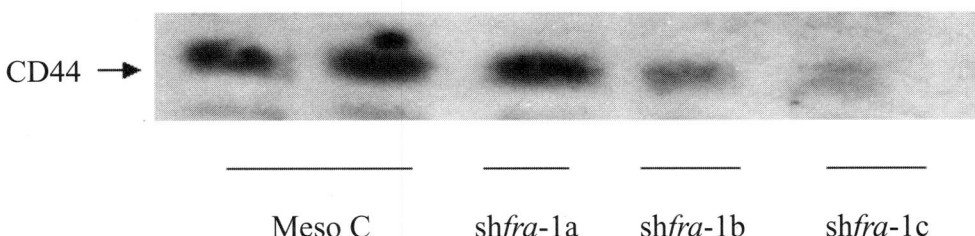

Figure 32.5. CD44 protein levels are related to levels of *fra-1* mRNA expression. A. RT-QPCR of mesothelioma cells with different levels of *fra-1* RNAi and B. Western blot showing different levels of CD44 as a result of different levels of *fra-1* RNAi. Meso C = Mesothelioma control transfected with empty vector.

c-met also is overexpressed in mesotheliomas and linked to growth, migration, and invasion of tumors (Klominek et al., 1998; Tolnay et al., 1998). Recent studies have confirmed that HGF is increased in bronchoalveolar and pleural lavage fluids after administration of crocidolite to rats, and causes mitogenesis in mesothelial cells (Adamson and Bakowska, 2001). Moreover, SV40 replication in human mesothelial cells induces HGF/Met receptor activation via an autocrine loop that causes cell cycle progression into S-phase (Cacciotti et al., 2001).

The link between *fra-1* gene expression and that of *cd44* was further demonstrated with the use of a vector-based rat *fra-1* shRNA, co-transfected with a neomycin resistant vector. After selection with neomycin, a set of colonies with different expression levels of RNAi was obtained. The use of permanent cell lines with different *fra*-1 RNAi capabilities allowed us to show by Western blot analysis the relationship between *fra-1* and CD44 expression (Figure 32.5).

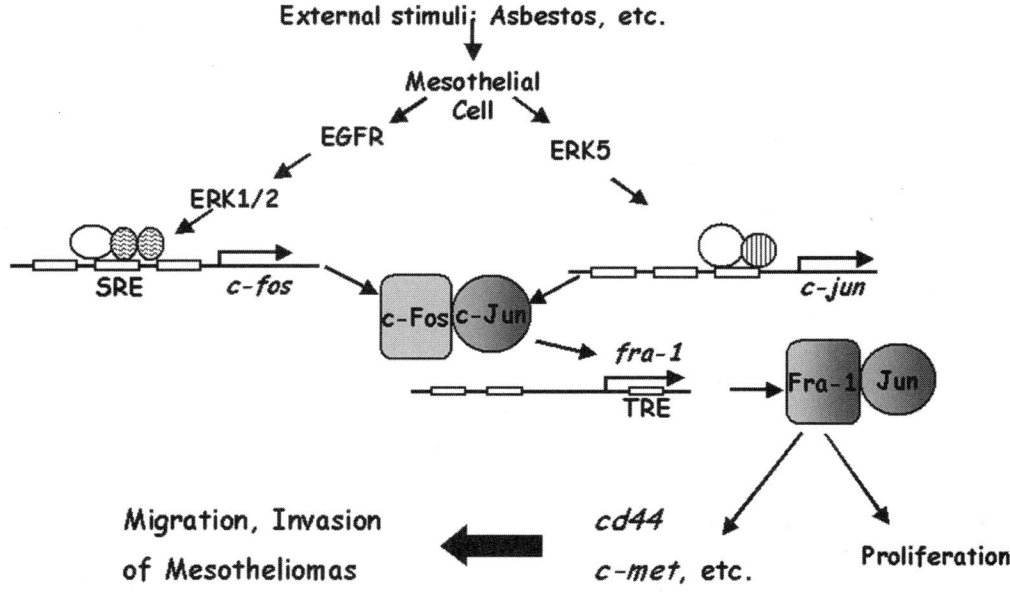

Figure 32.6. Hypothetical flow chart for the effect of asbestos in the MAPK pathways during the development of mesothelioma.

Recent studies using microarray and RNAi techniques also reveal the possible role of ERK5 in chemoresistance of human mesothelioma cells to the anticancer drugs, Carmustine and Doxorubicin (Ramos-Nino and Mossman, manuscript in preparation). Microarray analyses of rat mesotheliomas (see Table 32.1) and kinase activity assays in human mesothelioma cells have demonstrated that ERK5 is up-regulated in mesotheliomas. Moreover, a vector-based shRNA shows that knock-out of ERK5 sensitizes human mesothelioma cells to treatment with these anticancer drugs. The arrival of viral-mediated delivery mechanisms for RNAi will allow us to test our *in vitro* results in *in vivo* experiments. For example, Xia and colleagues (2002) recently demonstrated using a viral-mediated delivery mechanism, the specific silencing of β-glucuronidase in HeLa cells and in mouse brain and liver. Furthermore, they applied this strategy to a model system of a major class of neurodegenerative disorders, the polyglutamine diseases, to achieve reduced polyglutamine aggregation in cells. These results show the usefulness of the RNAi technique in therapy of human diseases.

In Figure 32.6, we present our hypothesis on how crocidolite asbestos contributes to the development and maintenance of mesothelioma through the MAPKs and the AP-1 transcription factor. Asbestos either directly or via formation of reactive oxygen species (ROS) causes phosphorylation of the EGFR and ERK1/2 phosphorylation and increased c-*fos* expression (Zanella et al., 1996; Pache et al., 1998; Zanella et al., 1999). Increases in ERK5 phosphorylation (Scapoli et al., 2004) result in increased expression of c-*jun*, thus forming an AP-1 complex and increased *fra*-1 expression, an AP-1 dependent gene. This transition and increased

Fra-1 in AP-1 complexes leads to increased proliferation and cd44/c-met expression (Ramos-Nino et al., 2002; Ramos-Nino et al., 2003).

Summary and conclusions

The mitogenic pathway is critically involved in human cancers, and cancer cells have reduced dependency on normal growth control (McCormick, 1999). This reduced dependency derives from the activity of oncogenes that generate constitutive mitogen signals (Hunter, 1991). An example of this is the K-ras mutation found in lung, pancreatic and colon cancers (Bos, 1989; Ellis and Clark, 2000). Alternatively, tumor cells can produce growth factors through MAPK cascades in an autocrine fashion, as has been demonstrated for human mesotheliomas (Mossman and Gruenert, 2002). Our data summarized above demonstrate the importance of the MAPK pathway as a possible target for therapy in human mesotheliomas. The combined technologies of microarray and RNAi demonstrate their usefulness in the determination of MAPK and AP-1 related gene function and drug targets for this tumor type.

The use of RNAi in cancer therapy is a future possibility; but questions need to be addressed before the technique is used as a therapeutic tool (Dillin, 2003). Some of these questions include: Can a specific siRNA be used in multiple cell types with the same efficiency? Is siRNA transferable to neighbor cells? In plants and worms, this mechanism exists. On the one hand, this may be a concern if the delivery of RNAi to tumor cells also affects normal neighboring cells. Alternatively, siRNA transfer may be an advantage in killing premalignant or remaining cells after removal of tumor masses. Can cells devoting much of their RNAi machinery to the process of mRNA degradation become defective in chromosome function or be more susceptible to viral infection? For example, worms and green algae with defects in the RNAi machinery have a higher incidence of transposon hopping, which in turn could lead to increased mutagenesis (Ketting et al., 1999; Ketting and Plasterk, 2000; Wu-Scharf et al., 2000). It is not known whether this function is conserved in humans, but if so, it could lead to genomic instability and the development of cancer later in life (Dillin, 2003).

As microarray data on malignant mesotheliomas accumulates in animal models and in human tumors (Rihn et al., 2000; Mohr et al., 2001; Gordon et al., 2002; Gordon et al., 2003; Ramos-Nino et al., 2003; Singhal et al., 2003) genes that can be targeted to reverse functional and phenotypic changes and/or kill tumor cells will be revealed. RNAi strategies will be required to assess gene function in the process of carcinogenesis and to exploit modifications in their functions as possible drug targets for cancer therapy. The arrival of RNAi technology provides hope in the battle to defeat and treat this often lethal disease.

REFERENCES

Abe, J., Kusuhara, M., Ulevitch, R. J., Berk, B. C. and Lee, J. D. (1996). Big mitogen-activated protein kinase 1 (BMK1) is a redox-sensitive kinase. *Journal of Biological Chemistry*, **271**, 16586–16590.

Adamson, I. and Bakowska, J. (2001). KGF and HGF are growth factors for mesothelial cells in pleural lavage fluid after intratracheal asbestos. *Experimental Lung Research*, **27**, 605–616.

Alizadeh, A. A., Eisen, M. B., Davis, R. E., Ma, C., Lossos, I. S., Rosenwald, A. et al. (2000). Distinct types of diffuse large B-cell lymphoma identified by gene expression profiling. *Nature*, **403**, 503–511.

Alon, U., Barkai, N., Notterman, D. A., Gish, K., Ybarra, S., Mack, D., et al. (1999). Broad patterns of gene expression revealed by clustering analysis of tumor and normal colon tissues probed by oligonucleotide arrays. *Proceedings of the National Academy of Sciences USA*, **96**, 6745–6750.

Battista, S., de Nigris, F., Fedele, M., Chiappetta, G., Scala, S., Vallone, D., et al. (1998). Increase in AP-1 activity is a general event in thyroid cell transformation *in vitro* and *in vivo*. *Oncogene*, **17**, 377–385.

Bennett, A. and Tonks, N. (1997). Regulation of distinct stages of skeletal muscle differentiation by mitogen-activated protein kinases. *Science*, **278**, 1288–1291.

Bergers, G., Graninger, P., Braselmann, S., Wrighton, C. and Busslinger, M. (1995). Transcriptional activation of the fra-1 gene by AP-1 is mediated by regulatory sequences in the first intron. *Molecular and Cellular Biology*, **15**, 3748–3758.

Bernstein, E., Denli, A. M. and Hannon, G. J. (2001). The rest is silence. *RNA*, **7**, 1509–1521.

Bhat, N. and Zhang, P. (1999). Hydrogen peroxide activation of multiple mitogen-activated protein kinases in an oligodendrocyte cell line: Role of extracellular signal-regulated kinase in hydrogen peroxide-induced cell death. *Journal of Neurochemistry*, **72**, 112–119.

Bos, J. L. (1989). ras oncogenes in human cancer: A review. *Cancer Research*, **49**, 4682–4689.

Braasch, D. A. and Corey, D. R. (2002). Novel antisense and peptide nucleic acid strategies for controlling gene expression. *Biochemistry*, **41**, 4503–4510.

Branch, A. D. (1998). A good antisense molecule is hard to find. *Trends in Biochemical Sciences*, **23**, 45–50.

Brummelkamp, T. R., Bernards, R. and Agami, R. (2002). A system for stable expression of short interfering RNAs in mammalian cells. *Science*, **296**, 550–553.

Butz, K., Ristriani, T., Hengstermann, A., Denk, C., Scheffner, M. and Hoppe-Seyler, F. (2003). siRNA targeting of the viral E6 oncogene efficiently kills human papillomavirus-positive cancer cells. *Oncogene*, **22**, 5938–5945.

Cacciotti, P., Libener, R., Betta, P., Martini, F., Porta, C., Procopio, A., et al. (2001). SV40 replication in human mesothelial cells induces HGF/Met receptor activation: A model for viral-related carcinogenesis of human malignant mesothelioma. *Proceedings of the National Academy of Sciences USA*, **98**, 12032–12037.

Caplen, N. J., Parrish, S., Imani, F., Fire, A. and Morgan, R. A. (2001). Specific inhibition of gene expression by small double-stranded RNAs in invertebrate and vertebrate systems. *Proceedings of the National Academy of Sciences USA*, **98**, 9742–9747.

Carbone, M., Kratzke, R. and Testa, J. (2002). The pathogenesis of mesothelioma. *Seminars in Oncology*, **29**, 2–17.

Carmichael, G. G. (2002). Medicine: Silencing viruses with RNA. *Nature*, **418**, 379–380.

Chase, D., Serafinas, C., Ashcroft, N., Kosinski, M., Longo, D., Ferris, D. K., et al. (2000). The polo-like kinase PLK-1 is required for nuclear envelope breakdown and the completion of meiosis in *Caenorhabditis elegans*. *Genetics*, **26**, 26–41.

Chen, Y., Stamatoyannopoulos, G. and Song, C. Z. (2003). Down-regulation of CXCR4 by inducible small interfering RNA inhibits breast cancer cell invasion *in vitro*. *Cancer Research*, **63**, 4801–4804.

Chi, J. T., Chang, H. Y., Wang, N. N., Chang, D. S., Dunphy, N. and Brown, P. O. (2003). Genomewide view of gene silencing by small interfering RNAs. *Proceedings of the National Academy of Sciences USA*, **100**, 6343–6346.

Chin, B., Choi, M., Burdick, M., Strieter, R., Risby, T. and Choi, A. (1998). Induction of apoptosis by particulate matter: Role of TNF-alpha and MAPK. *American Journal of Physiology (Lung Cellular and Molecular Physiology)*, **275**, L942–L949.

Cowley, S., Paterson, H., Kemp, P. and Marshall, C. (1994). Activation of MAP kinase kinase is necessary and sufficient for PC12 differentiation and for transformation of NIH 3T3 cells. *Cell*, **77**, 841–852.

Craighead, J., Akley, N., Gould, L. and Libbus, B. (1987). Characteristics of tumors and tumor cells cultured from experimental asbestos-induced mesothelioma in rats. *American Journal of Pathology*, **129**, 448–462.

De Schrijver, E., Brusselmans, K., Heyns, W., Verhoeven, G. and Swinnen, J. V. (2003). RNA interference-mediated silencing of the fatty acid synthase gene attenuates growth and induces morphological changes and apoptosis of LNCaP prostate cancer cells. *Cancer Research*, **63**, 3799–3804.

Denhardt, D. (1996). Signal transducing protein phosphorylation cascades mediated by Ras/Rho proteins in the mammalian cell: The potential for multiplex signaling. *Biochemical Journal*, **318**, 729–747.

Dillin, A. (2003). The specifics of small interfering RNA specificity. *Proceedings of the National Academy of Sciences USA*, **100**, 6289–6291.

Doench, J. G., Petersen, C. P. and Sharp, P. A. (2003). siRNAs can function as miRNAs. *Genes & Development*, **17**, 438–442.

Elbashir, S. M., Harborth, J., Lendeckel, W., Yalcin, A., Weber, K. and Tuschl, T. (2001). Duplexes of 21-nucleotide RNAs mediate RNA interference in cultured mammalian cells. *Nature*, **411**, 494–498.

Elbashir, S. M., Lendeckel, W. and Tuschl, T. (2001). RNA interference is mediated by 21- and 22-nucleotide RNAs. *Genes & Development*, **15**, 188–200.

Ellis, C. A. and Clark, G. (2000). The importance of being K-Ras. *Cellular Signaling*, **12**, 425–434.

English, J., Pearson, G., Hockenberry, T., Shivakumar, L., White, M. and Cobb, M. (1999). Contribution of the ERK5/MEK5 pathway to Ras/Raf signaling and growth control. *Journal of Biological Chemistry*, **274**, 31588–31592.

Esparis-Ogando, A., Diaz-Rodriguez, E., Montero, J., Yuste, L., Crespo, P. and Pandiella, A. (2002). Erk5 participates in neuregulin signal transduction and is constitutively active in breast cancer cells overexpressing ErbB2. *Molecular and Cellular Biology*, **22**, 270–285.

Faassen, A., Schrager, J., Klein, D., Oegema, T., Couchman, J. and McCarthy, J. (1992). A cell surface chondroitin sulfate proteoglycan, immunologically related to CD44, is involved in type I collagen-mediated melanoma cell motility and invasion. *Journal of Cell Biology*, **116**, 521–531.

Fire, A., Xu, S., Montgomery, M. K., Kostas, S. A., Driver, S. E. and Mello, C. C. (1998). Potent and specific genetic interference by double-stranded RNA in *Caenorhabditis elegans*. *Nature*, **391**, 806–811.

Gil, J. and Esteban, M. (2000). Induction of apoptosis by the dsRNA-dependent protein kinase (PKR): Mechanism of action. *Apoptosis*, **5**, 107–114.

Gille, H., Sharrocks, A. and Shaw, P. (1992). Phosphorylation of transcription factor p62TCF by MAP kinase stimulates ternary complex formation at c-fos promoter. *Nature*, **358**, 414–417.

Golub, T. R., Slonim, D. K., Tamayo, P., Huard, C., Gaasenbeek, M., Mesirov, J. P., et al. (1999). Molecular classification of cancer: Class discovery and class prediction by gene expression monitoring. *Science*, **286**, 531–537.

Gordon, G. J., Jensen, R. V., Hsiao, L. L., Gullans, S. R., Blumenstock, J. E., Ramaswamy, S., et al. (2002). Translation of microarray data into clinically relevant cancer diagnostic tests using gene expression ratios in lung cancer and mesothelioma. *Cancer Research*, **62**, 4963–4967.

Gordon, G. J., Jensen, R. V., Hsiao, L. L., Gullans, S. R., Blumenstock, J. E., Richards, W. G., et al. (2003). Using gene expression ratios to predict outcome among patients with mesothelioma. *Journal of the National Cancer Institute*, **95**, 598–605.

Grishok, A., Pasquinelli, A. E., Conte, D., Li, N., Parrish, S., Ha, I., Baillie, D. L., Fire, A., Ruvkun, G. and Mello, C. C. (2001). Genes and mechanisms related to RNA interference regulate expression of the small temporal RNAs that control *C. elegans* developmental timing. *Cell*, **106**, 23–34.

Gunthert, U., Hofmann, M., Rudy, W., Reber, S., Zoller, M., Haussmann, I., et al. (1991). A new variant of glycoprotein CD44 confers metastatic potential to rat carcinoma cells. *Cell*, **65**, 13–24.

Guo, Y., Ma, J., Wang, J., Che, X., Narula, J., Bigby, M., et al. (1994). Inhibition of human melanoma growth and metastasis *in vivo* by anti-CD44 monoclonal antibody. *Cancer Research*, **54**, 1561–1565.

Harvey, A. and Crompton, M. (2003). Use of RNA interference to validate Brk as novel therapeutic target in breast cancer: Brk promotes breast carcinoma cell proliferation. *Oncogene*, **22**, 5006–5010.

Harvey, P., Clark, I., Jaurand, M., Warn, R. and Edwards, D. (2000). Hepatocyte growth factor/scatter factor enhances the invasion of mesothelioma cell lines and the expression of matrix metalloproteinases. *British Journal of Cancer*, **83**, 1147–1153.

Hasuwa, H., Kaseda, K., Einarsdottir, T. and Okabe, M. (2002). Small interfering RNA and gene silencing in transgenic mice and rats. *Federation of European Biochemical Society Letters*, **532**, 227–230.

Heintz, N., Janssen, Y. and Mossman, B. (1993). Persistent induction of c-*fos* and c-*jun* expression by asbestos. *Proceedings of the National Academy of Sciences USA*, **90**, 3299–3303.

Hemann, M. T., Fridman, J. S., Zilfou, J. T., Hernando, E., Paddison, P. J., Cordon-Cardo, C., et al. (2003). An epi-allelic series of p53 hypomorphs created by stable RNAi produces distinct tumor phenotypes *in vivo*. *Nature Genetics*, **33**, 396–400.

Hofmann, M., Rudy, W., Gunthert, U., Zimmer, S., Zawadzki, V., Zoller, M., et al. (1993). A link between ras and metastatic behavior of tumor cells: Ras induces CD44 promoter activity and leads to low-level expression of metastasis-specific variants of CD44 in CREF cells. *Cancer Research*, **53**, 1516–1521.

Hunter, T. (1991). Cooperation between oncogenes. *Cell*, **64**, 249–270.

Hütvagner, G., McLachlan, J., Pasquinelli, A. E., Balint, E., Tuschl, T. and Zamore, P. D. (2001). A cellular function for the RNA-interference enzyme Dicer in the maturation of the let-7 small temporal RNA. *Science*, **293**, 834–838.

Ilves, H., Barske, C., Junker, U., Bohnlein, E. and Veres, G. (1996). Retroviral vectors designed for targeted expression of RNA polymerase III-driven transcripts: A comparative study. *Gene*, **171**, 203–208.

Jennings, P. A. and Molloy, P. L. (1987). Inhibition of SV40 replicon function by engineered antisense RNA transcribed by RNA polymerase III. *European Molecular Biology Organization Journal*, **6**, 3043–3047.

Jimenez, L., Zanella, C., Fung, H., Janssen, Y., Vacek, P., Charland, C., et al. (1997). Role of extracellular signal-regulated protein kinases in apoptosis by asbestos and H_2O_2. *American Journal of Physiology (Lung Cellular and Molecular Physiology)*, **273**, L1029–L1035.

Jochum, W., David, J., Elliott, C., Wutz, A., Plenk, H. J., Matsuo, K., et al. (2000). Increased bone formation and osteosclerosis in mice overexpressing the transcription factor Fra-1. *Nature Medicine*, **6**, 980–984.

Jorgensen, R. A., Cluster, P. D., English, J., Que, Q. and Napoli, C. A. (1996). Chalcone synthase cosuppression phenotypes in petunia flowers: Comparison of sense vs. antisense constructs and single-copy vs. complex T-DNA sequences. *Plant Molecular Biology*, **31**, 957–73.

Kamakura, S., Moriguchi, T. and Nishida, E. (1999). Activation of the protein kinase ERK5/BMK1 by receptor tyrosine kinases. Identification and characterization of a signaling pathway to the nucleus. *Journal of Biological Chemistry*, **274**, 26563–26571.

Kamath, R. S., Fraser, A. G., Dong, Y., Poulin, G., Durbin, R., Gotta, M., et al. (2003). Systematic functional analysis of the *Caenorhabditis elegans* genome using RNAi. *Nature*, **421**, 231–237.

Karin, M., Liu, Z. and Zandi, E. (1997). AP-1 function and regulation. *Current Opinion in Cell Biology*, **9**, 240–246.

Kato, Y., Tapping, R., Huang, S., Watson, M., Ulevitch, R. and Lee, J. (1998). BMK1/ERK5 is required for cell proliferation induced by epidermal growth factor. *Nature*, **395**, 713–716.

Ketting, R. F., Fischer, S. E., Bernstein, E., Sijen, T., Hannon, G. J. and Plasterk, R. H. (2001). Dicer functions in RNA interference and in synthesis of small RNA involved in developmental timing in *C. elegans*. *Genes & Development*, **15**, 2654–2659.

Ketting, R. F., Haverkamp, T. H., van Luenen, H. G. and Plasterk, R. H. (1999). Mut-7 of *C. elegans*, required for transposon silencing and RNA interference, is a homolog of Werner syndrome helicase and RNaseD. *Cell*, **99**, 133–141.

Ketting, R. F. and Plasterk, R. H. (2000). A genetic link between co-suppression and RNA interference in *C. elegans*. *Nature*, **404**, 296–298.

Klein, G., Powers, A. and Croce, C. (2002). Association of SV40 with human tumors. *Oncogene*, **21**, 1141–1149.

Klominek, J., Baskin, B., Liu, Z. and Hauzenberger, D. (1998). Hepatocyte growth factor/scatter factor stimulates chemotaxis and growth of malignant mesothelioma cells through c-met receptor. *International Journal of Cancer*, **76**, 240–249.

Knight, S. W. and Bass, B. L. (2001). A role for the RNase III enzyme DCR-1 in RNA interference and germ line development in *Caenorhabditis elegans*. *Science*, **293**, 2269–2271.

Lamb, R., Hennigan, R., Turnbull, K., Katsanakis, K., MacKenzie, E., Birnie, G., et al. (1997). AP-1 mediated invasion requires increased expression of the hyaluronan receptor CD44. *Molecular and Cellular Biology*, **17**, 963–976.

Lander, E. S., Linton, L. M., Birren, B., Nusbaum, C., Zody, M. C., Baldwin, J., et al. (2001). Initial sequencing and analysis of the human genome. *Nature*, **409**, 860–921.

Lebedeva, I. and Stein, C. A. (2001). Antisense oligonucleotides: Promise and reality. *Annual Review of Pharmacology and Toxicology*, **41**, 403–419.

Lee, N. S., Dohjima, T., Bauer, G., Li, H., Li, M. J., Ehsani, A., et al. (2002). Expression of small interfering RNAs targeted against HIV-1 rev transcripts in human cells. *Nature Biotechnology*, **20**, 500–505.

Lewis, C., Townsend, P. and Isacke, C. (2001). Ca(2+)/calmodulin-dependent protein kinase mediates the phosphorylation of CD44 required for cell migration on hyaluronan. *Biochemical Journal*, **357**, 843–850.

Li, Y. and Heldin, P. (2001). Hyaluronan production increases the malignant properties of mesothelioma cells. *British Journal of Cancer*, **85**, 600–607.

Liu, L., Liu, Z., Jiang, H., Zhang, W., Qi, S., Hu, J., et al. (2003). Gene expession profiles of hepatoma cell lines HLE. *World Journal of Gastroenterology*, **9**, 683–687.

Lloyd, A., Yancheva, N. and Wasylyk, B. (1991). Transformation suppressor activity of a Jun transcription factor lacking its activation domain. *Nature*, **352**, 635–638.

Malumbres, M. and Pellicer, A. (1998). RAS pathways to cell cycle control and cell transformation. *Frontiers of Bioscience*, **3**, 887–912.

Mansour, S., Matten, W., Hermann, A., Candia, J., Rong, S., Fukasawa, K., et al. (1994). Transformation of mammalian cells by constitutively active MAP kinase kinase. *Science*, **265**, 966–970.

Marais, R., Wynne, J. and Treisman, R. (1993). The SRF accessory protein Elk-1 contains a growth factor-regulated transcriptional activation domain. *Cell*, **73**, 381–393.

McCormick, F. (1999). Signalling networks that cause cancer. *Trends in Cellular Biology*, **9**, M53-M56.

McDonald, J. and Wagner, A. (1996). The epidemiology of mesothelioma in historical context. *European Respiratory Journal*, **9**, 1932–1942.

McManus, M. T. and Sharp, P. A. (2002). Gene silencing in mammals by small interfering RNAs. *Nature Reviews Genetics*, **3**, 737–747.

Mechta, F., Lallemand, D., Pfarr, C. and Yaniv, M. (1997). Transformation by ras modifies AP-1 composition and activity. *Oncogene*, **14**, 837–847.

Mehta, P., Jenkins, B., McCarthy, L., Thilak, L., Robson, C., Neal, D., et al. (2003). MEK5 overexpression is associated with metastatic prostate cancer, and stimulates proliferation, MMP-9 expression and invasion. *Oncogene*, **22**, 1381–1389.

Miyagishi, M. and Taira, K. (2002). U6 promoter-driven siRNAs with four uridine 3′ overhangs efficiently suppress targeted gene expression in mammalian cells. *Nature Biotechnology*, **20**, 497–500.

Mohr, S. and Rihn, B. (2001). Gene expression profiling in human mesothelioma cells using DNA microarray and high-density filter array technologies. *Bulletin du Cancer*, **88**, 305–313.

Mossman, B., Bignon, J., Corn, M., Seaton, A. and Gee, J. (1990). Asbestos: Scientific developments and implications for public policy. *Science*, **247**, 294–301.

Mossman, B. and Gee, J. (1989). Asbestos related disease. *New England Journal of Medicine*, **320**, 1721–1730.

Mossman, B. and Gruenert, D. (2002). SV40, growth factors, and mesothelioma – Another piece of the puzzle. *American Journal of Respiratory Cell and Molecular Biology*, **26**, 167–170.

Mulloy, R., Salinas, S., Philips, A. and Hipskind, R. A. (2003). Activation of cyclin D1 expression by the ERK5 cascade. *Oncogene*, **22**, 5387–5398.

Nagy, P., Arndt-Jovin, D. J. and Jovin, T. M. (2003). Small interfering RNAs suppress the expression of endogenous and GFP-fused epidermal growth factor receptor (erbB1) and induce apoptosis in erbB1-overexpressing cells. *Experimental Cell Research*, **285**, 39–49.

Napoli, C., Lemieux, C. and Jorgensen, R. (1990). Introduction of a chimeric chalcone synthase gene into petunia results in reversible co-suppression of homologous genes in trans. *Plant Cell*, **2**, 279–289.

Ojwang, J. O., Hampel, A., Looney, D. J., Wong-Staal, F. and Rappaport, J. (1992). Inhibition of human immunodeficiency virus type 1 expression by a hairpin ribozyme. *Proceedings of the National Academy of Sciences USA*, **89**, 10802–10806.

Pache, J., Janssen, Y., Walsh, E., Quinlan, T., Zanella, C., Low, R., et al. (1998). Increased epidermal growth factor-receptor (EGF-R) protein in a human mesothelial cell line in response to long asbestos fibers. *American Journal of Pathology*, **152**, 333–340.

Paddison, P. J., Caudy, A. A., Bernstein, E., Hannon, G. J. and Conklin, D. S. (2002). Short hairpin RNAs (shRNAs) induce sequence-specific silencing in mammalian cells. *Genes & Development*, **16**, 948–958.

Pass, H., Donington, J., Wu, P., Rizzo, P., Nishimura, M., Kennedy, R., et al. (1998). Human mesotheliomas contain the simian virus-40 regulatory region and large tumor antigen DNA sequences. *Journal of Thoracic and Cardiovascular Surgery*, **116**, 854–849.

Patterson, T., Vuong, H., Liaw, Y., Wu, R., Kalvakolanu, D. and Reddy, S. (2001). Mechanism of repression of squamous differentiation marker, SPRR1B, in malignant bronchial epithelial cells: Role of critical TRE-sites and its transacting factors. *Oncogene*, **20**, 634–644.

Paul, C. P., Good, P. D., Winer, I. and Engelke, D. R. (2002). Effective expression of small interfering RNA in human cells. *Nature Biotechnology*, **20**, 505–508.

Pearson, G., English, J. M., White, M. A. and Cobb, M. H. (2001). ERK5 and ERK2 cooperate to regulate NF-kappaB and cell transformation. *Journal of Biological Chemistry*, **276**, 7927–7931.

Perou, C. M., Sorlie, T., Eisen, M. B., van de Rijn, M., Jeffrey, S. S., Rees, C. A., et al. (2000). Molecular portraits of human breast tumours. *Nature*, **406**, 747–752.

Pollack, J. R., Perou, C. M., Alizadeh, A. A., Eisen, M. B., Pergamenschikov, A., Williams, C. F., et al. (1999). Genome-wide analysis of DNA copy-number changes using cDNA microarrays. *Nature Genetics*, **23**, 41–46.

Ramaswamy, S., Tamayo, P., Rifkin, R., Mukherjee, S., Yeang, C. H., Angelo, M., et al. (2001). Multiclass cancer diagnosis using tumor gene expression signatures. *Proceedings of the National Academy of Sciences USA*, **98**, 15149–15154.

Ramos-Nino, M., Timblin, C. and Mossman, B. (2002). Mesothelial cell transformation requires increased AP-1 binding activity and ERK-dependent Fra-1 expression. *Cancer Research*, **62**, 6065–6069.

Ramos-Nino, M. E., Haegens, A., Shukla, A. and Mossman, B. T. (2002). Role of mitogen-activated protein kinases (MAPK) in cell injury and proliferation by environmental particulates. *Molecular and Cellular Biochemistry*, **234–235**, 111–118.

Ramos-Nino, M. E., Scapoli, L., Martinelli, M., Land, S. and Mossman, B. T. (2003). Microarray analysis and RNA silencing link fra-1 to cd44 and c-met expression in mesothelioma. *Cancer Research*, **63**, 3539–3545.

Reddy, S. and Mossman, B. (2002). Role and regulation of activator protein-1 (AP-1) in toxicant-induced responses of the lung. *American Journal of Physiology (Lung Cellular and Molecular Physiology)*, **283**, L1161-L1178.

Regan, C. P., Li, W., Boucher, D. M., Spatz, S., Su, M. S. and Kuida, K. (2002). Erk5 null mice display multiple extraembryonic vascular and embryonic cardiovascular defects. *Proceedings of the National Academy of Sciences, USA*, **99**, 9248–9253.

Rihn, B. H., Mohr, S., McDowell, S.A., Binet, S., Loubinoux, J., Galateau, F., et al. (2000). Differential gene expression in mesothelioma. *Federation of European Biochemical Society Letters*, **480**, 95–100.

Rincón, M., Flavell, R. and Davis, R. (2000). The JNK and p38 MAP kinase signaling pathways in T-cell mediated immune responses. *Free Radical Biology and Medicine*, **28**, 1328–1337.

Risse-Hackl, G., Adamkiewicz, J., Wimmel, A. and Schuermann, M. (1998). Transition from SCLC to NSCLC phenotype is accompanied by an increased TRE-binding activity and recruitment of specific AP-1 proteins. *Oncogene*, **16**, 3057–3068.

Robinson, D., He, F., Pretlow, T. and Kung, H. J. (1996). A tyrosine kinase profile of prostate carcinoma. *Proceedings of the National Academy of Sciences USA*, **93**, 5958–5962.

Sandhu, H., Dehnen, W., Roller, M., Abel, J. and Unfried, K. (2000). mRNA expression patterns in different stages of asbestos-induced carcinogenesis in rats. *Carcinogenesis*, **21**, 1023–1029.

Scapoli, L., Ramos-Nino, M., Martinelli, M. and Mossman, B. (2004). Src-dependent ERK5 and Src/EGFR-dependent ERK1/2 activation is required for cell proliferation by asbestos. *Oncogene*, **23**, 805–813.

Schaeffer, H. and MJ, W. (1999). Mitogen-activated protein kinases: Specific messages from ubiquitous messengers. *Molecular and Cellular Biology*, **19**, 2435–2444.

Sebolt-Leopold, J., Dudley, D., Herrera, R., Van Becelaere, K., Wiland, A., Gowan, R., et al. (1999). Blockade of the MAP kinase pathway suppresses growth of colon tumors *in vivo*. *Nature Medicine*, **5**, 810–816.

Semizarov, D., Frost, L., Sarthy, A., Kroeger, P., Halbert, D. N. and Fesik, S. W. (2003). Specificity of short interfering RNA determined through gene expression signatures. *Proceedings of the National Academy of Sciences USA*, **100**, 6347–6352.

Sharp, P. A. (2001). RNA interference – 2001. *Genes & Development*, **15**, 485–490.

Shukla, A., Timblin, C., Hubbard, A., Bravman, J. and Mossman, B. (2001). Silica-induced activation of c-Jun-NH2-terminal amino kinases, protracted expression of the activator protein-1 proto-oncogene, fra-1, and S-phase alterations are mediated via oxidative stress. *Cancer Research*, **61**, 1791–1795.

Singhal, S., Wiewrodt, R. , Malden, L.D. , Amin, K.M., Matzie, K., Friedberg, J., et al. (2003). Gene expression profiling of malignant mesothelioma. *Clinical Cancer Research*, **9**, 3080–3097.

Sohn, S. J., Sarvis, B. K., Cado, D. and Winoto, A. (2002). ERK5 MAPK regulates embryonic angiogenesis and acts as a hypoxia-sensitive repressor of vascular endothelial growth factor expression. *Journal of Biological Chemistry*, **277**, 43344–43351.

Spankuch-Schmitt, B., Bereiter-Hahn, J., Kaufmann, M. and Strebhardt, K. (2002). Effect of RNA silencing of polo-like kinase-1 (PLK1) on apoptosis and spindle formation in human cancer cells. *Journal of the National Cancer Institute*, **94**, 1863–1877.

Sun, B., Nishihira, J., Suzuki, M., Fukushima, N., Ishibashi, T., Kondo, M., et al. (2003). Induction of macrophage migration inhibitory factor by lysophosphatidic acid: Relevance to tumor growth and angiogenesis. *International Journal of Molecular Medicine*, **12**, 633–641.

Thomas, L., Etoh, T., Stamenkovic, I., Mihm, M. J. and Byers, H. (1993). Migration of human melanoma cells on hyaluronate is related to CD44 expression. *Journal of Investigative Dermatology*, **100**, 115–120.

Timblin, C., Guthrie, G., Janssen, Y., Walsh, E., Vacek, P. and Mossman, B. (1998). Patterns of c-fos and c-jun protooncogene expression, apoptosis and proliferation in rat pleural mesothelial cells exposed to erionite or asbestos fibers. *Toxicology and Applied Pharmacology*, **151**, 88–97.

Tolnay, E., Kuhnen, C., Wiethege, T., Konig, J., Voss, B. and Muller, K. (1998). Hepatocyte growth factor/scatter factor and its receptor c-Met are overexpressed and associated with an increased microvessel density in malignant pleural mesothelioma. *Journal of Cancer Research and Clinical Oncology*, **124**, 291–296.

Treinies, I., Paterson, H., Hooper, S., Wilson, R. and Marshall, C. (1999). Activated MEK stimulates expression of AP-1 components independently of phosphatidylinositol 3-kinase (PI3-kinase) but requires a PI3-kinase signal To stimulate DNA synthesis. *Molecular and Cellular Biology*, **19**, 321–329.

Tuschl, T. and Borkhardt, A. (2002). Small interfering RNAs a revolutionary tool for the analyiss of gene function and gene therapy. *Molecular Interventions*, **2**, 158–167.

van Dartel, M., Cornelissen, P. W., Redeker, S., Tarkkanen, M., Knuutila, S., Hogendoorn, P. C., et al. (2002). Amplification of 17p11.2 approximately p12, including PMP22, TOP3A, and MAPK7, in high-grade osteosarcoma. *Cancer Genetics and Cytogenetics*, **139**, 91–96.

van der Krol, A., Mur, L., Beld, M., Mol, J. and Stuitje, A. (1990). Flavonoid genes in petunia: Addition of a limited number of gene copies may lead to a suppression of gene expression. *Plant Cell*, **2**, 291–299.

van 't Veer, L. J., Dai, H., van de Vijver, M. J., He, Y. D., Hart, A. A., Mao, M., et al. (2002). Gene expression profiling predicts clinical outcome of breast cancer. *Nature*, **415**, 530–536.

Venter, J. C., Adams, M. D., Myers, E. W., Li, P. W., Mural, R. J., Sutton, G. G., et al. (2001). The sequence of the human genome. *Science*, **291**, 1304–1351.

Vuong, H., Patterson, T., Adiseshaiah, P., Shapiro, P., Kalvakolanu, D. and Reddy, S. (2002). JNK1 and AP-1 regulate PMA-inducible squamous differentiation marker expression in Clara-like H441 cells. *American Journal of Physiology (Lung Cellular and Molecular Physiology)*, **282**, L215–L225.

Watts, R., Huang, C., Young, M., Li, J., Dong, Z., Pennie, W., et al. (1998). Expression of dominant negative ERK2 inhibits AP-1 transactivation and neoplastic transformation. *Oncogene*, **17**, 3493–3498.

Weg-Remers, S., Ponta, H., Herrlich, P. and Konig, H. (2001). Regulation of alternative pre-mRNA splicing by the ERK MAP-kinase pathway. *European Molecular Biology Organization Journal*, **20**, 4194–4203.

Whalen, A., Galasinski, S., Shapiro, P., Nahreini, T. and Ahn, N. (1997). Megakaryocytic differentiation induced by constitutive activation of mitogen-activated protein kinase kinase. *Molecular and Cellular Biology*, **17**, 1947–1958.

Williams, B. R. (1997). Role of the double-stranded RNA-activated protein kinase (PKR) in cell regulation. *Biochemical Society Transactions*, **25**, 509–513.

Williams, N. S., Gaynor, R. B., Scoggin, S., Verma, U., Gokaslan, T., Simmang, C., et al. (2003). Identification and validation of genes involved in the pathogenesis of colorectal cancer using cDNA microarrays and RNA interference. *Clinical Cancer Research*, **9**, 931–946.

Wisdom, R. and Verma, I. (1993). Transformation by Fos proteins requires a C-terminal transactivation domain. *Molecular and Cellular Biology*, **13**, 7429–7438.

Wu-Scharf, D., Jeong, B., Zhang, C. and Cerutti, H. (2000). Transgene and transposon silencing in Chlamydomonas reinhardtii by a DEAH-box RNA helicase. *Science*, **290**, 1159–1162.

Xia, H., Mao, Q., Paulson, H. L. and Davidson, B. L. (2002). siRNA-mediated gene silencing *in vitro* and *in vivo*. *Nature Biotechnology*, **20**, 1006–1010.

Yan, C., Takahashi, M., Okuda, M., Lee, J. D. and Berk, B. C. (1999). Fluid shear stress stimulates big mitogen-activated protein kinase 1 (BMK1) activity in endothelial cells. Dependence on tyrosine kinases and intracellular calcium. *Journal of Biological Chemistry*, **274**, 143–150.

Yu, J. Y., DeRuiter, S. L. and Turner, D. L. (2002). RNA interference by expression of short-interfering RNAs and hairpin RNAs in mammalian cells. *Proceedings of the National Academy of Sciences USA*, **99**, 6047–6052.

Zamore, P. D. (2001). RNA interference: Listening to the sound of silence. *Nature Structural Biology*, **8**, 746–750.

Zanella, C., Posada, J., Tritton, T. and Mossman, B. (1996). Asbestos causes stimulation of the ERK-1 mitogen-activated protein kinase cascade after phosphorylation of the epidermal growth factor receptor. *Cancer Research*, **56**, 5334–5338.

Zanella, C., Timblin, C., Cummins, A., Jung, M., Goldberg, J., Raabe, R., et al. (1999). Asbestos-induced phosphorylation of epidermal growth factor receptor is linked to c-fos expression and apoptosis. *American Journal of Physiology (Lung Cellular and Molecular Physiology)*, **277**, L684–L693.

Zhang, L., Yang, N., Mohamed-Hadley, A., Rubin, S. C. and Coukos, G. (2003). Vector-based RNAi, a novel tool for isoform-specific knock-down of VEGF and anti-angiogenesis gene therapy of cancer. *Biochemical and Biophysical Research Communications*, **303**, 1169–1178.

33 High-throughput RNA interference

Howard Y. Chang, Nancy N. Wang, and Jen-Tsan Chi

Introduction

RNA interference (RNAi) is an evolutionarily conserved pathway of gene silencing that identifies and destroys mRNA sequences derived from selfish repetitive or viral sequences. The mechanisms and physiologic functions of RNAi pathways are discussed in Section 1. Because RNAi can be triggered by exogenously supplied double-stranded RNA (dsRNA) molecules, in principle one can effectively silence the expression of a gene and infer its physiologic function given its sequence. RNAi analysis on a genome-wide scale is a versatile and powerful tool that holds great promise in functional annotation of the human genome and in acceleration of drug target discovery and validation. The rapid advances to date have illustrated that high-throughput RNAi can complement genetic studies in traditional model organisms to extend our understanding and probe the mechanisms of well-studied pathways. In this chapter, we will focus on emerging methodologies and issues surrounding high-throughput RNAi in mammalian systems, and highlight the potential advances and pitfalls with these new technologies.

Generation of genome-wide RNAi reagents

In order to silence each gene in the genome using RNAi, the first step is to create the appropriate library of dsRNA reagents. In *Caenorhabditis elegans* and *Drosophila melanogaster*, long segments of dsRNA are readily taken up by the animal or tissue culture cells, respectively, and genome-wide RNAi libraries can be constructed by annealing complementary RNAs synthesized from both strands of the same cDNA. In contrast, mammalian cells do not readily take up dsRNA sequences, and long dsRNAs (>500 bp) activate the protein kinase R pathway of translational shutdown and interferon production. Therefore, alternative strategies are needed to design and deliver dsRNA reagents for RNAi in mammalian cells. Short interfering RNAs (siRNAs), intermediates in the RNAi pathways, are 21-bp RNA duplexes derived from endonucleolytic cleavage of trigger dsRNAs and recognized by RISC complexes to target endogenous mRNAs for degradation. Tuschl and colleagues demonstrated that chemically synthesized siRNAs can be transfected into mammalian cells to induce gene silencing without triggering the translational shutdown incurred by long dsRNAs (Elbashir et al., 2001). Because the efficacy of each siRNA duplex is variable, several siRNA duplexes may need to be tested to identify an appropriate knockdown reagent for a target gene. To address this issue, several investigators have devised strategies that mimic the natural RNAi pathway: Long dsRNA corresponding to the target gene is synthesized and then fragmented into multiple 21-bp duplexes using recombinant Dicer enzyme or *E. coli* RNase III (Kawasaki et al., 2003; Myers et al., 2003; Yang et al., 2002). This pool of short dsRNAs, termed Dicer-generated short interfering RNA (d-siRNA) or endonuclease-generated short interfering RNA (esiRNA), can then be transfected into cells and generally has equivalent or improved efficacy in gene silencing as the best selected siRNAs. In the two above strategies, the RNAi trigger is not self-renewing and exists only transiently in the transfected cells. To allow regeneration of the dsRNA reagent and effect stable gene knockdown in cells and organisms, several investigators have constructed plasmid and viral vectors that encode short inverted repeats and thus express short hairpin RNAs (Brummelkamp et al., 2002; Devroe and Silver, 2002; Paddison et al., 2002; Qin et al., 2003; Rubinson et al., 2003). Analogous to endogenous microRNAs, the hairpin RNAs are thought to be processed by Dicer-like endonucleases and generate active antisense short RNA within cells. Ishii et al. reported that confinement of long dsRNA molecules to the nucleus by addition of ribozyme sequences (a strategy termed DECAP) can also be used to generate pools of gene-specific siRNAs from a RNA polymerase II-driven vector (Shinagawa and Ishii, 2003). Many of the above strategies can be applied to sequenced genomes or cDNA libraries to generate libraries of genome-wide RNAi reagents. As experience with single-gene RNAi experiments accumulates, it is likely that libraries of validated siRNAs against mammalian genes will also become commercially available and affordable.

Specificity of RNAi

To infer physiologic functions of genes based on their RNAi phenotype, it is essential to first understand and account for the specificity of RNAi-mediated silencing. The issue of specificity is especially important in high-throughput applications because, given a rare phenotype, off-target effects can produce more false positive and negative "hits" than true hits and obscure the biologic interpretation.

Initial studies indicated that siRNA-mediated mRNA cleavage can be abrogated by even a single base pair mismatch between the siRNA and cognate target sequence (Elbashir et al., 2001); recent studies have confirmed exquisite specificity of siRNA on a genomic scale but also highlighted possible sources of off-target effects. One effective strategy to identify possible off-target effects of RNAi is global gene expression profiling. To date, four studies have systematically examined the global gene expression pattern of mammalian cells after RNAi. Fesik and colleagues observed a common stress signature in cells exposed to high levels (100 nM) of siRNA which was irrespective of siRNA sequence; lowering the siRNA concentration to 20 nM or less allowed specific inactivation of the target gene without apparent off-target effects (Semizarov et al., 2003). Indeed, silencing of the same target gene with 5 different siRNAs gave indistinguishable global gene expression signatures while expression signatures of cells silenced for different genes were distinct and largely non-overlapping. Instead of endogenous genes, Chi et al. investigated the silencing of a model target gene, green fluorescent protein (GFP), the absence of which was not expected to perturb the transcriptome in 293 embryonic kidney cells (Chi et al., 2003). At 48 and 72 hours after GFP silencing, the global expression pattern of approximately 20,000 endogenous genes measured on the microarray was not altered compared to mock treated cells or cells treated with an irrelevant siRNAs. Chi et al. also employed a low siRNA concentration, confirming the exquisite sequence specificity of RNAi that can be achieved. In contrast, Jackson et al. noted the appearance of off-target effects in HeLa cells at >100 nM of siRNAs that could be diminished but not completely removed by lowering the siRNA concentration. Interestingly, some of the off-target genes affected at high siRNA concentration showed limited sequence similarity to the input siRNA, suggesting that a degree of sequence promiscuity may occur at high siRNA concentrations. The degree of mismatch between siRNA and affected mRNAs are reminiscent of interactions between endogenous microRNAs and their cognate targets, which lead normally to translational repression (Saxena et al., 2003).

The same RISC complexes that execute siRNA-mediated RNA degradation also mediate miRNA-directed translation repression (Doench et al., 2003), but the precise rules that dictate these two possible outcomes from the short RNA:mRNA duplex are currently not clear. Bridge et al. studied cells that have been transduced with lentiviral vectors that direct siRNA expression show activation of the interferon response (Bridge et al., 2003). While cells exposed to siRNAs showed specific silencing of the target gene, expression of the same shRNA from viral vectors was associated with activation of the interferon response, particularly in cells with high copies of viral integrants that likely facilitate tandem integration and transcriptional read-through events that can generate long dsRNA.

In addition, a phenomenon termed transitive RNAi may produce off-target effects, even on genes that do not demonstrate sequence similarity to the target sequence. In certain organisms such as *C. elegans*, a RNA-dependent RNA polymerase (RdRP) amplifies the RNAi effect; using the antisense siRNA as the primer and the mRNA as the template, the RdRP synthesizes antisense RNA and produces long dsRNA. The long dsRNA products can then be processed by Dicer and RISC complexes to form additional siRNAs. In effect, the sequence specificity of gene silencing spreads to include sequences 5′ to the initial designed target. In mammalian genomes, no RdRP has been discovered, and at least two lines of evidence indicate that transitive RNAi does not occur in mammalian cells: i) chemically modified siRNAs that contain bulky blocking groups on the 3′ end of the antisense strand still function to mediate mRNA degradation, indicating that siRNAs act as guides instead of primers (Schwarz et al., 2002); and ii) the success of exon-specific and allele-specific silencing of natural mRNAs and model substrates engineered to contain perfect sequence identities 5' of the cognate siRNA target site also demonstrates the lack of transitive RNAi (Chi et al., 2003; Lassus et al., 2002). In summary, off-target effects may derive from toxic effects of excessive loading of siRNAs, and from inadvertent activation of the long dsRNA response, and from the machinery of the RNAi mechanism itself.

Based on understanding of the above mechanisms, we can implement several strategies to enhance the specificity of the high-throughput RNAi screen. First of all, siRNAs should be filtered against genomic sequences to select for siRNAs that correspond to sequences most unique to the target gene. Secondly, it is preferable to generate and compare at least two distinct siRNAs to a target gene. This multiple-fold coverage minimizes the problem of unpredictable efficacy of any single siRNA and ensures that the phenotype is likely due to the gene in question rather than off-target effects from related sequences. Thirdly, nonspecific toxic effects or activation of the interferon response can be readily recognized by fine tuning the high-throughput RNAi screen first on a smaller scale (e.g. 20–100 siRNAs along with positive and negative controls). If the control siRNAs do not perform as expected or too many hits are observed, the design of the experiment can be changed to lower the amount of siRNA used or method of delivery. Finally, all RNAi experiments benefit from validation by additional approaches. In the case of genetically tractable model organisms, identification of genetic mutants that yield a similar phenotype as the RNAi knockdown is a powerful confirmation. In mammalian cells, the use of an unrelated siRNA sequence acting on the same gene or pinpointing the biochemical nature of the phenotypic effect can strengthen the biologic interpretation.

High-throughput RNAi screening

The rapidly expanding catalogue of eukaryotic genes from different sequencing projects present scientists with an enormous amount of information, but the effort of annotating and understanding gene function still lags far behind due to lack of effective genetic tools. The ability of RNAi to degrade specific genes with knowledge of the gene sequences provides a powerful addition to our arsenals of genetic tools to dissect different biological processes. RNAi-based genetic approaches are, in many aspects, superior to the traditional genetic screens because the prior knowledge of the targeted genes in RNAi bypasses the laborious mapping of the mutations while linking loss-of-function phenotypes to specific genes in the traditional forward genetic screening. Thus, RNA interference has the potential to combine the forward and reverse genetics to provide the direct connection between gene sequencing and biological functions in different model systems.

RNAi was first described in the nematode *C. elegans* in which injection or digestion of double-stranded RNA (dsRNA) can mediate gene-sepcific silencing (Fire et al., 1998; Timmons and Fire, 1998). But the dreams of genome-wide genetic screen cannot be realized until loss-of-function RNAi phenotypes can be generated efficiently by feeding nematodes with bacterial clones expressing dsRNAs that are identical to target genes (Kamath et al., 2001). Libraries of bacterial clones have been used to target genes on significant regions of chromosome I (Fraser et al., 2000), III (Gonczy et al., 2000) and 86% of all predicted genes in *C. elegans* genome (Kamath et al., 2003) to identify genes important for the viability and gross development of wild-type strains and RNAi hypersensitive *rrf-3* mutant nematodes (Simmer et al., 2003). These studies not only assigned loss-of-function phenotypes in many genes of *C. elegans* genome but also create an RNAi feeding library of bacterial clones that can be replicated and reused for other screening efforts. Indeed, these RNAi bacterial clones have also allowed comprehensive dissection of genes that regulate aging (Lee et al., 2003; Murphy et al., 2003), fat metabolism (Ashrafi et al., 2003), genomic mutation (Pothof et al., 2003), gonad morphogenesis (Cram et al., 2003), microtubule length (Le Bot et al., 2003) and transposon silencing (Vastenhouw et al., 2003) in the nematodes. These studies have shown the power of genome-wide RNAi in elucidating the connections between the gene sequences and gene functions.

Another model organism that has been extensively studied by traditional screens is the fruit fly *Drosophila*. Unlikely *C. elegans*, early attempts to inject interfering RNA were proven unfeasible for large-scale screening in the whole fly. Instead, it has been possible to introduce dsRNA into the cultured *Drosophila* S2 cells (Clemens et al., 2000). Large-scale pools of interfering dsRNAs can be generated by annealing *in vitro* transcribed RNA from PCR amplified products corresponding to the intended target genes. These dsRNA libraries have been instrumental in dissecting the *Drosophila* genes involved in the actin-based lamella formation (Rogers et al., 2003) cell morphogenesis (Kiger et al., 2003), Hedgehog pathways (Lum et al., 2003) and bacterial phagocytosis (Ramet et al., 2002). It is also interesting to note that very diverse phenotypic assays were used to perform high-throughput screening, including high-throughput microscopy (Kiger et al., 2003; Rogers et al., 2003), luciferase reporter activities (Lum et al., 2003) and bacterial ingestion (Ramet et al., 2002), highlights the adaptability of RNAi-based genetic screens of S2 cells to analyze components of investigated pathways.

To apply similar large scale genome-based RNAi to mammalian cells, it is essential to note that long dsRNAs (>500 bp) that were used in *Drosophila* screening will activate the protein kinase R pathway of translational shutdown and interferon production in mammalian cells. To circumvent this technical barrier, many solutions have been proposed to deliver siRNAs or plasmids vectors encoding siRNAs into mammalian cells as summarized above. But the costs of chemically synthesized RNA duplex and incomplete understandings of the targeting efficiency still present great financial barriers to the application of genome-wide siRNA screening in mammalian cells. The use of parallel and high-throughput screening techniques will greatly facilitate the annotation of gene functions on a large scale and make it possible to design genetic experiments to screen for cells with certain phenotypes. The recent development of high-throughput cDNA transfection on microarrays (Ziauddin and Sabatini, 2001) provides the technical impetus to print siRNAs on high-density microarrays and perform RNAi in mammalian cells in a highly parallel fashion.

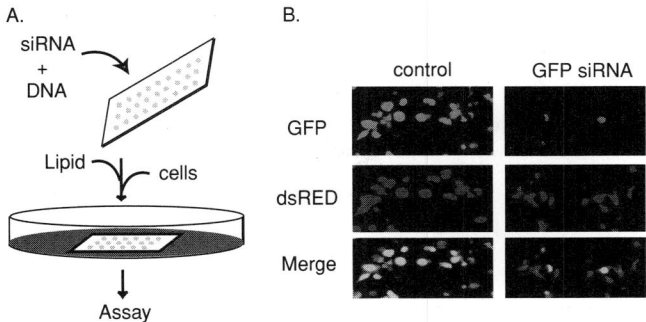

Figure 33.1. siRNA microarray for gene silencing. (A) Experimental strategy for siRNA microarrays. The desired cDNA and siRNAs are printed as individual spots on glass slides and exposed briefly to lipid before placing HEK293 cells on the printed slides in culture dish. Transfected cells are visualized using fluorescent microscopy and evaluated for the effect of RNAi. (B) Parallel RNAi on microarrays. Fluorescence photomicrograph of cells after reverse transfection of the indicated siRNA and cDNAs is shown. (See color section.)

We tested the feasibility of siRNAs-mediated RNAi on microarrays (Figure 33.1A). Expression constructs for GFP, dsRED, and siRNAs were spotted in the desired combinations on amine glass slides using a robotic arrayer. We hypothesized that in the presence of lipids, siRNA will complex with the DNA on the slide and form liposomes containing both reagents. Expression of dsRED served as an internal control for reverse transfection and localization of the printed spots. After air-drying, the printed arrays were exposed to lipids briefly and then placed in a tissue culture dish. HEK293 cells were then plated on the printed side of the arrays and allowed to attach and grow on the slides. The cells were examined with fluorescence microscopy 72 hours after transfection. As shown in Figure 33.1B, HEK293 expressed dsRED in all the cell clusters above the printed spots after reverse transfection. In contrast, GFP expression is robust in the control spots but is selectively decreased by the GFP siRNA E1 (Figure 33.1B) but not by control siRNA C1 (data not shown). The merged image allowed quick detection of the specific RNAi effect by the red shift of the affected cell clusters (Figure 33.1B). These results demonstrate that siRNA-mediated gene silencing can be adapted to microarray format. By arraying different siRNAs on microarrays, one can generate large panels of cells silenced for different genes for highly parallel tests of gene function. Similar results have also been demonstrated by two recent reports (Kumar et al., 2003; Mousses et al., 2003) to silence reporter genes and dissect pathways in TRAIL-induced apoptosis (Aza-Blanc et al., 2003) of human cells.

Discussion

The power of RNA interference lies in its ability to selectively inactivate specific genes with the knowledge of gene sequences. Its potential for setting high-throughput genetic screens is starting to be realized, and the quantity and quality of functional data derived from RNAi-based screens in different organisms will expand our understanding of how gene functions are linked with their corresponding loss-of-function phenotypes. It is interesting to note that both *Caenorhabditis elegans* and *Drosophila melanogaster* have been the target organism for extensive genetic screens over the years. With the published large scale RNAi-based screens, many previously unknown components of well-known pathways were identified. These results reveal that previous extensive screening efforts in worms and flies are not saturated and there are still many novel components to be discovered. Large-scale RNAi-based approaches show great promise in providing the novel windows to identify these essential components which were missed in the traditional forward genetic screens.

It is also noteworthy that understanding RNAi specificity is essential in setting up large-scale RNAi-based genetic screens since even small non-specific off-target effects during gene silencing might create false positive hits and affect the interpretation of such screening efforts. In all the published studies so far, the issue of false-positives seems to be minor and presents no significant backgrounds for the phenotype assays for different target genes. This suggests that RNAi is a robust and reliable tool in revealing the functions of targeted genes without significant "background noise" (false-positives) in these screens in worms, flies and human cells. But it is still essential for researchers to keep in mind the possibility of "off-target" effects of RNAi-mediated gene silencing during the design, execution and interpretation of large scale RNAi-based experiments. It also highlights the need to validate an identified phenotypes caused by loss of function of the genes through independent approaches.

We have demonstrated the feasibility of arraying siRNAs on microarrays and performing RNAi experiments by reverse transfection. Conceptually, this development allows the performance of highly parallel RNAi experiments in mammalian cells in a spatially addressable fashion. We estimate that 10,000 array elements can be accommodated on a glass slide in the fashion that we described. Thus, the limiting component for genome-wide RNAi experiments is the availability of siRNAs themselves. The rapid progress in siRNA technology has already provided several potential strategies such as plasmids encoding hairpin RNAs or d-siRNA/esiRNA as vehicles used in reverse transfection to achieve stable or transient gene silencing. As we have demonstrated, two reporter genes can be used to monitor transfection and gene silencing independently. With the same token, arrays of cells silenced for different genes can be screened for altered morphology, activation of signal transduction pathways using specific reporter genes, or expression of endogenous markers using immunofluorescence. The microarray format also lends itself to comprehensively test the effect of silencing various combinations of genes within a family and address the issues of redundancy and compensation that frequently arise in mammalian genetics. Furthermore, the ability to apply RNAi using siRNA on solid support suggests the possibility of coupling siRNAs to patterned substrates in order to test the geometric and localized requirements of gene function. Finally, it is also possible to couple RNAi with immobilized microenvironments to explore the roles of different genes during cellular activation and differentiation of overlayed cells triggered by the spotted microenvironments.

The implementation of high-throughput siRNA technology should open many doors in experimental biology and human therapeutics.

Acknowledgements

We are indebted to Dr. Patrick Brown and members of Brown lab for support and advices.

REFERENCES

Ashrafi, K., Chang, F. Y., Watts, J. L., Fraser, A. G., Kamath, R. S., Ahringer, J. and Ruvkun, G. (2003). Genome-wide RNAi analysis of *Caenorhabditis elegans* fat regulatory genes. *Nature*, **421**, 268–272.

Aza-Blanc, P., Cooper, C. L., Wagner, K., Batalov, S., Deveraux, Q. L. and Cooke, M. P. (2003). Identification of modulators of TRAIL-induced apoptosis via RNAi-based phenotypic screening. *Molecular Cell*, **12**, 627–637.

Bridge, A. J., Pebernard, S., Ducraux, A., Nicoulaz, A. L. and Iggo, R. (2003). Induction of an interferon response by RNAi vectors in mammalian cells. *Nature Genetics*, **34**, 263–264.

Brummelkamp, T. R., Bernards, R. and Agami, R. (2002). A system for stable expression of short interfering RNAs in mammalian cells. *Science*, **296**, 550–553.

Chi, J. T., Chang, H. Y., Wang, N. N., Chang, D. S., Dunphy, N. and Brown, P. O. (2003). Genomewide view of gene silencing by small interfering RNAs. *Proceedings of the National Academy of Sciences USA*, **100**, 6343–6346.

Clemens, J. C., Worby, C. A., Simonson-Leff, N., Muda, M., Maehama, T., Hemmings, B. A. and Dixon, J. E. (2000). Use of double-stranded RNA interference in *Drosophila* cell lines to dissect signal transduction pathways. *Proceedings of the National Academy of Sciences USA*, **97**, 6499–6503.

Cram, E. J., Clark, S. G. and Schwarzbauer, J. E. (2003). Talin loss-of-function uncovers roles in cell contractility and migration in *C. elegans*. *Journal of Cell Science*, **116**, 3871–3878.

Devroe, E. and Silver, P. A. (2002). Retrovirus-delivered siRNA. *BMC Biotechnology*, **2**, 15.

Doench, J. G., Petersen, C. P. and Sharp, P. A. (2003). siRNAs can function as miRNAs. *Genes & Development*, **17**, 438–442.

Elbashir, S. M., Harborth, J., Lendeckel, W., Yalcin, A., Weber, K. and Tuschl, T. (2001). Duplexes of 21-nucleotide RNAs mediate RNA interference in cultured mammalian cells. *Nature*, **411**, 494–498.

Fire, A., Xu, S., Montgomery, M. K., Kostas, S. A., Driver, S. E. and Mello, C. C. (1998). Potent and specific genetic interference by double-stranded RNA in *Caenorhabditis elegans*. *Nature*, **391**, 806–811.

Fraser, A. G., Kamath, R. S., Zipperlen, P., Martinez-Campos, M., Sohrmann, M. and Ahringer, J. (2000). Functional genomic analysis of *C. elegans* chromosome I by systematic RNA interference. *Nature*, **408**, 325–330.

Gonczy, P., Echeverri, C., Oegema, K., Coulson, A., Jones, S. J., Copley, R. R., Duperon, J., Oegema, J., Brehm, M., Cassin, E., Hannak, E., Kirkham, M., Pichler, S., Flohrs, K., Goessen, A., Leidel, S., Alleaume, A. M., Martin, C., Ozlu, N., Bork, P. and Hyman, A. A. (2000). Functional genomic analysis of cell division in *C. elegans* using RNAi of genes on chromosome III. *Nature*, **408**, 331–336.

Kamath, R. S., Fraser, A. G., Dong, Y., Poulin, G., Durbin, R., Gotta, M., Kanapin, A., Le Bot, N., Moreno, S., Sohrmann, M., et al. (2003). Systematic functional analysis of the *Caenorhabditis elegans* genome using RNAi. *Nature*, **421**, 231–237.

Kamath, R. S., Martinez-Campos, M., Zipperlen, P., Fraser, A. G. and Ahringer, J. (2001). Effectiveness of specific RNA-mediated interference through ingested double-stranded RNA in *Caenorhabditis elegans*. *Genome Biology*, **2**, RESEARCH0002.

Kawasaki, H., Suyama, E., Iyo, M. and Taira, K. (2003). siRNAs generated by recombinant human Dicer induce specific and significant but target site-independent gene silencing in human cells. *Nucleic Acids Reseach*, **31**, 981–987.

Kiger, A., Baum, B., Jones, S., Jones, M., Coulson, A., Echeverri, C. and Perrimon, N. (2003). A functional genomic analysis of cell morphology using RNA interference. *Journal of Biology*, **2**, 27.

Kumar, R., Conklin, D. S. and Mittal, V. (2003). High-throughput selection of effective RNAi probes for gene silencing. *Genome Research*, **13**, 2333–2340.

Lassus, P., Rodriguez, J. and Lazebnik, Y. (2002). Confirming specificity of RNAi in mammalian cells. *Science STKE*, **2002**, PL13.

Le Bot, N., Tsai, M. C., Andrews, R. K. and Ahringer, J. (2003). TAC-1, a regulator of microtubule length in the *C. elegans* embryo. *Current Biology*, **13**, 1499–1505.

Lee, S. S., Lee, R. Y., Fraser, A. G., Kamath, R. S., Ahringer, J. and Ruvkun, G. (2003). A systematic RNAi screen identifies a critical role for mitochondria in *C. elegans* longevity. *Nature Genetics*, **33**, 40–48.

Lum, L., Yao, S., Mozer, B., Rovescalli, A., Von Kessler, D., Nirenberg, M. and Beachy, P. A. (2003). Identification of Hedgehog pathway components by RNAi in *Drosophila* cultured cells. *Science*, **299**, 2039–2045.

Mousses, S., Caplen, N. J., Cornelison, R., Weaver, D., Basik, M., Hautaniemi, S., Elkahloun, A. G., Lotufo, R. A., Choudary, A., Dougherty, E. R., et al. (2003). RNAi microarray analysis in cultured mammalian cells. *Genome Research*, **13**, 2341–2347.

Murphy, C. T., McCarroll, S. A., Bargmann, C. I., Fraser, A., Kamath, R. S., Ahringer, J., Li, H. and Kenyon, C. (2003). Genes that act downstream of DAF-16 to influence the lifespan of *Caenorhabditis elegans*. *Nature*, **424**, 277–283.

Myers, J. W., Jones, J. T., Meyer, T. and Ferrell, J. E., Jr. (2003). Recombinant Dicer efficiently converts large dsRNAs into siRNAs suitable for gene silencing. *Nature Biotechnology*, **21**, 324–328.

Paddison, P. J., Caudy, A. A., Bernstein, E., Hannon, G. J. and Conklin, D. S. (2002). Short hairpin RNAs (shRNAs) induce sequence-specific silencing in mammalian cells. *Genes & Development*, **16**, 948–958.

Pothof, J., van Haaften, G., Thijssen, K., Kamath, R. S., Fraser, A. G., Ahringer, J., Plasterk, R. H. and Tijsterman, M. (2003). Identification of genes that protect the *C. elegans* genome against mutations by genome-wide RNAi. *Genes & Development*, **17**, 443–448.

Qin, X. F., An, D. S., Chen, I. S. and Baltimore, D. (2003). Inhibiting HIV-1 infection in human T cells by lentiviral-mediated delivery of small interfering RNA against CCR5. *Proceedings of the National Academy of Sciences USA*, **100**, 183–188.

Ramet, M., Manfruelli, P., Pearson, A., Mathey-Prevot, B. and Ezekowitz, R. A. (2002). Functional genomic analysis of phagocytosis and identification of a *Drosophila* receptor for *E. coli*. *Nature*, **416**, 644–648.

Rogers, S. L., Wiedemann, U., Stuurman, N. and Vale, R. D. (2003). Molecular requirements for actin-based lamella formation in *Drosophila* S2 cells. *Journal of Cell Biology*, **162**, 1079–1088.

Rubinson, D. A., Dillon, C. P., Kwiatkowski, A. V., Sievers, C., Yang, L., Kopinja, J., Rooney, D. L., Ihrig, M. M., McManus, M. T., Gertler, F. B., Scott, M. L. and Van Parijs, L. (2003). A lentivirus-based system to functionally silence genes in primary mammalian cells, stem cells and transgenic mice by RNA interference. *Nature Genetics*, **33**, 401–406.

Saxena, S., Jonsson, Z. O. and Dutta, A. (2003). Small RNAs with imperfect match to endogenous mRNA repress translation. Implications for off-target activity of small inhibitory RNA in mammalian cells. *Journal of Biological Chemistry*, **278**, 44312–44319.

Schwarz, D. S., Hütvagner, G., Haley, B. and Zamore, P. D. (2002). Evidence that siRNAs function as guides, not primers, in the *Drosophila* and human RNAi pathways. *Molecular Cell*, **10**, 537–548.

Semizarov, D., Frost, L., Sarthy, A., Kroeger, P., Halbert, D. N. and Fesik, S. W. (2003). Specificity of short interfering RNA determined through gene expression signatures. *Proceedings of the National Academy of Sciences USA*, **100**, 6347–6352.

Shinagawa, T. and Ishii, S. (2003). Generation of Ski-knockdown mice by expressing a long double-strand RNA from an RNA polymerase II promoter. *Genes & Development*, **17**, 1340–1345.

Simmer, F., Moorman, C., Van Der Linden, A. M., Kuijk, E., Van Den Berghe, P. V., Kamath, R., Fraser, A. G., Ahringer, J. and Plasterk, R. H. (2003). Genome-wide RNAi of *C. elegans* using the hypersensitive rrf-3 strain reveals novel gene functions. *Public Library of Science Biology*, **1**, E12.

Timmons, L. and Fire, A. (1998). Specific interference by ingested dsRNA. *Nature*, **395**, 854.

Vastenhouw, N. L., Fischer, S. E., Robert, V. J., Thijssen, K. L., Fraser, A. G., Kamath, R. S., Ahringer, J. and Plasterk, R. H. (2003). A genome-wide screen identifies 27 genes involved in transposon silencing in *C. elegans*. Current Biology, **13**, 1311–1316.

Yang, D., Buchholz, F., Huang, Z., Goga, A., Chen, C. Y., Brodsky, F. M. and Bishop, J. M. (2002). Short RNA duplexes produced by hydrolysis with *Escherichia coli* RNase III mediate effective RNA interference in mammalian cells. *Proceedings of the National Academy of Sciences USA*, **99**, 9942–9947.

Ziauddin, J. and Sabatini, D. M. (2001). Microarrays of cells expressing defined cDNAs. *Nature*, **411**, 107–110.

34 Generation of highly specific vector-based shRNAi libraries directed against the entire human genome

Makoto Miyagishi, Sahohime Matsumoto, Takashi Futami, Hideo Akashi, Krishnarao Appasani, Yasuomi Takagi, Shizuyo Sutou, Takashi Kadowaki, Ryozo Nagai, and Kazunari Taira

Introduction

The success of the Human Genome Project and the availability of the complete sequence of the human genome have revealed the existence of numerous genes whose functions are unknown. Although many functional genes were successfully identified using libraries of randomized ribozymes (Kruger et al., 2000; Li et al., 2000; Welch et al., 2000; Beger et al., 2001; Kawasaki et al., 2002; Kawasaki and Taira, 2002; Onuki et al., 2002; Nelson et al., 2003; Rhoades and Wong-Staal, 2003; Suyama et al., 2003a,b; Chatterton et al., 2004; Kuwabara et al., 2004; Onuki et al., 2004), the randomized library naturally contained many ribozymes that could not hybridize with any human gene transcripts, resulting in some false positives. While exploitation of RNA interference (RNAi) is hampered by the induction of the interferon response upon the introduction of double-stranded RNA (dsRNA) into mammalian cells (Elbashir et al., 2001), RNAi can be a powerful tool in gene analysis and, for example, the functions of a number of genes in the nematode *Caenorhabditis elegans* have been identified as being associated with particular mutant phenotypes (Fraser et al., 2000; Gonczy et al., 2000). Recent progress in mammalian cells, as discussed in this chapter, suggests that it should soon be possible to extend genome-wide approaches to mammalian cells, using libraries of small interfering RNA (siRNA) oligonucleotides or siRNA-expression libraries.

Efforts to exploit the phenomenon of RNA interference (RNAi) in mammalian cells have been hampered by the interferon response in cells into which foreign RNA is introduced. However, we have found that the use of vector-based short dsRNAs, as distinct from the widely used synthetic short dsRNAs, reduces the interferon response. We performed a series of studies to optimize vector-based siRNAs and determined that hairpin siRNAs, transcribed under the control of the U6 promoter, and including mismatches in the stem that prevent mutations during maintenance and amplification in *E. coli*, as well as a loop derived from the sequence of a human microRNA, had strong activity as suppressors of the activity of target genes.

We have been able to use our vector-based siRNA library, which we call an shRNAi (i.e., short-hairpin RNAi) library, for the identification of the functions of previously uncharacterized genes in the human genome. We have developed an algorithm for the identification of effective target sites and, since the sequence of the entire human genome is now available, we are able to predict the sites in individual genes that can be cleaved by shRNA, effectively eliminating the false-positive results obtained with other conventional libraries. Moreover, the suppressive effects of shRNAs are greater than those of other antisense molecules including ribozymes and, thus, detectable effects can be achieved at lower concentrations of the former than the latter.

Several groups, including our own, have already started to generate siRNA libraries directed against the entire human genome. In this chapter, we discuss the practical problems that are faced in efforts to construct siRNA-expression libraries and some methods for overcoming these problems. In addition, we have been able to demonstrate the utility of our library, by identifying novel genes that are involved in apoptosis in human cells.

Methodology

We shall begin with a brief overview of the methodology that we used in the studies described in this chapter.

Construction of vectors

For transient transfections, we generated both tandem-type and hairpin-type siRNA-expression vectors using the piGENEhU6 vector (Miyagishi and Taira, 2002b), which contains a human U6 promoter and two *Bsp*MI sites. For the construction of tandem-type siRNA-expression plasmids, we amplified, by PCR, DNA fragments that included the sequences of the relevant sense and antisense regions and the U6 promoter, using the piGENEhU6 vector as template and primers that included the sense or antisense sequence plus the terminator. After digestion of the products of PCR with *Bsp*MI, we ligated each fragment into the *Bsp*MI sites of the piGENEhU6 vector to yield a series of siRNA expression vectors. For the construction of hairpin-type siRNA-expression vectors, we synthesized oligonucleotides with the hairpin sequence, terminator sequence and overhanging sequences. Then we annealed the fragments and inserted them into the *Bsp*MI sites of the piGENEhU6. We generated H1 promoter-driven siRNA-expression vectors as described by Brummelkamp et al. (2002). Schematic representations of the constructs are shown in Figure 34.1. Details of relevant sequences are indicated in other figures and can be found in cited publications.

Culture and transfection of cells

We performed our experiments with HeLa S3 cells and transfected cells using the Lipofectamine™ 2000 reagent (Invitrogen Corp., Carlsbad, CA). We used increasing concentrations of siRNA expression plasmids and plasmids that encoded firefly luciferase and *Renilla* luciferase, respectively, so that we could monitor the

Tandem type

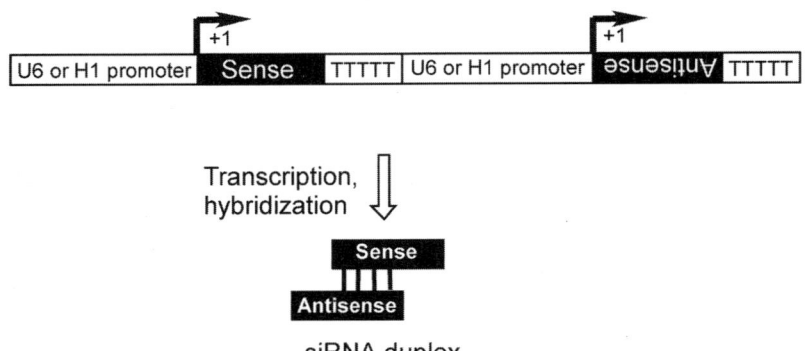

Hairpin type

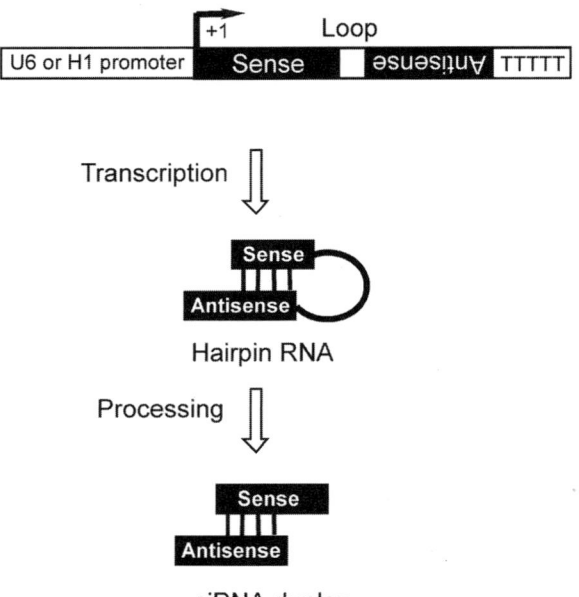

Figure 34.1. Schematic representation of the constructs used for expression of tandem-type and hairpin-type siRNAs.

activity of the firefly luciferase and normalize it against the activity of the *Renilla* luciferase. We added empty vectors to ensure that equal amounts of DNA were used in each transfection.

Algorithm for the identification of favorable target sites

We used a partial least-squares (PLS) method to generate models for the prediction of the effects of RNAi. We analyzed more than seven hundred sets of data which consisted of sequences of siRNAs and their respective suppressive effects, as well as factors that might be expected to be significantly correlated with suppressive activity, such as GC content and preference for nucleotide position (Yoshinari et al., 2004). We then made a PLS calibration model, using sequences of target sites, suppressive activities, correlated factors, and the algorithm PLS1 of our own making (Lindberg et al., 1983). We determined the optimum conditions for identification of a target site from the lowest standard errors in cross-validation analysis.

Preparation of duplexes of RNA oligonucleotides

Using T7 RNA polymerase, we prepared sense and antisense RNAs of 50 nucleotides (nt) in length for suppression of the firefly gene for luciferase by RNAi. We purified these RNAs by standard methods and then generated duplexes as described elsewhere (Akashi et al., 2002).

Construction of U6-based long-hairpin RNA-expression plasmids

For construction of U6-based long-hairpin RNA-expression plasmids, we used a plasmid that contains a human U6 promoter (Miyagishi and Taira, 2002a; Ohkawa and Taira, 2000). To generate the vector, pU6i50S, for expression of a 50-bp hairpin RNA, we inserted a sense sequence, with the mutations indicated in Figure 34.2C, between the U6 promoter and the corresponding antisense sequence. The sequences downstream of the U6 promoter were the same as those of synthetic RNAs that had been transcribed *in vitro* with the exception that the loop sequence (5′-TTCAAGAGA-3′) was different.

Identification of genes involved in apoptosis using the shRNAi library

We transfected HeLa S3 cells in multi-well plates with one short-hairpin siRNA-expression vector per well, using Lipofectamine™ 2000. Then, 36 h later, we treated cells with puromycin for 24 h. Surviving cells were resuspended in culture medium and, 18 h later, apoptosis was induced by treatment of cells with double-stranded RNA (dsRNA). Twenty-four hours later, cells were washed, fixed and stained with crystal violet.

Full details of all experimental methods can be found in recent publications from our laboratory (Miyagishi and Taira, 2002a, b; Kawasaki and Taira, 2003; Miyagishi and Taira, 2003; Miyagishi et al., 2004a, b; Yoshinari et al., 2004).

Results and discussion

The choice between tandem-type and hairpin-type siRNA-expression systems

The members of a useful siRNA-expression library must have strong suppressive activity and strong genetic stability. In our studies, since the level of transcripts generated from Pol III promoters is significantly higher than that of transcripts from pol II promoters (Koseki et al., 1999), we have worked with Pol III promoter-based siRNA-expression systems exclusively (Taira and Miyagishi, 2001).

The siRNA-expression systems that exploit the U6 and the H1 promoter can be divided into two groups, namely, tandem-type and hairpin-type (Figure 34.1). In

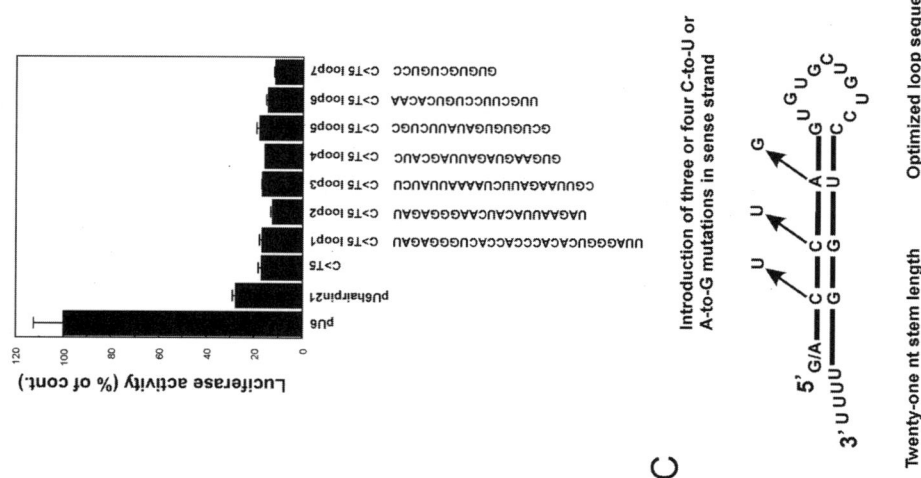

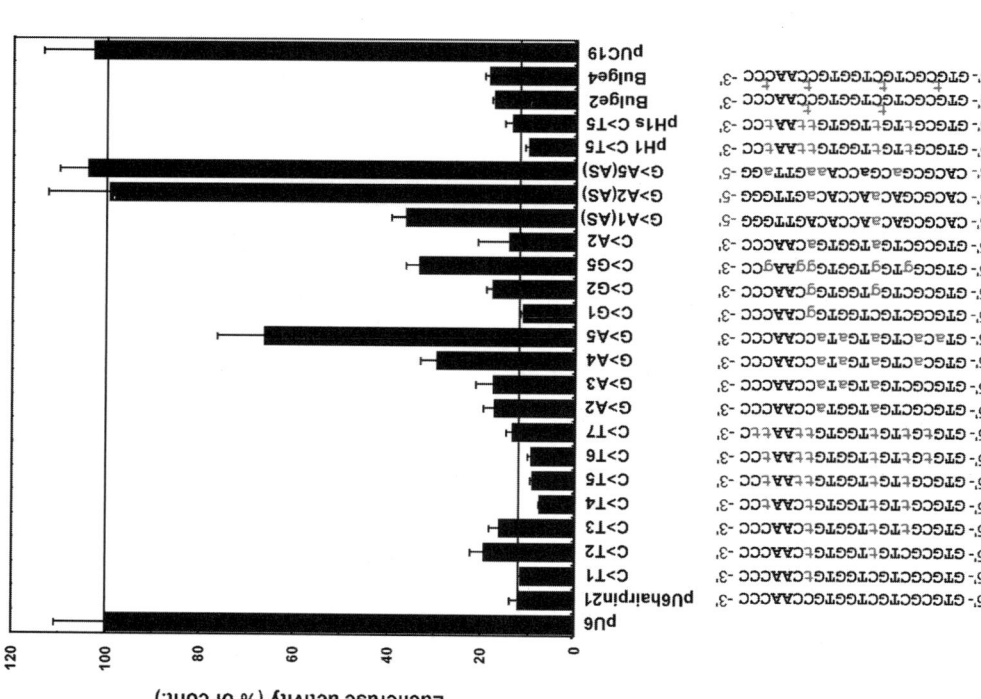

the case of tandem-type siRNA-expression vectors, transcription of the sense and antisense strands is driven by separate respective Pol III promoters. The transcripts anneal within cells and form siRNA duplexes with overhangs of approximately four nucleotides at each 3′ end (Lee et al., 2002, Miyagishi and Taira, 2002a). By contrast, in the hairpin type, sense and antisense nucleotides are connected by a loop and are transcribed as a single unit (Brummelkamp et al., 2002). The transcribed RNA appears to form a hairpin structure, with a stem and a loop, that is processed to siRNA by Dicer (Ketting et al., 2001; Kawasaki and Taira, 2003).

When we compared the activities of the two types of siRNA-expression vector that were directed against the same site in the transcript of the firefly gene for luciferase, we found that both had strong activity when the concentration of the siRNA expression vector was high. However, at lower concentrations, the hairpin-type siRNA-expression vector (with a stem that was 21 nt long) had significantly higher activity than the tandem-type vector (Figure 34.2A).

The choice of hairpin-type siRNA-expression vector (shRNAi vector)

The construction of hairpin-type siRNA-expression vectors, which we call shRNAi vectors, presents two serious technical problems. The first problem is that it is difficult to sequence constructs that contain a hairpin region, most likely because of the tight palindromic structure. The second, and potentially more serious, problem, is that mutations appear in approximately 20 to 40% of constructs when they are amplified and maintained in *E. coli*. When we sequenced such mutated constructs, we found that we were able to determine the sequences of constructs with two or more point mutations or insertions/deletions within the sense or the antisense region without any problems. Therefore, we examined whether such constructs might, themselves, have silencing activity.

The introduction of as many as seven C to T point mutations in the sense strand, which generated G:U base pairing, had no effect on the silencing activity of the construct (Figure 34.2A). Moreover, the introduction of as many as four "bulge" insertions in the sense strand also had no negative effect on silencing. By contrast, the introduction of five C to G mutations or more than four G to A mutations significantly reduced the silencing activity. Reduction and elimination of the suppressive activity were recognized upon introduction of a single G to A mutation in the antisense strand [G>A1(AS)], which recognizes the target

Figure 34.2. (A) A comparison of the effects of vectors that encode U6 promoter-driven tandem-type 21-nt siRNAs, U6 promoter-driven hairpin-type shRNAs with 21-nt stems, H1 promoter-driven hairpin-type shRNA with a 21-nt stem, and U6 promoter-driven hairpin-type shRNAs with various mutations in the stem, as indicated. HeLa S3 cells were transfected with plasmids that encoded the firefly gene for luciferase (the reporter gene), the *Renilla* gene for luciferase (for normalization of the activity of firefly luciferase) and an expression plasmid directed against the transcript of the firefly gene. Each column and bar indicates the mean and S.E.M. of results from triplicate assays. (B) The effects of various loop sequences on suppressive activity. Details are the same as those in the legend to Figure 34.2A. (C) Schematic representation of the introduction of mutations into the stem of the hairpin siRNA.

sequence in the target transcript, and upon introduction of two or five G to A mutations [G>A2(AS) or G>A5(AS), respectively] in the antisense strand.

Thus it appears that, in practice, it is impossible to introduce mutations or bulges into the antisense strand of a hairpin-type construct without significant or total loss of suppressive activity. By contrast, the introduction of three mutations into the sense (or antisense) strand markedly reduced rates of mutation in the hairpin region of constructs in *E. coli*. This reduction is obviously advantageous for the stable maintenance of plasmids in their bacterial host.

Our analysis of the effects of mutations in the stem of the hairpin-type siRNA-expression vector suggested that introduction of multiple C to T (or A to G) mutations in the sense strand not only prevented mutation of the vector in the bacterial host but also, in fact, enhanced the silencing activity of the siRNA (compare the results for pU6hairpin21 with C>T4, C>T5 and C>T6, respectively, in Figure 34.2A).

Choice of Pol III promoter

We compared the silencing activity of shRNAi directed by the U6 promoter and by the H1 promoter. These promoters contain consensus motifs, a distal sequence element (DSE), a proximal sequence element (PSE) and a TATA box. As shown in Figure 34.2A, the H1 promoter was slightly less effective than the U6 promoter when both were tested in an assay of suppression of the firefly gene for luciferase (for example, compare results for pH1C>T5 with those for pU6hairpin21). Therefore, we chose the U6 promoter for construction of our siRNA-expression libraries. It was demonstrated recently that tRNA-linked siRNA-expression vectors allow the constitutive transport of the hairpin RNA into the cytoplasm (Kawasaki and Taira, 2003). However, this system requires a hairpin stem of 30 nt in length and is, thus, much more costly than the U6 promoter-driven siRNA-expression system that we have adopted for the construction of libraries.

Choice of the loop sequence of the hairpin

Having characterized the optimal features of the stem and chosen the promoter for our hairpin-type siRNA-expression vector, we next examined the effects of changes in the loop of the hairpin sequence. When the stem was 19 nt long, a hairpin RNA with a loop of 9 nt had greater silencing activity than the same stem with a shorter loop sequence (Brummelkamp et al., 2002), and a 4-nt, 5'-UUCG-3', tetraloop sequence was also used (Paul et al., 2002).

Since natural microRNAs (miRNAs) include a loop, we examined the effects of hairpin RNAs with loop sequences derived from seven human miRNAs (Miyagishi and Taira, 2003; Miyagishi et al., 2004a,b). We found that hairpins that included each of the loops from human miRNAs had suppressive activity that was similar to or greater than that of the corresponding hairpin with the original 9-nt loop (Figure 34.2B). Loops 2 and 7 appeared to be particularly effective even though the effects of the loop sequence were relatively small. However, in practice, we found that several-fold higher concentrations of the C>T5 hairpin with the original 9-nt loop than of the C>T5 loop 7 hairpin were required for similar suppressive

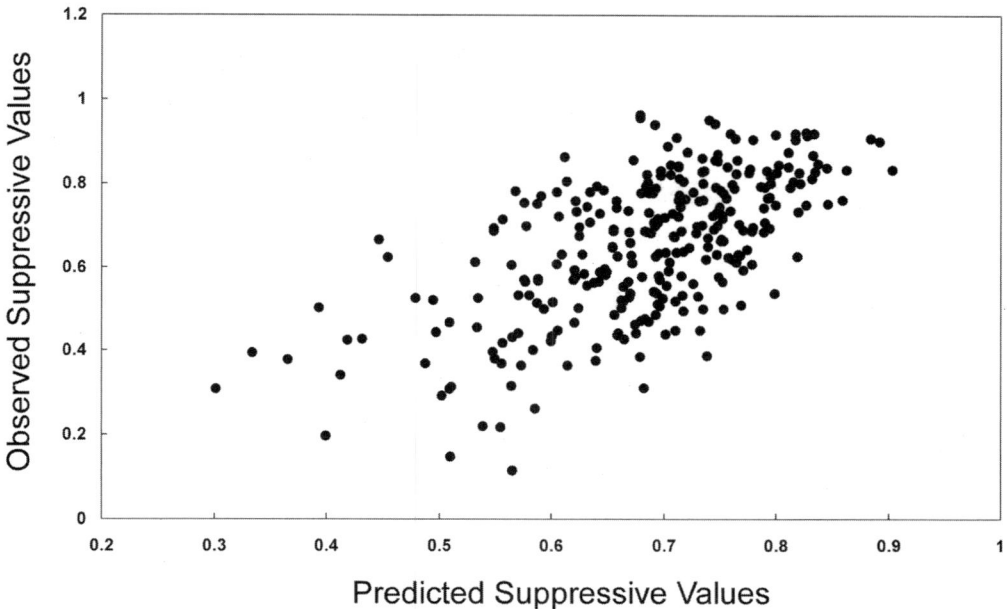

Figure 34.3. Development of the algorithm for prediction of effective target sites. The graph shows the leave-one-out cross validation result between values predicted by application of the algorithm and experimental results. Experiments were performed as described in the legend to Figure 34.2A.

activities. Therefore, the loop from a natural human miRNA was incorporated into the design of the hairpin for the siRNA library.

Choice of target sites in the firefly gene for luciferase and other genes

The choice of appropriate target sites is essential for the effective exploitation of gene silencing by siRNA. There is ample evidence that the effectiveness of siRNA depends on the target site within the target mRNA and that the probability of successful gene silencing at an arbitrarily selected target site ranges between zero and 40%. As part of our efforts to generate a library directed against the entire human genome, we developed an algorithm for the prediction of suitable target sites in mRNAs. We analyzed the data from about 1,000 attempts at silencing mRNAs by targeting specific sequences and we were able to identify a number of factors that were strongly correlated with strong suppressive activity. Then we used nonlinear regression methods to generate an algorithm that would allow us to predict the suppressive activity for a specific site within an mRNA.

Figure 34.3 shows the correlation between results predicted by application of our experimentally derived algorithm and the results that were actually obtained in an analysis of the effects of siRNAs directed against targets in the EGFP gene. The correlation coefficient for the results shown in Figure 34.3 is about 0.7. More recent fine-tuning of our algorithm has allowed us to achieve a correlation coefficient as high as 0.8.

For the construction of libraries, we chose target sites with predicted scores of more than 0.85.

Confirmation of the specificity of target sequences

When we select a particular target sequence within a specific gene, we need to make sure that there are no strongly homologous sequences in other genes in the same organism. Confirmation of specificity requires, however, that we also choose the stringency of the conditions under which we check for homology. When, for example, more than three base-mismatches are needed for discrimination between the chosen and homologous sequences, application of the BLAST search program (Altschul et al., 1990) reveals that significant numbers of apparently favorable target sites must be eliminated under such high-stringency conditions. When only one base mismatch is allowed, the designed siRNA might have the potential to disrupt not only expression of the target gene but also expression of other homologous genes, even if the efficiency of suppression of such homologous genes might be lower. We chose the latter conditions for our studies of target-site specificity because, if several genes are identified using a specific siRNA, more detailed analyses can be performed using new siRNAs directed against other specific sites within the candidate gene.

The existence of families of strongly homologous genes and the frequent alternative splicing of transcripts also complicate the selection of target sites, but it is obviously possible to design siRNAs directed against a specific isogene or against a specific form of spliced transcript. However, for the generation of the "first-draft" library, when the target gene yields multiple variously spliced transcripts, the target should be a sequence that is common to all the transcripts. Similarly, when the target is a member of a family of genes, a sequence that is common to all members of the family should be chosen as the initial target site. These strategies allow the economical construction of the first library. Additional specific siRNAs can be generated if the successfully targeted genes appear to be attractive candidates for further analysis.

An improved method for the construction of libraries

The practical problems associated with construction of libraries are those of efficiency, economy and speed. Our large-scale method for constructing libraries is shown in Figure 34.4. It allows us to generate between 3000 and 5000 shRNAi vectors per month at reasonable cost (Taira and Miyagishi, 2001).

In a typical experiment, 96 sets of oligonucleotides, including the hairpin sequences that correspond to 96 target sequences, were synthesized and annealed separately. Then the annealed oligonucleotides were mixed together and ligated into the *Bsp*MI site of the U6 promoter-driven shRNAi vector. After introduction of the vectors into *E. coli*, we picked up 384 clones and sequenced them for the identification of each clone. We generally recover about 90% of the clones of interest at this stage. The sequences of the clones that we fail to recover are included in the next 96 sets of oligonucleotides.

Our procedure allows us to generate shRNAi vectors very much more cheaply and rapidly than when each clone is prepared separately, as is customarily the case when the "traditional" procedure is used.

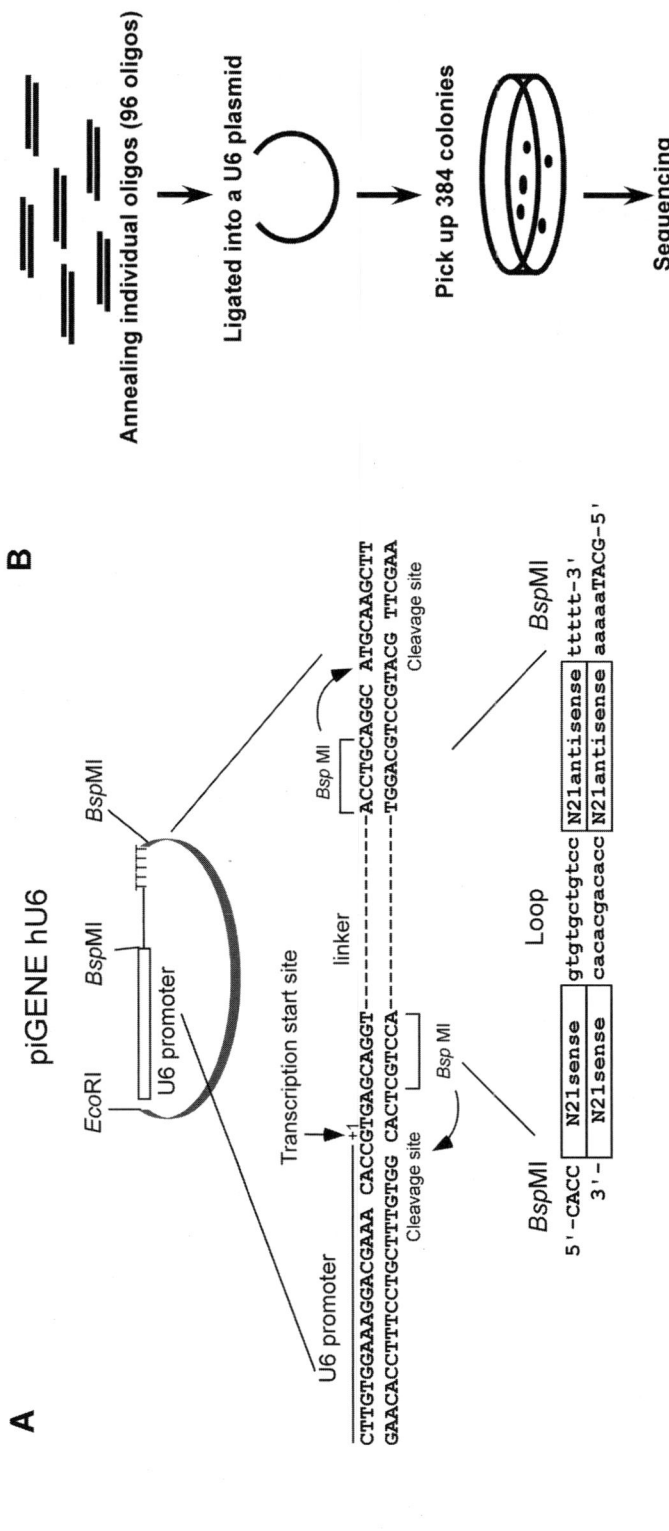

Figure 34.4. Construction of the shRNAi library. (A) Schematic representation of the plasmid that is used to construct the library. (B) Schematic representation of the procedure for generating clones in the library. Details of the procedure can be found elsewhere (Miyagishi and Taira, 2003).

489

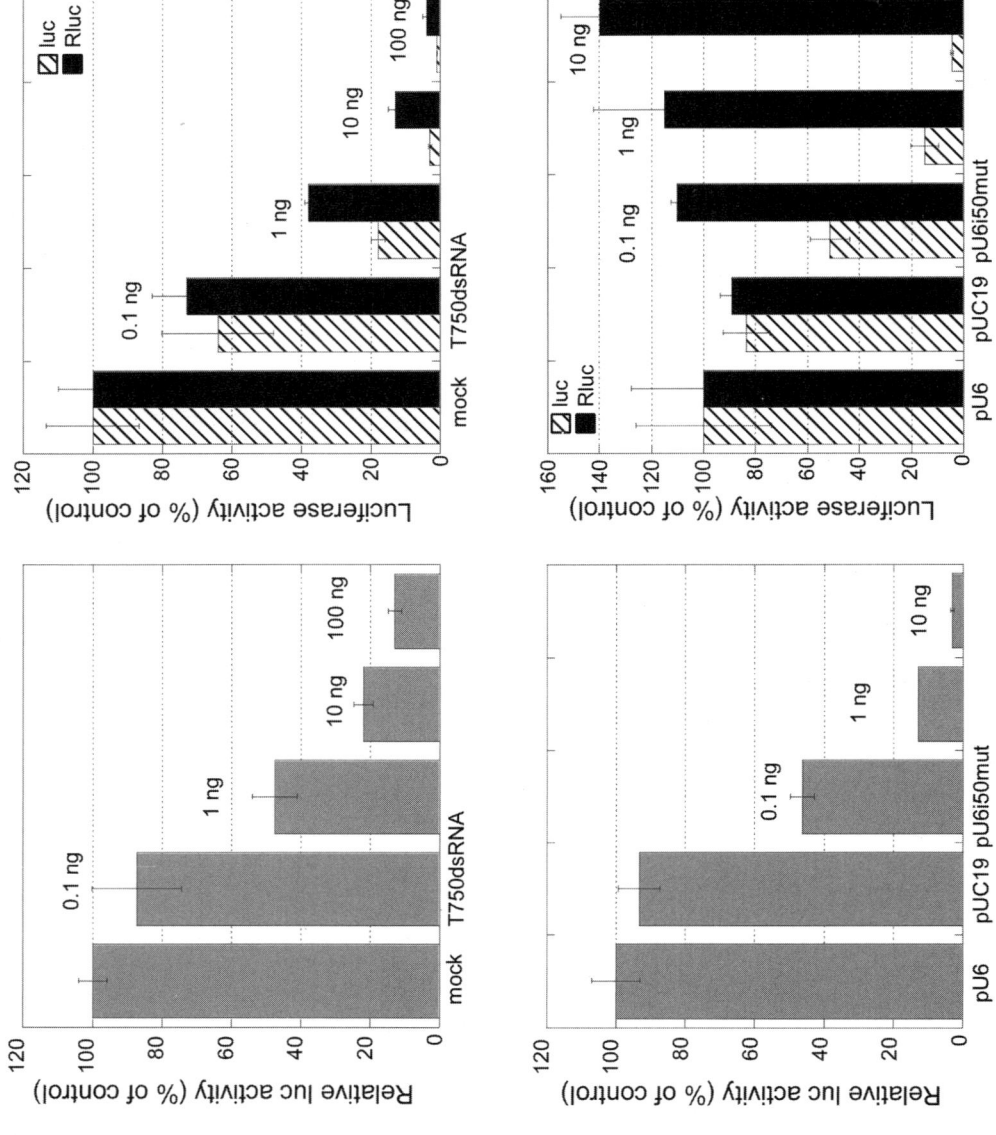

Some advantages and disadvantages of shRNAi libraries and siRNA oligonucleotide libraries

Plasmid-based shRNAi libraries and oligonucleotide-based siRNA libraries each have their own advantages and disadvantages. The efficiency of the introduction of siRNA into cells by lipofection depends on the cell type and the RNAi effect tends to be transient. When plasmid-based siRNA is used, it is easy to eliminate, by selection with the appropriate antibiotic, those cells that have not been transfected with a plasmid. Moreover, the effects of plasmid-based RNAi are of much longer duration than those observed when siRNA oligonucleotides are used. Viral vectors allow the delivery of siRNA-expression cassettes into cells with high efficiency and, when a lentivirus or retrovirus is used, integration of the viral vector into the host cell's genome facilitates the generation of stable lines of "knock-down" cells. In some cells, nonetheless, the efficiency of transfection with oligonucleotide siRNA exceeds 90%, which is greater than the efficiency that can be achieved with a plasmid. However, once plasmid-based siRNAs have been generated for a library, they can be amplified indefinitely, in particular when our construct (Figure 34.2C) is used. Our construct, unlike more conventional constructs, is not subject to mutation during its amplification in *E. coli*, and thus its use is particularly effective and economical.

The major advantage of the vector-based shRNAi system: Reduction of the interferon response

The most significant advantage of our shRNAi system, as compared to the use of synthetic siRNAs, is the reduction in the interferon response in transfected cells (Akashi et al., unpublished data). Figures 34.5A and B show the suppressive effects of synthetic and vector-derived siRNAs, respectively.

In the experiments for which results are shown in Figure 34.5, we used short dsRNA of 50 bp in order to sensitize cells. We included appropriate C to U (or A to G) mutations, as indicated in Figure 34.2C. The antisense strand was completely complementary to part of the firefly gene for luciferase and did not bind to the control transcript, Rluc, of the *Renilla* gene for luciferase. It appeared that the synthetic (i.e., *in vitro* transcribed) dsRNA suppressed the firefly luciferase activity in a dose-dependent manner. However, when we compared the effects on the firefly luciferase transcript with those on Rluc, it was clear, as shown in the right panel in Figure 34.5A, that levels of the control Rluc activity had been depressed

Figure 34.5. Demonstration of the non-specific effects of short (50-nt) dsRNA targeted to the firefly gene for luciferase. (A) HeLa S3 cells were transfected with 25 ng of firefly luciferase-expression plasmid, 2.5 ng of *Renilla* luciferase-expression plasmid and from 0.1 to 100 ng of dsRNA, as indicated. The left panel shows the activity of firefly luciferase (luc), normalized to that of *Renilla* luciferase (Rluc). The right panel shows the actual activities of the firefly and *Renilla* enzymes. (B) HeLa S3 cells were treated as in (A) with the exception that a U6 promoter-driven 50 bp hairpin RNA-expression vector was used instead of dsRNA. The left panel shows the reductions in relative luciferase activity and resembles the left panel in Figure 34.5A. The right panel shows that the dsRNA depressed the activity of the firefly luciferase but failed to depress the activity of the luciferase from *Renilla*. Each experiment was performed in triplicate and results are shown as means +/− S.E.M.

similarly to those of the firefly luciferase by the siRNA. Thus, while specific silencing of the firefly gene might have occurred, there were clearly non-specific effects that depressed levels of the Rluc transcript, which was not complementary to the siRNA that we had used.

When we performed a similar experiment using a U6 promoter-based 50 bp hairpin RNA-expression vector, with the same stem region as the synthetic dsRNA that we used in the first experiment, the effect on the expression of the firefly gene for luciferase was specific and the non-specific effect on the Rluc transcript was significantly reduced. The data in Figure 34.5B also indicated that, in transient-expression experiments, relatively few copies of shRNAi or of a 50 bp hairpin RNA-expression vector are needed for almost complete suppression of the target gene without, most importantly, any significant induction of the interferon response.

Some short siRNAs have been shown to induce the interferon response (Sledz et al., 2003). The extent of induction of the response depends on the sequence of such short siRNAs but it seems to be clearly advantageous to avoid any possibility of such a response by using shRNAi vectors, rather than synthetic siRNAs, for the construction of libraries.

Demonstration of the utility of shRNAi library

To confirm the utility of our shRNAi library in analyses of the human genome, we have begun efforts to identify the functions of genes of unknown function within the genome. Figure 34.6 shows the results of an attempt to identify genes involved in apoptosis, as an example. As positive control vectors, we used shRNAi vectors targeted against the transcript of the gene for dsRNA-dependent protein kinase (PKR), which is involved in the induction by dsRNA of apoptosis. These positive control vectors blocked dsRNA-induced apoptosis (Figure 34.6A). When we examined the effects of shRNAi vectors directed against genes of unknown function (Figure 34.6B; C1 through C11), we found one shRNAi vector (C3) that blocked apoptosis by interfering with the function of one of these uncharacterized genes. We anticipate that shRNAi libraries will be useful for the similar characterization of the functions of many genes that have been recognized within the human genome but whose functions remain to be determined.

Concluding remarks

Our work has allowed us to develop a method for the efficient and relatively inexpensive generation of libraries of plasmids that express short hairpin RNAs (shRNAi libraries). Introduction of mutations in the stem of the hairpin stabilizes the vectors during their maintenance and amplification in *E. coli*, and incorporation of the loop sequence from a naturally occurring human miRNA also enhances their activity. The use of these vectors minimizes the confounding and non-specific effects of the interferon response, which is induced by some short siRNAs and by longer dsRNAs. We have demonstrated the effectiveness of our library by identifying a gene of previously unknown function that is involved in apoptosis in human cells.

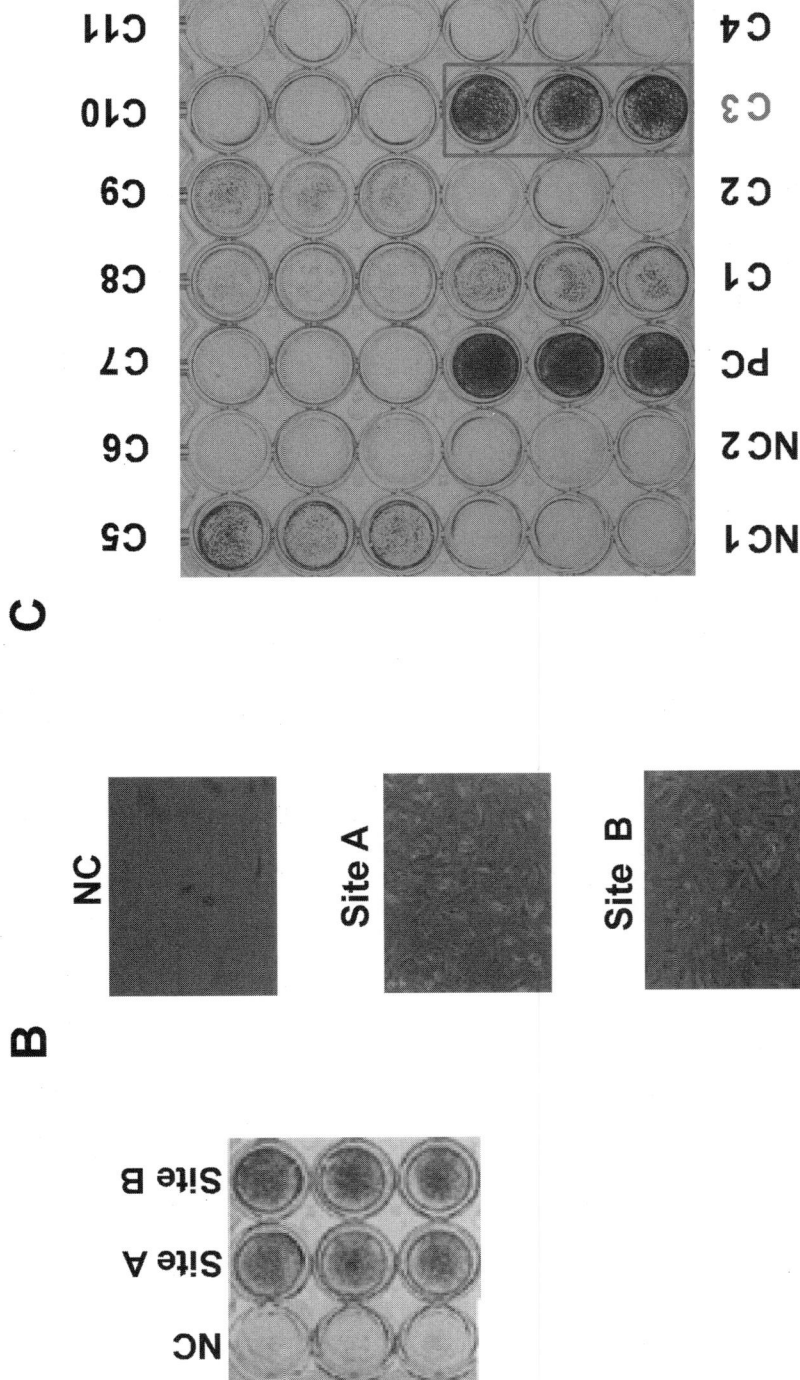

Figure 34.6. Identification of a previously unidentified gene that is involved in apoptosis in human cells. (A) shRNAi vectors directed against two sites (A and B) in the transcript of the gene for protein kinase R (PKR) prevented apoptosis in response to dsRNA. After the induction of apoptosis and fixation, HeLa S3 cells were stained with crystal violet, which only stains living cells. NC, Negative control. (B) Images of cells shown in (A) prior to staining. (C) The shRNAs designated C1 through C11 were tested for their ability to prevent apoptosis. These shRNAs were designed to target the transcripts of genes for kinases or transcription factors or genes that might be related to apoptosis. Each shRNA was tested in triplicate and cells that had been challenged with dsRNA were fixed and subjected to staining with crystal violet. PC, Positive control (shRNA directed against the gene for PKR); NC1 and NC2, negative controls; C1 through C11, shRNAs directed against specific genes. The shRNA designated C3 prevented apoptosis and the cells were stained purple with crystal violet. (See color section.)

493

The availability of our library of vectors, which encode short hairpin RNAs that disrupt the activity of specific target genes by inducing the cleavage of their transcripts at specific sites, should help in efforts to identify the functions of all the genes in the human genome.

Note: The first four authors contributed equally to this work.

REFERENCES

Akashi, H., Kawasaki, H., Kim, W. J., Akaike, T., Taira, K. and Maruyama, A. (2002). Enhancement in the cleavage activity of a hammerhead ribozyme by cationic comb-type polymers and an RNA helicase *in vitro*. *Journal of Biochemistry (Tokyo)*, **131**, 687–692.

Altschul, S. F., Gish, W., Miller, W., Myers, E. W. and Lipman, D. J. (1990). Basic local alignment search tool. *Journal of Molecular Biology*, **215**, 403–410.

Beger, C., Pierce, L. N., Kruger, M., Marcusson, E. G., Robbins, J. M., Welcsh, P., Welch, P. J., Welte, K., King, M. C., Barber, J. R. and Wong-Staal, F. (2001). Identification of Id4 as a regulator of BRCA1 expression by using a ribozyme-library-based inverse genomics approach. *Proceedings of the National Academy of Sciences USA*, **98**, 130–135.

Brummelkamp, T. R., Bernards, R. and Agami, R. (2002). A system for stable expression of short interfering RNAs in mammalian cells. *Science*, **296**, 550–553.

Chatterton, J., Hu, X. and Wong-Staal, F. (2004). Ribozymes in gene identification, target validation, and drug discovery. *Targets*, **3**, 10–17.

Elbashir, S. M., Harborth, J., Lendeckel, W., Yalcin, A., Weber, K. and Tuschl, T. (2001). Duplexes of 21-nucleotide RNAs mediate RNA interference in cultured mammalian cells. *Nature*, **411**, 494–498.

Fraser, A. G., Kamath, R. S., Zipperlen, P., Martinez-Campos, M., Sohrmann, M. and Ahringer, J. (2000). Functional genomic analysis of *C. elegans* chromosome 1 by systemic RNA interference. *Nature*, **408**, 225–230.

Gonczy, P., Echeverri, C., Oegema, K., Coulson, A., Jones, S. J., Copley, R. R., Duperon, J., Oegema, J., Brehm, M., Cassin, E., Hannak, E., Kirkham, M., Pichler, S., Flohrs, K., Goessen, A., Leidel, S., Alleaume, A. M., Martin, C., Ozlu, N., Bork, P. and Hyman, A. A. (2000). Functional genomic analysis of cell division in *C. elegans* using RNAi of genes on chromosome III. *Nature*, **408**, 331–336.

Kawasaki, H. and Taira, K. (2002). Identification of genes by hybrid ribozymes that couple cleavage activity with the unwinding activity of an endogenous RNA helicase. *European Molecular Biology Organization Reports*, **3**, 443–450.

Kawasaki, H. and Taira, K. (2003). Short hairpin type of dsRNAs that are controlled by tRNA^{Val} promoter significantly induce RNAi-mediated gene silencing in the cytoplasm of human cells. *Nucleic Acids Research*, **31**, 700–707.

Kawasaki, H., Onuki, R., Suyama, E. and Taira, K. (2002). Identification of genes that function in the TNF-alpha-mediated apoptotic pathway using randomized hybrid ribozyme libraries. *Nature Biotechnology*, **20**, 376–380.

Ketting, R. F., Fischer, S. E., Bernstein, E., Sijen, T., Hannon, G. J. and Plasterk, R. H. (2001). Dicer functions in RNA interference and in synthesis of small RNA involved in developmental timing in *Caenorhabditis elegans*. *Genes & Development*, **15**, 2654–2659.

Koseki, S., Tanabe, T., Tani, K., Asano, S., Shioda, T., Nagai, Y., Shimada, T., Ohkawa, J. and Taira, K. (1999). Factors governing the activity *in vivo* of ribozymes transcribed by RNA polymerase III. *Journal of Virology*, **73**, 1868–1877.

Kruger, M., Beger, C., Li, Q. X., Welch, P. J., Tritz, R., Leavitt, M., Barber, J. R. and Wong-Staal, F. (2000). Identification of eIF2Bgamma and eIF2gamma as cofactors of hepatitis C virus internal ribosome entry site-mediated translation using a functional genomics approach. *Proceedings of the National Academy of Sciences USA*, **97**, 8566–8571.

Kuwabara, T., Hsieh, J., Nakashima, K., Taira, K. and Gage, F. H. (2004). A small modulatory dsRNA specifies the fate of adult neural stem cells. *Cell*, **116**, 779–793.

Lee, N. S., Dohjima, T., Bauer, G., Li, H., Li, M. J., Ehsani, A., Salvaterra, P. and Rossi, J. (2002). Expression of small interfering RNAs targeted against HIV-1 rev transcripts in human cells. *Nature Biotechnology*, **20**, 500–505.

Li, Q. X., Robbins, J. M., Welch, P. J., Wong-Staal, F. and Barber, J. R. (2000). A novel functional genomics approach identifies mTERT as a suppressor of fibroblast transformation. *Nucleic Acids Research*, **28**, 2605–2612.

Lindberg, W., Persson, J. A. and Wold, S. (1983). Partial least-squares method for spectrofluorimetric analysis of mixtures of humic acid and lignin sulfonate. *Analytical Chemistry*, **55**, 643–648.

Miyagishi, M. and Taira, K. (2002a). U6 promoter-driven siRNAs with four uridine 3′ overhangs efficiently suppress targeted gene expression in mammalian cells. *Nature Biotechnology*, **20**, 497–500.

Miyagishi, M. and Taira, K. (2002b) Expression of siRNA from a Pol III promoter in mammalian cells. In *Perspectives in Gene Expression*, ed. K. Appasani, pp. 361–375. Westboro, MA: The Eaton Publishers.

Miyagishi, M. and Taira, K. (2003). Strategies for generation of an siRNA expression library directed against the human genome. *Oligonucleotides*, **13**, 325–333.

Miyagishi, M., Matsumoto, S. and Taira, K. (2004a). Generation of an shRNAi expression library against the whole human transcripts. *Virus Research*, **102**, 117–24.

Miyagishi, M., Sumimoto, H., Miyoshi, H., Kawakami, Y. and Taira, K. (2004b). Optimization of an siRNA-expression system with a mutated hairpin and its significant suppressive effects upon HIV vector-mediated transfer into mammalian cells. *The Journal of Gene Medicine*, in press.

Nelson, L., Suyama, E., Kawasaki, H. and Taira, K. (2003). Use of random ribozyme libraries for the rapid screening of apoptosis- and metastasis-related genes. *Targets*, **2**, 191–200.

Ohkawa, J. and Taira, K. (2000). Control of the functional activity of an antisense RNA by a tetracycline-responsive derivative of the human U6 snRNA promoter. *Human Gene Therapy*, **11**, 577–585.

Onuki, R., Bando, Y., Suyama, E., Katayama, T., Kawasaki, H., Baba, T., Tohyama, M. and Taira, K. (2004). An RNA-dependent protein kinase is involved in tunicamycin-induced apoptosis and Alzheimer's disease. *The European Molecular Biology Organization Journal*, **23**, 959–968.

Onuki, R., Nagasaki, A., Kawasaki, H., Baba, T., Uyeda, T. Q. and Taira, K. (2002). Confirmation by FRET in individual living cells of the absence of significant amyloid beta-mediated caspase 8 activation. *Proceedings of the National Academy of Sciences USA*, **99**, 14716–14721.

Paul, C. P., Good, P. D., Winer, I. and Engelke, D. R. (2002). Effective expression of small interfering RNA in human cells. *Nature Biotechnology*, **20**, 505–508.

Rhoades, K. and Wong-Staal, F. (2003). Inverse Genomics as a powerful tool to identify novel targets for the treatment of neurodegenerative diseases. *Mechanisms of Ageing and Development*, **124**, 125–132.

Sledz, C. A., Holko, M., de Veer, M. J., Silverman, R. H. and Williams, B. R. (2003). Activation of the interferon system by short-interfering RNAs. *Nature Cell Biology*, **5**, 834–839.

Suyama, E., Kawasaki, H., Kasaoka, T. and Taira, K. (2003a). Identification of genes responsible for cell migration by a library of randomized ribozymes. *Cancer Research*, **63**, 119–124.

Suyama, E., Kawasaki, H., Nakajima, M. and Taira, K. (2003b). Identification of genes involved in cell invasion by using a library of randomized hybrid ribozymes. *Proceedings of the National Academy of Sciences USA*, **100**, 5616–5621.

Taira, K. and Miyagishi, M. (2001). siRNA expression systems and methods for producing functional knock-down cells. *Japanese Patent Application*, Heisei-13–363385 (PCT/JP02/12447).

Welch, P. J., Marcusson, E. G., Li, Q. X., Beger, C., Kruger, M., Zhou, C., Leavitt, M., Wong-Staal, F. and Barber, J. R. (2000). Identification and validation of a gene involved in anchorage-independent cell growth control using a library of randomized hairpin ribozymes. *Genomics*, **66**, 74–83.

Yoshinari, K., Miyagishi, M. and Taira, K. (2004). Effects on RNAi of the tight structure, sequence and position of the targeted region. *Nucleic Acids Research*, **32**, 691–699.

Index